Plasma Electrolytic Oxidation – Fundamentals, Advances and Applications

Viswanathan S. Saji • Mikhail Zheludkevich
Editors

Plasma Electrolytic Oxidation – Fundamentals, Advances and Applications

Edition Details

Springer

Editors
Viswanathan S. Saji
Interdisciplinary Research Center for Advanced Materials
King Fahd University of Petroleum & Minerals
Dhahran, Saudi Arabia

Mikhail Zheludkevich
Insitute for Materials Research
Helmholtz-Zentrum Geesthacht
Geesthacht, Schleswig-Holstein, Germany

ISBN 978-3-031-92428-6 ISBN 978-3-031-92429-3 (eBook)
https://doi.org/10.1007/978-3-031-92429-3

This Springer imprint is published by the registered company Springer Nature Switzerland AG
The registered company address is: Gewerbestrasse 11, 6330 Cham, Switzerland

Preface

Plasma electrolytic oxidation (PEO) or micro-arc oxidation (MAO) is a pivotal surface modification/coating approach to create thick and adherent protective/functional oxide layers on metals and alloys. PEO is an electrochemical conversion coating approach conducted at high voltages where the formation mechanism includes complex electrochemical, thermal, and plasma chemical reactions. The process is accompanied by many short-lived micro-discharges and dielectric breakdowns at high voltages, producing an adherent porous conversion coating. The crystalline micro/nanoscale surface oxides formed by the plasma-assisted deposition could substantially enhance the wear and corrosion resistance. PEO treatment is well known for enhancing surface hardness and mechanical properties. PEO layers with incorporated active components are explored in diverse potential applications, including automobile, aerospace, machinery, biomedical, thermal protection, catalysis, and energy materials. Recent advancements have expanded the scope of PEO applications to include medical devices with antibacterial properties, electronics with enhanced thermal conductivity, and environmental technologies for pollutant degradation. These emergent applications demonstrate the versatility and potential of PEO coatings in addressing modern technological challenges. Despite these advancements, several key unresolved fundamental questions remain. Understanding the precise mechanisms of micro-discharge events and their influence on coating properties is crucial for optimizing the PEO process. Additionally, improving the scalability and cost-effectiveness of PEO for large-scale industrial applications remains a significant challenge. Further research is needed to enhance the adhesion and durability of PEO coatings under extreme conditions, such as high temperatures and aggressive chemical environments. A finer understanding and perception of the PEO process are essential to improve the process's efficacy and coating durability. In this book, we aim to provide a comprehensive account of the fundamentals and applications of PEO, addressing both the current state of the technology and the future directions for research and development.

The chapters are arranged in four parts. Part I of the book has five chapters on fundamental aspects of PEO. Chapter 1 provides a concise account of the basics of PEO, whereas Chap. 2 details the effect of electrolyte and electrode parameters on

PEO. Chapter 3 reviews sparking discharge regimes and the role of soft sparking in PEO. Sparking phenomena inherent to PEO are not fully understood yet, and several sparking regimes can occur during PEO. Chapter 4 introduces innovative PEO techniques, such as two-step/multistep PEO, laser-assisted PEO, molten-salt PEO, plasma electrolytic fluoridation, ultrasound-assisted PEO, PEO in magnetic fields, cathode PEO, and scanning PEO processes. These PEO techniques are based on the traditional anodic PEO theory but with different modifications of the technical routes. The last chapter of this part (Chap. 5) discusses the impact of PEO on the mechanical properties of metallic systems. The variation of mechanical properties of PEO-coated metals and alloys is a topic of significant recent research interest.

Part II of the book provides a comprehensive coverage of the growth mechanisms of PEO, with three chapters (Chaps. 6–8) dedicated to this topic. Chapter 6 focuses explicitly on the PEO growth mechanism of valve and non-valve metals (carbon steel, brass, and copper). Chapter 7 offers an interesting account of the growth mechanism of PEO on metal-matrix composites. The last chapter of this part (Chap. 8) accounts for cathodic PEO on valve and non-valve metals, a method that could enhance the oxide growth rate. Compared to the conventional anodic PEO, cathodic PEO utilizes PEO on the cathode electrode.

The book's third part (Part III) has three chapters (Chaps. 9–11) that deal with the effect of in situ particle incorporation on the developed PEO layers of various metals and alloys. The type and nature of the electrolyte additives (particularly nano/microparticles) used in PEO play a decisive role in shaping the morphology, compactness, porosity, composition, thickness, wear/corrosion resistance, mechanical properties, and other functionalities of the resulting layer. Chapter 9 provides a detailed account of the types, approaches, and mechanisms of different in situ particle incorporation, whereas Chap. 10 provides an account of the effect of particle incorporation on improving the corrosion resistance of PEO coatings. The last chapter of this part (Chap. 11) comprehends the research progress of particle-incorporated PEO coatings on metallic materials for biomedical implant applications.

Part IV has 12 chapters on different application sectors of PEO. The first four chapters (Chaps. 12–15) address research advancements in anticorrosion and anti-wear PEO coatings on Al, Ti, and Mg alloys and steel. Chapters 12 and 13 provide interesting accounts of ongoing research and development efforts on anticorrosion and anti-wear PEO coatings on Al and Mg alloys. Chapter 14 addresses advances in anticorrosion and anti-wear PEO coatings for Ti and its alloys, whereas Chap. 15 explains anticorrosion and anti-wear PEO coatings on steel.

The next chapter of Part IV (Chap. 16) focuses on multilayer and self-healing coatings based on PEO in different applications. Chapters 17 and 18 offer comprehensive accounts of high-temperature thermal protection coatings based on PEO. Chapter 17 provides explicitly the recent advancements in PEO-derived thermal protection coatings on Ti and Nb alloys, with particular emphasis on their thermal conductivity, emissivity, and multiple protection characteristics for applications on the hot-end components in industries such as rockets, spacecraft, missiles, and high-speed aircraft. Chapter 18 accounts for the mechanisms underlying spectrally selective absorption and emission within PEO coatings and strategies for optimizing their composition, microstructure, and functional properties.

Chapter 19 accounts for PEO coatings in catalytic and battery materials applications. The chapter discusses PEO-derived electro, photo, and photoelectrocatalysts in different applications. PEO-derived anode electrode materials for lithium-ion batteries are also discussed. Chapter 20 summarizes the photoluminescent studies of PEO of metals such as Zr, Al, Ti, Zn, Nb, Hf, Ta, Y, and Gd in electrolytes containing Ln_2O_3 (Ln = Eu, Sm, Pr, Tm, Dy, Er, Ho), CeO_2, and Tb_3O_4 particles. Rare earths are known for their outstanding luminescence characteristics and intense emission.

Chapter 21 provides an interesting account of antibacterial and antibiofouling PEO coatings in biomedical implants, surgical tools, and marine structures. Chapter 22 focuses on the latest advances in using PEO coatings for various biomedical applications, including biocompatibility improvement, biodegradability control, and antimicrobial applications. The book's last chapter (Chap. 23) covers various posttreatment sealing strategies for PEO coatings aiming to enhance the adhesion, appearance, and overall functionality of PEO coatings.

This book results from our genuine effort to collectively present research advancements in this critical area. The book will be a handy reference tool for students and researchers in materials science, electrochemistry, corrosion science, surface modifications, and coatings.

Dhahran, Saudi Arabia — Viswanathan S. Saji
Geesthacht, Schleswig-Holstein, Germany — Mikhail Zheludkevich

Acknowledgments

We are thankful to all the authors for their valued contribution to this book. We would also like to express our gratitude to all those who have granted us the copyright permissions for reproducing illustrations. We acknowledge the support received from King Fahd University of Petroleum & Minerals (KFUPM) during the writing phase of the book. Our sincere thanks to the Springer Nature team for evolving this book into its final shape.

Contents

Part II Growth Mechanism

About the Editors

Viswanathan S. Saji is a Research Scientist-II at the Interdisciplinary Research Center for Advanced Materials, King Fahd University of Petroleum & Minerals (KFUPM), Saudi Arabia. Before that, he worked as an Endeavour Research Fellow at the University of Adelaide (2014), Australia; Research Professor at Korea University (2010–2013); Senior Research Scientist at Ulsan National Institute of Science and Technology (2009–2010); Postdoctoral Researcher at Sunchon National University (2009); Research Professor at Chosun University (2008–2009); Postdoctoral Researcher at Yonsei University (2007–2008); and Research Associate at the Indian Institute of Science (2005–2007) and the Indian Institute of Technology Bombay (2004–2005). He is a Graduate Advisor to Tongren University, China, under Sino-Foreign joint training program. He has authored over 100 journal publications and contributed 15 books and 25 book chapters. He has given over 50 international/national conference presentations and invited talks and seminars; he has also been awarded several fellowships and funding. His current research interest lies in electrochemistry and energy materials.

Mikhail Zheludkevich currently serves as the Director of the Institute of Surface Science at Helmholtz-Zentrum Hereon, Germany. Additionally, he holds a full professorship at the University of Kiel. His research interests revolve around electrochemistry, multifunctional surfaces, and active protection of lightweight materials and multi-material systems. In 2002, Prof. Zheludkevich earned his PhD in Physical Chemistry from the Belarusian State University, specializing in the interaction of atomic gases with metallic surfaces. Following a decade of diverse roles at the University of Aveiro, ranging from postdoctoral researcher to group leader, he joined Helmholtz-Zentrum, Geesthacht, as the head of the Corrosion and Surface Technology Department. In 2021, he established the Institute of Surface Science, a position he continues to lead. Prof. Zheludkevich's contributions have garnered recognition, including the Tajima Prize from the International Society of Electrochemistry in 2016, the distinction of Person of the Year 2023 by the International Mg Society, and Cavallaro Medal from the European Federation of Corrosion (EFC) in 2024. Furthermore, he serves as the Chairman of the Aerospace Working Party at the EFC.

Part I
Fundamental

Chapter 1
Fundamental Aspects of Plasma Electrolytic Oxidation

Aleksey Yerokhin

1.1 Introduction

Surface engineering of light alloys (Mg, Al, Ti) is vital for broadening their applications in structural engineering components, functional devices and consumer products. Plasma electrolytic oxidation (PEO) is an advanced electrochemical technique that attracts significant attention for deposition of oxide ceramic coatings on light alloys that are sought after in aerospace, automotive, manufacturing, chemical engineering, medicine and electronics. PEO originates from conventional anodising wherein field-assisted growth of surface oxide films is facilitated by anodic bias applied to the metal component immersed in an aqueous electrolyte solution. Unlike acid anodising however, PEO coatings are grown above breakdown voltage, which promotes their crystallisation and formation of specific fused morphology. This provides coatings with superior properties but makes PEO more energy intense than anodising, although the coating process can be simplified thanks to using non-aggressive and environmentally electrolytes such as mildly alkaline solutions of phosphates, silicates, borates or aluminates.

Controlling morphology and phase composition of anodic oxide films grown under conditions of dielectric breakdown requires careful selection of polarisation signal and imposes stringent requirements to electrical power supplies and process control algorithms. The fundamental understanding of PEO coating formation process relies upon advancing knowledge in electrochemistry, plasma physics and materials science, whereas experimental studies are often hindered by the lack of non-invasive process diagnostics techniques. So far, this caused heuristic approaches to prevail in process development, with scientific understanding often lagging

A. Yerokhin (✉)
Department of Materials, Henry Royce Institute, University of Manchester, Manchester, UK
e-mail: Aleksey.Yerokhin@manchester.ac.uk

V. S. Saji, M. Zheludkevich (eds.), *Plasma Electrolytic Oxidation – Fundamentals, Advances and Applications*,
https://doi.org/10.1007/978-3-031-92429-3_1

behind. Here, we revisit earlier studies to mark milestones on the road leading to contemporary PEO technology, discuss its current trends and recent advancements in fundamental understanding achieved due to the development of new *in-operando* process diagnostics techniques.

1.2 Historical Retrospective

1.2.1 Early Studies

Light-emitting electrical discharges at the electrode-electrolyte interface have probably been observed in the earliest electrochemical experiments [1]. However, the first scientific description of these phenomena is often credited to the work of Fizeau and Foucault (1844), who reported on the measurements of characteristic light emission from arcs observed during electrolysis of aqueous solutions on carbon and metal electrodes [2] (Fig. 1.1). Later on, Sluginov has resolved two types of electrolytic luminosity [3] that can be identified with two phenomena commonly observed in anodic treatments nowadays. The first is a uniform glow inherent to processes such as conventional anodising, where it is termed 'galvanoluminescence', and electrolytic plasma polishing, where a more violent but still uniform discharge is established [4], often termed 'contact glow discharge'. The second type of luminosity was attributed to distinct discrete microdischarges with clearly resolved spectral lines. This is characteristic to microdischarges observed in modern PEO treatments and sometimes also present in related non-oxidising processes, such as electrolytic plasma hardening or thermodiffusion treatments. The latter processes have been developed by Yasnogorodsky and practically implemented in the USSR during World War II [5], whereas the basis for fundamental understanding of underlying mechanisms was laid out in later works on Contact Glow Discharge Electrolysis by Kellogg [6] and Hickling [7] and further developed in more recent studies by Duradzhi [8], Sen Gupta [9], Belkin [10, 11] and others [12].

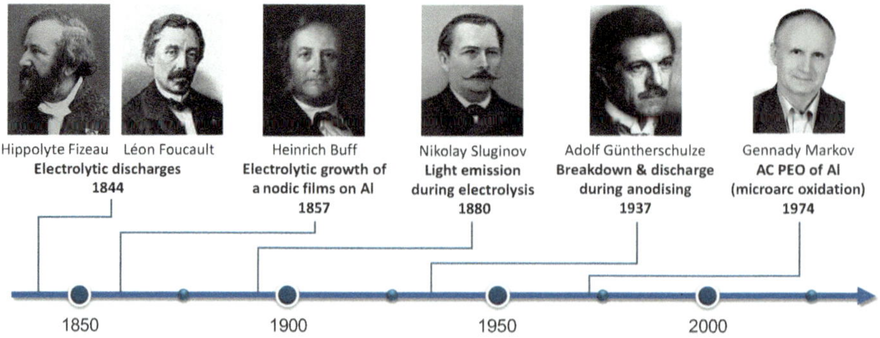

Fig. 1.1 Key contributors and developments that shaped contemporary PEO technology

The anodising technique stems from the work by Buff in 1857 [13], who has found that current passing through Al anode under constant polarisation in a solution of sulphuric acid drops sharply, which he attributed to the growth of surface oxide film. Subsequent studies have shown that the film remained stable and conducted well in the opposite direction [14]. This was probably the first observation of the valve effect, which has triggered interest in anodic treatments of Al for electrical engineering applications and laid the basis of modern anodising technology. First systematic studies of the valve effect, dielectric breakdown phenomena and electrolytic discharges during anodising of valve metals in various electrolytes—aqueous, non-aqueous solutions and molten salts—have been carried out early in twentieth century by Güntherschulze and summarised in his prominent monograph on 'Electrolytic Capacitors' [15]. Despite processes being patented at that time claiming to produce highly insulating oxide films under sparking conditions [16], the main practical driver for that research was to simultaneously increase both breakdown voltage and capacitance, the properties that oppositely depend on the film thickness, while sparking was generally deemed to be adverse phenomenon which should be avoided. This paradigm prevailed for a long time during which various acid anodising processes were developed for protection of Al alloys from corrosion [17] and wear [18].

Important prerequisites leading to the departure from the above paradigm were provided around 1960 when first reports on the synthesis of solid compounds by anodic spark reaction between metal electrodes and oxyanions from alkaline solutions started to emerge. McNeil [19] produced $Cd_2Nb_2O_7$ on Cd anode treated under sparking conditions in a potassium niobate solution. Hradcovsky and Belohradsky [20] claimed a method for production of silicate coatings by high-voltage anodising of Al in silicate-molybdate solutions. Anodic spark deposition (ASD) of solid compounds from silicate and aluminate solutions on Cu, Ni and Al was later studied by Brown et al. [21]. However, those deposits were often rough, porous and weakly adhered to metal substrates, which significantly limited their practical applications.

1.2.2 First Generation of PEO Technology

The breakthrough occurred in the early 1980 due to a series of works by Markov et al. [22–24], who developed electrolytic methods for discharge-assisted deposition of oxide coatings on Al driven by pulse reverse and asymmetric AC polarisation as well as a simple capacitive power supply device for the latter process (Fig. 1.2). The 50 Hz AC process termed microarc oxidation (MAO) has turned out to be particularly attractive thanks to its simplicity, excellent throwing power, utilisation of a low-cost environmentally benign electrolyte based on a dilute alkaline solution of technical liquid glass and some other advantages. But most importantly, it provided an unmatched combination of coating mechanical properties, including extremely high hardness (up to 23–25 GPa) and adhesion to the metal substrate. This was achieved due to AC polarisation promoting formation of the inner dense coating

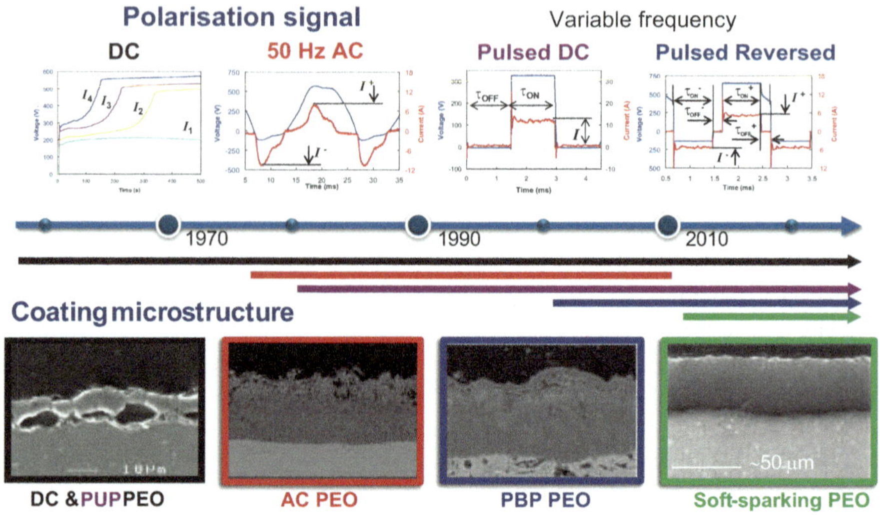

Fig. 1.2 Historical evolution of polarisation signal leading to morphological refinement of PEO coatings on Al. PUP and PBP stand for pulsed unipolar and pulsed bipolar polarisation modes, respectively

region enriched with hard α-Al_2O_3 phase, which normally requires temperatures above 1000 °C to occur. Initial tribological evaluations carried out by the group of Fedorov in Moscow Oil and Gas Institute have been very promising [25], which ignited interest in these coatings in many sectors beyond foundation industries, including manufacturing, aerospace, defence, transportation, biomedical and electronic devices [26].

Extensive research was subsequently initiated in several institutions in the Soviet Union, and following its collapse in the early 1990s, these studies have propagated quickly into Europe, Asia and North America. Whilst practical benefits of the MAO process have been clear, much of basic information was missing mainly due to a lack of appropriate experimental tools for diagnostics of high-voltage electrochemical processes. Applied research and development therefore prevailed over fundamental studies, which was reflected in the terminology coined during that period. Some terms such as MAO, ASD or ANOF (anodische Oxidation unter Funkenentladung, adopted in Germany [27]) were merely a result of phenomenological observations and seeming analogies with sparks and arcs observed elsewhere rather than reflecting the true physics of microdischarge phenomena, whereas many others were simply tradenames and marketing terms. Often, this caused confusion and hindered acceptance of the new research area by plasma physics community. To disambiguate nomenclature, more generic terms 'Plasma Electrolytic Oxidation' and 'microdischarge' introduced by the author have been broadly adopted although some earlier terms are still being used, especially in China, to refer to essentially the same group of processes.

1.3 State of the Art

1.3.1 Pulsed Bipolar PEO Technology

The success of Markov's development is reflected in the fact that it remained the main PEO technology commercially available for the treatment of Al alloys up to mid-2000s, with a number of proprietary versions of the 50 Hz process having been developed during that time and some lingering till nowadays. However, its main drawbacks have quickly become apparent. Firstly, a dependence of the impedance of capacitive power supply on the electrical load caused difficulties in process control and scaling up and down. Secondly, a possibility to control coating morphology by adjusting characteristics of discharge events was limited by the AC waveform linked to the mains frequency. An obvious solution to these problems would be an application of polarisation waveforms with explicit control over pulse magnitudes and durations for both positive and negative bias, in analogy with pulse and pulse reverse electrodeposition, which has been developed since the 1950s [28]. However, no off-the-shelf power supply solutions covering the required ranges of voltage, current and frequency existed at the time, which hindered research progress in that direction.

In the early 2000s, advancements in solid-state electronics enabled using H bridges with appropriate specifications to synchronise outputs from two high-voltage DC sources combining them into a bipolar pulsed waveform with ±1 kV magnitude and the minimum period of 20 μs corresponding to the maximum 50 kHz pulse frequency [29, 30]. These were utilised in adopting power supplies previously used for electroplating and vacuum coatings to the needs of PEO processing. Initial studies of the second generation of PEO technology have shown that increasing pulse frequency from 50 Hz to 1 to 3 kHz provided major advantages to the characteristics of both the process and resulting coatings [31]. It was possible to increase coating growth rate from 1 to 2…3 μm/min while maintaining a similar specific energy consumption ~10 kW h/(μm m^2) and expand the fraction of the inner dense region to about 90% of the total coating thickness while retaining sufficiently high hardness and adhesion. This has significantly reduced the need for surface finishing essential for tribological applications and made PEO coatings more competitive against both hard anodic oxide films and plasma-sprayed alumina coatings [32].

Since the second half of 2000s, research and development in the field of PEO have grown significantly, spreading from Al to other valve metals, primarily Mg [33], where it has become the main corrosion protection technique and Ti [34] where it has proved to be viable in biomedical applications. Studies have started to emerge on PEO treatments of Zr, Ta and Nb alloys [35–37]. Electrolytic plasma processing for coating of non-valve metals such as steels, Cu and Zn alloys has also begun to be explored [38]. The number of research articles in archival journals increased steadily from dozens to several hundred per annum [39], with accelerating trend sustaining till nowadays. Reports on practical applications of PEO coatings have also grown significantly [40], testifying commercial feasibility of the advanced process.

At the same time, heuristic approaches prevailing in experimental studies highlighted problems associated with complexity of the pulsed bipolar PEO technology. Despite the possibility of explicit control over polarisation waveform, a large number of variables embedded in the pulsed bipolar signal rendered process optimisation based on factorial experimental design impractical. Further improvements in power supply hardware and control algorithms enabled yet more complex waveforms containing pulse trains to be used for PEO treatments, which yielded some interesting results [41]. However, these attempts were limited by insufficient understanding of the behaviour of anodic oxides under such conditions. Due to high specific energy consumption per mole of oxide formed and demanding power density requirements the second generation of PEO treatments of Al remained to be seen less competitive than acid anodising processes in a broad range of applications beyond hard coatings. For Mg though this appeared to be less of a problem, especially when using pulsed current for short-term 'flash-PEO' treatments that showed promise in replacing hazardous chromate conversion coatings [42].

1.3.2 Soft Sparking PEO

One of the most interesting and practically important phenomena observed during PEO treatments of Al is a spontaneous transition to a so called 'soft sparking' [43], which occurs under AC or pulsed bipolar polarisation with a high ratio of cathodic to anodic current, $R > 1$. Galvanostatic growth of PEO coatings is usually accompanied with reduction in the number of discharge events that become larger and more powerful. This may coarsen coating morphology, increase surface roughness and defect density including porosity and thermal cracks, which is detrimental to its protective properties. Under soft sparking however, anodic voltage drops, and powerful discharges are replaced with faint glow or periodic light flashes featuring diffuse emission spectra with occasional lines of elements with high electron transition probabilities (K, Na, H). The coating growth rate increases at the expense of the inner dense region, which becomes enriched in hard α-alumina without significant changes in surface morphology. Together with the reduction of anodic voltage this creates important prerequisites for development of the new generation of energy-efficient PEO processes. Similar phenomena have been observed on other valve metals [44], and research is currently underway to understand the fundamental physics underlying soft sparking transition and how it can be facilitated.

A feasible way was proposed recently [45] to accelerate the transition to soft sparking and ultimately eliminate discharge at all, by gradually reducing initially high current ratio R as the PEO treatment progresses. Relatively thin (15 to 30 μm) yet dense and fully crystalline alumina layers featuring hard α-Al_2O_3 phase formed without apparent contribution of plasma discharge can be produced using this approach. The formation of high-temperature alpha alumina under such conditions is rather intriguing and can be explained by field-assisted crystallisation enhanced by Joule heating of the oxide layer. Coating growth under such conditions appears to follow a laminar pattern described by a simple plug-flow rector model, which makes it more predictable compared to the reformative growth under the influence of localised microdischarges in conventional PEO.

1.3.3 Digitalisation of PEO Processes

Recent studies highlighted the increasing role of complex polarisation waveforms and adaptive process control algorithms in contemporary PEO processes [46]. This makes digitalisation instrumental in the development of the next generation PEO technology, which will increasingly rely on digital twins incorporating advanced process diagnostic and control methods based on physics informed models of PEO coating formation and elements of machine learning [47]. With smart arbitrary waveform generators being developed and adopted to high-voltage electrochemistry [48], one of the most attractive approaches to digitalisation of PEO processes consists in application of intelligent polarisation signals to simultaneously drive and diagnose coating formation process in real time (Fig. 1.3).

Non-invasive methods of electrical identification including pulse transient analysis [49], impedance spectroscopy [50] and voltammetric techniques [51] have shown to be successful in obtaining new information on charge transfer processes underpinning growth of PEO coatings. This data can be linked through appropriate electrical circuit models to interfacial mass transfer processes described by existing theories of anodic film growth adopted to PEO conditions. In domains where implementation of physically justified models is hindered by fundamental uncertainties

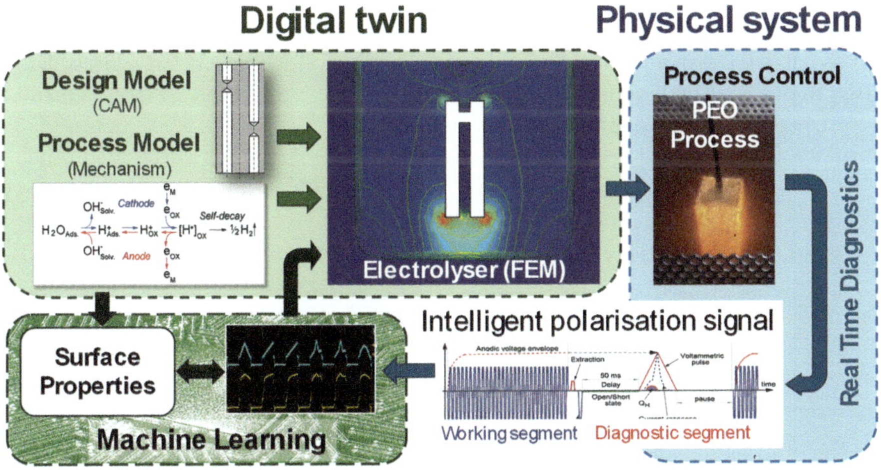

Fig. 1.3 Digitalisation of PEO technology based on application of intelligent polarisation signal

and missing information, elements of artificial intelligence and machine learning could be used to simulate surface layer microstructure and predict properties relevant to intended applications. However, practical implementation of this approach faces challenges associated with the lack of fundamental understanding of the electrochemical aspect of PEO processes. A brief snapshot of current understanding is provided in Sect. 1.4 of this chapter.

1.4 Electrochemical Aspects

Currently, the largest cluster of missing information is related to the behaviour of anodic oxide films on valve metals under pulsed reversed polarisation conditions. Existing theories of passivation and high-field growth of anodic oxides often assume steady-state conditions under constant electrode potential or constant electric field across the oxide film [52], whereas transient phenomena are usually taken into account via capacitive effects associated with charging of electrical double layer and dielectric relaxation of oxide material [53]. Processes occurring in aqueous electrolytes under reverse bias are often reduced to hydrogen evolution on oxidised metal electrodes without proper account of their reactivity and amphoteric properties.

1.4.1 Anodic Oxide Growth

Reactions of valve metals with oxygen are usually accompanied with large negative changes of Gibbs free energy, i.e. oxide formation occurs spontaneously. External polarisation in anodising treatments is therefore required exclusively to maintain appropriate kinetics of oxide film growth. If the oxidation of a valve metal, such as Al, is carried out in aqueous alkaline solutions, the following set of reactions can be used to describe the anodic process:

At the metal-oxide interface

$$\text{Electrochemical step}: \mathrm{Al}(met) \rightarrow \mathrm{Al}^{3+}(ox) + 3\mathrm{e}^- \tag{1.1}$$

At the oxide electrolyte interface

$$\text{Adsorption}: \mathrm{OH}^-(aq) \rightarrow \mathrm{O}^{2-}(surf) + \mathrm{H}^+(aq) \tag{1.2}$$

$$\text{Electrochemical step}: \mathrm{O}^{2-}(surf) \rightarrow \mathrm{O}(surf) + 2\mathrm{e}^-(ox) \tag{1.3}$$

Recombination and desorption steps leading to:

$$\text{Oxygen evolution}: 2\mathrm{O}(surf) \rightarrow \mathrm{O}_2(g)\uparrow \tag{1.4}$$

$$\text{Metal dissolution}: Al^{3+}(surf) + 4OH^-(aq) \rightarrow Al(OH)_4^-(aq) \tag{1.5}$$

Within the oxide layer

$$\text{New oxide formation}: 2Al^{3+}(ox) + 3O^{2-}(ox) \rightarrow Al_2O_3(ox) \tag{1.6}$$

Relative kinetics of these reactions define a competition between three main processes resulting in the growth of anodic oxide (1.1, 1.2 → 1.6), metal dissolution (1.1 → 1.5), and oxygen evolution (1.2 → 1.3 → 1.4).

The dissolution reaction (1.5) is promoted in situations where the formation of continuous surface oxide film is compromised, e.g. under highly alkaline conditions in the vicinity of Al electrodes (as well as in aggressive acidic solutions, although yielding Al^{3+} rather than oxyanion). In alkaline solutions containing oxyanions of metals the dissolution reaction (1.5) often results in formation of gel deposits [54]. The adsorption of hydroxyl anions (1.2) can also be locally affected by presence of halide anions leading to pitting corrosion. Anodic dissolution can be significant on other metals, in particular Mg, whose oxide is not protective and the main native passivating agent, $Mg(OH)_2$, becomes unstable in non-passivating electrolytes at pH < 13. Additionally, chemical dissolution, where the redox process occurs with direct electron exchange between metal and electrolyte species, e.g. Mg(met) + $2H^+$ (*aq*) → Mg^{2+}(*aq*) + H_2(*g*), can potentially increase the apparent yield of anodic dissolution reaction. However, the contribution of such processes at high anodic potentials is usually considered to be rather small for most practically important valve metals. Under such conditions, field-assisted ejection (1.5), which relies upon a supply of metal cations (driven by electric field) at the oxide–electrolyte interface is deemed to play the key role [55].

Oxygen evolution requires electronic conductivity through the oxide film (1.3), e.g. in TiO_2 or in anodic films formed on other metals above breakdown voltage, and a weaker bonding in the oxide than in O_2. These attributes are characteristic of noble metal oxides, whereas on valve metal electrodes the oxygen evolution reaction is normally inhibited up to breakdown voltages. Above those however, it becomes the main anodic electrochemical process [56].

In general, the growth of anodic oxide films can be controlled by interfacial steps (1.1), (1.2) as pustulated respectively in Cabrerra-Mott [57] and Dewald [58] theories, or by the bulk reaction (1.6) itself, Verwey theory [59]. The latter is commonly accepted to be the case for valve metal oxides and treated in terms of the point defect model [60] which assumes that the rate-determining step of anodic oxide film growth is related to the transport of non-reacting vacancies of metal cations and oxide anions across the film under high field conditions. This model has found the most convincing experimental confirmations and is in agreement with the empirical Eq. (1.7) proposed by Güntherschulze [15], providing exponential dependence of anodic current on the electric field across the barrier anodic oxide:

$$i = i_0 \exp\left(\frac{\beta \, \Delta U}{\delta}\right), \tag{1.7}$$

where i_0 and β are constants and ΔU is the voltage drop across the barrier oxide with thickness δ.

A competition between the kinetics of reactions (1.5) and (1.6) determines three possible scenarios of morphological evolution in the growing anodic oxide film (Fig. 1.4). If dissolution prevails, a thin (few nm) oxide film does not grow, and the oxide-electrolyte interface remains stable with increasing cell voltage up to the point of dielectric instability leading to the development of interfacial glow discharge. In the case of negligible dissolution, stable growth of the barrier anodic oxide is sustained up to the point of dielectric breakdown leading to the appearance of localised microdischarges at the film surface. The nature of dielectric instabilities and breakdown resides outside the scope of this chapter and will be discussed elsewhere in this book. Here, it will be sufficient to note that for typical values of

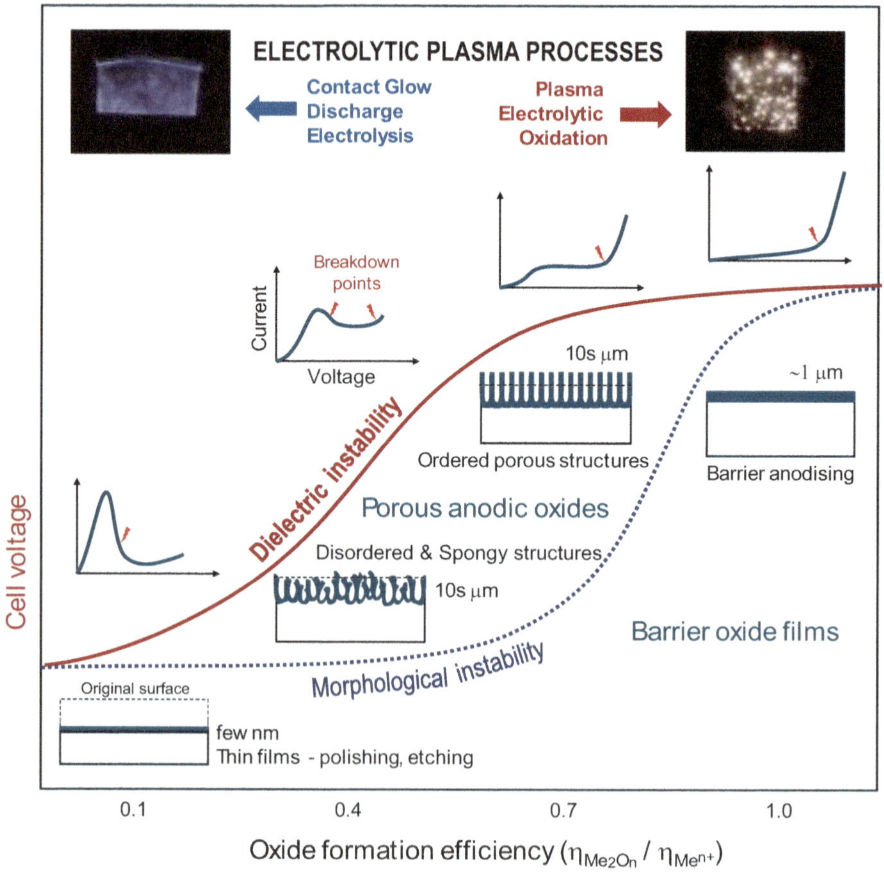

Fig. 1.4 Relationships between morphological evolution and dielectric instabilities in anodic oxides during high filed growth depending on the ratio of current yields $\eta_{Me^{n+}}$ and $\eta_{Me_2O_n}$ of reactions (1.5) and (1.6), respectively, leading to different types of electrolytic plasma processes

breakdown voltage (400–600 V), the maximum barrier film thickness under high-field conditions ($\frac{\Delta U}{\delta}$ = 0.6–0.7 V/nm) does not exceed 1 μm.

If the kinetics of oxide formation and dissolution are comparable the films tend to experience morphological instabilities at the oxide-electrolyte interface resulting in the formation of nanoscale porous structures. In most cases, these structures are disordered or spongy. However in the narrow range of oxide formation efficiencies, η_{Me2On}~0.7, the films may self-organise into ordered linear porous structures typically associated with acid anodising [61]. With proper cooling, these porous layers can grow much thicker than barrier films, up to hundreds of microns, without experiencing dielectric instabilities. However, such growth mode is not typically observed in the films formed in alkaline solutions used for PEO treatments as the reduction of surface pH associated with reaction (1.2) would normally stabilise valve metal oxides.

1.4.2 Processes Under Reverse Bias

Anodic reactions on valve metals are usually irreversible, and subsequent cathodic polarisation of oxide-coated electrodes triggers a different set of reactions. In aqueous alkaline solutions, transition metal cations are hydrated and cannot be directly reduced to metals; the main cathodic reaction is therefore hydrogen evolution:

$$2H_2O + 2e^- \rightarrow H_2(g) + 2OH^- \tag{1.8}$$

Reaction (1.8) is a two-step process involving formation oxide-bonded H intermediates and hydroxyl ions. The latter may cause surface alkalisation and hydrate the outer region of the barrier layer in the oxide film. This in turn would promote intercalation of hydrogen intermediates in the amorphous oxo-hydroxide structure (Fig. 1.5 [62]). An important consequence of these processes is that it changes electrical properties of the barrier oxide [63] and reduces the thickness of the dielectric part of the barrier layer. Recent studies using dynamic voltammetry of the PEO process show that relaxation processes during pause and subsequent H deintercalation have significant influence on anodic behaviour of the oxide film in the following on polarisation cycle [64], which may provide plausible explanation to several unusual phenomena observed during PEO treatments of valve metals, including transition to the soft sparking regime.

1.5 Conclusions and Outlook

A brief overview of past developments and current trends in the field of electrolytic plasma processing has highlighted main practical drivers, fundamental challenges and some unique capabilities of Plasma Electrolytic Oxidation. Despite many

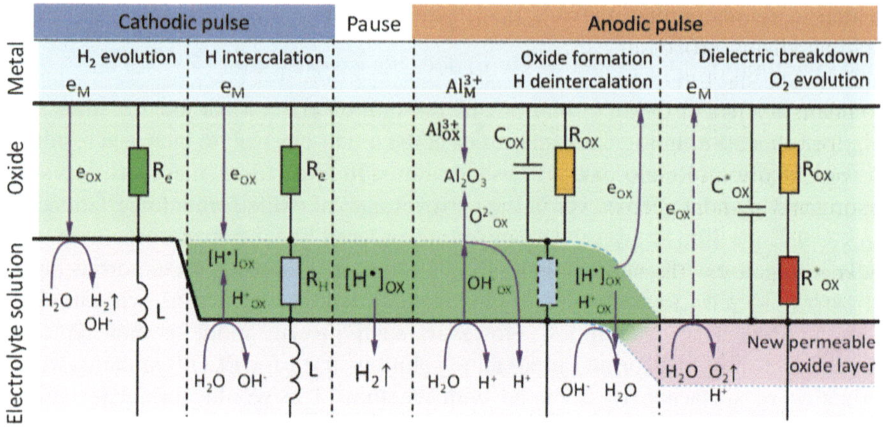

Fig. 1.5 Main electrochemical reactions taking place during bipolar PEO treatment of Al and corresponding equivalent circuit elements [62]. *Reproduced with permission from [62] © 2022 The Authors. Published by Elsevier B.V.*

prerequisites that existed in the 1970s, the development of the first generation of 50 Hz AC microarc oxidation process by Markov was a result of speculative rather than physics informed studies. Physics and chemistry of PEO process are very complex, and fundamental understanding of the mechanisms underpinning formation of PEO coatings is hindered by a lack of experimental tools available for probing electrochemical behaviour of anodic oxides under high-voltage polarisation by transient pulsed reverse waveforms. Current technological developments are mostly informed by phenomenological observations, progress in power supply hardware and process control algorithms.

Future progress in PEO processing is inseparably linked to digital technology combining physics-informed experimentation, advanced instrumentation and modelling with elements of artificial intelligence and machine learning. Recent progress in fundamental understanding of the behaviour of anodic oxide films during transition to the soft sparking PEO modes has been achieved due to application of intelligent polarisation signals incorporating diagnostic segments into working pulsed bipolar waveforms. This behaviour appears to be influenced by dynamic rearrangements in the interfacial barrier layer of anodic oxide caused by periodic intercalation and deintercalation of H species under alternating cathodic and anodic polarisation. Although the barrier layer occupies a small portion of PEO coating situated at the interface with metal substrate its behaviour seems to influence both structural and morphological transformations in the whole PEO coating. The fact that these rearrangements under soft sparking conditions can be described by a relatively simple plug flow reactor model allows more hope for mechanistic understanding to be incorporated into feedback loops used for process control in the future.

Adaptive control modes promoting earlier transition to the soft sparking allow reducing specific energy and power density requirements for coating formation, which increases competitiveness of PEO treatments compared with conventional acid anodising processes used for corrosion protection of Al. It could therefore be anticipated that future generations of digitally controlled energy efficient and environmentally friendly PEO processes would shape up the mainstream technology platform for development of high-performance, smart and multifunctional coatings on practically important light alloys to successfully address future industrial, economic and societal needs.

Acknowledgments The author is grateful to all former and current PhD students, postdoctoral researchers, academic colleagues and industrial collaborators who contributed to the development of ideas and thoughts discussed in this publication. Current research has been supported by UK EPSRC (grant EP/T024607/1, 'CoatIN') and is acknowledged with thanks.

References

1. Trasatti, S. (1999). *Journal of Electroanalytical Chemistry., 476*, 90–91.
2. Fizeau, H., & Foucault, L. (1844). *Annales de Chimie et de Physique, 11*, 370–383.
3. Sluginov, N. (1880). *Journal of Russian Physical-Chemical Society, 12*, 193–203.
4. Yerokhin, A., Mukaeva, V. R., Parfenov, E. V., Laugel, N., & Matthews, A. (2019). *Electrochimica Acta, 312*, 441–456.
5. Yasnogorodsky, I. Z. (1949). Heating metals and alloys in electrolyte, Mashgiz, Moscow.
6. Kellogg, H. H. (1950). *Journal of The Electrochemical Society, 97*, 133.
7. Hickling, A., & Ingram, M. D. (1964). *Transactions of the Faraday Society., 60*, 783–793.
8. Duradzhi, V., & Parsadanyan, A. (1988). *Nagrev metallov v elektrolitnoi plazme (Heating of metals in Electrolyte Plasma)*. Shtiintsa.
9. Sen Gupta, S. K. (2017). *Plasma Chemistry and Plasma Processing, 37*, 897–945.
10. Belkin, P. N., Yerokhin, A., & Kusmanov, S. A. (2016). *Surface and Coatings Technology, 307*, 1194–1218.
11. Belkin, P. N., Kusmanov, S. A., & Parfenov, E. V. (2020). *Applied Surface Science Advances, 1*, 100016.
12. Nestler, K., Böttger-Hiller, F., Adamitzki, W., Glowa, G., Zeidler, H., & Schubert, A. (2016). *Procedia CIRP, 42*, 503–507.
13. Buff, H. (1857). *Annalen der Chemie und Pharmacie, 102*, 265–284.
14. Wohler, F., & Buff, H. (1857). *Annalen der Chemie und Pharmacie, 103*, 218–229.
15. Güntherschulze, A., & Betz, H. (1937). Elektrolytkondensatoren. In *Ihre Entwicklung, Wissenschaftliche Grundlage, Herstellung, Messung und Verwendung* (p. 178). Krayn, Moritz, Technischer Verlag.
16. Elektrotechnische Industrie, assignee. (1910). Verfahren, Draht aus Aluminium oder verwandten Metällen mit einer Isolationsschicht von hoher mechanischer,thrmischer und elektrischer Widerstandsfähigkeit zu versehen. Germany patent DE263603C. 28 June 1910.
17. Bengough, G. D., & Stuart, J. M. (1924). Improved process of protecting surfaces of aluminium of aluminium alloys. UK patent GB223994A. 2 August, 1923.
18. Tomashov, N. D.(1946). Anodic oxidation and its application in the construction of Engines. Vestnik Inzhenerov i Tekhniskov. p. 59.
19. McNeill, W. (1958). *Journal of The Electrochemical Society, 105*, 544.
20. Hradčovský, R., & Bělohradský, Z. (1962). Způsob výroby isolační křemičitanové vrstvy na hliníkové destičce pro zkratové můstky světelných. Československo patent 104927. 15 March 1962.

21. Brown, S. D., Kuna, K. J., & Van, T. B. (1971). *Journal of the American Ceramic Society, 54*, 384–390.
22. Markov, G. A., Gizatulin, B.S., & Rychazhkova, I. B., & Inventors. (1982). Method for electrolytic deposition of silicate coatings. USSR patent SU 926083. 1982.
23. Markov, G. A., Shulepko, E. K., Terleeva, O. P., Kirillov, V. I., Fedorov, V. A., & Kan, A.G., et al. (1987). Device for conversion of alternating voltage into asymmetric alternating one. USSR patent SU 1339818. 23 September 1987.
24. Markov, G. A., Shulepko, E. K., & Terleeva, O. P. (1989). Method for coating deposition on metals and alloys. USSR patent SU 1200591. 7 April 1989.
25. Petrosyants, A. A., Malyshev, V. N., Fedorov, V. A., & Markov, G. A. (1985). *Trenie Iznos, 101*, 360.
26. Kurze, P., Krysmann, W., & Schneider, H. G. (1986). *Crystal Research and Technology, 21*, 1603–1609.
27. Dittrich, K.-H., Krysmann, W., Kurze, P., & Schneider, H. G. (1984). *Crystal Research and Technology, 19*, 93–99.
28. Popov, K. I., & Maksimovic, M. D. (1989). Theory of the effect of electrodeposition at a periodically changing rate on the morphology of metal deposits. In B. E. Conway, J. O. M. Bockris, & R. E. White (Eds.), *Modern aspects of electrochemistry. 19* (pp. 193–250). Premium Press.
29. Mark, G., & MELEC, GMbH. (2000). Current supply apparatus for bipolar current supply of installation of plasma or surface technology, has DC power packs connected via series circuit of power switches, for separate bridge circuits. Germany patent DE10018879A1. 17 April 2000.
30. Canacsinh, H., Redondo, L. M., & Silva, J. F. (2012). *IEEE Transactions on Plasma Science, 40*, 2603–2610.
31. Yerokhin, A. L., Shatrov, A., Samsonov, V., Shashkov, P., Pilkington, A., Leyland, A., et al. (2005). *Surface and Coatings Technology, 199*, 150–157.
32. Wheeler, J. M., Curran, J. A., & Shrestha, S. (2012). *Surface and Coatings Technology, 207*, 480–488.
33. Blawert, C., & Bala Srinivasan, P. (2010). Plasma electrolytic oxidation treatment of magnesium alloys. In H. Dong (Ed.), *Surface engineering of light alloys* (pp. 155–183). Woodhead Publishing.
34. Paital, S. R., & Dahotre, N. B. (2009). *Materials Science and Engineering: R: Reports, 66*, 1–70.
35. Wang, C., Wang, F., & Han, Y. (2016). *Applied Surface Science, 361*, 190–198.
36. Stojadinović, S., Tadić, N., Radić, N., Stefanov, P., Grbić, B., & Vasilić, R. (2015). *Applied Surface Science, 355*, 912–920.
37. Yan, Y., Han, Y., & Huang, J. (2008). *Scripta Materialia, 59*, 203–206.
38. Nie, X., Cai, R., Zhao, C., Sun, J., Zhang, J., & Matthews, D. T. A. (2022). *Surface and Coatings Technology, 442*, 128403.
39. Clyne, T. W., & Troughton, S. C. (2019). *International Materials Reviews, 64*, 127–162.
40. Part III: Applications for surface engineered light alloys. In: Dong H, editor. Surface engineering of light alloys: aluminium, magnesium and titanium alloys. Metals and surface engineering. : Woodhead Publishing; 2010. p. 549-641.
41. Rogov, A. B., Yerokhin, A., & Matthews, A. (2018). *Journal of Physics D: Applied Physics, 51*, 405303.
42. Wierzbicka, E., Vaghefinazari, B., Lamaka, S. V., Zheludkevich, M. L., Mohedano, M., Moreno, L., et al. (2021). *Corrosion Science, 180*, 109189.
43. Jaspard-Mécuson, F., Czerwiec, T., Henrion, G., Belmonte, T., Dujardin, L., Viola, A., et al. (2007). *Surface and Coatings Technology, 201*, 8677–8682.
44. Rogov, A. B., Huang, Y., Shore, D., Matthews, A., & Yerokhin, A. (2021). *Ceramics International, 47*, 34137–34158.
45. Bousser, E., Rogov, A., Shashkov, P., Gholinia, A., Laugel, N., Slater, T. J. A., et al. (2023). *Acta Materialia, 244*, 118587.

46. Parfenov, E. V., Yerokhin, A., Nevyantseva, R. R., Gorbatkov, M. V., Liang, C. J., & Matthews, A. (2015). *Surface and Coatings Technology, 269*, 2–22.
47. Fernández-López, P., Alves, S. A., Rogov, A., Yerokhin, A., Quintana, I., Duo, A., et al. (2024). *Coatings., 14*, 979.
48. Rogov, A. B. (2019). *Instruments, 3*, 13.
49. Gao, Y., Yerokhin, A., Parfenov, E., & Matthews, A. (2014). *Electrochimica Acta, 149*, 218–330.
50. Yerokhin, A., Parfenov, E. V., & Matthews, A. (2016). *Surface and Coatings Technology, 301*, 54–62.
51. Rogov, A. B., Matthews, A., & Yerokhin, A. (2019). *Electrochimica Acta, 317*, 221–231.
52. Lee, W., & Park, S.-J. (2014). *Chemical Reviews, 114*, 7487–7556.
53. Lohrengel, M. M. (1993). *Materials Science and Engineering: R: Reports, 11*, 243–294.
54. Morlidge, J. R., Skeldon, P., Thompson, G. E., Habazaki, H., Shimizu, K., & Wood, G. C. (1999). *Electrochimica Acta, 44*, 2423–2435.
55. Li, C., Ni, Y., Gong, J., Song, Y., Gong, T., & Zhu, X. (2022). *Nanoscale Advances, 4*, 322–333.
56. Snizhko, L. O., Yerokhin, A. L., Pilkington, A., Gurevina, N. L., Misnyankin, D. O., Leyland, A., et al. (2004). *Electrochimica Acta, 49*, 2085–2095.
57. Cabrera, N., & Mott, N. F. (1949). *Reports on Progress in Physics, 12*, 163.
58. Dewald, J. F. (1955). *Journal of The Electrochemical Society, 102*, 1.
59. Verwey, E. J. W. (1935). *Physica, 2*, 1059–1063.
60. Macdonald, D. D. (2011). *Electrochimica Acta, 56*, 1761–1772.
61. Hebert, K. R., Albu, S. P., Paramasivam, I., & Schmuki, P. (2012). *Nature Materials, 11*, 162–166.
62. Rogov, A. B., Nemcova, A., Hashimoto, T., Matthews, A., & Yerokhin, A. (2022). *Surface and Coatings Technology, 442*, 128142.
63. Ivker, M., & Agranat, A. J. (2004). *Journal of Applied Physics, 96*, 6405–6408.
64. Rogov, A. B., Matthews, A., & Yerokhin, A. (2020). *The Journal of Physical Chemistry C, 124*, 23784–23797.

Chapter 2
Effects of Electrolyte and Electrode Parameters on Plasma Electrolytic Oxidation

Shuqi Wang and Yaming Wang

2.1 Introduction

The plasma electrolytic oxidation (PEO, also known as microarc oxidation [1]) is the preparation of protective ceramic coatings on various valve metals, using plasma discharge-assisted oxidation, that improves surface performances, such as structural reliability and functional properties [2, 3]. During the PEO process, a high voltage is applied to the metal surface to cause breakdown discharge, forming the plasma discharge state accompanying high pressure and temperature that be imparted to the anode substrate surface in the electrolyte, leading to the occurrence of functional ceramic coating with good adhesive strength to the substrate [4]. The coating composition is mainly composed of the substrate element oxides and the elements contained in the electrolyte participating in the reaction to form non-substrate oxides. The composition, structure and properties of the PEO coating can be optimized by the selection of electrolyte components and the matching of electrical parameters, to obtain a series of special functional properties (such as wear resistance, corrosion resistance, thermal protection, thermal control, dielectric insulation, catalysis, biological performances, etc.) [5]. Therefore, it is the key to design and prepare excellent functional coatings to determine the influencing factors of PEO process parameters. The growth rules and microstructure of PEO coating are mainly divided into internal factors (substrate natural characteristics, electrolyte composition, pH, and conductivity) [6] and external factors (power supply types (AC, DC, pulsed DC), electrical parameters and solution temperature) [7]. Among them, the

S. Wang (✉) · Y. Wang (✉)
Institute for Advanced Ceramics, Harbin Institute of Technology, Harbin, China

Key Laboratory of Advanced Structure-Function Integrated Materials and Green Manufacturing Technology, Harbin Institute of Technology, Harbin, China
e-mail: wang-shuqi@hit.edu.cn; wangyaming@hit.edu.cn

V. S. Saji, M. Zheludkevich (eds.), *Plasma Electrolytic Oxidation – Fundamentals, Advances and Applications*,
https://doi.org/10.1007/978-3-031-92429-3_2

composition and proportioning ratio of the electrolyte and the applied electrical parameters play a decisive role in the PEO process. This means that the coating with special phase and structure can be obtained through the design of the above parameters to meet the customized demand for a variety of functional coatings. On the other hand, the natural characteristics of different substrates (such as electrical conductivity, microstructure, defects, second phase, and relative phase content, etc.) also have a great influence on the PEO process, and sometimes it is necessary to use PEO pretreatment, special solutions and electrical parameters to ensure the coating uniform growth.

Hence, this chapter mainly focuses on the growth process and microstructure evolution of the coating from the aspects of power supply regime, electrical parameters, electrolyte composition, and substrate materials. Finally, the existing issues for uniform growth of customized coatings are presented, and their future perspectives are prospected.

2.2 Effect of Power Supply Output Regime on Coating Microstructure

2.2.1 Power Supply Output Mode

The investigation of effect of power supply regime on the structure and properties of coating is an important direction in the development of PEO process. In 1932, *Guntershculzte* and *Betz* [8] first reported the discharge spark phenomenon on the metal surface under a high electric field, and forming an oxide coating. However, due to the backward power supply, the discharge characteristics were difficult to adjust, which limited the controllability of PEO process, so it was only used in the preparation of simple parts with thin coating. In the 1970s, Van et al. [9] used unipolar pulse DC power sources to study this technology and named as spark discharge anodizing. Intermittent discharge can be controlled by pulsed DC power supply, so as to tailor the duration of the arc discharge, and the pulse form can also be changed. However, due to the limited development of the power module with a low pulse frequency, the strong spark discharges caused burning damage of the ceramic layer, resulting in that prepared coating being particularly loose without dense structure. In 1977, Ikonopisov et al. [10, 11] adopted the AC power supply mode with a discharge voltage higher than the anodic oxidation voltage, which can form a relatively dense ceramic coating. As well as it is found that the additional polarization of the electrode can be avoided by the AC power supply, and the discharge process can be improved by the pulse discharge interruption. However, the stability and design of the initial pulse power supply were still limited, and the frequency and duty cycle of the power supply cannot be adjusted independently. Since 2000, researchers have continued to study the power mode in more detail, mainly focusing on the investigation of single pulse waveform and bipolar pulse, aiming to improve the peak power, ceramic layer processing efficiency and processing area, and ability to customize different functional coatings. Meanwhile, some works have also been done to restrain the strong discharge, improve the service life of the power supply, and reduce the energy consumption, etc.

Hence, this section summarizes typical power supply regime and their effects on coating growth:

1. DC power supply mode

   The output voltage or current of the DC power supply does not change with the change of time. The discharge voltage generated by DC power supply with high discharge energy is obviously higher than that of AC power supply, which is not conducive to uniform dense growth of coating. The discharge spark generated by DC power supply is large and dispersed, that is easy to form strong arc discharge, which adversely affects the quality and compactness of the coating. In a word, under DC power supply, although the coating grows fast and thick, the coating has many defects, large roughness, and other problems.

2. Pulse power mode

   Compared with the traditional DC power supply, the output waveform, frequency, duty cycle and average current density of the pulse power supply can be adjusted and set separately. It can be divided into unipolar pulse, bipolar symmetrical pulse, and bipolar asymmetric pulse mode.

   (a) Unipolar pulse power supply mode

      The unipolar square wave pulse is obtained by chopping the DC voltage or current, which has the advantages of a simple circuit, operability and low cost. The controllable rectifier bridge can rectify the power frequency voltage to obtain a DC voltage with adjustable voltage amplitude, and filter the voltage ripple through the large capacity capacitor to obtain a stable DC voltage. Meanwhile, the frequency and duty cycle are adjustable, which greatly improves the matching design freedom of electrical parameter. The unipolar square wave pulse is shown in Fig. 2.1a.

   (b) Bipolar symmetrical pulse power supply mode

      Generally, in the unipolar pulse discharge reaction process, the direction of charge accumulation on the workpiece surface and ion movement in the electrolyte is always unidirectional polarity, which is easy to form a polarization effect on the substrate metal, resulting in the substrate unable to release excess charge. Hence, the DC source with adjustable amplitude can be obtained through the rectifier circuit and filter. The symmetrical bipolar pulse with positive and negative bipolar is obtained through the inverter circuit to achieve bipolar symmetrical pulse power supply mode. The waveform is shown in Fig. 2.1b.

   (c) Bipolar asymmetric pulse power supply mode

      A bipolar symmetric pulse can realize the reversal of the polarity of the electric field applied to the load to eliminate the charge polarization effect, further improving the coating quality. However, its pulse output of positive and negative polarity is symmetrical, while the discharge reaction often has

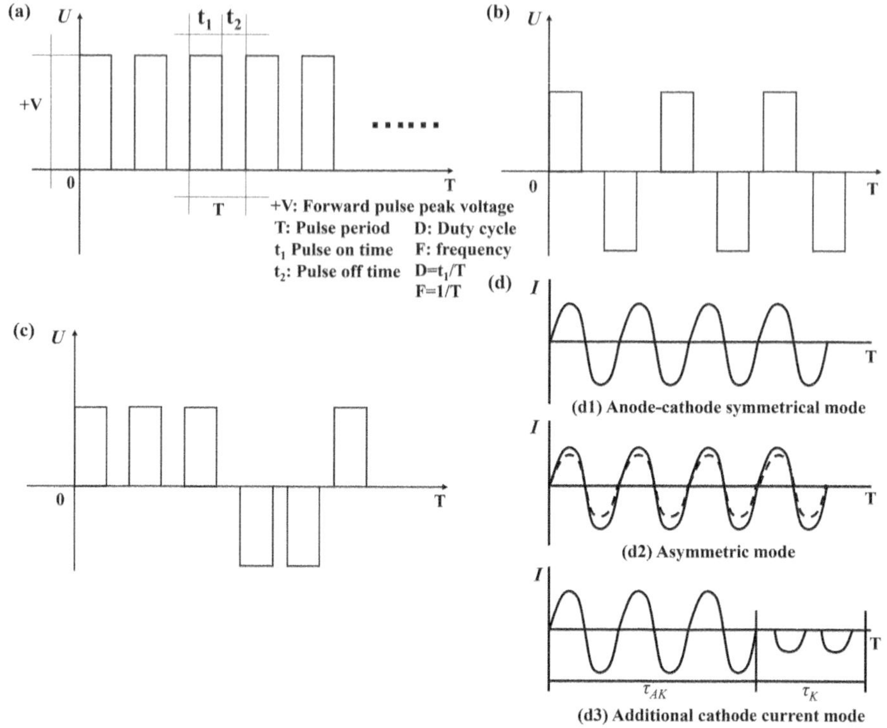

Fig. 2.1 Typical PEO power supply regime:(**a**) unipolar pulse waveform and electrical parameter definition including voltage, frequency, duty cycle; (**b**) bipolar symmetrical pulse waveform; (**c**) bipolar asymmetric pulse waveform; (**d**) pulse power supply with sine waveform

different requirements for the electric field of positive and negative polarity, which restricts the application range of bipolar symmetric pulse, so the bipolar asymmetric pulse mode emerged. The mode has two forms: one is that the square wave pulse amplitude is equal, but the conduction time and the number of positive and negative pulse proportions are inconsistent; Second, the pulse amplitude, duty cycle, frequency, and number of positive and negative pulses are not equal, as shown in Fig. 2.1c.

In short, the current density generated in the pulse power supply mode is stable, accompanied by a long-term microarc discharge stage, which is conducive to the even growth and densification of the coating. At the same time, the white sparks with small and uniform are produced by the pulse power mode, which is easy to form a coating with low porosity. The prepared coating has fewer defects and low roughness. The pulse power supply with lower consumption is conducive to green energy saving and environmental protection.

In addition to square wave (mentioned above), sine wave pulse power supply is always used to prepare PEO coating, where the ratio of anode current to cathode current density (I_a/I_k) is used to represent the basic characteristics of the waveform. The ratio of cathode current to anode current (I_k/I_a) is 1.0 and 1.2, corresponding to Fig. 2.1d1 (bipolar symmetry) and Fig. 2.1d2 (bipolar asymmetry), respectively. The cathode current in the PEO process enables the production of coatings with new characteristics. That is, the growth pattern, rate and quality of the coating can be changed by adjusting the polarity of the current, the cathode and anode currents (Fig. 2.1d2, d3), particularly, the coating quality is closely related to the ratio of anode current to cathode current density (I_a/I_k), final voltage (U) and pulse waveform. Of which, according to the coating process and performance requirements, the anode-cathode type additional cathode current mode (Fig. 2.1d3) can be used to create a self-regulating discharge mode, that is, different amounts, strengths and waveforms of additional cathode current are inserted at different times and positions so that the coating with high quality can grow on demand. It is worth emphasizing that the self-regulation and self-feedback of the power output mode is a development trend of digital and intelligent PEO in the future, and the best waveform, discharge intensity, duration, etc. can be optimized to make the coating grow on demand, so as to truly achieve customization and scale.

2.2.2 *Effect of Electrical Parameters*

The power supply and its output mode determine the energy distribution and duration of a single microarc discharge, whose process significantly affects the rate of oxide formation, and the structural characteristics of the coating surface and interface.

Taking Ti6Al4V alloy as material and bipolar pulse power supply as an example, the effects of process parameters (voltage, current density, frequency, duty cycle, and time) on the coating structure are discussed. Figure 2.2 shows the applied bipolar pulse parameters and waveform. The pulse parameters (such as voltage or current, pulse on and pulse off time, positive and negative pulse working time) can be adjusted independently, which provides a lot of freedom to choose the discharge intensity to grow a coating with specific structure. Generally, the positive pulse of the power supply plays a key role in tailoring the coating structure and thickness. The few negative pulses distributed in the positive pulse beam can interrupt the continuous discharge, allowing the discharge micro-region to cool rapidly, and inducing the soluble component to reconvert to oxide. As shown in Fig. 2.2, we set 20 positive pulses with additional 2 negative pulses, of which the negative pulses distributed in the positive pulse can reduce the discharge energy to form uniform dense plasma discharge.

Combined with the process test results, the effect law of pulse electrical parameters on the growth and microstructure of the PEO coating on Ti alloy can be shown in Table 2.1.

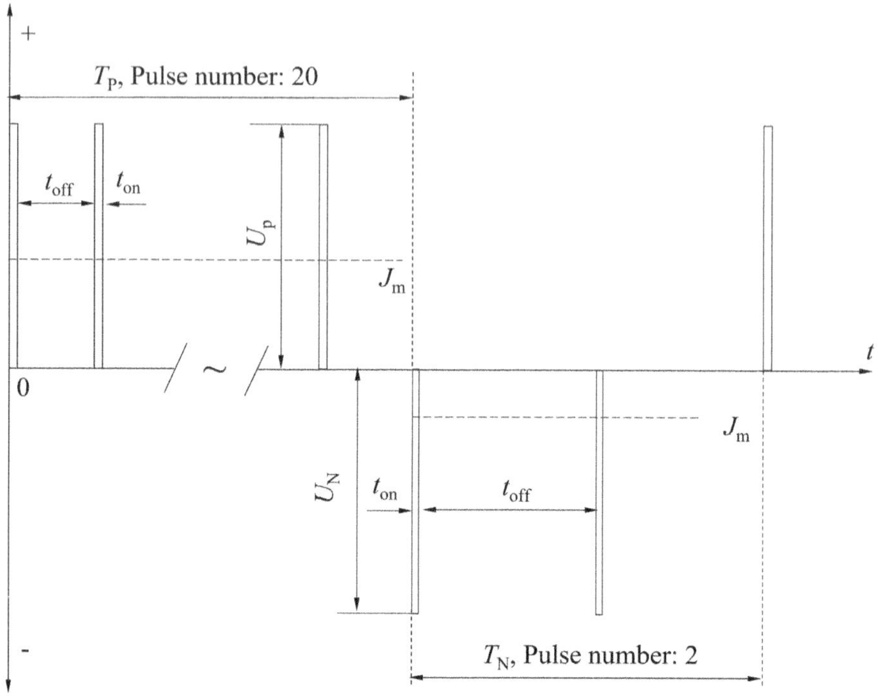

Fig. 2.2 Schematic diagram of pulse output waveform and parameters of PEO bipolar pulse power supply. t_{on} positive pulse on time, t_{off} positive pulse off time, t_{on} negative pulse on time, t_{off} negative pulse off time, T_p positive pulse number, T_N negative pulse number, J_m positive mean current density, J_m negative mean current density, U_p and U_N positive and negative voltage amplitudes, respectively

The coating undergoes repetitive cycles of component synthesis, grain nucleation, growth, and remelting under the influence of cyclic electric pulses during the PEO process. Consequently, the growth rate and surface quality of the coating are mainly determined by the single pulse energy (E_p, Eq. 2.1) [12]:

$$E_p = \int_0^{t_p} U_p I_p \mathrm{d}t, \tag{2.1}$$

where U_p and I_p represent the voltage and the current of the pulse, respectively, and t_p is the pulse on-time.

Increasing pulse voltage, duty cycle, current density or decreasing frequency can enhance the monopulse discharge energy. In this case, the plasma discharge zone temperature increases rapidly, with strong discharge phenomenon, and the amount of molten products of the substrate and coating in the discharge zone increases, which leads to an increase of coating thickness and roughness simultaneously. Moreover, the increase of single pulse discharge energy is conducive to the conversion of anatase to rutile phase TiO_2 in titanium alloy [13]. It is reported that similar

Table 2.1 Effects of pulse electrical parameters on the growth and microstructure of coated Ti alloy

Electrical parameters	Coating growth	Surface morphology	Phase
Voltage, V 350, 400, 450, 500, 550	Voltage↑, coating growth velocity↑ (4–33 μm)	Porosity↓, pore size↑, roughness↑	As the voltage increases, anatase converts to a rutile phase
Frequency, Hz 1800, 1400, 1000, 600, 200	Frequency ↑, coating growth velocity ↓ (6–36 μm)		
Duty cycle 4%, 8%, 12%, 16%, 20%	Duty cycle ↑, coating growth velocity ↑ (10–35 μm)		
Oxidation time, min 15, 30, 60, 120, 180	Coating growth velocity ↑ (10–35 μm)		Rutile is the dominant phase, The phase transition from anatase to rutile is not obvious
Current density, mA/cm^2 40, 60, 80, 100, 120	Current density ↑, coating growth velocity ↑ (17–21 μm)		Rutile is the dominant phase, The phase transition from anatase to rutile is not obvious

results occur in the PEO process of aluminum alloy, with the increase of voltage and current density, the discharge intensity is significantly enhanced, resulting in an increase in the coating thickness and the α-Al_2O_3 content [14].

Further, the above-mentioned power supply, output mode and electrical parameters are summarized as follows: For unipolar pulse power supply, the ceramic coating with dense structure can be prepared by selecting reasonable electrical parameters. Meanwhile, the pore structure, porosity and growth rate can be adjusted within a certain range, whose range is relatively narrow; on the basis of unipolar pulse power supply, each pulse period increases a certain proportion of negative current in addition to the forward current to form a bipolar pulse mode. In this mode, the negative current will affect the coating growth, namely, the oxide film as the cathode can be dissolved through the reduction reaction, and the dissolved ions are randomly distributed in the micropores and micro-cracks. When the anode current passes the next time, these randomly distributed ions form oxide or complex compounds through electrochemical anode reaction and fill the micropores and microcracks to achieve the improvement of coating compactness. In order to further improve the coating compactness, the discharge intensity staged "attenuation" mode is adopted [15]. Specifically speaking, in constant current mode (60 mA/cm^2), the coating grows rapidly under the action of relatively strong discharge spark (high single pulse discharge energy), and a relatively thick and dense layer is formed in the early stage of PEO (Duty cycle of 8%, Frequency of 300–500 Hz). At the late stage of PEO (Duty cycle of 2%, Frequency of 500–800 Hz), the porous structure formed in the early stage is repaired to a certain extent under relatively weak discharge intensity. The staged "attenuation" mode greatly improves the coating compactness, compared with the traditional constant current density, which is related to the change of discharge behavior. That is, the diminished current density in the late discharge period can heal the previously formed micropores. It is worth noting that the "soft" spark discharge regime (see Sect. 2.2.3 for details) is regulated in a bipolar pulse mode to form a uniform and dense discharge spark, thereby simultaneously increasing the thickness and density of the coating.

In addition, based on the nature of plasma discharge, the influence of the relative position between the two electrodes on the PEO process can be attributed to discharge intensity. Results show that the larger the spacing between positive and negative electrodes, the smaller the anode current. In contrast, the current flowing through the front surface of the sample is higher than the back surface, and the high current density induces high discharge intensity. So, the high current density leads to the growth of thicker and harder coatings, which is consistent with the higher wear and corrosion resistance of the front surface of the sample than the back surface.

2.2.3 "Soft" Spark Discharge Regime

The fabrication of a relatively compact and homogeneous coating with good performance in terms of mechanical and corrosion properties is needed for numerous industrial applications. Therefore, much attention has been paid in recent years to suppress the strong destructive discharges by the development of the so-called "soft" plasma discharges at the later stages of the PEO process. The condition would lead to an increase in not only the uniformity and compactness of the coating, but also the hardness and adhesion strength between the coating and its substrate [16–19]. Soft plasma discharges were usually characterized by decrement of the anodic voltage, hysteresis in the transient current–voltage curves, a reduction of the acoustic emissions, and a more uniform distribution of the plasma discharges over the substrate surface. The transition from the strong destructive discharges to the soft plasma discharges was generally discovered at an optimized AC or bipolar PEO conditions having higher cathodic current with respect to the anodic current [20]. Namely, increasing the cathodic charge quantity with respect to the anodic one, while appropriately tuning the duration of the positive and negative current pulses might result in a PEO regime for which no detrimental arc appears anymore, resulting in more homogeneous coatings. Within these conditions, though the first two steps of the process (sparks and next micro-arcs) remain unchanged, the process turns into a phase for which very tiny sparks appear, only observable in the dark. Usually, the "soft" spark discharge transition mode can be determined by the ratio of anode charge to cathode charge (called charge ratio R), the charge transfer rate of the large cathode than the anode ($R < 1$) can lead to the formation of soft discharge.

Moreover, in the "soft" spark discharge mode, the voltage changes with the increase of time, and showing a stable and declining trend (Fig. 2.3) [21, 22], accompanied by a fine discharge spark, which avoids the phenomenon of strong discharge accumulation in the late stage, making the coating uniform and dense growth. At the same time, the anode layer (including the interface layer) is thick with low porosity, and there are no large holes and cracks significantly throughout the coating. After the soft plasma discharge regime is established, the diameter of the molten pool formed around the discharge channel remains almost constant, while under standard conditions (non-soft spark mode), these diameters tend to increase linearly with the processing time. Besides, soft sparks are caused by a large number of fine discharges occurring at the metal/oxide interface, and these discharges do not develop in the usual way, therefore do not reach the coating surface, almost without light emission. And the "soft" spark may represent a reversal of the dominant growth mechanism during anodizing. In this mode, the coating growth efficiency is improved because of the uniform distribution of electric field, but how the electric field redistribution affects the charge transfer mechanism needs to be further explored.

In addition, the formation of the "soft" spark mode is also affected by other factors: higher current density, higher AC power supply frequency, electrolyte composition and concentration (including conductivity, pH, ion species content), and pre-anodized layers (coating thickness playing an important role in the transition process), which will accelerate the speed of the transition to the "soft" spark mode. During the transition to the "soft" spark, the reducing voltage indicates that the power consumed is decreasing. In other words, "soft" spark discharge intensity is

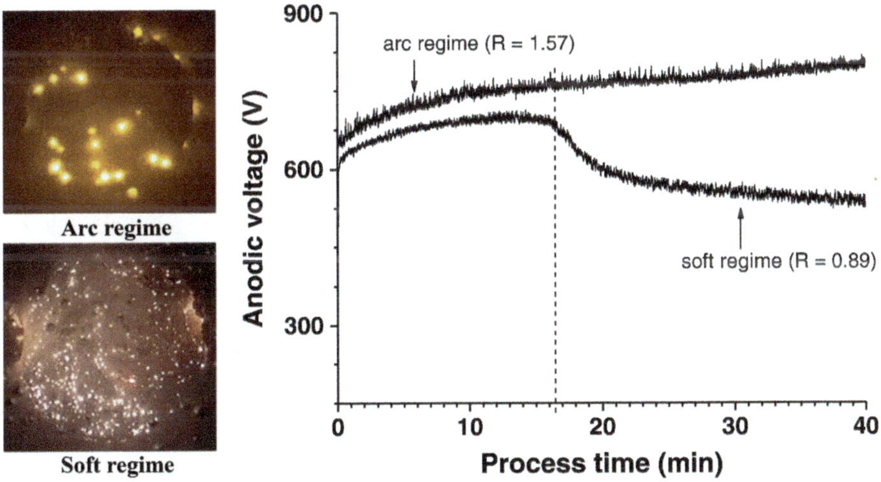

Fig. 2.3 (Left) Morphology of discharge sparks under different modes [21], and (Right) variation of the anodic voltage over the process duration. Within the present conditions, the transition to soft regime appears around 15 min process (vertical dash line) [22]. *Reproduced with permission from* [21] © *2022 Elsevier B.V. and* [22] © *2011 Elsevier B.V.*

relatively small (reduced current density), which is expected to solve the problem of high energy consumption.

2.3 Effect of Electrolyte Composition on Coating Microstructure

The selection of electrolyte composition plays a decisive role in the PEO process. Different components and concentrations of electrolyte show different characteristics (such as pH, conductivity, viscosity, stability, dispersion, film-forming ability, etc.), which have different degrees of influence on ion migration, charge change, oxygen supply, film passivation, dielectric breakdown, arc starting time, discharge voltage, discharge current, spark characteristics (size, number, life, and distribution), and further affecting the growth behavior and properties of ceramic coatings.

2.3.1 Function and Classification of Electrolyte

As an important factor affecting the PEO process, the role of electrolyte can be summarized as follows: ① Promoting metal surface passivation to form a thin dielectric insulating film; ② Forming a conductive solution medium, that transfers the energy required for oxidation reactions occurring at the substrate metal/electrolyte interface; ③ Providing the "O" source required for metal oxidation; ④ Special components of the electrolyte can enter the coating during the PEO process, providing a way to design novel coatings; ⑤ Some of the electrolyte components can selectively dissolve the coating, and improve the coating porosity to form a hierarchical micro-nano structure; ⑥ The non-hydrolytic liquid system can provide the required electrolyte composition for the preparation of non-oxide coatings; ⑦ The development of ionic liquid electrolyte systems will provide greater freedom for the structure/composition regulation of PEO coatings.

Besides, to meet the prerequisites for dielectric breakdown, additives that can promote strong metal passivation (such as silicates, aluminates and phosphates) are widely used in the basic composition of the electrolyte. The three types of strong passivators have the following advantages: ① In the PEO pre-discharge stage, strong passivation additives can make the discharge voltage easily reach the breakdown voltage of the insulation film on the metal surface; ② The components in the electrolyte (such as SiO_3^{2-}, AlO_2^-, and PO_4^{3-}) are easy to enter the coating through polypolymerization or deposition, increasing the growth rate of the coating; ③ Above electrolyte without heavy metal ions is environment-friendly, and can prepare anti-wear and anti-corrosion coatings; ④ Colloidal sodium silicate and potassium silicate solutions as well as sodium silicate-based multi-component electrolytes have been mostly used in PEO industrial production. In fact, it has been proven that even in a single alkaline NaOH solution (acting as passivation and conduction), a discharge can be generated to grow the coating, but a single or simple solution system leads to a slow coating growth rate and high-energy consumption, which is not conducive to the industrial application of the process. Hence, the exploration of composite electrolyte systems and formulations has great absorption for both academic researchers and engineering applications.

To optimize and tailor the coating microstructure, the effects and optimization of the special electrolyte on the microstructure and properties of the coating are discussed in the following sections, respectively, from the basic electrolyte system, electrolyte system with special ions/particles, self-sealing component, and other electrolyte system.

2.3.2 Basic Electrolyte System

The electrolyte system has developed from the early acidic (sulfuric acid, phosphoric acid, or mixed with phosphate, etc.) to the current alkaline, and is widely used. This is due to the fact that in the alkaline electrolyte, metal ions and other metal ions generated by the anode reaction are easily transformed into negatively charged colloidal particles and reenter the coating, and new functional properties and be obtained by adjusting and improving the composition of the coating. Whether acidic or alkaline electrolytes, the main mechanism of oxygen transport to the metal involves the movement of OH^- ions through the electrolyte under the influence of an electric field. In the anodic oxidation process, ions only flow steadily along the electrolyte channel, while the PEO process is more dynamic and transient. As well as the conductivity of the electrolyte is important because it affects the current and the voltage drop through the electrolyte, thus affecting the arcing ability of the plasma discharge.

On the other hand, the alkaline electrolyte systems are mostly composed of the solution corresponding to the anion with strong adsorption capacity. The results of selective adsorption of anions in solution by oxide film show that the order of anion adsorption capacity from strong to weak is as follows: $SiO_3^{2-} > PO_4^{3-} > VO_4^{3-} > MoO_4^{3-} > WO_4^{3-} > B_4O_7^{2-} > CrO_4^{3-}$. Among them, silicate and phosphate are the two main systems of alkaline electrolyte, as well as aluminate and hydroxide systems are also usually selected objects.

Table 2.2 summarizes the effect of typical electrolyte on the coating structure: ① Silicate can increase the solution conductivity to reduce the breakdown voltage, further promoting the metal to form passivation film quickly. Meanwhile, SiO_3^{2-} synergies with other anions in the electrolyte, which can promote the combustion and extinguishing of discharge sparks alternately on the coating surface, accelerating the movement of discharge sparks on the coating surface, so as to avoid the production of macro small pits and micro-cracks caused by local heat accumulation; ② Phosphate can increase the coating thickness and reduce the surface roughness, making the coating smoother and denser; ③ The addition of aluminate and

Table 2.2 Effect of typical electrolyte on the growth of PEO coating

Electrolyte	Film-forming rate	Coating structure	Thickness	Porosity
Na_2SiO_3	Fast	Thick coating; high density; high roughness; less cracks	Thick	Dense inner layer and porous outer layer
Phosphate	Moderate	Thin and dense coating; small roughness; surface smoothing; less cracks	Thin	Low porosity and thin inner dense layer; small pore size
$NaAlO_2$, hydroxide	Fast	Thick and dense coating; high density; less cracks	Thick	Low porosity; dense inner layer
Tetraborate	Moderate; strong selective dissolution	Micro-nano structure, "cortex-like" structure	Unevenness	High porosity; micro/nanohole; uneven coating thickness
Sodium carbonate	Moderate	Micro-nano pore structure	Unevenness	Reaming effect, with high porosity

hydroxide can greatly improve the electrolyte conductivity, to reduce the breakdown voltage and electric arc size, so as to promote the uniform increase of coating thickness, forming a relatively dense structure; ④ The addition of tetraborate can dissolve the oxide formed in the plasma discharge process in the electrolyte, with little or no deposition outside the discharge channels, which can build a two-scale "cortex-like" structure on the coating surface; ⑤ Sodium carbonate has a good reaming effect, which can significantly improve the porosity and pore size of the coating surface to construct the multi-scale pore structures.

Generally, to improve the coating structure and its overall performance, composite electrolyte systems (containing two or more electrolytes) are widely explored and have received great attention. In particular, some additives can provide electrolyte components into the coating, meanwhile improve discharge spark and extend the microarc discharge period, etc., thereby achieving the multifunctional characteristics of coating. For example: adding Na_2WO_4, Na_2SnO_3, Na_2MnO_4, etc., and reduce the breakdown voltage, to increase the coating thickness and surface roughness simultaneously; Adding glycerin, $C_4H_4Na_2O_6$, etc. can stabilize the microarc discharge spark and reduce the coating porosity; Adding hydroxide, fluoride, etc. can adjust the pH value of the electrolyte, and tailor the growth rate and thickness of the coating; The addition of $Na_3C_6H_5O_7{\cdot}2H_2O$ can improve the stability and conductivity of the electrolyte, so as to increase the growth rate and thickness of the coating, which is contribute to the improvement of corrosion resistance; Adding EDTA can prolong the service life of electrolyte. it is worth highlighting that the higher the pH value or the higher the conductivity, the lower the breakdown voltage, which is conducive to the rapid formation of the dense structure of the coating, so as to improve the coating thickness and compactness. However, if the pH or conductivity is too large, the coating porosity will increase, accompanied by large roughness and poor quality. Therefore, it is necessary to adjust the appropriate pH value, generally in the range of 9–13.

Additionally, the growth rate and reaction degree can be enhanced with the increase in electrolyte concentration. On this basis, the size, distribution, and number of micropores of the coating can be greatly regulated by the proportion of different components of the electrolyte, simultaneously the coating thickness can also be increased to different degrees to optimize the coating surface characteristics [23]. Usually, the decrease of electrolyte concentration is conducive to preventing the formation of large pores in the coating. However, the electrolyte concentration is too small (the reaction is difficult to carry out, and not conducive to film-formation) or too large (the reaction is too violent, and not conducive to the formation of a dense film, even has a corrosive effect on the substrate and coating), which both is not conducive to the coating growth and the regulation of microstructure.

In summary, for the electrolyte, the prerequisite for coating growth is dielectric breakdown. In order to optimize the coating growth and tailor the microstructure, additives that improve the metal passivation can be used to reduce the breakdown voltage, accelerating the coating reaction and growth. Further, by adjusting the electrolyte formula and considering the influence of a single electrolyte component on the coating growth, two or more electrolyte components are mixed and matched, which can better stabilize the microarc discharge of the coating surface, so as to achieve the purpose of improving the coating thickness and compactness simultaneously.

2.3.3 Electrolyte System with Special Ions/Particles

In order to construct a series of functional PEO composite coatings (such as antibacterial, catalytic, electromagnetic, optical, biological, ferroelectric, decorative properties), a variety of unique formulations have been successfully developed. The addition of special ions, molecules, micro-nano metal/ceramic/polymer particles to the electrolyte will have a great impact on the structure and performance of the coating. The ionic additives (taking NH_4VO_3 as an example) are added to the basic electrolyte, and NH_4VO_3 dissolves in the solution and ionizes to produce VO_3^-. During the coating growth process, the temperature of the microarc discharge zone is about 3000–6000 K, or even higher, and the instantaneous high temperature of heat-affected zone also reaches 1600–2000 K [1]. Such high temperature promotes the adsorption and reaction of VO_3^- to convert into vanadium oxides, forming V_2O_3 and V_2O_5. As a result, ionic additives can participate in coating reactions and form corresponding oxides or other compounds, which is of great value for designing special functional coatings.

Moreover, micro-nano ceramic particles can be added to the electrolyte [24], at which time a large number of particles will preferentially accumulate, deposit and sintering near large holes and defects in the coating. Most of the particles are mainly located in the outer and middle layers of the coating, indicating that the added particles can flow back with the molten ceramic layer oxide through the discharge channels, as well as grow together with the oxide ceramic layer and agglomerate around the holes [25]. Besides, a large number of particles will aggregate to form large-size particles on the coating surface [26] or form a multilayer coating with different micro-nano structures [27], thus achieving functional characteristics to meet the needs of different environmental services.

It is worth noting that organic micro-nano particles (such as PTFE) added to the electrolyte can fill and seal the pores, which is conducive to improving the corrosion resistance and electrical insulation of the coating. Meanwhile, the hydrophobicity of the coating can be realized by adjusting the electrical parameters and combining with the low surface energy of PTFE, which can meet the requirements of functional application [28].

In order to design and prepare the special functional coating, vanadium salt, copper salt, nickel salt, iron salt, zinc salt, molybdenum salt, cobalt salt, chromium salt, cerium salt, zirconate, fluorozirconate, and some rare earth salts are added to electrolyte, so as to improve the optical, thermal control, catalytic, electromagnetic, decorative, and biomedical properties. For instance, to improve wear resistance or corrosion resistance, ionic additives (such as Na_2MoO_4, K_4ZrF_6) or particle additives with high hardness and high melting point (Al_2O_3, $MoSi_2$, SiC, ZrO_2), or dry lubricants (graphite, PTFE, CNTs, diamond) are added into the electrolyte. Additionally, colorants (NH_4VO_3, Na_3VO_4, $Cu(CH_3COO)_2$, Na_2WO_4, CeO_2, Ce_2O_3, Cr_2O_3, V_2O_3), bioactive hydroxyapatite, etc. can be added to achieve different functions. It can be seen that the special ions/particles addition of the electrolyte system provides a new way for the construction of functional coatings and opens up new application fields.

2.3.4 Electrolyte System with Self-Sealing Component

PEO coating surface typically exhibits a porous structure, which is unfavorable to corrosion protection in corrosive media, especially for highly corrosion-sensitive materials like magnesium and aluminum alloys. The formation of these pores is attributed to the fact that the volume of metal oxide generated during the coating growth process, relative to the volume of metal consumed (the Pilling–Bedworth ratio, or PBR) is less than 1 (e.g., MgO). Therefore, it is necessary for the coatings prepared in the PEO process to have self-sealing ability. There are four primary approaches to achieve self-sealing PEO coatings: (1) Incorporation of low-melting-point ion-forming components in the electrolyte, such as fluorine salts, that participate in the coating formation process; (2) Addition of highly oxidative ionic components (e.g., potassium per-manganate) into the electrolyte to enhance coating integrity; (3) Micro-nano particles are added into the electrolyte to fill the pores through plasma-assisted deposition and sintering or to form a self-sealing double-layer structure; (4) Incorporation of self-healing refractory salts within the electrolyte.

2.3.4.1 Ionic Components

Ionic components (such as fluorate) in the electrolyte participate in film formation and can form a low melting point phase. A new fluotitanate electrolyte system (2 g/L $Na_5P_3O_{10}$, 6 g/L $(NaPO_3)_6$, 3 g/L NaOH and 10 g/L K_2TiF_6) was developed based on the principle that the metal compound salts in the electrolyte decompose and participate in film formation at high temperature [29]. The electrical parameters were set to: current density of 40 mA/cm^2, voltage of 420 V, oxidation time of 2 min, frequency of 1000 Hz, duty cycle of 30%. Since MgF_2, which has the lowest melting point during the melt condensation process, can be condensed and deposited in the micro-pores, completing the sealing treatment of the discharge channel. So the micro-pores on the coating surface have self-sealing characteristics (Fig. 2.4). In addition to the sealing effect, the formed coating contains a large number of titanium oxides with high chemical stability, whose corrosion resistance is 5–6 times higher than that of conventional PEO coatings on the Mg alloys. Except for fluotitanate, self-sealing PEO coating can also be formed in the electrolyte containing fluorozirconate, which has high compactness, less defects and good corrosion resistance [30].

Self-sealing mechanism: The primary components of the electrolyte system include fluorotitanate or fluorozirconate salts, which leverage the significant melting point differences among the resulting phases to achieve self-sealing of the PEO coating through a sequence of solidification. Namely, the internal pressure in the discharge channels is markedly lower than the external pressure during the PEO process, which promotes the deposition of oxidation products on the inner channel walls rather than on the coating surface. The melting point of each component of the coating formed in the fluorate electrolyte is different, and MgF_2 with the lowest

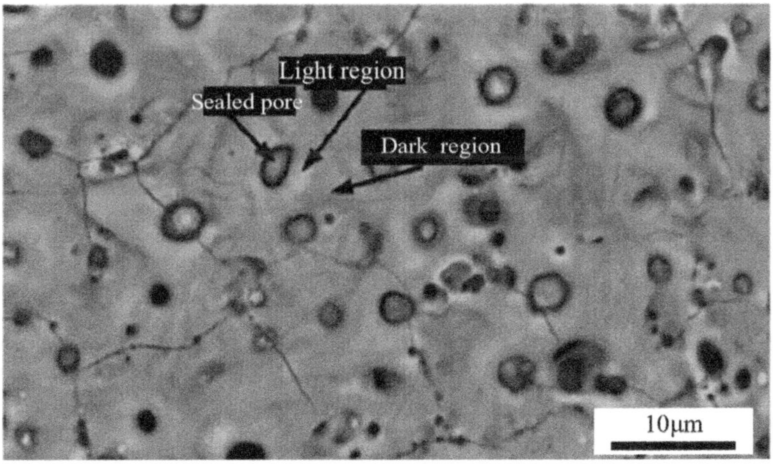

Fig. 2.4 Morphologies of the self-sealing pore coating [29]. *Reproduced with permission from [29] © 2013 National Engineering Research Center for Magnesium Alloys of China, Chongqing University. Production and hosting by Elsevier B.V.*

freezing point is finally condensed and deposited in the micro-pores to achieve sealing. As well as stable titanium or zirconium oxides with a PBR greater than 1 are formed, also enhancing coating compactness. The self-sealing coating combined with highly dense inner and outer layers both prevents the intrusion of corrosive media to improve salt spray corrosion resistance for a long time.

2.3.4.2 Micro-Nano Particles

The sealed layer can be formed by plasma-assisted deposition and sintering of micro-nano particles. For instance, the magnesium alloy was treated by PEO in the electrolyte containing Al_2O_3 particles, that participated in the film formation reaction, which increased the Al_2O_3 content in the coating to improve the coating compactness [31]. When a certain amount of SiC nanoparticles was added to the electrolyte, it was found that part of the SiC was oxidized to nonconductive SiO_2, which inhibited the strong spark generation, and further promoted the coating growth and in-situ sealing [32]. The author et al. [33] proposed a new technology of PEO synchronous deposition and sintering of nanoparticles (metal/ceramic/polymer particles) to efficiently construct double-layer composite coatings with large thickness. Under the action of strong electric field, a large number of nanoparticles were oriented migration and synchronous deposition and sintering to form a double-layer composite coating composed of PEO bottom layer and nanoparticle rich outer layer, realizing the elimination and sealing of pores (Fig. 2.5).

2.3.4.3 Strong Oxidizing Ionic Components

Strong oxidizing ionic components (such as potassium permanganate) involved in film formation are added to the electrolyte to achieve self-sealing pore. Specifically, with the increase of potassium permanganate concentration, the applied voltage and current of the PEO process are significantly reduced to avoid generating strong discharge spark. Meanwhile, the strong oxidation of potassium permanganate causes the increase of the thickness and compactness of the PEO coating simultaneously. As well as potassium permanganate has a good in-situ sealing effect on porous coatings, so as to form a uniform self-sealing coating.

2.3.4.4 Self-Healing Salt with Low Solubility

The destructive corrosion of PEO coating mainly originates from the infiltration of corrosive liquid along the pores. Therefore, the corrosion inhibition salts such as insoluble phosphates are used to achieve self-healing pores. It is reported that the salt spray corrosion resistance time of the prepared super-corrosion resistant coating can reach 1800 h, which is expected to be applied to a variety of corrosion environment and long-life aluminum alloy corrosion protection products. For example, a

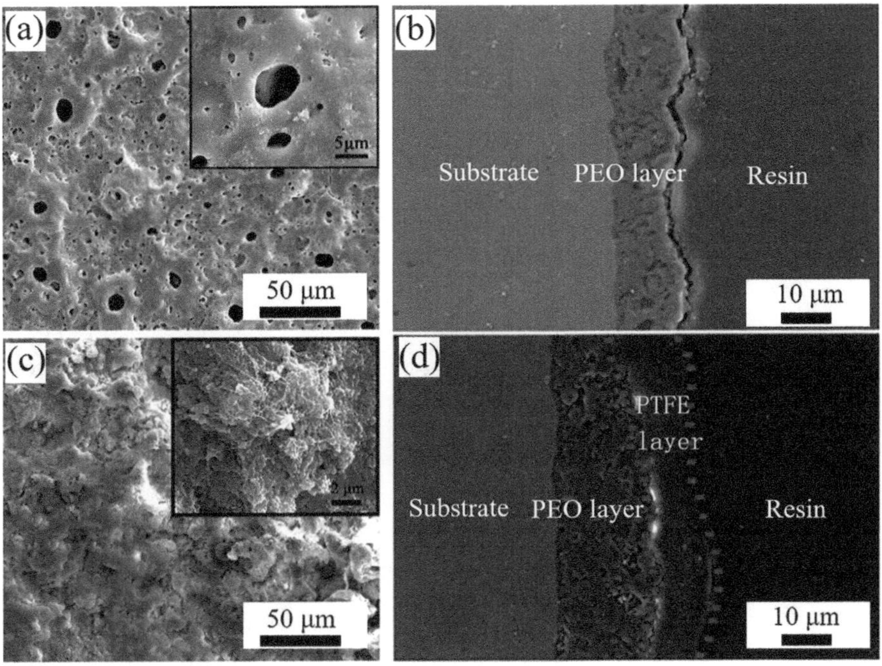

Fig. 2.5 Surface and cross sectional morphologies of traditional PEO coating and PEO-synchronous deposition & sintering nanoparticle double-layer coating [33]. *Reproduced with permission from* [33] *© 2020 Elsevier B.V.*

compact ceramic coating containing stable Fe and P phases was prepared by PEO on the aluminum alloy surface in an electrolyte containing sodium hexametaphosphate and ammonium ferric oxalate hydrate. The coating surface has low roughness and high hardness of 1200 HV. Meanwhile, the insoluble salt can achieve the coating sealing effect, showing a good corrosion resistance [34].

2.3.5 Other Electrolyte Systems

Nonaqueous electrolyte systems for plasma electrolytic fluoridation or oxidation involve the use of fluoride-containing molten salts, nitrate melts, or alcohol-fluoride solutions as electrolytes. Table 2.3 summarizes the structures, properties, and advantages/disadvantages of coatings prepared in nonaqueous electrolyte systems.

1. Low-temperature fluoride molten salt electrolytes for plasma electrolytic fluoridation

   - **Preparation process**: In an ultra-high concentration NH_4HF_2 molten fluoride system at 150 °C, applying a voltage of 100–140 V for 30 s results in fluoride coating formation.

Table 2.3 Structure, properties, and advantages/disadvantages of coatings prepared in nonaqueous electrolyte

Nonaqueous electrolyte	Preparation process	Structural features	Properties	Advantages/disadvantages
Low-temperature fluoride molten salt system	Ultra-high concentration NH_4HF_2 molten salt at 150 °C	Pore size: 600–900 nm, thickness: 1–14 μm, MgF_2	Good corrosion resistance	**Advantages:** No cooling requirements, high coating growth rates, uniform coating structure, simple composition, and the potential to form non-oxide coatings. **Disadvantages:** Increased energy consumption due to complex heating systems, high temperature requirements, molten salt corrosion issues, material compatibility, and high costs
Nitrate melt system	$NaNO_3$-KNO_3 (45.7:54.3) eutectic nitrate at 280 °C	Dense γ-Al_2O_3 inner layer and α-Al_2O_3 outer layer, composition: No contaminants	Corrosion resistance in molten salt-produced coatings is 3 times that of coatings prepared in silicate solution	
Alcohol-fluoride solution	An EG-NH_4F electrolyte (8 wt% NH_4F in ethylene glycol, 25 °C)	MgF_2 (magnesium) or AlF_3 (aluminum), uniform composition, porous structure	Good protection in neutral/acidic media	Alcohol–fluoride solution facilitates non-oxide coating formation at room temperature without heating, have broad substrate adaptability, and present a safer and nonpolluting experimental environment **Limitations:** Coating thickness and compactness

- **Structural features**: Coating pore size ranges from 600 to 900 nm, with a thickness of 1 to14 μm, and the coating mainly consists of MgF_2.
- **Performance**: The coating exhibits markedly improved corrosion resistance in simulated body fluid compared to magnesium alloys [35].

2. Nitrate melt as an electrolyte for plasma electrolytic oxidation

- **Preparation process**: An aluminum alloy was treated in a $NaNO_3$–KNO_3 eutectic nitrate melt system (mass ratio 45.7:54.3) at 280 °C, with current density of 70 mA/cm^2, and frequency of 50 Hz for 10 min.
- **Structural characteristics**: The coating consists of a dense γ-Al_2O_3 inner layer and an α-Al_2O_3 outer layer, free from any electrolyte contaminants of aqueous systems. With slower cooling rates, the coating shows no through-holes or cracks. As well as the growth rate in molten salt systems is three times higher than in silicate electrolytes, with nearly six times greater energy efficiency.

- **Performance**: Corrosion resistance in molten salt-produced coatings is three times that of coatings prepared in silicate solution and nearly two orders of magnitude greater than untreated alloys, attributed to the high compactness of coating [36].

3. Alcohol-fluoride as electrolyte for plasma electrolytic fluoridation

- **Preparation process**: Plasma electrolytic fluoridation of magnesium or aluminum alloys is conducted in an EG-NH_4F electrolyte (8 wt% NH_4F in ethylene glycol, 25 °C), with a current density of 8 A/dm^2, frequency of 100 Hz, duty cycle of 8%.
- **Structural characteristics**: The coatings primarily comprise MgF_2 (magnesium substrate) or AlF_3 (aluminum substrate) with uniform composition and porous structures.
- **Performance**: MgF_2 and AlF_3 coatings provide effective corrosion protection in neutral and acidic media for magnesium and aluminum alloys, respectively [37].

Molten salt systems offer advantages such as no cooling requirements, high coating growth rates, uniform coating structure, simple composition, and the potential to form nonoxide coatings. However, some challenges include increased energy consumption due to complex heating systems, high temperature requirements, molten salt corrosion issues, material compatibility, safety considerations, and high costs. Alcohol–fluoride solution facilitates non-oxide coating formation at room temperature without heating, have broad substrate adaptability, and present a safer and nonpolluting experimental environment. Nonetheless, limitations remain in coating thickness and density, and further process refinement is needed.

The selection of electrolyte components, concentrations, and additive proportions significantly influences coating structure and performance. The composition and structure of coatings can be precisely controlled by adding nanoparticles, ions, or dopants, providing pathways for the development of novel functional coatings. Continued research on new electrolytic systems (including self-sealing, nonaqueous, and ionic liquid, etc.), reaction mechanisms, coating formation processes, and impacts on structure and performance is essential.

2.4 Effect of Substrate Metal Composition on Coating Microstructure

For different substrate materials, due to their different elemental compositions and metallurgical states, that will present different film-forming characteristics. Among them, the metal alloy composition, second phase, intermetallic compounds, metallurgical treatment and pretreatment will affect the plasma discharge state, further affecting the coating growth process, structure uniformity and surface quality. Hence, on the basis of exploring the influence of the above factors on the plasma discharge process, it is particularly important to select the appropriate process parameters to optimize the coating structure.

2.4.1 The Difficulty and Thermodynamic Analysis of PEO Reaction of Metals

The change in Gibbs free energy associated with metal oxidation is relevant to the PEO process. At ambient temperature, this energy change is negative for all metals (excluding gold), indicating that metal oxidation is energetically favorable, at least at temperatures where the metal remains solid or liquid. However, due to plasma collapse and cooling during the PEO process, there may be competitive oxidation reactions with higher driving forces. Notably, the plasma likely contains hydrogen, presenting a competition between the driving forces for metal oxidation and hydrogen conversion to water. Metal oxidation is thus more likely to occur if its energetic favorability exceeds that of hydrogen oxidation.

For metals with low energy changes (Cu, Ni, Fe, Sn, Mo, and W), the oxidation driving force is similar to or even lower than that of hydrogen forming water, making oxidation less likely during plasma collapse and cooling in the PEO process. This characteristic renders certain metals (such as copper alloys, steels, and nickel-based superalloys) unsuitable for PEO treatment. In most cases, oxide formation on steel and copper surfaces likely results from the deposition and sintering of electrolyte constituents, producing brittle coatings with poor adhesion. For metals in the second category (e.g., Cr, Nb, Ta, and V), oxidation is energetically possible despite being marginally above hydrogen oxidation. Lastly, the third category comprises metals with significantly high oxidation energies (Ti, Al, Mg, Be, Hf, Zr, Y, and Sc), providing a robust thermodynamic driving force for oxidation. Notably, Al, Mg, and Ti are commonly treated with PEO, while Nb, Ta, Zr, and Be (toxic) are receiving growing attention, and Y and Sc are potential candidates for future PEO applications [38].

It may be noted at this point that there is also the option, for metals unsuited to PEO processing (including steels and alloys of Ni, Cu, Sn, etc.), of first plating them with a layer of a metal (such as Al) that is suitable and then producing the PEO coating (possibly consuming all of the plated metal). This has been explored in some detail, mostly for Al layers on steels [39–41]. In principle, this is a viable and attractive option, but there are problems in practice. One is that while electroplating is the most convenient option for producing such layers, it is inherently difficult for the metals best suited to PEO. Nevertheless, the main problem is that there is a tendency to leave a residual (incomplete oxidation) aluminum layer and/or poorly bonded interface between the substrate and the coating. This is unfortunate since one of the inherent attractions of PEO is usually having very good substrate/coating adhesion. In particular, the combination of pretreatment and PEO is still the direction of optimization, with many variants to be explored [38].

2.4.2 Effect of Metal Alloy Composition on Coating Growth

The difficulty of initiating a PEO reaction depends on the elemental composition of the metallic substrate and the properties of its oxides. The presence of other alloying elements significantly affects PEO coating growth process. Typically, solute elements in the metal can be incorporated into the coating as: (a) The coating is mainly composed of substrate oxides, supplemented by other alloying element oxides; (b) Mixed oxides; (c) Other inclusions. (d) These elements may also enter the electrolyte in ionic or solid suspension forms. Consequently, studying the influence of secondary phases in alloys on the PEO process provides a foundation for optimizing PEO processes for multiphase metallic materials and controlling oxide film microstructures and properties.

2.4.2.1 Influence of Alloy Composition on Coating Growth in Aluminum Alloys

Table 2.4 summarizes the effects of other alloying elements on the PEO coating growth, structure, and properties in aluminum alloy substrates. Traditional anodizing is difficult to prepare uniform films on high silicon cast aluminum alloy or high copper aluminum alloy. PEO is able to form a thick coating on its surface to improve wear and corrosion resistance, albeit with localized defects in Si- or Cu-rich areas. For instance:

1. For Al–Si alloys, the coating surface uniformity decreases as Si content increases, but coatings of acceptable quality can still be formed at up to 30% Si by PEO treatment. Although the porosity of the coating increases due to the non-conductivity of the Si component, causing hardness of the coating decreases, the wear resistance and corrosion resistance of Al-Si alloy can be greatly improved.
2. For Al–Cu alloys, copper exists primarily as precipitates. It is found that there are Cu residues and Cu_2O produced in PEO coating, and Cu is unfavorable to α-Al_2O_3 phase transition. According to the evolution of copper-rich precipitation in PEO-treated AA2214 aluminum alloy, it is founded that dealloying produces copper nanoparticles. This shows that copper nanoparticles increase the conductivity of oxides and delay the delaying dielectric breakdown, which is not conducive to uniform coating growth.
3. For Al–Fe alloys, Fe influences dielectric breakdown, discharge occurrence, and coating growth, particularly in the early stages. Namely, Fe resists passive film formation, leading to current leakage and uneven coating growth. Besides, Zinc and magnesium in aluminum alloys also inhibit the formation of α-Al_2O_3 in the PEO coating, with magnesium exerting a more substantial inhibitory effect.

To better understand the coating growth process with different second phases in metal substrates, $AlSi_9Cu_3$ alloy was treated with PEO for various times (15–480 s) in a phosphate electrolyte, forming coatings composed mainly of γ-Al_2O_3, mullite,

Table 2.4 Effects of other alloying elements in aluminum alloy substrates on the PEO coating growth process, structure, and performance

Substrates	Alloying element	Com-position	Coating growth process	Coating structure	Performance
Al alloy	Si	<20%	Si-enriched areas exhibiting low discharge activity with high breakdown voltage, affecting coating growth (minor effect)	Porous structure, unique morphology in some areas	Anti-wear, anti-corrosion
		>20% High Si cast aluminum	High-Si zones showing local defects and high energy consumption, leading to uneven coating growth and increased porosity	Loose and porous structure, unique morphologies in multiple areas	Anti-wear, anti-corrosion
	Cu	High Cu-Al alloy	High-Cu areas displaying local defects and high energy consumption, resulting in uneven coating growth and high porosity	Porous structure with relatively low compactness	Anti-wear, anti-corrosion
	Fe	Al-Fe	Difficult to form passivation film in Fe region, causing current leakage and high energy consumption	Porous structure with relatively low compactness	Anti-wear, anti-corrosion
	Zn	3–7.5%	Inhibiting coating growth	Smooth surface with high porosity	Anti-wear, anti-corrosion
	Mn	1–8%	Mn can replace Al in the Al_2O_3 crystal structure, accelerating coating growth, with high porosity and roughness	Increased surface roughness, thickness, and porosity	Anti-wear, anti-corrosion
	Mg		Inhibiting α-Al_2O_3 formation	Rough surface	Anti-wear, anti-corrosion
	$AlSi_9Cu_3$		Different elements causing different discharge order	Rough surface, high porosity, unique morphology in some areas	Anti-wear, anti-corrosion
	Rare earth modification		The Er addition can refine the grain and improve the corrosion resistance	Increasing compactness, hard-ness, and adhesion strength	Anti-wear, anti-corrosion

and amorphous phases. Results showed that plasma discharge involves phases in the following sequence: initial discharge from the Al substrate, simultaneous discharge on intermetallic compounds $A_{12}Cu$ and β-Al_5FeSi, and finally on eutectic Si. Most initial discharges appear preferentially around α-Al near the Al_2Cu compound, indicating more intense plasma discharge in these areas. The presence of intermetallic compound Al_2Cu significantly influences early dissolution, conversion product deposition, and breakdown initiation. Eutectic Si in the substrate demonstrates the highest electrochemical stability throughout all stages, forming a unique coating morphology. As well as extending treatment time results in more uniform PEO coatings with a bilayer structure [42].

Moreover, rare earth elements can also impact coating structure and performance. For instance, the Al_3Er phase generated by the reaction of Er and Al can produce a large number of heterogeneous nucleation, further achieving the effect of refining grains. Hence, the PEO coating on aluminum alloy with Er will result in higher compactness, lower roughness, and increased hardness. Meanwhile, the addition of Er reduces the probability of corrosion between the coating grains to improve the corrosion resistance.

2.4.2.2 Influence of Alloy Composition on Coating Growth in Magnesium Alloys

Compared to aluminum alloys, ceramic coatings on magnesium alloys are generally denser but more challenging to thicken. In acidic and weakly basic solutions, magnesium exists in theory as metallic magnesium at cathode potential below −2.5 V, as ions at anode potential above −2.5 V. It can be seen that the magnesium surface in a strong alkali electrolyte (pH > 12) is prone to forming passivation film, which can promote the formation of relatively dense oxide film. When the solution pH < 12, magnesium alloy is not prone to passivation in solution, and plasma discharge does not occur for a long time. Therefore, tailoring the pH and composition of the electrolyte is the key to preparing highly dense coatings on the magnesium alloy surface.

Table 2.5 summarizes the effects of other elements in the magnesium alloy on the growth process, structure and properties of the PEO coating.

1. **Different phase components**: During the PEO process, the α phase in the magnesium alloy exhibits a higher growth rate compared to the β phase. The coating on the β phase predominantly develops around the oxide layer formed on the α phase. This difference is likely due to the higher reactivity of the α phase, which can release more Mg^{2+} ions for the film-forming reaction under the influence of a high electric field and Joule heating. While the presence of the β phase causes slight non-uniformity in the coating growth.
2. **Other alloying elements**: The other elements on magnesium alloy actively participate in the film-forming reaction, which directly impacts the composition and properties of the coating. In the AZ and AM series magnesium alloys, aluminum

Table 2.5 Effects of other alloying elements in magnesium alloy substrates on coating growth process, structure, and performance

Substrate	Alloying element	Composition	Coating growth process	Coating structure	Performance
AZ91D	α, β phases		The α-phase has a higher growth rate than the β-phase; coating on β-phase grows around α-phase oxide film, due to the higher reactivity of the α-phase	Porous structure, controllable porosity	Anti-corrosion and anti-wear
AZ/AM	Al		Forming magnesia–alumina spinel, allowing uniform coating growth	Porous structure with controllable porosity; AZ series has better compactness, adhesion, and uniformity than AM series	Anti-corrosion and anti-wear
AZ1/AZ91			Lower Al content in AZ31 resulting in lower oxide film stability and higher dissolution rate. Large horizontal pores formed on AZ31, whereas AZ91 only showing micro-discharge and having dense structure	The coated AZ91 is denser, with no large pores, offering higher corrosion resistance, than that of AZ31	Anti-corrosion
Mg-Mn	Mn	0–4%	Increasing Mn content reducing initial, breakdown, and stable voltages	As Mn content in-creases, micropore size decreases, porosity increases, and thickness initially decreases then rises	Anti-corrosion and anti-wear
Mg-Zn-Ga	Zn	1–5%	Uniform coating growth	Porous structure, controllable porosity	Anti-corrosion and anti-wear
Mg-Si	Si		Si-enriched areas showing low discharge activity and high breakdown voltage, affecting uniform coating growth and increasing porosity	Loose and porous structure and unique morphology in some areas	Anti-corrosion and anti-wear
Mg-Cu	Cu		High-energy consumption in cu-enriched areas and delaying dielectric breakdown, resulting in uneven coating growth and high porosity	Loose and porous structure	Anti-corrosion and anti-wear

participates in reactions in the electrolyte to form magnesium–aluminum spinel. The coating on AZ series alloys demonstrates superior compactness, adhesion strength, compared to the AM series. Further comparison of the coating growth on AZ31 and AZ91 alloys shows that the coating formed on AZ91 is denser than that on AZ31, attributed to differences in the rates of inner layer growth and oxide dissolution. Due to its lower aluminum content, the oxide layer on AZ31 has lower stability and a higher dissolution rate [43]. Additionally, the PEO process on AZ31 generates higher-energy discharges, which tend to produce large pores. In contrast, AZ91 exhibits only micro discharges, resulting in a coating with no large through-pores and better corrosion resistance. This indicates that alloying elements influence discharge intensity and pore structures during coating formation.

It is noteworthy that the substrate's impact on PEO coating is largely attributed to variations in the Pilling–Bedworth Ratio (PBR) of the oxide layers on different alloys, where a higher PBR correlates with denser PEO coatings. Comparisons of PBR values across different alloys reveal that Mg-Gd-Y > AZ91D > ZM6, suggesting that Mg-Gd-Y surfaces can form a denser and more uniform coating, with a good corrosion resistance.

2.4.2.3 Influence of Alloying Elements on Coating Growth in Titanium Alloys

For titanium alloys, alloying elements participate in the coating reaction, forming corresponding oxides, mixed oxides, or complex glassy phases. For instance, although nickel alloys are unsuitable for PEO treatment, in high-titanium-content alloys, such as nickel-titanium shape memory alloys, anatase-phase coatings can be prepared on NiTi substrates through aqua regia dealloying combined with PEO. Studies show that the manufacturability of PEO coatings on NiTi alloys is comparable to that on industrial titanium, producing porous coatings with fewer cracks [44].

2.4.3 *Effect of Substrate in Different Metallurgical States on Coating Growth*

The substrate's metallurgical state and forming processes also impact the PEO coating growth. For example, after PEO treatment of ultrafine-grained AZ91D magnesium alloy, the as-prepared coating porosity is reduced to 8.3%, showing a superior corrosion resistance compared to coatings on the commonly used AZ91D magnesium alloy. After solution treatment of Mg–Al alloy, the uniformity of elements alleviates the nonuniformity of PEO selective film formation, further reducing the micro-cracks number, and increasing the coating growth rate. For cast aluminum alloys, appropriate heat treatment refines the substrate grain structure, and evenly distributes Si elements, enhancing the uniformity of the coating structure. Age hardening also releases residual stresses within the PEO coating, reducing microcracks. For β titanium alloys, solution treatment modifies the phase composition and element distribution, facilitating a uniform discharge. Compared with the untreated sample, the coating prepared on the solution treated sample is more compact.

2.4.4 *Effect of Substrate Surface Pretreatment on Coating Growth*

In addition to the substrate's metallurgical state, pretreatment can also influence the coating growth behavior and properties. Techniques such as shot peening and surface mechanical attrition treatment (SMAT) can notably enhance the fatigue life and corrosion resistance of coated alloys [45, 46]; Preanodizing, laser processing, chemical etching, and electrochemical etching can create PEO coatings with larger specific surface areas, suited for energy material applications. As well as preanodizing cannot only accelerate the coating growth, but also make it easier to achieve "soft" spark conditions, further improving the compactness and thickness of the coating. The niobium alloy siliconizing pretreatment markedly enhances thermal protection, making the PEO coated Nb alloys potentially viable for service in high-temperature environments [47].

2.5 Conclusions and Outlook

PEO coating was concluded to be promising and competitive to modify surface structures of metallic materials that were suitable for both structural and functional applications in many industrial fields. Therein the coating structure and quality are determined by many factors such as power supply regime, electrical parameters, electrolyte and substrate materials, etc. Particularly, the micropores and discharge channels are inescapable during the evolution of the coating structure, due to complex processes involving thermal, electrochemical and plasma chemical reactions, resulting in a great challenge on the design and preparation of customized PEO coatings.

Herein, future research should address the following critical areas:

1. For the power supply output regime, with the rapid development of current new power modules, exploiting low-energy power supplies (such as double-pulse inverter power supplies) and high-efficiency long-life electrolytic liquid systems can reduce coating manufacturing costs, which is one of the future development trends. Meanwhile, to apply PEO technology to large-scale complex workpieces and mass production line, it is urgent to develop high-power power supply, automatic production line and online monitoring and testing system that are compat-

ible with industrial automated production, so as to greatly enhance the processing capacity of PEO power and the ability to match and adjust electrical parameters. In the future, there is still great space for development of high value-added special functional PEO coating technology.
2. For the electrolyte composition design, it is necessary to further understand the internal relationship between plasma discharge characteristics and film forming mechanism in different electrolyte systems in the future, so as to enrich and develop the theory of anode plasma electrolysis. The electrolyte composition has a great influence on the structure and performance of the coating, so the best concentration of the electrolyte and the appropriate proportion of additives should be selected. The composition and structure of the coating can be coordinated by adding micro-nano particles/ions/dopants, which provides a way for the construction of new functional coatings. Simultaneously, the development of new electrolyte systems, including self-sealing, nonaqueous and ionic liquid, etc., whose reaction mechanism, film formation process and effects on coating structure and properties need to be further studied in the future.
3. The PEO process of substrate metal is different due to different material properties, heat treatment state, parts structure and quantity, film thickness, film-forming speed, equipment power, and so on. Particularly, the properties of different substrates (such as electrical conductivity, microstructure and defects, second phase doping and content) have a great influence on the PEO process; thus, special solutions and electrical parameters need to be matched to ensure uniform coating growth.

All in all, corrosion dissolution, oxidation passivation, impedance change, dielectric breakdown, plasma discharge, and electrochemical reaction of different phases in the substrate metals during the PEO process have important effects on coating growth. Hence, it is necessary to strengthen the discharge mechanism research to provide theoretical guidance for process optimization and practical application. In other words, the precise matching of the metal materials to the power, electrical parameters and electrolytes can truly enable the customization of high-quality coatings.

Acknowledgments The partial supports from the NSFC Grant nos. 52301084, U21B2053, China Postdoctoral Science Foundation (2023 M730843), Aviation Science Foundation (ASFC-2023 M045077001), Open Project Funding of State Key Laboratory for High Performance Tools (GXNGJSKL 2024), Postdoctoral Fellowship Program of CPSF (GZB20230945), and Research start-up Fund by HIT are gratefully acknowledged.

References

1. Yerokhin, A. L., Nie, X., Leyland, A., Matthews, A., & Dowey, S. J. (1999). *Surface and Coatings Technology, 122*, 73–93.
2. Kaseem, M., Fatimah, S., Nashrah, N., & Ko, Y. G. (2020). *Progress in Materials Science, 117*, 100735.

3. Jumbad, V. R., Chel, A., Verma, U., & Kaushik, G. (2020). *Letters in Applied NanoBioScience, 9*, 1249–1262.
4. Jiang, B. L., & Wang, Y. M. (2010). *Surface Engineering of Light Alloys*, 110–154.
5. Aliofkhazraei, M., Macdonald, D., Matykina, E., Parfenov, E., Egorkin, V., Curran, J., Troughton, S., Sinebryukhov, S., Gnedenkov, S., & Lampke, T. (2021). *Applied Surface Science Advances, 5*, 100121.
6. Wang, L., Chen, L., Yan, Z. C., & Fu, W. (2010). *Surface and Coatings Technology, 205*, 1651–1658.
7. Sikdar, S., Menezes, P. V., Maccione, R., Jacob, T., & Menezes, P. L. (2021). *Nanomaterials, 11*, 1375.
8. Güntherschulze, A., & Betz, H. (1934). *Zeitschrift für Physik, 91*, 70–96.
9. Van, T. B., Brown, S. D., & Wirtz, G. P. (1977). Mechanism of Anodic Spark Deposition. *American Ceramic Society Bulletin (United States)*.
10. Ikonopisov, S. (1977). *Electrochimica Acta, 22*, 1077–1082.
11. Ikonopisov, S., Girginov, A., & Machkova, M. (1977). *Electrochimica Acta, 22*, 1283–1286.
12. Zou, Y., Wang, Y., Sun, Z., Cui, Y., Jin, T., Wei, D., Ouyang, J., Jia, D., & Zhou, Y. (2018). *Surface and Coating Technology, 342*, 198–208.
13. Wang, Y., Jiang, B., Lei, T., & Guo, L. (2004). *Materials Letters, 58*, 1907–1911.
14. Li, Y., Chen, M., Li, W., Wang, Q., Wang, Y., & You, C. (2019). *Surface and Coating Technology, 358*, 637–645.
15. Wang, M. Y., Jia, C. D., & Guo, X. L. (2005). *Materials Chemical Physics, 90*, 128–133.
16. Cheng, Y., Wang, T., Li, S., Cheng, Y., Cao, J., & Xie, H. (2017). *Electrochimica Acta, 225*, 47–68.
17. Sah, S. P., Tsuji, E., Aoki, Y., & Habazaki, H. (2012). *Corrosion Science, 55*, 90–96.
18. Tsai, D. S., Chen, G. W., & Chou, C. C. (2019). *Surface and Coating Technology, 357*, 235–243.
19. Rogov, A. B., Yerokhin, A., & Matthews, A. (2017). *Langmuir, 33*, 11059–11069.
20. Tsai, D. S., & Chou, C. C. (2018). *Metals, 8*, 105.
21. Jiang, C., Wang, Y., Wang, S., Li, Y., Zou, Y., Ouyang, J., Jia, D., & Zhou, Y. (2022). *Surface and Coating Technology, 444*, 128692.
22. Melhem, A., Henrion, G., Czerwiec, T., Briançon, J., Duchanoy, T., Brochard, F., & Belmonte, T. (2011). *Surface and Coating Technology, 205*, 133–136.
23. Curran, J., & Clyne, T. (2006). *Acta Materialia, 54*, 1985–1993.
24. Lu, X., Mohedano, M., Blawert, C., Matykina, E., Arrabal, R., Kainer, K. U., & Zheludkevich, M. L. (2016). *Surface and Coating Technology, 307*, 1165–1182.
25. O'Hara, M., Troughton, S. C., Francis, R., & Clyne, T. W. (2021). *Surface and Coating Technology, 385*, 125354.
26. Borisov, A. M., Krit, B. L., Lyudin, V. B., Morozova, N. V., Suminov, I. V., & Apelfeld, A. V. (2016). *Surface Engineering and Applied Electrochemistry, 52*, 50–78.
27. Wang, S., Wen, L., Wang, Y., Cheng, Y., Cheng, Y., Zou, Y., Zhu, Y., Chen, G., Ouyang, J., & Jia, D. (2022). *Applied Surface Science, 592*, 153043.
28. Wang, S., Wang, Y., Zou, Y., Chen, G., Ouyang, J., Jia, D., & Zhou, Y. (2020). *ACS Applied Materials & Interfaces, 12*, 35502–35512.
29. Song, Y., Dong, K., Shan, D., & Han, E.-H. (2013). *Journal of Magnesium and Alloys, 1*, 82–87.
30. Liu, F., Shan, D. Y., Song, Y. W., & Han, E. H. (2011). *Transactions of Nonferrous Metals Society of China, 21*, 943–948.
31. Wang, Y., Wei, D., Yu, J., & Di, S. (2014). *Journal of Materials Science and Technology, 30*, 984–990.
32. Zou, Y., Wang, Y., Wei, D., Ge, Y., Ouyang, J., Jia, D., Zhou, Y., & Appl, A. C. S. (2019). *Journal of Electronic Materials., 1*, 1527–1537.
33. Wang, S., Wang, Y., Zou, Y., Wu, Y., Chen, G., Ouyang, J., Jia, D., & Zhou, Y. (2020). *Chemical Engineering Journal, 402*, 126116.

34. Ji, S., Weng, Y., Wu, Z., Ma, Z., Tian, X., Fu, R. K. Y., Lin, H., Wu, G., Chu, P. K., & Pan, F. (2017). *Journal of Alloys and Compounds, 710*, 452–459.
35. Jiang, H. B., Wu, G., Lee, S.-B., & Kim, K.-M. (2017). *Surface and Coating Technology, 313*, 282–287.
36. Sobolev, A., Kossenko, A., Zinigrad, M., & Borodianskiy, K. (2018). *Surface and Coating Technology, 344*, 590–595.
37. Qi, Y., Peng, Z., Wang, L., Zhou, J., Wang, P., & Liang, J. (2021). *Surface Engineering, 37*, 360–364.
38. Clyne, T. W., & Troughton, S. C. (2019). *International Materials Review, 64*, 127–162.
39. Kharanagh, V. J., Sani, M. A. F., & Rafizadeh, E. (2014). *Surface Engineering, 30*, 224–228.
40. Gu, W. C., Lv, G. H., Chen, H., Chen, G. L., Feng, W. R., Zhang, G. L., & Yang, S. Z. (2007). *Journal of Alloys and Compounds, 430*, 308–312.
41. Zhao, J., Zhao, G., Li, T., Liu, X., Li, J., & Han, E. (2012). *Transmission Materials and Heat Treatment, 33*, 129–132.
42. Wu, T., Blawert, C., & Zheludkevich, M. L. (2020). *Journal of Materials Science and Technology, 50*, 75–85.
43. Rakoch, A., Monakhova, E., Khabibullina, Z., Serdechnova, M., Blawert, C., Zheludkevich, M., & Gladkova, A. (2020). *Journal of Magnesium and Alloys, 8*, 587–600.
44. Lu, L. H., Zhang, J. W., & Shen, D. J. (2012). *Applied Mechanics and Materials, 121–126*, 3837–3841.
45. Lei, W., Yaming, W., Yu, Z., Changli, W., Lixin, G., & Jiahu, O. (2009). *Rare Metal Materials and Engineering, 38*, 747–750.
46. Wen, L., Wang, Y., Zhou, Y., Guo, L., & Ouyang, J. H. (2011). *Corrosion Science, 53*, 473–480.
47. Wang, Z., Wang, Y., Wang, S., Zou, Y., Chen, G., Wen, L., Zhang, G., Zhao, L., Ouyang, J., & Jia, D. (2022). *Corrosion Science, 209*, 110733.

Chapter 3
Plasma Electrolytic Oxidation: Sparking Discharge Regimes and the Role of Soft Sparking

J. Martin, T. Czerwiec, and G. Henrion

3.1 Introduction

The phenomena of electric discharges occurring in electrolytes were discovered more than 180 years ago and should be attributed to Foucault and Fizeau [1] according to Wüthrich and Mandin [2]. However, the exploitation of electrolytic plasmas in the field of surface treatments was not really undertaken until the 1960s when Mc Neill and Gruss used a spark discharge to develop a layer of cadmium niobate on a cadmium anode immersed in an electrolyte containing a niobium salt [3, 4]. They showed that the composition of the electrolyte as well as the treated material has an influence on the discharge and in particular on the value of the breakdown voltage of the insulating oxide layer [5]. In 1967, Wood and Pearson [6] established the link between the breakdown voltage, the type of alloying elements for a given metal, and the concentration, of the electrolyte constituents. In 1970, Yahalom and Hoar showed that the increase in voltage observed at the beginning of treatment was all the more rapid as the imposed current density was high [7]. On the other hand, they established that the breakdown voltage was independent of the current density. By cross-reviewing previous results from various authors, Yahalom and Zahavi deduced that the breakdown was controlled at the film/electrolyte interface rather than at the heart of the oxide layer [8]. During the same period, Brown et al. showed that, if a constant voltage was maintained during treatment, the current intensity decreased as the resistivity of the oxide layer increased until the discharge was quenched [9]. Therefore, in 1976, Markov et al. [10] increased the voltage during treatment to maintain the current density at a sufficient level, avoid quenching the discharge and thus allow the oxide film to continue growing. Deeper studies have been carried out

J. Martin · T. Czerwiec · G. Henrion (✉)
Université de Lorraine, CNRS, IJL, Nancy, France
e-mail: gerard.henrion@univ-lorraine.fr

V. S. Saji, M. Zheludkevich (eds.), *Plasma Electrolytic Oxidation – Fundamentals, Advances and Applications*,
https://doi.org/10.1007/978-3-031-92429-3_3

since the end of the 1970s. Some authors proposed theoretical models [10–13], but most works on PEO dealt with experimental results concerning the properties of the resulting oxide layer (in terms of thickness, mechanical properties, corrosion resistance, etc.) as a function of various treatment parameters (current density, AC or DC signal, composition and concentration of the electrolyte) [14]. Plasma electrolytic oxidation (PEO), also known as micro-arc oxidation [10], spark anodic oxidation [15] or spark anodizing [16], has attracted increasing interest in recent years for the treatment of light metal alloys (Ti, Al, Mg alloys) [17].

Depending on the current or voltage supply used to power the electrodes, the PEO process can exhibit different working regimes, especially for what concerns the PEO discharge ignition and behaviour. This chapter deals with the physics that supports these regimes. A focus is brought to the so-called soft sparking regime and the parameters that influence its occurrence. Finally, trends and new strategies involving this soft sparking regime to improve the PEO process will be presented.

3.2 Electrolyte–Metal Interaction in the Context of PEO

In order to describe the electrical phenomena occurring in PEO processes, it is essential to describe the phenomena observed when a metal is in contact with an electrolyte.

3.2.1 Space Charges in Semiconductor–Electrolyte Contact: Electrolyte Side

In the case of aluminium, a layer of alumina is always present on the surface. When such a surface is immersed in water, it binds water molecules in the form of hydroxyl groups. The hydrated oxide surface is amphoteric (it is able to adsorb or release protons) and this interaction can be described by Eqs. (3.1) and (3.2) where ≡AlOH is an aluminol group [18, 19]

$$\equiv \mathrm{AlOH} + \mathrm{H}^{+} \rightleftharpoons \mathrm{AlOH_2}^{+} \tag{3.1}$$

$$\equiv \mathrm{AlOH} \rightleftharpoons \equiv \mathrm{AlO}^{-} + \mathrm{H}^{+}_{\mathrm{aqu}} \tag{3.2}$$

These reactions produce an electrical charge on the oxide surface that depends on both the type of oxide and the pH of the solution (Fig. 3.1a). The isoelectric point (IEP) of α-Al_2O_3 is about 9.1 in pure water [18]. This means that, when immersed in a solution with pH = IEP, the surface carries no net charge and is electrically neutral. For pH values lower than IEP, the aluminium oxide carries a positive surface charge, whereas the charge is negative for pH higher than IEP. The surface charge is also affected by the specific adsorption of anions or cations present in the

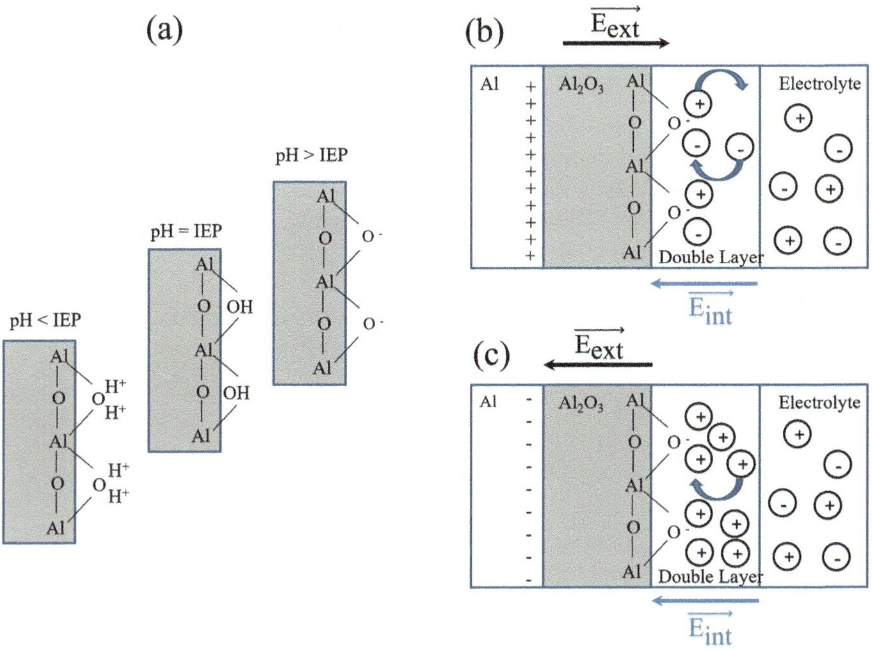

Fig. 3.1 Schematic description of the microscopic phenomena at the interface between alumina and electrolyte. (**a**) Surface charge of alumina at different pH values compared to the isoelectric point IEP, (**b**) ion displacement in the double layer of the electrolyte induced by an applied external field with a positive bias of the sample (anodizing) for pH < IEP, and (**c**) same as (b) with a negative bias. *Adapted with permission from* [18] © *2009 WILEY-VCH Verlag GmbH & Co. KGaA)*

solution other than H^+/OH^- ions, resulting in an IEP shift. In alkaline solutions used in PEO (pH usually around 12), the surface of the growing alumina is negatively charged (Fig. 3.1a). This charge distribution produces the so-called double layer (DL or electric double layer, EDL) that appears on the surface of an object when exposed to a fluid. There are two parallel layers of charges surrounding the object. In the first layer, the surface charge (positive or negative) consists of ions adsorbed on the surface due to chemical interactions such as those we have described for alumina. The second layer consists of ions attracted to the surface charge via the Coulomb force, which screens the first layer. Unlike the ions in the first layer, the ions in the second layer can move in the fluid under the influence of an electric field or temperature. Figure 3.1b, c depicts the movements of ions in the double layer when the sample is positively or negatively polarized. For positive polarization, the external field repels the positive ions towards the electrolyte and attracts the negative ions towards the double layer. As a consequence of this movement of charges, an internal field is formed in the double layer with a direction opposite to the applied external field. When the surface is negatively polarized, the internal field has the same direction as the external field.

3.2.2 Space Charges in Semiconductor–Electrolyte Contact: Material Side

The presence of charges at the surface of the oxide film modifies the distribution of charges inside the surrounding electrolyte. We will assume that alumina behaves like an n-doped semiconductor. In the case of pH = IEP, the positive charges of the immobile donor state are compensated by the negative charges of the mobile electrons. The case corresponding to IEP > pH with a positive bias (anodic situation) is presented in Fig. 3.2a, d. Since the surface of the semiconductor is negatively charged and also due to the fact that the metal is positively biased, the electrons are repelled towards the bulk of the semiconductor. The immobile positive charges of the donors remaining in place result in the formation of a depletion zone (Fig. 3.2d). As a consequence, there is a bending of the band edges downwards from the interface (Fig. 3.2a).

With the same IEP value and under a negative bias (cathodic situation), the electrons are not efficiently repelled from the region at the surface of the semiconductor because of the negative bias of the metal (Fig. 3.2e), which leads to a week bending of the band edges upwards from the interface (Fig. 3.2b). When the negative applied

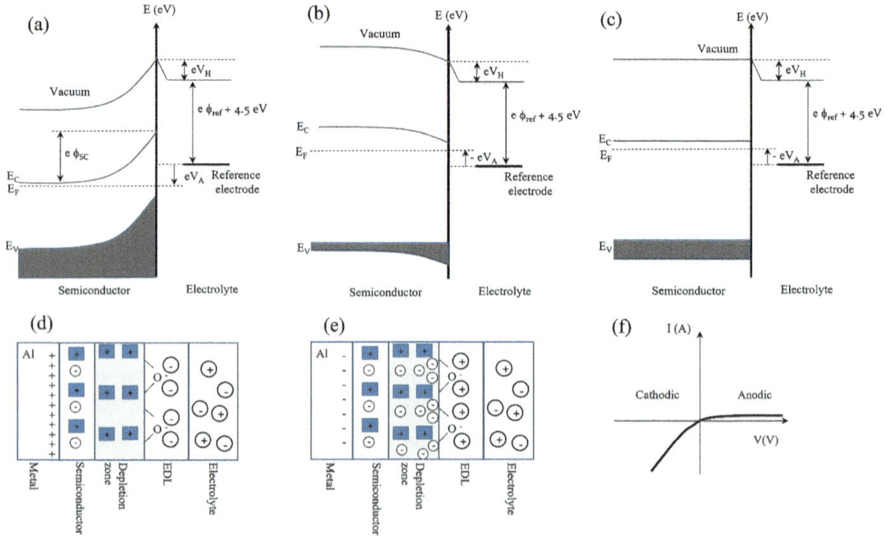

Fig. 3.2 Energy representation of the interface between an n-doped semiconductor and an electrolyte when a bias (V_A) is applied. (**a**) V_A voltage is positive, (**b**) V_A voltage is negative, and (c) V_A voltage is sufficiently negative to reduce the band bending to zero. (**d**) Space charge layer in n-doped semiconductor and in electrolyte corresponding to situation (**a**), (**e**) same for situation (**b**), and (**f**) corresponding current-voltage curve. V_H is the potential drop across the double layer, Φ_{sc} is the potential drop across the space charge region, and $e\Phi_{ref}$ is the work function of the electrolyte. *Adapted with permission from* [19] © *2012 Springer Science + Business Media and* [20] © *1990 Elsevier*

voltage is high enough, the electrons are not efficiently repelled from the region at the surface of the semiconductor, and the bands are not bent (flat band situation) (Fig. 3.2c). As a result, the electrolyte–semiconductor interface is non-conductive when positively biased (anodic region) and can be conductive when negatively biased. This is illustrated in the current–voltage characteristic of Fig. 3.2f. Passivated aluminium demonstrated noticeable differences in current response to positive and negative polarization, such as rectifying properties of the metal-oxide–electrolyte junctions [21].

3.2.3 Oxide Film Formation, Stability, and Breakdown

In the previous paragraphs, the origin of the presence of an oxide film when a metal is exposed to an electrolyte is not explained. Providing explanations for this phenomenon is beyond the scope of this chapter, but some elements can be found in the literature [22, 23]. We will simply show that it is possible to form a passive oxide film in the alkaline electrolytes generally used in PEO. Figure 3.3A shows a transmission electron microscope (TEM) micrograph of the oxide layer formed on aluminium after anodizing in a KOH (2.5 10^{-2} M, pH = 12.4) electrolyte under an increasing voltage of 40 V with a voltage ramp of 10 V·min^{-1}. The layer can be subdivided into a dense inner layer of 50 nm (passive oxide film) and a porous layer

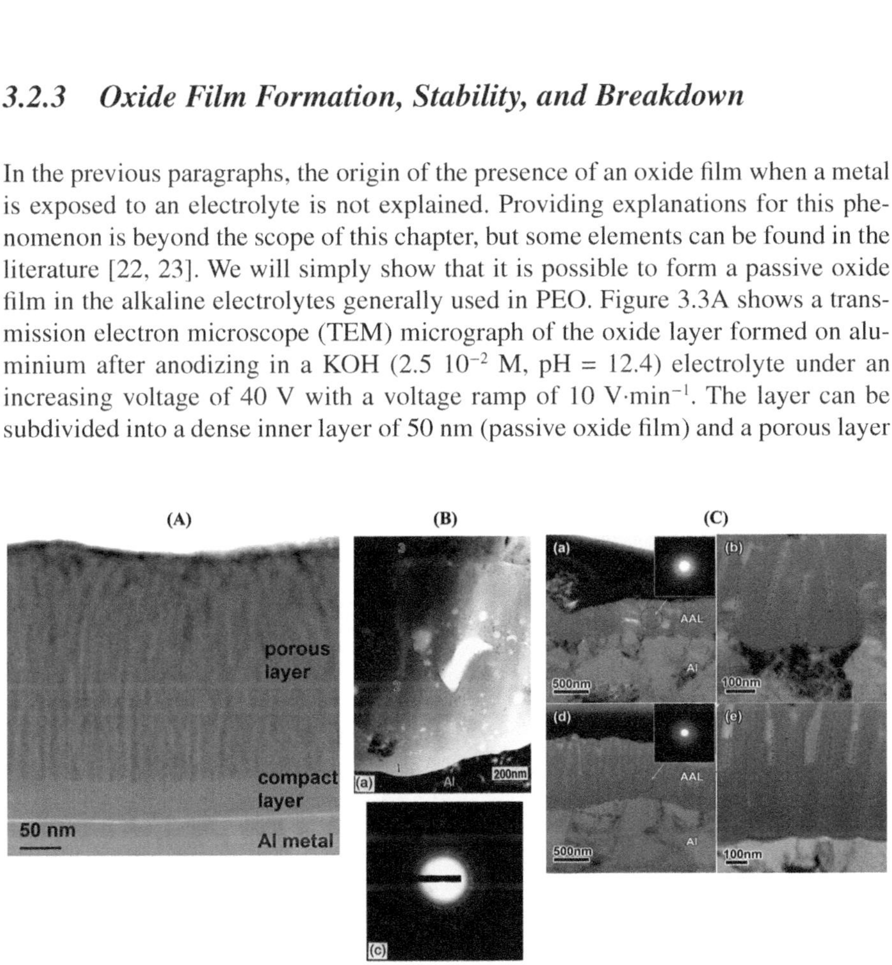

Fig. 3.3 (*A*) STEM cross-section observations of aluminium anodized layer in KOH electrolyte up to 40 V (scan rate: 10 V·min^{-1}). *Reproduced with permission from* [24] *© 2021 Elsevier.* (*B*) Cross-section TEM images of the inner layer near coating-substrate interface (**a**), SAED patterns taken from sublayer 2 of a 1 to 2 h PEO treatment in silicate electrolyte (**b**). *Reproduced with permission from* [25] *© 2002 Elsevier.* (*C*) Cross-section TEM images and the corresponding selected area electron diffraction of the circled amorphous alumina layer (AAL) after a PEO treatment of 4 s (**a**, **b**) and 8 s (**d**, **e**). *Reproduced with permission from* [26] *© 2021 Elsevier*

of about 300 nm. Hence, under electric conditions that do not allow discharges to occur, the formation of a passive oxide film at a rate of about 10 nm·min^{-1} is possible.

The formation of a dense amorphous layer at the interface between the metal and the oxide layer was also observed after PEO treatments. Fig. 3.3B shows an amorphous layer of about 140 nm for a treatment of 1–2 h in a solution of sodium silicate (2–10 g·L^{-1}) in distilled water with addition of KOH (1–2 g·L^{-1}) with bipolar AC voltage pulsed at 50 Hz in the range of 400–600 V in the positive half cycle and −200 V in the negative half cycle and a current density of 100 mA·cm^{-2} [25]. Similar observations were reported in reference [27]. Recently, Zhu et al. [26] showed the same oxide layer morphologies for very short PEO treatments performed in a solution of sodium silicate (12 g·L^{-1}) in distilled water with the addition of KOH (5 g·L^{-1}) with a unipolar pulsed current generator (100 Hz, 40% duty cycle, 0.1 A·cm^{-2}). For treatment times of 4 and 8 s, under conditions for which no discharge occurs (respectively, 150 and 190 V), the thickness of anodic amorphous oxide layers are, respectively, 500 and 1000 nm (Fig. 3.3C). According to the authors, such an increase in thickness is explained by an increase in temperature estimated at 250°C (see explanations below). These observations validate the growth mode of the amorphous oxide layer by migration of Al^{3+} ions from the metal–oxide interface and O^{2-} ions from the electrolyte–oxide interface under a strong electric field. The formation of nanopores in the upper oxide layer would be due to deformations induced by the formation of the oxide and by plasticity induced by the electric field [26]. In this situation, and for oxide with low electronic conductivity, the transport is thermally activated, and the relation between current density (j) and electric field in the oxide (E) is given by Eq. (3.3) where β and A are constants, and V and d are the potential differences between the oxide layer and the thickness of the oxide film, respectively [22, 23]

$$j = A\,e^{\beta E} \text{ and } E = \frac{V}{d} \tag{3.3}$$

Using the explanations provided in Fig. 3.1b, for positively polarized surfaces, the decrease in electric field acts as an additional energy barrier that hinders the anions migration from the electrolyte, thus leading to an increase in temperature in the oxide layer [26]. According to Zhu et al. [26], this increase in temperature constitutes the main difference between the start of a PEO treatment and an anodization treatment.

The concept of breakdown in electrolytes is rather poorly defined and depends on the physical or chemical point of view adopted. From a strict physical point of view, electric breakdown of an insulation oxide layer or the depletion region of a semiconductive layer occurs if extra charge carriers are generated either by tunnelling (Zener breakdown) or by collisions (avalanche breakdown) [22]. In chemistry, the breakdown is defined as an immediate increase in local conductivity or passivity breakdown [23]. Local destruction of the passive film can be induced by a dielectric breakdown, such as defined above or by a corrosion process that weakens the passive film locally. Aggressive ions like chloride may also penetrate into the film, or hydrogen incorporation into the film with hydrogen gas evolution can produce film reduction, dissolution or flaking off. In conclusion, the passive film produced during positive polarizations (anodic regime) will likely undergo a dielectric breakdown in the physical sense, while it will likely undergo a breakdown in the chemical sense during negative polarizations (cathodic regime).

Veys-Renaux and Rocca [28] performed potentiodynamic measurements of the current density as a function of voltage (scan rate: 15 $V{\cdot}min^{-1}$) on aluminium alloys treated in a solution of sodium silicate (9.3 $g{\cdot}L^{-1}$) in distilled water with the addition of KOH (1.9 $g{\cdot}L^{-1}$) with DC voltage. The current density values are very low until the anodic alumina layer (estimated to be between 200 and 300 nm thick) undergoes an electric breakdown for a voltage of 360 V.

The oxide film can also be damaged when interacting with the electrolyte. Following Hassel and Lohrengel [29], cathodic processes can be subdivided into two mechanisms: charge transport by electrons and hydrogen evolution at the oxide/electrolyte interface or charge transport by protons and hydrogen evolution at the metal/oxide interface. By applying successive anodic and cathodic polarization with various durations on pre-anodized aluminium samples, Sah et al. [30] showed that a previous electric breakdown during the anodic pulse favours a local destruction of the passive alumina layer, which in turn favours a chemical breakdown associated to a high current density during the following cathodic pulse. During the second anodic pulse, the anodic current is much lower than that of the first anodic pulse. In contrast, without an intermittent cathodic pulse, breakdowns occur hardly during the second anodic pulse. According to the authors, these ruptures would be of a physical nature because they produce an enlargement of the discharge channels. The type of breakdown (electrical discharge or passivity breakdown) remains however an open question. Although anodic breakdown sites are less protective, they turn out to be more protective after the cathodic pulse. Anodic breakdown in the succeeding anodic pulse occurs preferentially in the regions where no prior breakdown has occurred.

3.3 Type and Mode of Electrical Discharge in PEO

After having shown that an oxide layer having a more or less dielectric behaviour following the applied anodic–cathodic sequence forms fairly quickly on a valve metal placed in an alkaline electrolyte, we will discuss the different modes of application of an electric field to achieve dielectric breakdown.

3.3.1 Direct Voltage and Direct Current

As displayed in Fig. 3.4, a typical PEO treatment process can be powered by an AC or DC generator. Originally, most generators operated in DC mode, with the sample to be treated placed as an anode, making difficult the control of the discharges. Two different types of electrical generators exist: voltage-controlled generators (Fig. 3.4a) and current-controlled generators (Fig. 3.4b) [31, 32]. DC modes at constant voltage or current (Fig. 3.4c) are only viable for short treatment times [32].

3.3.2 Alternating, Unipolar-Pulsed Regimes

It appears interesting to periodically cut the electrical signal in order to limit the lifetime of the discharges. This can be done using an AC (Fig. 3.4d) or still better, unipolar (Fig. 3.4e), bipolar (Fig. 3.4f) pulsed signals or bipolar signals with pause times (Fig. 3.4g) [14].

Figure 3.5 summarizes very well the different behaviours that can be observed when applying different forms of electrical signal to an oxide layer formed by the PEO process in sodium hydroxide (2 g·L^{-1}) and tri-polyphosphate (4 g·L^{-1}). In all cases, a preliminary period of formation of an oxide layer is carried out. When the treatment is done with AC voltage (Fig. 3.5a), a shift between the voltage and the current is observed. Non-zero values of the current are observed only when the breakdown voltage is established. We thus have current–voltage characteristics such as those presented in Fig. 3.6a. Such behaviour is consistent with the explanations given in Sect. 3.2.3. The oxide layer behaves as an electrical insulator as long as the breakdown voltage is not reached.

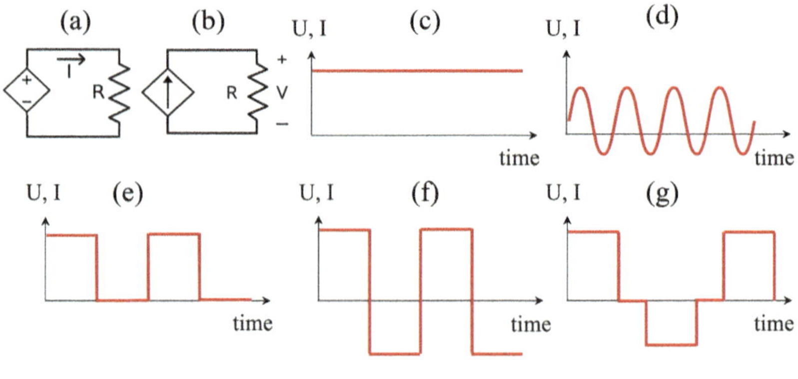

Fig. 3.4 Simplified electrical circuits for (**a**) voltage control, (**b**) current control, and mode of PEO processing: (**c**) DC, (**d**) AC, (**e**) pulsed unipolar, (**f**) pulsed bipolar, and (**g**) pulsed bipolar with pauses

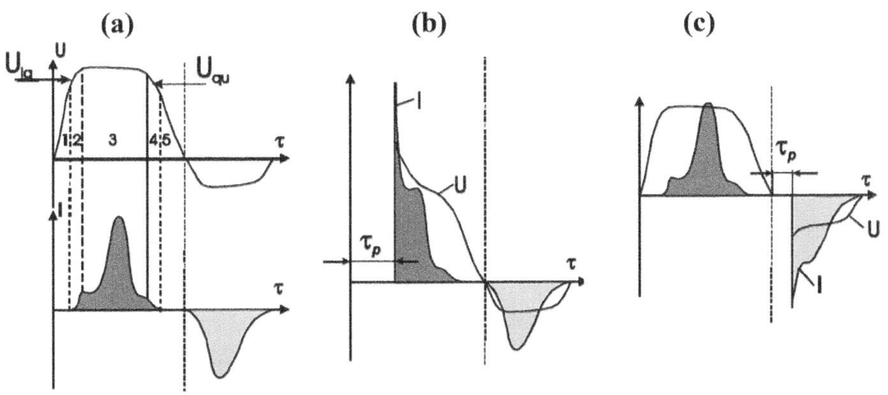

Fig. 3.5 Typical voltage (U) and current (I) curves over one period of polarization in (**a**) AC, (**b**) positive pulses polarization (PP) mode (AC + positive voltage pulses), and (**c**) negative pulse polarization (NP) mode (AC + negative voltage pulses). *Reproduced with permission from* [33] © *2005 Elsevier*

An interesting phenomenon occurs when a positive or negative pulse is superimposed on the AC signal. A very rapid rise in current is observed as soon as the voltage is applied. This phenomenon results in a very different shape of the current-voltage characteristic as shown in Fig. 3.6c. It was obtained in an electrolyte composed of sodium silicate (10–20 $g{\cdot}L^{-1}$) in distilled water with the addition of KOH (1–5 $g{\cdot}L^{-1}$) with an asymmetric AC pulsed electric signal. AC power sources are not so much used given the limitations in the power (max. 10 kW) and current frequency.

This phenomenon is also commonly observed in PEO experiments performed with a unipolar voltage power supply [36–38]. In the past, these current spikes have been interpreted as being due to electrical discharges. Hermanns et al. [36] showed that during these electrical current pulse virtually no discharge emission is detected. This initial pulse might be associated with charge transport before the avalanche breakdown. This initial pulse is associated with the charging process of a capacitor (the oxide layer). The time evolution of the displacement current can be described by a simple capacitive–resistive equivalent circuit. It is only when point B (see Fig. 3.6c) is reached that dielectric breakdown occurs. From this point onwards, the current–voltage characteristics are identical to those observed in AC (Fig. 3.6a).

The current decay during the capacitor discharge phase produces an even more peculiar shape of the current–voltage characteristic with a unipolar voltage generator. Thus, the current can decrease while the voltage increases. The voltage increases linearly with the current for a voltage higher than 340 V, which would correspond to the dielectric breakdown voltage. The current oscillations observed during time t_s would be produced by electric discharges. For voltages higher than 340 V, the electric discharges appear. The shape of the current–voltage characteristic is very dependent on the type of electrolyte used [39].

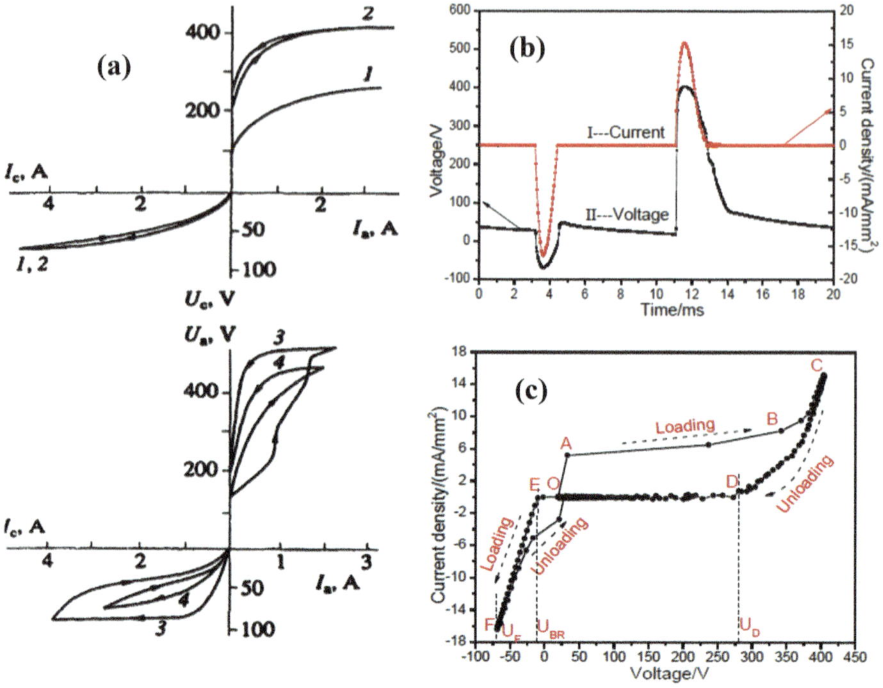

Fig. 3.6 (**a**) Transient V–I characteristics at different processing times 0.5 min (1), 5 min (2), 60 min (3), and 150 min (4). *Reproduced with permission from* [34] © *2000 Consejo Superior de Investigaciones Científicas (CSIC).* (**b**) Evolution of current (I) and voltage (II) as a function time in the first minute of a PEO process with AC pulses, and (**c**) corresponding transient V–I characteristics. *Reproduced with permission from* [35] © *2012 Published by ESG*

3.3.3 *Bipolar-Pulsed Regimes*

In our experiments, the electrolytic solution consists of KOH and Na_2SiO_3 diluted in distilled water; the resulting electrolyte conductivity ranges from 2.3 to 2.5 $mS \cdot cm^{-1}$. A cooling system maintains the electrolyte temperature below 30 °C during the process. Two bipolar pulse current generators (CERATRONIC® process [40]) are used, which deliver current to the substrate with amplitude in the range 0–36 A and 75–150 A, respectively. Process parameters, such as the period (T), the different duration of each step in the current waveform (T_i), and the amplitude of the positive (I_P) and negative (I_N) current can be adjusted over a wide range (Fig. 3.7). We define the parameter R_{CQ} as being the positive to negative charge quantity ratio:

$$R_{CQ} = \frac{Q_P}{Q_N}, \tag{3.4}$$

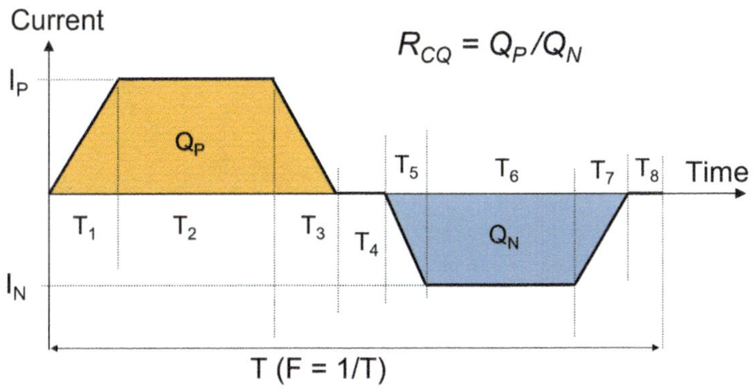

Fig. 3.7 Schematic representation of a typical bipolar current waveform. The current waveform can be finely adjusted by tuning the current frequency (F), the anodic and cathodic current amplitude (I_P and I_N, respectively), and the duration of each time step (T_i)

where Q_P and Q_N are the anodic (positive) and cathodic (negative) charge quantity respectively defined as:

$$Q_P = \int_0^{T_1+T_2+T_3} I(t)dt$$
$$Q_N = \int_{T_1+T_2+T_3}^{T} I(t)dt \tag{3.5}$$

Figure 3.8a shows the voltage–time response of PEO treatments performed under different R_{CQ} values. In current control mode, as the PEO process starts, a sharp increase in voltage occurs during the first milliseconds, which is then followed by a much-reduced voltage increase rate. Figure 3.8a shows the corresponding evolutions of the light emission recorded over 10 ms for different R_{CQ}.

Microdischarges (MDs) appear only during the anodic polarization. For treatment times higher than a few minutes, a delay is observed between light emission and the rising edge of the anodic current. It appears that the delay in the MDs appearance is longer for lower R_{CQ} values. Figure 3.8b shows the exponential behaviour of the anodic voltage–time response (Eq. 3.4).

$$V(t) = V_{max}\left(1 - e^{-\frac{t}{\tau}}\right) \tag{3.6}$$

The corresponding τ values evolution as a function of treatment time is shown in Fig. 3.8c. From this figure, it can be seen that the time constant increases gradually with the processing time and remains constant for longer treatment times. A mechanism of charge accumulation taking place at the oxide/electrolyte interface and arising before the occurrence of dielectric breakdown is proposed to explain the ignition of MDs during pulsed bipolar PEO of aluminium [41].

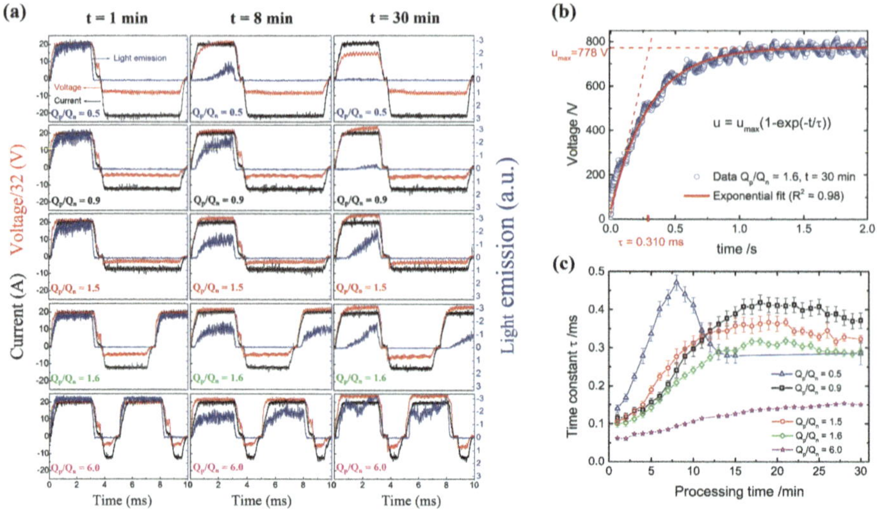

Fig. 3.8 (**a**) Time evolution of the light emission (blue lines), the corresponding current (black lines) and voltage (red lines) signals recorded over 10 ms for different values of the charge quantity ratio R_{CQ} and at various processing times, (**b**) example of the curve fitting for the transient anodic voltage–time response ($R_{CQ} = 1.5$, $t = 30$ min) using an exponential function and (**c**) evolution of the time constant τ (deduced from exponential fitting) with the processing time for various values of the charge quantity ratio R_{CQ}. *Reproduced with permission from* [41] © *2017 Elsevier*

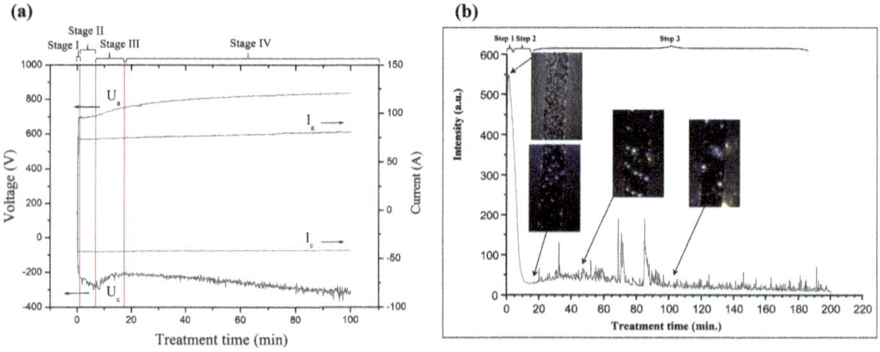

Fig. 3.9 (**a**) Evolution of the maximum current and voltage in the cathodic and anodic part of the bipolar current as a function of time during a PEO process [42]. (**b**)Typical time variation of the Al (396.2 nm) emission line intensity along a conventional PEO process (positive current density 38 A·dm^{-2} and $R_{CQ} = 1.57$) with side-view pictures of aluminium alloy samples at different times of the PEO process (1, 15, 45, and 100 min). The integration time of the camera is set between 8 and 10 ms. *Figure (**b**) is reproduced with permission from* [43] © *2006 Elsevier*

Figure 3.9a shows the evolution of the maximum current and voltage in the cathodic and anodic part of the bipolar pulse as a function of processing time during a PEO process. The typical time variation of the Al (396.2 nm) emission line intensity along a conventional PEO process is shown in Fig. 3.9b with side-view pictures of aluminium alloy samples at different times of the PEO process.

In the case of a current-regulated treatment, the voltage increases by several hundred volts per minute in the first moments of treatment (stage I) before reaching the breakdown voltage of the barrier layer of the native oxide (stage II). The first micro-discharges, small in size and very numerous, begin to appear sporadically on the surface of the sample. The appearance of discharges is marked by a break in the increase rate of the voltage; the voltage stops increasing so quickly, and the curve exhibits a concave shape.

After several minutes of treatment, the voltage evolution increases more slowly. The size, number and colour of the visible micro-discharges evolve according to the treatment time and the electrical parameters (stage III). In general, the micro-discharges reduce in number but increase in size. Their colour change from a white, slightly bluish tint to an orange tint. For long treatment times (stage IV), the visible micro-discharges become much less numerous but spatially more extensive.

Hussein et al. proposed a phenomenological model comprising three types of discharges [44]. They distinguish discharges occurring in the gas trapped in the micropores of the external layer (type C) and discharges occurring at the interface with the electrolyte (type A) and promoting reactions involving the elements of the electrolyte. Finally, discharges crossing the oxide and interacting directly with the metal substrate are proposed (type B). The latter are typically more intense for long treatment times and are associated with the "pancake" or "volcano" structures characteristic of PEO coatings.

Liu et al. also proposed a model distinguishing only two types of discharges according to their intensity (weak and strong) [45]. A model with 5 types of discharges was put forward by Cheng et al. based on the model of Hussein et al. and adding two additional types of discharges [46]. Considering the morphology of the layers, which often presents a large separation between the external sublayer and the internal sublayer, they proposed the existence of discharges that only partially cross the oxide: a type of discharge visible from the surface and that only crosses the external part of the layer (type E), and another type of discharge, typically undetectable from the surface and crossing only the internal part of the layer in contact with the metal substrate (type D). These discharges help to explain the different growth of the external layer and the internal layer and in the case of treatments under soft regime, it is suspected that the discharges not visible from the external surface help continuing the growth and densification of the layer.

3.4 Different Strategies to Optimize Layer Quality and Minimize Power Consumption

In addition to the so-called soft regime conditions, which are going to be developed in the next sections of this chapter, different techniques have been developed to improve the quality of the layers and reduce the electrical consumption of the PEO process. It should be noted that the conditions for obtaining the soft regime are strictly fulfilled only in the case of the current-controlled bipolar electrical regime for $R_{CQ} < 1$.

When working in a unipolar voltage-controlled regime, three strategies can be implemented. The first one consists of controlling the mean microdischarge size based on an in situ discharge detection process, consisting of CCD-camera measurements followed by fast image processing algorithms as reported by Hermanns et al. [47]. The duty cycle and total current are adjusted over treated substrate area for stabilizing the mean discharge size. However, this control method requires, in addition to a voltage-controlled unipolar generator, an optical measuring device and analysis and control software. Furthermore, the electrolyte must be transparent.

The second method consists of using intelligent self-adaptive control technology based on real-time active electrical diagnostics and the adjustment of these electrical parameters. Thus, a voltage slightly higher than the breakdown voltage value is achieved to be located just above the current corresponding to the breakdown [38]. According to Yang et al. [38], it is thus possible to suppress the current pulse that generally occurs when the PEO process operates in voltage-controlled unipolar mode. A trapezoidal or sawtooth current waveform can be used to control and gradually ramp the current up or down [48]. This allows to control transitions between positive and negative pulses in order to avoid current spikes that can have detrimental effects on coating formation or lead to unwanted discharge behaviour. It was shown that a negative ramp in a bipolar mode reduces porosity, enhances uniformity, and thickens the inner barrier layer, which consequently improves coating quality [48].

The third promising approach consists of working on oxide layers made prior to a PEO treatment. Basically, this amounts to combining treatments in unipolar and bipolar modes [49, 50].

3.5 The Soft Sparking Mode

3.5.1 General Features of the Soft Sparking Mode

PEO coatings usually exhibit a high porosity level, which is often detrimental to the usage properties, especially for what concern corrosion protection. This porosity results from the PEO discharges that become bigger and stronger as the process goes on and when the growing oxide layer gets thicker. This particularly occurs when PEO is operated under DC current (galvanostatic mode), under unipolar anodic current, or to a lesser extend under pure AC current mode (sine current waveform). The use of a bipolar current (as schematically represented in Fig. 3.7) to supply the electrodes offers the possibility to adjust independently the anodic part and the cathodic part of the current waveform, which allows tuning the anodic to cathodic charge quantity ratio R_{CQ} (see Sect. 3.3.3).

F. Mécuson-Jaspard et al. investigated the effect of the different current waveform parameters on the resulting morphology of the produced PEO coating in terms of thickness homogeneity and the presence of defects in the PEO layers [42]. The qualitative results collected and reported in Fig. 3.10 showed that there exists a certain correlation between the coating "quality" and the current waveform parameters, especially the R_{CQ} parameter.

This was the premise of what was later on reported as the soft sparking regime in PEO by the same author group [43].

The soft sparking regime is manifested through three noticeable aspects:

- A decrease in PEO discharge intensity (Fig. 3.11a)
- A decrease in acoustic emissions
- A drop in anodic voltage amplitude (when working under constant current amplitude mode) (Fig. 3.11b)

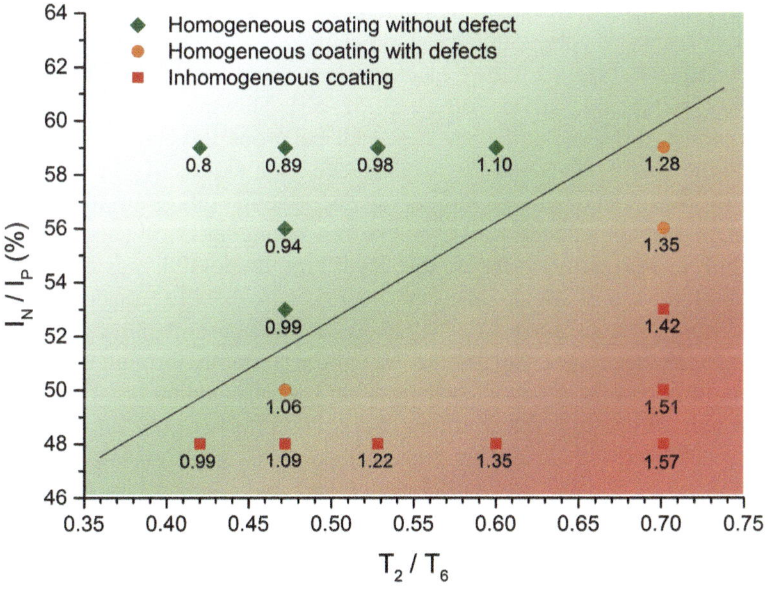

Fig. 3.10 Schematic representation of the PEO coating morphology (in terms of thickness and defects) on Al2214 as a function of the waveform parameters I_N/I_P and T_2/T_6. Figures under each data point represent the values of R_{CQ}. $J_P = 38$ A·dm^{-2}, $F = 100$ Hz, process time = 80 min

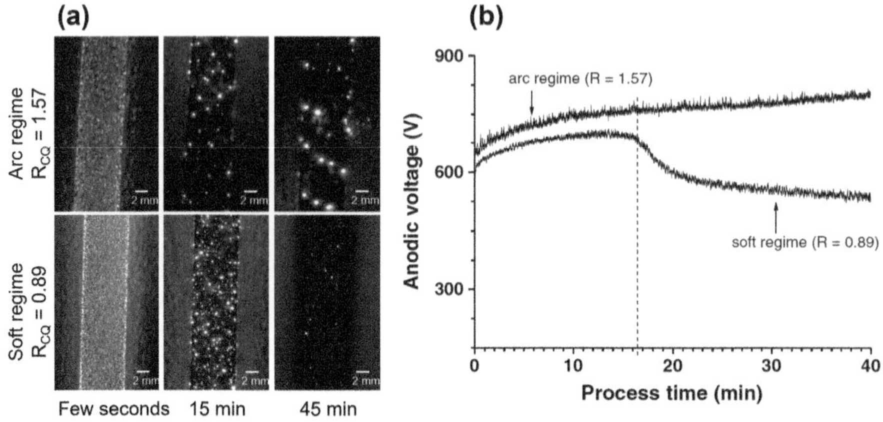

Fig. 3.11 (**a**) Comparison of the discharge appearance at different processing times under arc regime (R_{CQ} = 1.57, upper row) and under soft regime conditions (R_{CQ} = 0.89, lower row). J_P = 38 A·dm^{-2}, F = 100 Hz. Integration time of the camera is set between 8 ms and 10 ms except for pictures at 45 min under soft conditions for which it was set at 2000 ms. *Reproduced with permission from* [43] © *2006 Elsevier.* (**b**) Variation of the anodic voltage amplitude for arc regime (R_{CQ} = 1.57) and under soft regime conditions (R_{CQ} = 0.89). J_P = 92 A·dm^{-2}, F = 100 Hz. Transition to the soft regime occurs when the anodic voltage starts dropping (dash line). *Reproduced with permission from* [51] © *2011 Elsevier*

Although it is still not that clear what mechanisms govern the transition to soft regime, it is however well established that the transition only starts after the growing oxide layer reaches a certain thickness, typically ~30–50 μm depending on the processing conditions (grade of alloy, electrolyte composition, current density, relative position of anode and cathode, etc.) [43, 52–57]. This means that, as the coating reaches such a "threshold" thickness, its electrical properties change, which is likely related to the coating morphology together with its thickness. As the process is running under constant current amplitude, the decrease in voltage amplitude implies a decrease in the impedance of the substrate-coating–electrolyte system and all related interfaces, including the double layer and the possible gas envelope at the solid–liquid interface. Meanwhile, the decrease in the PEO discharge intensity while keeping the same current density shows that the current pathways through the coating also change, favouring the current to flow smoothly through the oxide layer, including through inner D-type discharges [46]. Consequently, ionic species better flow through the coating, which allows the oxide to further grow while densifying.

The mechanisms that support these phenomenological observations are still unclear. In the following, we will consider only the conductivity of the growing oxide and that of the surrounding electric double layer (EDL). Indeed, provided that the volume of the electrolyte is greater than the sample one, the conductivity of the bulk electrolyte can be assumed as constant. Similarly, the conductivity of the bulk metal beneath the coating is also constant, irrespective of the polarization of the working electrode.

3.6 Influence of Process Parameters on the Transition to Soft Regime

As mentioned above, the transition to soft sparking regime requires the oxide layer reaches some "threshold" thickness to occur, together with other conditions related to the current waveform applied to the working electrode, especially the presence of a cathodic part of the current. It can thus be expected that the PEO processing parameters that influence the coating growth rate also affect the transition to soft sparking, namely, the time at which the transition to soft regime starts. Except for the current density, the effect of other process parameters on the transition to the soft regime has not been deeply investigated.

For a long time, it has been established that increasing the current density results in an increase in the PEO oxide growth rate [58, 59]. In many cases, the growth rate of the PEO oxide layer increases linearly with the current density [43, 60, 61]. Although the measured values of the growth rate most often correspond to average values determined after a given processing time, it can be reasonably considered that the early steps of the coating growth are also accelerated as the current density increases. It can thus be expected that the transition from arc to soft regime occurs earlier as the current density increases, and the processing time to soft transition decreases linearly with the applied current density. This was confirmed by Martin et al. [61] and Melhem et al. [51, 62], who used a bipolar current generator to supply the electrodes that were placed in a PEO cell containing an alkaline electrolyte ([KOH] = 1 $g{\cdot}L^{-1}$ + [Na_2SiO_3] = 1.65 $g{\cdot}L^{-1}$). The working electrode (anode) was placed in between two counter-electrodes (cathodes); the distance $D_{A\text{–}C}$ of the anode to each cathode was set at 55 mm. Figure 3.12a clearly evidences a linear relationship between the current density and the variation of the processing time at which the transition to soft regime occurs, and with the variation of the average growth rate of the PEO coating as well.

Another process parameter that can influence the coating growth, and hence the transition to soft regime, is the relative position of the anode (working electrode) and the cathode, namely the distance $D_{A\text{–}C}$ in between both electrodes. This parameter has been scarcely studied despite its importance in the distribution of the applied electric field and, consequently, in the mechanisms that sustain the coating growth through the distribution of charges in the vicinity of the sample surface. Moreover, it is still unclear whether $D_{A\text{–}C}$ has an influence on the growth or not. Indeed, Rama Krishna et al. [58] studied this parameter for $D_{A\text{–}C}$ ranging from 60 mm to 120 mm in the case of PEO processing of Al7075 alloy under AC voltage power supply with a current density of 20 $A{\cdot}dm^{-2}$ in a usual electrolyte (KOH + Na_2SiO_3). From their results, they concluded that the gap distance between the electrodes does not affect the coating growth. On the other hand, Wei et al. [63] used a unipolar DC pulsed current supply to process Al2024 samples with $D_{A\text{–}C}$ values in the range of 50 mm to 250 mm, and they concluded that the anode current is influenced by the distance between the electrodes and has a critical effect on the oxidation efficiency.

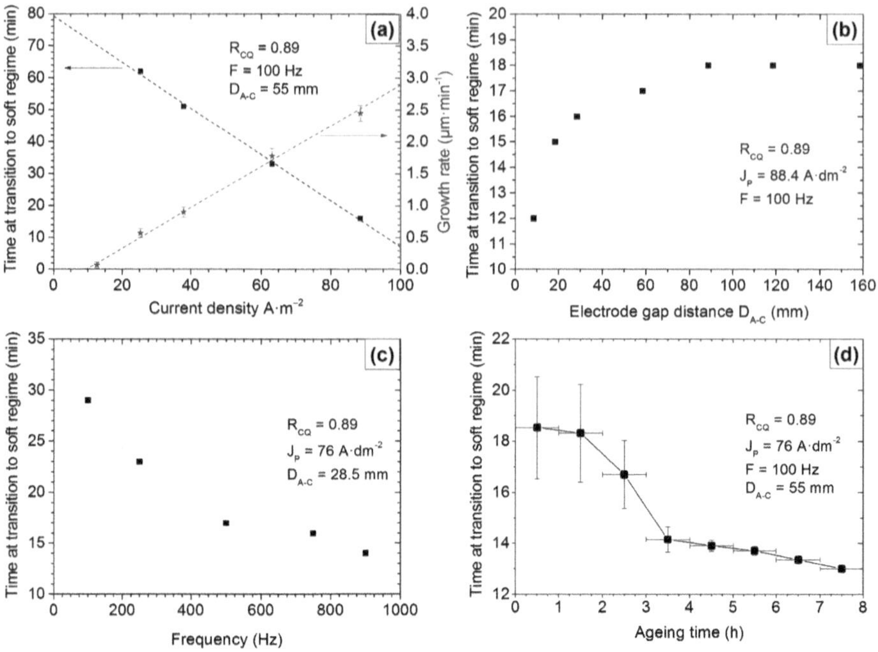

Fig. 3.12 Time at transition to soft regime as a function of (**a**) the current frequency, (**b**) the electrode gap distance, (**c**) the bipolar current pulse frequency [62], and (**d**) the age of the electrolyte. *Figure (**d**) is adapted with permission from* [64] *© 2014 Elsevier*

Regarding the effect of D_{A-C} on the soft regime transition, Melhem et al. [62] performed PEO on Al2214 samples by using bipolar current pulses in KOH (1 g·L^{-1}) + Na_2SiO_3 (1.65 g·L^{-1}) electrolyte. The gap between the electrodes varied from 8.5 mm to 158.5 mm. The R_{CQ} was set at 0.89, which is a suitable value for the soft regime to occur. The results reported in Fig. 3.12b clearly show an effect of the distance between the electrodes on the transition to soft regime. Meanwhile, A. Melhem et al. [62] noticed a slight increase in the average growth rate as D_{A-C} increases, which would either be inconsistent with the above-mentioned considerations on the threshold thickness required to start the transition to soft regime or be significant of some unexplained mechanisms regarding the relationship between the growth mechanisms and the soft regime. They also mentioned a strong thickness gradient from the centre to the edge of the processed sample, which could also affect the values of the average growth rate.

The same author also studied the effect of the current pulse frequency on the delay the process requires before the transition to soft sparking regime. The experimental conditions were similar to those mentioned above except the frequency that served as studying parameter and the electrode gap that was set at 28.5 mm. The results showed that the higher the current frequency, the earlier the transition to soft regime (Fig. 3.12c). A. Melhem et al. also noticed that the average growth rate increased with the current frequency while the denser inner part of the coating increased as well [62]. The latter point was explained by the fact that because the transition occurred earlier at high frequency, the process ran a longer time under soft regime (11 min at 100 Hz, 17 min at 250 Hz, 23 min at 500 Hz, 24 min at 750 Hz, and 26 min at 900 Hz).

Finally, Martin et al. [64] questioned about the electrolyte ageing and its role in the PEO process. It is indeed well known that due to the electrical discharges and the chemical reactions occurring during the PEO process, the electrolyte bath gets progressively polluted with particles from the material and compounds such as silicate, phosphates, metal oxides, and others, depending on the electrolyte and material composition. The physicochemical properties of the electrolyte can thus change as the electrolyte gets more used, which can influence the PEO process as well. Regarding the transition to soft regime, the results point out the influence of the current frequency on the transition to soft regime (Fig. 3.12d). The earlier transition to soft regime was ascribed to a decrease in both the electrolyte conductivity and the discharge strength as the electrolyte gets older.

3.7 Mechanisms at the Interfaces and the Concept of Active Zone

We are considering here PEO conditions of electrolyte that are suitable for growing thick oxide layers. This means that the pH of electrolyte is higher than the isoelectric point (IEP) of the oxide (in the case of alumina, IEP = 9.1), which is usually the case for the PEO of aluminium in an alkaline electrolyte with pH > 12.

A. Rogov et al. studied the behaviour of charges (anions, cations) within the EDL and at the interface between the metal and the oxide layer, which they called the active zone [65]. This active zone (AZ) is created at the early beginning of the PEO process. It corresponds to the initial oxide layer and supports the main voltage drop during the sharp voltage change over the first stage of the process (Fig. 3.9). According to this model, the major part of the inner PEO sublayer grows due to the migration of (i) the metal cations from the underneath metal and (ii) the transportation of anion from the electrolyte through the interconnected network of pores [66] assisted by the applied electric field (Fig. 3.13). The AZ conductivity depends on the shape of the current waveform (whether there is a cathodic part of the current or not). Indeed, during cathodic polarization, hydrogen radicals (likely hydrated) are formed in the EDL, which enriches the AZ in these radicals close to the oxide–EDL interface. As a consequence, the conductivity within that part of the AZ increases and the electric field is hence concentrated to the remaining narrow part of the AZ. Consequently, the subsequent anodic polarization faces a lowest potential barrier in the EDL which suppresses the electrical breakdown. All these processes are well described in Rogov's paper [65] and schematically depicted in Fig. 3.14.

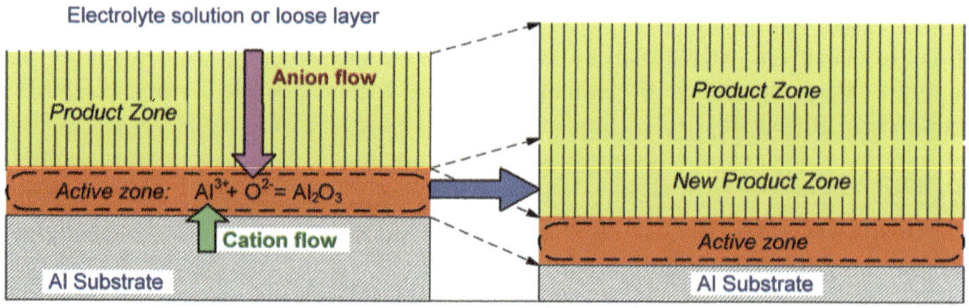

Fig. 3.13 Schematic representation of the concept of active zone in the case of PEO processing of aluminium. *Reproduced with permission from* [65] © *2017 American Chemical Society*

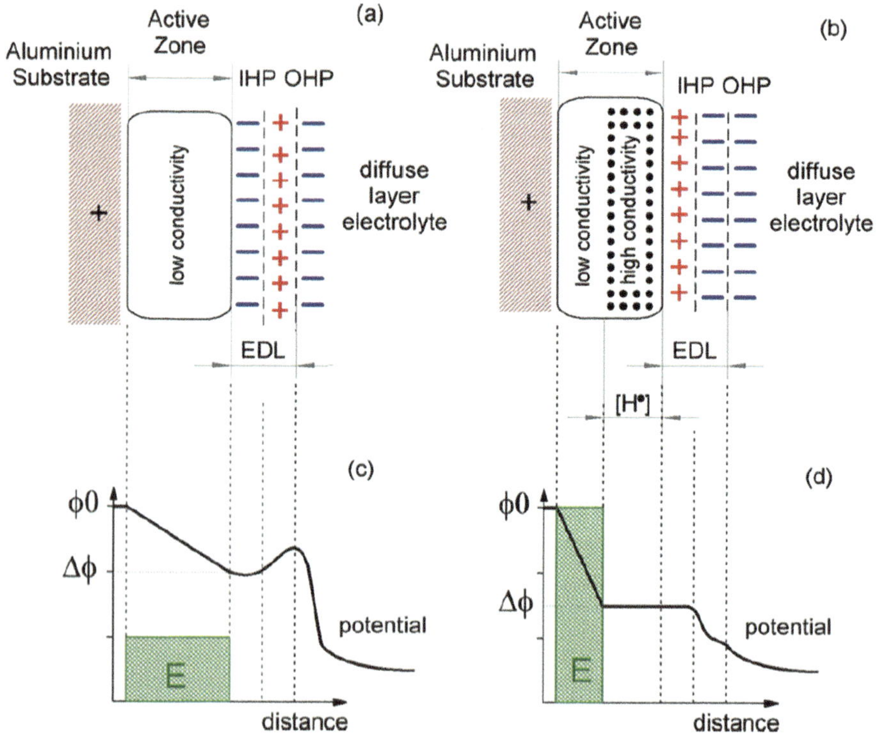

Fig. 3.14 Schematic representation of charge (**a**, **b**), potential and electric field distribution (**c**, **d**) in the active zone and EDL of PEO coating under anodic polarization without (**a**, **c**) and with (**b**, **d**) prior cathodic polarization. The symbol [H•] denotes hydrogen enriched area. *Reproduced with permission from* [65] © *2017 American Chemical Society*

3.8 New Strategies Involving Soft Sparking Regime

3.8.1 Pre-anodized Samples

As previously explained, high electrical energy consumption is currently limiting the use of PEO to niche applications in high-end technologies such as aerospace, gas and oil industries, where the outstanding range of mechanical, corrosion and wear resistance properties of PEO coatings can be exploited. During the last years, with the objective of broadening PEO for mass production applications, there has been an increasing number of works addressing practical strategies in order to improve the energy efficiency of the PEO process. One approach consists in considering the specific soft sparking regime as a potential lever to overcome the cost limitations of the PEO technique.

In pioneer studies, Matykina et al. applied the PEO process on a pre-anodized Al substrate [52, 67]. The possibility that the presence of a porous alumina precursor film (27-μm thick with a 360 nm thick barrier layer elaborated in sulphuric acid) may influence the establishment of the soft sparking regime was particularly examined. As shown on the voltage–time responses for AC PEO of a pure Al, with and without pre-anodizing, displayed in Fig. 3.15a, authors clearly evidenced that a conventional anodizing pre-treatment leads to an earlier establishment of the soft sparking regime. The time needed for the achievement of a relatively low voltage (~220 V) was clearly much reduced using pre-anodized Al (~300 s) than without pre-treatment (~1500 s). For the pre-anodized samples, an earlier disappearance of noisy and large

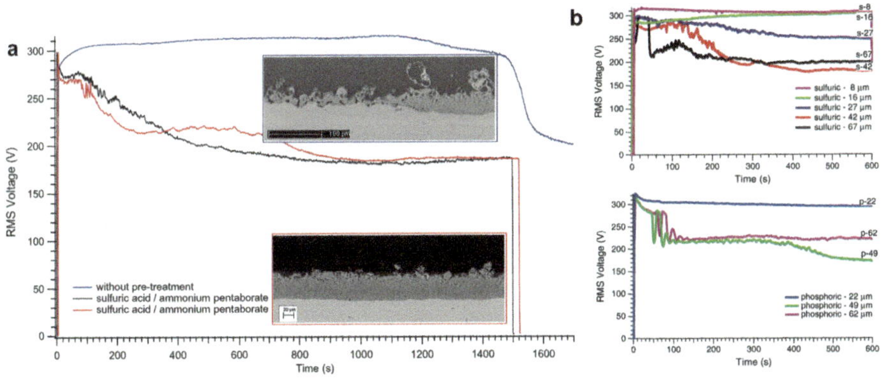

Fig. 3.15 (**a**) Voltage-time responses for AC PEO of pure Al (55 A·dm^{-2} rms in 10.5 g·L^{-1} Na_2SiO_3/2.8 g·L^{-1} KOH), with and without porous alumina precursor films, and the resulting SEM cross-section micrographs of the PEO coatings formed on Al without (for 600 s) and with pre-anodizing (for 470 s). *Reproduced with permission from* [52] © *2009 Elsevier.* (**b**) Voltage-time responses for AC PEO of pure Al (55 A·dm^{-2} rms in 10.5 g·L^{-1} Na_2SiO_3/2.8 g·L^{-1} KOH) with porous alumina precursor films of various thicknesses and pore diameters prepared in different acids (sulfuric, oxalic, and phosphoric acids). *Reproduced with permission from* [53] © *2010 Elsevier*

sparks concomitant with an earlier appearance of muted and finer bluish-white micro-discharges was also observed, confirming an earlier transition from the "arcs" to the soft sparking regime. As observed in Fig. 3.15a, authors also showed that the homogeneity and compactness of the elaborated PEO coatings can be improved, and the growth kinetic increased by pre-anodizing the Al substrate. In these conditions, authors concluded that due to the earlier achievement of a low voltage condition under the soft sparking regime, the specific electrical energy needed to produce the final PEO coating can be significantly reduced by a factor of about 4 or more. For example, they established that a specific energy of about 3 $kWh \cdot m^{-2} \cdot \mu m^{-1}$ is required to produce a PEO coating with a 27 μm thick pre-formed anodic film while it is more than about 12 $kWh \cdot m^{-2} \cdot \mu m^{-1}$ without precursor anodic film.

The role played by pre-anodizing on the onset of the soft sparking regime was explored further by the same group. For that, the influence of the initial morphology of the porous anodic films on the subsequent PEO treatment was closely examined by preparing alumina films in various acid media (sulfuric, oxalic and phosphoric acids) [53]. As shown in Fig. 3.15b, the transition to the soft sparking regime seems to appear when a sufficient thickness of the pre-anodized layer is achieved (up to 16 μm when prepared in sulfuric acid and up to 22 μm when prepared in phosphoric acid). In addition, this transition occurs sooner when anodic film is prepared in oxalic acid (at ~10 s for a thickness of ~44 μm) and phosphoric acid (at ~40 s for a thickness of ~49 μm) than in sulfuric (at ~140 s for a thickness of ~42 μm). As oxalic and phosphoric acids usually result in larger pore diameter of the alumina anodic films (~70 μm) than those produced in sulfuric acid (~15 μm), authors claim that a certain combination of pore length and diameter promotes the soft sparking regime due to a rapid increase of gas pressure within pores, locally resulting in a hollow cathode effect and contraction of the positive column of the arc discharge. Under this soft discharge regime, involving local melting and rapid cooling, the porous alumina film is gradually transformed in composition, structure and morphology and is finally incorporated into the growing PEO coating. It was also evidenced that an inward growth of the PEO coating mainly takes place from the base of the anodic film to the substrate, as usually encountered for the growth of PEO coatings within the soft sparking regime. Same authors also show enhanced mechanical properties of the PEO coatings when grown on pre-anodized Al substrate that was explained by the authors by considering the greater amount of α-Al_2O_3 phase that preferentially forms under the soft sparking regime, the earlier the soft sparking occurs and the higher microhardness is achieved [68].

More recently, Mohedano et al. combined different strategies based on pre-anodizing in different acids [53, 68], substrates [68, 69] and electrolytes selections [70], and electrical parameter adjustment [49] in order to reduce the specific energy consumption of PEO. Particularly, and using a bipolar voltage-controlled mode to supply the electrodes, authors observed a marked current drop in the different rms current–time transients (Fig. 3.16a). Authors assigned this common feature to the uniform establishment of the soft sparking regime over the entire treated surface, although this specific PEO regime was initially and usually observed under alternating current-controlled mode. Authors attributed such gradual current drop to an

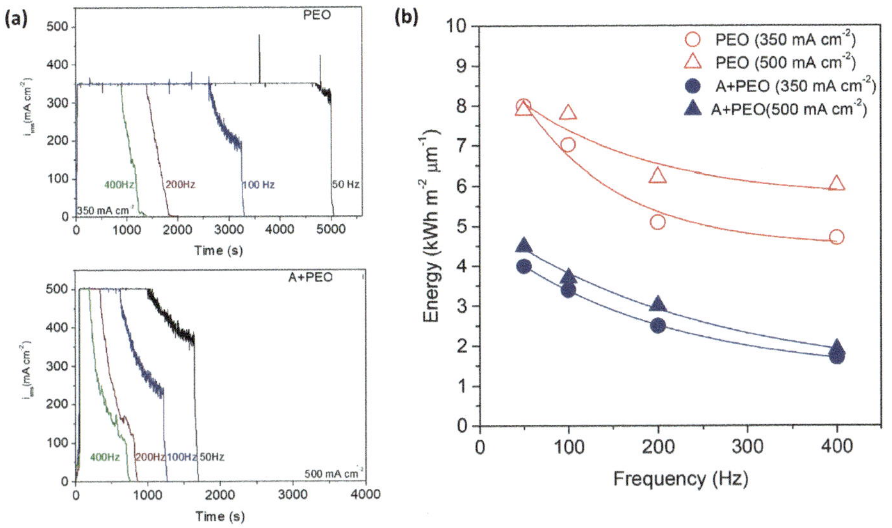

Fig. 3.16 (**a**) Current–time responses for PEO of a 6082 Al alloy with (noted A + PEO) and without (noted PEO) precursor anodic films at different frequencies (50, 100, 200, and 400 Hz) and current densities (350 and 500 mA·cm^{-2}) and, (**b**) the associated variations of the specific energy consumption of the PEO treatments. *Reproduced with permission from* [49] *© 2022 The Author(s). Published by Elsevier*

increased thickness and density of the PEO coating and consequently the higher impedance of the oxide material to mass transfer. As shown in Fig. 3.16a, authors observed that the time at which the current drop decreases with increasing frequency and current density, this tendency is even more pronounced in the case of pre-anodized Al substrate. Consequently, and as illustrated in Fig. 3.16b, they evidenced that both the use of precursor anodic film and well-designed electrical waveform parameters highly influence the energy consumption of the PEO process. In the range of the process parameters investigated, Mohedano et al. succeeded in establishing that the combination of a 20-µm thick anodic film and high current density (500 mA·cm^{-2}) and frequency (400 Hz) during PEO enables up to 76% energy saving compared to direct PEO treatment [49].

3.8.2 *Duplex Treatments: Cold-Spray + PEO*

As presented previously, the PEO process is well controlled when applied on Al alloys to achieve long-term protection in terms of wear and corrosion resistance. Especially, the optimization of the electrical conditions in parallel with the chemistry of the electrolytes, allows now to better manage the occurrence of the beneficial soft sparking regime for rapidly form thick and compact protective alumina coatings, involving hard and stable α-Al_2O_3 corundum phase, with a reduced electrical energy consumption. However, for other strategic lightweight metallic alloys such as titanium and magnesium alloys, it should be mentioned that PEO oxide coatings are not as protective as those produced on Al alloys and are not sufficient in many harsh service conditions. This is mainly due to the fact that PEO layers grown on magnesium and titanium are generally thin, porous, poorly adherent to the substrate and often need additional sealing post-treatments. In addition, although some studies report the occurrence of a soft sparking regime on Mg- and Ti-based alloys, the benefits of coating microstructures and functional properties remain marginal with regard to Al-based alloys, mainly due to the absence of crystalline alumina phases in the produced oxide coatings. To overcome these limitations and widen the use of the PEO process to various metallic materials, some combined methods were developed in which the substrate is first coated with an Al-based coating—mainly using spraying technologies [71–73]—and then converted into alumina by the use of the PEO process in a second step.

In pioneer works, Martin et al. demonstrated the feasibility of applying a duplex treatment combining cold-spray deposition and PEO under the specific soft regime electrical conditions ($R = 0.9$) with the aim of efficiently forming thick and dense PEO alumina coatings on both magnesium (EV31) and low-carbon steel (S235) substrates [55, 56]. Particularly, and as displayed in Fig. 3.17a, authors evidenced that the presence of a cold-spray Al coating on substrates has a significant influence on the voltage amplitude and promotes an earlier appearance of the soft sparking regime. This is particularly obvious for thick sprayed coatings (~200 μm thick) for which the gradual drop in the voltage amplitude occurs at about 12 min while it takes about 24 min for direct PEO of bulk Al. Alongside these observations, authors evidenced that the duplex surface treatment results in thicker and more compact PEO oxide layers (~130 μm thick after 20 min PEO) than those grown by direct PEO on a bulk Al substrate (~60 μm thick after 35 min PEO). Particularly, in the

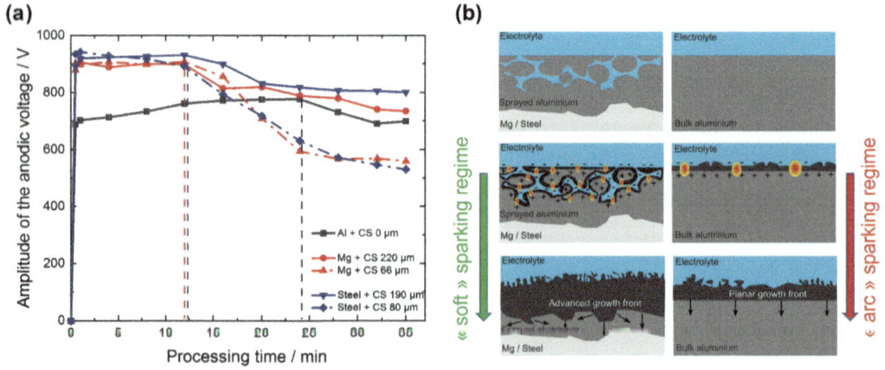

Fig. 3.17 (**a**) Voltage–time responses for pulsed bipolar current-controlled PEO of an Al 1050 substrate without cold-sprayed coating and Mg (EV31) and steel (S235) substrates with Al cold-sprayed coatings in various thicknesses. (**b**) Schematics of the influence of the porosities in a cold-sprayed Al coating on the gradual growth of a PEO oxide layer. *Reproduced with permission from* [55] © *2020 Elsevier*

case of substrates with cold-sprayed coatings, it was observed that the typical sponge-like structure known to incorporate elements from the electrolyte (e.g. Si, Na, and K) is more developed over the top-surface, thereby confirming an earlier occurrence of the soft regime. Interestingly, in the case of thinner pre-deposited coatings on a bulk magnesium (<80 μm thick) associated with longer PEO processing time (>20 min), it was also observed by Martin et al. that magnesium is PEO converted in magnesium oxide beyond the total conversion of the cold-sprayed Al coating while maintaining the soft sparking regime. In this specific case, authors demonstrated for the first time the feasibility to rapidly produce ceramic-based composite coating (Al_2O_3/MgO) on magnesium by combining cold-spray and PEO methods, this new opportunity having been exploited in a more recent study [57]. Finally, an enhancement of the growth kinetic of the PEO oxide layer, by a factor 3 or more, was established when applying duplex treatments (~6 $\mu m \cdot min^{-1}$) than using single-step PEO (~1.5 $\mu m \cdot min^{-1}$). In their preliminary study, Martin et al. proposed a potential explanation of the role played by the intrinsic porosities of a cold-sprayed Al coating on its subsequent PEO treatment [55]. While porosities in cold-sprayed coatings are generally unwanted, they have a positive effect on the PEO growth kinetics since they promote an earlier appearance of the soft sparking regime. As illustrated in Fig. 3.17b, when the PEO processing starts, the electrolyte penetrates into the interconnected porosity of the cold-sprayed coating and a thin insulating oxide layer grows up deeply into the pre-deposited Al coating. Arcs ignite deeply inside the sprayed coating and they turn earlier into tiny or even not visible micro-discharges whose behaviour is comparable to the inner D-type micro-discharges that are usually associated with the soft sparking regime. These inner micro-discharges rapidly convert the surrounding aluminium into alumina, and the advanced fronts of oxidation overlap with the adjacent ones forming a thicker PEO coating compared to those observed for the PEO process of bulk aluminium.

In a more recent study, Maizeray et al. applied the combination of cold-spray and PEO methods to develop Al-based metal matrix composite (MMC) coatings, including dispersed α-Al_2O_3 particles up to 14 vol%, with improved tribological properties [57]. At the same time, the authors investigated the effects of these dispersed particles on the subsequent PEO process, within the arc and soft sparking regimes, on the produced morphology and composition of the PEO oxide layers, and finally, on the wearing properties of the produced duplex coatings. As displayed in Fig. 3.18a, it was observed that the presence of dispersed α-Al_2O_3 particles in the pre-deposited Al-MMC coatings triggers the transition to the soft sparking regime earlier (~22 min) compared to sprayed Al coatings without particles (~27 min), thus explaining a higher amount of alumina phase in the final PEO coatings as well as improved tribological properties. Applying the same experimental approach on a Ti6Al4V substrate, Shao et al. also reported that PEO coatings produced under the soft sparking regime exhibited a more compact structure, a relatively higher alumina content, and a higher micro-hardness than that formed under the conventional arc regime [74]. Interestingly, and as observed in SEM micrographs displayed in Fig. 3.18b, Maizeray et al. also demonstrated that the dispersed α-Al_2O_3 particles remain unaffected by the PEO process within the arcs sparking regime and keep

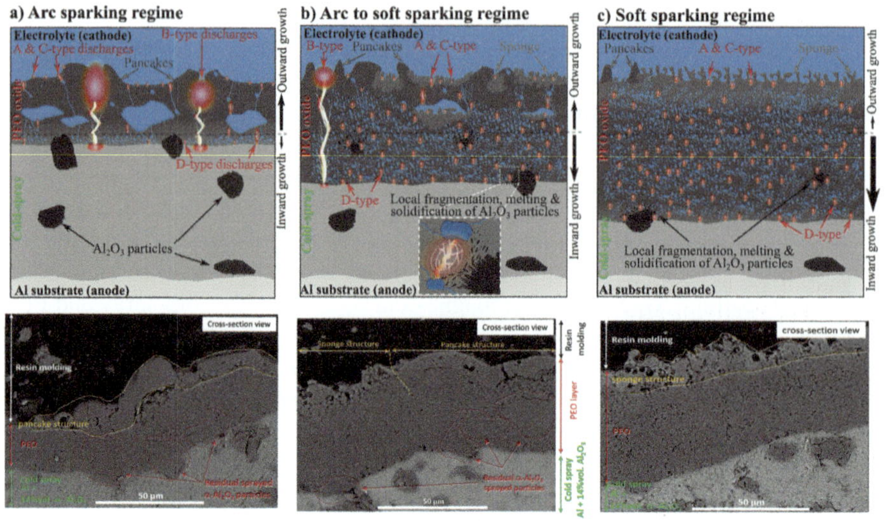

Fig. 3.18 Schematic illustration of the gradual transformation of the dispersed α-Al_2O_3 particles into an Al cold-sprayed coating within the initial arc regime (**a**), during the transition to soft regime (**b**), and after fully established soft sparking (**c**) together with the corresponding cross-section SEM micrographs. *Reproduced with permission from* [57] *© 2024 Elsevier*

their structural integrity [57]. Oppositely, they are gradually transformed to finely porous morphology within the soft regime, making them unrecognizable throughout the PEO inner sublayer. As illustrated in Fig. 3.18b, the authors proposed explanations that are mainly based on the existing types of discharges appearing during each specific sparking regime. First, under the arc sparking regime, strong B-type discharges develop throughout the overall oxide layer mainly resulting in an outward growth of the PEO coating due to the ejection and solidification of molten alumina. Although these B-type discharges dominate the arc regime, it is suggested that they occur few in number since a large current leakage is expected in each discharge, thus making the dispersed particles less exposed to a potential morphological transformation. With the transition to the soft sparking regime, the intense and visible B-type discharges are gradually replaced by tiny and more numerous inner D-type micro-discharges that preferentially initiate in defects, pores or cracks of the growing oxide layer than through the denser sprayed α-Al_2O_3 particles. As a consequence, these dispersed particles may locally undergo high mechanical stress due to shock-wave propagation as well as a significant increase in temperature, thus resulting in their local fragmentation, heating, and melting and finally, making them transformed in the PEO coating formed within the soft sparking regime.

3.8.3 Incorporation of Particles

As previously explained, the intrinsic high porosity of PEO coatings is one of the main restrictions to achieve a wider range of applications and desirable properties such as long-term protection or multifunctional surfaces. The incorporation of suspended micro- and nanometre-size particles from the electrolyte during the PEO oxide growth is a common strategy to reduce porosities as well as modify some microstructural features and composition of the coating [75–78]. It results in the formation of an oxide-based composite coating with an improvement of its specific functional properties such as tribological properties, corrosion resistance properties, catalytic activities as well as biomedical functionalities. A wide range of particles can be chosen to modify oxide coatings produced on various substrates (Al, Mg Ti, and Zr), and they can be classified into oxide (Al_2O_3, SiO_2, TiO_2, CeO_2, ZrO_2), non-oxide (SiC, Si_3N_4, TiN, TiC) and carbonaceous particles (graphene, fullerene, carbon nanotubes, carbon black). The mechanisms of incorporation of these particles into PEO oxide layers are still unclear, although it is generally proposed that they migrate toward the anode due to mechanical/electrophoretic forces, undergo inert or reactive incorporation depending on the applied conditions and on the nature of the particles, and remain finally entrapped into molten oxide during coating growth. Although the incorporation of particles into the growing PEO oxide layer has been extensively documented, with a large focus on the final performances of the produced composite coatings, only a few studies report on the influence of these particles on the process itself, particularly on the occurrence of the soft sparking regime.

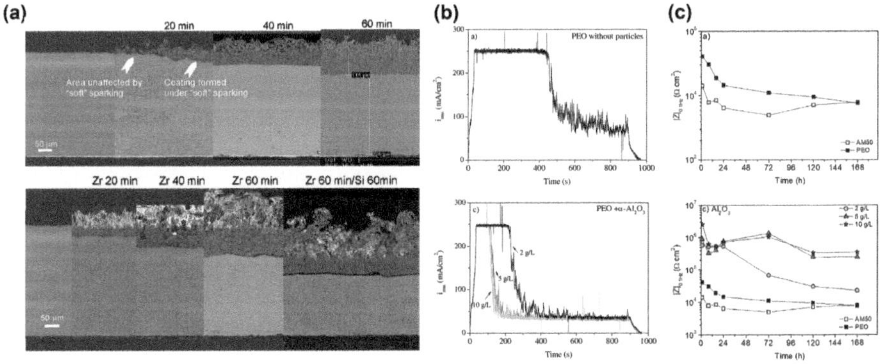

Fig. 3.19 (**a**) Cross-section SEM micrographs of PEO coatings formed at different processing times on an AM50 Mg alloy with dispersed m-ZrO_2 particles (noted Zr in bottom micrographs) and without particles. *Reproduced with permission from* [79] © *2009 Elsevier.* (**b**) Current–time responses for PEO of an AM50 Mg alloy with dispersed α-Al_2O_3 particles in the electrolyte in various concentrations (2, 5 and 10 g·L^{-1}) and without particles. (**c**) Evolution of the impedance modulus with an immersion time of PEO coatings produced an AM50 Mg alloy with dispersed α-Al_2O_3 particles in the electrolyte in various concentrations (2, 5, and 10 g·L^{-1}) and without particles (at 0.1 Hz in 0.5 wt% NaCl). *Reproduced with permission from* [80] © *2017 Elsevier*

Matykina et al. reported for the first time the incorporation of monoclinic zirconia nanoparticles during AC PEO of aluminium within the specific soft sparking conditions [79]. As illustrated in Fig. 3.19a, it was found that during the first stage of the PEO process—i.e. within the preliminary arcs sparking regime—zirconia particles are extensively incorporated into the porous outer sublayer of the PEO coatings which grows up outwardly with respect to the original aluminium surface. With the gradual transition to the soft sparking regime, authors suggested that this porous outer layer acts as a reservoir of zirconia particles for the subsequent inward growth of the denser inner sublayer. It is suspected that zirconia particles are transported from the outer to the inner sublayer through pores, cracks and channels formed during inner micro-discharges or possibly by previous micro-discharges. It was also not excluded that local melting of the particles and coating materials could also take part in the inward transport of zirconia phases. However, Matykina et al. explained the absence of zirconia in the deeper regions of the PEO coatings by considering a limited inward mass transfer, possibly related to the small diameter of the discharge channels during the soft sparking regime [79]. Mohedano et al. investigated further the role of different types of metallic oxide nanoparticles (α-Al_2O_3, m-ZrO_2, CeO_2) and concentration on the PEO process of an AM50 Mg alloy [80]. Using a voltage-controlled mode (square voltage waveform), and whatever the nature of the suspended particles, they observed an early drop of the rms current amplitude during the PEO process—that they considered as the transition to the soft sparking regime—with the concentration in particles (see example with dispersed α-Al_2O_3 nanoparticles in Fig. 3.19b). Authors suggested that the presence of dispersed particles contribute to a better sealing effect of the coating porosities which in turn results in a higher resistance of the growing oxide layer to mass transfer, and consequently an earlier appearance of the current drop and soft sparking regime occurrence. Although particles remain preferentially incorporated in the outer oxide sublayer and partially in the inner sublayer, as observed by Matykina et al. [79], Mohedano et al. [80] evidenced superior corrosion resistance properties of PEO coatings formed in particles containing electrolytes (see example with dispersed α-Al_2O_3 nanoparticles in Fig. 3.19c).

In a more recent study, with the aim of modifying the electrical resistivity of the PEO coatings on aluminium alloys, Da Silva Tousch et al. investigated the possibility to incorporate multi-walled carbon nanotubes (MWCNT) into PEO oxide layers, and their influence on the PEO process performed under the soft sparking conditions was discussed [81, 82]. Meanwhile, some similar experiments were performed by Magniez et al. using carbon black nanoparticles (CB), and common trends emerged [82, 83]. First, and as displayed in Fig. 3.20a, authors observed that the addition of carbonaceous particles into the electrolyte lowers the breakdown voltage and also promotes an earlier transition to the soft sparking regime. For example, in the range of the investigated conditions, the transition starts after 12 min for the sample processed without MWCNT, while it occurs after only 4 min for an MWCNT concentration of 1.5 $g \cdot L^{-1}$. In addition, they also evidenced that the complete establishment of the soft sparking regime—characterized by the gradual voltage drop in current-controlled mode—requires less time as the concentrations in MWCNT get

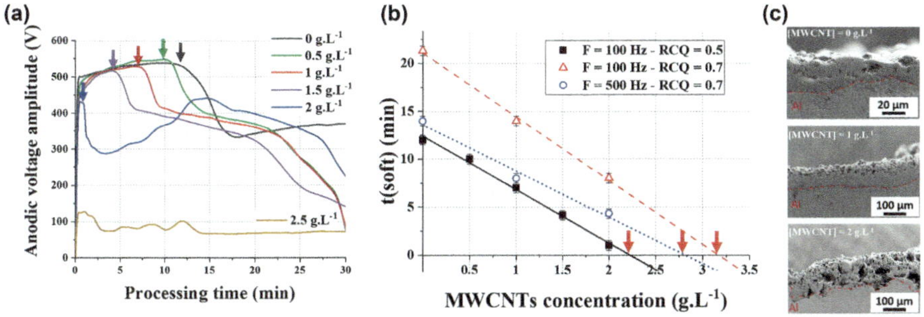

Fig. 3.20 (**a**) Evolution of the anodic voltage amplitude as a function of the processing time (**a**) for different concentrations in carbon nanotubes (MWCNT, $F = 100$ Hz, $R_{CQ} = 0.5$). (**b**) Variations of the time to soft sparking regime transition as a function of the concentrations in MWCNT for different current pulse frequencies and different charge quantity ratios. (Arrows indicate the MWCNT concentration threshold beyond which the PEO process is inhibited for each set of electrical conditions applied.) *Reproduced with permission from* [82] © *2023 Elsevier.* (***c***) Cross-section SEM micrographs of samples processed with different concentrations in MWCNTS ($F = 100$ Hz, $R_{CQ} = 0.7$). *Reproduced with permission from* [81] © *2022 Elsevier*

higher (~6 min and ~2 min without and with MWCNT, respectively). As shown in Fig. 3.20b, the same authors established a linear relationship between the time at which the transition to soft sparking appears and the concentration in carbonaceous particles, clearly evidencing a direct effect of the quantity of dispersed particles into the electrolyte on the transition to soft regime. Up to now, although the phenomena that sustain this effect are not clearly understood, authors proposed that the particles charge under the applied electric field, which would therefore affect the potential distribution at the electrolyte/coating interface, favouring the transition to the soft sparking regime. Moreover, the linear fit of the experimental data points in Fig. 3.20b also suggests a critical concentration in particles for which the occurrence of the soft sparking regime tends to 0 s and beyond which the PEO process should be inhibited. This was experimentally confirmed when using a MWCNT concentration of 2.5 $g{\cdot}L^{-1}$ while the concentration threshold—with a current frequency of 100 Hz and a charge quantity ratio of 0.5—was expected at 2.25 $g{\cdot}L^{-1}$. Finally, and surprisingly, Da Silva Tousch et al. [82] and Magniez et al. [84] reached the same conclusion: the benefits from an earlier transition to the soft sparking regime for higher concentration of dispersed carbonaceous particles should be mitigated. Indeed, and as illustrated in Fig. 3.20c, although the soft sparking regime is usually beneficial for improving the density and phase composition of PEO coatings, a too early transition to this regime due to the addition of excessive amounts of particles leads to the formation of highly heterogeneous and porous coatings with numerous defects and also a decrease in the proportion of α-Al_2O_3 relative to η-Al_2O_3. They suggested that, for PEO processing of aluminium, the transition to the soft sparking regime, as indicated by an anodic voltage drop, is not necessarily the relevant criterion for dense and high-quality PEO coatings.

3.9 Conclusions and Outlook

Through this chapter, some insights into the sparking mechanisms in PEO have been presented, with a particular focus on the so-called soft sparking regime. So far, the general model has not been developed despite the detailed practical understanding of the influence of the process parameters and electrolyte composition directly defining the coatings' morphology and properties. Advanced characterization techniques of the PEO process, including voltage/current pulse transient analysis, optical emission spectroscopy, in connection with the semiconductor and electrochemical charge transfer theories, can produce significant breakthroughs in the PEO fundamentals.

During the PEO process, cascades of high-energy discharges are generated to promote coating growth. However, the same discharges can also contribute to the formation of structural defects. Therefore, controlling and mitigating discharges during the PEO process is crucial. One promising approach to discharge control may involve the use of optimally selected current or voltage waveforms.

However, selecting the appropriate waveform parameters is challenging. Despite these challenges, the limitations in our current understanding also provide clear directions for future research.

The relationship between waveform parameters such as frequency, duty cycle, and their combined effects on the properties of coatings grown by PEO remains insufficiently explored. This is largely because many studies focus on analysing the effects of only one parameter at a time, leading to an incomplete understanding of their complex interactions.

Furthermore, there is a notable lack of research investigating the effects of sawtooth and trapezoidal waveforms, even though these waveforms could potentially reduce and mitigate discharge energy. This gap highlights the need for further investigation into their potential benefits.

Over the last 10 years, progress has been achieved regarding the mechanisms that support the soft sparking regime. However, some aspects are still unclear and need further investigations. For instance, if it is commonly admitted that the soft regime can only occur under bipolar or AC polarization with $R_{CQ} < 1$, the poor results, in terms of coating growth and properties, achieved with very low R_{CQ} values (typ. $R_{CQ} < 0.5$) do not comply with the above-mentioned theoretical considerations.

Moreover, the best advances in understanding the soft regime concern the PEO processing of aluminium alloys. Results obtained on aluminium cannot be directly applied to other valve metals like titanium or magnesium. Indeed, for those metals, the criteria to achieve soft sparking regime are different and mastering the PEO of these alloys requires also deep investigation despite huge efforts that have been brought on this topic over the last decades.

References

1. Fizeau, H., & Foucault, L. (1844). *Annales de Chimie et de Physique XI (3rd série)* (pp. 370–383).
2. Wüthrich, R., & Mandin, P. (2009). *Electrochimica Acta, 54*, 4031–4035.
3. McNeill, W. (1958). *Journal of the Electrochemical Society, 105*, 544–547.
4. McNeill, W., & Gruss, L. L. (1966). Anodic spark reaction processes and articles. US Patent 3.293.158.
5. Gruss, L., & Mackus, T. (1974). *Rare Earth Metal., 121*, 1402.
6. Wood, G. C., & Pearson, C. (1967). *Corrosion Science, 7*, 119–125.
7. Yahalom, J., & Hoar, T. (1970). *Electrochimica Acta, 15*, 877–884.
8. Yahalom, J., & Zahavi, J. (1971). *Electrochimica Acta, 16*, 603–607.
9. Brown, S. D., Kuna, K. J., & Van, T. B. (1971). *Journal of the American Ceramic Society, 54*, 384–390.
10. Markov, G., & Markova, G. (1976) USSR Patent 526 961, Bulletin of Inventions, 32.
11. Ikonopisov, S. (1977). *Electrochimica Acta, 22*, 1077–1082.
12. Klapkiv, M. D., Nykyforchyn, H. M., & Posuvailo, V. M. (1994). *Materials Science, 30*, 333–343.
13. Klapkiv, M. D. (1995). *Materials Science, 31*, 494–499.
14. Yerokhin, A. L., Nie, X., Leyland, A., Matthews, A., & Dowey, S. J. (1999). *Surface and Coating Technology, 122*, 73–93.
15. Kurze, P., Krysmann, W., & Schneider, H. G. (1986). *Crystal Research and Technology, 21*, 1603.
16. Monfort, F., Berkani, V., Matykina, E., Skeldon, P., Thompson, G. E., Habazaki, H., & Shimizu, K. (2005). *Journal of the Electrochemical Society, 152*, C382.
17. A. Yerokhin (ed), Special issue on plasma electrolysis, Surface and Coating Technology, 199, 2005, 119-242.
18. Lukes, P., Clupek, M., Babicky, V., & Sunka, P. (2009). *Plasma Processes and Polymers, 6*, 719–728.
19. Van de Krol, R., & Grätzel, M. (2012). *Photoelectrochemical hydrogen production*. Springer.
20. Gerischer, H. (1990). *Electrochimica Acta, 35*, 1677–1699.
21. Rogov, A. B., Huang, Y., Shore, D., Matthews, A., & Yerokhin, A. (2021). *Ceramics International, 47*, 34137–34158.
22. Schmuki, P. (2002). *Journal of Solid State Electrochemistry, 6*, 145–164.
23. Schultze, J. W., & Lohrengel, M. M. (2000). *Electrochimica Acta, 45*, 2499–2513.
24. Romdhane, A. B., Veys-Renaux, D., NDiaye, M. M., Bruyère, S., Elleuch, K., & Rocca, E. (2022). *Applied Surface Science, 572*, 151436.
25. Nie, X., Meletis, E. I., Jiang, J. C., Leyland, A., Yerokhin, A. L., & Matthews, A. (2002). *Surface and Coating Technology, 149*, 245–251.
26. Zhu, L., Ke, X., Li, J., Zhang, Y., Zhang, Z., & Sui, M. (2021). *Surfaces and Interfaces, 25*, 101186.
27. Monfort, F., Berkani, V., Matykina, E., Skeldon, P., Thompson, G. E., Habazaki, H., & Shimizu, K. (2007). *Corrosion Science, 49*, 672–693.
28. Veys-Renaux, D., & Rocca, E. (2015). *Journal of Solid State Electrochemistry, 19*, 3121–3129.
29. Hassel, A. W., & Lohrengel, M. M. (1995). *Electrochimica Acta, 40*, 433–437.
30. Sah, S. P., Tsuji, E., Aoki, Y., & Habazaki, H. (2012). *Corrosion Science, 55*, 90–96.
31. Matykina, E., Arrabal, R., Mohedano, M., Mingo, B., Gonzalez, J., Pardo, A., & Merino, M. C. (2017). *Transactions of the Nonferrous Metals Society of China, 27*, 1439–1454.
32. Sisarwal, V., Dong, S., Toh, R. J., Gamaleldin, K., Kulkarni, S., Li, H., Cole, I. S., Dong, J., & Chen, X. (2022). In V. S. Saji, T. S. N. S. Narayanan, & X. B. Chen (Eds.), *Conversion coatings for magnesium and its alloys*. Springer.

33. Timoshenko, A. V., & Magurova, Y. V. (2005). *Surface and Coating Technology, 199*, 135–140.
34. Timoshenko, A. V., & Magurova, Y. V. (2000). *Revista de Metalurgia, 36*, 323–330.
35. Duan, H., Li, Y., Xia, Y., & Chen, S. (2012). *International Journal of Electrochemical Science, 7*, 7619–7630.
36. Hermanns, P., Boeddeker, S., Bracht, V., Bibinov, N., & Awakowicz, P. (2020). *Journal of Physics D: Applied Physics, 53*, 435204.
37. Yang, K., & Cao, B. (2015). *Materials Letters, 143*, 177–180.
38. Yang, K., Huang, H., Chen, J., & Cao, B. (2019). *Materials, 12*, 2286.
39. Meshreghi, H., Gorbatkov, M., Yerokhin, A., & Matthews, A. The Third Conference for Engineering Sciences and Technology (CEST-2020) 01–03 December 2020 Alkhoms—Libya.
40. Beauvir, J. (2001) Oxidising electrolytic method for obtaining a ceramic coating at the surface of a metal. Patent WO 01/81658 A1.
41. Martin, J., Nominé, A., Brochard, F., Briançon, J. L., Noël, C., Belmonte, T., & Henrion, G. (2017). *Applied Surface Science, 410*, 29–41.
42. Mécuson-Jaspard, F. (2005). Plasma electrolytic oxidation of aeronautical aluminium alloys: Optical investigation of the process and study of the growth kinetics. PhD thesis, University of Lorraine, Nancy (France) (in French).
43. Jaspard-Mécuson, F., Czerwiec, T., Henrion, G., Belmonte, T., Dujardin, L., Viola, A., & Beauvir, J. (2007). *Surface and Coating Technology, 201*, 8677–8682.
44. Hussein, R. O., Nie, X., Northwood, D. O., Yerokhin, A., & Matthews, A. (2010). *Journal of Physics D: Applied Physics, 43*, 105203.
45. Liu, R., Wu, J., Xue, W., Qu, Y., Yang, C., Wang, B., & Wu, X. (2014). *Materials Chemistry and Physics, 148*, 284–292.
46. Cheng, Y. L., Xue, Z. G., Wang, Q., Wu, X. Q., Matykina, E., Skeldon, P., & Thompson, G. (2013). *Electrochimica Acta, 107*, 358–378.
47. Hermanns, P., Boeddeker, S., Bracht, V., Bibinov, N., Grundmeier, G., & Awakowicz, P. (2021). *Journal of Physics D: Applied Physics, 54*, 045205.
48. Guo, Y., Rogov, A., Hird, A., Mingo, B., Matthews, A., & Yerokhin, A. (2022). *Surface and Coating Technology, 429*, 127938.
49. Mohedano, M., Lopez, E., Mingo, B., Moon, S., Matykina, E., & Arrabal, R. (2022). *Journal of Materials Research and Technology, 21*, 2061–2075.
50. Oh, G. H., Yoon, J. K., Huh, J. Y., & Doh, J. M. (2024). *Corrosion Science, 233*, 112123.
51. Melhem, A., Henrion, G., Czerwiec, T., Briançon, J. L., Duchanoy, T., Brochard, F., & Belmonte, T. (2011). *Surface and Coating Technology, 205*, S133–S136.
52. Matykina, E., Arrabal, R., Mohamed, A., Skeldon, P., & Thompson, G. E. (2009). *Corrosion Science, 51*, 2897–2905.
53. Matykina, E., Arrabal, R., Skeldon, P., Thompson, G. E., & Belenguer, P. (2010). *Surface and Coating Technology, 205*, 1668–1678.
54. Gebarowski, W., & Pietrzyk, S. (2014). *Archives of Metallurgy and Materials, 59*, 407–411.
55. J. Martin, K. Akoda, V. Ntomprougkidis, O. Ferry, A; Maizeray, A. Bastien, P. Brenot, G. Ezo'o, G. Henrion. (2020). *Surface and Coating Technology*, 392, 125756.
56. Martin, J., Maizeray, A., Da Silva Tousch, C., Marcos, G., Czerwiec, T., & Henrion, G. (2023). *Materials Today Communications, 36*, 106676.
57. Maizeray, A., Marcos, G., Cappella, A., Planche, M.-P., Liao, H., Henrion, G., Czerwiec, T., & Martin, J. (2024). *Surface and Coating Technology, 482*, 130713.
58. Rama Krishna, L., Somaraju, K. R. C., & Sundararajan, G. (2003). *Surface and Coating Technology, 163–164*, 484.
59. Khan, R. H. U., Yerokhin, A., Li, X., Dong, H., & Matthews, A. (2010). *Surface and Coating Technology, 205*, 1679–1688.
60. Pillai, A. M., Rajendra, A., & Sharma, A. K. (2018). *Journal of Applied Electrochemistry, 48*, 543–557.
61. Martin, J., Melhem, A., Shchedrina, I., Duchanoy, T., Nominé, A., Henrion, G., Czerwiec, T., & Belmonte, T. (2013). *Surface and Coating Technology, 221*, 70–76.

62. Melhem, A. (2011). Plasma electrolytic oxidation: Influence of process parameters on the behaviour of micro-discharges, and effect on the oxide coatings. PhD thesis. University of Lorraine, Nancy (France) (in French).
63. Wei, C. B., Tian, X. B., Yang, S. Q., Wang, X. B., Fu, R. K. Y., & Chu, P. K. (2007). *Surface and Coating Technology, 201*, 5021–5024.
64. Rogov, A. B., Yerokhin, A., & Matthews, A. (2017). *Langmuir, 33*, 11059–11069.
65. Matykina, E., Arrabal, R., Scurr, D. J., Baron, A., Skeldon, P., & Thompson, G. E. (2010). *Corrosion Science, 52*, 1070–1076.
66. Matykina, E., Arrabal, R., Skeldon, P., & Thompson, G. E. (2010). *Surface and Interface Analysis, 42*, 221–226.
67. Matykina, E., Arrabal, R., Pardo, A., Mohedano, M., Mingo, B., Rodríguez, I., & González, J. (2014). *Materials Letters, 127*, 13–16.
68. Mohedano, M., Matykina, E., Arrabal, R., Mingo, B., & Zheludkevitch, M. L. (2016). *Surface and Interface Analysis, 48*, 953–959.
69. Mohedano, M., Mingo, B., Mora-Sánchez, H., Matykina, E., & Arrabal, R. (2021). *Surface and Coating Technology, 409*, 126892.
70. Koshuro, V., Formin, A., & Rodionov, I. (2018). *Ceramics International, 44*, 12593–12599.
71. Huang, C. C., Li, H. M., Li, D. H., & Lin, S. Y. (2021). *Composites and Advanced Materials, 30*, 1–11.
72. Pokhmurskii, V., Nykyforchyn, H., Student, M., Klapkiv, M., Pokhmurska, H., Wielage, B., Grund, T., & Wank, A. (2007). *Journal of Thermal Spray Technology, 16*, 998–1004.
73. Yürektürk, Y. (2021). *Journal of Thermal Spray Technology, 30*, 1274–1283.
74. Shao, M., Wang, W., Yang, H., & He, X. (2021). *Coatings, 11*, 1288–1301.
75. Borisov, A. M., Krit, B. L., Lyudin, V. B., Morozova, N. V., Suminov, I. V., & Apelfeld, A. V. (2016). *Surface Engineering and Applied Electrochemistry, 52*, 50–78.
76. Lu, X., Mohedano, M., Blawert, C., Matykina, E., Arrabal, R., Kainer, K. U., & Zheludkevich, M. L. (2016). *Surface and Coating Technology, 307*, 1165–1182.
77. Fattah-Alhosseini, A., Chaharmahali, R., & Babai, K. (2020). *Journal of Magnesium and Alloys, 8*, 799–818.
78. Tsai, D.-S., & Chou, C.-C. (2021). *Coatings, 11*, 270.
79. Matykina, E., Arrabal, R., Skeldon, P., & Thompson, G. E. (2009). *Electrochimica Acta, 54*, 6767–6778.
80. Mohedano, M., Arrabal, R., Mingo, B., Pardo, A., & Matykina, E. (2018). *Surface and Coating Technology, 334*, 328–335.
81. Da Silva Tousch, C., Martin, J., Marcos, G., Czerwiec, T., & Henrion, G. (2022). *Surface and Coating Technology, 440*, 128489.
82. Da Silva Tousch, C., Magniez, L., Fontana, S., Marcos, G., Henrion, G., Czerwiec, T., & Martin, J. (2023). *Surface and Coating Technology, 468*, 129779.
83. Magniez, L., Da Silva Tousch, C., Fontana, S., Cahen, S., Martin, J., Hérold, C., & Henrion, G. (2023). *Surface and Coating Technology, 473*, 129990.
84. Magniez, L., Da Silva Tousch, C., Fontana, S., Martin, J., Czerwiec, T., Hérold, C., & Henrion, G. (2023). *Materials Letters, 350*, 134960.

Chapter 4
Innovative Plasma Electrolytic Oxidation Approaches

Chi Xu and Wenbin Xue

4.1 Introduction

An adequate protective oxide coating on valve metals should have a compact microstructure. Although the PEO process can effectively improve the corrosion resistance of valve metals, the defects and cracks of PEO coatings may lead to their limited corrosion protection [1–4]. Various efforts have been made to improve the coating qualities or compactness through optimization of the PEO treatment parameters, such as an optimized current density waveform (a freely decaying current density [5] or a stepped current regime [6], etc.), or through a combination of different preparation technologies (such as sol-gel [7], mechanical attrition [8], anodization [9], etc. pre-treatments). As one of the optimization of the PEO treatments, two-step or more step PEO processes have been researched, which involve the consequent PEO treatments in different electrolytes or the use of different electrical parameters. Under proper experimental designs, the two-step coating may improve the coating quality in terms of corrosion resistance, coating compactness, new functional phases/ingredients, hydrophilicity, etc., through formation of a bi-layered composite structure.

So far, researchers have designed and developed various PEO composite coatings in order to further enhance the corrosion protection of valve metals, which can be roughly classified as direct composite approaches (direct doping of nanoparticles, etc.) [4, 10, 11] and secondary composite approaches (laser pre/posttreatments, inorganic/organic coating posttreatment, etc.) [4, 12, 13]. As a combinational approach, laser techniques have been shown to significantly improve the quality of the PEO coating either through a reorganization or readjustment of the surface

C. Xu · W. Xue (✉)
Beijing Normal University, Beijing, China
e-mail: xuchi@bnu.edu.cn; xuewb@bnu.edu.cn

V. S. Saji, M. Zheludkevich (eds.), *Plasma Electrolytic Oxidation – Fundamentals, Advances and Applications*,
https://doi.org/10.1007/978-3-031-92429-3_4

structures of the metal substrate in a laser pretreatment process or through remelting and reorganization of the PEO coating in a laser posttreatment process. An advantage for both the laser techniques and the PEO process is that they have little effect on the microstructure of the bulk substrate material and only provide structural modifications to the surface coating.

It is well known that the conventional PEO processes are carried out in aqueous electrolytes, mostly water-containing, including silicate-based, phosphate-based and aluminate-based solutions. The electrolyte functions as an electric conducting medium where the frequent micro-discharges happen. In this perspective, the molten-salt is considered a viable electrolyte comparable to the aqueous solution based electrolytes. Compared to the PEO treatment in aqueous solution electrolyte, the PEO process in molten-salt electrolyte is claimed to overcome the system heating problem of aqueous solutions based PEO, provide higher energy efficiency and produce denser coatings with less pores. The molten-salt PEO is also advantageous in limiting the undesirable incorporations of electrolyte components in the case of aqueous solution-based PEO processes. In replacing the aqueous solution electrolytes, it is possible to produce a coating comprised primarily of fluorides instead of oxides in molten fluoride salts or fluorine-containing organic solvents, which is named plasma electrolytic fluoridation. The produced fluoride coatings significantly improve the corrosion resistance in harsh environments.

In addition to the PEO enhancements with auxiliary methods, efforts have also been made to add extra influences into the PEO process to improve the coating quality. These include methods such as ultrasound, magnetic fields, etc. With the rapid development of power ultrasound technology, ultrasonic treatment is being widely used for material preparations and property improvement including assistance in electroless plating, electrochemical plating, and melts solidifying to refine the grains, which increases the homogeneity or reduce the micro-segregation of the processed materials [14–20]. The imposition of ultrasound on electroplating systems has been shown to give rise to an increase in effective current density and change the chemical and physical properties of deposited metal film [21]. Similarly, the ultrasound shock has been applied to the PEO process, which was aimed at solving the problem of nonuniform microstructure and low growth rate of ceramic coatings in the mid-to-late PEO processes. So far, the ultrasound-assisted PEO has been shown effective in promoting the coating growth and its compactness. Following this logic, varying magnetic fields have been applied to the PEO process and were found to increase the electron current density due to the magneto-hydrodynamics effect.

Different from the conventional PEO treatments with the work pieces as the anode, the cathodic PEO fabricates an oxide coating depending on the oxidation of the cathode metal. The growth of cathodic PEO coatings relies on the breakdown of vapor gaseous envelopes around the cathode. Therefore, the cathodic PEO has no restrictions on matrix materials, and it can fabricate the oxide coatings on valve or non-valve metals.

One of the outstanding advantages of PEO treatments is the capability of growing ceramic coatings on intricate valve metal parts or workpieces. However, the size of the to-be-treated compartments is largely limited by the electrolyte-containing cell, as well as the current limitations of the power source in the conventional PEO treatments. In order to overcome the size limitation of the to-be-treated metal parts, scanning plasma electrolytic oxidation treatments have been developed. The scanning PEO typically involves a stationary nozzle cathode with a flowing electrolyte and a moving valve metal part as the anode. The scanning PEO significantly reduces the total treating current for large metal parts and also relieves the size limitation for the to-be-treated metal parts.

The objective of this chapter is to provide basic introductions of technical details and coating characteristics for the abovementioned innovative PEO techniques. The following text elaborates these contents through specific examples of two-step PEO, laser-assisted PEO, molten-salt PEO, plasma-electrolytic fluoridation, ultrasound-assisted PEO, the magnetic fields influenced PEO, cathodic PEO, and scanning PEO.

4.2 Two-Step PEO

A two-step PEO treatment normally involves consecutive first-step and second-step PEO treatments in different electrolytes or voltage conditions. Zhang et al. [22] claimed that the second-step PEO process was a competition between old film dissolution and new film formation. The transition from the process dominated by old-film dissolution to that by new film formation happens when the new film has completely covered the old film. The formation of the new film next to the old film results in a bi-layered structure with different compositions or structures. Practically, two-step PEO treatment exists in multiple forms depending on the parameters. The most common type uses another electrolyte in the subsequent step. Others may also involve the change in electrical parameters, etc.

In the first type of two-step treatment, two different electrolytes (typically one basic and one acidic) are used subsequently. For Ti alloys, a two-step PEO treatment reported by Tsunekawa et al. [23] involves a first step in alkaline electrolyte (0.15 mol/L $K_2Al_2O_4$, 0.02 mol/L Na_3PO_4, 0.015 mol/L NaOH) and a second step in acid electrolyte (0.1 mol/L H_2SO_4, 0.1 mol/L H_3PO_4). It was claimed to have improved the adhesion strength of PEO coating by two-step PEO treatment. The subsequent PEO treatment in acid electrolyte formed an amorphous structure next to the alloy substrate (as shown in Fig. 4.1) and was compact and adherent, while the coating formed in the alkaline electrolyte alone consisted predominantly of a highly crystalline Al_2TiO_5 phase with fine pores and low adherence. Through the second-step PEO treatment, the adhesion of the coating is markedly improved without deteriorating the high wear resistance of the PEO coating.

In a similar practice, Pereira et al. [24] performed a two-step PEO process on Ti-25Nb-25Ta alloy through a first-step PEO treatment in phosphate electrolyte (80 vol.% H_3PO_4+ 20 vol.% H_2O_2), and a second-step PEO treatment in an electrolyte of calcium + phosphorus (80 vol.% $Ca(CH_3CO_2) \cdot H_2O$ + 20 vol.% H_3PO_4) solution under the direct-current mode. The first-step PEO in the phosphate solution is

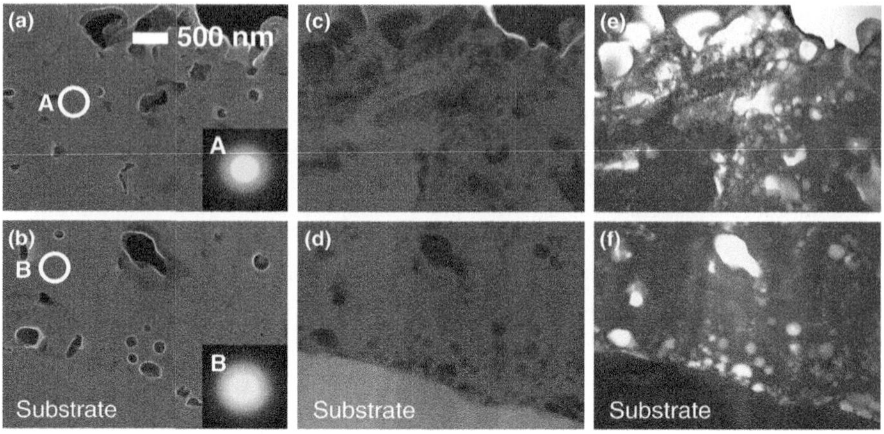

Fig. 4.1 Scanning transmission electron micrographs of the Ti–15 V–3Al specimen after two-step PEO coating in alkaline aluminate and following mixed sulfuric and phosphoric acid electrolytes. (**a**, **b**) secondary electron images, (**c**, **d**) Z-contrast images, and (**e**, **f**) transmission electron images [23]. *Reproduced with permission from* [23] *© 2011 Elsevier B.V.*

considered a pretreatment to the second-step PEO in the calcium/phosphorous solution. Through the two-step treatment, two distinctive layers with different Ca/P ratios are formed, and the existence of crystalline hydroxyapatite formation is also confirmed. The formation of an adherent and porous coating with inter-dispersed apatite is positive for osseointegration and the bio-applications.

On the other hand, Liu et al. [25] performed a two-step PEO treatment on 2A70 aluminum alloy to study the growth mechanism of PEO coating. The process included a first-step PEO in silicate electrolyte (Na_2SiO_3, NaF, $C_6H_8O_7$, and NaOH) and a second step in phosphate electrolyte (Na_3PO_4, zirconic salt, TEA, and NaOH). Through the two-step process, it was observed that the P and Si elements revealed the layered distributions in the as-formed PEO coating. The later-formed P-rich layer covers the Si-rich layer from the first-step process, which increases the coating thickness. As shown in Fig. 4.2, discharge channels can be found in the coating by the P distribution, which connects alloy substrate and the outside of coating. The inner newly formed P-rich layer increases its thickness as the oxidation time increases. These results lead to the conclusion that the PEO coating process followed an inward growth mechanism.

In another case, Rehman et al. [26] fabricated a compact coating on Al6061 alloy via a two-step PEO process with a first-step in 8 g/L Na_3PO_4, 2 g/L NaOH and 0.5 g/L Na_2SiF_6, and a second-step in the same electrolyte with 5 g/L K_2ZrF_6 additive. The two-step process coating showed a significant improvement in corrosion and hardness properties compared with the one-step process. It was found that the addition of K_2ZrF_6 resulted in the highly rich ZrO_2 and pseudo-craters-based coatings, however with damaged inner layer structure. By using a combined solution of K_2ZrF_6 and Na_3PO_4, the coating growth and structural density were highly improved.

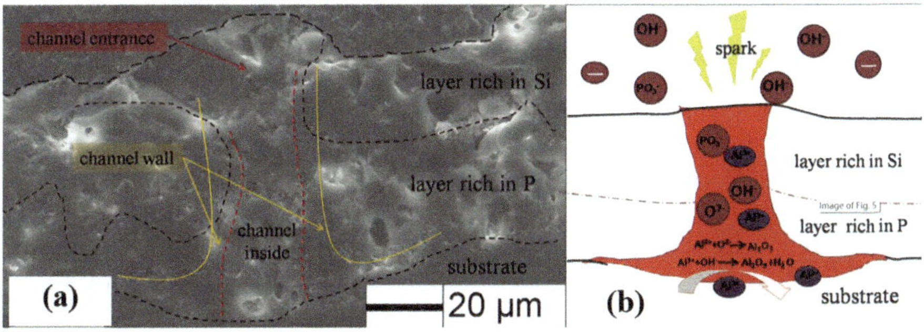

Fig. 4.2 (**a**) SEM image of cross-sectional morphology of a two-step PEO coating on a 2A70 Al alloy. (**b**) Schematic illustration of the PEO reaction process [25]. *Reproduced with permission from* [25] © *2013 Elsevier B.V.*

Wu et al. [27] performed both single-step and two-step PEO treatments on aluminum using electrolytes of alkaline silicate solution (2 g/L NaOH+8 g/L Na_2SiO_3 $9H_2O$) and alkaline silicate electrolyte with tungstate (Na_2WO_4) addition (2 g/L NaOH+8 g/L Na_2SiO_3 $9H_2O$ +8 g/L Na2WO4 · $2H_2O$), respectively. The W species can fill up the surface craters of coatings and the thickness of the inner barrier layer in the two-step PEO coatings is determined by the second-step PEO process. Furthermore, the two-step treatment has also provided new insights into the growth model of the PEO coating on Al. It was found that the structure in the primary coating prepared in the first-step PEO affects the final microstructure obtained after the second-step PEO. The open-hole surface craters formed in the primary coatings provide preferential access to reoccurrence of the penetrating discharges during the follow-up PEO process.

For the two-step treatments of Mg alloys, Einkah et al. [28] applied the two-step PEO on AZ31 Mg alloy in an alkaline electrolyte (0.02 mol/L KOH, 0.03 mol/L Na3PO4 $12H_2O$) and an acid electrolyte (0.02 mol/L H_3PO_4, 0.02 mol/L K_2ZrF_6) subsequently. The pretreatment with alkaline electrolyte prevented the dissolution of Mg substrate in comparison to the direct reaction with acidic electrolyte. The micro-pores were filled with ZrO_2 during the two-step PEO treatment. The as-formed dense oxide layer with $Mg_2Zr_5O_{12}$ and ZrO_2 phases on the outer region significantly improved the corrosion resistance of the Mg alloy substrate. Lee et al. [29] performed a similar two-step PEO on Mg alloy with the same alkaline electrolyte and K_2ZrF_6 containing acidic electrolyte. The results suggest that the composite PEO coating includes three distinctive layers (inner, intermediate, and outer layers) where the Zr content is increased from inner to outer layers and Mg content is the opposite. The outer oxide layer was mainly comprised of ZrO_2 compound, while both the ZrO_2 and $Mg_2Zr_5O_{12}$ compounds existed together in the intermediate oxide layer.

Rehman et al. [30] adopted a different electrolyte recipe for the two-step PEO treatment on an AZ91D Mg alloy, where a first-step electrolyte of 12 g/L N_2SiO_3 · $10H_2O$, 3.5 g/L NaOH, 0.3 g/L Na_2SiF_6, and a second-step electrolyte of the same electrolyte with additional 6 g/L K_2ZrF_6 were used. It was found that the microstructural features of the primary coating significantly influenced the properties of the second-step PEO coating. Unlike the primary coating, the discharge channels in the secondary coatings are directional in nature (as schematically presented in Fig. 4.3). The major phases in the coating were detected as ZrO_2, MgO, and Mg_2SiO_4, along with minor other phases. Rehman et al. [31, 32] also evaluated the effect of the concentrations of Na_2SiO_3 and K_2ZrF_6 on the two-step PEO process on the AZ91 Mg alloy. Both the Na_2SiO_3 and K_2ZrF_6 concentrations have pronounced impacts on the growth mechanics during the two-step PEO process on the AZ91 Mg alloy. For the same substrate material, Zhang et al. [22] performed a two-step PEO on a AZ91 HP Mg alloy with a first-step electrolyte of alkaline silicate solution (18 g/L $Na_2SiO_3 \cdot 9H_2O$, 8 g/L NaOH, 9 g/L $KF \cdot 2H_2O$) and a second-step electrolyte of a neutral solution (5 g/L $C_6H_{18}O_{24}P_6$, 20 g/L HF, 35 g/L H_3BO_3, 58 g/L H_3PO_4, 360 g/L hexamethylenetetramine and neutralized by ammonia). Their results showed that the second step PEO was a competition process between old film dissolution and new film formation. The transition from the process dominated by the old film dissolution to that by the new film formation took place after 12 min when the new film had completely covered the old film.

The two-step PEO treatment was also carried out on Nb alloy to improve its surface properties in applications such as medical implantation. Pereira et al. [33] performed a two-step PEO treatment on Nb with a first-step PEO in phosphorus acid (0.8 mol/L phosphoric acid +3 mol/L H_2O_2) and second-step PEO in different

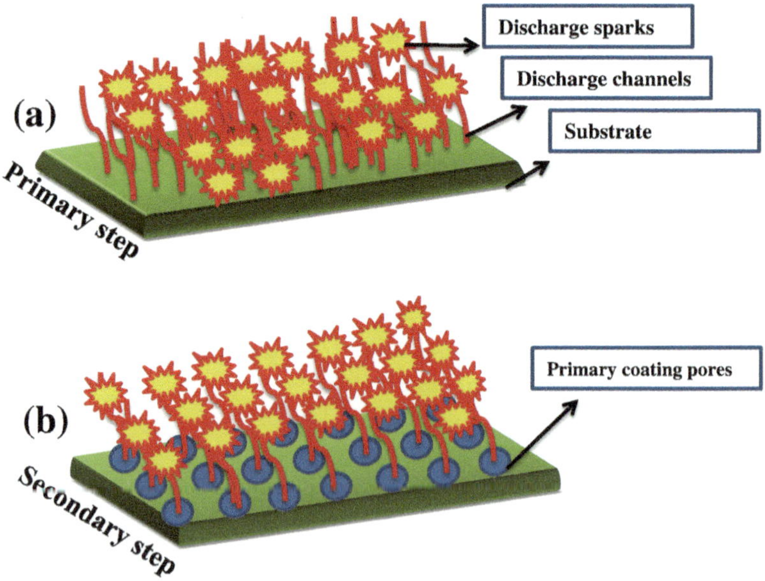

Fig. 4.3 Schematic illustration of the discharges in (**a**) primary step and (**b**) secondary step coatings during a two-step PEO treatment on an AZ91 Mg alloy [30]. *Reproduced with permission from* [30] *© 2016 Elsevier B.V.*

calcium acetate solutions in order to grow an apatite containing surface layer. Through the two-step PEO treatments, the porous surface containing calcium and phosphorus was obtained. The orthorhombic Nb_2O_5 phase was detected in the PEO coating. The two-step PEO results in crystalline apatite phase and improves hydrophilicity of coating, which is considered favorable for application situations where rapid osseointegration processes are necessary.

On the other hand, a second type of two-step method involves subsequent PEO treatments at two different voltage or current modes. Wang et al. [34] conducted the two-step PEO coating on Ti_2AlNb alloy using a one-step voltage-controlled treatment (550 V for 20 min) and a subsequent two-step treatment (550 V for 10 min and 400 V for 10 min) in an alkaline electrolyte (25 g/L $NaAlO_2$). The two-step PEO coating with a relatively compact microstructure was composed of Al_2TiO_5, R-TiO_2, α-Al_2O_3, and γ-Al_2O_3 phases. Due to the compact microstructure and a high content of Al_2TiO_5 phase, the two-step PEO coating exhibits better high-temperature properties, such as good oxidation resistance, high infrared emissivity, low friction coefficient, and small wear rate. The cross-sectional morphologies of bare alloy, single-step PEO, and two-step PEO after isothermal oxidation tests at 800 °C for 150 h are shown in Fig. 4.4. It can be seen that the oxidation layer formed between the coating and substrate is thinner (~4 μm) in the two-step PEO coated sample than in the single-step PEO coated sample (~6 μm), which indicates a better oxidation resistance for the former.

Two-step PEO treatments in different current modes have also been reported. Liang et al. [5] performed a PEO treatment on AM60B Mg alloy using two different current densities consecutively in a silicate electrolyte (10 g/L Na_2SiO_3, 1 g/L KOH). Two improved waveforms in current density (waveforms B and C), were applied on the samples in contrast to the constant current density mode (waveform A), as illustrated in Fig. 4.5. It was found that higher current densities and voltages in the later PEO process led to the enhanced sparking discharge intensity, which contributed to the formation of the coarse microstructure with large micro-pores. In contrast with the constant current density mode, the decrease of current densities in the later stages (waveforms B and C) readily tended to form the fine and less porous oxide coatings in the structure. The waveform B was found to be the optimum

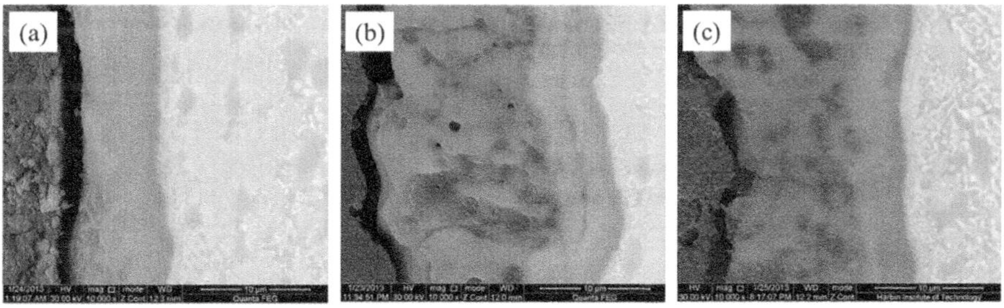

Fig. 4.4 Cross-section morphologies of the bared Ti_2AlNb alloy and different PEO coatings after isothermal oxidation tests at 800 °C for 150 h: (**a**) Ti_2AlNb alloy; (**b**) single-step PEO coating; (**c**) two-step PEO coating [34]. *Reproduced with permission from* [34] © *2014 Elsevier B.V.*

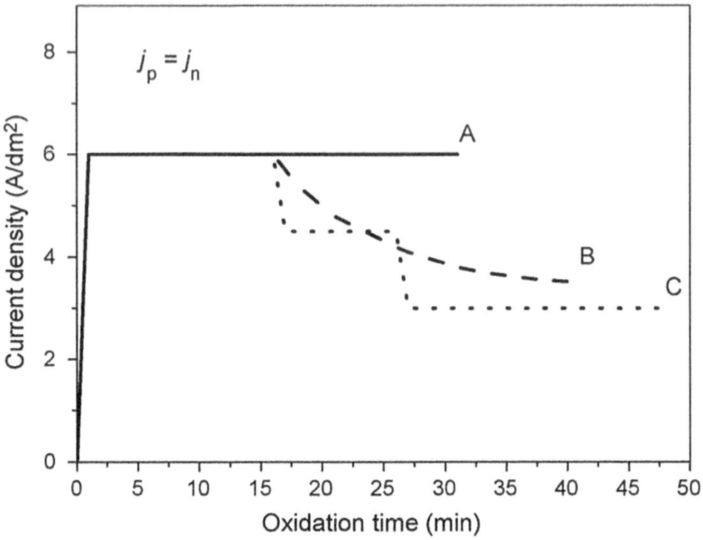

Fig. 4.5 Waveforms of current density employed in AM60B alloy during the PEO process: (**a**) constant current density; (**b**) decaying freely current density in the later stages; (**c**) stepped decreasing of current density [5]. *Reproduced with permission from* [5] © *2007 Elsevier B.V.*

current density waveform with freely decaying current density in the later stage of the PEO process, which fabricated the most intact and homogeneous microstructure of oxide coating. The corresponding corrosion tests also show that the coating prepared in the optimized current density mode exhibits the highest corrosion resistance.

In addition to the two-step PEO treatments, multiple-step (over two steps) PEO processes have also been reported. Ma et al. [35] performed various three-step PEO treatments on a ZL109 aluminum alloy as a self-lubricating composite coating for diesel engine piston skirts. Different PEO electrical parameters, including positive duty cycles, frequency and positive voltages have been applied on the three subsequent stages of PEO treatments in an electrolyte containing Na_2SiO_3, Na_2WO_3, KOH, EDTA-2Na, and DI water. In addition, a total of 16 variants for three-step PEO treatments with varying action times at identical step-positive voltages of 340 V, 380 V, and 420 V have been performed. It was found that the positive voltage has a more significant effect on the surface morphologies than positive duty cycle and frequency. The most suitable protocol of positive voltage and action time is the sequential of 340 V/4 min, 380 V/4 min, and 420 V/4 min. The friction-wear test shows that the friction coefficients of the composite coating were reduced by about 60%.

4.3 Laser-Assisted PEO

The laser-based surface modification techniques, such as laser surface melting (LSM) (Fig. 4.6a), laser surface alloying (LSA), laser shock adhesion (LSA), laser surface texturing (LST) (Fig. 4.6b), and laser shock peening (LSP), have been investigated to improve the corrosion properties of PEO coatings on Mg, Ti alloys and other alloys. A detailed review of functionalizing PEO coatings on titanium substrate with laser surface treatment can be found in Ref. [36]. Another in-depth review on the amelioration of corrosion behavior of Mg-based alloys through the PEO process combined with the surface laser process is referred to Ref. [37]. In practice, there are two types of combination between the laser techniques and the PEO treatment, either by laser-assisted pretreatment following PEO, or PEO pretreatment followed by laser-assisted posttreatment. In the laser pretreatment PEO process, the laser preprocessing causes a redistribution of alloying elements and precipitates on the metal substrate surface. Thus, it will produce improvements for the substrate material and the subsequent PEO coating as well. Meanwhile, some researchers have adopted laser-assisted postprocessing to PEO treatment, in order to achieve a rapid, high precision and powerless processing of metal materials.

The laser treatment can effectively reduce the nanopores or microcracks in the outer layer of PEO coating through laser remelting, which results in a compact coating and an improvement in corrosion resistance. The laser-assisted treatment can also improve the process efficiency of PEO coatings. For example, Wang et al. [38] confirmed that the efficiency of laser and electrochemical processing based on internal total reflection increased by 118.8%, which was attributed to the increase in current density in the processing area caused by laser-induced temperature rise. The

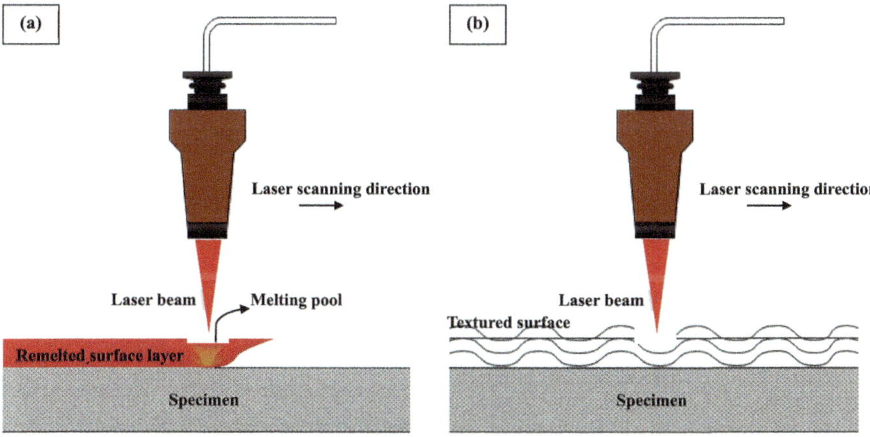

Fig. 4.6 The schematic illustrations of (**a**) LSM and (**b**) LST treatments [36]. *Reproduced with permission from* [36] *© 2023 The Author(s). Published by Elsevier B.V.*

following text deals with laser pre-processing and postprocessing PEO treatments separately.

4.3.1 Laser Pretreated PEO Processes

The laser pretreatment can improve the properties of PEO coating by affecting the surface state of the metal substrate. The microstructural changes induced by laser preprocessing include the refined grain structures, production of evenly distributed precipitates, redistribution of alloying elements, etc. Wang et al. [39] applied LSM pretreatment and PEO subsequently on an AZ91D Mg alloy, and the LSM pretreatment resulted in the microstructure refinement, the Al element enrichment, and the redistribution of β-phase (Mg17Al12) on the matrix. The PEO coating produced on LSM pretreated magnesium alloy showed slight difference in microstructure and compositions with the PEO coating on as-received magnesium alloy, including more compact cross-sectional microstructure, less cracks on the surface and higher content of $MgAl_2O_4$ spinel phase in the coating. The corrosion resistance of the LSM–PEO treated magnesium alloy shows remarkable improvement compared to that of the single PEO treated magnesium alloy.

Liu et al. [40] employed LSM pretreatment and PEO process on AZ31 magnesium alloy to prepare ceramic coatings with better corrosion resistance. The LSM pretreatment was performed using a 10 kW transverse-flow continuous-wave CO_2 laser with a beam diameter of 3 mm and track overlap ratio of 50%, which could be attributed to the modified layer and uniformly distributed intermetallic compounds after LSM treatment. The LSM treatment produced a homogeneous modified layer with redistributed intermetallic compounds. It was found that the LSM treatment had no obvious influence on the surface and cross-sectional microstructures of the PEO coatings on AZ31 alloy, and MgO was the main constituent for PEO coatings, regardless of LSM pretreatment. However, the LSM pretreatment revealed large enhancement for the long-term corrosion property of the PEO-coated AZ31 alloy after 587 h immersion test in 3.5 wt.% NaCl solution. The better corrosion resistance for the LSM pretreated PEO coated alloy was ascribed to Al enrichment in the Mg alloy matrix which decreased the corrosion susceptibility of α-Mg matrix, and also the refinement of Al_8Mn_5 intermetallic phase. That lowered the micro-galvanic corrosion in the α-Mg matrix where the intermetallic phase serves as micro-cathodes.

Nisar et al. [41] performed femtosecond (Fs) laser texturing treatment on a Ti-6Al-4 V alloy prior to PEO treatment. The extremely brief pulse duration and the substantial laser peak fluence of the Fs laser create the so-called laser-induced periodic surface structures (LIPSSs), which consist of parallel trench-shaped structures spaced ~1 μm on the surface (as shown in Fig. 4.7). It was deduced that the Fs texturing with and without PEO coating can be considered to be the innovative surface modification alternatives for biomaterials due to improved hydrophilicity, corrosion

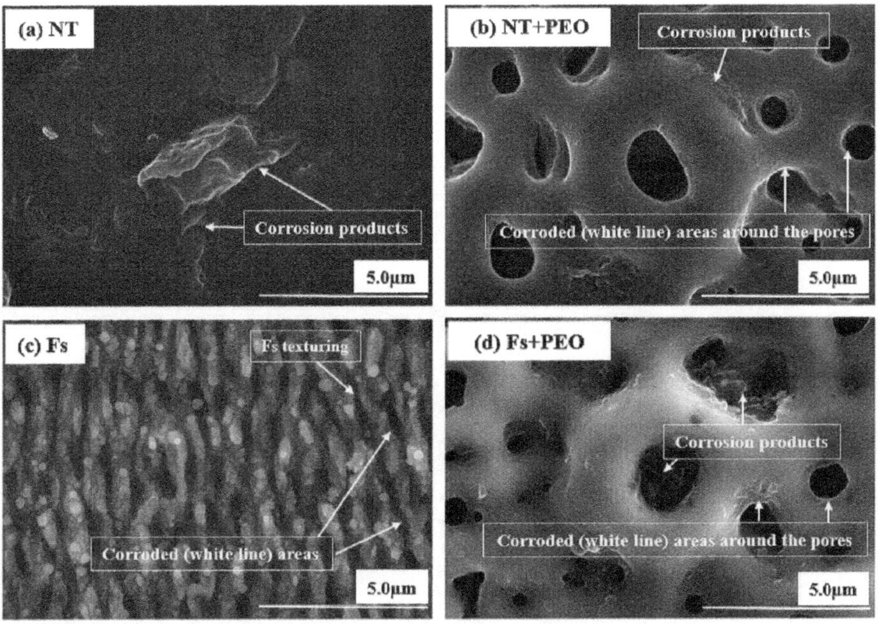

Fig. 4.7 Corrosion morphologies of (**a**) no-treatment (NT), (**b**) NT + PEO, (**c**) femtosecond laser (Fs) treated, and (**d**) Fs + PEO in 0.9% NaCl recorded by FESEM [41]. *Reproduced with permission from* [41] *© 2023 Elsevier B.V.*

resistance, and surface roughness properties with hardness comparable to that of cortical bone thereby enhancing osseointegration capacity of bone to the implant. A cell culture test indicated that the combined Fs texturing and PEO-coating could promote high cellular proliferation. A similar treatment on the Ti-6Al-4 V (TC4) alloy was also performed by Wu et al. [42], in which three different surface textures (dimple, groove and mesh) were constructed on the TC4 alloy prior to the PEO treatment. It was shown that the laser texturing improved the wear properties of the TC4 alloy by retaining abrasive debris in the textured structures during wear.

4.3.2 Laser Posttreated PEO Processes

The laser-assisted postprocessing can enhance the surface structure of PEO coating through surface remelting and consolidation, which is effective in removing the pores or microcracks that typically exist in the outer layer of PEO coating. Rapheal et al. [43] compared the wear performance of PEO coatings prepared in phosphate and silicate solutions, respectively on MRI 230D Mg alloy, and hybrid coatings prepared by first PEO treatment and followed by laser surface alloying (LSA) treatment with Al and Al_2O_3. The hybrid coating with post LSA treatment shows much

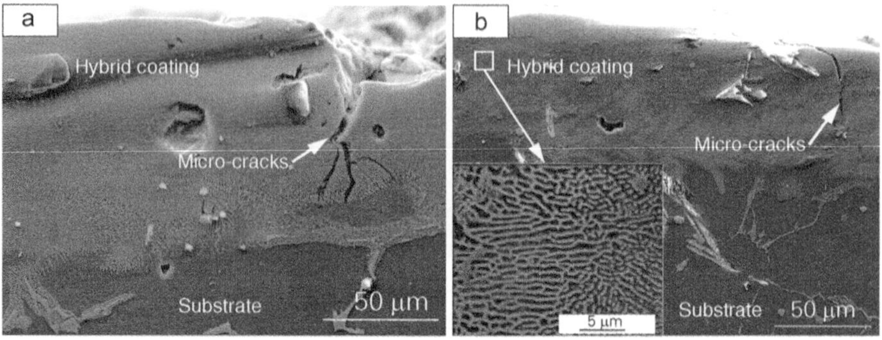

Fig. 4.8 Cross-sectional SEM images of hybrid coatings formed by PEO treatments in (**a**) silicate-based electrolyte and (**b**) phosphate-based electrolytes and followed by 5 J laser-surface alloying (LSA) treatments with Al and Al_2O_3 [43]. *Reproduced with permission from* [43] © *2011 Elsevier B.V.*

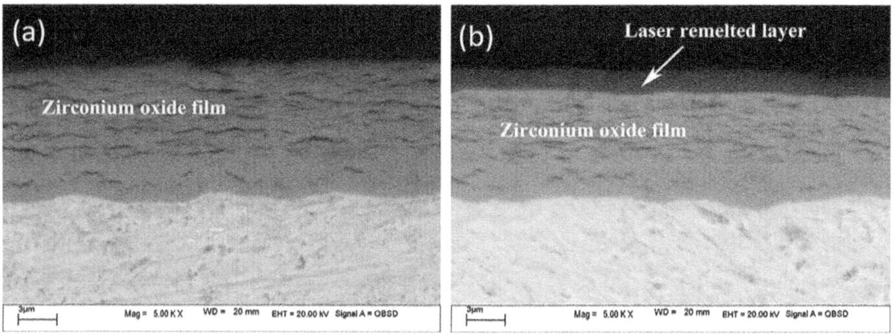

Fig. 4.9 The backscattered SEM images of the cross-sectional morphology of the cladding tubes after the autoclave test: (**a**) PEO-coated Zr-1Nb alloy; (**b**) PEO-coated and excimer laser-treated Zr-1Nb alloy [44]. *Reproduced with permission from* [44] © *2015 Elsevier B.V.*

less porous structures as compared with the PEO coatings, as shown in Fig. 4.8. Both the PEO coatings melted/decomposed on laser irradiation and all the hybrid coatings exhibited similar microstructure and wear behavior irrespective of the nature of the primary PEO coating or laser energies. The wear track depth measurements revealed improved wear resistance for both the PEO and hybrid coatings as compared to the as-cast material at 2 N load. However, PEO coatings indicated better wear resistance than the hybrid coatings due to the higher hardness value for the former, and the PEO coating prepared in silicate solution exhibited better wear resistance than that in phosphate solution [43].

Yang et al. [44] performed excimer laser postprocessing on the PEO film of Zr-1Nb alloy as a cladding tube material for commercial light-water nuclear power reactors. The excimer laser (248 nm, FWHM 30 ns, max. Pulse energy 1.2 J) formed a compact remelting layer on the surface of the PEO coating, as shown in Fig. 4.9b. The laser remelting also reduced surface roughness and thickness of PEO coating.

The composite coating had a strong bonding with the substrate and the change in thermal expansion coefficient was found to have no significant effect on the performance of the Zr-1Nb cladding tubes. After 94 days of autoclave corrosion in reactor water environment (360 °C, 18.6 MPa, 0.01 mol/L LiOH aqueous solution), the corrosion weight gains in samples processed by excimer laser (~ 10 mg/dm^2) were significantly lower than in PEO-coated Zr-1Nb (~19.5 mg/dm^2) and the bare Zr-1Nb alloy (~22 mg/dm^2).

Sampatirao et al. [45] fabricated PEO coatings with and without the incorporation of ZrO_2 particles on ZM21 Mg alloy as a biodegradable implant material. The PEO coating was followed by a laser texturing process to further improve its corrosion resistance. The fabricated samples exhibited better contact and response with the fibroblast cells, compared to the metal substrate. The addition of ZrO_2 nanoparticles in the PEO process resulted in a higher coating thickness and improved corrosion resistance of the samples than the PEO-coated samples without ZrO_2 nanoparticles addition. The laser surface textured PEO-coated and PEO-coated with ZrO_2 incorporation samples enhance the corrosion resistance and cell growth and are considered as a suitable prototype for biodegradable implant applications.

As a summary, the laser pre-/post- treatments change the coating properties through influencing the surface properties of either the substrate materials or the final PEO coating. It is reported that laser pretreatments have slight or trivial impacts on the microstructure of PEO coatings [39, 40]. Nonetheless, the pretreatments show significant improvements in long-term corrosion resistance [40]. On the other hand, the laser posttreatments actually create double-layered structures, and the outer layer has significant improvement in surface smoothness or compactness. Posttreatment may also introduce phase transitions in the coating and modify the coating properties such as the thermal expansion coefficients, etc. [44]. Challenges might exist in the application of the laser treatments, which may produce undesirable modifications to the metal substrate or the PEO coating, such as the introduction of micro-cracks or the secondary phases. Furthermore, laser treatments also add up energy consumption and economic cost. However, the undesirable effects may be minimized through proper selections of laser processing parameters, and the significant improvements of coating properties will balance over the increase in costs in cases where the property enhancements are strongly requested.

4.4 Molten-Salt PEO

The molten-salt PEO treatments are PEO treatments that utilize the molten salt as the working electrolyte instead of an aqueous solution. Sobolev et al. [46] compared the PEO coating on AA1050 Al alloy prepared in silicate-based aqueous solution and nitrate-based molten salt electrolytes. For the molten-salt-based PEO treatment, an eutectic nitrite consisting of 45.7 wt.% $NaNO_3$ and 54.3 wt.% KNO_3 at 280 °C was used as the electrolyte in a Ni crucible as the cathode and Al sample as the anode. The coating formed in PEO in aqueous electrolyte (2 g/L KOH,

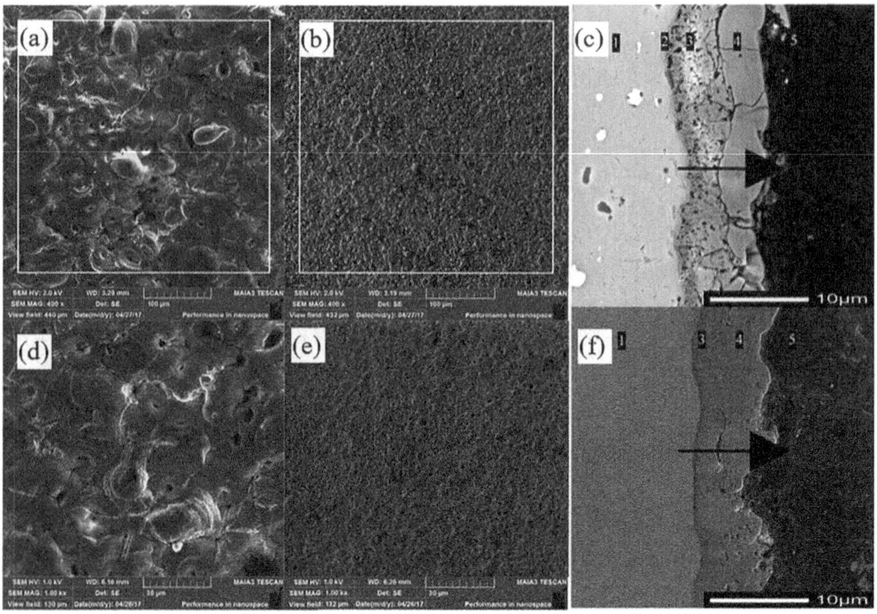

Fig. 4.10 Surface morphology of the Al 1050 alloy treated by the PEO approach in different electrolytes: (**a–c**) in aqueous electrolyte solution; (**d–f**) in molten salt [46]. *Reproduced with permission from* [46] © *2018 Elsevier B.V.*

12.72 g/L $Na_2SiO_3 \cdot 5H_2O$) solution (Fig. 4.10a, b) consists of η-Al_2O_3 and amorphous SiO_2 in addition to hard α-Al_2O_3 and γ-Al_2O_3 phases, has through pores and shows lower corrosion resistance. Meanwhile, the molten-salt prepared coating (Fig. 4.10c, d) consists of hard α-Al_2O_3 and γ-Al_2O_3 and has no additional compounds or through pores and shows higher corrosion resistance than the coating formed in silicate electrolyte. During the PEO process, it was found that the micro-arc discharges in molten-salt electrolyte begin at 30 V, which is four times lower than in silicate electrolyte. The coating growth rate is about 3 times higher, and the energy efficiency is about 6 times higher in the molten-salt electrolyte than in the silicate electrolyte. Moreover, the presence of electrolyte components is inevitable in the coating obtained in silicate aqueous solution, while these components are not common to appear in the coating obtained in molten salt. The corrosion resistance and electrochemical analysis of the pure oxide coating prepared in molten-salt PEO process were also investigated in the NaCl solution [47]. The obtained ceramic coating contains two sublayers: a porous outer sublayer and a dense inner sublayer. It is revealed that the progress of the corrosion attack by chloride ions affects both sublayers of the PEO coating: the thickness of the outer sublayer decreased, and the inner sublayer became more compact, resulting in high resistance properties. A more in-depth analysis of the molten-salt fabricated PEO coating on AA1050 Al alloy was reported in [48].

Sobolev et al. [49] also investigated the influence of frequency (200, 300, and 400 Hz, duty cycle 50%) on the PEO process in molten salt (280 °C, eutectic composition of KNO_3–$NaNO_3$) on Al 7075 alloy. It was found that higher frequency resulted in more dense coating with higher corrosion resistance.

In addition to the Al alloy, the molten-salt-based PEO treatment has also been applied for the Ti-6Al-4 V alloy [50] and an Al-Cu alloy [47], where the same molten-salt electrolyte was used (280 °C, 45.7 wt.% $NaNO_3$ and 54.3 wt.% KNO_3). For the molten-salt PEO treatment on Ti-6Al-4 V alloy, thin titanium oxide layer of 2–2.5 μm thick consisting of rutile and anatase was formed on the Ti-6Al-4 V alloys. The PEO frequency increased from 50 Hz to 150 Hz, leading to the formation of a thick and uniform ceramic coating with sub-micron pores. Both morphology and the hydrophobicity were found to significantly affect the corrosion resistance of the coating, and the highest corrosion resistance was observed for the coating obtained using 150 Hz. However, the application of an even higher current frequency (200 Hz) led to the coating degradation resulting in the non-uniform layered structure. For the molten-salt PEO treatment on Al–Cu alloy (Al 2024, Cu 3.9–4.9 wt.%), a double-layered structure composed of a porous outer layer and a compact inner layer was produced [51]. It was found that intermediate current frequencies created a relatively thick compact inner layer that protected the alloy substrate against corrosion. High and low current frequencies produce thin and nonuniform inner layers, respectively, which typically contain some defects such as cavities and pores.

The molten-salt PEO technique is an innovative approach to the formation of oxide coatings through the PEO method. The challenges of this technique may include the high temperature of the electrolyte, which significantly affects the coating growth rate, and a much lower discharge voltage (~30 V for Al alloy treatment [46]), which means that the fine adjustment of process parameters is necessary for achieving best coating quality. The current research is more focused on the PEO treatments for Al-based materials, which has resulted with improved coating properties over conventional PEO coating, such as the higher compactness, growth rate, corrosion resistance and disappearance of electrolyte components in the coating. It is expected that the molten-salt PEO technique can be applied to a wider variety of valve metals and may show more different coating characteristics than the coatings formed in aqueous electrolyte.

4.5 Plasma Electrolytic Fluoridation

Through the development of the PEO technique, it has been found that the fluoride additives (NaF, KF, NH_4F, etc.) in the electrolytes may improve the PEO coating properties or growth rate. For example, KF addition was found to enhance the corrosion resistance of PEO coating on Mg alloys in a basic silicate solution. Guo et al. [52] found that the KF addition in electrolytes significantly improved the growth rate of PEO coating on an AZ91D Mg alloy. In the traditional PEO process, the coatings formed in the aqueous electrolytes are mostly comprised of oxide compounds, which limits the composition and functions of the PEO coatings. However, the PEO processes in molten salt or organic solutions introduce new possibilities for the compositions or functions of PEO coatings. In these nonaqueous solutions electrolytes, it is possible to form fluoride coatings where fluoride rather than oxide is the major component and hence termed plasma electrolytic fluoridation (PEF).

PEF process has been performed in organic solvent solutions. Qi et al. [53] fabricated fluoride (MgF_2) dominated coating on AZ91 Mg alloy using the PEF technique in nonaqueous NH_4F-ethylene glycol (EG) electrolyte (NH_4F 8 wt.%). The fluoride coating had homogeneous composition and porous microstructure, which displayed a marked improvement of corrosion resistance in 3.5 wt.% NaCl solution (acidic pH = 3). As shown in Fig. 4.11, the surface morphology of the PEF coating features a fine granular feature. The interior of the coating was found to be more compact than the outer coating. A similar experiment was reported by Luo et al. [54], where the effect of the NH_4F concentration was studied on the PEF of an AZ31 Mg alloy in the EG-NH_4F electrolyte. The results showed that the working voltage and breakdown voltage of the PEF process decreased gradually with the increase of the NH_4F concentration. The pore size and surface roughness of the

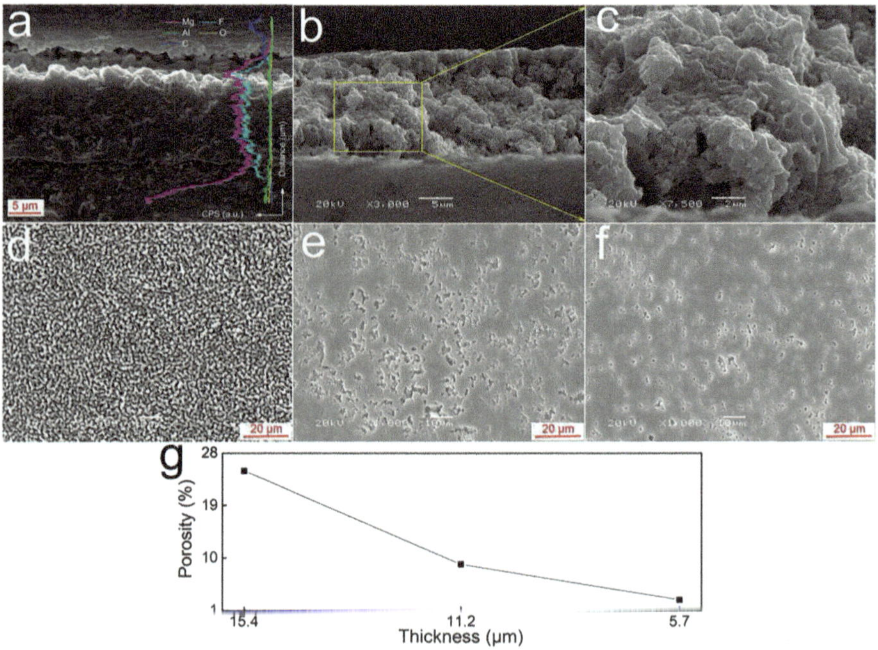

Fig. 4.11 Internal characteristics of the coating: (**a**) polished morphology and EDS line-scanning, (**b**) fractured morphology, (**c**) enlarged view of (**b**); morphology of as-coated surface (**d**), the coating polished to 11.2 μm (**e**) and 5.7 μm (**f**); (**g**) porosity statistics [53]. *Reproduced with permission from* [53] © 2020 Institute of Materials, Minerals and Mining Published by Taylor & Francis on behalf of the Institute and © 2021, Sage Publications

coating decreased with the increase in concentration. The MgF_2 coating was found to significantly improve the corrosion resistance of AZ31 magnesium alloy, and the coating prepared at a concentration of 100 g/L NH_4F had optimal corrosion protection properties.

4.6 Ultrasound-Assisted PEO

The imposition of ultrasound on PEO is typically realized through introducing an ultrasound source to the reaction cell during the PEO process. Shen et al. [21] applied the ultrasonic technique to the PEO treatment of a 6061 Al alloy. The experiment sketch is shown in Fig. 4.12, where an ultrasonic device is placed under a normal PEO electrolyte cell with a maximum power of 100 W and frequency of 130 kHz. The ultrasonic processing promoted ceramic growth, especially in the mid-to-late period of the PEO process. The ultrasonic action was considered to increase the discharge breakdown points by producing flaws in the ceramic coating. In addition, the introduction of ultrasonic is claimed to have promoted the formation of α-Al_2O_3 instead of γ-Al_2O_3 and raised the compactness of the coatings.

A systematic comparison between the PEO process with and without ultrasound effect was also performed by Shen et al. [55] on the 6061 Al alloy in an aqueous silicate containing electrolyte. Figure 4.13 shows the evolution of surface morphology between the conventional PEO and ultrasound-assisted PEO (UPEO) at 5 min, 15 min, and 30 min durations. It was observed that the ultrasound effect became significant in the later stage of the PEO process. After 30 min treatment, the pores in the UPEO coatings were much smaller in number and size (about 4.1 um) than

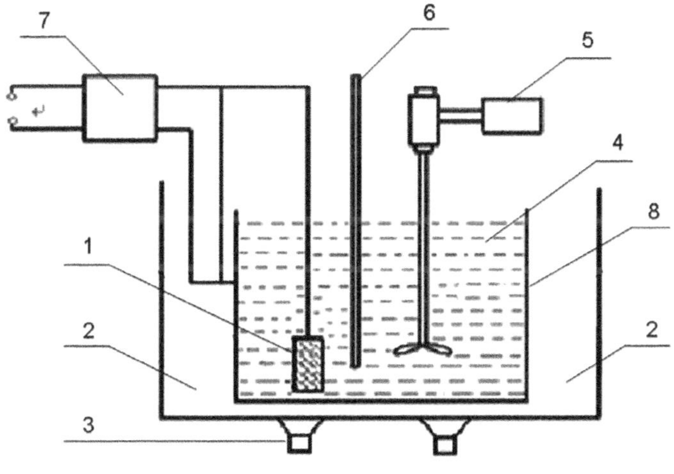

Fig. 4.12 Plasma-electrolysis depositing apparatus sketch. 1-specimen; 2-cooling water; 3-ultrasonic device; 4-electrolyte; 5-stirrer; 6-thermocouple; 7-power supply; 8-stainless steel cathode as the electrolytic cell [21]. *Reproduced with permission from* [21] © *2013 Elsevier Ltd.*

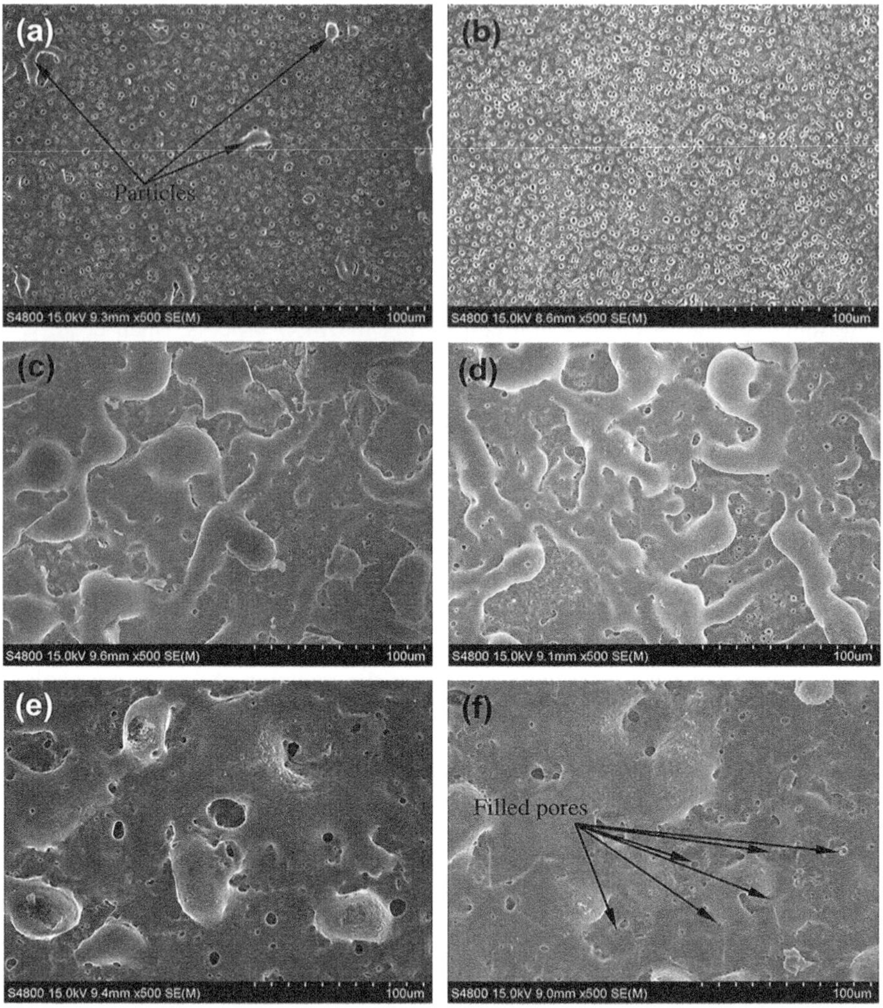

Fig. 4.13 Surface morphologies of the coatings, (**a**, **c**, **e**) in conventional PEO system for 5 min, 15 min, and 30 min, respectively, and (**b**, **d**, **f**) in ultrasound-assisted PEO system for 5 min, 15 min, 30 min, respectively [55]. *Reproduced with permission from* [55] © *2014 Elsevier B.V.*

those in the conventional PEO coatings (about 6.2 um). The ultrasound effect was concluded as promoting the dielectric breakdown of PEO coatings during the UPEO process. The dielectric breakdown voltage of UPEO and traditional PEO coatings are 225 V and 270 V, respectively. Ultrasound increases the number of micro-discharges while decreasing their average size. The number of residual discharge pores in the coatings decreased after 30 min with the assistance of ultrasound. In addition, ultrasound was found to promote the formation of α-Al_2O_3 and mullite and result in a homogeneous distribution of elements in the coatings. Similar results were also reported in [56] for the ultrasound-assisted PEO on an AA1060 Al alloy.

More about the UPEO theory, Fatimah et al. [57] investigated the effect of ultrasonic frequency (0, 40 kHz, and 135 kHz) on the coating structure and corrosion properties of AZ31 Mg alloy. It was found that in the presence of 40 kHz ultrasonic waves, the formation of a uniform and dense oxide layer on Mg alloy was facilitated, which was attributed to the plasma softening, acoustic streaming, and improved mass transport for successful deposition and continuous reforming of the oxide layer during UPEO processes. However, increasing ultrasonic frequency from 40 kHz to 135 kHz suppressed the optimum growth of the oxide layer due to the occurrence of super-soft plasma swarms, which resulted in a low coating thickness.

In addition to Al alloys, UPEO has been applied to other metals as well. Zhang et al. [56] applied the UPEO on pure Ti to fabricate a Cu-incorporated uniform TiO_2 coating as an implanted biomaterial. It was found that a uniform brownish appearance was achieved after the introduction of ultrasonic vibration instead of the coexistence of central brown area and peripheral black area in the Cu-incorporated TiO_2 coating fabricated by the conventional PEO process. The introduction of ultrasonic vibration also significantly modified the elemental composition, phase component, and chemical configuration of the coating, which exhibited a bi-layered structure with the porous TiO_2 as the inner layer and the amorphous Ca, P, O, and Cu components as the outer layer. Ajiriyanto et al. [58] applied UPEO to accelerate the growth of coating formed on a zircaloy-4 alloy through the PEO process at low electric power. The UPEO process accelerated the coating growth by lowering the breakdown voltage, generating more plasma discharge, and avoiding the gas shielding effect. It was considered that the vibrational energy from sonication stimulated the breakdown of the dielectric layer, hence generating more sparks on the surface.

Ultrasonic effects have also been combined with the PEO process in multiple other forms. A short-duration (5 s) ultrasonic shot peening (USP) has also been used as a posttreatment to PEO coatings produced on 2017 Al alloy under various sparking regimes (arc or soft spark) and time durations [59]. The USP postprocessing was used for the purpose of eliminating the outer sublayer with high porosity. The results show that the multidirectional USP is effective in removing the loose outer sublayer. It is especially beneficial for specific soft sparking regime PEO coatings which keep the structural integrity while experiencing a significant densification and a resurfacing of the porous top surface. Ultrasound has also been applied to improve the attrition properties of Ti-6Al-4 V [60] and SiC single crystal [61] through combination with PEO and plasma electrochemical oxidation, respectively. Moreover, acoustic detection techniques have been shown to be an effective method in analyzing the mechanics and dynamic processes of PEO on various metals [62–64].

4.7 PEO Treatments in Magnetic Fields

The PEO discharge process is known to be related to the electron avalanche breakdown as in the anodic oxidation process [65–67]. Recently, studies have shown that electronic charges contribute to electron avalanche, whereas ionic charges contribute to the growth of coatings [68]. In addition, the incorporation ratio of ionic charges has a significant effect on the growth of PEO coating and can be adjusted using the current density and electrolyte conductivity. Generally, a low electronic charge density can be used to obtain a high-density PEO coating. However, investigations on the growth of PEO coatings by changing the incorporation ratio of electronic charge are still scarce. Zhai et al. [69] proposed a method of adjusting the incorporation ratio of electronic charge by applying a magnetic field based on the theory of collision ionization. The experiment setup is shown in Fig. 4.14. A varying magnetic field of 0.3–0.35 T was formed when the Ti-6Al-4 V alloy samples rotated with the copper rod (the rotation axis) between two square neodymium magnets in the electrolyte. The results show that the magneto-hydrodynamics (MHD) effect increased the electron current density during the PEO process. The higher the rotation speed of the sample in the magnetic field, the more significant the MHD effect obtained (the effect of rotation speed is displayed in Fig. 4.15). The MHD effect significantly increased the electron current density during the PEO process, resulting in a higher breakdown voltage of the anodic oxide film, faster growth of the PEO coatings, and stronger and more concentrated micro-arcs. As a result, thick, uneven coatings were obtained on the samples rotating at a lower speed in the magnetic field, whereas thin, uniform coatings were obtained on the samples rotating at a higher speed in the magnetic field.

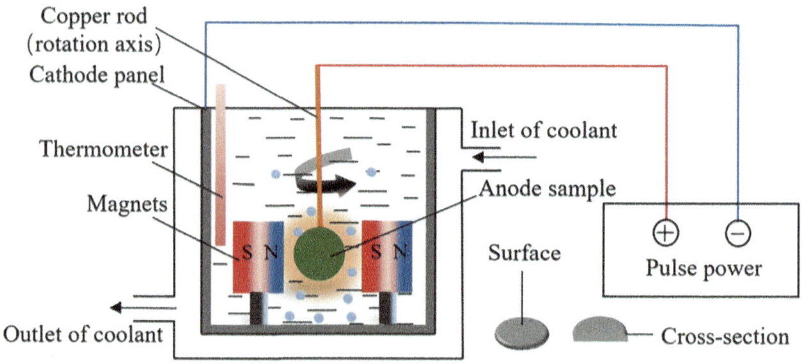

Fig. 4.14 Schematic diagram of PEO equipment with incorporation of magnetic field [69]. *Reproduced with permission from* [69] © *2022, The Minerals, Metals & Materials Society and ASM International*

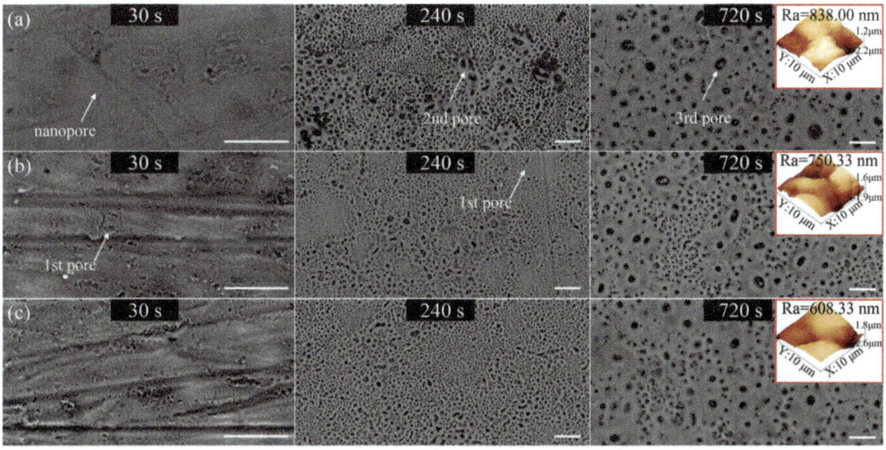

Fig. 4.15 Surface SEM (BSD) morphology and roughness of the MAO coatings obtained on samples with different rotation speeds: (**a**) R500 (500 r/min), (**b**) R1500 (1500 r/min), and (**c**) R3500 (3500 r/min) in a magnetic field. White scale bar, 10 μm [69]. *Reproduced with permission from* [69] © *2022, The Minerals, Metals & Materials Society and ASM International*

4.8 Cathodic PEO Treatments

Conventionally, plasma electrolytic oxidation (PEO) is applied to valve metals such as Al, Mg, Ti, and Zr [70], where the workpieces are set as the anode. It is believed that the cathode metal cannot be directly oxidized by the PEO process based on the electrochemical theory. However, Jin and Xue et al. found that an oxide film could be quickly formed on the cathode of 304 stainless steel [71, 72], Q235 low carbon steel [73], and T8 high carbon steel [74] by plasma discharge in solution. This technique was named cathodic plasma electrolytic oxidation (CPEO). Furthermore, Stojadinovic et al. [75] found that the CPEO process could be carried out on molybdenum. Recently, Gao et al. [76] reported that the CPEO process might also be applied to valve metals such as TiAl alloy.

The formation mechanisms of CPEO and PEO coatings have significant differences. The conventional PEO process depends on the dielectric breakdown of passive film accompanied with spark discharge on anodes [70], which should be more specifically named as anodic plasma electrolytic oxidation (APEO). However, in the CPEO process, the oxide coating is quickly formed due to the breakdown discharge of the vapor gaseous envelope around the metal cathode [73, 76]. During the CPEO process, the surface temperature of the discharge cathode is up to 1082 °C [60]. More descriptions about the characterization, properties and formation mechanism of CPEO coatings will be introduced in Chap. 8.

4.9 Scanning PEO Treatments

Through time, lots of efforts have been tried for widening the application of PEO process and breaking through the restrictions on the size of the workpieces [77–85]. For example, Pogrebnjak et al. [82] developed an apparatus for PEO treating extended carbon composite articles with metalized (Al-coated) surfaces, as shown in Fig. 4.16a. During the PEO treatment, article 5 (tube with a diameter of 5 cm) is driven by mechanism 6 through a closed ring 7 attached to cathode 10, where the article is the anode and electrolyte is filled between 5 and 7. The authors managed to PEO treat a carbon composite tube with a length of 65 cm and were able to obtain high-quality local oxide coating in low-accessible sites of articles.

Lv et al. [77] developed a method to break through the area restrictions by using fine cathode scanning to complete the PEO treating of the whole area of a ruled surface workpiece. The method is hence termed scanning PEO (SPEO) or scanning microarc oxidation (SMAO). Similarly, Wang et al. [85] have applied an apparatus as schematically shown in Fig. 4.16b, c to perform a SPEO on Ti6Al4V alloy using

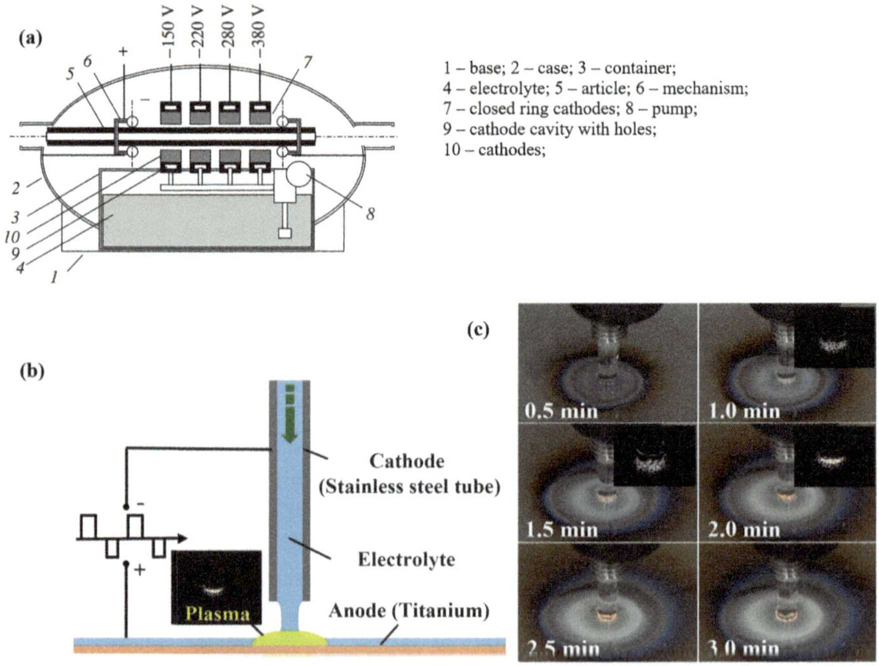

Fig. 4.16 (**a**) Schematic of an apparatus for PEO treating extended carbon composite articles with metalized (Al-coated) surface [69]; (**b**) schematic of a scanning PEO (SPEO) apparatus; (**c**) images showing the SPEO characteristics at different time intervals [72]. *Reproduced (**a**) with permission from* [69] *© 1969, MAIK Nauka/Interperiodica and (**b**, **c**) from* [72] *© 2023 Institute of Materials, Minerals and Mining Published by Taylor & Francis on behalf of the Institute and © 2023, Sage Publications*

an electrolyte of 10 g/L Na_2SiO_3, 5 g/L Na_2HPO_4, and 4 g/L NaOH. The result shows that a higher working voltage is necessary during SPEO treatment owing to additional resistance from the electrolyte column. During the SPEO process, the spark discharge region moves with the movement of the electrolyte ejecting nozzle and exhibits an annular shape (Fig. 4.16c). The current density with Gaussian distribution is observed between the stainless-steel tube (cathode) and titanium substrate (anode), which causes the SPEO coating to show a higher thickness in the middle and lower on both sides. Xia et al. [83] performed an in-depth comparison between SPEO with stationary nozzle (conventional PEO) and moving nozzle (SPEO) and studied the effect of scanning speed and the inter-electrode gap on SPEO coating formation and growth using a 2-mm-thick A356 aluminum sheet and a 15-mm-diameter nozzle. According to their conclusions, micro-hardness distributions and cross-sectional morphology are similar to MAO for a stationary nozzle, but a denser outer layer develops when a moving nozzle is used. This is attributed to a high density of discharge occurring in micropores of the oxide film and remelting, which results from the moving electrolyte column. The SPEO coating thickness and growth rate are linear and are inversely proportional to the inter-electrode gap and the scanning speed when a moving nozzle is used. Analysis of voltage–current characteristic curves shows that the resistance of the electrolyte column is essentially linear over the range considered and results indicate that it can be modeled as a variable length resistor. The results confirm that SPEO is suitable for coating large, planar parts and for repairing worn surfaces.

4.10 Conclusions and Outlook

As shown in this chapter, the different innovative PEO techniques have shown new capabilities that may improve the quality and properties of PEO coatings either through modifications of the parameters of conventional PEO processes or through adding external influences during PEO processes. The two-step or more step PEO treatments have provided more variants for adjusting the conventional PEO treatments. Prospects lie in the proper combination of multiple PEO treatment steps that produce composite structured coatings with more desirable properties. The laser pre/posttreatments, in addition to the PEO treatment, can change the surface structure of the metal substrate or final coating to improve the coating properties such as corrosion resistance. The molten-salt PEO technique can significantly increase the coating efficiencies of PEO treatments and also provide the benefit of reduced inclusions of electrolyte elements in the final coatings. Additionally, the plasma electrolytic processes in molten fluoride salts or fluorine-containing organic solvents can produce fluoride coatings that possess extraordinary anti-corrosion properties. The imposition of ultrasound and magnetic fields to PEO treatments is effective in improving the growth rates as well as the coating quality including uniformity, compactness, etc. The cathodic PEO treatments enable the rapid formation of oxide coatings on both valve and non-valve metals, which has greatly expanded the applicability of the PEO technique. The scanning PEO enables the treatment of large-area workpieces and reduces the instant power needed for the PEO treatments.

The abovementioned techniques are examples of innovative PEO techniques, which are driven by either application needs or fundamental mechanistic studies. Through the application of these innovative methods, the potential application fields of PEO processes will also expand and may include a wide range of application areas, including the surface modifications of mechanical parts, medical implants, nuclear materials, new energy materials, etc. Driven by these applications, it is expected that more innovative PEO approaches will appear in the near future.

References

1. Guan, H. H., Zhou, Q., Xu, C., Jin, X. Y., Du, J. C., & Xue, W. B. (2024). *Journal of Nuclear Materials, 597*, 155091.
2. Sankara Narayanan, T. S. N., Park, I. S., & Lee, M. H. (2014). *Progress in Materials Science, 60*, 1–71.
3. Yang, S., Wang, C., Li, F., Liu, N., Shi, P., Wang, B., & Sun, R. (2022). *Journal of Alloys and Compounds, 909*, 164710.
4. Farshid, S., & Kharaziha, M. (2021). *Journal of Magnesium and Alloys, 9*, 1487–1504.
5. Liang, J., Hu, L., & Hao, J. (2007). *Applied Surface Science, 253*, 6939–6945.
6. Wang, Y., Lei, T., Jiang, B., & Guo, L. (2004). *Applied Surface Science, 233*, 258–267.
7. Shi, P., Ng, W. F., Wong, M. H., & Cheng, F. T. (2009). *Journal of Alloys and Compounds, 469*, 286–292.
8. Wen, L., Wang, Y., Zhou, Y., Guo, L., & Ouyang, J. H. (2011). *Corrosion Science, 53*, 473–480.
9. Cui, W. F., Jin, L., & Zhou, L. (2013). *Materials Science and Engineering: C, 33*, 3775–3779.
10. Fatimah, S., Kamil, M. P., Kwon, J. H., Kaseem, M., & Ko, Y. G. (2017). *Journal of Alloys and Compounds, 707*, 358–364.
11. Bordbar-Khiabani, A., Yarmand, B., & Mozafari, M. (2019). *Surface and Coating Technology, 360*, 153–171.
12. Zehra, T., & Kaseem, M. (2022). *Journal of Molecular Liquids, 365*, 120091.
13. Liu, Y., Wen, S., Liang, G., & Tian, G. (2021). *Journal of Materials Research and Technology, 15*, 6956–6973.
14. Lu, Y. (2010). *Applied Surface Science, 256*, 3554–3558.
15. Namgoong, E., & Chun, J. S. (1984). *Thin Solid Films, 120*, 153–159.
16. Touyeras, F., Hihn, J. Y., Bourgoin, X., Jacques, B., Hallez, L., & Branger, V. (2005). *Ultrasonics Sonochemistry, 12*, 13–19.
17. Nie, K. B., Wang, X. J., Wu, K., Zheng, M. Y., & Hu, X. S. (2011). *Materials Science and Engineering A, 528*, 7484–7487.
18. Jian, X., Xu, H., Meek, T. T., & Han, Q. (2005). *Materials Letters, 59*, 190–193.
19. Xu, H., Jian, X., Meek, T. T., & Han, Q. (2004). *Materials Letters, 58*, 3669–3673.
20. Li, J., Momono, T., Tayu, Y., & Fu, Y. (2008). *Materials Letters, 62*, 4152–4154.
21. Shen, D., Cai, J., Li, G., He, D., Wu, L., Ma, H., Xia, Y., Chen, H., & Yang, Y. (2014). *Vacuum, 99*, 143–148.
22. Zhang, R. F. (2010). *Corrosion Science, 52*, 1285–1290.
23. Tsunekawa, S., Aoki, Y., & Habazaki, H. (2011). *Surface and Coating Technology, 205*, 4732–4740.
24. Pereira, B. L., Beilner, G., Lepienski, C. M., Szameitat, E. S., Chee, B. S., Kuromoto, N. K., dos Santos, L. L., Mazzaro, I., Claro, A. P. R. A., & Nugent, M. J. D. (2020). *Surface and Coating Technology, 382*, 125224.

25. Liu, X., Zhu, L., Liu, H., & Li, W. (2014). *Applied Surface Science, 293*, 12–17.
26. Rehman, Z. U., Shin, S. H., Kaseem, M., Uzair, M., & Koo, B. H. (2017). *Surface and Coating Technology, 328*, 355–360.
27. Wu, Y. K., Han, H. P., Wang, R. Q., Wu, G. R., Zhang, X. Z., Su, Y., Li, D. L., Guo, C. H., Yu, S. X., Shen, D. J., & Nash, P. (2019). *International Journal of Electrochemical Science, 14*, 4730–4753.
28. Einkhah, F., Lee, K. M., Sani, M. A. F., Yoo, B., & Shin, D. H. (2014). *Surface and Coating Technology, 238*, 75–79.
29. Lee, K. M., Ko, Y. G., & Shin, D. H. (2015). *Journal of Alloys and Compounds, 615*, S418–S422.
30. Rehman, Z. U., Shin, S. H., Hussain, I., & Koo, B. H. (2016). *Surface and Coating Technology, 307*, 484–490.
31. Ur Rehman, Z., & Koo, B. H. (2017). *Surface and Coating Technology, 317*, 117–124.
32. Rehman, Z. U., Koo, B. H., Jung, Y. G., Lee, J. H., & Choi, D. (2020). *Materials, 13*, 499.
33. Pereira, B. L., Lepienski, C. M., Mazzaro, I., & Kuromoto, N. K. (2017). *Materials Science and Engineering: C, 77*, 1235–1241.
34. Wang, Y. H., Ouyang, J. H., Liu, Z. G., Wang, Y. M., & Wang, Y. J. (2014). *Applied Surface Science, 307*, 62–68.
35. Ma, C., Liu, J., Zhu, X., Yan, Z., Cheng, D., Xue, W., Fu, J., & Ma, S. (2021). *Wear, 466–467*, 203557.
36. Fattah-alhosseini, A., & Molaei, M. (2023). *Applied Surface Science Advances, 18*, 100506.
37. Fattah-alhosseini, A., & Chaharmahali, R. (2023). *Journal of Magnesium and Alloys., 11*, 4390–4406.
38. Wang, Y., Yang, F., & Zhang, W. (2019). *Journal of the Electrochemical Society, 166*, E481–E488.
39. Wang, L., Zhou, J., Liang, J., & Chen, J. (2012). *Surface and Coating Technology, 206*, 3109–3115.
40. Liu, C., Liang, J., Zhou, J., Li, Q., & Wang, L. (2016). *Applied Surface Science, 382*, 47–55.
41. Nisar, S. S., Arun, S., & Choe, H. C. (2023). *Surface and Coating Technology, 464*, 129553.
42. Wu, G., Zhu, L., Chen, X., Li, L., Zhang, S., Wang, Y., Wen, C., & Yao, J. (2023). *Surface and Coating Technology, 475*, 130108.
43. Rapheal, G., Kumar, S., Blawert, C., & Dahotre, N. B. (2011). *Wear, 271*, 1987–1997.
44. Yang, J., Wang, X., Wen, Q., Wang, X., Wang, R., Zhang, Y., & Xue, W. (2015). *Journal of Nuclear Materials, 467*, 186–193.
45. Sampatirao, H., Amruthaluru, S., Chennampalli, P., Lingamaneni, R. K., & Nagumothu, R. (2021). *Journal of Magnesium and Alloys, 9*, 910–926.
46. Sobolev, A., Kossenko, A., Zinigrad, M., & Borodianskiy, K. (2018). *Surface and Coating Technology, 344*, 590–595.
47. Sobolev, A., Peretz, T., & Borodianskiy, K. (2021). *Journal of Alloys and Compounds, 869*, 159309.
48. Sobolev, A., Bograchev, D., Borodianskiy, K., & Zinigrad, M. (2022). *Corrosion Science, 208*, 110604.
49. Sobolev, A., Peretz, T., & Borodianskiy, K. (2020). *Coatings., 10*, 1–12.
50. Sobolev, A., Zinigrad, M., & Borodianskiy, K. (2021). *Surface and Coating Technology, 408*, 126847.
51. Sobolev, A., Bograchev, D., Zinigrad, M., & Borodianskiy, K. (2022). *Ceramics International, 48*, 10990–10998.
52. Guo, H. F., & An, M. Z. (2005). *Applied Surface Science, 246*, 229–238.
53. Qi, Y., Peng, Z., Wang, L., Zhou, J., Wang, P., & Liang, J. (2021). *Surface Engineering, 37*, 360–364.
54. Lan, L., Zhen-jun, P., Jian-song, Z., & Jun, L. (2023). *Surface Technology, 52*, 2023.
55. Shen, D., He, D., Liu, F., Guo, C., Cai, J., Li, G., & Ma, H. (2014). *Ultrasonics, 54*, 1065–1070.
56. Zhang, X., Zhang, T., Lv, Y., Zhang, Y., Lu, X., Xiao, J., Ma, C., Li, Z., & Dong, Z. (2021). *Applied Surface Science, 569*, 150932.

57. Fatimah, S., Hazmatulhaq, F., Sheng, Y., Suhartono, T., Oh, J. M., Nashrah, N., Kang, J. H., & Ko, Y. G. (2023). *Materials, 16*, 5424.
58. Ajiriyanto, M. K., & Anawati, A. (2023). *Surface and Coating Technology, 456*, 129261.
59. Martin, J., Teriokhina, E., Marcos, G., Czerwiec, T., Henrion, G., Novelli, M., & Grosdidier, T. (2024). *Materials Letters, 370*, 136840.
60. Li, S., Wu, Y., Yamamura, K., Nomura, M., & Fujii, T. (2017). *CIRP Annals, 66*, 345–348.
61. Yin, X., Wang, Y., Liu, J., Deng, L., & Li, S. (2024). *Arabian Journal for Science and Engineering, 49*, 599–611.
62. Boinet, M., Verdier, S., Maximovitch, S., & Dalard, F. (2005). *Surface and Coating Technology, 199*, 141–149.
63. Liao, Y., Zhou, Q., Gao, C., Li, H., Xu, C., Jin, X., Du, J., Xue, W., & Zhang, Y. (2024). *Ultrasonics, 138*, 107213.
64. Liao, Y., Zhou, Q., Gao, C., Xu, C., Jin, X., Du, J., Xue, W., & Zhang, Y. (2023). *The Review of Scientific Instruments, 94*, 063905.
65. Vijh, A. K. (1971). *Corrosion Science, 11*, 161–167.
66. Albella, J. M., Montero, I., & Martinez-Duart, J. M. (1987). *Electrochimica Acta, 32*, 255–258.
67. Ikonopisov, S., Girginov, A., & Machkova, M. (1977). *Electrochimica Acta, 22*, 1283–1286.
68. Mortazavi, G., Jiang, J., & Meletis, E. I. (2019). *Applied Surface Science, 488*, 370–382.
69. Zhai, D., Li, X., Shen, J., & Feng, K. (2022). *Metallurgical and Materials Transactions A: Physical Metallurgy and Materials Science, 53*, 1200–1207.
70. Clyne, T. W., & Troughton, S. C. (2019). *International Materials Review, 64*, 127–162.
71. Jin, X., Wang, B., Xue, W., Du, J., Wu, X., & Wu, J. (2013). *Surface and Coating Technology, 236*, 22–28.
72. Jin, X., Zhang, Y., Chen, L., Yu, J., & Xue, W. (2018). *Surface and Coating Technology, 338*, 38–44.
73. Yu, J., Zhang, Y., Jin, X., Chen, L., Du, J., & Xue, W. (2019). *Surface and Coating Technology, 363*, 411–418.
74. Wu, J., Dong, L., Deng, J., Hou, D., Li, G., Li, D., & Xue, W. (2018). *Surface and Coating Technology, 338*, 63–68.
75. Stojadinović, S., Tadić, N., Šišović, N. M., & Vasilić, R. (2015). *Journal of Applied Physics, 117*, 233304.
76. Gao, C., Zhu, M., Wang, B., Liao, Y., Li, H., Jin, X., Xu, C., Du, J., & Xue, W. (2022). *Surface and Coating Technology, 444*, 128666.
77. Lv, P. X., Chi, G. X., Wei, D. B., & Di, S. C. (2011). *Advances in Materials Research, 189–193*, 1296–1300.
78. Lu, J., Zhan, S., Liu, B., & Zhao, Y. (2022). *International Journal of Extreme Manufacturing, 4*, 045101.
79. Shen, X., Nie, X., & Tjong, J. (2019). *Heat and Mass Transfer und Stoffuebertragung, 55*, 2503–2515.
80. Clare, A. T., Speidel, A., Mitchell-Smith, J., Bisterov, I., & Murray, J. W. (2019). *CIRP Annals, 68*, 177–180.
81. Gu, W. C., Lv, G. H., Chen, H., Chen, G. L., Feng, W. R., & Yang, S. Z. (2007). *Surface and Coating Technology, 201*, 6619–6622.
82. Pogrebnjak, A. D., & Tyurin, Y. N. (2004). *Technical Physics, 49*, 1064–1067.
83. Xia, L., Han, J., Domblesky, J. P., Yang, Z., & Li, W. (2017). *Journal of Materials Engineering and Performance, 26*, 5323–5332.
84. Wen, L., Wang, Y., Jin, Y., Liu, B., Zhou, Y., & Sun, D. (2013). *Surface and Coating Technology, 228*, 92–99.
85. Wang, Y., Shen, J., Wu, G., & Yao, J. (2023). *Surface Engineering, 39*, 218–228.

Chapter 5
The Effect of Plasma Electrolytic Oxidation on the Mechanical Properties of Different Metallic Systems

Mikhail Zheludkevich and Carsten Blawert

5.1 Introduction

Plasma Electrolytic Oxidation (PEO) is a surface treatment process where a high voltage is applied to an electrolytic solution, resulting in the formation of a hard, ceramic-like oxide layer on the metal surface. This process is characterized by the generation of plasma discharges in the electrolyte, which lead to the formation of the oxide layer. The resulting coatings are typically thick and dense, exhibiting excellent mechanical properties such as increased hardness, wear, and corrosion resistance. The thickness and properties of the oxide layer can be controlled by adjusting process parameters like applied voltage, current density, treatment time, and electrolyte composition. Depending on the composition of the electrolytes used and the nature of the metal substrates, the formed layers can also consist of more complex compounds such as phosphates, silicates, or aluminates. This method is particularly well suited for metals like aluminum, magnesium, and titanium, which are widely used in aerospace, automotive, and biomedical industries. The PEO process is highly versatile and can be tailored to produce coatings with specific properties for a wide range of applications.

For example, PEO coatings on aluminum alloys can significantly enhance their wear and corrosion resistance, making them suitable for use in aerospace and automotive components. Similarly, PEO coatings on magnesium alloys can improve their surface hardness and corrosion resistance, which is beneficial for applications in the automotive and biomedical industries. Titanium alloys treated with PEO exhibit enhanced biocompatibility and wear resistance, making them ideal for use in medical implants and orthopedic devices. The microstructure of the PEO coatings can be

M. Zheludkevich (✉) · C. Blawert
Institute of Surface Science, Helmholtz-Zentrum Hereon, Geesthacht, Germany
e-mail: mikhail.zheludkevich@hereon.de

V. S. Saji, M. Zheludkevich (eds.), *Plasma Electrolytic Oxidation – Fundamentals, Advances and Applications*,
https://doi.org/10.1007/978-3-031-92429-3_5

further modified by incorporating nanoparticles or other additives into the coating, which can enhance the coating's properties and expand its range of applications.

The ceramic-like layers formed during the PEO process significantly enhance the mechanical properties of treated surfaces, often increasing their hardness and wear resistance. However, the PEO process has its limitations. One major challenge is the formation of microcracks and pores within the oxide layer, which can serve as initiation points for defects under mechanical loads or in corrosive environments. These defects can compromise the coating's overall mechanical properties and effectiveness. Additionally, the hard surface layer can induce a notching effect, further serving as initiation points for defects under mechanical or combined mechanical and corrosive loads. To mitigate these issues, various posttreatment processes have been developed to improve the quality and performance of PEO coatings. For instance, sealing treatments can fill the pores and cracks within the oxide layer, enhancing its protective capabilities and overall mechanical properties.

The impact of PEO on the mechanical properties of metallic systems is a topic of significant interest and ongoing research. This chapter briefly addresses the effects of PEO on the mechanical properties of metallic systems, including surface hardness, wear resistance, stress corrosion cracking, and corrosion fatigue. Additionally, the impact of posttreatment processes on the mechanical properties of PEO-coated substrates is discussed. By understanding the mechanisms and factors that influence the performance of PEO coatings, researchers and engineers can optimize the process to develop advanced materials with superior properties for a wide range of applications.

5.2 Surface Hardness and Wear Resistance of PEO-Coated Metals

PEO significantly enhances the surface hardness of metallic substrates by forming a dense oxide layer. The microhardness of pure metals like aluminum, magnesium, and titanium typically ranges from 20 to 150 HV (Vickers Hardness). Metal alloys, such as steel or aluminum alloys, generally exhibit higher microhardness values, ranging from 150 to 600 HV, depending on their composition and heat treatment. At the same time, the application of a PEO layer can lead to a substantial increase in microhardness primarily due to the formation of hard ceramic phases during the oxidation process, which significantly enhances the surface properties of the underlying metal. The typical range of microhardness values of PEO layers ranges from 800 to 2000 HV.

5.2.1 Aluminum Alloys

PEO coatings on aluminum alloys are reported to achieve microhardness values between 1000 and 2000 HV. PEO-treated aluminum alloys demonstrate remarkable hardness and wear resistance. The coatings typically feature a three-layered structure: a thin barrier layer adjacent to the substrate, a dense intermediate layer, and a loose outer layer. The barrier layer is crucial for its anticorrosion properties, while the intermediate layer significantly enhances hardness and wear resistance [1]. Additionally, to the morphology of the layer, the phase composition plays an important role in the mechanical properties of the coating. The outer layer was demonstrated to be predominantly constituted by γ-Al_2O_3 and Al-Si-O compounds when the coating is applied in the silicate-containing electrolyte. Moreover, the increase in the mullite ($3Al_2O_3$ $2SiO_2$) content in the PEO layer was reported to decrease the microhardness of the coating [2]. However, these phases still possess significantly higher hardness than the substrate and oxide coatings produced by anodizing or hard-anodizing.

The intermediate layer in turn is richer in α-Al_2O_3 which confers a significantly higher hardness [1]. One of the strategies reported to significantly improve the properties of the outer layer is the incorporation of hard particles during the PEO synthesis. During AC PEO treatment, α-Al_2O_3 particles were integrated into the outer coating layer. PEO coatings with embedded alumina particles exhibited a higher α-Al_2O_3 phase content in the outer layer, resulting in a denser and harder coating compared to those formed in a particle-free electrolyte on 6082-T6 alloy, as shown in Fig. 5.1 [3]. The incorporation of these particles likely occurs through deposition, electrophoresis, or mechanical entrapment by material expelled from discharge channels. Some particles undergo solid-state sintering, fusing their boundaries with the surrounding oxide material rather than melting.

The composition of the electrolyte used for PEO treatment of Al alloys is an essential factor defining the composition and morphology of the obtained coatings. A systematic study of the most common alkaline electrolytes including those containing phosphates and silicates on AA2024 demonstrated a significant variation in the coating thickness, phase composition, and morphology [4]. For example, a mixed electrolyte with low hydroxide concentration, high phosphate concentration, and increasing silicate levels significantly enhances the coating thickness, while also improving the uniformity and density of the coatings as shown in Fig. 5.2. The coating growth mechanism exhibited two primary directions. In Si-based electrolytes, the coating thickens primarily due to the rapid deposition of electrolyte compounds, with growth occurring mainly toward the electrolyte/coating interface. Conversely, in OH- and P-based electrolytes, inward growth is predominantly driven by substrate oxidation. Various phases were observed depending on the electrolyte composition and final voltages. At high final voltages (over 470 V) in mixed electrolytes with low concentrations of hydroxide, silicate, or phosphate (2 and 6 g/L), the γ-Al_2O_3 crystalline phase predominates in the PEO layer composition.

The composition of the aluminum alloy can also affect the PEO layers. The presence of elements like Zn, Cu, or Mg in the alloy was shown to inhibit the transition from γ- to α-alumina, resulting in reduced hardness [5]. However, despite certain differences in the phase composition, the wear properties in the rubber-wheel test show no evident dependence on the substrate material [6].

The PEO process parameters also strongly affect the hardness of the formed layers. When a pulsed PEO regime is used, the increase in the duty cycle is reported to

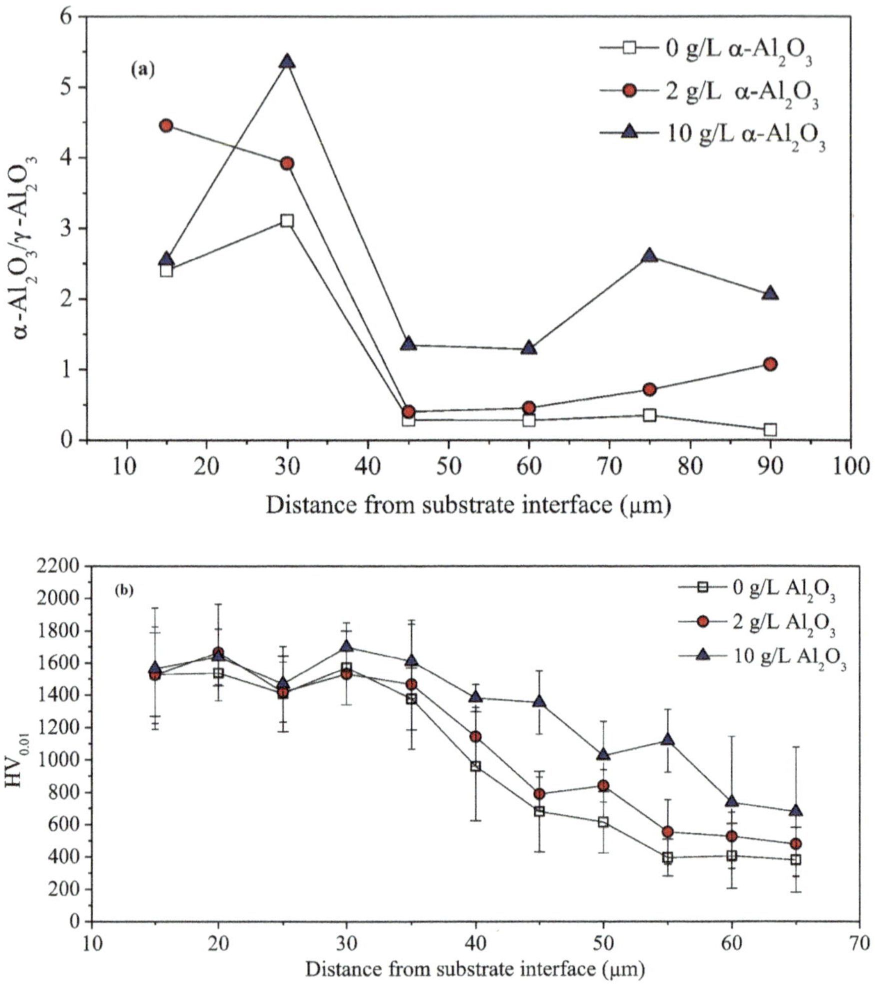

Fig. 5.1 α-Al2O3/γ-Al2O3 ratio (**a**) and hardness (**b**) as a function of distance from coating/substrate (AA6082-T6) interface for the PEO coatings with and without α-Al_2O_3 particles [3]. *Reproduced with permission from [3] © 2014 Elsevier B.V.*

increase the hardness of the PEO layers [7]. As expected, the higher amount of the injected energy is reflected in the formation of thicker PEO, which in turn improves tribological properties.

The exceptional hardness, residual compressive stresses, and compact microstructure of PEO-produced oxide coatings contribute to their high wear resistance. However, the wear resistance was demonstrated to be dependent not only on the PEO process itself but also on the aluminum alloy substrate, applied loads, and type of abrasive materials. For example, notable differences were observed in the phase composition, microstructure, and hardness of the PEO layers on AA7075 wrought

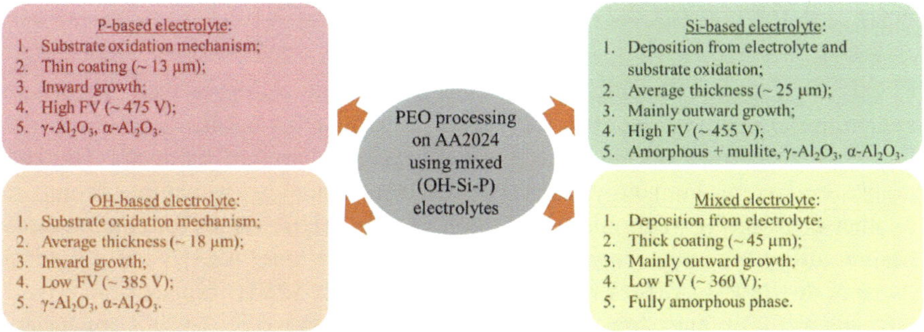

Fig. 5.2 Schematic illustration of PEO coatings characteristics obtained in different alkaline electrolytes on AA2024 [4]. *Reproduced with permission from [4] © 2023 The Author(s). Published by Elsevier B.V.*

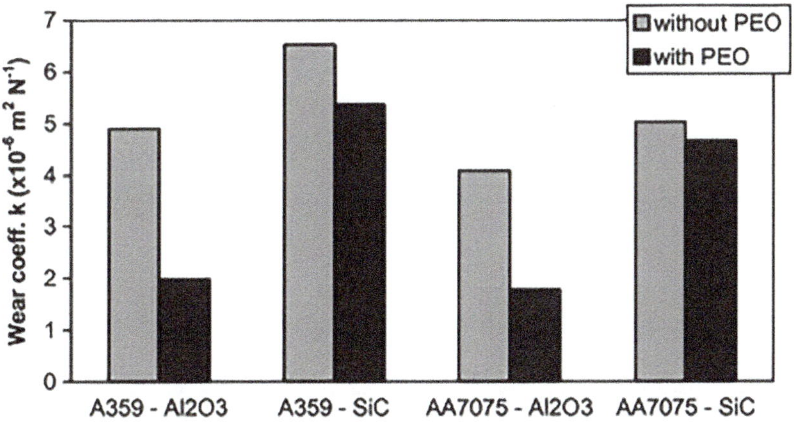

Fig. 5.3 Wear coefficients measured by micro-scale abrasion tests on uncoated and PEO-treated A359 and AA7075 aluminum alloys [8]. *Reproduced with permission from [8] © 2009 Elsevier Ltd.*

alloy and A359 cast alloy, being driven by the distinct compositions and microstructures of the substrates [8]. Use of alumina particles as abrasive material demonstrated a significant difference between the bare and PEO-coated substrates, while this difference becomes much smaller when more aggressive SiC is employed for the ball-cratering test, as shown in Fig. 5.3. In unlubricated sliding conditions, the PEO treatment notably enhanced the wear resistance of both aluminum alloys at low applied loads. However, under high applied loads, the hardness of the PEO layer and the load-bearing capacity of the substrate are the primary factors affecting wear resistance.

The application of PEO coatings, known for their extreme hardness, reduces the impact toughness of the aluminum alloy substrate. Consequently, these coatings may not be ideal for environments where the substrate is subjected to impact [1].

5.2.2 Magnesium Alloys

PEO is often applied to modify the surface of magnesium alloys as well. Magnesium and aluminum alloys are both widely used in engineering applications due to their favorable properties, but they differ significantly in terms of hardness. The commonly used magnesium alloys are known for their relatively low hardness compared to other engineering materials. The Brinell hardness (HB) values for common magnesium alloys typically range from 42 to 95. For example, AZ31B has a hardness range of 46–73 HB, AZ61A ranges from 50 to 60 HB, AZ91D ranges from 60 to 77 HB, and WE43 ranges from 75 to 95 HB. When comparing the two, magnesium alloys have a hardness range of 42–95 HB, while aluminum alloys have a broader range of 21–160 HB. Typical hardness values for magnesium alloys like AZ31B and AZ91D are around 50–77 HB, whereas aluminum alloys like 6061-T6 and 7075-T6 have hardness values around 95–150 HB.

Due to their lower hardness, magnesium alloys are often not selected as materials of choice in applications requiring better wear resistance and structural integrity, such as aerospace and automotive components. On the other hand, magnesium alloys, being much lighter, are very attractive in applications where weight reduction is critical. Therefore, the use of the PEO layers to improve the surface hardness and wear resistance of Mg alloys becomes a very attractive option. The hardness of PEO coatings on magnesium alloys is predominantly influenced by the presence of crystalline phases. These phases can be meticulously controlled through the composition of electrolytes, the substrate, and the process parameters [9]. PEO of magnesium alloys is typically conducted in alkaline electrolytes containing different concentrations of aluminate, silicate, or phosphate. The mixed electrolytes or those containing fluoride ions are also often considered. The resulting coatings generally comprise a mixture of amorphous and crystalline phases, including MgO, $MgAl_2O_4$, Mg_2SiO_4, and $Mg_3(PO_4)_2$ [10] and exhibit a two- or three-layer structure, characterized by a thin barrier layer at the substrate/coating interface, typically only a few hundred nanometers thick. Current PEO techniques favor AC, pulsed DC, and bipolar electrical regimes, as they provide superior control over plasma chemical processes compared to traditional DC regimes. This enhanced control minimizes the adverse effects associated with DC discharges, resulting in more uniform and less porous coatings [11].

PEO coatings developed in silicate electrolytes demonstrate significantly better wear resistance than those formed in phosphate electrolytes, as shown in Fig. 5.4. This superior wear resistance is attributed to the higher hardness of the silicate-based coatings, as noted in [12]. The silicate-based PEO coating primarily comprised Mg_2SiO_4, whereas the phosphate-based PEO coating mainly consisted of $Mg_3(PO_4)_2$, both along with some MgO. Both types of PEO coatings exhibited high hardness values, with Si-PEO achieving 1660 ± 367 Hv and P-PEO reaching 560 ± 153 Hv due to their ceramic nature. Aluminate-based PEO coatings generally exhibit even higher hardness values than silicate-based coatings. This difference is primarily due to the presence of the $MgAl_2O_4$ phase in aluminate-based coatings,

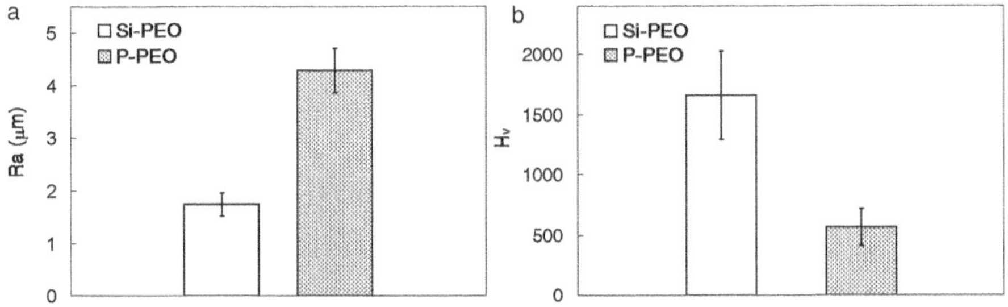

Fig. 5.4 (**a**) Average surface roughness and (**b**) Vickers hardness of silicate-based PEO (Si-PEO) and phosphate-based PEO (P-PEO) coatings [12]. *Reproduced with permission from [12] © 2011 Elsevier B.V.*

which is harder than the Mg_2SiO_4 phase found in silicate-based coatings. The denser and more compact microstructure of aluminate-based coatings also contributes to their superior hardness and wear resistance.

The composition of Mg alloys can also be an important factor affecting the hardness of formed PEO layers. In an earlier work [11], a set of Mg alloys with different principal alloying elements was coated by silicate/phosphate PEO, and the mechanical properties of the formed layers were studied. The hardness of the intermediate layer of the PEO coatings varied considerably based on the alloy composition, with measurements spanning from 270 to 450 $HV_{0.05}$. Coatings on AZ61, AZ91D, and WE43 alloys exhibited the highest hardness levels. In contrast, coatings on ZC71, ZE41, and AZ31 alloys demonstrated moderate hardness, while those on pure magnesium had the lowest hardness values. Thus, even a small variation of the PEO layer composition caused by the inclusion of the alloying elements into the composition can remarkably affect the hardness.

PEO coatings on magnesium and its alloys often struggle to provide long-term wear protection due to their high porosity and limited range of phase compositions. Therefore, a promising pathway is to modify the PEO coatings with addition of hard particles or those, which can form hard phases as a result of a reactive incorporation. Such a particle addition to the electrolyte can lead to coatings with significantly enhanced wear resistance [13]. For example, the hardness of PEO coatings on AZ91D magnesium alloy was boosted more than twice by the reactive incorporation of Al_2O_3 nanoparticles with higher particle concentrations in the electrolyte, which was linked to a significant rise in the $MgAl_2O_4$ phase within the layer [14]. However, the inert incorporation of such particles can also have a strong effect. For instance, the hardness of PEO-coated AZ31 magnesium alloy was also significantly increased from 130 HV to 358 HV with the inert incorporation of Al_2O_3 nanoparticles [15]. The significant increase of the hardness and improved wear resistance was also reported after incorporation of SiC and TiO_2 thanks to formation of denser and thicker layers [16–18].

The wear resistance of the PEO layers on Mg can be improved not only by the incorporation or formation of harder phases but also by a decrease in the friction coefficient. The introduction of solid lubricant PTFE particles into PEO coatings led to a stable and significantly lower (below 0.2) friction coefficient compared to particle-free coatings (0.5–0.7), effectively preventing surface wear [19].

The wear resistance of the magnesium alloys usually is dramatically enhanced by the application of PEO coatings. However, the coating thickness plays a pivotal role in further boosting wear resistance. At elevated initial stress levels, substrate deformation can cause the coating to crack and flake off, especially when the coating is thin. In such scenarios, a thicker PEO coating offers superior load-bearing capacity, resulting in markedly improved wear resistance [20].

5.2.3 *Titanium Alloys*

When comparing the hardness of PEO layers on titanium and its alloys to the hardness of the alloys themselves, it is evident that PEO coatings significantly enhance surface hardness. PEO coatings on pure titanium achieve hardness values of 600 to 1200 HV, while those on titanium alloys like Ti6Al4V can reach 800 to 1500 HV. In contrast, the hardness of titanium alloys without PEO coatings typically ranges from 330 to 450 HV. This substantial increase in hardness due to PEO coatings is primarily attributed to the formation of hard oxide phases and a denser microstructure [21, 22]. The specific hardness value can be influenced by the electrolyte composition and the electrical parameters used during the PEO process.

The typical electrolyte systems for Ti are often similar to the conventional solutions also used for Al and Mg alloys such as phosphate [23], aluminate [22], silicate [21], or their mixtures. Often, the electrolytes are modified by addition of fluorides [24] or acetates [25]. The phase composition of plasma electrolytic oxidation (PEO) coatings on titanium is significantly influenced by the composition of the electrolyte used during the PEO process. Different electrolytes can lead to the formation of various crystalline and amorphous phases, which in turn affect the mechanical and chemical properties of the coatings. Very often, the composition of such layers is tuned toward improved biocompatibility through the introduction of phosphate and calcium. Electrolytes containing calcium and phosphorus ions can result in the formation of hydroxyapatite ($Ca_{10}(PO_4)_6(OH)_2$) on the titanium surface. This phase is particularly beneficial for biomedical applications due to its excellent biocompatibility and ability to promote bone growth. The incorporation of hydroxyapatite enhances the bioactivity of the PEO coatings. Although, the effect of such additives on the mechanical properties is not always positive.

Nano-scale indentation, scratch, and multiple impact testing have been employed to quantitatively assess the wear performance of PEO coatings for Ti–6Al–4 V obtained in aluminate, silicate, phosphate, and mixed electrolytes [26]. Among the tested systems, an aluminate-based coating demonstrated superior performance, even surpassing the bare substrate. Coatings containing Al_2O_3 and Al_2TiO_5 exhibited excellent erosion resistance due to the high hardness of these phases. However, the aluminate-based coating showed low interfacial adhesion strength, as evidenced by delamination during metallographic preparation and chipping failure during erosion. This issue might be mitigated by adjusting processing parameters, such as increasing substrate surface roughness through shot peening before coating.

The use of pure aluminate electrolytes without addition of sodium hydroxide was reported to result in the formation of mainly α-Al_2O_3-containing layer on pure titanium [22]. At the same time, a mixture of sodium aluminate and sodium hydroxide leads to the formation of a predominantly aluminum titanate structure, which exhibits inferior mechanical properties compared to the α-Al_2O_3 enriched coatings [27].

The formation of titania phases is dominating when phosphate/acetate-based electrolytes are used on Ti alloys. The ratio between rutile and anatase is one of the factors dominating the hardness of the resulting coatings. This ratio can be adjusted by the PEO process parameters. At the higher voltage (300 V), rutile was found to dominate, leading to a PEO coating with a higher hardness value. However, at lower voltage (200 V), the film has a reduced hardness due to anatase oxide and tiny pores on the surface [28]. It was also found that an increase in the PEO treatment time can lead not only to an increased coating thickness but also to a significant change in the phase composition [29]. As the PEO surface treatment time increased, there was a notable promotion of the rutile phase and a corresponding decrease in the anatase phase. These coatings also possessed a micrometric porous characteristic, which resulted in a lower concentration per unit area when subjected to longer PEO treatment times. The formation of rutile contributed to an increase in the hardness of the PEO coating with extended treatment time.

As in the case of PEO layers on Al and Mg alloys, the addition of particles to the electrolyte can be considered as one of the pathways for improvement of the hardness and wear resistance of the coatings on Ti alloys. Numerous electrolyte systems with different particles have been recently considered, including hydroxyapatite [30], zirconia [31], alumina [27], and silicon nitride [32]. In many cases, presence of hard particles incorporated inertly was linked to the improved mechanical properties.

In the work of Lederer et al. [33], the increase in mechanical properties was achieved by the addition of zirconia nanoparticles into the electrolyte. The uncoated titanium substrate exhibited Young's modulus of 96 ± 15 GPa. Coatings, both with and without ZrO_2 nanoparticles (NPs), demonstrated a higher Young's modulus compared to the substrate. However, the coating containing ZrO_2 NPs showed a significantly higher Young's modulus of 205 ± 70 GPa, in contrast to the 156 ± 63 GPa observed in the coating without ZrO_2, as can be seen in Fig. 5.5.

Incorporating solid lubricant particles such as MoS_2 into the electrolyte has been shown to significantly reduce the friction coefficient of PEO coatings [34]. An alternative is to use graphite as an additive. Despite resulting in a thinner coating, the friction coefficient was significantly lowered from 0.6–0.8 to 0.15 with the inert addition of graphite particles [35]. Incorporating graphene oxide (GO) additives into the base silicate-hypophosphite electrolyte was found to significantly enhance the hardness of PEO coatings, increasing from 331 HV to 366 HV as the GO concentration rises from 0.1 to 0.5 kg/m^3. Additionally, the friction coefficient and wear rate of the coatings decrease from 0.73 to 0.69 and from 8.04 to 5.2 mm^3/N·m, respectively, with the increase in GO concentration from 0 to 0.5 kg/m^3. This improvement is attributed to the formation of a lubricating trilayer containing GO, which forms upon contact between the coating and the counter-body in the friction pair [36].

In summary, the influence of electrolyte composition on the hardness and wear resistance of PEO coatings is significant. For example, the addition of nanoparticles to the electrolyte has been shown to further enhance the hardness of PEO coatings on Mg alloys. The presence of these nanoparticles within the oxide layer helps to fill the micropores, resulting in a denser and more homogeneous coating. This, in turn, increases the surface hardness and wear resistance of the treated metal. Additionally, the use of different electrolyte compositions can lead to the formation of various oxide phases, each with distinct hardness properties.

Process parameters such as current density, treatment time, and voltage also play a crucial role in determining the wear resistance of PEO coatings. Higher current densities and longer treatment times generally result in thicker and more wear-resistant oxide layers. However, there is an optimal range for these parameters, beyond which the coating may become brittle and prone to cracking. For example, excessively high current densities can lead to the formation of large pores and cracks within the oxide layer, which can reduce the overall wear resistance of the coating. Therefore, optimizing these parameters is essential to achieve the best possible wear resistance and mechanical properties of PEO-coated substrates.

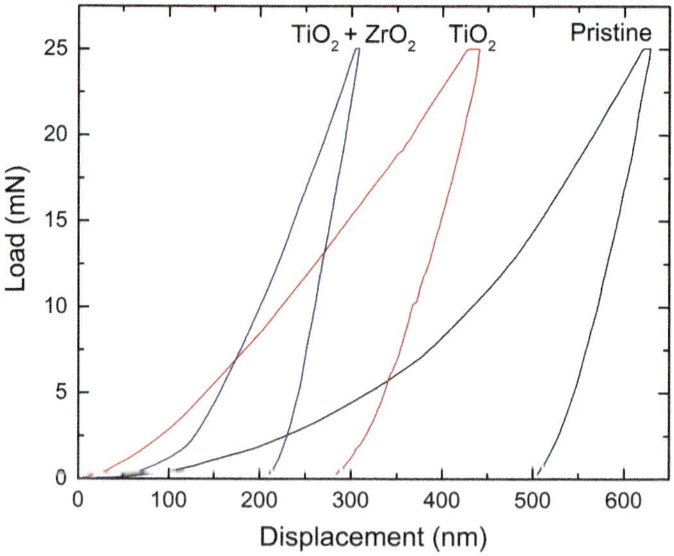

Fig. 5.5 Load versus indenter displacement for the pristine material and PEO-coated samples with and without zirconia particle addition [33]. *Reproduced with permission from [33] © 2017 Elsevier B.V.*

5.3 Effect of PEO on Stress Corrosion Cracking and Corrosion Fatigue

There are no doubts about the positive influence of PEO coatings on general corrosion [37–39] and wear [39, 40] performance as explained above. However, PEO coatings are ceramic-based coatings with relative high strength, but low ductility applied on light metals with much higher ductility and lower strength. As the interfacial bonding strength of PEO coatings to the substrates is generally considered as good, the coating must follow the deformations of the substrate and even tolerable elastic deformations of the substrate may cause high stresses and overloads for the coating leading to new or growing cracks in the coating. Also, cyclic loads much lower than the yield strength of the substrate or coating can cause problems as the PEO coatings by nature are brittle and full of defects (cracks and pores) and have rough surfaces and interfaces. All those aspects are considered to have negative effects on the fatigue behavior as already existing cracks can grow or new cracks can easily form and progress from the coating into the substrate. If additionally, an aggressive surrounding interacts with the external and internal surfaces of a PEO coating/light metal substrate system the good and sufficient protectiveness can quickly diminish, and performance might be even worse than without a PEO coating applied. Fatigue of PEO coatings in air is already a concern, and a corrosive atmosphere will not make it better. While in many corrosion cases, an increasing coating thickness is a solution to reduce the corrosion of the substrate, thicker PEO coatings may be even more prone to stress because of their inherent stiffness and the higher likelihood of internal stress build-up, thus offering no proper solution for improving SCC and CF.

The following sub-chapters address the main points and mechanisms of PEO coating performance on light metals (Mg, Al, Ti) in conditions allowing SCC or CF. In spite of the obvious substrate material influences on the coating's behavior under stress and corrosion (e.g., the native oxide film on Ti is overall much more stable than on Al or Mg; on the other hand, Mg becomes much more stable under strong alkaline conditions) the focus of the chapter should be more on the general PEO coating influence as a thick, porous ceramic layer on a ductile metal substrate.

5.3.1 *Stress Corrosion Cracking*

Stress corrosion cracking (SCC) requires the combined influence of tensile stress and a corrosive environment. The problem in applications is that the failure may occur already at stress levels much lower than the yield stress of the materials, suddenly without warning, sometimes even after only short exposures to the corrosive environment. SCC goes together with a loss of ductility, which is often associated with hydrogen embrittlement. Hydrogen can contribute to SCC by promoting crack initiation and propagation. When hydrogen atoms are present in a corrosive environment, they can enter the metal and create local internal stresses that contribute to crack formation. The source of hydrogen is quite often related to the corrosion processes themselves [41] or to the coating process in which hydrogen can be a reaction product. Latter appears to be not so relevant in the anodization of light metals in alkaline electrolytes.

The bare PEO coatings with their internal residual stress levels, inherent process-related defects and brittle ceramic character are not able to prevent SCC efficiently. The micro-cracks and pores act as pathways for corrosive media, leading to localized corrosion and consequently quite often to SCC. Localized corrosion under the PEO coating can cause stress (due to volume increase due to solid corrosion products) and subsequent delamination [42] generating new starting points for cracks in the interface. However, the already existing cracks in the coating are also critical, as they can grow under the action of stress and corrosive media, progressing across the interface from the coating into the substrate, where new quite active surfaces (plastically deformed) are exposed to the electrolyte and prone to corrosion, with the risk of hydrogen formation as a by-product. Thus, new cracks can form, or existing ones can grow, allowing even more corrosive media to enter and accelerate the SCC processes.

SCC does not need externally applied tensile stress. Internal residual stress due to a mismatch of thermal expansion coefficients and different mechanical properties of either the various coating phases themselves or between the ceramic coating and the metallic substrate can be sufficient for causing SCC. Compressive stress levels of 40–50 MPa were measured for PEO coatings on Al, 130–150 MPa for PEO coatings on Mg, and 95–113 MPa for coating on titanium [41–44]. This is positive for the SCC performance of the coating, because the internal coating stresses are firstly compensating applied external tensile stresses. However, the substrate is even more under tensile load and more prone to SCC in the case the corrosive media reaches the substrate. Therefore, it is often found that the PEO coating is retarding the onset of SCC (at least by the time it needs that the electrolyte reaches the interface and damages the barrier layer) but not being able to prevent SCC to a certain point that the electrolyte is reaching the interface/substrate causing localized corrosion as a possible starting point of crack growth in the substrate [41]. A common approach to determine the susceptibility to SCC is slow to strain rate tests comparing the performance in air and in the respective corrosive media of concern (Fig. 5.6). Those tests sometimes demonstrate the advantage of PEO-coated specimens compared to the uncoated reference specimen and sometimes not [45–49]. However, it is disputable if such tests are the best choice for brittle ceramic-based coatings that cannot follow the plastic deformation of the substrate, but at least they can be considered as the worst-case scenario when the coating is damaged and cracked. A more realistic but much slower and time-demanding approach would be testing under constant loads below the yield strength of the coating/substrate.

Environmental factors such as temperature, humidity, and the presence of specific ions, can influence SCC behavior of PEO-coated light metals. PEO coatings may be particularly vulnerable to SCC in chloride ion-containing environments because of the localized breakdown of the oxide layer due to chemical reactions

Fig. 5.6 Typical setup for slow strain rate testing—SCC experiment with CT specimen

with chloride ions [41]. Thus, proper corrosion inhibitors and smart corrosion protection concepts may delay the SCC of PEO coatings. Even more effectively applied are sealing coatings (quite often polymer-based paints), which prevent the access of corrosive species into the coating and migration toward the coating–substrate interface [40]. The combination of such additional (better not too brittle) sealing barrier coatings with smart corrosion protection concepts might be the solution for much better SCC resistance of PEO coatings.

In summary, SCC of PEO coatings is a complex process that depends on coating (phase composition, thickness, etc.), substrate material, and environmental factors. By controlling the coating's microstructure (reduce defects, increase amount of corrosion-resistant phases), improving its adhesion to the substrate, select a less SCC-sensitive substrate material, and managing the stresses and environmental exposure, it is possible to reduce the risk of SCC.

5.3.2 *Corrosion Fatigue*

Corrosion fatigue is the mechanical degradation of a material under the combined action of cyclic loading and a corrosive environment. It involves the initiation and propagation of cracks due to the synergistic effects of mechanical stress and corrosion. The corrosive environment accelerates crack growth, reducing the fatigue life of the material. Similar to SCC, it can lead to unexpected failures even under stress levels much lower than the yield strength of the material. The severity of corrosion fatigue depends on factors such as the type of material, the corrosive environment, the frequency and amplitude of cyclic loading, and the presence of stress concentrators [50], [51]. Different substrate materials in combination with a PEO coating, such as magnesium, aluminum, and titanium alloys, exhibit different fatigue behaviors, but this is not the focus of this sub-chapter.

Coatings can significantly influence the fatigue behavior of those metals, both positively and negatively [50, 52]. Compared with hard anodization, the PEO coatings have fewer negative effects on the fatigue properties [53]. However, to mitigate CF, it is important to select a substrate that is less susceptible to fatigue and has good compatibility with PEO coatings. The coating thickness, adhesion quality, surface roughness, and the presence of defects like micro-cracks are the main influencing parameters. On the one hand, coatings can introduce stress concentrations that may accelerate fatigue crack propagation, but on the other hand, they can retard the access of corrosive species toward the substrate. The mechanisms are overall quite complex, and what is dominating depends on the application and environment.

Overall, the majority of fatigue studies of PEO coatings in air see a reduction of the fatigue strength with increasing coating thickness/roughness, and only thin coatings might have negligible negative effect [39, 43, 50, 52, 54, 55] or even positive effect [52]. This is not surprising, as the damage tolerance of PEO coatings is decreasing with increasing coating thickness [56]. Interestingly the fatigue life can be improved by shot peening before PEO processing while the plain PEO shows the typical negative effect of thickness [57]. The improvement is related to large compressive stresses in the substrate due to the shot peening, removing the critical residual tensile stresses in the substrate [51, 58, 59]. The other critical points for PEO coatings are the adhesion strength in the interface and the presence of defects such as micro-cracks and pores, which are sometimes even agglomerated in specific regions forming pore bands reducing the cohesive strength within the PEO coatings. Furthermore, corrosion can weaken the bond between the PEO coating and the substrate, leading to delamination or detachment of the coating [42]. Another issue is the typically high surface roughness of PEO coatings. Higher roughness can lead to more defects and stress concentrations, reducing fatigue life [59]. However, in corrosive environments, the fatigue limit normally decreases further (Fig. 5.7).

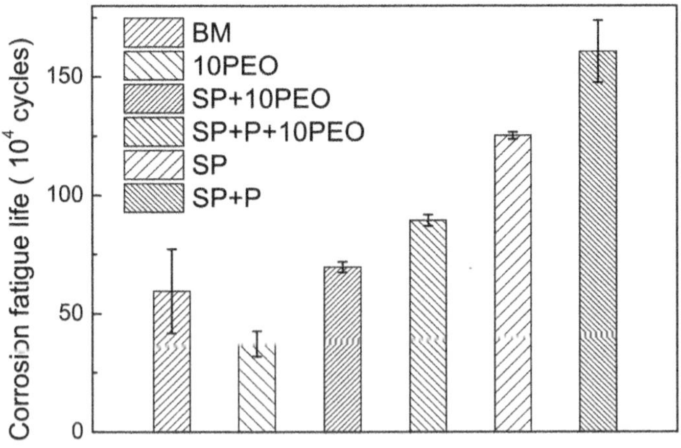

Fig. 5.7 Corrosion fatigue life of samples with different surface conditions. (BM: uncoated, 10PEO: PEO coated (10 min), SP: shot peening, P: polished substrates) [59]. *Reproduced with permission from [59] © 2019 Elsevier B.V.*

However, under cyclic stress, the PEO coating may degrade over time, especially at interfaces between the coating and the substrate. Corrosive environments (like saltwater, acidic solutions, or high-humidity conditions) can accelerate the degradation by breaking down the oxide layer. Thus, the coating can suffer from localized corrosion (like pitting) under cyclic loading, leading to crack formation. These cracks may propagate, eventually causing coating failure. Once cracks initiate, corrosion at the crack tip may accelerate the propagation under repeated loading, leading to faster failure of the system. Additionally, as soon as cracks/electrolytes reach the interface, electrochemical reactions with the metal can result in hydrogen formation and hydrogen embrittlement making the metal even more susceptible to cracking under cyclic loading. Hydrogen can enhance localized plastic deformation (hydrogen-enhanced localized plasticity), which further accelerates crack initiation and propagation. The overall outcome is a further reduced fatigue strength of PEO-coated substrates in corrosive environments [60–62].

Thus, to enhance the fatigue performance of PEO coatings the coating quality can be controlled by reducing the number of micro-cracks and pores (by optimized PEO parameters), which can act as stress concentrators and reduce fatigue life. A uniform thickness avoids weak points that could initiate fatigue cracks. Surface preparation before PEO coating, which reduces roughness before applying the PEO coating can lead to better adhesion and fewer stress concentrations. Shoot peening before PEO treatment is successfully applied to introduce compressive stresses to the substrate surface with positive effects on the CF [59]. Especially for corrosion fatigue the use of sealing methods to fill micro-pores and cracks can have twice positive effects, reducing the porosity and blocking pathways for corrosive media and thus improving overall durability. Combining PEO coatings with polymer-based top coats can improve their overall resistance to corrosion fatigue. A comprehensive review of the quite complex fatigue-resistant design of PEO-coated Al alloys is given by Zhang et al. [51]. In summary, understanding the complex interaction between the coating, the substrate, and the environment is crucial to improving the corrosion fatigue resistance of these coatings. The result depends quite often on the overall barrier properties of the PEO-coating system. If the positive impact of preventing fast interaction of the electrolyte with the substrate on the fatigue strength is higher than the negative effect due to the defects and loss of ductility by the coating application, positive effects can be seen. Here, also the base metal's susceptibility to corrosion will affect the overall durability of the coating.

5.4 Effect of Posttreatment of PEO on the Mechanical Properties of PEO-Coated Metallic Substrates

The PEO coatings often exhibit porosity and micro-cracks, which can compromise their protective properties. Playing with the electrolyte composition and the process parameters can help to partially control the defect formed in the PEO layers.

Moreover, addition of the particles to the PEO electrolytes can affect not only the composition of the obtained coatings but also the morphology of the pores and defects. However, still the critical pores and defects might remain causing underperformance in terms of corrosion protection and mechanical properties. To address these issues, various posttreatment and sealing methods have been developed to enhance the performance of PEO-coated substrates. Posttreatment processes can further enhance the mechanical properties of PEO-coated substrates by addressing the microstructural defects within the oxide layer. For example, sealing treatments can fill the pores and cracks within the oxide layer, improving its protective capabilities and enhancing the overall mechanical properties of the coating. Studies have shown that posttreatment processes such as immersion in sealing solutions or the application of additional coatings can significantly improve the corrosion resistance, wear resistance, and fatigue properties of PEO-coated substrates. However, the effectiveness of posttreatment processes depends on several factors, including the type of sealing solution used and the specific conditions of the treatment. Optimizing these factors is crucial for maximizing the benefits of posttreatment processes and enhancing the overall performance of PEO-coated substrates.

One common posttreatment method is the application of sealing treatments, which aim to fill the pores and cracks in the PEO coating, thereby improving its barrier properties. Sealing treatments can be broadly categorized into organic and inorganic methods. Organic sealing involves the use of polymers, such as epoxy resins, which penetrate the porous structure of the PEO coating and provide a protective layer. Inorganic sealing, on the other hand, involves the use of chemical solutions, such as silicates or phosphates, which react with the oxide layer to form a more compact and less porous structure.

A two-step PEO approach can be used to perform a partial sealing of the coatings. For instance, a wear-resistant 10 μm thick PEO layer formed on Ti–15–3 alloy using an alkaline aluminate electrolyte was shown to lack adhesion to the substrate due to voids and pores near the substrate. To enhance adhesion, a subsequent PEO process is conducted in an acidic electrolyte containing phosphoric and sulfuric acids. This step forms a new amorphous oxide layer next to the substrate, filling the voids and reducing porosity. Consequently, the coating's adhesion is significantly improved without compromising its high wear resistance [63].

Hydrothermal treatment is another effective posttreatment method for PEO coatings. This process involves immersing the PEO-coated substrate in hot water or steam, which promotes the formation of additional oxide layers and the hydration of existing oxides. The hydrothermal treatment can significantly reduce the porosity of the PEO coating and enhance its corrosion resistance. One of the hydrothermal processes recently reported for PEO is Layered Double Hydroxide (LDH) sealing, which demonstrates considerable promise in improving the mechanical properties of PEO-coated metallic substrates. The influence of LDH sealing on attributes such as hardness, wear resistance, and fatigue performance is especially significant. The imperfections within the PEO film act as reservoirs for LDHs. These LDHs effectively seal the corrosive pathways in the PEO film, blocking corrosive ions from penetrating the film and damaging the alloys. At present, composite coatings made up of oxide ceramics and LDHs have been successfully developed on a range of aluminum and magnesium alloys.

Wear resistance is a vital mechanical property that significantly benefits from LDH sealing. The layered structure of LDH forms a strong barrier against abrasive forces, thereby reducing the wear rate of PEO-coated substrates. For instance, studies on PEO-coated aluminum alloys with LDH sealing have demonstrated a substantial enhancement in wear resistance. The LDH layer functions as a protective shield, effectively minimizing material loss during wear tests. Beyond corrosion resistance, the composite coating also shows remarkable wear resistance. Specifically, the wear rate of samples with PEO-LDH coatings decreased by 84% at room temperature and by 98.9% at a high temperature of 250 °C compared to the aged bare alloy [64]. The application of MgAl–LDH posttreatment clearly improved wear resistance compared to untreated PEO coatings as can be observed in Fig. 5.8. This enhancement is due to the LDH layer's ability to better distribute and absorb mechanical stresses than the unsealed PEO coating.

However, the hydrothermal LDH treatment can lead to a partial conversion of the phases constituting PEO into layered double hydroxides. The dissolution of PEO phases can also result in the drop of the mechanical properties of the coating. For example, the scratch resistance of PEO layers on AZ31 Mg alloy was partially compromised [65].

There are other hydrothermal approaches based on the partial conversion of the PEO coating materials into new often more voluminous phases, which allow partial sealing of the pores leading to improvement of mechanical and protective properties. For example, exposing the PEO-coated substrate to hot water or steam. The high temperature promotes the formation of additional oxide/hydroxide layers and the hydration of existing oxides, resulting in a denser and more uniform coating. Hydrothermal treatment is particularly effective for Mg alloys, where it can form a protective magnesium hydroxide layer that significantly improves protection properties.

Sol-gel sealing significantly improved the performance of both PEO coatings, as confirmed by the wear test results. The recent study has evaluated the tribological performance of several sol-gel sealed PEO layers on AA2024 alloy. Overall, sol-gel-treated PEO layers showed significant improvement in tribological properties compared to untreated samples. Importantly, the average coefficient of friction (COF) showed significantly lower values compared to PEO especially for the systems sealed with combination of tetraethoxysilane (TEOS) with (3-glycidyloxypropyl)trimethoxysilane (GPTMS) as well as the system having methacryloxypropyltrimethoxysilane and zirconium(IV) propoxide in the sealing formulation. Specific wear rates were significantly lower for these sealed systems compared to PEO, indicating better wear resistance. The presence of GPTMS and TEOS created a dense sol-gel network, enhancing the sealing and reducing COF and wear loss volume. The system showed lesser wear under higher loads due to the alignment of the sol-gel layer, providing lubricating characteristics. [66].

Combination of several posttreatment methods can also be considered as a viable approach often offering synergistic effect originated from combination of various

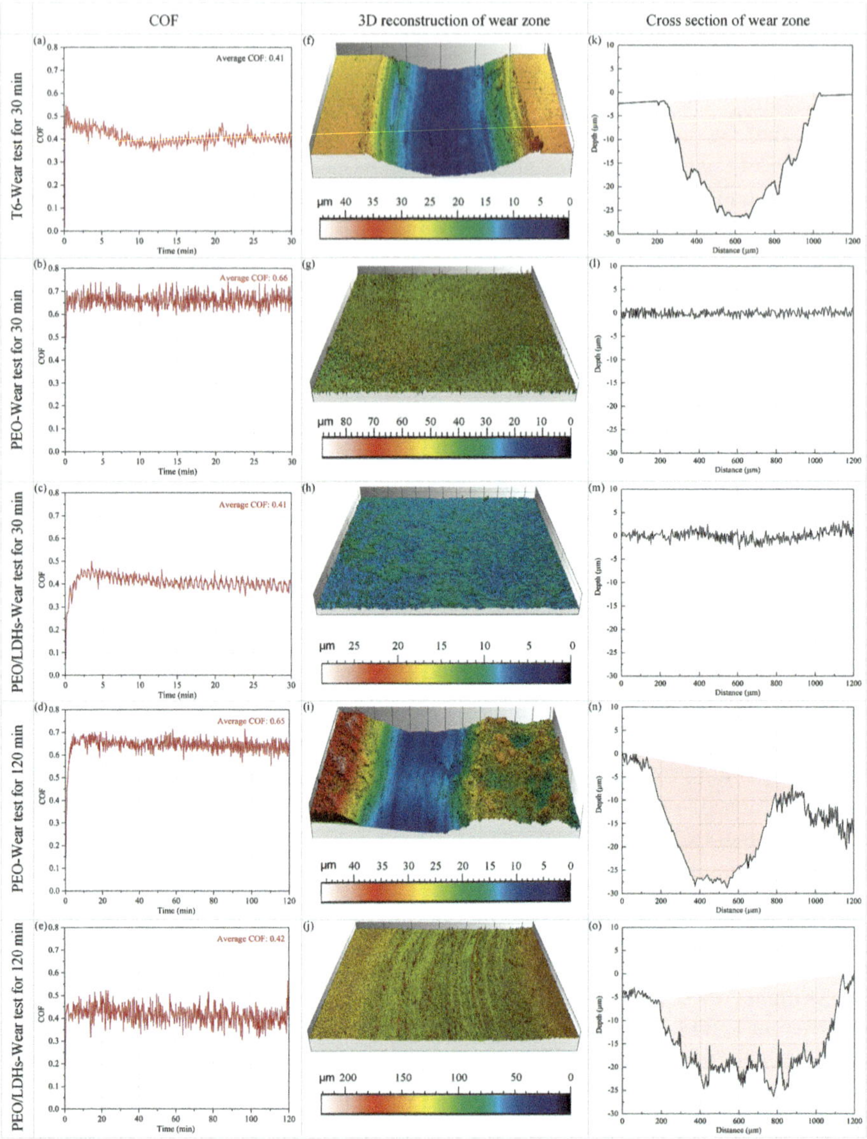

Fig. 5.8 Wear test results at room temperature for bare 2195-T6 Al–Li alloy, PEO, and PEO/LDHs samples: (**a–e**) COF curves; (**f–j**) 3D reconstruction of wear zone; (**k–o**) Cross-sectional profile of the wear zone [64]. *Reproduced with permission from [64] © 2023 The Authors. Published by Elsevier B.V.*

protection mechanisms. As an example, the combination of the hydrothermal treatment with a polymer layer can be mentioned here. In the recent work focused on improvement of protective properties of PEO coating on Mg-based biodegradable

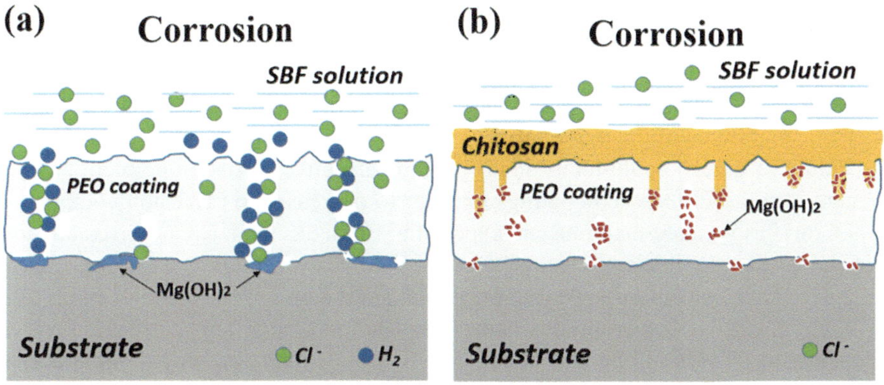

Fig. 5.9 Schematic illustrations of the proposed tribocorrosion protection mechanisms of (**a**) PEO-treated AZ31B samples and (**b**) CS-posttreated samples [67]. *Reproduced with permission from [67] © 2022 Elsevier B.V.*

implants, the hydrothermal posttreatment was followed by application of a chitosan biopolymer layer. The chitosan layer itself provides an additional barrier against the corrosive species and also lubricates the surface decreasing the friction coefficient as shown in Fig. 5.9. When comparing to the blank PEO and one with only chitosan sealing, the combined posttreatment has provided the lowest COF and wear rate as well as the most stable tribocorrosion performance [67].

In addition to sealing and hydrothermal treatments, laser treatment is another posttreatment method that has been explored for PEO coatings. Laser treatment involves the use of a high-energy laser beam to melt and re-solidify the surface of the PEO coating, which can reduce porosity and improve the coating's adhesion to the substrate. This method has been shown to enhance the wear resistance and fatigue performance of PEO-coated Ti alloys [68].

In conclusion, posttreatment methods such as sealing, hydrothermal treatment, sol-gel coatings, and laser treatment play a crucial role in enhancing the mechanical properties of PEO-coated metallic substrates. These treatments help to reduce porosity, improve adhesion, and enhance the wear resistance and corrosion resistance of the PEO coatings. The successful application of these post-treatment methods on Al, Mg, and Ti alloys demonstrates their potential for improving the performance of PEO-coated substrates in various industrial applications.

5.5 Conclusions and Outlook

PEO is a promising surface treatment technique that can significantly enhance the mechanical properties of metallic substrates by forming a hard, wear-resistant, and corrosion-resistant oxide layer. The effectiveness of PEO in improving surface hardness, wear resistance, stress corrosion cracking resistance, and corrosion fatigue resistance depends on several factors, including the composition of the electrolyte, process parameters, and the presence of microstructural defects within the oxide layer. Posttreatment processes can further enhance the mechanical properties of PEO-coated substrates by addressing these defects and improving the overall protective capabilities of the oxide layer.

Future research should focus on introducing an active corrosion protection component into the PEO-based coatings in order to decrease the corrosion component when the PEO-coated materials are exposed to tribocorrosion, stress corrosion, and corrosion fatigue loads. Moreover, a new strategy of finding environmentally friendly electrolyte additives or posttreatment approaches to form composite layers with enhanced barrier properties and lower friction coefficient, but having lower tendency to form hard phases causing notching effect and abrading particle debris are essential.

References

1. Wei, T., Yan, F., & Tian, J. (2018). *Corrosion Science, 130*, 95–102.
2. Kolomeichenko, A. V., & Kravchenko, I. N. (2019). *Russian Metallurgy*, 1410–1413.
3. Arrabal, R., Mohedano, M., Matykina, E., Pardo, A., Mingo, B., & Merino, M. C. (2015). *Surface and Coating Technology, 269*, 64–73.
4. Yeshmanova, G., Blawert, C., Serdechnova, M., Florian Wieland, D. C., Starykevich, M., Gazenbiller, E., Höche, D., Smagulov, D., & Zheludkevich, M. L. (2024). *Surfaces and Interfaces, 44*, 103,797.
5. Oh, Y.-J., Mun, J.-I., & Kim, J.-H. (2009). *Surface and Coating Technology, 204*, 141–148.
6. Sieber, M., Simchen, F., Morgenstern, R., Scharf, I., & Lampke, T. (2018). *Metals, 8*, 356.
7. Vyaliy, I. E., Egorkin, V. S., Sinebryukhov, S. L., Minaev, A. N., & Gnedenkov, S. V. (1874). *AIP Conference Proceedings, 2017*, 040056.
8. Sabatini, G., Ceschini, L., Martini, C., Williams, J. A., & Hutchings, I. M. (2010). *Materials and Design, 31*, 816–828.
9. Blawert, C., Sah, S. P., Scharnagl, N., & Kannan, M. B. (2015). Plasma electrolytic oxidation/micro-arc oxidation of magnesium and its alloys. In *Surface modification of magnesium and its alloys for biomedical applications* (pp. 193–234). Woodhead Publishing.
10. Blawert, C., Dietzel, W., Ghali, E., & Song, G. (2006). *Advanced Engineering Materials, 8*, 51.
11. Arrabal, R., Matykina, E., Hashimoto, T., Skeldon, P., & Thompson, G. E. (2009). *Surface and Coating Technology, 203*, 2207–2220.
12. Rapheal, G., Kumar, S., Blawert, C., & Dahotre, N. B. (2011). *Wear, 271*, 1987–1997.
13. Lu, X., Mohedano, M., Blawert, C., Matykina, E., Arrabal, R., Kainer, K. U., & Zheludkevich, M. L. (2016). *Surface and Coating Technology, 307*, 1165–1182.
14. Wang, Y., Wei, D., Yu, J., & Di, S. (2014). *Journal of Materials Science and Technology, 30*, 984–990.
15. Zhang, D., Gou, Y., Liu, Y., & Guo, X. (2013). *Surface and Coating Technology, 236*, 52–57.
16. Daroonparvar, M., Mat Yajid, M. A., Yusof, N. M., & Bakhsheshi-Rad, H. R. (2016). *Journal of Alloys and Compounds, 688*, 841–857.
17. Yu, L., Cao, J., & Cheng, Y. (2015). *Surface and Coating Technology, 276*, 266–278.
18. Mingo, B., Arrabal, R., Mohedano, M., Pardo, A., & Matykina, E. (2017). *Surface and Coating Technology, 309*, 1023–1032.
19. Guo, J., Wang, L., Wang, S. C., Liang, J., Xue, Q., & Yan, F. (2009). *Journal of Materials Science, 44*, 1998–2006.

20. Srinivasan, P. B., Blawert, C., & Dietzel, W. (2009). *Wear, 266*, 1241–1247.
21. Shao, M., Wang, W., Yang, H., Zhang, X., & He, X. (2021). *Coatings, 11*, 1288.
22. Malinovschi, V., Marin, A. H., Ducu, C., Moga, S., Andrei, V., Coaca, E., Craciun, V., Lungu, M., & Lungu, C. P. (2022). *Coatings, 12*, 29.
23. Sowa, M., Piotrowska, M., Widziołek, M., Dercz, G., Tylko, G., Gorewoda, T., Osyczka, A. M., & Simka, W. (2015). *Materials Science and Engineering: C, 49*, 159–173.
24. Khanmohammadi, H., Allahkaram, S. R., Towhidi, N., & Rashidfarokhi, A. M. (2016). *Surface Engineering, 32*, 448–456.
25. Zhang, Z.-Y., Huang, T.-Y., Zhai, D.-J., Wang, H.-B., Feng, K.-Q., & Xiang, L. (2022). *Surface and Coating Technology, 451*, 129045.
26. Wheeler, J. M., Collier, C. A., Paillard, J. M., & Curran, J. A. (2010). *Surface and Coating Technology, 204*, 3399–3409.
27. Malinovschi, V., Marin, A., Ducu, C., Andrei, V., Coaca, E., Craciun, V., & Lungu, M. (2021). *Surface and Coating Technology, 418*, 127240.
28. Kuroda, P. A. B., Montes, T. H., Rossi, M. C., Grandini, C. R., & Afonso, C. R. M. (2024). *Materials Letters, 356*, 135607.
29. Kuroda, P. A. B., Cardoso, G. C., Rossi, M. C., Afonso, C. R. M., & Grandini, C. R. (2024). *Materials Letters, 363*, 136,226.
30. Yeung, W. K., Sukhorukova, I. V., Shtansky, D. V., Levashov, E. A., Zhitnyak, I. Y., Gloushankova, N. A., Kiryukhantsev-Korneev, P. V., Petrzhik, M. I., Matthews, A., & Yerokhin, A. (2016). *RSC Advances, 6*, 12688–12698.
31. Lederer, S., Arat, S., & Fuerbeth, W. (2021). *Materials., 14*, 5364.
32. Aliofkhazraei, M., & Rouhaghdam, A. S. (2012). *Applied Surface Science, 258*, 2093–2097.
33. Lederer, S., Lutz, P., & Fürbeth, W. (2018). *Surface and Coating Technology, 335*, 62–71.
34. Mu, M., Liang, J., Zhou, X., & Xiao, Q. (2013). *Surface and Coating Technology, 214*, 124–130.
35. Mu, M., Zhou, X., Xiao, Q., Liang, J., & Huo, X. (2012). *Applied Surface Science, 258*, 8570–8576.
36. Grigoriev, S., Peretyagin, N., Apelfeld, A., Smirnov, A., Morozov, A., Torskaya, E., Volosova, M., Yanushevich, O., Yarygin, N., Krikheli, N., & Peretyagin, P. (2023). *Materials., 16*, 3928.
37. Yang, C., Chen, P., Wu, W., Sheng, L., Zheng, Y., & Chu, P. K. (2024). *Coatings, 14*, 451.
38. Blawert, C., Dietzel, W., Ghali, E., & Song, G. (2006). *Advanced Engineering Materials, 8*, 511–533.
39. Sikdar, S., Menezes, P. V., Maccione, R., Jacob, T., & Menezes, P. L. (2021). *Nanomaterials., 11*, 1375.
40. Gnedenkov, S. V., Sinebryukhov, S. L., Sergienko, V. I., & Gnedenkov, A. S. (2019). *Corrosion Science and Technology, 18*, 212–219.
41. Qi, X., Huang, Y., Zhao, Y., Liu, Y., Dai, W., & Tian, Y. (2022). *Journal of the Electrochemical Society, 169*, 020559.
42. Liang, J., Srinivasan, P. B., Blawert, C., Störmer, M., & Dietzel, W. (2009). *Electrochimica Acta, 54*, 3842–3850.
43. Xi, F., Huang, Y., Zhao, Y., Liu, Y., Dai, W., & Tian, Y. (2022). *Materials., 15*, 4256.
44. Dean, J., Gu, T., & Clyne, T. W. (2015). *Surface and Coating Technology, 269*, 47–53.
45. Venugopal, A., Panda, R., Manwatkar, S., Sreekumar, S., & Ramakrishna, L. (2012). *International Journal of Corrosion, 823*, 967.
46. Ghosh, R., Venugopal, A., Pradeep, P. I., Ramakrishna, L., Narayanan, P. R., Pant, B., & Cherian, R. M. (2018). *Corrosion Science and Technology., 17*, 101–108.
47. Srinivasan, P. B., Blawert, C., & Dietzel, W. (2008). *Corrosion Science, 50*, 2415–2418.
48. Srinivasan, P. B., Blawert, C., & Dietzel, W. (2008). *Materials Science and Engineering A, 494*, 401–406.
49. Nachtsheim, J., Ma, S., Burja, J., & Markert, B. (2024). *Surface and Coating Technology, 477*, 130,391.
50. Winter, L., et al. (2016). *IOP Conference Series: Materials Science and Engineering, 118*, 012033.

51. Zhang, J., Dai, W., Wang, X., Wang, Y., Yue, H., Li, Q., Yang, X., Guo, C., & Li, C. (2023). *Journal of Materials Research and Technology, 23*, 4307–4333.
52. Dai, W., Li, C., He, D., Jia, D., Zhang, Y., & Tan, Z. (2019). *Surface and Coating Technology, 360*, 347–357.
53. Lonyuk, B., Apachitei, I., & Duszczyk, J. (2007). *Surface and Coating Technology, 201*, 8688–8694.
54. Zakharchenko, K., et al. (2020). *Journal of Physics: Conference Series, 1666*, 012019.
55. Morri, A., Ceschini, L., Martini, C., & Bernardi, A. (2020). *Coatings, 10*, 1180.
56. Rojas, C. A., Simchen, F., Winter, L., Rymer, L.-M., & Lampke, T. (2023). *Surface and Coating Technology, 475*, 130125.
57. Madhavi, Y., Krishna, L. R., & Narasaiah, N. (2019). *International Journal of Fatigue, 126*, 297–305.
58. Asquith, D. T., Yerokhin, A. L., Yates, J. R., & Matthews, A. (2006). *Thin Solid Films, 515*, 1187–1191.
59. Ye, Z., Liu, D., Zhang, X., Wu, Z., & Long, F. (2019). *Applied Surface Science, 486*, 72–79.
60. Klein, M., Lu, X., Blawert, C., Kainer, K. U., Zheludkevich, M. L., & Walther, F. (2017). *Materials and Corrosion, 68*, 50–57.
61. Němcová, A., Skeldon, P., Thompson, G. E., Morse, S., Čížek, J., & Pacal, B. (2014). *Corrosion Science, 82*, 58–66.
62. Apachitei, I., Lonyuk, B., Fratila-Apachitei, L. E., Zhou, J., & Duszczyk, J. (2009). *Scripta Materialia, 61*, 113–116.
63. Tsunekawa, S., Aoki, Y., & Habazaki, H. (2011). *Surface and Coating Technology, 206*, 4732–4740.
64. Wang, Y., Pan, Y., Xu, X., Zhang, W., Feng, R., Li, H., & Gao, Z. (2024). *Journal of Materials Research and Technology, 28*, 1044–1061.
65. Shulha, T., Serdechnova, M., Wu, T., Naacke, T., Wiese, G., Blawert, C., & Zheludkevich, M. L. (2024). *Nano Materials Science, 6*, 428–442.
66. Khalid, H. A., Akbarzadeh, S., Paint, Y., Vitry, V., & Olivier, M.-G. (2023). *Coatings, 13*, 871.
67. Xu, L., Fu, X., Su, H., Sun, H., Li, R., & Wan, Y. (2022). *Progress in Organic Coating, 170*, 107,002.
68. Rahmadani, S., & Anawati, A. (2024). *Indonesian Journal of Applied Physics, 14*, 399–416.

Part II
Growth Mechanism

Chapter 6
Growth Mechanism of Plasma Electrolytic Oxidation on Valve Metals and Non-valve Metals

Yingliang Cheng

6.1 Introduction

According to electrochemistry, thick and porous anodic oxide films can be formed on certain metals such as Al and Ti in specific electrolytes at high potentials up to 300–700 V. This is the underpinning scientific principle for the technique with the name "plasma electrolytic oxidation (PEO)". PEO is evolved from the traditional anodizing technique. In 1857, Buff first found that aluminum can be electrochemically oxidized in aqueous solution to form an oxide layer that is thicker than the native one [1]. This phenomenon is called "anodization" because the aluminum being processed is the anode in the electrolytic cell. The anodized Al demonstrate noticeable difference in current response to positive and negative polarization, i.e., current is more easily passing through the electrode system when it was under cathodic polarization. Therefore, Güntherschulze and Betz introduced the term "valve action" [2] as a semi-quantitative characteristic of the current blocking effect exhibited by an oxide covered electrode immersed in a suitable solution. Later, this rectifying property was also reported for several other metals, which found practical applications as both electrolytic rectifiers and electrolytic capacitors [3]. Up to now, it is commonly quoted that Al, Mg, Ti, Ta, Nb, Zr, and Be are valve metals, while steel, nickel, copper, and its alloys are non-valve metals [4]. Traditional anodization is usually performed below the breakdown potentials of the passive films being formed (10–100 V) since polarization above the breakdown voltage terminates the growth of a dense barrier-type oxide layer which is required for applications in rectifiers and capacitors [3]. Nowadays, the anodized products can be easily found in

Y. Cheng (✉)
College of Materials Science and Engineering, State Key Laboratory of Cemented Carbide, Hunan University, Changsha, China
e-mail: chengyingliang@hnu.edu.cn

V. S. Saji, M. Zheludkevich (eds.), *Plasma Electrolytic Oxidation – Fundamentals, Advances and Applications*,
https://doi.org/10.1007/978-3-031-92429-3_6

electronic gadgets, electrolytic capacitors, cookware, outdoor products, plasma equipment, vehicles, architectural materials, machine parts, etc. [1]. The growth mechanism of anodic oxides on valve metals has been studied extensively, and it is well established that oxide films grow irreversibly by a field-assisted migration of ions through the film following the high-field model [5, 6].

PEO is distinguished from conventional anodizing by the occurrence of plasma discharges, which are a result of the breakdown of the oxide layer on the metals being processed. Therefore, unlike the continuous growth of oxide layer by the outward and inward migration of cations and anions, respectively, in conventional anodizing, PEO coatings grow discontinuously at the location of plasma discharges. The plasma discharges make the PEO process more complex than anodizing, and indeed than virtually all other coating procedures [4]. Historically, PEO has been termed with different names, such as anodic oxidation under spark discharge, spark anodizing, micro-plasma oxidation, and microarc oxidation (MAO), with the last name still being widely used by current researchers. The pioneer in the evolution of plasma-assisted anodic oxidation technology may be the invention by McNeill and coworkers, who used a spark discharge to deposit cadmium niobate onto a cadmium anode in a Nb-containing electrolyte in the 1960s as well as the identification of a "micro-arcing" phenomenon on Al by Markov and Markova in the 1970s [3, 7]. In the last two decades of the twentieth century, scientists from Russia, Germany, the United States, and China have also become involved in the field of plasma electrolytic oxidation [3, 7, 8]. As time enters this century, PEO technology has received continuous attention, with worldwide participation from more countries such as the United Kingdom, Spain, Serbia, and Iran. The early PEO studies mainly focused on anti-wear and corrosion-resistant coatings. Since 2005, new functional coatings related to biology, catalysis, thermal control, optical and luminescence properties have gradually attracted the attention of researchers [9–16]. For example, Zhang et al. [17–19] have developed coatings on Ti and Mg–Ag alloys with antibacterial capability and cyto-compatibility. Early PEO research mainly focused on Al [8], Mg [20] and their alloys, and now it has been extended to a broad scope of metals such as Ti [21], Zr [22], and Ta [23].

Up to now, PEO has been reviewed considerably by scholars in this field. The initial review was made as early as 1999 by Yerokhin and his coworkers [7], focusing on the physical and chemical fundamentals of plasma electrolysis. In 2019, Clyne and Troughton [4] described the recent progress on understanding and quantification of the various phenomena that take place during plasma electrolytic oxidation, with their main focus on the dielectric breakdown and its relationship with a number of influencing factors. A recent review by Kaseem et al. [24] reports various aspects of PEO from the scientific principles of discharge behavior, coating growth, and inorganic particle incorporation to coating properties such as tribological and corrosion performances, biomedical, catalysis, light, and energy performances. An abundant of review papers covering some specific metals such as Mg [25–27] and Ti [28] or certain special aspects of PEO, e.g., particles incorporation [29] and soft sparking [30], are also present.

Although the PEO process has been extensively reviewed and progress has been made, some fundamental mechanisms of the plasma discharges and coating growth remain controversial and unresolved. The discharge is initially viewed as a result of the dielectric breakdown of the oxide layer being formed under the mechanism of electron avalanching [31, 32]. However, it seems that the dielectric breakdown model cannot explain the fact that the polarizing potential does not increase proportionally with the oxide thickness and the lifetime of spark discharges is much longer than the conventional electron avalanche events [33]. Therefore, gaseous breakdown models, such as discharge in pore and contact glow discharge models are proposed and very popular in recent years [33, 34]. However, recent research on the PEO of carbon steel shows that the formation of an initial dielectric layer seems to be a prerequisite for the initiation of spark discharge and thus challenges the gaseous phase breakdown models [35].

Initially, it was thought that the non-valve metals could not be PEO-treated [36]. This impression may come from the fact that thick, porous anodic films are mostly obtained on valve metals such as Al, Ti, Zr, Nb, Ta, and W, although a recent work reported the formation of barrier-type and nanoporous-type anodic films on iron by potentiodynamic anodizing in an ethylene glycol electrolyte [37]. According to some researchers, non-valve metals normally do not form stable oxides, and high anodic potential to non-valve metals usually results in VGE (vapor–gaseous envelope) formation, promoting non-oxidizing treatments, such as polishing [38]. Moreover, Clyne et al. [4] regarded that Ni, Fe, Cu, Sn, Mo, and W cannot normally be processed by PEO since the driving force for oxidation of these metals is similar to, or lower than, that of hydrogen being converted to water. However, these years witness the gradually expanding of PEO to the treatment of a various non-valve metals such as steels [35], brass [39], and copper [40]. The applicability of PEO to non-valve metals not only broadened the application scope of the technique but also led to the reevaluation of the existing views on the mechanism of oxide breakdown and plasma occurrence and coating formation mechanisms. Actually, the patent issued in 1966 by McNeill and Gruss [41] reported that spark reactions in aqueous solutions of $NaAlO_2$, Na_3WO_4, and Na_2SiO_3 can deposit coatings on a variety of metals, including some typical non-valve metals such as Fe, Ni, and Cu. However, the method described in [41] is more like a deposition of electrolyte anion species on target metal and the mechanism was not discussed. Unlike the PEO of valve metals, the problem in PEO of non-valve metals is the difficulty in building a barrier layer, which is essential for converting the process into stable spark regime [42]. Due to the important role of non-valve metals such as steel and copper in industry and human life, researchers have been endeavored to developing suitable plasma electrolytic oxidation regimes to grow hard ceramics on them. Initially, Al layers were applied on steels by sputtering [43], arc spraying [36], or hot dipping [44] and then the Al-coated steel was PEO-treated. However, these methods are essentially the PEO of Al. In the early years of the first decade of this century, some researchers from Harbin Institute of Technology in China begun to pay attention to the plasma electrolytic oxidation of steels itself (without plated layers), confirming that PEO coatings can be grown in-situ on steels in a variety of aqueous solutions such as

$NaAlO_2$, Na_2SiO_3, and $(NaPO_3)_6$, etc. [45–47]. Afterward, Wang et al. [48–52] published a series of papers on the PEO of carbon steel, focusing on the microstructure, phase composition, and corrosion and wear performance of the coatings. Similar works on PEO of low carbon steels [53–56] and 39NiCrMo$_3$ steel [57] have also been reported by others. In 2018, Li et al. [35] published a paper on the mechanism of PEO of carbon steel in aluminate electrolyte with the addition of NaH_2PO_4 and proposed that the formation of a thin dielectric layer of $AlPO_4$ or alumina aluminum phosphate (AAP) is critical to the plasma generation and coating formation, and therefore, the previous models that dielectric breakdown in PEO is preceded by the breakdown of a bubble layer need to be reconsidered. Afterward, PEO has been expanding to brass [39] and copper [40], signifying the importance of formation an initial dielectric layer of silica or AAP in building stable plasma discharges for PEO treatment. However, it seems that less substrate metal has participated in the coating formation process on non-valve metals and the coatings normally do not have good adhesion to the substrate. At present, PEO of non-valve metals is still in its infancy and much effort should be made to develop suitable processing regime and understand its mechanism.

The author of this chapter has been engaged in PEO study for nearly 20 years. We have carried out PEO research on a variety of valve metals (Al, Mg, Ti, Zr, and their alloys) as well as some non-valve metals (carbon steel, brass, and copper). This chapter aims to summarize the author's current understanding of the PEO mechanisms of valve metals as well as non-valve metals. This chapter begins with the PEO of some typical valve metals, concentrating on various phenomena during PEO and the growth mechanism of PEO coatings. After that, PEO of some non-valve metals (steel, copper, and brass) is discussed. A comparison of the coating growth mechanism between valve and non-valve metals has been made.

6.2 PEO Growth Mechanism on Valve Metals

6.2.1 Electrical Regimes

Historically, PEO had been developed from conventional anodizing, where DC operation is the most prevalent. Therefore, galvanostatic DC regimes of PEO were used in the early stage of the development of the technique. However, the DC regimes usually resulted in the large sparks (also called "arcs") at the later stage of treatment, leading to the appearance of coarse structural defect due to local overheating [33]. Afterward, it has been repeatedly found that the process runs more effectively and the coatings are of higher quality when AC power supplies are used. Currently, AC (symmetric or asymmetric), pulsed DC (unipolar and/or bipolar) regimes are commonly encountered in industrial and experimental applications. The frequencies of the AC and pulsed DC regimes can range from 50 Hz to several kHz. It was previously thought that the positive pulse is more important than the negative pulse since the positive one is directly related to coating formation and the negative pulse performs only as a means to interrupt the spark discharges, permit the surface to cool, and induce the re-conversion of soluble components into metal oxide [58]. However, practice and investigation in these years found that the introduction of cathodic polarization (negative pulse) can improve significantly the current efficiency and coating quality [59–62]. Especially, the application of specific cathodic-to-anodic current or charge quantity ratio (*R*) at values $R > 1$ can trigger the so-called "soft sparking" phenomenon, which can greatly improve coating quality in terms of thickness and homogeneity, and therefore, is a very favorable research topic in recent years [63–66]. The underlying mechanism for the function of negative pulse (cathodic polarization) and soft sparking will be discussed in detail in Sect. 6.2.9. The frequency also influences the PEO processes. Although the low frequencies in 50 or 100 Hz are usually used in PEO treatment, it is reported in some cases that bipolar current in the kHz frequency range resulted in improved coating growth rate and reduced the volume fraction of the outer porous layer [67].

PEO can be controlled under the electrical modes of constant current/voltage/power, respectively. The constant current (galvanostatic) mode is most commonly used in PEO treatment, in which the cell potential increases to maintain the preset current as the coating thickens. The benefit of the constant current mode is that a reasonable coating growth rate can be maintained during the whole PEO process. The constant voltage (potentiostatic) mode shows large current values at the initial stage, while the current decreases quickly as the coating thickens, signifying much-reduced coating growth rates at prolonged treatment time. The constant power mode is seldom reported in the PEO study.

6.2.2 *Cell Potential-Time Responses and PEO Stages*

Figure 6.1 shows the cell potential-time responses during PEO of an Al–Cu–Li alloy in a moderately concentrated aluminate electrolyte of 24 g L^{-1} $NaAlO_2$ + 1 g L^{-1} KOH under the galvanostatic pulsed bipolar mode [68]. The current is employed in square waveforms, where the magnitudes of the positive and negative current are kept at constant levels. Both the positive and negative potentials in peak value are displayed in Fig. 6.1. For the sample immersed for 0 min in the aluminate electrolyte, the curve of the positive potential (black curve) can be divided into three stages. In stage I, the positive potential rises rapidly with a rate of ~30 V/s to ~130 V at ~4.2 s. This stage corresponds to the formation of a thin barrier layer (conventional anodizing) on the metal. The conventional anodizing stage is found with the PEO of all valve metals. Uniform light emission, which is called galvanoluminescence (GL), can be observed on the sample surface as shown in Fig. 6.1. GL had been found with the anodization of a variety of valve metals such as Al, Mg, Nb, Ti, Zr, and W [69]. It should be noted that the brightness of GL is very low, which needs to be clearly observed in darkness (The GL in Fig. 6.1 was recorded with a long exposure time of the camera). The positive potential in stage II shows a reduced rising

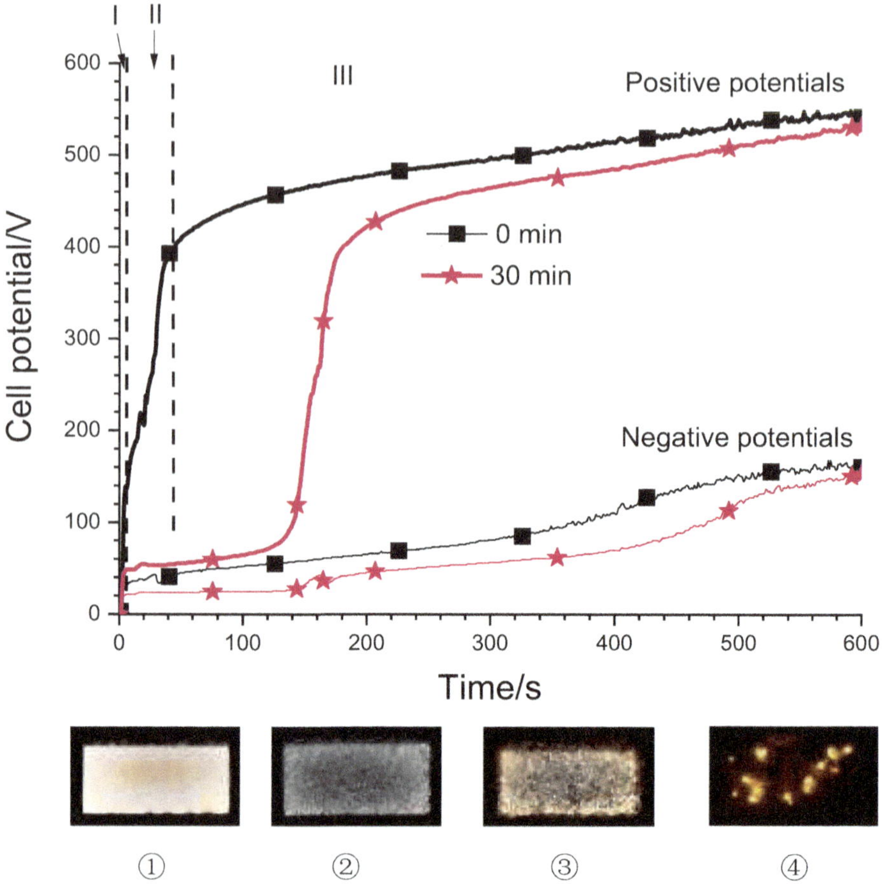

Fig. 6.1 Cell potential-time responses and discharge appearance during PEO (pulsed bipolar regime) of an Al–Cu–Li alloy after being immersed for 0 and 30 min in 24 g L^{-1} $NaAlO_2$ + 1 g L^{-1} KOH [68]. Discharge appearances recorded at different stages: ① GL recorded at ~100 V in stage I; ② stage II; ③ the early stage in III; ④ later stage in III. *Reproduced with permission from* [68] © *2014 Elsevier Ltd.*

rate (~7 V/s). In stage II, another type of light emission, the short-lived discrete sparks begin to appear on the sample surface. The potential at which the first spark appears on the electrode is called the "first-spark voltage"(U_{FS}) [31]. Numerous fine white sparks are observed as the potential approaches the higher values in stage II. At the end of stage II, an inflexion appears on the cell potential curve and the PEO enters stage III. The potential at the inflexion point is called the breakdown potential/voltage (U_B) in literature [31]. The potential rises very slowly in stage III, showing a value of ~0.26 V/s. The sparks in stage III exhibit a yellow color, and the number of sparks is significant reduced, but their intensities are increased at the later stage. The color of sparks is related to the species within the plasma discharges. In this case, the yellow or orange color can be attributed to Na species, which usually are detected by optical emission spectroscopy (OES) [59]. Green discharges have been observed during the PEO of ML5 magnesium alloy in NH_4F solution [70], which is possibly due to the presence of Mg species in the plasma. Compared with the positive cell potential response in Fig. 6.1, the negative potential shows a similar rapid rise at the initial time, and then it rises slowly to the end. However, the magnitude of the negative potential is much smaller than the anodic one.

PEO of other valve metals normally shows similar potential-time responses and discharge behavior as that of PEO of the Al–Cu–Li alloy with the immersion time of 0 min (the black curve in Fig. 6.1). However, PEO of the Al–Cu–Li alloy with a delayed immersion time of 30 min shows different cell potential responses (the magenta curves in Fig. 6.1). The positive potential lingers at lower values (~50–70 V) for a relative long period before it rises sharply and displays the same PEO behavior as the sample without immersion. The period of the lingering positive potential is believed to be associated with the balance between the thinning and repairing of the initial anodic oxide film [68]. It is known that alumina, the composition of the thin anodic films on Al alloy, is an amphoteric oxide, which can be attacked by an alkaline electrolyte. PEO of other valve metals that form stable oxide films, for example, Zr, does not show the lingering potential at the initial stage.

In some literature work [7, 58], the PEO process was also classified into four stages according to the appearance of discharges: Conventional anodizing stage; Sparking discharge stage; Micro arcing stage; Powerful arcing stage. However, it must be pointed out that the classification is somewhat based on subjective impression. In some circumstance, powerful arcs at the later stage of PEO may be developed on the sample surface, leading to destruction such as thermal cracking or very porous coating. However, it is not easy to distinguish what type of discharges is the "powerful arcing," since big-sized discharges always occur at the later stage of PEO and they do not always lead to coating destruction.

6.2.3 *Growth of the Conventional Anodizing Film*

In the PEO process, a conventional anodic film grows to a critical thickness above which the dielectric breakdown occurs. The critical thickness may depend on the nature of the passive oxide film formed. Growth of conventional anodic films has been investigated for at least more than 100 years. Conventional anodic film is known to be formed by a continuous process with the migration of cations and anions under high electrical field. In the case of Al anodization, Al^{3+} ions migrate outwardly toward the oxide/electrolyte interface, while O^{2-} or OH^- anions move inwardly toward the metal/oxide interface [1]. Therefore, one can consider both the metal/oxide and the oxide/electrolyte interfaces as the growth front of anodic oxide during anodization of a valve-metal [1]. Accurately, anodic oxide film growth on aluminum in alkaline electrolytes can be described with the following processes [71]:

1. Oxide formation due to opposite migration of aluminum ionized at the metal/oxide interface and anions formed at the oxide/electrolyte interface, as a result of adsorbed water dissociation

$$Al \rightarrow Al^{3+} + 3e^- \tag{6.1}$$

$$2H_2O \rightarrow O^{2-} + OH^- + 3H^+ \tag{6.2}$$

$$2Al^{3+} + 3O^{2-} \rightarrow Al_2O_3 \tag{6.3}$$

2. Chemical dissolution of aluminum oxide caused by removal of the protective oxide film from its surface

$$Al_2O_3 + 2OH^- \rightarrow 2AlO_2^- + H_2O \tag{6.4}$$

$$2AlO_2^- + 2H_3O^+ + (n-3)H_2O \leftrightarrow Al_2O_3 \cdot nH_2O \tag{6.5}$$

3. Direct ejection of Al into the electrolyte

$$Al_{ej} + 3OH^- \rightarrow Al(OH)_3 + 3e^- \tag{6.6}$$

During the PEO process, the growth of the initial anodic film would not proceed at 100% efficiency, which can be manifested by the evolution of a large amount of anodic gas during the process. The total current i passing through the anode surface is considered to comprise the sum of the ionic i_i and the electronic i_e current components. The latter arises due to the increased electronic conductivity of the oxide film at high electrode potentials [71]. The electronic i_e current accounts for the liberation of gaseous oxygen, according to reaction (6.7):

$$4OH^- \rightarrow 2H_2O + O_2 \uparrow + 4e^- \tag{6.7}$$

Although Eqs. (6.1)–(6.7) describe the reaction processes for the growth of anodic film on Al, similar procedures will be anticipated for other valve metals. The oxygen gas evolution during the stage of growth conventional anodic film is governed by electrochemical mechanism, obeying the Faradaic law. However, anomalous gas evolution will occur under the plasma discharges, which exceeds the volume of gas evolution based on Faradaic yield by more than 10 times. The mechanism of gas evolution under plasma discharges will be discussed in Sect. 6.2.7.

6.2.4 Breakdown Theory

Actually, the breakdown of barrier anodic films has long been investigated in the field of conventional anodizing. After reaching the breakdown voltage, sparks leads to local thickening, cracking, blistering, or burning of oxide film, which degrades the dielectric properties of the oxide. An empirical equation indicates that the breakdown potential increases linearly with the logarithm of the electrolyte resistivity with the following equation:

$$U_B = A + B \log \rho_e \tag{6.8}$$

where *A* and *B* are the constants depending on the electrolyte composition and the nature of the anodized metal [72–75].

Initially, suggestion has been made that breakdowns are triggered by conduction through electrolyte-filled fissures and flaws of the oxide [73, 76–78]. However, it cannot explain why U_B does not depend on the history of film formation, the presence of forgoing breakdowns and topography of the electrode surface since all these factors should affect the number of flaws and fissures [31]. The dependence of U_B upon the nature of the anodized metal, as well as on the composition and the resistivity of the electrolyte, corroborates the idea that breakdowns are controlled by the electrochemical reactions at the electrolyte/oxide interface and depend on the solid-state properties of the oxides. Based on this recognition, Ikonopisov [31] developed a model that is capable of explaining qualitatively most of the observed features of the breakdown. The breakdown is viewed as an electron avalanche breakdown in the bulk of the anodic film, with the initial electrons being injected from the electrolyte into the oxide conduction band (CB). It is believed that the high field within the anodic film layer accelerates the electrons to critical energy, which can free other (secondary) electrons by impact ionization, so that an avalanche multiplication occurs. Although Ikonopisov's model has been cited widely for breakdown mechanism, it has been criticized by others. For example, Shimizu reexamined the validity of the theory presented by Ikonopisov et al. and pointed out that the data clearly indicate the invalidity of the theory instead of supporting it [79]. Moreover, in Ikonopisov's model, the role of ionic current is viewed as it only ensures the gradual increasing in thickness and does not take an immediate part in breakdown effects. The validity of this assumption is unknown. Therefore, electrolytic breakdown is a rather complex phenomenon, and the understanding of its mechanism is far from complete.

In conventional anodization, it is generally regarded as a rule that the U_B is independent of the current density used for the anodizing process [31]. However, it is not always the case in PEO. Dehavani et al. [80] investigated the PEO of Al alloy under a pulsed unipolar DC regime in a typical silicate electrolyte. Figure 6.2 is the result of the relationship between the PEO anodizing current density and maximum and breakdown potential obtained during the PEO processes. It is clearly shown that the breakdown potential increases with increase of the current density, especially at high frequency of 1000 Hz and lower duty cycle of 20%. Therefore, PEO technology, with a greater number of adjustable variables by the employment of AC or pulsed waveforms, did show some differences to the conventional anodizing, where DC is usually predominant.

In PEO study, the theory for breakdown and plasma occurrence was initially attributed to the same model for the dielectric breakdown of conventional barrier

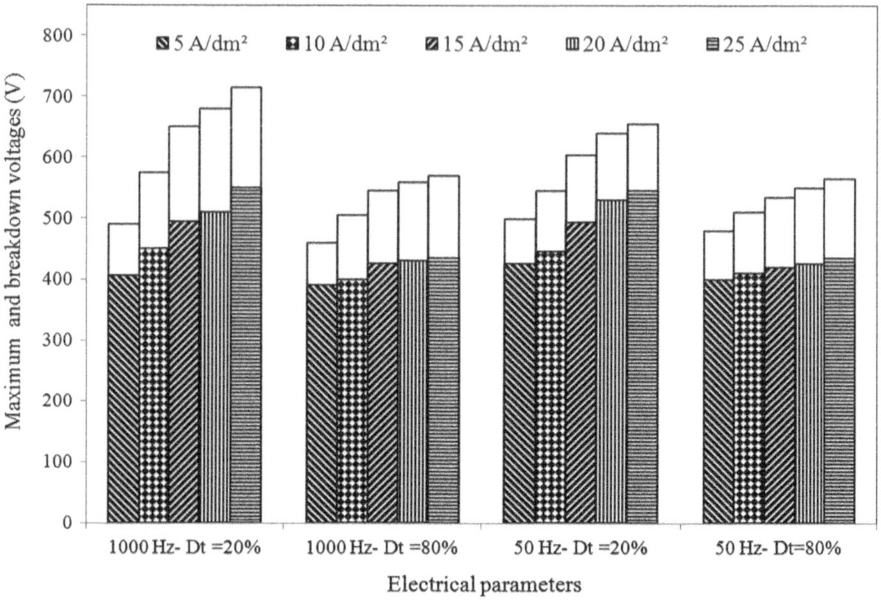

Fig. 6.2 Maximum and breakdown voltages reached during PEO of 6061 aluminum alloy samples using different electrical parameters (in each column, the height represents the maximum voltage and the inner column represents the breakdown voltage) [80]. *Reproduced with permission from* [80] *Crown copyright © 2014 Published by Elsevier B.V.*

anodic films, i.e., the breakdown is treated as a 'streamer propagation' due to the electron avalanche effects induced by film dopants and structural defects [33]. However, in the PEO process, the rate of the increase of polarizing potential is significantly reduced after the breakdown voltage (see Fig. 6.1), whereas the thickness of the PEO coating grows continuously. This contradicts the point of view that the polarizing voltage should be in a direct relationship with the thickness of the oxide layer in order to maintain a sufficient electric field for breakdown. Second, high-speed photographing [81] shows that the microdischarge lifetime could be at least two orders of magnitude larger than that expected during conventional electron avalanche events. Therefore, the oxide film dielectric breakdown model was regarded as obsolete and new models based on gas phase breakdown are proposed for explaining the microdischarge occurrence during PEO. There are mainly two categories of gas phase breakdown models [33]: the discharge-in-pore model and the model of contact glow discharge electrolysis. The discharge-in-pore model considers each microdischarge as a gas discharge occurring in a micropore of the oxide film. However, Yerokhin et al. [33] believed that there exists a substantial discrepancy between the minimum resolvable diameter of a microdischarge (≈100 μm) and the average pore size in the outer region of the surface oxide layer (5–25 μm), which raises major concerns regarding the applicability of "discharge-in-pore" models. The model of contact glow discharge electrolysis was proposed by Yerokhin et al.

based on analogy with the contact glow discharge studied by Hickling and Ingram [82]. In this model, discharge was triggered by electron emission from the electrolyte surface (partial cathode) into a gaseous phase (gas bubbles above the oxide) rather than dielectric breakdown of the growing oxide film.

In recent years, it has become favorable to classify plasma discharges into several types according to their position of occurrence and related coating microstructures. In 2010, Hussein et al. [83] investigated the PEO of Al using OES spectroscopy and surface morphological characterization methods. They believe that there are three types of discharge, which originate either at the substrate/coating interface (type B), within the upper layer (type C), or at the coating top (type A). Type B is the strongest discharge, resulting in surface morphology (for example, the pancake structure) with high content from the substrate. The type C discharge occurred inside the relatively deep surface holes in the oxide layer, corresponding to the feature with composition from both the electrolyte and the substrate. The features caused by type A discharge consist of composition mostly from electrolytes. Afterward, Cheng et al. [84] found that in some cases, empty voids (big pores) are found beneath some pancake structures of the PEO coatings on Al alloy. Obviously, this kind of microstructure cannot be attributed to the type B discharge in Hussein's model. Moreover, sign of discharge in internal pores has also been detected in the coating microstructure. Therefore, additional discharge types of D and E are proposed for the discharge in pore and the pancake with underlying big pores, respectively.

Although the specific coating features can be easily attributed to a certain type of discharge in the PEO process, there is a lack of physical meanings for those types of discharges. We do not know the physics of these different types of discharges. Moreover, Tu et al. [85] have found that the pancake structure on the PEO coatings of Mg alloy formed in an aluminate-tungstate electrolyte contains a large amount of Al and W species, which come from the electrolyte. Therefore, the B-type discharges, which penetrate the whole coating layer, also lead to the decomposition of electrolyte species, which is a role previously attributed to the type A and C discharges.

Due to the complexity of the coating breakdown and plasma occurrence during the PEO process, although it has been studied for a long time, a theory that can explain satisfactorily all aspects of the PEO phenomena has not been reached yet. Initially, the microdischarges are regarded as a result of the dielectric breakdown of oxide film due to electron avalanche. Afterward, the gas phase breakdown models are more predominant. However, the work of Li et al. [35] found that a thin dielectric layer is a prerequisite for discharge initiation, and therefore, questioned the gas phase breakdown model. The research group of Clyne has studied the PEO dielectric breakdown based on the small area sample method [86–89]. They found that the cascades of discharge are prone to occur in the same location of a deep pore [89]. Therefore, it must be local microstructural features that promote repeated discharges in the same place, and the actual dielectric breakdown occurs across a relatively thin oxide layer at the base of a deep pore in the coating [4]. The fact that discharge occurs across a thin oxide in the pore can also explain the observed phenomenon that polarizing potential does not increase linearly with the increase of PEO coating thickness.

6.2.5 Features of Discharges

The most distinctive feature of the PEO process is the discharges, which can take on different forms according to the PEO stages and the electrical or electrolytic parameters. The nature of the discharges and their size, lifetime, and spatial distribution are central to understanding and control of the process. The size, appearance, and spatial distribution of plasma discharges during the PEO process can be studied by digital video imaging, as reported in [33, 90]. The appearance of discharges during the PEO process usually evolves in the pattern as shown in Fig. 6.1. Numerous sparks appear in the early stage, whereas only a few large sparks are observed in the later stage. The form of discharge is also influenced by other factors, such as the electrolyte concentration [59] and electrical regime [60]. It is noteworthy that in some circumstances, the plasma discharge would be very weak or nearly invisible, but significantly improved coating efficiency and quality are obtained [60]. This is the soft sparking phenomenon and will be discussed in Sect. 6.2.9. Due to the short-lived nature of the discharges, high-speed video imaging method has been used to study the details of the discharges. The work of Dunleavy et al. [86] and Nomine et al. [91] indicated that the individual discharges (light emission events) tend to occur in sequences ("cascades"), with lifetimes from ~30 to 300 μs and ("incubation") periods between them from ~0.1 to 1 ms.

Compared to conventional anodizing, which normally forms amorphous films, one of the advantages of PEO is that crystalline materials, such as α–Al_2O_3 and γ–Al_2O_3, can be formed within the coatings. The crystallization of coating materials was induced by the high temperature within the discharge channels where melting and quenching of the oxide occur. Due to the localized nature of the microdischarges, the plasma temperature cannot be measured by conventional temperature measurement methods. However, the plasma electron temperature (T_e) can be calculated based on optical emission spectroscopy (OES).

Taking the assumption of at least local thermal thermodynamic equilibrium (LTE) [83], the spectral lines of OES spectra can be used to calculate T_e. Plasma temperature can be calculated based on the two-line method or the multi-line method, respectively. For the two-line method [34, 83, 86, 92, 93], the relative intensity of two spectral lines of the same atomic or ionic species at the same ionization stage is used to calculate T_e according to:

$$T_e = \frac{E_2 - E_1}{k_B}\left[\ln\left(\frac{I_1 \times \lambda_1 \times g_2 \times A_2}{I_2 \times \lambda_2 \times g_1 \times A_1}\right)\right]^{-1}, \tag{6.9}$$

where k_B is the Boltzmann constant (1.38×10^{-23} J K^{-1}), E_i is the energy of the upper level of the lines, I_i is the measured intensity of line I, λ_i is the wavelength, g_i is the statistical weight, A_i is the transition probability of the active species.

The multi-line method (also called the Boltzmann plot) based on multiple spectral lines from the same species, which is considered to be more reliable, uses the following equation to calculate plasma temperature [94, 95]:

$$\ln\left(\frac{I\lambda}{gA}\right) = \frac{E}{k_B T}, \tag{6.10}$$

where I, λ, g, A, E, and k_B are the relative line intensity, wavelength, upper-level statistical weight, transition probability, and Boltzmann constant, respectively. Plotting $\ln[I\lambda/(gA)]$ as a function of E, the electron temperature will be obtained from the slope of the straight line, provided the system is in the state of LTE.

T_e of the PEO discharges is normally reported to be in the range ~3000–10,000 K. Hussein et al. [61] demonstrated that the values of T_e were in the ranges of 4000–7000 K and 4000–5500 K for the coating formed on pure Al under unipolar and bipolar current modes, respectively. Similar T_e values (4500–5800 K) have also been reported for the PEO of AJ62 Mg alloy by the same authors [96]. Dunleavy et al. [86] studied the T_e of the discharges during PEO of an AA6082 alloy, leading to the conclusion that the plasma consists of a central core of high temperature (~16,000 ± 3500 K) and a cooler peripheral region (~3000–4000 K). Stojadinović et al. [97] analyzed Ti I lines of the OES spectra collected from the PEO of pure T_i in a silicate electrolyte, leading to a T_e value of ~3700 K. The T_e *of the plasma generated from the PEO of a Zr alloy in aluminate electrolyte reported by Cheng* et al. *was mostly around ~6000 K* [59]. *Plasma electron temperature of the PEO of brass* [39], *a non-valve metal in silicate electrolyte has been estimated by two-line method and also the Boltzmann plot with multiple lines of Cu I. The temperature lies between ~4393 and ~6911 K. However, the spectral lines of Cu I in the ultraviolet region are found to be not suitable for temperature estimation, possibly due to the self-absorption by the electrolyte.*

It should be pointed out that although the plasma electron temperature reaches thousands Kelvin, PEO never causes detrimental thermal effect on the substrate alloy. This is due to the fact that the plasma discharge is only confined to a very small micro-region within the PEO coating. Yang et al. has used a thermocouple to measure the temperature of the underlying Al substrate during PEO in a silicate electrolyte, indicating that temperature of substrate was kept below 120 °C during 3600 s PEO process [98]. The low temperature of the substrate can be attributed to the effective cooling effect of the aqueous electrolyte and the good thermal conduction of the metal substrate.

6.2.6 *Surface Characteristics of the PEO Coatings on Valve Metals*

6.2.6.1 The Pancake Structure

One of the typical morphological features on the surface of PEO coatings is the pancake structure, sometimes called the volcanic crater structure in literature. The structure may be formed by the solidification of the molten discharge channels. Sometimes, a pinhole or pore is found at the center of the pancake. The size of the pancake increases with extended PEO time [99]. Figure 6.3 shows the surfaces of PEO coatings on several valve metals. It can be seen that pancakes are formed on the PEO coatings of Al–Cu–Li alloy [84], Zr alloy [59], Ta [100], and the coating on Mg alloy formed in aluminate electrolyte [85]. However, PEO of the Mg alloy in silicate electrolyte [101] does not form pancakes. Pancake structure is also absent on the surface of the coating on Ti6Al4V alloy [102].

It is hypothesized here that the formation of the pancake structure is related to the insulating properties of the formed oxide. Those coatings consisting of oxides with wider band gap usually have high resistance, which helps to confine the anodic current within the discrete discharge channels on the sample surface, leading to the formation of pancake structure after extinguishment of the discharges. However, for semiconductor coatings such as TiO_2, anodic current may be more diffusive within the coating, with less probability to form the pancakes.

Figure 6.4 shows the relative position of the band gap values of various oxides that may be found with the PEO coatings. It can be seen that Al_2O_3 and ZrO_2 have large band gap values, which may be the reason that pancakes are usually found on the PEO coatings on Al and Zr alloys. In the case of PEO of Mg alloys in aluminate electrolyte, $MgAl_2O_4$ is the main composition of the resulting coatings [85]. $MgAl_2O_4$ has a large band gap value of 7.8 eV [103], which supports the above hypothesis for the formation of pancake structure in this condition. However, in silicate electrolytes, the main composition of PEO coatings on Mg alloy is Mg_2SiO_4 [101], which has a lower band gap value of 5.60 eV [104]. Moreover, Duan et al. have demonstrated that the coatings formed on AZ91 magnesium alloy by PEO in silicate electrolyte show n-type semiconducting behavior [105, 106]. Therefore, it is not unexpected that pancake structure is absent from the surface morphology of the PEO coatings on Mg alloys formed in silicate electrolyte.

It is noted that Ta_2O_5 has a band gap value of 4–4.6 eV [5], which is lower than that of Mg_2SiO_4. However, pancakes are found with the PEO coatings on Ta instead of the coating on Mg alloy with Mg_2SiO_4. This fact suggests that band gap value may not be the sole factor that determines the formation of the pancake structure on PEO coatings. However, the electronic property of the band gap of the formed oxides should be an important factor in determining the microstructure of the PEO coatings.

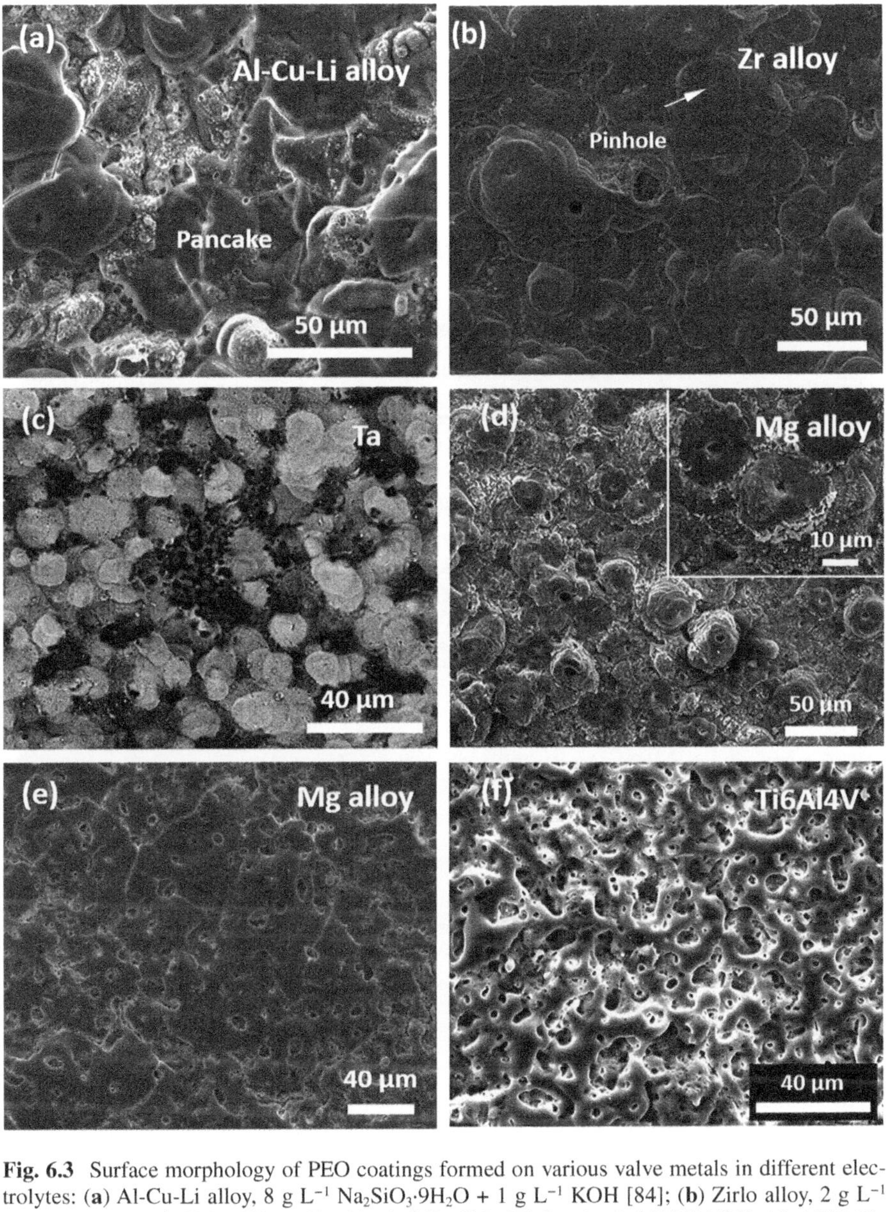

Fig. 6.3 Surface morphology of PEO coatings formed on various valve metals in different electrolytes: (**a**) Al-Cu-Li alloy, 8 g L^{-1} $Na_2SiO_3{\cdot}9H_2O$ + 1 g L^{-1} KOH [84]; (**b**) Zirlo alloy, 2 g L^{-1} Na_2AlO_2 + 1 g L^{-1} KOH [59]; (**c**) Ta, 10 g L^{-1} $Na_2SiO_3{\cdot}9H_2O$ + 1 g L^{-1} KOH [100]; (**d**) AZ31 Mg alloy, 10 g L^{-1} Na_2AlO_2 + 3 g L^{-1} $C_6H_8O_7$ + 2 g L^{-1} KOH [85]; (**e**) AZ31 Mg alloy, 12 g L^{-1} $Na_2SiO_3{\cdot}9H_2O$ + 15 g L^{-1} $(NaPO_3)_6$ [101]; (**f**) Ti6Al4V, 12 g L^{-1} $Na_2SiO_3{\cdot}9H_2O$ + 15 g L^{-1} $(NaPO_3)_6$ [102]. Pancakes are dominant in (**a–d**), but not found in (**e**) and (**f**). *Reproduced (**a**) with permission from* [84] *© 2013 Elsevier Ltd; (**b**) from* [59] *© 2016 Elsevier Ltd; (**c**) from* [100] *© 2020 Elsevier Ltd and Techna Group S.r.l.; (**d**) from* [85] *© 2017 Elsevier B.V.; (**e**) from* [101] *© 2015 Elsevier B.V.; and (**f**) from* [102] *© 2015 Elsevier Ltd.*

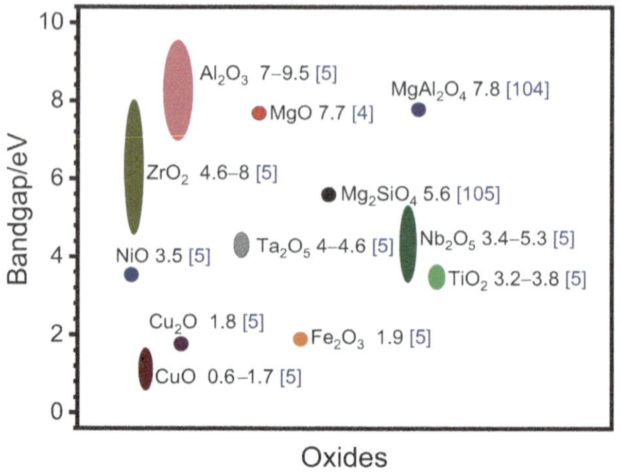

Fig. 6.4 Band gap values of various oxides. Data references are indicated in the plot

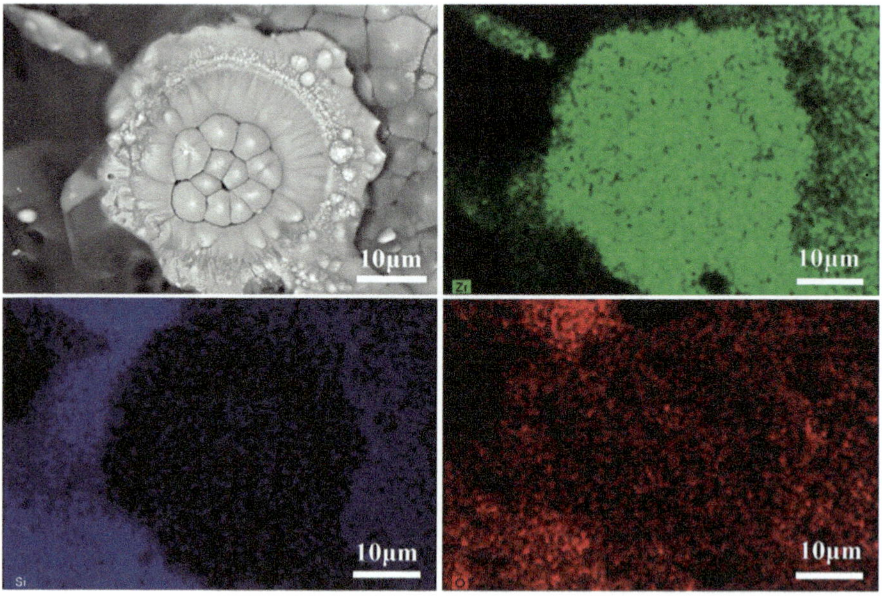

Fig. 6.5 A characteristic solidification structure on the PEO coating of a Zircaloy-4 alloy formed in silicate electrolyte for 3600 s [22]. *Reproduced with permission from* [22] © *2011 Elsevier Ltd.*

6.2.6.2 Characteristic Solidification Structure

The characteristic solidification structure was first found with the PEO of Zr alloy. Figure 6.5 shows the structure formed on the surface of a Zircaloy-4 alloy by PEO in an electrolyte of 30 g L^{-1} $Na_2SiO_3{\cdot}5H_2O$ + 4.88 g L^{-1} KOH for 3600 s [22]. The characteristic solidification structure consists of clusters of equiaxed grain in the central region, surrounded by a ring of radially orientated, elongated grains. The elemental mapping analysis confirmed a composition based on zirconium oxide, with silicon being enriched in the adjacent material.

According to [22], the characteristic solidification structure may be formed by the formation of melting pools on the coating surface by repeated discharges at the same location. The different ratios of *G*/*R*, where *G* and *R* are the temperature gradients and crystal growth rates, at the center and periphery of a melting pool, respectively, determine the formation of crystals with different morphologies at different locations. In a subsequent investigation [107], it is suggested that the low thermal conductivity of ZrO_2 (3.5 W m^{-1} K^{-1}) favors the formation of the melting pools on the PEO coatings of Zr alloys. Up to now, the characteristic solidification structure is only found with PEO of Zr alloys in silicate electrolytes and PEO of Mg alloy in aluminate–tungstate electrolytes [85]. The formation of the solidification structure indicates that the microstructure on the PEO coatings may be formed by a relatively long process in time span, in contrast to the previous understanding that the surface features originate from the rapid quenching of erupted molten oxides.

6.2.6.3 Features Derived from Electrolyte Species

Electrolyte species also take part in the PEO coating formation process. Electrolyte species can be incorporated inside the PEO coatings. However, features enriched with electrolyte components are more easily found on the coating surface. Figure 6.6 shows the typical silicon-enriched morphologies on the coatings formed by PEO of an Al–Cu–Li alloy [84] and Ta [100] in silicate electrolyte. In Fig. 6.6a, fine nodules are found on the surface of the PEO coating on the Al–Cu–Li alloy. These fine

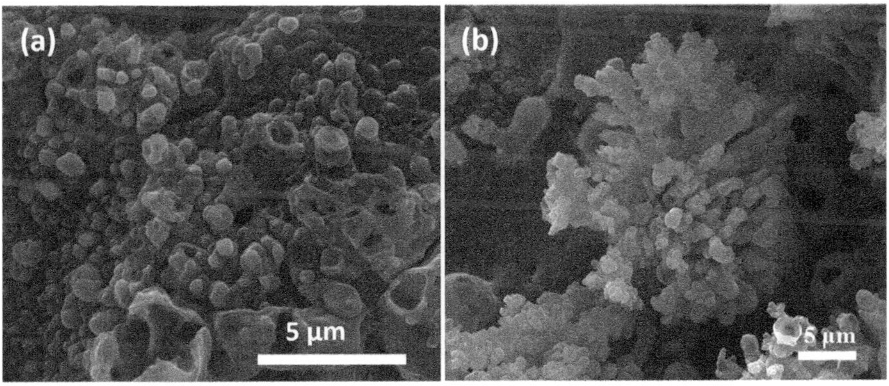

Fig. 6.6 The silicon-enriched features formed on Al–Cu–Li alloy [84] and Ta [100] by PEO in 8 g L^{-1} $Na_2SiO_3{\cdot}9H_2O$ + 1 g L^{-1}KOH and 10 g L^{-1} $Na_2SiO_3{\cdot}9H_2O$ + 1 g L^{-1}KOH, respectively. *Reproduced with permission from* [84, 100] *© 2013 Elsevier Ltd. and © 2020 Elsevier Ltd and Techna Group S.r.l.*

nodules are enriched with Si from the electrolyte species. According to EDS analyses, the ratio of Si/Al of the fine nodules reaches ~1.6–2.8, compared with a low value of ~0.09 for the adjacent pancake structure. Figure 6.6b shows a "coral reef" feature on the surface of the PEO coating on Ta. The "coral reef" feature consists of 72.9% O, 3.7% Na, 21.5% Si, 1.3% K, 0.6 Ta% (at. %), indicating that it was silicon-rich.

The features enriched with electrolyte species on the coating surface were attributed to the types A and C discharges in Hussein's model, since these types of discharges are believed to occur on the coating surface or the shallow pores below the surface. However, in the work of Tu et al. [85], the electrolyte species W are clearly presented within the pancakes, which is the structure caused by the strongest discharges originated from the substrate (the discharge channels). Therefore, the authors suspected that the penetrating discharges (type B) can also decompose the electrolyte species, forming the fine nodules or coral reef features on the coating surface. At the later stage of most of the PEO process, only a few numbers of big sparks are present on the coating surface. Observation shows that each spark shines below a gas bubble, which is termed as the plasma bubble [108]. The structure of plasma bubbles imply that the discharges initiate from deep within the coating (type B discharges). Therefore, there seems to be no evidence to support the presence of types of A and C discharges, which is thought to be weak discharges in Hussein's model.

6.2.7 *Efficiency and Non-Faradaic Processes*

PEO is normally considered to be a high-energy consumption, low efficiency technology so that it has only been exploited for some highly demanding and niche applications [67]. Snizhko et al. [109] had analyzed the basic electrochemical processes (such as oxide film growth, anodic dissolution and oxygen liberation) for the galvanostatic DC PEO of an aluminum anode in a model alkaline solution (KOH). The overall current efficiency of the oxide film formation was estimated to be in the 10–30% range. This low efficiency of the PEO coating formation was considered due to the side reactions of gas generation and dissolution processes [109–111]; however, it was found later that the coating growth rate is strongly dependent on the electrolyte concentration and current regime. In some circumstances, the apparent current efficiency can be very high, even exceed 100%.

Unlike conventional anodizing, electrolyte species play an important role in the PEO coating formation. Increasing the concentration of electrolyte species can significantly improve the coating growth rate, and therefore, the current efficiency. Cheng et al. [112] have investigated the PEO of an Al_Cu–Li alloy in 5, 32 and 56 g L^{-1} $NaAlO_2$ + 1 g L^{-1} KOH, respectively, using a bipolar regime with average positive and negative current densities of ~0.22 and 0.18 A cm^{-2}. It was found that the coating growth rates are ~1.5 μm min^{-1}, 4.6 μm min^{-1} and 11.3 μm min^{-1} as the electrolyte concentration increases from 5, 32 to 56 g L^{-1} $NaAlO_2$, respectively. For the highest concentration of 56 g L^{-1} $NaAlO_2$ + 1 g L^{-1} KOH, a single-layered ~45 μm coating was formed for short period of 4 min. The authors have estimated the anodic current efficiency based on the weight of the stripped coatings, yielding efficiencies of ~16, 60 and 111% for electrolytes of 5, 32 and 56 g L^{-1} $NaAlO_2$, respectively. The efficiency >100% reveals the non-electrochemical aspect of PEO technology. It was proposed that the high-temperature plasma has caused thermal decomposition of the aluminum-bearing species in the concentrated sodium aluminate electrolyte, contributing to the non-Faradaic formation of alumina within the discharge channels:

$$2Al(OH)_4^-(l) \xrightarrow{\text{Plasma}} Al_2O_3(s) + 3H_2O(g) + 2OH^-(l) \qquad (6.11)$$

The influence of electrolyte concentration on the coating growth rate has also been confirmed in the investigations of the PEO of a Zr alloy in aluminate electrolyte [59] and PEO of the Al-Cu-Li alloy in silicate electrolyte [84]. Besides the electrolyte concentration, the coating growth efficiency is also significantly affected by the application of negative current (cathodic polarization), which will be discussed in detail in sect. 6.2.9 ("cathodic polarization and soft sparking").

Regardless of the electrochemical nature of the PEO technology, non-Faradaic processes take part in the PEO process. Apart from the thermal decomposition of the electrolyte species, the gas evolution under the plasma discharges is another phenomenon that does not obey Faraday's law. The underlying mechanism for the PEO anodic gas evolution has not been fully understood yet. Snizhko et al. [109] investigated the evolution rate and composition of anodic gaseous products during PEO of a 6082 aluminum alloy in dilute KOH solutions, detecting a substantial surplus of gaseous products over the expected Faradaic yield. The authors have analyzed a few possible mechanisms for anomalous gas evolution and attributed it to the radiolytic effect of plasma discharge events on the adjacent electrolyte volume. Afterward, the same authors hypothesized that peroxide decomposition taking place at the discharge–electrolyte interface caused the anomalous gas evolution [110]. Cheng et al. [102] have collected the gas evolution volumes under different conditions during PEO of a Zr alloy and Ti_6Al_4V alloy in a silicate–hexametaphosphate electrolyte. It was found that volume of gas liberation on Zr alloy under the strong discharges is ~16 times greater than that of the Faradaic yield of the anodic oxygen. It should be noted that only the strong discharges can cause the liberation of large amount of gas. The weak discharges such as soft sparking do not induce the anomalous gas evolution, despite intensive gas liberation in the form of fine bubbles was observed for the PEO of Ti_6Al_4V at a relatively high potential of 600 V. According to the observation by naked eye, when strong discharges are established, large bubbles are ejected out of the discharging points with high velocity and loud acoustic emission. Therefore, the anomalous gas liberation should be generated within the discharge channels. At the later stage of PEO, the number of discharges is fewer and the anodic current is almost exclusively passing through the discharge channels. Tu et al. [85] estimated an extremely high anodic current density of ~10^4 A cm^{-2} within the discharge channels. Based on this evaluation, Tu et al. proposed that the thermal decomposition of water within the discharge channels is the main reason for the anomalous gas evolution. This effect has also been evaluated by Snizhko et al., but they thought that this effect was unimportant due to the extremely strong bonds of water molecules [109]. However, the temperature of the plasma can reach as high as ~6000 K [59], which is much higher than the temperatures for the direct thermal decomposition of water (the thermal decomposition of water begins at 2000 K [113, 114] and nearly completed at ~4000 K [115]).

According to the mechanism of thermal decomposition of water, the PEO gases should contain both oxygen and hydrogen. However, the composition of the gas evolved during PEO was initially a subject of considerable debate. Snizhko et al. [109, 110] first reported that the majority of the gas that evolved during PEO is oxygen, while the same authors [116] demonstrated in a later publication that the composition of the anodic gas consists mainly of hydrogen. In a more recent publication, Rogov et al. [117] analyzed the composition of gas evolved during the bipolar PEO of Al at the average anodic current density of 100 mA cm^{-2} and two values of cathodic current density, 80 and 120 mA cm^{-2}, which constitute cathodic-to-anodic current ratios $R = 0.8$ and 1.2, corresponding to the conventional microarcing and soft sparking regimes, respectively. Figure 6.7 shows the cell potential responses and analysis of the evolution of gas composition during the PEO processes. It can be seen that the proportion of hydrogen is greater than that of oxygen during the whole process. This result is not unexpected, since both the thermal decomposition of water and the cathodic polarization contribute to the release of hydrogen. A higher fraction of O_2 is detected at the early PEO stage, which indicates that oxygen is the main gas product during conventional anodizing stage (hydrogen at this stage

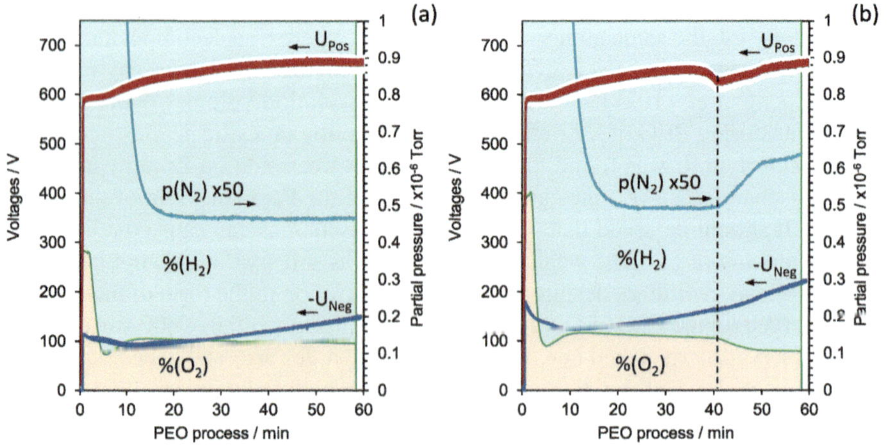

Fig. 6.7 Evolution of cell potentials, partial pressures of nitrogen and hydrogen-to-oxygen proportion during PEO treatments at various current ratios: (**a**) R = 0.8 and (**b**) R = 1.2 [117]. *Reproduced with permission from* [117] *© 2022 The Authors. Published by Elsevier B.V.*

is due to application of cathodic current). They also noticed that the oxygen fraction decreases as the PEO enters the soft sparking regime.

The viewpoint that the PEO anodic gas originates from thermal decomposition of water is further supported by the work of Wei et al. [118]. The authors collected the gas released from Zirlo alloy electrode during PEO process in KF solution and phosphate solution. It was found that the content of H_2 reaches up to 81.6 vol.% under strong plasma discharge in phosphate solution. A little CO was also detected while the glycerol was added into electrolyte, providing evidence for the direct thermal decomposition of the electrolyte component.

From the above discussion, it is evident that good ventilation should be ensured during the PEO industrial practice to prevent the potential explosion risk caused by hydrogen release.

6.2.8 Effect of Electrolyte Concentration

Since electrolyte species take part in the PEO coating formation process, the electrolyte concentration influences the growth rate and microstructure of the coatings. It is a general rule that thicker coatings are formed in a concentrated electrolyte. Cheng et al. [112] have carried out the PEO of an Al–Cu–Li alloy in different concentrations of aluminate electrolyte, using average positive and negative current densities of ~0.22 and ~0.18 A cm^{-2}, respectively. Figure 6.8 shows the cross sections of the coatings formed in 5, 32 and 56 g L^{-1} $NaAlO_2$ + 1g L^{-1} KOH with the treatment time of 20, 8 and 4 min, respectively. It can be seen that a non-uniform, bi-layer structure coating is formed in the dilute electrolyte of 5 g L^{-1} $NaAlO_2$. The typical thickness is ~30 μm, with thicker regions being due to blistering of the outer layer. Big pores, with dimension up to ~50 μm, are found at the region of blistering. It will be shown later that these big pores are formed due to the strong discharges. Each big pore corresponds to a pancake structure on the surface [84]. The big pores are eliminated as the electrolyte concentration increases to 32 and 56 g L^{-1} $NaAlO_2$ + 1 g L^{-1} KOH. The coating growth rate is also significantly improved with the increasing concentration, being ~1.5 μm min^{-1}, 4.6 μm min^{-1}, and 11.3 μm min^{-1} for the coatings formed in 5, 32, and 56 g L^{-1} $NaAlO_2$, respectively.

PEO in other electrolytes and using different substrate metals shows the same regularity for the influence of electrolyte concentration. In the study of the PEO of an Al-Cu-Li alloy in silicate electrolyte using average positive and negative current densities of 150 and100 mA cm^{-2}, an ~145 μm coating was formed for 60 min in 36 g L^{-1} $Na_2SiO_3{\cdot}9H_2O$ + 1 g L^{-1} KOH, compared with an ~90 μm coating formed for the same treatment time in 8 g L^{-1} $Na_2SiO_3{\cdot}9H_2O$ + 1 g L^{-1} KOH [84]. PEO of a Zircaloy-2 [119] and a zirlo alloy [59] also confirms that thicker and more uniform coatings are formed in concentrated electrolyte. Therefore, increasing the electrolyte concentration can improve the coating growth rate and uniformity. However, the coatings obtained in electrolytes with excessively high concentration are not of the highest quality, as reported in [84, 112].

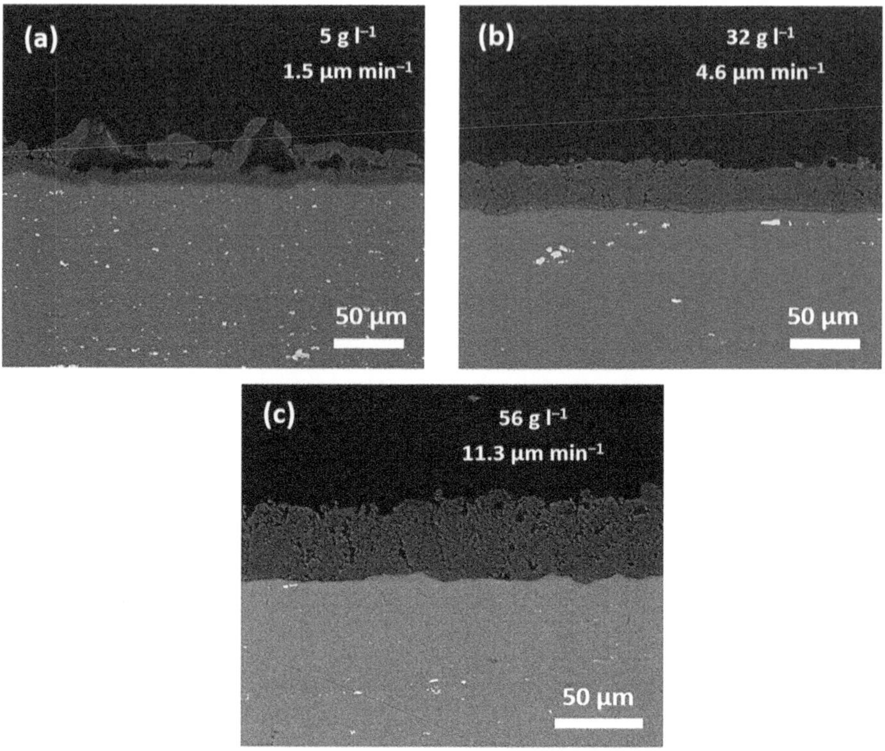

Fig. 6.8 Scanning electron micrographs showing the cross sections of the PEO coatings formed on an Al–Cu–Li alloy in (**a**) 5 g L^{-1} $NaAlO_2$ + 1 g L^{-1} KOH for 20 min; (**b**) 32 g L^{-1} $NaAlO_2$ + 1 g L^{-1} KOH for 8 min; (**c**) 56 g L^{-1} $NaAlO_2$ + 1 g L^{-1} KOH for 4 min [112]. The growth rate is indicated in each image. *Reproduced with permission from* [112] © *2015 Elsevier B.V.*

6.2.9 Cathodic Polarization and Soft Sparking

The application of cathodic polarization (negative current) is another factor that strongly influences the microstructure and quality of the PEO coatings. Especially, in the case of Al alloy, a specific cathodic-to-anodic current or charge ratio (R) can trigger the so-called soft sparking phenomenon, which is a PEO regime associated with high efficiency and high-quality coatings and has been the focus of many researchers.

6.2.9.1 Cathodic Polarization

Although PEO is regarded as a technique based on anodic polarization, i.e., coatings only grow under the anodic current cycles, the application of an additional cathodic polarization improves the efficiency and quality of the resulting coatings [33, 60,

61, 67]. The work of Cheng et al. [62] clearly demonstrated the influence of cathodic polarization on the microstructure of the PEO coatings formed on an Al–Cu–Li alloy in aluminate electrolytes with different concentrations. In their work, the alloy was treated without (unipolar) and with (bipolar) the application of cathodic polarization in electrolytes of 5, 32 and 56 g L^{-1} $NaAlO_2$ with the addition of 1 g L^{-1} KOH. Figure 6.9 shows the surface and cross sectional morphologies of the unipolar and bipolar coatings formed in 5 and 32 g L^{-1} $NaAlO_2$. For the unipolar coating formed in 5 g L^{-1} $NaAlO_2$, a wide gap is observed between the outer and the inner layers of the coating (Fig. 6.9a). This kind of coating microstructure has also been observed in the PEO treatment of a Zr alloy in dilute aluminate electrolyte and is seriously detrimental to the wear performance of the coating [119]. The addition of cathodic polarization eliminated the gap structure, although big pores are still observed in the cross section (Fig. 6.9b). Despite the different microstructures in the cross section, the insets show that both of the coatings formed with and without

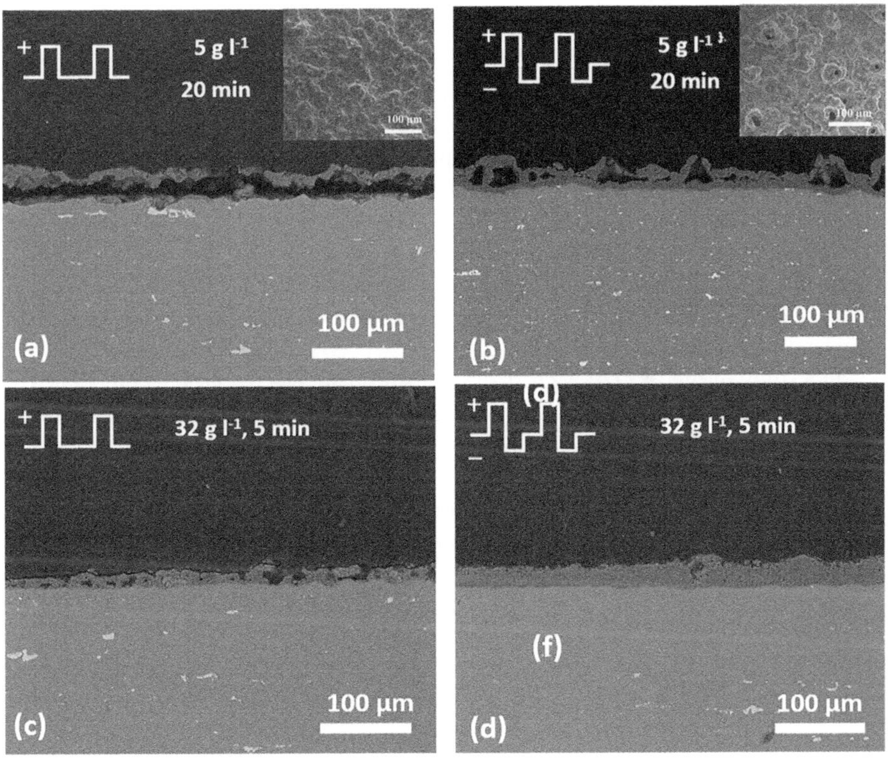

Fig. 6.9 Effect of an electrical regime (unipolar and bipolar) on the cross-sectional microstructure of the PEO coatings formed on an Al–Cu–Li alloy in (**a**, **b**) 5 g L^{-1} $NaAlO_2$ + 1 g L^{-1} KOH and (**c**, **d**) 32 g L^{-1} $NaAlO_2$ [62]. Coatings were formed for 20 min and 5 min in the 5 g L^{-1} and 32 g L^{-1} electrolytes, respectively. The insets in (**a**) and (**b**) are surface morphology. The average current densities are 0.22 A cm^{-2} and 0.22/−0.13 A cm^{-2} for the unipolar and bipolar regimes, respectively. *Reproduced with permission from* [62] © *2016 Elsevier B.V.*

cathodic polarization show a similar surface morphology dominated by the pancake structures. For the coating formed in 32 g L^{-1} $NaAlO_2$, the bipolar regime results in a thicker and more homogenous coating cross section (see Fig. 6.9d), whereas internal pores are still occasionally presented within the cross section of the unipolar coating (see Fig. 6.9c). Regardless of the application of cathodic polarization or not, the coatings formed within 56 g L^{-1} $NaAlO_2$ are single-layered; however, the surface of the coatings formed under the bipolar regime is smoother than its counterpart [62]. The findings demonstrate that both electrolyte concentration and cathodic polarization are important in determining the coating microstructure. Homogenous and dense coatings are obtained in a more concentrated electrolyte with the application of cathodic polarization.

The role that cathodic polarization played in PEO process is very complex and has been intensively studied. Initially, the main effect of a reversal of sample polarity is thought to be the discharge interruption, which prevents the development of long-lived, very large microdischarges [33]. However, the effect of discharge interruption cannot explain the increased growth rate with cathodic polarization. Some other factors have been mentioned to affect the discharge characteristics and the formation of an oxide layer. For example, the hydrogen liberation during negative biasing may provide a plasma-supporting gas medium that is favorable for the immediate establishment of a gas phase discharge during positive biasing [33]. The occurrence of a discharge during cathodic polarization may be also possible, particularly for the late stages of the layer formation; however, in the study of Yerokhin et al. [33], the authors did not find sufficient evidence for its actual presence. Moreover, Rogov et al. [66] have summarized several hypotheses about the role of cathodic polarization such as electrostatic charge accumulation [120, 121], hydration [122] or intercalation [123, 124] on the surface or in the bulk of the oxide layer, and heating of the coating [125].

In a study of the PEO of a Zirlo alloy in aluminate electrolytes, Cheng et al. [59] proposed a hypothesis on the role of cathodic polarization. Figure 6.10 presents the main ideas of the hypothesis. It is widely known that for any electrode reaction, there is always a diffusion layer near the electrode surface, through which reaction species migrate to the surface. In an aqueous solution of sodium aluminate, the anionic species normally exist in the form of $Al(OH)_4^-$ [68]. Figure 6.10a shows that $Al(OH)_4^-$ ions migrate across the diffusion layer to form the coating with big pores under the unipolar regime. In contrast, the cathodic hydrogen evolution during the negative pulse exerts an additional stirring effect to the liquid of the diffusion layer. Therefore, more $Al(OH)_4^-$ ions (reaction species) are brought into the diffusion layer and take part in the coating formation process (Fig. 6.10b). As a result, a thicker and more homogeneous coating is formed under the bipolar regime.

It should be noted that excessive cathodic polarization is detrimental to the coating formation. Gebarowski and Pietrzyk [126] found that a high cathodic-to-anodic current value of $R = 1.43$ produced non-uniform layer. Martin et al. [120] also found that using a higher R ratio of 2, the surface of the processed sample of an Al alloy appears "as if shortly processed." In a PEO study of an Al-Cu-Li alloy, Cheng et al. [127] systematically varied the R-value from 0 to 3.3, and it was found that the

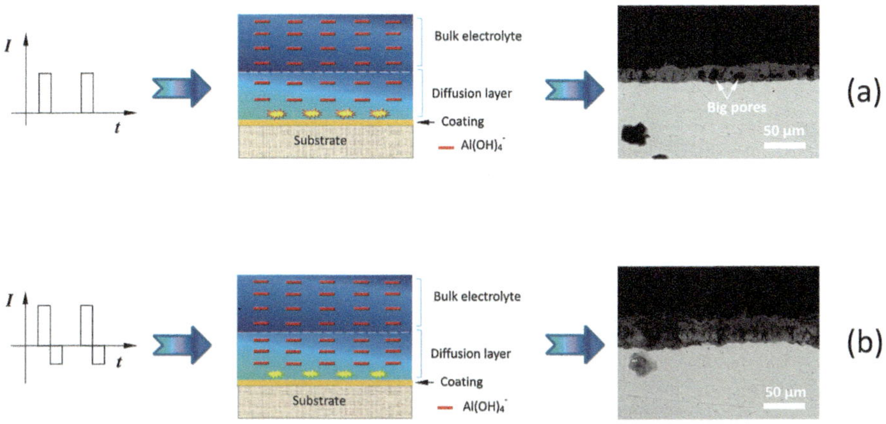

Fig. 6.10 Schematic illustration of the role of negative pulse (cathodic polarization) during the PEO of a Zirlo alloy in 32 g L^{-1} $NaAlO_2$: (**a**) unipolar regime, (**b**) bipolar regime [59]. *Reproduced with permission from* [59] *© 2016 Elsevier Ltd.*

coatings formed under the high *R* values of 2 and 3.3 was damaged at the later stages in the form of blistering or flaking-off. It is believed that the stress generated by excessive hydrogen evolution within the interior of the coatings damages the coatings formed under high *R* values [127]. The same regularity has also been found in the PEO of pure Mg and different Mg alloys [128]. However, the tolerance of the substrates to cathodic polarization is different. For pure Mg and AZ31, damages to the coating, such as blistering and detachment, occur at $R = 0.9$, while in the case of AZ91, the formed coatings are relatively intact even at the highest *R* value of 3.2.

6.2.9.2 Soft Sparking

Soft sparking was first discovered in PEO study on aluminum. It was found that when the applied cathodic current density (j_c) or charge (Q_c) exceeds the anodic current density (j_a) or charge (Q_a) in AC or pulsed bipolar regimes, a discharge phenomenon called soft sparking may be triggered [60, 63, 129]. The most significant feature of soft sparking is the decrease in anodic potential and the significant reduction in discharge intensity. Rogov et al. [124] has made a schematic illustration (Fig. 6.11) of the behavior of anodic potential and coating thickness evolution during PEO under the conventional and soft sparking conditions, respectively. In Fig. 6.11, the cathodic-to-anodic current density ratio of $R < 1$ and $R > 1$ are associated with the conventional sparks (arcs) and soft sparking, respectively. However, more precisely, soft sparking occurs at *R* range between ~1.1 and 1.2 [28, 60, 126]. Higher *R* values, as mentioned previously, lead to thinner or damaged coatings [127]. For the conventional PEO regime under $R < 1$, the anodic potential does not decrease during the whole process, and big and destructive sparks/arcs may occur at a later stage, leading to damaged surface morphology. In the case of a soft sparking

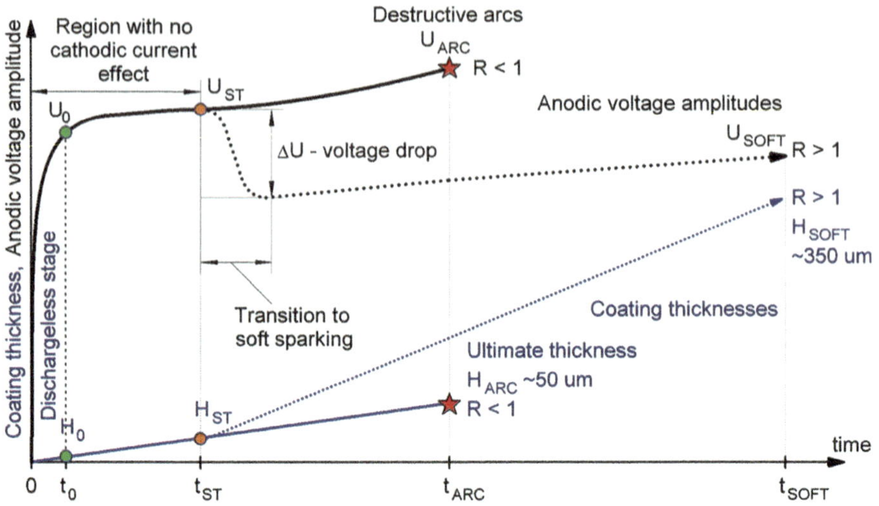

Fig. 6.11 Schematic representation of anodic potential behavior and coating thickness evolution during PEO of aluminum alloy under conventional and soft sparking regimes, respectively [124]. *Reproduced with permission from* [124] © *2017, American Chemical Society*

regime, after reaching a critical thickness (H_{ST}), the anodic potential drops to a lower level (up to 30% of the previous potential) and increases slowly to the final potential. Accompanying the potential drop, the discharge behavior is gradually converted into a state with a total elimination of the strong sparks. In this duration, the light emission is significantly reduced, although close observation reveals the scintillation of numerous tiny sparks. As shown in Fig. 6.11, the coating exhibits a significant improved growth rate during the soft sparking duration, and the coating thickness may approach ~350 μm after a certain time of treatment.

The mechanism of soft sparking has not been fully understood yet. For the drop of the anodic potential, Rogov et al. [124] have made a hypothesis that after reaching the critical thickness, the electron transfer becomes obstructed and field-assisted acidification takes place. The electrolyte acidification modified the EDL (electrical double layer) structure as well as the thickness of low-conductive part within active zone; therefore a transition to the soft sparking mode occurs. However, in a subsequent research of Cheng et al. [127], the authors found that anodic potential drop did not always occur at the PEO practice at *R* ratio of 1.2, which is the optimum electrical regime leading to soft sparking. The authors guessed that the phenomenon of potential drop may be partly associated with the damage to the mounting resin at the sample edge due to repeated spark discharges. Therefore, various complicated factors may need to be considered in explaining the soft sparking phenomenon. Electrochemical impedance spectroscopy (EIS) measurements of the coatings formed before and after the occurrence soft sparking indicate a clearly decrease in the impedance of the coating when soft sparking occurs [126]. Rogov et al. [124] first made an assumption that neutral hydrogen complexes (denoted as $[H^{\bullet}]_{OX}$) are formed during the PEO with cathodic polarization and these $[H^{\bullet}]_{OX}$ complexes lead to improved conductivity of the resulting coating. Cheng et al. [127] hold the same view that the incorporated hydrogen species may contributed to the decrease of the coating impedance. However, hydrogen may exist in various forms, such as proton, neutral hydrogen species and even negatively charged hydrogen species—hydride ions, in oxides [130]. At present, the presence of hydrogen species in oxides is far from well understood. Therefore, this topic would be the focus in future study.

Another advantage of soft sparking is the ability of formation of large amount of α-Al_2O_3 within the coatings. At the first look, this phenomenon is against our common understanding of the process since the suppressed discharges and lowered energy input under the soft sparking condition seem to be unfavorable to the formation of the high-temperature α-Al_2O_3 phase. In order to explain the α-Al_2O_3 formation, Rogov et al. [66] proposed a hypothesis that neutral hydrogen complexes $[H^{\bullet}]$ provide the seeds for the $\gamma \rightarrow \alpha$ phase transition of alumina. However, Bousser et al. [131] suggests that the α phase formation after disappearance of sparking is due to the lowered critical temperature for the phase transition, which is caused by the "ionic migration within the oxide and hydrothermal dissolution/precipitation in the associated microporous network that facilitate species mobility at the grain boundaries."

Based on a careful study of the PEO of an Al–Cu–Li alloy and measuring the interface temperatures at the electrolyte/coating and coating/substrate, He et al. pointed out that the formation of α–Al_2O_3 under soft sparking should be attributed to the conventional thermal activation mechanism. According to Cheng et al. [127] and He et al. [132], soft sparking coatings are different from those coatings formed under conventional sparking both in appearance and microstructure. Figure 6.12 shows the cell potential-time responses, interface temperatures, and the surface and cross sections of the coatings formed on an Al–Cu–Li alloy using the conventional bipolar ($R = 0.8$) and soft sparking ($R = 1.2$) regimes. For PEO under $R = 1.2$, three stages, namely, the stage with conventional discharges, transition stage and soft sparking stage are discerned from the cell potential-time curve. Interestingly, despite the lowered anodic potential, the temperature at both the electrolyte/coating and coating/substrate interfaces detected by the thermocouple is higher for the soft sparking regime. Therefore, it could be deduced that the temperature inside the coating should be higher in the soft sparking process. Moreover, high-temperature species, such as CuO and cristobalite, are also detected within the soft sparking coating, confirming the presence of high-temperature environment during its formation. Due to the oxidation of the alloying element Cu, the coating formed under the conventional sparks shows a brownish-red appearance (Fig. 6.12a). However, a white layer is formed on the top of the soft sparking coating (Fig. 6.12f). The SEM of the cross section shows that this outer white layer has a thickness of ~49 μm and is very porous (Fig. 6.12h). The outer layer consists of Al, O, Si, Na, K and a little Cu, and is highly amorphous according to XRD [132]. It is also widely known that there is a thin amorphous alumina layer at the coating/substrate interface, i.e., the barrier layer [63, 133–136]. Therefore, the soft sparking coating has a sandwiched structure of amorphous outer layer/crystalline Al_2O_3/amorphous alumina (barrier

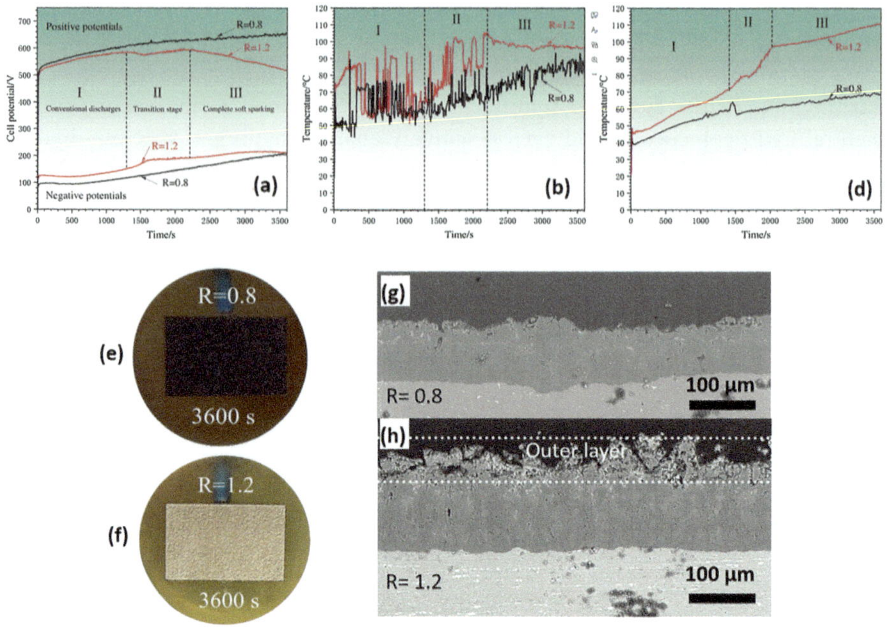

Fig. 6.12 Cell potential-time responses, temperature at interfaces measured by thermocouple, coating appearance and cross section of the coatings for the PEO of an Al–Cu–Li alloy under the conventional sparks ($j_a = 0.17$ A cm^{-2}, $j_c = 0.13$ A cm^{-2}, $R = 0.8$) and soft sparking ($j_a = 0.17$ A cm^{-2}, $j_c = 0.21$ A cm^{-2}, $R = 1.2$) regimes. (**a**) cell potential-time responses; (**b**) temperature at the electrolyte/coating interface; (**c**) temperature at the coating/substrate interface; (**e** and **f**) the appearance of the coatings formed under $R = 0.8$ and 1.2 for 3600 s; (**g** and **h**) the cross sections of the coatings formed under $R = 0.8$ and 1.2 for 3600 s [132]. *Reproduced with permission from* [132] © *2023 The American Ceramic Society*

layer). Based on the fact that amorphous materials have very low thermal conductivity, the authors have proposed a hypothesis that the amorphous structure of the outer and barrier layers effectively blocks the heat dissipation, facilitating the formation of α–Al_2O_3 under soft sparking [132].

It should be pointed out that at present, only the soft sparking in the PEO of Al and its alloys in silicate electrolyte results in the formation of high-quality coating with improved efficiency. Many authors have devoted to the development of same effect of soft sparking on other types of alloys (Mg, Ti, etc.), but it was turned out to be not very successful: "The soft sparking phenomena can be reproduced, but the acquired structures are inferior to those on aluminum alloys" [30]. Tjiang et al. [137] have reproduced soft sparking on the high-purity Mg substrate in a solution of K_2ZrF_6, NaOH, and $Na_6P_6O_{18}$, under a pulse bipolar regime with $R = 1.16$. The transition to soft sparking occurred at 21 min, featured by sudden decreases in light and acoustic emissions and a potential drop under constant current. The resultant coatings comprise a porous top layer and an inner layer of lower porosity, but it still contains cracks and pores. Arrabal et al. [129] have produced soft sparking transitions on a number of Mg-based alloys, specifically 99.9% Mg, AZ31, AZ61, AZ91D, ZC71, ZE41, and WE43-T6. Their PEO experiments were performed with a 50 Hz AC current of square waveform and $R = 1.2$ in an electrolytic solution of Na_2SiO_3 and $Na_2P_2O_7$. The coating microstructure is similar to that of Tjiang et al. [137], again with considerable cracks and pores in the inner layer. Le et al. [128] have investigated the influence of cathodic polarization in the PEO of Mg, AZ31 and AZ91alloys in 5 g L^{-1} $Na_2SiO_3 \cdot 9H_2O$ + 1 g L^{-1} KOH, which is a typical electrolyte used for the soft sparking study of Al. It was found that the optimal R ratio is 0.6, which forms the most homogenous and thickest coatings without the occurrence of soft sparking. Soft sparking did occur when higher R ratios were employed, but leading to thinner or damaged (blistering, flake-off) coatings. Rogov et al. [3] have compared the characteristics of PEO processes on different valve metals (Al, Mg, Zr, Ta, Ti, and Nb) in a dilute silicate electrolyte, with the aim of finding the influence of the nature of the valve metal on the transition to the soft sparking regime. In their study, the negative current was higher than positive for most of the valve metals, i.e., $R > 1$. However, an R ratio of 0.6 is used for Ti since it was impossible to initiate the PEO process at $R > 1$. They found that soft sparking is a characteristic feature on Mg, Al, Zr, and Ta that form insulating surface oxides with negligible electronic conductivity, whereas Ti and Nb forming semiconducting oxides do not exhibit a characteristic voltage drop, although some features of spark softening were detected. An examination of the cross sections of the coatings shows that coatings formed on non-Al metals with soft sparking phenomenon do not possess the same microstructure with that on Al.

Liu et al. [23] have carried out the PEO of tantalum in 10 g L^{-1} $NaAlO_2$ + 2 g L^{-1} KOH under bipolar regime with different R ratios, focusing on the soft sparking phenomenon and influence of frequency in the electrical regime. Rather complex PEO behaviors are observed in the research. For PEO at 1000 Hz, an excessive R ratio of 2.3 has caused early extinguishment of the plasma discharges with potential drops, which is the typical characteristic of soft sparking. However, the soft sparking coating is much thinner and enriched with a new crystalline phase of $NaTaO_3$. In contrast, in the PEO of tantalum at 100 Hz, patches of soft sparking occur at the R ratio of 1.3 and 2.3, leading to non-uniform coatings. Unlike the soft sparking coating at 1000 Hz, the soft sparking coatings developed at 100 Hz are much thicker at the region that had experienced soft sparking (up to 207 μm).

Since the reproduce of the soft sparking effect (homogeneity in the inner layer and improved coating formation efficiency) on metals other than Al and in electrolytes other than silicate is largely unsuccessful, the author of this chapter believes that the underlying reason lies in that the porous outer layer with amorphous materials, which is shown in the PEO of Al in silicate electrolyte (see Fig. 6.12f), is not formed in these cases. The softening of the discharges is an effect of the application of a due extent of cathodic polarization; however, it does not guarantee the formation of thick and compact coatings. Only in the suitable case of the PEO of Al and its alloys in silicate electrolyte, the sandwiched structure of amorphous outer layer/crystalline Al_2O_3/amorphous alumina (barrier layer) is formed, which promotes the fast formation of an inner layer by sintering effect.

It should be pointed out that there are different types of soft sparking in PEO studies. The traditional soft sparking is those observed in the PEO of Al and its alloys, as discussed above [60, 124, 127]. In this type of soft sparking, numerous tiny sparks distribute homogeneously on the whole sample surface. The sparks initiate and extinguish rapidly, and their luminescence intensity is very weak, accompanied by an anodic potential drop. The other type of soft sparking has a localized and transient nature. Cheng et al. [119] first reported this type of soft sparking behavior in PEO of a Zr alloy in 8 g L^{-1} $NaAlO_2$ + 1 g L^{-1} KOH. The PEO was carried out under a bipolar regime, with 20% duty cycle, 1000 Hz frequency, and average positive and negative current densities of 150 and 100 mA cm^{-2}, respectively. Figure 6.13 shows the process of the initiation and termination of the localized soft sparking after the conventional PEO treatment for ~3108 s. The image marked with "0 s" indicates the appearance of the sample surface before the occurrence of soft sparking. A patch of fine sparks first emerges at the left-hand corner of the specimen and then moves across to the right-hand corner. The "soft" sparking" lasts ~100 s, and strong discharges re-appear at ~108 s. The same as the conventional soft sparking behavior in Al, the localized soft sparking also featured a reduction in acoustic emission and potential drop [119].

The localized soft sparking in Fig. 6.13 has also caused a significant change in the coating microstructure. The soft sparking did not alter the surface morphology of the coatings but have caused significant changes in the cross-section. Figure 6.14a shows the cross-section of a coating formed for 1035 s before transition into soft sparking state. The coating shows the presence of a large cavity between the outer and inner part of the coating. Figure 6.14b shows the cross section of a coating formed for 1414 s, with a final period of transient soft sparking. It is found that the previous cavity is filled with dark material, which was alumina-rich according to EDS analysis [119]. Evidently, large amount of electrolyte species has been drawn and incorporated into the inner cavities of the coating during the localized soft sparking period.

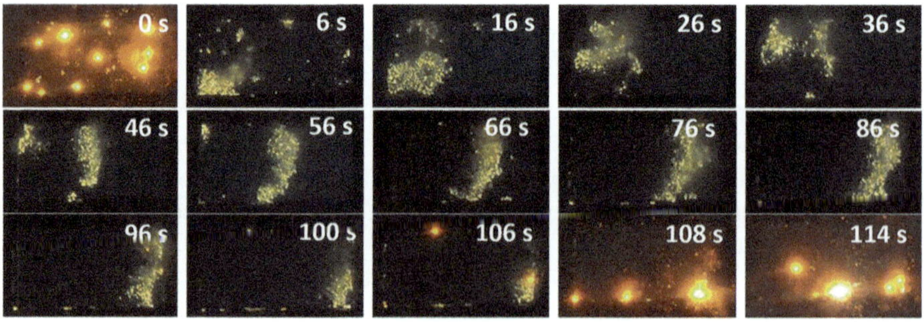

Fig. 6.13 Localized soft sparking behavior was observed during PEO of Zircaloy-2 alloy in 8 g L^{-1} $NaAlO_2$ + 1 g L^{-1} KOH. The discharges were recorded with a camera shutter speed of 1/200 s [119]. *Reproduced with permission from* [119] © *2013 Elsevier Ltd.*

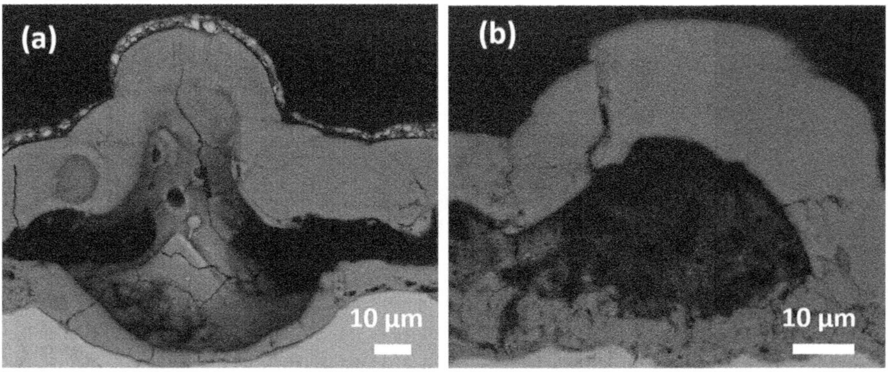

Fig. 6.14 Scanning electron micrographs showing cross-sections of coatings on Zircaloy-2 formed before and after a period of soft sparking in 8 g L^{-1} $NaAlO_2$ + 1 g L^{-1} KOH: (**a**) coating formed for 1350 s, before the commencement of soft sparking; (**b**) coating formed for 1414 s, including a final period of soft sparking with a duration of 94 s [119]. *Reproduced with permission from* [119] © *2013 Elsevier Ltd.*

6.2.10 Growth Models

6.2.10.1 Growth under Conventional Discharges

From the PEO study in the past 20 years, the coating growth mode under the strong plasma discharges is somewhat clear. It is known that coating grows discontinuously under the conventional discharge mode. The anodic current is almost exclusively passing through the locations of the discharge channels, leading to the thickening of coating at those regions experiencing the discharges. The early work of Sundararajan and Rama Krishna [99] had found that the size of the pancake diameter on Al increases linearly as a function of PEO treatment time, and they proposed a crude coating growth model with pancakes as the basic building unit. In the work of Yerokhin et al. [33], a cylindrical zone is depicted to evaluate the thermal effect in the oxide layer caused by the microdischarges. However, subsequent research shows that the discharge channels may have complex shapes. Gao et al. [138] used the Ti tracer method to investigate the microstructure of individual discharge channel of the PEO coating on aluminized Ti. They found that the individual discharge channel consists of a trumpet-shaped region in the outer porous coating layer, a cylinder-shaped region in the compact inner coating layer and a groove-like oxidation region near the substrate.

In our research group's previous study of the PEO of AZ31 magnesium alloy [85], sequential PEO treatment in aluminate and tungsten-containing electrolytes has been adopted to investigate the coating formation mechanisms by tracing W and Al species in the coatings. The AZ31 alloy was first treated in W-free electrolyte (10 g L^{-1} $NaAlO_2$ + 3 g L^{-1} $C_6H_8O_7{\cdot}H_2O$ + 2 g L^{-1} KOH) for 300 s to form a precursor coating, and then the precursor coating was treated in W-containing electrolyte (W-free electrolyte +10 g L^{-1} $Na_2WO_4{\cdot}H_2O$) for 60 s. The SEM micrograph of the surface following the two-step PEO treatment shows the irregular distribution of patches of white material on pancakes, which is due to the incorporation of W species during the second step [85]. In early PEO work, the pancake is viewed to be formed as ejection of molten oxide onto the top surface [99]. This viewpoint is debatable since electrolyte species are also incorporated into the pancake structure. Figure 6.15 shows the cross section of the sample following the two-step PEO treatment. It can be seen that the discharges have caused a significant inward growth of the PEO coating at the discharging points. The magnification of the boxed area at left-hand side shows the part of the coating that has not experienced the discharges in the second electrolyte. This part is thin and free from the white material. The adjacent area (the boxed area in the right-hand side) is the region affected by plasma discharges in the second electrolyte, showing much improved thickness and the white W species near the coating/substrate interface. With reference to the region without W incorporation, we can see that the region with W incorporation extends into the side of the AZ31 substrate, indicating an inward growth of the coating.

Based on the results of the above study, Tu et al. proposed a model for the growth of PEO coating under strong discharges (Fig. 6.16). Although this model is proposed for the coating growth on Mg alloy, it can be extended to other valve metals. As shown in the figure, the coating consists of an outer layer and an inner layer, which is usually found in the PEO of valve metals, such as Al [80, 84, 134], Mg [85, 128], and Zr [22, 119]. Two discharge events are described in the figure. A big

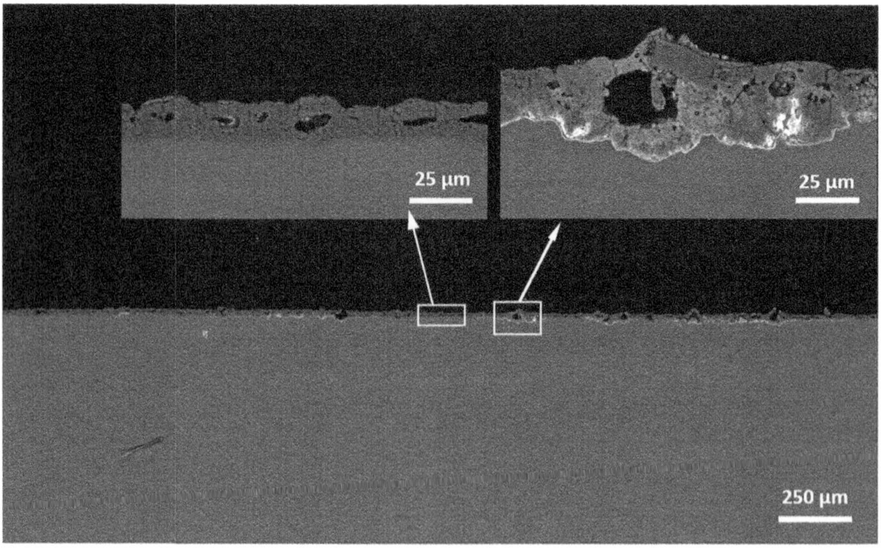

Fig. 6.15 Backscattered electron micrographs showing the cross section of the coating formed on AZ31 Mg alloy by a two-step PEO treatment. The alloy is first treated in W-free electrolyte (10 g L^{-1} $NaAlO_2$ + 3 g L^{-1} $C_6H_8O_7$ H_2O + 2 g L^{-1} KOH) for 300 s, and then in the W-containing electrolyte (W-free electrolyte +10 g L^{-1} Na_2WO_4 H_2O) for 60 s [85]. *Reproduced with permission from* [85] © *2017 Elsevier B.V.*

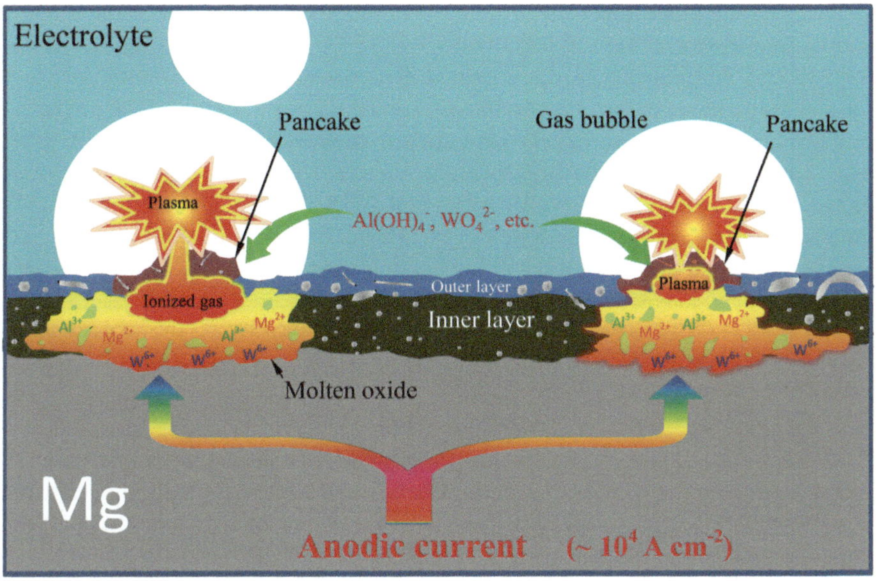

Fig. 6.16 Model for the PEO of Mg alloy under strong discharges [85]. *Reproduced with permission from* [85] © *2017 Elsevier B.V.*

bubble is shown above each discharging point, which has been called the "plasma bubble." The process of development of these bubbles has been studied by using fast-speed video [108]. Since the coating materials are dielectric oxides with high resistivity, the anodic current is almost concentrated at the discharging points. The anodic current density within the discharge channels has been estimated as high as the order ~10^4 A cm^{-2}. This extremely high anodic current density leads to a series of high-temperature processes within the discharge channels, such as the melting of the substrate metal, formation of molten oxides, and ionization of gas species into plasma. Besides the visible spark above a pancake structure, plasma may also be present within the coating. The collapse of the plasma inside the coating after a discharge event forms the big pores beneath the pancake structure (see Figs. 6.8a, 6.14a, and 6.15). During a discharge event, anionic species of the electrolyte are also drawn into the discharge channels to take part in the coating formation. It is seen that the discharge channels are irregularly shaped and protrude toward the substrate. The substrate metals, such as Al and Mg, usually have low melting points. Therefore, the substrate metal is preferentially melted at the frontier of a discharge channel and then reacts with oxygen species to form the molten oxides. As a result, the interface of coating/substrate is pushed forward into the substrate, leading to the inward growth of the oxide layers. This mechanism explains why the PEO coatings formed by the strong discharges usually have a wavy coating/substrate interface.

A subsequent study by Hu et al. [139] revealed more detail about the microstructure of discharge channels. In the work, PEO of AZ31 magnesium alloy covered with a ~10 μm magnetron sputtered Al layer was carried out in a silicate–hexametaphosphate electrolyte. After the Al-layered was consumed by oxidation, the substrate Mg will participate in the PEO coating formation. By tracing the Mg and Al elements in the coatings, it is found that new coating is mainly formed at the lower part of the discharge channels. The old oxides in the upper part of discharge channels are largely kept at their original position, which means that they did not melt during the subsequent discharge events. The high melting points of the formed oxides (2054 °C for Al_2O_3 and 2825 °C for MgO) help to remain in their solid state during the next discharges. Only a small amount of the molten oxide flows out through the micropores of the coating, reaching the top layer. These findings are largely consistent with the model in Fig. 6.16.

6.2.10.2 Growth Under Soft Sparking

From Sect. 6.2.9.2, we can see that coating grows in a different manner under the soft sparking regime. Unlike the coatings developed under strong discharges, which usually show a wavy coating/substrate interface due to the formation of bulges at the coating/substrate caused by the strong discharges, the soft sparking coatings show a smooth and flat coating/substrate interface. Figure 6.17 is the description of the growth of PEO coating on Al alloy under the soft sparking mode [127]. As shown in the figure, a PEO coating with the porous outer layer, dense inner layer and barrier layer has been developed under the soft sparking regime. Anionic

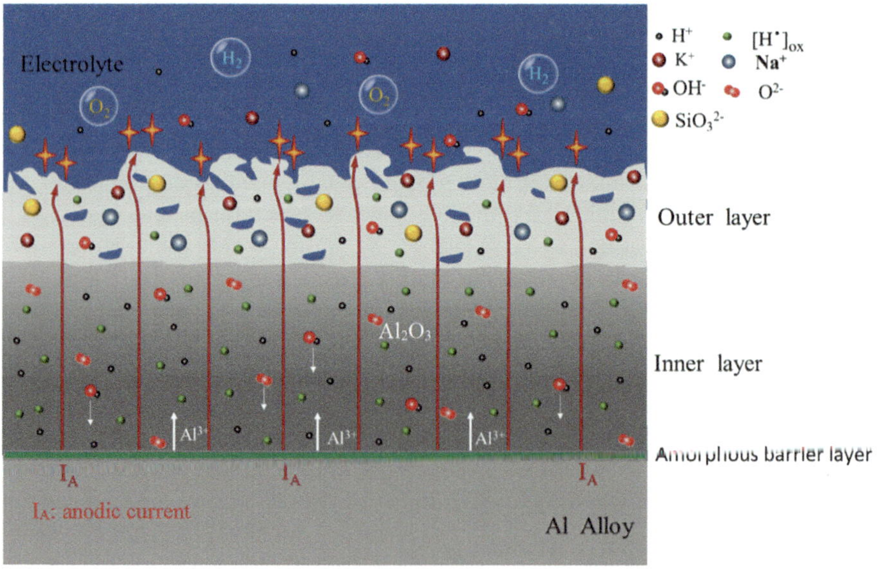

Fig. 6.17 Growth of PEO coating on Al alloy under the soft sparking mode [127]. *Reproduced with permission from* [127] *© 2022 The Electrochemical Society ("ECS"). Published on behalf of ECS by IOP Publishing Limited*

species of SiO_3^{2-} and cations of Na^+ and K^+ are incorporated into the outer layer, contributing to the formation of a highly amorphous layer with possible composition of silica, natrolite, and Lucite [127]. The transition to this soft sparking mode is due to the improved coating conductivity by cathodic hydrogen intercalation ($[H^{\bullet}]_{OX}$). The improved conductivity also makes the uniform distribution of anodic current on numerous sites of weak sparks. The inner layer is dense and is enriched with the high-temperature phase of α-Al_2O_3. In the case of Al–Cu–Li alloy, the inner layer appears black due to the oxidation of Cu species into CuO [132]. The thin barrier layer consists of amorphous alumina. As discussed previously, the Joule heat generated during PEO is effectively conserved by the sandwiched structure with the two amorphous layers, facilitating the fast formation of α-Al_2O_3 and other high-temperature species. The model in Fig. 6.17 is only suitable for PEO of Al and its alloys in silicate electrolyte. Softening of sparks may be observed for PEO of other valve metals when applies high cathodic polarization, but does not lead to the effect of improved efficiency and formation of dense coatings.

6.3 PEO Growth Mechanism on Non-valve Metals

Non-valve metals are less studied in the field of PEO compared with valve metals. Actually, the classification of metals into valve and non-valve metals is not based on rigid scientific principles. The term "valve action" represents the current blocking effect exhibited by an oxide-covered electrode immersed in a suitable solution [2]. Therefore, the classification of Fe, Cu, and Ni into non-valve metals implies the inability of forming thick oxides for these metals. It is seldom discussed why thick oxide layers are not formed on these non-valve metals. Some researchers believed that valve metals form n-type semiconductor oxide films, whereas the non-valve metals, such as Fe, Cr, Ni, W, and Cu, either form p-type oxide films or do not form stable oxides at all [38]. However, the relationship between the valve effect of a metal and the type of semiconductor of its oxide has not been validated. PEO of non-valve metals first began with steels and has been extended to other non-valve metals such as brass [39] and copper [40]. Compared with that of valve metals, PEO of non-valve metals is less studied and the present understanding of the coating formation mechanism is limited. However, a few papers on the coating growth mechanism have been published in recent years.

6.3.1 PEO of Steels

It is known that the electrolyte composition plays an important role in plasma electrolytic oxidation. The practical realization of PEO requires a careful matching of the metal-electrolyte composition. For PEO of valve metals, the so-called inorganic polymers (e.g., silicates, aluminates, tungstates, molybdates) and phosphates of alkaline metals are widely used [7]. However, the selection of suitable electrolytes for PEO of non-valve metals is more critical since plasma discharges are difficult to be built on non-valve metals. Table 6.1 listed the specific electrolyte compositions used for PEO of steels from the literature published between 2009 and 2018. It can be seen that most of the works use sodium aluminate or sodium silicate, two common electrolyte compositions used for PEO of valve metals, as the basic composition of the electrolyte. However, it is noticed that additives such as NaH_2PO_4 and NaH_2PO_2 are usually added to the PEO electrolytes. Among these works, most of the researchers focused on the microstructure (morphology, phase, and composition), and corrosion and mechanical (hardness, thermal shock resistance and wear) properties of the coatings [48–57]; fewer have paid attention to the coating formation mechanism.

In 2018, Li et al. [35] of our research group had published a paper on the PEO of carbon steel, focusing on the plasma discharge initiation and coating formation mechanism. In the work, Q345 carbon steel was treated in 8 g L^{-1} $NaAlO_2$ + 2 g L^{-1} NaH_2PO_4 using a constant current mode under the pulsed unipolar regime at 2000 Hz. Various influencing factors, such as the surface roughness of the steel substrate, agitating speed of the stirrer, current density, and duty cycle of the power supply on the PEO behavior have been investigated. The results show that although the appearance of plasma discharges and the microstructure of the PEO coatings are similar to those on the valve metals, the PEO of the Q345 steel did show some unique characteristics belonging to non-valve metals. Figure 6.18 shows the cell potential-time responses, the appearance of discharges, and SEM morphologies of surface for the PEO of the Q345 steel. Figure 6.18a shows the cell potential-time responses for the PEO of the carbon steel at 5% duty cycle under the conditions with and without agitation. The samples have been polished to achieve a mirror surface. It shows that the electrolyte stirring has an effect on the potential curves. For the PEO in the stagnant electrolyte, the cell potential rises quickly to the inflexion point (breakdown potential), which is similar to that of the PEO of valve metals. However, in the electrolyte with magnetic stirring, the cell potential shows a reducing trend after an initial rise to 633 Vat 20 s, before it rises again at delayed time. Correspondingly, the occurrence of plasma discharges is much delayed in the electrolyte with stirring (see Fig. 6.18b, c). Moreover, it was also found that discharges occurred earlier for the unpolished sample with a rougher surface. These behaviors are not observed in the PEO of valve metals (the breakdown potential of valve metals is independent of various factors such as the rate of electrolyte stirring and the surface topography [31]). Apart from the effect of stirring, it was found that a rougher surface, an increased current density, and a low duty cycle are beneficial for the coating formation (the effect of duty cycle may be due to an increased peak current density of the waveform at low duty cycle). Figure 6.18d shows the formation of a thin cracked film on the Q345 carbon steel at the early PEO stage. It was shown that this thin film is a thin layer of aluminum phosphates ($AlPO_4$ and $AlPO_4{\cdot}2H_2O$) or alumina aluminum phosphate (AAP) which was formed due to a series of the following reactions during the initial PEO process [35]:

Table 6.1 Specific compositions of the electrolytes used for PEO of steels in recent publications

Sample	Electrolyte	Publishing year	Authors
Q235 carbon steel	1. $NaAlO_2 + NaH_2PO_4$ 2. $Na_2CO_3 + NaH_2PO_2$	2009	Wang et al. [50]
Q235 carbon steel	1. $Na_2SiO_3{\cdot}9H_2O + Na_2SO_4 + NaH_2PO_4{\cdot}H_2O + NaH_2PO_2{\cdot}H_2O$ 2. $Na_2CO_3 + NaH_2PO_4{\cdot}H_2O$	2009	Wang et al. [49]
Q235 carbon steel	10 g L^{-1} $NaAlO_2$ + 1.5 g L^{-1} NaH_2PO_4	2009	Wang et al. [48]
Q235 carbon steel	1. 10 g L^{-1} $NaAlO_2$ + 1.5 g L^{-1} NaH_2PO_4 2. 10 g L^{-1} $NaAlO_2$ + 1.5 g L^{-1} NaH_2PO_4 + 2 g L^{-1} $Na_2WO_4{\cdot}2H_2O$ 3. 10 g L^{-1} $NaAlO_2$ + 1.5 g L^{-1} NaH_2PO_4 + 0.5 g $L^{-1}$$Na_2SiO_3{\cdot}9H_2O$	2009	Wang et al. [51]
Q235 carbon steel	$NaAlO_2 + NaH_2PO_4$	2010	Wang et al. [52]
Iron	0.1 mol $L^{-1}$$NaAlO_2$ + 0.05 mol L^{-1} NaOH	2010	Karpushenkov et al. [42]
Low carbon steel	7 g L^{-1} $NaAlO_2$ + 2 g L^{-1} NaOH	2014	Malinovschi et al. [53]
Carbon steel S235JR	1. 25 g L^{-1} Na_2SiO_3 + 15 g L^{-1} Na_2CO_3 2. 18 g L^{-1} Na_2SiO_3 + 11 g L^{-1} Na_2CO_3	2016	Malinovschi et al. [54]
39NiCrMo3 steel	1. 10 g L^{-1} $NaAlO_2$ + 10 g L^{-1} Na_2SiO_3 + 10 g L^{-1} NaOH 2. 10 g L^{-1} $NaAlO_2$ + 100 g L^{-1} Na_2SiO_3 + 10 g L^{-1} NaOH 3. 100 g L^{-1} $NaAlO_2$ + 10 g L^{-1} Na_2SiO_3 + 10 g L^{-1} NaOH	2016	Pezzato, et al. [57]
Q235 carbon steel	8 g L^{-1} $Na_2SiO_3{\cdot}9H_2O$ + 1.2 g L^{-1} $NaH_2PO_2{\cdot}H_2O$	2016	Wang et al. [143]
Q235 carbon steel	10 g L^{-1} Na_2SO_4 + 2.65 g L^{-1} $Na_5P_3O_{10}$ + 13.35 g L^{-1} Na_6P6O_{18} and different content of K2ZrF6 (0.0 g, 0.5 g, 1.0 g, 1.5 g, 2.0 g)	2017	Wang et al. [144]
Low carbon steel	20 g L^{-1} $NaAlO_2$ + 1.2 g $L^{-1}$$NaH_2PO_4$	2017	Yang et al. [55]
Low carbon steel	1. 30 g L^{-1} $Na_2SiO_3{\cdot}9H_2O$ + 1.2 g L^{-1} $NaH_2PO_2{\cdot}H_2O$ 2. 30 g L^{-1} $Na_2SiO_3{\cdot}9H_2O$ + 1.2 g L^{-1} $NaH_2PO_2{\cdot}H_2O$ + 2 g L^{-1} Al nanoparticles	2017	Yang et al. [56]
Low carbon steel	1.2 g $L^{-1}$$NaH_2PO_2{\cdot}H_2O$ + 10, 20, 30, 40, 50 g L^{-1} $Na_2SiO_3{\cdot}9H_2O$	2018	Yang et al. [145]
Low carbon steel	1. 28 g L^{-1} $Na_2SiO_3{\cdot}9H_2O$ + 1.2 g L^{-1} $NaH_2PO_2{\cdot}H_2O$ 2. 10 g L^{-1} $NaAlO_2$ + 1.5 g L^{-1} $NaH_2PO_4{\cdot}2H_2O$	2018	Yang et al. [146]
Q345 carbon steel	8 g L^{-1} $NaAlO_2$ + 2 g L^{-1} NaH_2PO_4	2018	Li et al. [35]

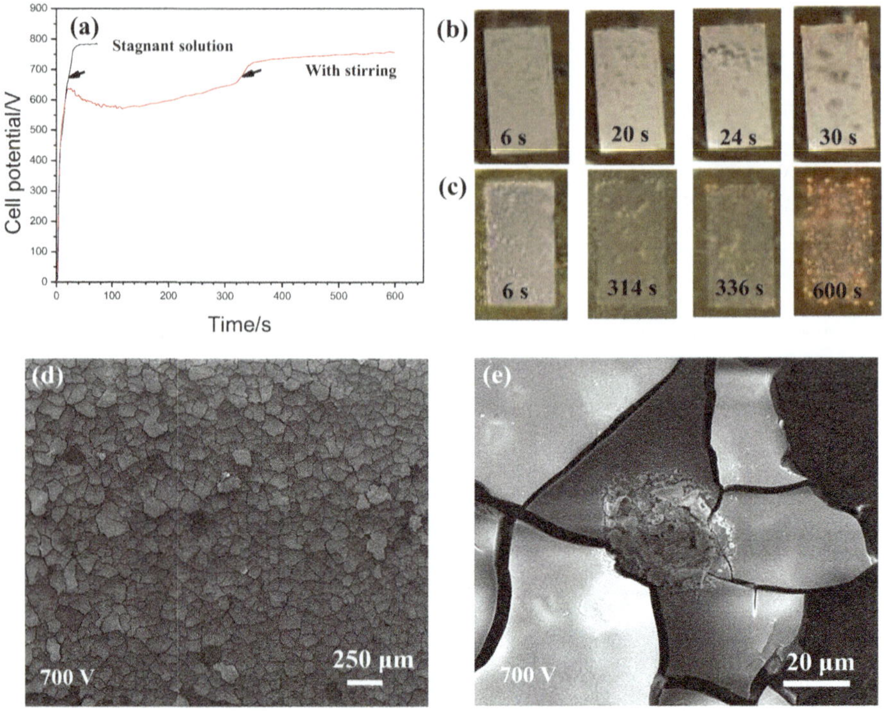

Fig. 6.18 PEO of Q345 carbon steel in 8 g L^{-1} $NaAlO_2$ + 2 g L^{-1} NaH_2PO_4 using pulsed unipolar regime at 2000 Hz, 5% duty cycle and ~0.16 A cm^{-2}: (**a**) cell potential-time responses of a polished sample under the conditions with and without electrolyte stirring, respectively (the black arrows in the curves indicate the potential at which sparks begin to appear); (**b** and **c**) are the corresponding sample appearance during the PEO in the stagnate electrolyte and the agitated electrolyte, respectively. (**d**) SEM image of the thin cracked film deposited on the unpolished sample at the early PEO stage with a cell potential of 700 V, (**e**) magnified image showing a discharge point on the film in (**d**) [35]. *Reproduced with permission from* [35] © *2018 Elsevier Ltd.*

$$2H_2O \rightarrow O_2 \uparrow + 4H^+ + 4e^- \tag{6.12}$$

$$4H^+ + Al(OH)_4^- \rightarrow Al^{3+} + 4H_2O \tag{6.13}$$

$$Al^{3+} + PO_4^{3-} \rightarrow AlPO_4 \tag{6.14}$$

$$Al^{3+} + H_2O \rightarrow Al(OH)^{2+} + H^+ \tag{6.15}$$

$$H_2PO_4^- + Al(OH)^{2+} + H_2O \rightarrow AlPO_4 \cdot 2H_2O + H^+ \tag{6.16}$$

$$3Al(OH)_4^- + HPO_4^{2-} \rightarrow Al_2O_3 \cdot AlPO_4 + 5OH^- + 4H_2O \tag{6.17}$$

Although the deposited layer is full of cracks as shown in Fig. 6.18d, the steel surface should be uniformly covered during the PEO process, since $AlPO_4$ and variscite ($AlPO_4 \cdot 2H_2O$) usually exist in the forms of sol [140] or colloidal state [141]. $AlPO_4$ shows close similarities to silica [142], which is electrically neutral and highly covalent. Therefore, a dielectric layer on the surface of carbon steel was built for the subsequent breakdown events. In the case of valve metals, the role of the dielectric layer was played by the original passive films. Figure 6.18e shows a feature on the deposited layer which is clearly an evidence of a discharging event. Despite the formation of an initial deposited layer, the morphology of PEO coating formed at the later stage, e.g., 1800 s on the carbon steel is similar to those on the conventional PEO coatings on valve metals, showing crater-like features and pores.

A model (Fig. 6.19) has been proposed to explain the coating formation mechanism for PEO of carbon steel in the aluminate electrolyte with phosphate additive

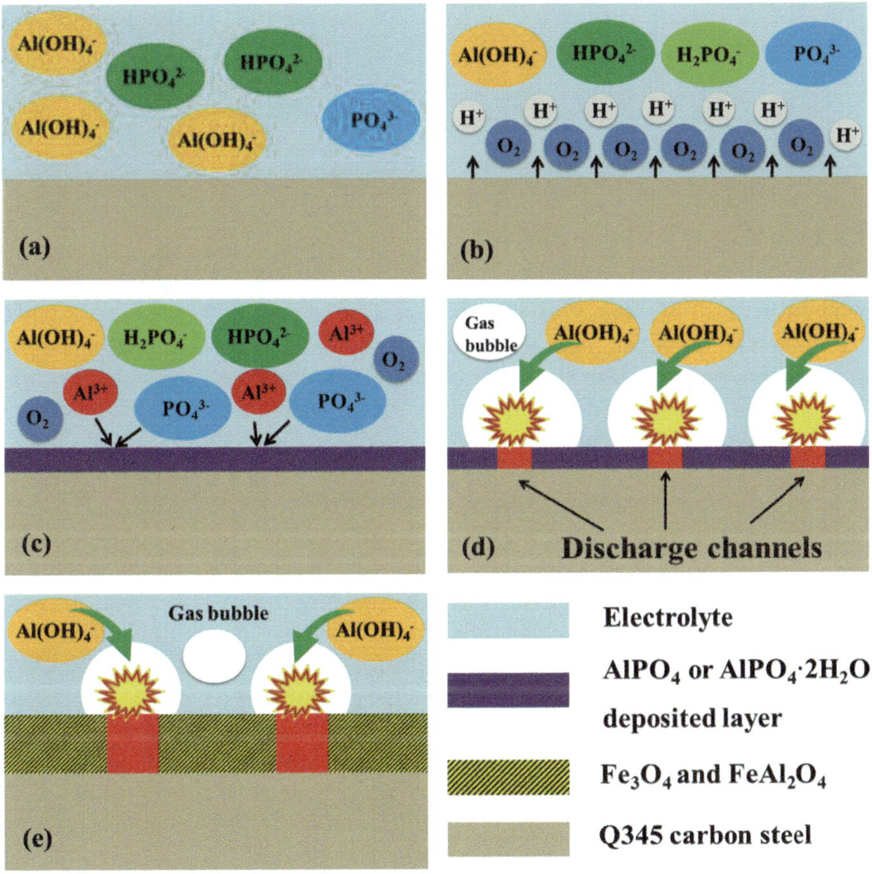

Fig. 6.19 Schematic illustration of the PEO of the carbon steel in 8 g L^{-1} $NaAlO_2$ + 2 g L^{-1} NaH_2PO_4: (**a**) Before the application of anodic current. (**b**) After the application of anodic current. (**c**) Deposition of an aluminum phosphates or AAP layer on the steel surface. (**d**) Breakdown of the dielectric layer. (**e**) Stable discharges at later stage of PEO [35]. *Reproduced with permission from* [35] *© 2018 Elsevier Ltd.*

[35]. In this electrolyte, the main anionic species in the electrolyte are $Al(OH)_4^-$, HPO_4^{2-}, and a few PO_4^{3-}(Fig. 6.19a). After the application of the anodic current, large numbers of oxygen bubbles are generated according to Eq. (6.12). This reaction will lead to a local acidification (generation of protons) of the thin electrolyte layer above the carbon steel surface (Fig. 6.19b). After that, a thin dielectric layer of aluminum phosphates ($AlPO_4$ and $AlPO_4{\cdot}2H_2O$) or alumina aluminum phosphate (AAP) (see Fig. 6.19c) is formed as discussed above. Figure 6.19d represents the breakdown of the thin dielectric layer with the incorporation of aluminate anions. The thin dielectric layer formed during the initial PEO stage helps the building of the stable plasma discharges at subsequent stages, and finally, a thick PEO coating with the composition of Fe_3O_4 and $FeAl_2O_4$ is formed (Fig. 6.19e). The oxides may be formed within the high-temperature discharge channels according to the following equations:

$$3Fe + 2O_2 \rightarrow Fe_3O_4 \tag{6.18}$$

$$Fe + 2AlO_2^- \rightarrow FeAl_2O_4 + 2e^- \tag{6.19}$$

6.3.2 *PEO of Copper and Brass*

Inspired by the PEO works on carbon steel, Cheng et al. [39, 40] have extended PEO to brass and copper. Since the additive of NaH_2PO_4 in the aluminate electrolyte was found to be critical for the oxide coating formation on carbon steel [35], PEO of copper was investigated in the aluminate-based electrolytes with the addition of 3 different types of phosphate: (1) 8 g L^{-1} $NaAlO_2$ + 2 g L^{-1} $Na_3PO_4{\cdot}12H_2O$ (trisodium phosphate, denoted as "TP"); (2) 8 g L^{-1} $NaAlO_2$ + 2 g L^{-1} $NaH_2PO_4{\cdot}2H_2O$ (sodium dihydrogen phosphate, denoted as "DP"); (3) (8 g L^{-1} $NaAlO_2$ + 2 g L^{-1} $NaH_2PO_2{\cdot}H_2O$ (sodium hypophosphite, denoted as "HP") [40]. Similar to carbon steel, PEO coatings were successfully developed on copper in the DP electrolyte. Again, a layer of aluminum phosphates or AAP was believed to have played an important role in the establishment of stable discharges. The morphologies of copper after PEO treatment in DP electrolyte for different times are displayed in Fig. 6.20, showing the AAP layer, the discharge point on the AAP layer, the surface, and the cross section of the coating formed for 1200 s. The morphologies of the deposited coatings are very similar to those on carbon steel, implying the same formation mechanism. According to XRD, the PEO coatings on copper formed in DP electrolyte consist of γ–Al_2O_3, α-Al_2O_3, CuO and Cu_2O. The alumina phases are formed by plasma-induced thermal decomposition of electrolyte, whereas the copper oxides come from the oxidation of the substrate. Coatings cannot be formed in the TP electrolyte; however, the presence of an initial thin AAP layer and the signs of coating formation at increased current density have been found in this electrolyte. The fact that plasma discharge and coating formation cannot be realized in TP electrolytes may be attributed to the insufficient phosphate ions in the electrolyte. Some

Fig. 6.20 PEO of copper in 8 g L^{-1} $NaAlO_2$ + 2 g L^{-1} NaH_2PO_4 using pulsed unipolar regime at 1000 Hz, 5% duty cycle and current density of ~0.28 A cm^{-2}: (**a**) the AAP layer formed at 20 s; (**b**) the boxed region in (**a**) showing the site of breakdown; (**c**) surface of the coating formed for 1200 s; (**d**) cross section of the coating in (**c**) [40]. *Reproduced with permission from* [40] © *2021 Elsevier B.V.*

peculiar behaviors have been observed during the PEO of copper in the HP electrolyte. A layer of loose materials with blue or green colors, which turned out to be $Cu(OH)_2$, are deposited on the copper surface at the initial PEO stages. SEM examination of the copper surface indicates that pitting corrosion has occurred at this stage. However, a black coating was ultimately formed on the surface after prolonged PEO treatment time. The pitting was attributed to a dual effect of local acidification and the attack of the passive layer by the hypophosphite ions with strong reducing capability. However, the deposited $Cu(OH)_2$ and the formation of $AlPO_4$ or Al_2O_3•$AlPO_4$ at later stage repassivated the copper surface and promoted the establishment of stable discharges.

The growth of the coatings with reference to the original metal surface has been investigated by comparing the cross sections of the copper foil and PEO-treated foil [40]. Figure 6.21a, b shows the backscattered electron micrographs of the aligned cross sections of the original copper foil (left-hand side) and the foils (right-hand side) after being PEO-treated in DP and HP for 1200 s, respectively. A comparison of the part of the samples above the white dotted lines shows clearly that there is only slight consumption of copper substrate during PEO treatment in the two

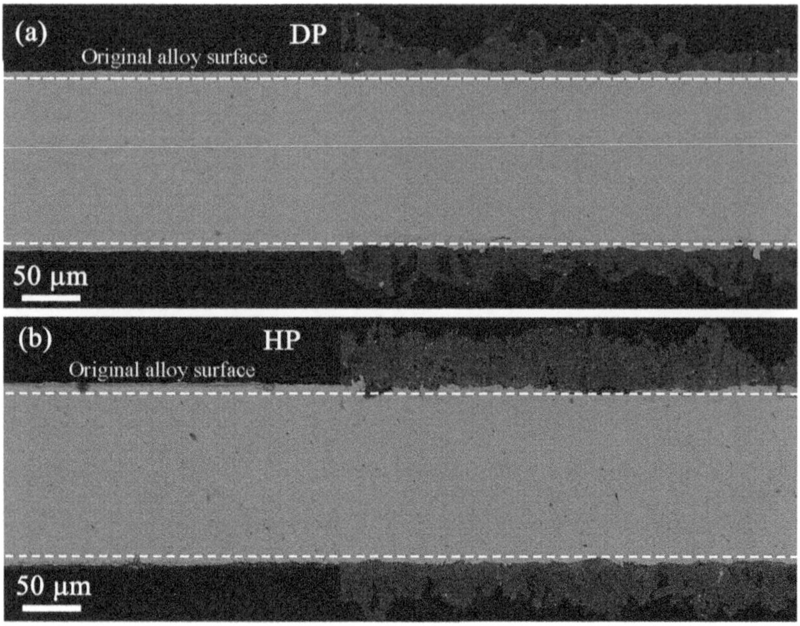

Fig. 6.21 Aligned cross sections of the untreated copper foil (left-hand side) and the PEO-treated foil in (**a**) DP for 20 min and (**b**) HP for 20 min, respectively [40]. *Reproduced with permission from* [40] © *2021 Elsevier B.V.*

electrolytes, indicating that the coatings are dominated by outward growth which is mainly contributed from the deposition of electrolyte species. The less participation of substrate metal in the coating formation implies a lower bonding force between the coating and substrate, which is an issue to be addressed in further research in the area of PEO of non-valve metals.

Apart from the aluminate–phosphate electrolyte system, PEO of non-valve metals can also be implemented in other electrolytes, such as silicate. For example, PEO coatings have been successfully formed on brass in 8 g L^{-1} $Na_2SiO_3·9H_2O$ + 1 g L^{-1} NaOH, using a pulsed unipolar mode under the frequency of 2000 Hz, duty cycle of 5% and average current density of ~0.36 A cm^{-2} [39]. It takes a relative long time (~590 s) to establish stable discharges on the brass surface. However, a black coating with the composition of CuO, Cu_2O, ZnO and amorphous SiO_2 was formed after 1200 s PEO treatment. The successful building of stable plasma charges was attributed to the formation of a dielectric silica layer on the brass surface, which is possibly formed at the early PEO stages according to Eqs. (6.20) and (6.21):

$$SiO_3^{2-} \rightarrow SiO_2 + \frac{1}{2}O_2 \uparrow 2e^- \quad (6.20)$$

$$SiO_3^{2-}(aq) + H_2O(aq) \rightarrow SiO_2(s) + 2OH^-(aq) \quad (6.21)$$

After that, oxides of substrate metal are formed under the sintering effect of plasma discharges:

$$2Cu + O_2 \rightarrow 2CuO \tag{6.22}$$

$$2Cu + \frac{1}{2}O_2 \rightarrow Cu_2O \tag{6.23}$$

$$Zn + \frac{1}{2}O_2 \rightarrow ZnO \tag{6.24}$$

The plasma discharges during the PEO of brass have been studied by OES [39]. The plasma electron temperature calculated based on the relative intensities of the lines of Cu, Zn, and H yielded electron temperature values of ~6806, 4393, and 5705 K, respectively. These values are consistent with those on valve metals. However, it is found that the spectral lines of Cu I at the ultraviolet region cannot be used for the calculation of the electron temperature, which may be due to the self-absorption of the light emission by the electrolyte [39].

6.4 Differences and Similarities Between the PEO Behaviors of Valve and Non-valve Metals

This book chapter discusses the PEO coating growth mechanism on valve and non-valve metals. PEO of valve metals has been extensively studied and abundant information has been obtained with regard to the growth mechanism. However, PEO of non-valve metals is less studied, and the coating growth mechanism is not as clear as that of valve metals. For growth of PEO coatings on both valve and non-valve metal, stable plasma discharges must be established on the sample surface. For PEO of valve metals, a variety of electrolytes, such as inorganic polymers (e.g., silicate, aluminates, tungstates, molybodates), phosphates of alkaline metals, and fluoride electrolytes (e.g., NaF, KF) can be used [7]. Thick coatings can even be formed on a 2024 aluminum alloy in a simple electrolyte of 5 g L^{-1} NaOH [8]. In the PEO of valve metals, the ease in building stable surface plasma discharges may originate from their ability to grow thick and protective passive films. The passive films provide a high resistant dielectric layer through which anodic current passes in the form of local breakdown. However, it must be careful to select the appropriate electrolyte system for PEO treatment of non-valve metals. For example, PEO of Q345 carbon steel can be realized in 8 g L^{-1} $NaAlO_2$ + 2 g L^{-1} NaH_2PO_4, but the stable discharge cannot be built in the same aluminate electrolyte without the phosphate additive [35]. PEO coatings on brass can also be obtained in the silicate electrolyte, but it takes a very long time to achieve a stable plasma discharge state [39]. It seems that regardless of the nature of a metal (valve or non-valve metals), the establishment of stable plasma discharges, therefore successful PEO treatment, depends on the formation of a dielectric layer on its surface. However, in the case of non-valve metals, the formation of the dielectric layer was assisted by the deposition of electrolyte species. The influence of electrolyte stirring and sample surface roughness on the PEO of non-valve metals may be implemented through their effect on the formation of the deposited layer.

According to the OES study [39], the electron temperature of the discharges on non-valve metals is similar to those on valve metals. The high electron temperature indicates that the plasma discharges are energy-intensive events on both valve and non-valve metals, which involve complex physical and chemical processes, such as high-temperature sintering, rapid melting and solidification, and electrochemical reactions. In PEO of valve metals, the discharge behavior is highly dependent on the electrical regimes (e.g., the application of cathodic current), the type of electrolyte and its concentration. Soft sparking may be produced under some circumstance, especially in the PEO of Al alloys. However, due to the limited reports on the PEO of non-valve metals, the effect of electrical regime parameters, such as cathodic polarization, has not been well understood. Moreover, there is no report of soft sparking on the PEO of non-valve metals.

According to the model in Fig. 6.16 of the PEO of valve metals, plasma discharges usually cause a local melting region, leading to an inward growth toward the substrate. Therefore, substrate metal participates significantly in the coating formation, although electrolyte species are also incorporated into the coating. The inward growth often results in a well-adherent coating with a high bonding force. However, the substrate metal of non-valve metals less participates in the coating formation in the PEO process. Therefore, the adherence of PEO coatings on non-valve metals is not so good, which may limit their potential application in industry.

6.5 Conclusions and Outlook

PEO evolves from traditional anodizing and has become a very attractive technique with industrial application and commercial use. This technique has the ability of forming hard, thick, uniform and strongly adherent ceramic coatings on metallic components of complex shape in eco-friendly alkaline electrolytes. The coatings usually endow the treated work-piece with good corrosion resistance, excellent wear performance, as well as functional properties such as biocompatibility and catalysis. At first, it was considered that only valve metals can be PEO-treated, however, recent years witness the expanding of PEO to a number of non-valve metals, such as steel, brass and copper. The PEO process is featured by repeated plasma discharges on the sample surface. The discharges usually take the form of visible sparks, which are large and violent at later stage. However, soft sparking is also reported and due to its ability to form homogenous coatings with high efficiency, becomes the focus of many researchers. The plasma discharges make the coating formation mechanism very complicated and has not been fully understood yet. At present, PEO practices are largely dependent on empirical selection of parameters;

however, an in-depth understanding of the growth mechanism is the key to the rational design of the PEO process.

Dielectric breakdown and occurrence of plasma discharge are central to the PEO coating formation mechanism. Dielectric breakdown has long been investigated in traditional anodizing theories, which is typically viewed as an electron avalanching event after the polarization potential exceeds the breakdown voltage of a dielectric layer, i.e., the oxide films on the treated metals. However, gas phase discharge models have been favorable in recent years in PEO studies. Based on the fact that plasma discharges could only be established after a highly resistant dielectric layer had been formed on a metal surface (e.g., passive films on valve metals, the $AlPO_4$ or AAP layer on carbon steel and copper), it is logically to assume that a discharge event should be preceded by the breakdown of the dielectric film. The discharge mechanism is one of the most challenging problems in PEO and physics, and its full revelation may depend on the development of sophisticated and advanced experimental methods in the future.

PEO is a versatile technique that is affected by many variables. Electrical parameters and the composition and concentration of the PEO electrolyte are most influential to the PEO coating formation. Although PEO is considered to be an electrochemically based technique, non-Faradaic processes are dominant in the PEO process. The non-Faradaic processes include the oxide formation by plasma-assisted thermal decomposition of electrolyte species and the anomalous gas evolution. It is clear that the strong plasma discharges have caused the thermal decomposition of water, leading to the production of large amounts of hydrogen and oxygen.

PEO of valve metals usually leads to the formation of pancake structures in the coating microstructure. The pancakes are believed to be formed on those metals that form oxides with large band gap values. Pancake structure is not found on the PEO coatings of Ti and its alloys which form semiconductor oxides. The pancakes are possibly formed by the melting-quenching process of a discharge channel. Electrolyte-derived features such as fine nodules and the coral reef-like feature are also found on the surface of the PEO coatings. In some circumstances, characteristic solidification structures are formed on the PEO coatings of certain metals, such as zirconium alloys.

Increase the electrolyte concentration is beneficial for the formation of thick and homogeneous coatings, thus greatly improves the efficiency of PEO process. However, highly concentrated electrolyte leads to the deposition of too much electrolyte species, which decreases the fraction of the oxides of substrate metal, e.g., Al_2O_3, in the resultant coatings. Cathodic polarization is another factor that greatly influences the coating formation. Suitable cathodic polarization can greatly improve the efficiency, and minimize the porosity, especially the big pores, leading to high-quality coatings. It is believed that the cathodic hydrogen evolution facilitates the transportation of more electrolyte species to the coating surface, where they take part in the subsequent coating formation reactions under anodic cycles. However, excessive cathodic polarization is detrimental to PEO coating formation, which may be due to the stress generated by hydrogen evolution.

When a suitable cathodic-to-anodic current /charge ratio (e.g., $R = 1.2$) is employed for the PEO of Al and its alloys, soft sparking is created after a certain coating thickness is reached. The soft sparking regime greatly improves the growth rate and homogeneity of the coating, with the formation of a large amount of α–Al_2O_3 within the inner dense layer. Soft sparking is first found with the PEO of Al alloys in silicate electrolyte. Researchers have tried to expand the soft sparking effect to other valve alloys (Mg, Ti, etc.), but most of the endeavors failed, although some softening of the discharging behavior has been observed. The drop of anodic potential during soft sparking indicates improved coating conductivity, which is believed to be associated with the incorporation of hydrogen species. However, little about the detail and kinetics of hydrogen incorporation reactions during the PEO process is known at present, which will be the topic of future research. The soft sparking coatings on Al show a sandwich structure of porous outer layer/dense inner layer/barrier layer. The outer layer and barrier layer consist of amorphous materials with very low thermal conductivity, which maintains a high-temperature environment within the inner coating, facilitating the $\alpha \rightarrow \gamma$ phase transition for alumina. The unsuccessful soft sparking practice on other valve metals may be due to the incapability of forming a similar coating structure.

There are two growth models for PEO of valve metals. In the first model, PEO coating is formed with the discrete plasma discharges. An individual discharge event causes a significant melting towards the substrate, leading to inward growth that is beneficial for the formation of high adherent coatings. Anionic species are also drawn into the discharge channels to take part in coating formation. An individual plasma discharge usually leads to the formation of a pancake structure and a big pore may be formed beneath the pancake after the collapse of plasma. The soft sparking coating grows under a different model. Under the soft sparking, the anodic current is uniformly distributed on the whole sample area; therefore, the soft sparking coating develops a flat coating/substrate interface.

At present, PEO of non-valve metals is still in the stage of laboratory research. Most of the work is about the PEO of carbon steels, with a few reports on brass and copper. The difficulty in PEO of non-valve metals may lie in the fact that anodic polarization does not lead to the growth of the original passive films on these metals. Therefore, it seems that the successful PEO treatment of non-valve metals reported so far is assisted by the forming of a dielectric layer via the deposition of electrolyte species, like the $AlPO_4$ and APP layers formed in the aluminate–phosphate electrolyte system, and the silica layer on brass in the silicate electrolyte. However, the present study shows that the substrate of non-valve metals participated less in the PEO coating formation, leading to inferior bonding compared with those coatings on valve metals. Compared with valve metals, research of the PEO of non-valve metals is still in its infancy, and the mechanism of the coating formation is less understood. Due to the enormous potential application prospects of the PEO coatings on non-valve metals, future research should focus on exploring new electrolyte systems and improving the coating properties, such as its adhesion, resistance to corrosion and wear, and others. It is still a long way to develop high-performance coatings on non-valve metals and gain a deeper understanding of their growth mechanisms.

Acknowledgments The author thanks the support from the China Scholarship Council, which allowed the author to carry out research on the PEO of Zr alloys as a visiting scholar at the University of Manchester from 2008 to 2009. After return to China, PEO research on a wide range of metals was continued in the author's laboratory under the financial support from the National Natural Science Foundation of China (Grant Nos: 51071066 and 51671084). Special gratitude is addressed to Prof. Peter Skeldon for his collaboration in many research works. The author thanks all the postgraduate students who had participated in the works leading to this publication.

References

1. Lee, W., & Park, S. (2014). *Chemical Reviews, 114*, 7487–7556.
2. Guntherschulze, A. (1928). *Electric rectifiers and valves*. John Wiley and Sons.
3. Rogov, A. B., Huang, Y., Shore, D., Matthews, A., & Yerokhin, A. (2021). *Ceramics International, 47*, 34137–34158.
4. Clyne, T. W., & Troughton, S. C. (2019). *International Materials Review, 64*, 127–162.
5. Schultze, J. W., & Lohrengel, M. M. (2000). *Electrochimica Acta, 45*, 2499–2513.
6. Patrito, E. M., Torresi, R. M., Leiva, E. P. M., & Macagno, V. A. (1990). *Journal of the Electrochemical Society, 137*, 524–530.
7. Yerokhin, A. L., Nie, X., Leyland, A., Matthews, A., & Dowey, S. J. (1999). *Surface and Coating Technology, 122*, 73–93.
8. Xue, W., Deng, Z., Lai, Y., & Chen, R. (1998). *Journal of the American Ceramic Society, 81*, 1365–1368.
9. Wang, Y., Zou, Y., Wang, S., Chen, G., Ouyang, J., Wei, D., Jia, D., & Zhou, Y. (2018). *China Surface Engineering, 31*, 20–45.
10. Qin, J., Shi, X., Li, H., Zhao, R., Li, G., Zhang, S., Ding, L., Cui, X., Zhao, Y., & Zhang, R. (2022). *Green Chemistry, 24*, 8113–8130.
11. Shi, X., Wang, Y., Li, H., Zhang, S., Zhao, R., Li, G., Zhang, R., Sheng, Y., Cao, S., Zhao, Y., Xu, L., & Zhao, Y. (2020). *Journal of Alloys and Compounds, 823*, 153721.
12. Coto, M., Troughton, S. C., Knight, P., Joshi, R., Francis, R., Kumar, R. V., & Clyne, T. W. (2021). *Surface and Coating Technology, 411*, 127000.
13. Kim, Y. S., Yang, H. W., Shin, K. R., Ko, Y. G., & Shin, D. H. (2015). *Surface and Coating Technology, 269*, 114–118.
14. Yao, Z., Xia, Q., Shen, Q., Ju, P., Su, P., Hu, B., & Jiang, Z. (2015). *Solar Energy Materials & Solar Cells, 143*, 236–241.
15. Stojadinović, S., Tadić, N., Radić, N., Grbić, B., & Vasilić, R. (2017). *Surface and Coating Technology, 310*, 98–105.
16. Ćirić, A., & Stojadinović, S. (2024). *Journal of Luminescence, 266*, 120268.
17. Zhang, X., Zhang, T., Lv, Y., Zhang, Y., Lu, X., Xiao, J., Ma, C., Li, Z., & Dong, Z. (2021). *Applied Surface Science, 569*, 150932.
18. Zhang, X., Peng, Z., Lu, X., Lv, Y., Cai, G., Yang, L., & Dong, Z. (2020). *Applied Surface Science, 508*, 144766.
19. Zhang, X., Zhang, Y., Lv, Y., Dong, Z., Yang, L., Zhang, E., Hashimoto, T., Zhou, X., (2023). *Journal of Magnesium and Alloys, 11*, 2182–2195.
20. Blawert, C., Dietzel, W., Ghali, E., & Song, G. (2006). *Advanced Engineering Materials, 8*, 511–533.
21. Qin, J., Chen, Y., Chen, C., Zhong, S., Yan, Z., Liu, W., Wang, Y., Lai, X., Zhao, Y., Zhao, R., & Zhang, R. (2023). *Surface and Coating Technology, 466*, 129655.

22. Cheng, Y., Matykina, E., Skeldon, P., & Thompson, G. (2011). *Electrochimica Acta, 56*, 8467–8476.
23. Liu, Z., Le, Z., He, X., Cheng, Y., Hu, P., & Cheng, Y. (2023). *Ceramics International, 46*, 13385–13396.
24. Kaseem, M., Fatimah, S., Nashrah, N., & Ko, Y. (2021). *Progress in Materials Science, 117*, 100735.
25. Sankara Narayanan, T. S. N., Park, I. S., & Lee, M. H. (2014). *Progress in Materials Science, 60*, 1–71.
26. Zhang, L., Zhang, J., Chen, C., & Gu, Y. (2015). *Corrosion Science, 91*, 7–28.
27. Barati Darband, G., Aliofkhazraei, M., Hamghalam, P., & Valizade, N. (2017). *Journal of Magnesium and Alloys, 5*, 74–132.
28. Aliofkhazraei, M., Macdonald, D. D., Matykina, E., Parfenov, E. V., Egorkin, V. S., Curran, J. A., Troughton, S. C., Sinebryukhov, S. L., Gnedenkov, S. V., Lampke, T., Simchen, F., & Nabavi, H. F. (2021). *Applied Surface Science Advances, 5*, 100121.
29. Lu, X., Mohedano, M., Blawert, C., Matykina, E., Arrabal, R., Kainer, K. U., & Zheludkevich, M. (2016). *Surface and Coating Technology, 307*, 1165–1182.
30. Tsai, D., & Chou, C. (2018). *Metals, 8*, 105.
31. Ikonopisov, S. (1977). *Electrochimica Acta, 22*, 1077–1082.
32. Albella, J. M., Montero, I., & Martinez-Duart, J. M. (1984). *Journal of the Electrochemical Society, 131*, 1101–1104.
33. Yerokhin, A. L., Snizhko, L. O., Gurevina, N. L., Leyland, A., Pilkington, A., & Matthews, A. (2003). *Journal of Physics D: Applied Physics, 36*, 2110–2120.
34. Wang, L., Chen, L., Yan, Z., & Fu, W. (2010). *Surface and Coating Technology, 205*, 1651–1658.
35. Li, Z., Cheng, Y., Kang, S., Tu, W., & Cheng, Y. (2018). *Electrochimica Acta, 284*, 681–695.
36. Gu, W., Shen, D., Wang, Y., Chen, G., Feng, W., Zhang, G., Fan, S., Liu, C., & Yang, S. (2006). *Applied Surface Science, 252*, 2927–2932.
37. Shahzad, K., Tsuji, E., Aoki, Y., Nagata, S., & Habazaki, H. (2015). *Electrochimica Acta, 151*, 363–369.
38. Parfenov, V., Yerokhin, A. L., Nevyantseva, R. R., Gorbatkov, M. V., Liang, C. J., & Matthews, A. (2015). *Surface and Coating Technology, 269*, 2–22.
39. Cheng, Y., Zhu, Z., Zhang, Q., Zhuang, X., & Cheng, Y. (2020). *Surface and Coating Technology, 385*, 125366.
40. Cheng, Y., Wei, B., Liu, Y., & Cheng, Y. (2021). *Applied Surface Science, 565*, 150477.
41. McNiell, W., & Gruss, L. (1966). Anodic Spark Reaction Processes and Articles. US Patent: 3293158, 1966212220.
42. Karpushenkov, S. A., Shchukin, G. L., Belanovich, A. L., Savenko, V. P., & Kulak, A. I. (2010). *Journal of Applied Electrochemistry, 40*, 365–374.
43. Efremov, A. P. (1989). *Protection of Metals, 25*, 176–180.
44. Wu, Z., Xia, Y., Li, G., & Xu, F. (2007). *Applied Surface Science, 253*, 8398–8403.
45. He, J. (July, 2006). The research on microarc oxidation on steel (dissertation). Harbin Institute of Technology.
46. Sun, J. (July, 2007). Plasma oxidation on steel (dissertation). Harbin Institute of Technology.
47. Di, S., & Pan, M. (2006). Micro arc oxidation treatment on the surface of iron and steels. Chinese Patent, No: CN1721578.
48. Wang, Y., Jiang, Z., & Yao, Z. (2009). *Current Applied Physics, 9*, 1067–1071.
49. Wang, Y., & Jiang, Z. (2009). *Applied Surface Science, 255*, 6240–6243.
50. Wang, Y., Jiang, Z., & Yao, Z. (2009). *Applied Surface Science, 256*, 650–656.
51. Wang, Y., Jiang, Z., & Yao, Z. (2009). *Journal of Alloys and Compounds, 481*, 725–729.
52. Wang, Y., Jiang, Z., Yao, Z., & Tang, H. (2010). *Surface and Coating Technology, 204*, 1685–1688.
53. Malinovschi, V., Marin, A., Moga, S., & Negrea, D. (2014). *Surface and Coating Technology, 253*, 194–198.

54. Malinovschi, V., Marin, A., Mihalache, M., & Iosub, I. (2016). *Surface and Coating Technology, 296*, 96–103.
55. Yang, W., Li, Q., Liu, W., Liang, J., Peng, Z., & Liu, B. (2017). *Vacuum, 144*, 207–216.
56. Yang, W., Liu, W., Peng, Z., Liu, B., & Liang, J. (2017). *Ceramics International, 43*, 16851–16858.
57. Pezzato, L., Brunelli, K., Dolcet, P., & Dabala, M. (2016). *Surface and Coating Technology, 307*, 73–80.
58. Jiang, B., & Wang, Y. (2010). *Plasma electrolytic oxidation treatment of aluminium and titanium alloys* (pp. 110–154). Woodhead Publishing.
59. Cheng, Y., Wang, T., Li, S., Cheng, Y., Cao, J., & Xie, H. (2017). *Electrochimica Acta, 225*, 47–68.
60. Jaspard-Mecuson, F., Czerwiec, T., Henrion, G., Belmonte, T., Dujardin, L., Viola, A., & Beauvir, J. (2007). *Surface and Coating Technology, 201*, 8677–8682.
61. Hussein, R. O., Nie, X., & Northwood, D. O. (2010). *Surface and Coating Technology, 205*, 1659–1667.
62. Cheng, Y., Cao, J., Mao, M., Xie, H., & Skeldon, P. (2016). *Surface and Coating Technology, 291*, 239–249.
63. Matykina, E., Arrabal, R., Skeldon, P., & Thompson, G. (2009). *Electrochimica Acta, 54*, 6767–6778.
64. Gębarowski, W., & Pietrzyk, S. (2013). *Archives of Metallurgy and Materials, 58*, 241–245.
65. Hakimizad, A., Raeissi, K., Santamaria, M., & Asghari, M. (2018). *Electrochimica Acta, 284*, 618–629.
66. Rogov, A. B., Matthews, A., & Yerokhin, A. (2019). *Electrochimica Acta, 317*, 221–231.
67. Yerokhin, A. L., Shatrov, A., Samsonov, V., Shashkov, P., Pilkington, A., Leyland, A., & Matthews, A. (2005). *Surface and Coating Technology, 199*, 150–157.
68. Cheng, Y., Mao, M., Cao, J., & Peng, Z. (2014). *Electrochimica Acta, 138*, 417–429.
69. Ikonopisov, S. (1975). *Electrochimica Acta, 20*, 783–793.
70. Rakoch, A. G., Gladkova, A. A., Linn, Z., & Strekalina, D. M. (269, 2015). The evidence of cathodic micro-discharges during plasma electrolytic oxidation of light metallic alloys and micro-discharge intensity depending on pH of the electrolyte. *Surface and Coatings Technology*, 138–144.
71. Snizhko, L. O., Yerokhin, A. L., Gurevina, N. L., Misnyankin, D. O., Pilkington, A., Leyland, A., & Matthews, A. (2005). *Electrochimica Acta, 50*, 5458–5464.
72. Ikonopisov, S., Girginov, A., & Machkova, M. (1979). *Electrochimica Acta, 24*, 451–456.
73. Burger, F. J., & Wu, J. C. (1971). *Journal of the Electrochemical Society, 118*, 2039–2042.
74. Di Quarto, F., Piazza, S., & Sunseri, C. (1988). *Journal of Electroanalytical Chemistry, 248*, 99–115.
75. Quarto, F. D., Piazza, S., & Sunseri, C. (1984). *Journal of the Electrochemical Society, 131*, 2901–2906.
76. Young, L., & Anodnye oksidnye plenki, Energiya. Leningrad. (1967). *Russian translation of anodic oxide films* (Vol. 1961). Academic Press.
77. Burger, F. J., & Young, L. (1962). Electrolytic capacitors. In L. B. Birks (Ed.), *Progress in dielectrics* (Vol. 5).
78. Young, L. (1955). *Transactions of the Faraday Society, 51*, 1250–1260.
79. Shimizu, K. (1981). *Electrochimica Acta, 26*, 1691–1692.
80. Dehnavi, V., Luan, B., Liu, X., Shoesmith, D., & Rohani, S. (2015). *Surface and Coating Technology, 269*, 91–99.
81. Golovanova, O. A., Sizikov, A. M., & Vuzov, I. (1996). *Series: Chemistry and Chemical Technology, 39*, 85. (in Russian) (ISSN 0579−2991).
82. Hickling, A., & Ingram, M. D. (1964). *Transactions of the Faraday Society, 60*, 783–793.
83. Hussein, R. O., Nie, X., Northwood, D. O., Yerokhin, A., & Matthews, A. (2010). *Journal of Physics D: Applied Physics, 43*, 105203.

84. Cheng, Y., Xue, Z., Wang, Q., Wu, X., Matykina, E., Skeldon, P., & Thompson, G. E. (2013). *Electrochimica Acta, 107*, 358–378.
85. Tu, W., Cheng, Y., Wang, X., Zhan, T., Han, J., & Cheng, Y. (2017). *Journal of Alloys and Compounds, 725*, 199–216.
86. Dunleavy, C. S., Golosnoy, I. O., Curran, J. A., & Clyne, T. W. (2009). *Surface and Coating Technology, 203*, 3410–3419.
87. Dunleavy, C. S., Curran, J. A., & Clyne, T. W. (2011). *Surface and Coating Technology, 206*, 1051–1061.
88. Dunleavy, C. S., Curran, J. A., & Clyne, T. W. (2013). *Applied Surface Science, 268*, 397–409.
89. Troughton, S. C., Nomine, A., Dean, J., & Clyne, T. W. (2016). *Applied Surface Science, 389*, 260–269.
90. Matykina, E., Berkani, A., Skeldon, P., & Thompson, G. E. (2007). *Electrochimica Acta, 53*, 1987–1994.
91. Nomine, A., Troughton, S. C., Nomine, A. V., Henrion, G., & Clyne, T. W. (2015). *Surface and Coating Technology, 269*, 125–130.
92. Klapkiv, M., Nykyforchyn, H., & Posuvailo, V. (1994). *Materials Science, 30*, 333–343.
93. Zhan, T., Tu, W., Cheng, Y., Han, J., Su, B., & Cheng, Y. (2018). *Ceramics International, 44*, 10402–10411.
94. Jovović, J., Stojadinović, S., Šišović, N., & Konjević, N. (2011). *Surface and Coating Technology, 206*, 24–28.
95. Stojadinović, S., Radic-Perić, J., Vasilić, R., & Perić, M. (2014). *Applied Spectroscopy, 68*, 101–112.
96. Hussein, R. O., Nie, X., & Northwood, D. O. (2013). *Electrochimica Acta, 112*, 111–119.
97. Stojadinović, S., Vasilić, R., Petković, M., Kasalica, B., Belča, I., Žekić, A., & Lj. Zeković, Appl. (2013). *Surface Science, 265*, 226–233.
98. Yang, X., Chen, L., Qu, Y., Liu, R., Wei, K., & Xue, W. (2017). *Surface and Coating Technology, 324*, 18–25.
99. Sundararajan, G., Rama Krishna L. (2003). *Surface and Coating Technology, 167*, 269–277.
100. Cheng, Y., Zhang, Q., Zhu, Z., Tu, W., Cheng, Y., & Skeldon, P. (2020). *Ceramics International, 46*, 13385–13396.
101. Yu, L., Cao, J., & Cheng, Y. (2015). *Surface and Coating Technology, 276*, 266–278.
102. Cheng, Y., Peng, Z., Wu, X., Cao, J., Skeldon, P., & Thompson, G. E. (2015). *Electrochimica Acta, 165*, 301–313.
103. Reshak, A. H., Khan, S. A., & Alahmed, Z. A. (2014). *Optical Materials, 37*, 322–326.
104. Karthikeyan, B., & Vettumperumal, R. (2022) *Optical Materials, 123*, 111878.
105. Duan, H., Yan, C., & Wang, F. (2007). *Electrochimica Acta, 52*, 3785–3793.
106. Duan, H., Yan, C., & Wang, F. (2007). *Electrochimica Acta, 52*, 5002–5009.
107. Cheng, Y., Wu, F., Matykina, E., Skeldon, P., & Thompson, G. E. (2012). *Corrosion Science, 59*, 307–315.
108. Troughton, S. C., Nominé, A., Nominé, A. V., Henrion, G., & Clyne, T. W. (2015). *Applied Surface Science, 359*, 405–411.
109. Snizhko, L. O., Yerokhin, A. L., Pilkington, A., Gurevina, N. L., Misnyankin, D. O., Leyland, A., & Matthews, A. (2004). *Electrochimica Acta, 49*, 2085–2095.
110. Snizhko, L. O., Yerokhin, A. L., Gurevina, N. L., Patalakha, V. A., & Matthews, A. (2007). *Thin Solid Films, 516*, 460–464.
111. Matykina, E., Arrabal, R., Pardo, A., Mohedano, M., Mingo, B., Rodríguez, I., & González, J. (2014). *Materials Letters, 127*, 13–16.
112. Cheng, Y., Cao, J., Mao, M., Peng, Z., Skeldon, P., & Thompson, G. E. (2015). *Surface and Coating Technology, 269*, 74–82.
113. Baykara, S. Z. (2004). *International Journal of Hydrogen Energy, 29*, 1451–1458.
114. Lede, J., Lapicque, F., & Villermaux, J. (1983). *International Journal of Hydrogen Energy, 8*, 675–679.
115. Zhai, X., Liu, K., & Han, Q. (2010). *New energy technology* (2nd ed.). Chemical Industry Press. (In Chinese).

116. Snezhko, L. A., Erokhin, A. L., Kalinichenko, O. A., & Misnyankin, D. A. (2016). *Materials Science, 52*, 421–430.
117. Rogov, A. B., Nemcova, A., Hashimoto, T., Matthews, A., & Yerokhin, A. (2022). *Surface and Coating Technology, 442*, 128142.
118. Wei, K., Zhang, Y., Yu, J., Liu, R., Du, J., Jiang, F., & Xue, W. (2020). *Materials Chemistry and Physics, 251*, 123054.
119. Cheng, Y., Cao, J., Peng, Z., Wang, Q., Matykina, E., Skeldon, P., & Thompson, G. E. (2014). *Electrochimica Acta, 116*, 453–466.
120. Martin, J., Nominé, A., Brochard, F., Briançon, J. L., Noël, C., Belmonte, T., Czerwiec, T., & Henrion, G. (2017). *Applied Surface Science, 410*, 29–41.
121. Nominé, A., Martin, J., Noël, C., Henrion, G., Belmonte, T., Bardin, I. V., & Lukeš, P. (2016). *Langmuir, 32*, 1405–1409.
122. Terleeva, O. P., Oh, Y., Slonova, A. I., Kireenko, I. B., Ok, M.-R., & Ha, H.-P. (2005). *Materials Transactions, 46*, 2077–2082.
123. Rogov, A. B., & Shayapov, V. R. (2017). *Applied Surface Science, 394*, 323–332.
124. Rogov, A. B., Yerokhin, A., & Matthews, A. (2017). *Langmuir, 33*, 11059–11069.
125. Shashkov, P., Khomutov, G., Yerokhin, A., & Usov, S. (2017). US 9551082 insulated metal substrate.
126. Gębarowski, W., & Pietrzyk, S. (2014). *Archives of Metallurgy and Materials, 59*, 407–411.
127. Cheng, Y., Feng, T., & Cheng, Y. (2022). *Journal of the Electrochemical Society, 169*, 071505.
128. Le, Z., Liu, Z., He, X., Cheng, Y., Hu, P., & Cheng, Y. (2023). *Coatings, 13*, 1736.
129. Arrabal, R., Matykina, E., Hashimoto, T., Skeldon, P., & Thompson, G. E. (2009). *Surface and Coating Technology, 203*, 2207–2220.
130. Norby, T., Widerøe, M., Glöckner, R., & Larring, Y. (2004). *Dalton Transactions, 19*, 3012–3018.
131. Bousser, E., Rogov, A., Shashkov, P., Gholinia, A., Laugel, N., Slater, T. J. A., Withers, P. J., Matthews, A., & Yerokhin, A. (2023). *Acta Materialia, 244*, 118587.
132. He, X., Feng, T., Cheng, Y., Hu, P., Le, Z., Liu, Z., & Cheng, Y. (2023). *Journal of the American Ceramic Society, 106*, 7019–7042.
133. Tillous, K., Toll-Duchanoy, T., Bauer-Grosse, E., Hericher, L., & Geandier, G. (2009). *Surface and Coating Technology, 203*, 2969–2973.
134. Zhu, L., Guo, Z., Zhang, Y., Li, Z., & Sui, M. (2016). *Electrochimica Acta, 208*, 296–303.
135. Nie, X., Meletis, E. I., Jiang, J. C., Leyland, A., Yerokhin, A. L., & Matthews, A. (2002). *Surface and Coating Technology, 149*, 245–251.
136. Oh, Y. J., Mun, J. I., & Kim, J. H. (2009). *Surface and Coating Technology, 204*, 141–148.
137. Tjiang, F., Ye, L. W., Huang, Y. J., Chou, C. C., & Tsai, D. S. (2017). *Ceramics International, 43*, S567–S572.
138. Gao, F., Hao, L., Li, G., & Xia, Y. (2018). *Applied Surface Science, 431*, 13–16.
139. Hu, P., Wei, B., Cheng, Y., & Cheng, Y. (2024). *Transactions of the Nonferrous Metals Society of China, 34*, 139–156.
140. Bao, Y., & Nicholson, P. S. (2006). *Journal of the American Ceramic Society, 89*, 465–470.
141. Morales, J. G., Clemente, R. R., & Matijevic, E. (1992). *Journal of Colloid and Interface Science, 151*, 555–562.
142. Bothe, J. V., & Brown, P. W. (1993). *Journal of the American Ceramic Society, 76*, 362–368.
143. Wang, J., Li, C., Yao, Z., Yang, M., Wang, Y., Xia, Q., & Jiang, Z. (2016). *Surface and Coating Technology, 307*, 1315–1321.
144. Wang, J., Yao, Z., Xia, Q., Wang, Y., & Jiang, Z. (2017). *Journal of Alloys and Compounds, 711*, 278–286.
145. Yang, W., Peng, Z., Liu, B., Liu, W., & Liang, J. (2018). *Journal of Materials Engineering and Performance, 27*, 2345–2353.
146. Yang, W., Li, Q., Liu, C., Liang, J., Peng, Z., & Liu, B. (2018). *Surface Engineering, 34*, 54–62.

Chapter 7
Growth Mechanism of Plasma Electrolytic Oxidation on Metal–Matrix Composites

Yongchun Zou and Yaming Wang

7.1 Introduction

Metal matrix composites (MMCs), including those based on aluminum (Al), magnesium (Mg), and titanium (Ti), have emerged as promising structural materials across industries such as automotive, aerospace, and biomedical engineering. Their appeal lies in the remarkable strength, stiffness, ductility, and fracture toughness they provide [1]. The broad applicability of metal matrix composites (MMCs) can be attributed to the wide range of fabrication techniques available. These techniques span solid-state methods, such as forging, rolling, and extrusion, to liquid-state processes, such as casting, and even include additive manufacturing approaches like 3D printing. In these production techniques, micro- or nano-sized particles or fibers (typically ceramic, carbon, or metallic) are integrated into the metal or alloy matrix to improve material properties. Each method has unique benefits, with the choice of fabrication approach being driven by the final product's specific requirements [2]. For instance, in electronic packaging, Al-MMCs are essential, meeting critical needs for mechanical support, thermal stability, electrical insulation, and interference shielding for chips [3]. Among these, aluminum composites reinforced with silicon carbide particles (SiCp/Al) are recognized as ideal materials for heat sinks in

Y. Zou (✉) · Y. Wang (✉)
Institute for Advanced Ceramics, Harbin Institute of Technology, Harbin, China

Key Laboratory of Advanced Structure-Function Integrated Materials and Green Manufacturing Technology, Harbin Institute of Technology, Harbin, China
e-mail: zouyongchun@hit.edu.cn; wangyaming@hit.edu.cn

V. S. Saji, M. Zheludkevich (eds.), *Plasma Electrolytic Oxidation – Fundamentals, Advances and Applications*,
https://doi.org/10.1007/978-3-031-92429-3_7

power devices, owing to their excellent thermal conductivity, adaptable coefficient of thermal expansion, and low density [4, 5]. Moreover, in biomedical applications, reinforcing Mg matrices with hard ceramic nanomaterials addresses challenges like limited mechanical properties and insufficient creep resistance. Such reinforcements also enhance corrosion and wear resistance.

Metal matrix composites (MMCs) offer distinct advantages by effectively overcoming the inherent limitations of the matrix material through the incorporation of reinforcements, typically ceramic materials at the micro- or nano-scale. To optimize the performance of MMCs, it is crucial to control the interactions at the matrix-reinforcement interface. These interactions may manifest as nonreactive and insoluble, nonreactive but mutually soluble, or as a reaction product layer formed through chemical processes [3, 6].

Although incorporating reinforcements enhances the mechanical properties of MMCs, it can negatively impact their corrosion behavior, primarily due to alterations in the interface interactions, particularly in the case of galvanic corrosion. When galvanic corrosion occurs between the matrix (acting as the anode) and the reinforcement (serving as the cathode), the material's corrosion sensitivity is markedly increased, leading to the dissolution of the matrix and detachment of the reinforcement. Furthermore, surface imperfections such as pores and microcracks can accelerate the corrosion process. The extent of this susceptibility is closely associated with both the material's processing techniques and its composition [2, 7].

Plasma electrolytic oxidation (PEO), also known as microarc oxidation (MAO), is a surface modification technique recognized for its simplicity, environmental sustainability, and high efficiency. PEO allows for the rapid formation of a ceramic layer with excellent adhesion to the metal substrate, eliminating the complex bonding steps and additional layers typically required in traditional packaging methods. During the PEO process, high voltage induces localized plasma discharges at the metal-electrolyte interface, transforming the metal surface into a robust, densely bonded oxide or compound layer. This protective layer forms as a result of reactions between the base metal and electrolyte components during the discharge phase [8–12]. The characteristics of PEO coatings are strongly influenced by discharge conditions, breakdown voltage, and electrolyte composition. Due to its ease of application, cost-effectiveness, and environmental benefits, PEO has gained significant traction in the aerospace, mechanical, and electronics industries, with increasing market demand [12].

While numerous studies have examined PEO on aluminum, magnesium, titanium, and their alloys, research on PEO applications for metal matrix composites remains limited. The influences of reinforcement materials, electrolyte composition, and electrolytic parameters on ceramic coating development have not yet been systematically reviewed. Consequently, further investigation into the formation mechanisms of ceramic coatings on aluminum, magnesium, and titanium-based composites through PEO is essential to deepen understanding of the PEO coating growth process.

7.2 Growth Mechanism of PEO on Aluminum-Matrix Composites

7.2.1 Effect of Electrolyte on Spark Discharge

A key feature of the PEO process is the generation of intense spark discharges, which are significantly influenced by the electrolyte composition. This factor plays a crucial role in determining the microstructure and phase composition of the coating. Figure 7.1a illustrates the current density–time (j–t) response of the SiCp/Al composite during PEO in various electrolyte systems, including phosphate (PEO-P), silicate (PEO-Si), and aluminate (PEO-Al) [13]. In constant voltage mode, the variation in current density during different stages of the coating process was recorded. The j-t curve of the PEO process can be divided into three distinct regions: Region I, Region II, and Region III. The figure also includes representative images of spark discharges and optical photographs of the coated samples (Fig. 7.1b) [14].

In the initial stage, Region I (0–90 s), current density increases linearly over time, indicating this phase as the conventional anodization phase. During this period, gas evolution on the surfaces of PEO-Al and PEO-Si coatings is more pronounced than on the PEO-P coating. In Region II (90–150 s), dielectric breakdown occurs in the oxide layer, triggering spark discharges. At this stage, the current density in PEO-Al and PEO-Si systems gradually reaches a peak, and the spark discharges transition from white to bright yellow light (Fig. 7.1b). In contrast, only isolated white sparks appear on the PEO-P coating. The final stage, Region III (after

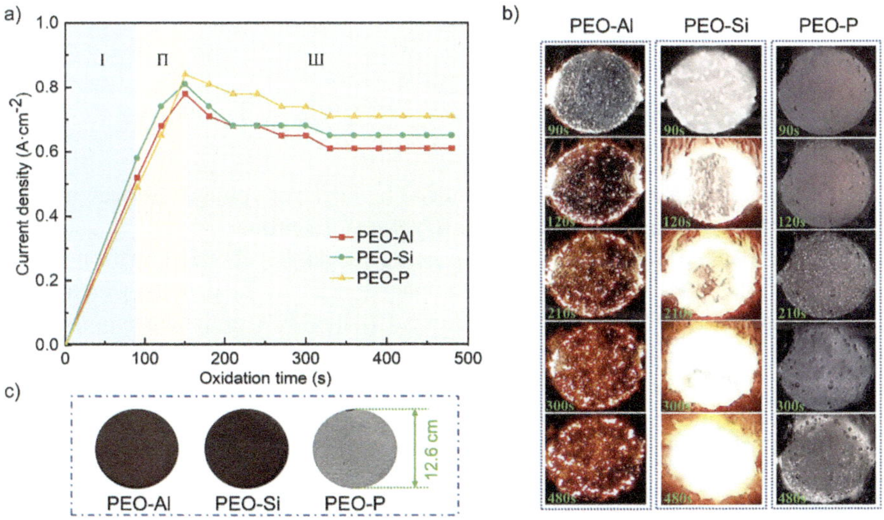

Fig. 7.1 PEO treatment in different basic electrolytes (**a**) applied current density as a function of time, (**b**) spark evolution, and (**c**) optical photographs of the coated specimens [14]. *Reproduced with permission from* [14] © *2023 Elsevier Inc.*

150 s), represents the microarc oxidation phase, during which current density decreases and stabilizes. At this point, bright yellow light spreads across the entire surface of PEO-Al and PEO-Si coatings, accompanied by a faint popping sound. Meanwhile, the PEO-P coating exhibits a uniform but faint white light without noticeable explosions [15].

While the overall pattern of the j–t curves is similar across the different electrolyte systems, the characteristics of the spark discharges vary notably. The initiation of the spark discharge phase is marked by the emergence of numerous short-lived arc sparks on the surface. Driven by high-temperature gradients and pulsed voltage, the coating forms through a "discharge-quenching" cycle [16]. Consequently, the behavior of the spark discharges is closely tied to the coating's growth and microstructural development. In phosphate electrolytes, the discharge intensity is the weakest, producing only faint white sparks on the surface. Although these weak discharges cause minimal damage to the coating, they may prevent a complete transition into the typical microarc oxidation stage, potentially resulting in a flocculent and discontinuous coating structure. In contrast, silicate electrolytes produce the most intense discharges, characterized by bright yellow light that covers the entire surface with vigorous gas evolution. This intense discharge activity can lead to the formation of large pores and cracks within the coating. The discharge intensity in aluminate electrolytes is moderate, promoting the development of a dense coating with high crystallinity. In the plasma discharge zone, local temperatures reach between 6000 K and 12,000 K in silicate electrolytes, 6000 K–10,000 K in aluminate electrolytes, and 4000 K–5000 K in phosphate electrolytes [14]. Figure 7.1c provides optical images of the coatings after the PEO process, where differences in surface color reflect the significant influence of electrolyte type on coating characteristics.

7.2.2 *Effect of Negative Voltage on Soft Spark Discharge*

The effect of soft sparking discharge induced by negative voltage has been investigated. The high-intensity arc sparking discharge is the primary cause of generating macropores on the PEO layer. The soft sparking discharge generated by the negative pulses can efficiently reduce the hostile impact and repair the defects, which could develop a dense PEO layer [3, 17–20]. For SiC/Al MMCs, an increase in negative voltage accelerates the transition from arc discharge to soft sparking. At low negative voltage (500 V positive/40 V negative), the transition from large arc sparks to dense, small white soft sparks occurs gradually. In contrast, at a higher negative voltage (500 V positive/80 V negative), the arc discharge phase is shortened, the spark intensity decreases rapidly, and the transition to soft sparking occurs much faster [17].

It is important to recognize that surpassing the optimal negative voltage range can have negative consequences. As the negative voltage increases from 0 V to 80 V, the coating thickness grows from 34.6 μm to 45.1 μm, while the porosity decreases from 21.8% to 5.1%. However, when the negative voltage is further raised to 120 V, both the coating thickness decreases and porosity increase [19]. A comparison between unipolar pulsed (U-P) and bipolar pulsed (B-P) PEO processes applied to SiC/Al MMCs (Fig. 7.2a) indicates that the B-P mode, characterized by soft sparking, results in thicker and denser coatings. The current density–time curve (Fig. 7.2b) supports the idea that negative voltage plays a beneficial role in modifying spark characteristics. In the conventional anodization phase (Stage I), a barrier-type oxide layer is formed. In the second stage, the breakdown voltage is reached, and an increasing number of low-brightness, small sparks appear. Subsequently, orange sparks emerge widely and evenly across the surface, accompanied by strong light and sound emissions (arc discharge, Stage III). As oxidation progresses, the micro-discharges in the B-P mode transition to soft sparking, while in the U-P mode, the micro-discharges remain in the arc sparking state.

The current density observed in the B-P samples was notably higher than that in the U-P samples, primarily due to the negative pulse's role in promoting electron accumulation, which in turn increases the current density. Scanning electron microscopy (SEM) images reveal that the PEO layer produced under the B-P mode (Fig. 7.2f–h) is more compact and contains fewer large pores when compared to the layer formed under the U-P mode (Fig. 7.2c–e) [3]. Furthermore, as the negative voltage rises from 0 to 120 V, the number of large pores (5–6 μm) within the PEO

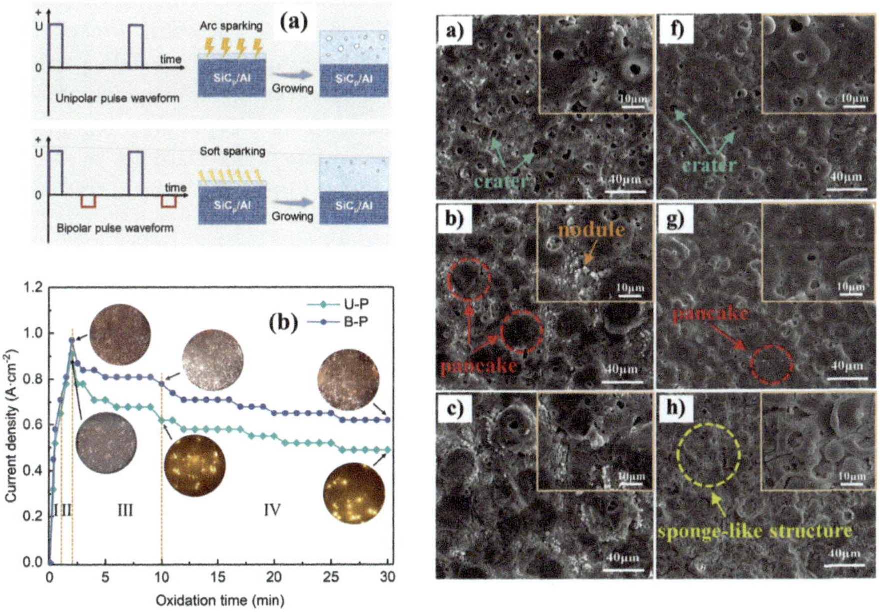

Fig. 7.2 (Left) (**a**) Scheme showing PEO under unipolar (U-P) and bipolar (B-P) pulse modes and (**b**) the associated current density-time plots. (Right) Surface morphologies of PEO layers after different PEO times under U-P (**a**–**c**) and B-P (**f**–**h**) modes: (**a, f**) 5, (**b, g**) 10, and (**c, h**) 15 min [3]. *Reproduced with permission from* [3] © *2022 Elsevier B.V.*

layer diminishes. However, the discharge intensity at these levels remains too weak to facilitate the phase transformation from γ-Al_2O_3 to α-Al_2O_3 [20]. This indicates that while negative voltage affects the characteristics of spark discharges, it also plays a crucial role in determining the microstructure and phase composition of the coatings.

7.2.3 *The Type of Enhancer Affects the Growth of the Coating*

The impact of reinforcements in metal matrix composites (MMCs) on the PEO process is substantial, as these particles often function as semiconductors or insulators, which decelerates the PEO layer's growth rate. Specifically, SiC, with its low electrical conductivity, restricts oxide layer growth but does not compromise film continuity. Research indicates that SiC particles can obstruct the expulsion of molten material, resulting in uneven growth of the coating [21, 22]. Conversely, conductive reinforcements like TiB_2 permit some current leakage in the early PEO stages. As oxidation advances, however, the formation of semiconducting TiO_2 on the surface quickly minimizes this leakage (Fig. 7.3). TiB_2 particles at the interface between the matrix and the PEO layer persist in restricting the inward growth of the PEO layer [23].

The inclusion of reinforcements raises the energy demand during oxidation, leading to a prolonged PEO process. Figure 7.4 depicts the microstructural changes that occur during PEO, such as the oxidation and melting of surface TiB_2 particles, dielectric breakdown of TiO_2 and the matrix, injection of electronegative particles into discharge channels followed by their reactions, melt crystallization, and the oxide layer's formation. Over several discharges, large TiB_2 particles eventually vanish. Zhu et al.'s study on B_4C/Al MMCs provides further evidence of this effect. B_4C, a semiconductor with a melting point near 2600 °C, shows behavior similar to SiC. During the initial anodic oxidation phase of PEO, B_4C particles impede the

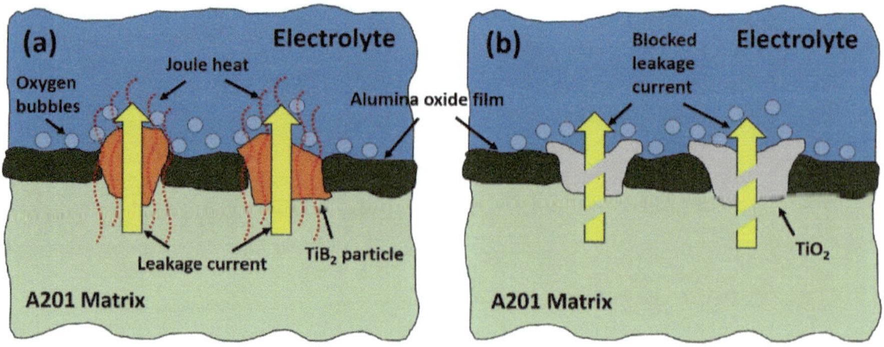

Fig. 7.3 Schematic diagram of the melting and oxidation process of TiB_2 particles: (**a**) before oxidation, and (**b**) after oxidation [23]. *Reproduced with permission from* [23] © *2017 Elsevier B.V.*

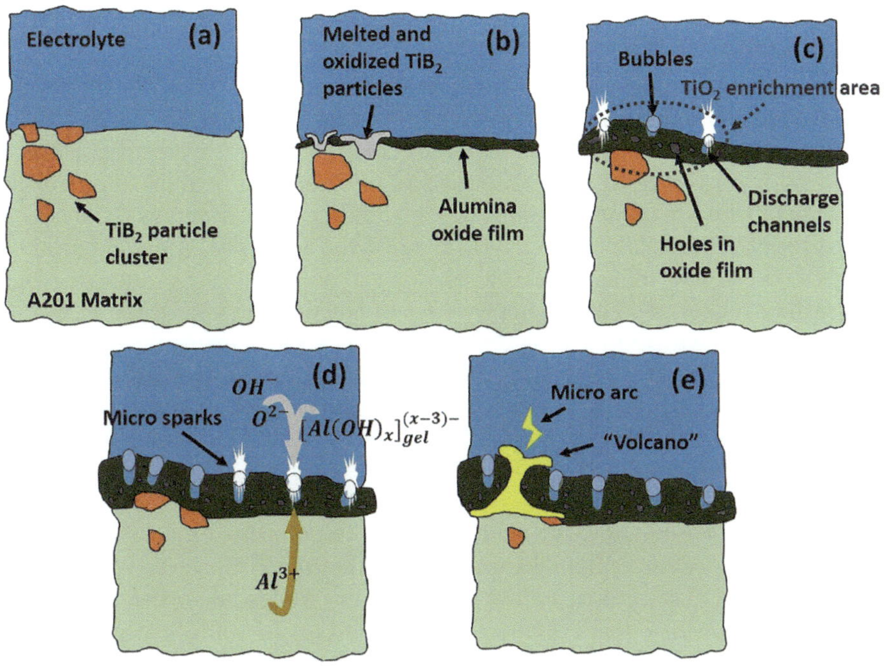

Fig. 7.4 Schematic diagram of microscopic growth of PEO film and the transformation of TiB_2 particles [23]. *Reproduced with permission from* [23] © *2017 Elsevier B.V.*

development of a dense oxide layer on the MMC surface. At the spark discharge stage (260 V, 160 s), the size of B_4C particles slightly decreases, suggesting that the temperature remains too low for complete particle breakdown. In the early microarc discharge phase (200 seconds), B_4C begins to oxidize into B_2O_3 and C, and after 10 min, it is fully converted, with C oxidizing to CO or CO_2 [24].

Comparative analyses of various reinforcement materials have indicated that Al_2O_3-reinforced MMCs do not experience substantial reductions in PEO layer thickness or hardness, which is attributed to the electrical insulating nature of Al_2O_3. In contrast, MMCs reinforced with SiC exhibit noticeable increases in porosity and a reduction in coating thickness. This behavior occurs because, during the micro-arc discharge process, SiC continuously transforms into oxide phases, which are accompanied by gas evolution, resulting in thinner coatings on SiC/MMC substrates [25].

7.3 Growth Mechanism of PEO on Magnesium-Matrix Composites

Magnesium metal matrix composites (Mg MMCs) have been identified as potential substitutes for Al MMCs in applications that demand a high strength-to-weight ratio. Considerable recent research has focused on Mg MMCs, which remain in the development phase [26, 27].

7.3.1 The Type of Enhancer Affects the Growth of the Coating

In magnesium metal matrix composites (MMCs), a lower volume fraction of SiC (1 vol%) has no significant effect on the PEO process, whereas higher fractions (3% and 10%) lead to thinner and more porous coatings [28]. Comparative studies of AZ91 alloy, $Al_{18}B_4O_{33}$/AZ91 MMC, and SiC/AZ91 MMC reveal that the PEO layers of the first two exhibit uniform pore sizes and distributions, primarily due to the consistency of fine spark discharges. In contrast, the SiC/AZ91 MMC demonstrates larger microarcs, resulting in uneven porosity and microcracks, as well as the highest surface roughness. This discrepancy is attributed to the differences in melting points between $Al_{18}B_4O_{33}$ and SiC, with the order of PEO coating thickness reduction being AZ91 alloy > $Al_{18}B_4O_{33}$/AZ91 MMC > SiC/AZ91 MMC. The lower melting point of $Al_{18}B_4O_{33}$ (1440 °C) allows it to integrate well into the coating, while the presence of SiC increases defects, weakens spark discharge, and promotes current leakage [29].

Additionally, research indicates that the PEO layer of glass microspheres (GM)-reinforced magnesium MMCs exhibits a more uniform microporous structure as the current density increases from 80 mA/cm^2 to 160 mA/cm^2. However, when the current density is further increased to 200 mA/cm^2, larger micropores and cracks appear due to intense spark discharges as shown Fig. 7.5 [30]. Studies on hollow glass microspheres (HGM)/Mg MMC reveal that the PEO coating consists of a loose outer layer and a dense inner layer, with pore sizes ranging from 0.3 μm to 5 μm, uniformly distributed across the surface. The matrix/PEO interface displays a serrated structure, attributed to the uneven dissolution of the matrix [31].

Additionally, Aydin et al. investigated the effect of graphene (GR) on the PEO coatings of MWCNT/Mg composites. The results indicated that the incorporation of GR led to denser and thicker coatings, while also introducing microcracks and small pores [26]. Comparative studies revealed that the presence of the reinforcement phase in the metal matrix composite (MMC) did not significantly affect the phase composition of the PEO layer, which primarily consisted of MgO, MgF_2, and Mg_2SiO_4 [29]. However, Xue et al. demonstrated that the PEO layer formed on SiC/AZ31 MMC using a phosphate electrolyte containing $Mg_3(PO_4)_2$, suggesting the involvement of the reinforcement phase [32]. Similar studies have confirmed the presence of MgO and Mg_2SiO_4 in SiC/AZ92 and MWCNT/Mg composites, with Arrabal et al. proposing that the SiC reinforcement phase contributed to the growth of the coating, resulting in a silicon-rich layer [5]. These findings underscore the importance of optimizing the content of reinforcements in MMCs to achieve dense and uniform PEO coatings.

7.3.2 Breakdown Voltage and Coating Thickness

During the PEO process, the occurrence of spark discharge signifies the breakdown of the barrier film and the initiation of PEO coating growth. Consequently, the breakdown voltage is a critical parameter in the PEO process, directly

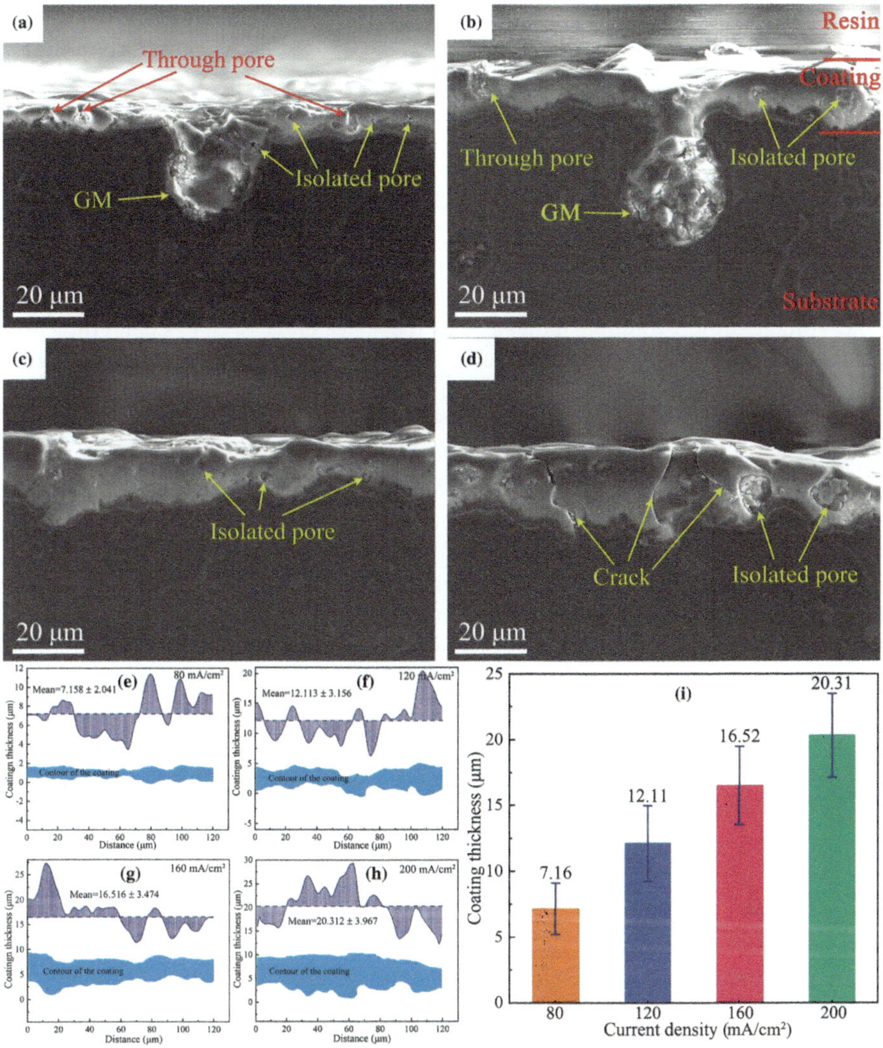

Fig. 7.5 Cross-sectional morphologies of the PEO coatings formed at different current densities: (**a**) 80 mA/cm^2, (**d**) 120 mA/cm^2, (**e**) 160 mA/cm^2, (**f**) 200 mA/cm^2. Panels (**e–h**) show the average thicknesses of the coatings measured from the contour of the cross-section. Panel (**i**) illustrates the influence of current density on the average thickness [30]. *Reproduced with permission from* [30] *© 2021, The Author(s), under exclusive license to Springer Science Business Media, LLC, part of Springer Nature*

corresponding to the appearance of the first visible spark; its magnitude reflects the difficulty of forming the PEO coating on the substrate. Generally, a lower breakdown voltage indicates an easier coating formation, while a higher breakdown voltage corresponds to increased difficulty in coating formation [33].

Figure 7.6 illustrates the relationship between breakdown voltage and coating thickness for SiC_p/AZ91 composites with varying SiC content [34]. As the volume fraction of SiC particles increases, the breakdown voltage rises significantly, while the coating thickness gradually decreases. This indicates that SiC particles exert a suppressive effect on the spark discharge during PEO coating formation, thereby reducing the growth efficiency of the coating. Specifically, as the SiC volume fraction increases from 0% to 2%, the breakdown voltage rises markedly, followed by an approximately linear increase with further increments in SiC content.

The thickness of the PEO coating is closely related to the discharge voltage range and the voltage–time evolution trend. A wider discharge voltage range and a favorable voltage–time evolution indicate that the PEO process can occur efficiently, promoting coating growth on the substrate. The AZ91 alloy exhibits the widest spark discharge voltage range and the best voltage–time curve, resulting in the maximum coating thickness. The presence of SiC particles not only increases the breakdown voltage but also narrows the discharge voltage range, reducing the rate of voltage increment. Consequently, as the SiC content increases, the coating thickness gradually decreases, indicating that SiC particles, as the reinforcement phase in the composite, adversely affect the growth of the PEO coating. Furthermore, the oxidation of SiC particles occurs not only during intense spark discharge but also in the early stages of PEO. Research indicates that within the voltage range of 150–200 V, SiC particles undergo the most vigorous oxidation reactions, even though only weak and invisible spark discharges are present at this time. This phenomenon suggests that the oxidation of SiC initiates before spark discharge, and by transforming

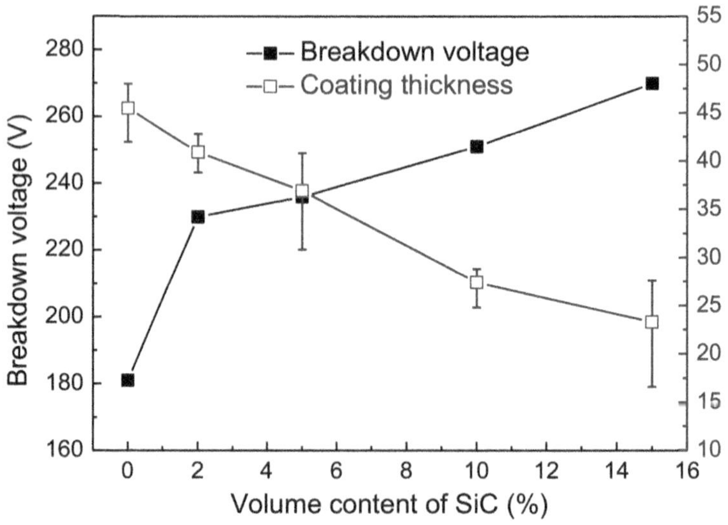

Fig. 7.6 Breakdown voltage and coating thickness of AZ91 Mg alloy and SiCp/AZ91 composites for 30 min PEO treatment at a current density of 15A/dm^2 [34]. *Reproduced with permission from* [34] © *2013 Elsevier B.V.*

semiconductive SiC into insulating SiO_2, it suppresses current leakage at the original SiC locations, thereby leading to increased voltage and spark discharge. Therefore, ensuring the integrity and electrical insulation of the barrier film is essential for the occurrence of spark discharge and coating growth. Compared to the oxidation of SiC during the spark discharge process, the oxidation of SiC particles in the early stages has a more significant impact on coating growth.

7.3.3 Plasma Electron Temperature

Under conditions of local thermal equilibrium (PLTE), the electron temperature (T_e) of the plasma can be determined by the intensity ratio of the Hα (656.2 nm) to Hβ (486.1 nm) spectral lines [35, 36]. In the localized discharge region, the electron temperature is approximately equal to the ion temperature. Nemcova et al. noted that the formation of MgO ceramic films on magnesium alloys is attributed to increased heat generated by larger discharges, which leads to an expanded melting range during the PEO process [37]. The surface morphology of Mg alloy PEO coatings confirms this melting process, as the melting point of MgO is 3125 K, indicating that the temperature in the local discharge zone is likely well above 3000 K.

In phosphate electrolytes, the average T_e for SiCP/AZ31 and Al_2O_3-SiO_2 (sf)/AZ91 magnesium-based composites during plasma discharge is approximately 5500 K, which aligns with the T_e values measured from the H-I line [35, 36] and Mg-I line [38] for magnesium alloys. However, due to self-absorption effects associated with the H-I line, the T_e value obtained from this line is slightly higher. Specifically, the SiC_P/AZ31 MMC exhibits a T_e exceeding 7600 K during the initial discharge phase, indicating a very high spark discharge density; after about 2 min, the average T_e stabilizes at 5500 K, as shown in Fig. 7.7a. For the Al_2O_3–$SiO_{2\ (sf)}$/AZ91 MMC, as the ceramic coating grows, the discharge sparks nearly vanish after 12 min of oxidation. The curve shown in Fig. 7.7b reflects changes in T_e only for the initial 12 min, with an average T_e of approximately 6000 K, which is higher than that for the SiC_P/AZ31 MMC. Moreover, both SiC_P/AZ31 and Al_2O_3-$SiO_{2(sf)}$/AZ91 MMCs exhibit numerous closely spaced temperature peaks, particularly with some peaks in the Al_2O_3–$SiO_{2(sf)}$/AZ91 MMC exceeding 10,000 K, corresponding to intense discharge sparks.

In summary, the T_e results in Fig. 7.7 indicate that the instantaneous temperature in the local plasma discharge zone during the PEO process for SiC_P/AZ31 and Al_2O_3–$SiO_{2(sf)}$/AZ91 MMCs is significantly higher than 5000 K. Given that the melting points of the SiC and Al_2O_3–SiO_2 reinforcements are 3003 K and 2123 K, respectively, both are considerably lower than the plasma temperature in the local discharge zone. Consequently, the reinforcements near the discharge channel may partially melt and participate in the plasma reactions.

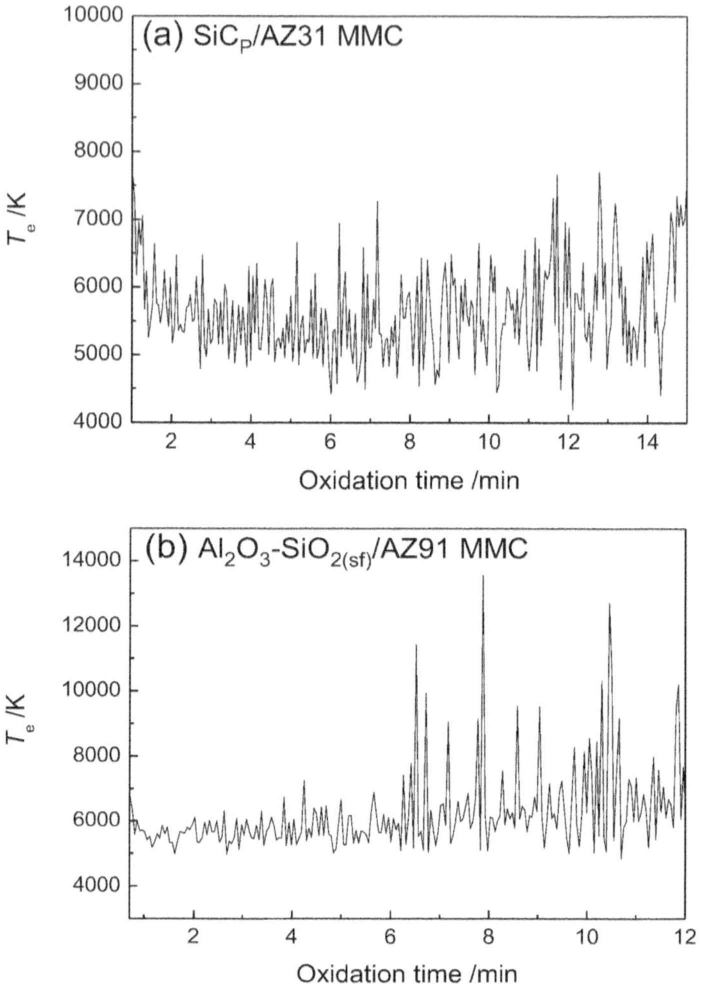

Fig. 7.7 Plasma electron temperature (T_e) as a function of oxidation time. (**a**) SiC_P/AZ31 MMC and (**b**) Al_2O_3–$SiO_{2(sf)}$/AZ91 MMC [39]. *Reproduced with permission from* [39] © *2014 Elsevier B.V.*

7.3.4 PEO Discharge Processes and Growth Kinetics

The growth kinetics of PEO layers on magnesium-based composites (MMCs) have been shown to depend on the material phases [30, 31]. Research on GM-enhanced Mg MMCs indicates that the PEO growth rate for the β-$Mg_{17}Al_{12}$ phase is higher than that for the **α**-Mg phase. During the PEO process, interfacial reaction products (Mg_2Si) and the reinforcement (GM) gradually melt and transform into Mg and Si oxides. As the coating thickens, the depressions caused by the reinforcements are completely covered. The voltage–time curve can be divided into three stages: the first stage involves the electric field promoting substrate dissolution, forming a thin, nonconductive oxide layer; the second stage sees the oxide layer unable to withstand the high-intensity electric field, initiating slight spark discharges; and the third stage is characterized by intense spark discharges and rapid growth of the oxide layer. The inward growth of the PEO layer is typically attributed to the electrochemical dissolution of the matrix and the diffusion of anions to the matrix/PEO interface, while outward growth is associated with the ejection and solidification of molten oxides due to oxide layer breakdown [36].

Liu et al. found that in silicate electrolytes, HGM and Mg_2Si hinder the growth of the initial PEO layer. However, as the PEO duration increases, the reinforcement phase and Mg_2Si are completely covered by the growing PEO layer, as shown Fig. 7.8. During the initial approximately 10 min, the PEO growth rate is low, primarily exhibiting outward growth with silicate oxide deposition as an auxiliary factor. After 10 min, both outward and inward growth rates increase rapidly, resulting in a linear rise in layer thickness. After 30 min, the growth rate of the PEO layer decreases as the increasing coating thickness impedes ion transfer, leading to growth kinetics dominated by inward growth. The layer thickness on the α phase is greater than that on the β phase, as the passivation layer formed by the β phase is more stable, thus hindering dielectric breakdown [31].

Comparative studies of AZ91 magnesium alloy and its MMCs ($Al_{18}B_4O_{33}$ and SiC reinforcements) indicate that the formation of PEO is influenced by the substrate and the reinforcement. The minimum voltages required to initiate PEO for AZ91 alloy, SiC/AZ91 MMC, and $Al_{18}B_4O_{33}$/AZ91 MMC are 150 V, 250 V, and 170 V, respectively. The AZ91 alloy exhibits the largest spark discharge voltage range and the highest mass gain. This suggests that the presence of reinforcements in the MMC reduces the integrity of the barrier layer, thereby delaying the onset of microarc discharge [29]. Wang et al. showed that the SiC reinforcement

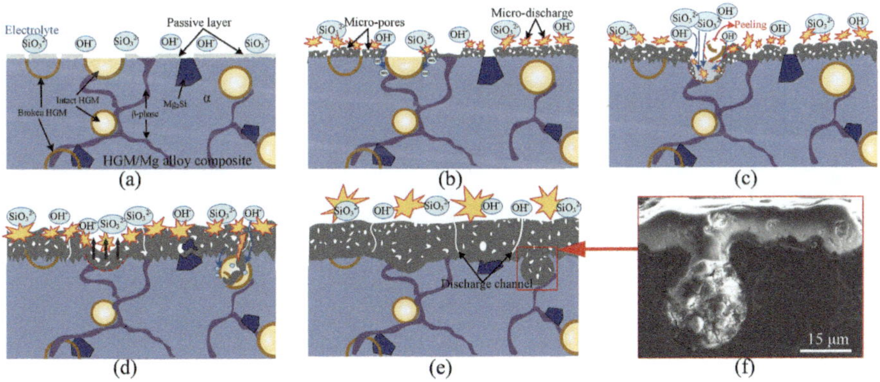

Fig. 7.8 Schematic of the growth process of the PEO coating on the HGM/Mg alloy composite: (**a**) ~10 min (**b**) 15 min (**c**) 20 min (**d**) 25 min (**e**) 30 min [31]. *Reproduced with permission from* [31] © *2021 Elsevier B.V.*

compromises the integrity and electrical insulation properties of the PEO barrier film, thus inhibiting early spark discharges. As the SiC volume fraction increases, the growth efficiency of the PEO layer decreases [34]. At higher concentrations, the high melting point of SiC may hinder the formation of the PEO layer, resulting in thicker coatings that develop micro-cracks and fine pores, as shown Fig. 7.9 [28].

Under the high temperatures within the discharge channel, SiC particles may melt and dissolve into Mg_2SiO_4 at optimal SiC levels, with some smaller particles retained in the PEO layer [32]. Studies on SiC and $Al_2O_3-SiO_2$ reinforced MMCs in phosphate electrolytes indicate that $Al_2O_3-SiO_2$ remains mostly in a molten state at high temperatures and does not oxidize, while it can react with MgO to form composite oxides. After oxidation/melting, reinforcements like SiC react directly with MgO [39]. Research on the formation of PEO coatings on Mg MMCs has identified several distinct process steps: the formation of a surface passivation layer, the dissolution, and release of matrix Mg^{2+} ions, the formation of a $Mg(OH)_2$ layer on the matrix, the outward migration of $Mg2^+$, the inward migration of SiO_32^-, and OH^- to form Mg_2SiO_4, the conversion of $Mg(OH)_2$ to MgO, and the melting and ejection of oxides from the discharge channel, which are subsequently cooled rapidly by the electrolyte, resulting in a porous structure with volcanic features [31].

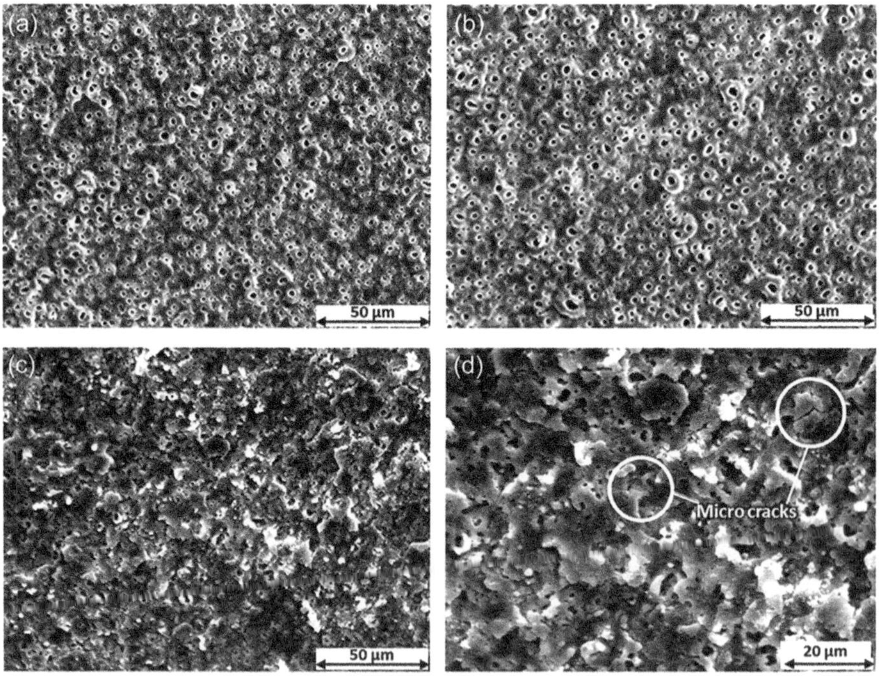

Fig. 7.9 Surface morphology of the plasma electrolytic oxidation (PEO) coatings on nanocomposite samples: (**a**) MM, (**b**) M1Sn, (**c**, **d**) M10Sn in different magnifications [28]. *Reproduced with permission from* [28] © *2022 The Authors. Materials and Corrosion published by Wiley-VCH GmbH*

7.4 Growth Mechanism of PEO on Titanium-Matrix Composites

Titanium-matrix composites (Ti MMCs) have the potential to enhance bioactivity, corrosion resistance, and wear resistance when compared to Ti alloys, with these properties further optimized by PEO under suitable conditions. However, research on the PEO of Ti MMCs remains limited [40–42].

Sousa et al. conducted a study where they fabricated biofunctionalized Al_2O_3/Ti MMCs via PEO for a brief duration (1 min) in a Ca- and P-based electrolyte (a mixture of 0.35 M of calcium acetate monohydrate and 0.02 M of β-glycerophosphate disodium salt pentahydrate). The MMCs, hot-pressed and sintered at two different temperatures (1000 and 1100 °C), were treated with PEO, resulting in the typical volcano-like structure (Fig. 7.10b). Due to the short PEO duration, the reinforcements were still visible on the coated surface. A reaction zone of approximately 6–11 μm thickness was observed around the reinforcement (Fig. 7.10d), and the PEO layer formed on this zone was more heterogeneous than that on the matrix. XRD spectra, Fig. 7.10e, revealed a similar crystalline structure consisting of Ti and TiO_2 in the form of anatase and rutile phases.

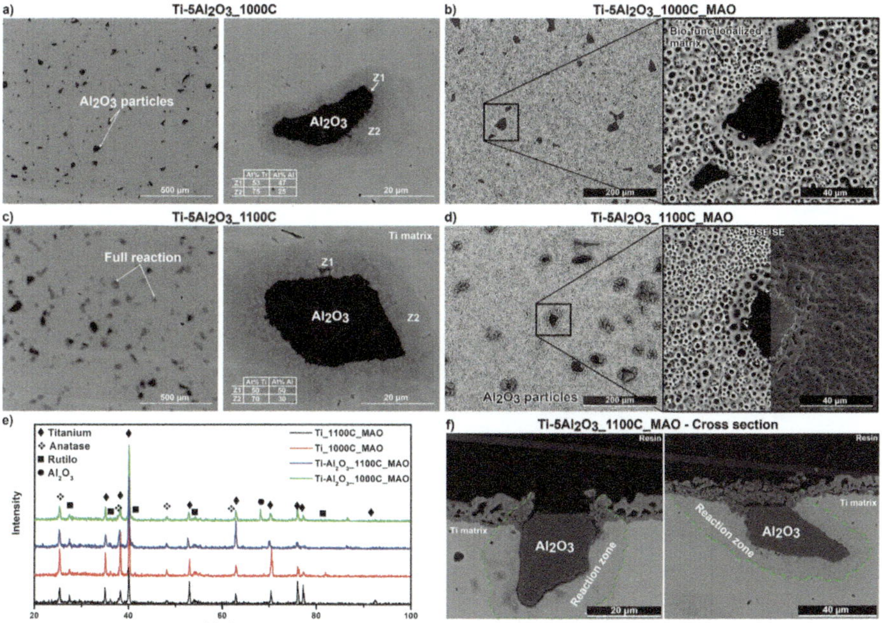

Fig. 7.10 Overall microstructure of the as-processed and biofunctionalized composites, cross-section BSE images from Ti-Al_2O_3_1100C group and XRD spectra for PEO treated groups [40]. *Reproduced with permission from* [40] © *2021 Elsevier B.V.*

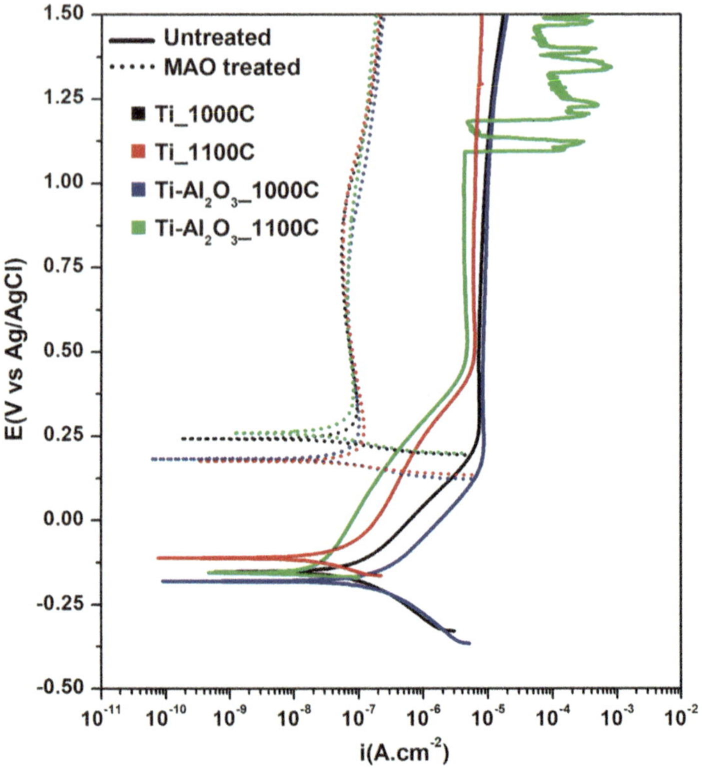

Fig. 7.11 Potentiodynamic polarization curves for all the conditions [40]. *Reproduced with permission from* [40] © *2021 Elsevier B.V.*

Despite this, the PEO coating significantly improved the corrosion resistance, as shown by a nearly two-decade reduction in passive current density (i_{pass}) in Fig. 7.11. The crystalline phases found in the PEO layer included anatase TiO_2, rutile TiO_2, and Al_2O_3, with EDS spectra confirming the incorporation of bioactive Ca and P components. The inherent passive oxides likely contributed to the improved corrosion resistance following the PEO treatment. Additionally, the PEO layer's protective properties were shown to enhance tribocorrosion resistance, as evidenced in Fig. 7.12, where the open circuit potential (OCP) showed minimal fluctuations, indicating negligible mechanical damage during sliding [40].

The same group presents a work aimed to explore the influence of PEO and TO treatments on the surfaces of in situ Ti-TiB-TiC and ex situ Ti-B_4C composites, and to assess their corrosion and tribocorrosion performance. Results indicate that B_4C-depleted surfaces were observed on the ex situ composites, where the typical micro-porous oxide layer formed by PEO was interrupted. A more detailed view of these areas can be observed in the cross-sectional optical OM image presented in Fig. 7.13e. These images illustrate that, during the PEO treatment, certain reinforcing particles were dislodged from the surface. Furthermore, the PEO layer extended

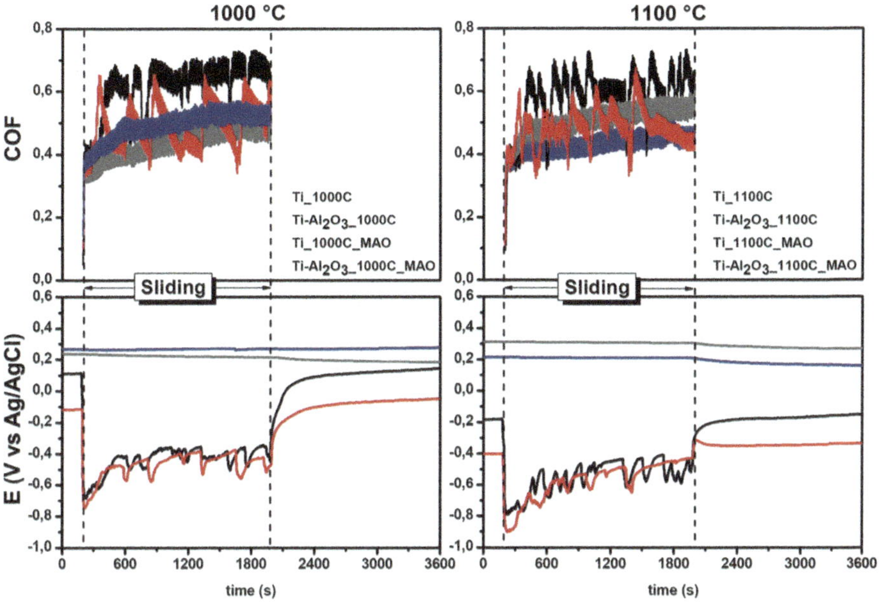

Fig. 7.12 Evolution of OCP before, during, and after sliding, together with the respective COF values registered during sliding [40]. *Reproduced with permission from* [40] © *2021 Elsevier B.V.*

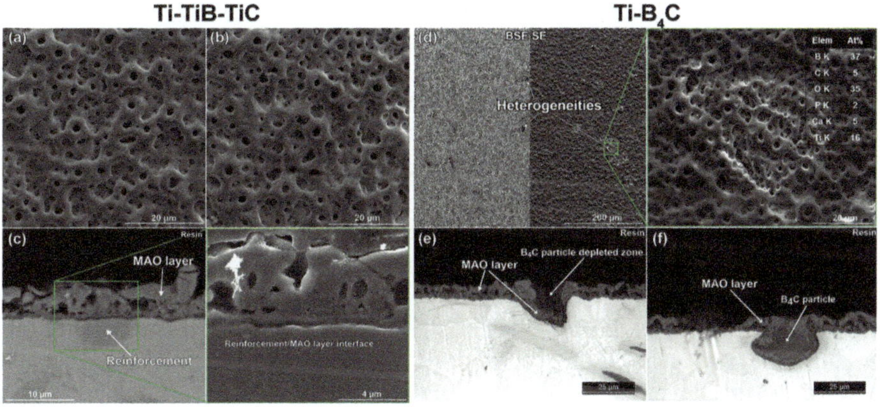

Fig. 7.13 The representative microstructures after PEO treatment: SE images displaying the PEO-treated unreinforced (**a**) Ti and (**b**) in situ composite; (**c**) cross-sectional BSE and SE images of the PEO-treated in situ composite; (**d**) BSE/SE images showing the surfaces of the ex situ composite after PEO; and cross-sectional OM images of the ex situ composites following PEO revealing (**e**) B_4C depleted zone and (**f**) B_4C particle enveloped by the PEO layer [42]. *Reproduced with permission from* [42] © *2024 Elsevier B.V.*

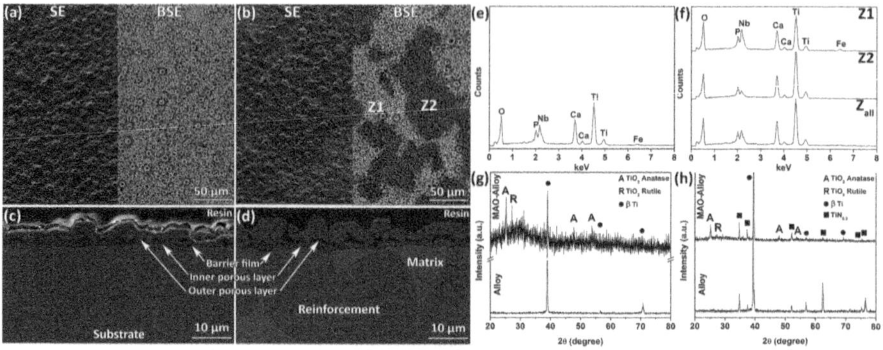

Fig. 7.14 Top view SE/BSE SEM, SE cross-section image, EDS spectra, and XRD spectra of PEO–Alloy in (**a**), (**c**), (**e**), and (**g**), and PEO–MMC in (**b**), (**d**), (**f**), and (**h**), respectively [41]. *Reproduced with permission from* [41] *© 2021 Elsevier B.V.*

over these areas. In addition to these zones, some B_4C particles encapsulated by the micro-porous oxide film were also observed (as shown in Fig. 7.13f). In these instances, the micro-porous oxide layer spread over the reaction zones and enclosed the reinforcing particles.

In a subsequent study, Ihsan Çaha demonstrated that Ti-25Nb-5Fe alloy and TiN-reinforced MMCs also exhibited a similar microporous volcano-like structure after PEO (Fig. 7.14). Similar morphologies were found on both the matrix and reinforcing phase of PEO-coated TiN/MMC (Fig. 7.14a and b). The resulting PEO layer consisted of a dense barrier film, an inner porous layer with fine pores, and an outer porous layer with larger pores (Fig. 7.14c and d). The average overall thickness of anodic layers for PEO–alloy and PEO–MMC samples was approximately 6 and 8 μm, respectively. Moreover, the anodic layer formed on both phases of PEO–MMC samples showed the same triplex structure, where no visible differences were observed. Regarding XRD results (Fig. 7.14e~h), while the XRD pattern of Alloy presented fully β-Ti phase, the MMC composed of β–Ti phase and TiN0.3 phase as in accordance with the EBSD results. In the case of bio-functionalized samples, in addition to their substrate peaks, both PEO-treated samples presented anatase and rutile TiO_2 peaks. This PEO treatment significantly enhanced both corrosion and tribocorrosion resistance [41]. More recent investigations have indicated that reinforcement detachment can occur during PEO treatment of Ti MMCs, likely due to the growth of the oxide layer at the reinforcement/reaction zone interface [42].

Xu et al. utilized PEO treatment on Ti–Mg metal–metal composites to mitigate degradation in simulated body fluid (SBF). Their findings revealed minor differences in the morphologies of the PEO layers formed on the Ti and Mg regions (Fig. 7.15). The PEO coating consisted primarily of amorphous MgO and SiO_2, along with crystalline TiO_2 phases. The PEO layer effectively impeded hydrogen evolution, significantly lowering the degradation rate. The optimal voltage condition for achieving enhanced corrosion resistance through PEO treatment was determined to be 300 V [43].

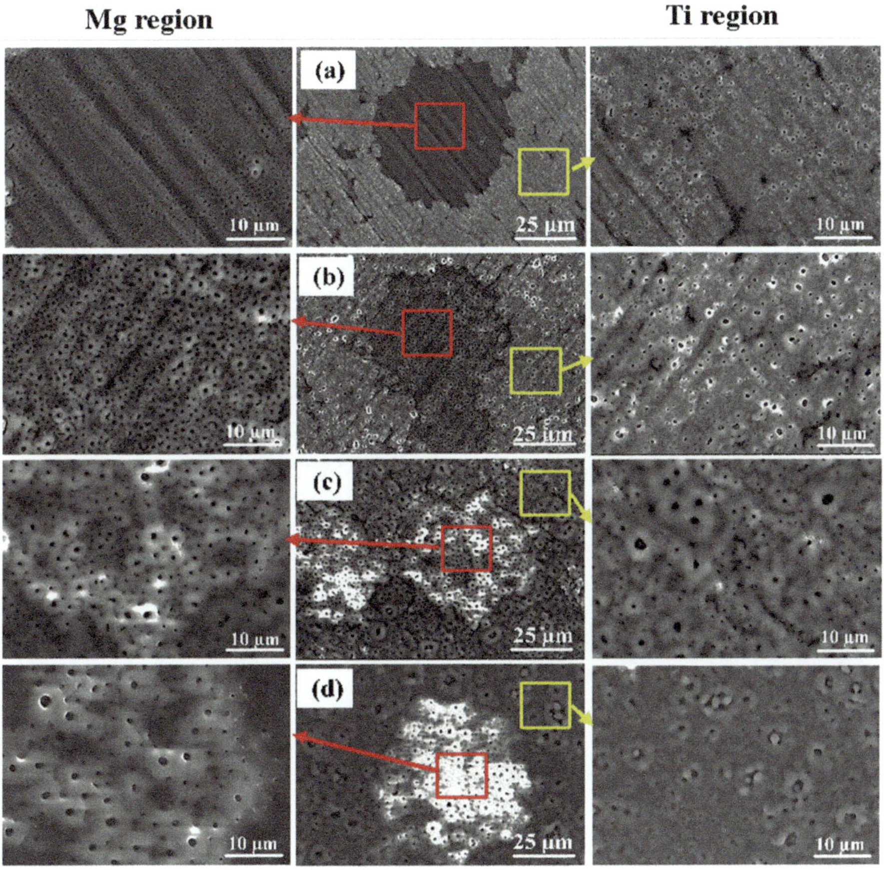

Fig. 7.15 Surface morphologies of the PEO-treated Ti-15 Mg MMC samples under different applied voltages: (**a**) 200 V; (**b**) 250 V; (**c**) 300 V; (**d**) 350 V [43]. *Reproduced with permission from* [43] © *2021 The Authors. Published by Elsevier B.V.*

7.5 Conclusions and Outlook

Despite extensive studies on the PEO growth mechanism in metal-matrix composites, further exploration is essential to clarify the specific influences of electrolyte composition, negative voltage application, and types of reinforcement materials. There is a pressing need for more studies on MMCs of other metals and alloys, beyond the current focus on Al, Mg, and Ti alloys.

1. Firstly, the electrolyte composition plays a critical role in shaping the spark discharge process. Changes in the electrolyte can alter the intensity and frequency of discharges, which in turn impacts the density and structural uniformity of the PEO coating. By optimizing electrolyte formulation, discharge-induced defects can be minimized, thereby enhancing the coating's overall properties.

2. Secondly, the role of negative voltage in promoting soft spark discharges has attracted substantial interest. Unlike positive voltage, applying negative voltage reduces spark intensity and discharge temperature, leading to more uniform and denser coatings. This "soft spark" mode effectively mitigates cracks and defects caused by high temperatures, improving the coating's wear and corrosion resistance.
3. Finally, the reinforcement material type is crucial for PEO coating growth, as different reinforcements exhibit varying levels of reactivity and melting behavior during PEO. Reinforcements with lower melting points or higher reactivity are more likely to contribute to coating growth, enhancing both thickness and density.

In summary, understanding the growth mechanism of plasma electrolytic oxidation is crucial to design and fabricate multifunctional coatings on the surface of metal matrix composites. Incorporating functional nanoparticles from the electrolyte into the PEO layer presents a promising direction for MMC development. Further investigation is also necessary to assess the long-term durability of PEO coatings under diverse environmental conditions.

Acknowledgments The partial support from the Outstanding Youth Project of Natural Science Foundation of Heilongjiang Province (No. YQ2023E008), Young Elite Scientists Sponsorship Program by CAST (2021QNRC001), NSFC Grant nos. 52001100, U21B2053, and Heilongjiang Touyan Team Program is gratefully acknowledged.

References

1. Saji, V. S. (2024). *Chemical Engineering Journal, 498*, 155066.
2. Mortensen, A., & Llorca, J. (2010). *Annual Review of Materials Research, 40*, 243–270.
3. Jiang, C. Y., Wang, Y. M., Wang, S. Q., Li, Y. F., Zou, Y. C., Ouyang, J. H., Jia, D. C., & Zhou, Y. (2022). *Surface and Coating Technology, 444*, 128692.
4. Zou, S. Y., Zhou, X. L., Rao, Y. H., Hua, X. Z., & Cui, X. (2019). *The Journal of Alloys and Compounds, 780*, 937–947.
5. Arrabal, R., Matykina, E., Skeldon, P., & Thompson, G. (2009). *Applied Surface Science, 255*, 5071–5078.
6. Yu, Z. Y., Zheng, K. H., Li, X. T., Xia, P., Xu, J., Sun, J. X., Zhou, N., & Pan, F. S. (2023). *The Journal of Materials Research and Technology, 26*, 7395–7411.
7. Hihara, L., & Latanision, R. (1994). *International Materials Review, 39*, 245–264.
8. Yerokhin, A., Mukaeva, V. R., Parfenov, E. V., Laugel, N., & Matthews, A. (2019). *Electrochimica Acta, 312*, 441–456.
9. Curran, J. A., & Clyne, T. W. (2005). *Surface and Coating Technology, 199*, 168–176.
10. Aliakseyeu, Y. G., Korolyov, A. Y., Niss, V. S., Parshuto, A. E., & Budnitskiy, A. S. (2018). *Science and Technology, 17*, 211–219.
11. Darband, G. B., Aliofkhazraei, M., Hamghalam, P., & Valizade, N. (2017). *Journal of Magnesium and Alloys, 5*, 74–132.
12. Kaseem, M., Fatimah, S., Nashrah, N., & Ko, Y. G. (2021). *Progress in Materials Science, 117*, 100735.

13. Wang, S. X., Liu, X. H., Yin, X. L., & Du, N. (2020). *Surface and Coating Technology, 381*, 125214.
14. Jiang, C. Y., Wang, Y. M., Wang, S. Q., Zou, Y. C., Ouyang, J. H., Jia, D. C., & Zhou, Y. (2023). *Materials Characterization, 205*, 113344.
15. Clyne, T. W., & Troughton, S. C. (2019). *International Materials Review, 64*, 127–162.
16. Hussein, R. O., Nie, X., Northwood, D. O., Yerokhin, A., & Matthews, A. (2010). *Journal of Physics D, 43*, 105203.
17. Jiang, C. Y., Wang, Y. M., Wang, S. Q., Zou, Y. C., Ouyang, J. H., Jia, D. C., & Zhou, Y. (2023). *Surface and Coating Technology, 466*, 129578.
18. Bahador, R., Hosseinabadi, N., & Yaghtin, A. (2021). *Journal of Materials Engineering and Performance, 30*, 2586–2604.
19. Jiang, C. Y., Wang, Y. M., Wang, S. Q., Zou, Y. C., Ouyang, J. H., Jia, D. C., & Zhou, Y. (2023). *Applied Surface Science, 636*, 157789.
20. Jaspard-Mecuson, F., Czerwiec, T., Henrion, G., Belmonte, T., Dujardin, L., Viola, A., & Beauvir, J. (2007). *Surface and Coating Technology, 201*, 8677–8682.
21. Wang, Y., Sheng, L., Xiong, R., & Li, B. (1999). *Surface Engineering, 15*, 112–114.
22. Du, C. Y., Huang, S. T., Yu, X. L., Wang, Q. Z., & Zhao, H. (2021). *Surface and Coating Technology, 420*, 127321.
23. Zhang, H. B., Geng, J. W., Li, X. F., Chen, Z., Wang, M. L., Ma, N. H., & Wang, H. W. (2017). *Applied Surface Science, 422*, 359–371.
24. Zhu, M. Y., Song, Y. W., Xu, J. N., Dong, K. H., & Han, E. H. (2023). *Surface and Coating Technology, 457*, 129334.
25. Morgenstern, R., Sieber, M., Lampke, T., & Iop. (2016). *Plasma electrolytic oxidation of AMCs* (Vol. 118, p. 012031). Iop Publishing Ltd.
26. Aydin, F., Ayday, A., Turan, M. E., & Zengin, H. (2020). *Surface Engineering, 36*, 791–799.
27. Wang, Y. Q., Zheng, M. Y., & Wu, K. (2005). *Materials Letters, 59*, 1727–1731.
28. Golhin, A. P., Kamrani, S., Fleck, C., & Ghasemi, A. (2022). *Werkstoffe und Korrosion, 73*, 1813–1825.
29. Wang, Y. Q., Wu, K., & Zheng, M. Y. (2006). *Surface and Coating Technology, 201*, 353–360.
30. Liu, L., Yu, S. R., Zhu, G., Li, Q., Liu, E. Y., Xiong, W., Wang, B. Y., & Yang, X. Z. (2021). *Journal of Materials Engineering and Performance, 56*, 15379–15396.
31. Liu, L., Yu, S. R., Liu, E. Y., Zhao, Y., Wang, B. Y., Niu, Y. F., Zhang, K., Zhu, G., & Li, Q. (2021). *Materials Chemistry and Physics, 271*, 124935.
32. Xue, W., Jin, Q., Zhu, Q., Hua, M., & Ma, Y. (2009). *Journal of Alloys and Compounds, 482*, 208–212.
33. Verdier, S., Boinet, M., Maximovitch, S., & Dalard, F. (2005). *Corrosion Science, 47*, 1429–1444.
34. Wang, Y. Q., Wang, X. J., Gong, W. X., Wu, K., & Wang, F. H. (2013). *Applied Surface Science, 283*, 906–913.
35. Wang, L., Chen, L., Yan, Z. C., & Fu, W. (2010). *Surface and Coating Technology, 205*, 1651–1658.
36. Hussein, R. O., Nie, X., & Northwood, D. O. (2013). *Electrochimica Acta, 112*, 111–119.
37. Nemcová, A., Skeldon, P., Thompson, G. E., & Pacal, B. (2013). *Surface and Coating Technology, 232*, 827–838.
38. Jovovic, J., Stojadinovic, S., Sisovic, N. M., & Konjevic, N. (2012). *Journal of Quantitative Spectroscopy & Radiative Transfer, 113*, 1928–1937.
39. Liu, R., Weng, N., Xue, W. B., Hua, M., Liu, G. J., & Li, W. F. (2015). *Surface and Coating Technology, 269*, 212–219.
40. Sousa, L., Basilio, L., Alves, A. C., & Toptan, F. (2021). *Surface and Coating Technology, 420*, 127329.
41. Çaha, I., Alves, A. C., Chirico, C., Pinto, A. M., Tsipas, S., Gordo, E., & Toptan, F. (2021). *Surface and Coating Technology, 415*, 127122.

42. Sousa, L., Costa, N. A., Rossi, A., Simoes, S., Toptan, F., & Alves, A. C. (2024). *Surface and Coating Technology, 485*, 130854.
43. Xu, J. L., Tang, J., Hu, J., Zhang, J. L., Luo, J. M., Hao, G. D., Han, S. L., & Liu, X. (2021). *Journal of Materials Research and Technology, 11*, 1654–1664.

Chapter 8
Cathodic Plasma Electrolytic Oxidation on Valve and Non-valve Metals

Xiaoyue Jin and Wenbin Xue

8.1 Introduction

Plasma electrolysis is a green and highly efficient surface modification method on metals [1–4]. It includes various technologies, such as plasma electrolytic oxidation (PEO) or microarc oxidation (MAO) [5, 6], cathodic microarc electrodeposition [7], plasma electrolytic saturation [8–11], and plasma electrolytic cleaning [12, 13].

In recent years, researchers have found that iron-based materials could be oxidized by the cathodic vapor gaseous envelope discharge to form dense oxide coatings under certain conditions. These coatings possess high hardness, thereby improving the wear and corrosion resistance of iron-based materials [14]. This innovative technique, which has revolutionized conventional plasma electrolysis, is referred to as cathodic plasma electrolytic oxidation (CPEO) to differentiate it from the traditional anodic plasma electrolytic oxidation (PEO) technology. This chapter reviewed the basic principles of the CPEO process, analyzed the morphology, structure, and properties of CPEO coatings on carbon steel, stainless steel, molybdenum (Mo), and titanium aluminide (TiAl) alloys, and discussed the growth mechanisms of CPEO coatings. It also described the synthesis of diamond-like carbon (DLC) components in composite oxide coatings and provided an outlook on the development prospects and future research directions of CPEO technology.

X. Jin
Beijing Academy of Science and Technology, Beijing, China
e-mail: jinxiaoyue@bjast.ac.cn

W. Xue (✉)
Beijing Normal University, Beijing, China
e-mail: xuewb@bnu.edu.cn

V. S. Saji, M. Zheludkevich (eds.), *Plasma Electrolytic Oxidation – Fundamentals, Advances and Applications*,
https://doi.org/10.1007/978-3-031-92429-3_8

8.2 Cathodic Plasma Discharge, Voltage, and Current

The discharge process of CPEO process is similar to that of cathodic plasma electrolytic carburizing and nitriding, and the reactions in CPEO process are more vigorous than those in PEO process [15]. Aqueous solutions of glycerin, urea, and borax are commonly used as electrolytes, with the addition of a small amount of inorganic salts, such as KCl and NH_4Cl, to adjust the electrical conductivity and maintain an appropriate voltage range [14, 16, 17]. CPEO process can be performed with a direct current (DC) or a pulsed power supply. When a pulsed power supply is used, the pulse frequency, duty ratio, and pulse width can be modulated, making the discharge process more suitable for the CPEO process [14, 18, 19].

Stojadinović et al. [19] investigated the CPEO process of Mo in a glycerin + borax solution. The current density–voltage curve of the Mo cathode and real-time photographs of the cathode at several typical voltages are shown in Fig. 8.1. During the initial rise in the applied voltage, the current increases linearly with an increase in voltage, which is consistent with Ohm's law (see Fig. 8.1a). With a further increase in voltage, the Joule heating effect causes local vaporization of the electrolyte solution around the cathode, leading to the formation of a large amount of water vapor bubbles. The electrochemical reaction also releases a certain amount of hydrogen bubbles from the cathode. Accumulation of these small vapor bubbles leads to the formation of a continuous vapor gaseous envelope (VGE) (typical thickness: approximately 0.1 mm) to isolate the Mo cathode from the electrolyte solution (see Fig. 8.1bii). When the current reaches a maximum value, the high electric field of the VGE ionizes the cathode and causes discharge at the edge of the cathode (see Fig. 8.1biii), and the corresponding voltage is the breakdown voltage of the vapor gaseous envelope. Subsequently, the current decreases rapidly and then stabilizes at a lower value. At this time, the Mo cathode surface is covered by red arcs (Fig. 8.1bv),

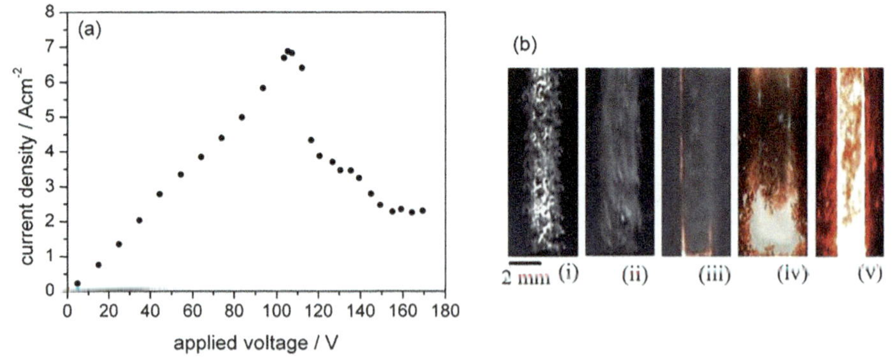

Fig. 8.1 (**a**) Correlation between current density and voltage during CPEO process of molybdenum, (**b**) Real-time images of CPEO process at an applied voltage of (i) 5 V, (ii) 30 V, (iii) 100 V, (iv) 115 V, and (v) 160 V [19]. (Reproduced with permission from [19] Rights managed by AIP Publishing)

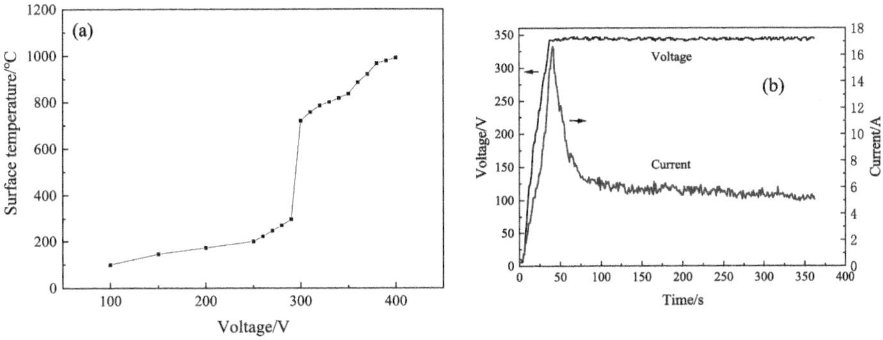

Fig. 8.2 (**a**) Near-surface temperature of TiAl cathode under different applied voltages, (**b**) Time-dependent applied voltage and current in CPEO process [18]. (Reproduced with permission from [18] © 2022 Elsevier B.V.)

and stable CPEO discharge can be achieved by maintaining this stage until the end of the CPEO process. Figure 8.1bv shows that the stable discharge occurs at an applied voltage of 160 V. The suitable CPEO voltage depends on various factors, such as electrolyte composition, electrolyte concentration, voltage waveform, and cathode material. More exploration is required to determine the suitable voltage range of the CPEO process.

Gao et al. [18] inserted a thermocouple into a cylindrical TiAl cathode to a depth of 0.5 mm from the bottom to measure the near-surface temperature when the cathode was placed in a 50% glycerin solution at different applied voltages. Figure 8.2a shows that the near-surface temperature of the cathode gradually rises when the voltage is increased up to 300 V. At 300 V, the TiAl cathode temperature increases sharply from 300 to 700 °C. Under the high voltage, the vapor gaseous envelope is broken down to cause surface discharge around the TiAl cathode, which heats the sample surface and forms a temperature gradient from the surface to the interior of the TiAl sample, thereby resulting in a tremendous increase in the near-surface temperature. At 400 V, the rate of temperature rise declines, whereas a near-surface temperature of up to 1000 °C is observed. A larger temperature gradient exists within the CPEO coating, and the interior temperature at 2 mm away from the surface reduces to 720 °C. Therefore, the near-surface temperature is much higher than the interior temperature inside the sample [20]. According to Fig. 8.2a, 300–360 V might be a suitable applied voltage range for TiAl during the CPEO process.

Figure 8.2b shows the trends of applied voltage and current with oxidation time. During the initial stage, the current of the TiAl cathode increases rapidly with an increase in voltage, which fulfills Ohm's law. Breakdown and discharge occur in the VGE on the cathode surface at approximately 30 s, and both the voltage and current reach their maximum values at approximately 40 s. Subsequently, the voltage remains relatively constant while the current decreases rapidly. Stable discharge occurs after 80 s, when the current remains relatively unchanged until the end of CPEO process.

8.3 CPEO on Non-valve Metals

8.3.1 Growth and Properties of CPEO Coatings on Low-Carbon Steels

Jin and Xue [14, 17, 20] reported that the outer layer of CPEO coatings on different Fe-based materials exhibited columnar crystal characteristics, whose diameter increased with an increase in discharge time and applied voltage.

As shown in Fig. 8.3a, the intact crystal grain structures are clearly observed on the surface of the CPEO coating after 3 min oxidation at 420 V on Q235 low-carbon steel in glycerin solution [20]. The average grain size of columnar crystals is about 20 μm after 3 min oxidation treatment and increases to about 30 μm after 5 min treatment. The grain boundaries in the oxide coating are completely covered by iron oxide after 9-min treatment. Figure 8.3b shows strong coating/substrate binding in the cross-sectional microstructure of the CPEO coating after 3 min oxidation treatment. It is also observed that carbon ions were produced from glycerin degradation and aggregated in the oxide coating to form carbon clusters (black areas indicated by white arrows). The thickness of the CPEO coating on Q235 low-carbon steel follows a nearly linear growth trend (45 μm, 70 μm, and 110 μm after 3 min, 5 min, and 9 min treatment, respectively). This indicates that rapid growth occurred in the CPEO coating under the prevailing discharge conditions and electrolytic environment [20]. The CPEO coating is mainly composed of the FeO phase with minor randomly distributed carbon particles. After a longer treatment, the Fe_2O_3 phase is also detected in the CPEO oxide coating.

Figure 8.4 shows the microhardness distribution and the electrochemical impedance spectra of Q235 low-carbon steel after CPEO treatment for 3, 5, and 9 min [20]. The microhardness values of the CPEO coatings are significantly higher than that of the Q235 low-carbon steel substrate. Meanwhile, they show minimal changes after different treatment durations. Their average microhardness values at 3 min,

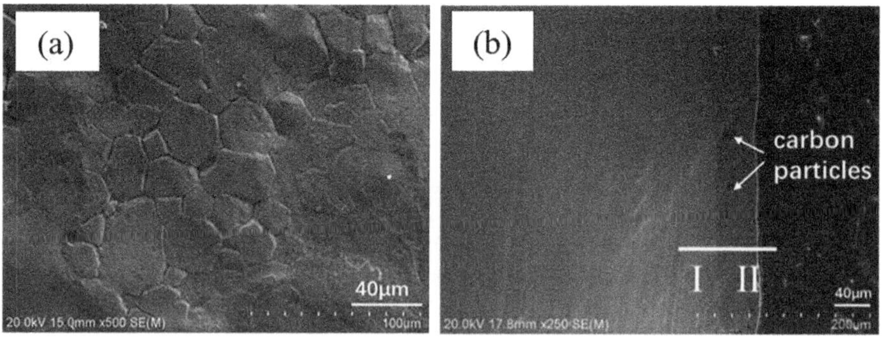

Fig. 8.3 Morphologies of CPEO coating on Q235 low-carbon steel after 420 V/3 min treatment in glycerin electrolyte solution. (**a**) Surface morphology, (**b**) cross-sectional microstructure [20]. (Reproduced with permission from [20] © 2019 Elsevier B.V.)

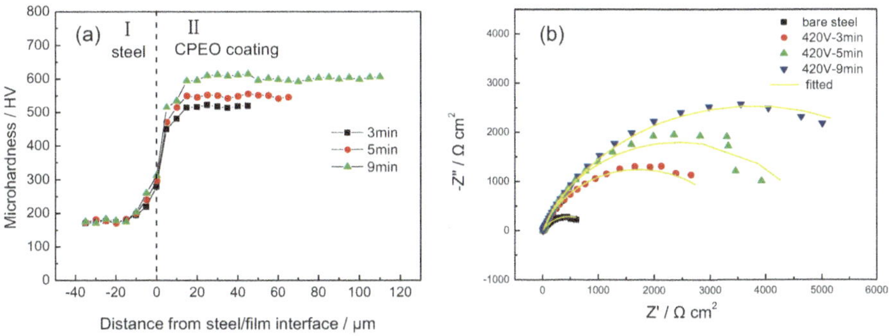

Fig. 8.4 (**a**) Microhardness distributions and (**b**) electrochemical impedance spectra of CPEO-treated Q235 low-carbon steel at different discharge times [20]. (Reproduced with permission from [20] © 2019 Elsevier B.V.)

5 min, and 9 min are 520 HV, 550 HV, and 605 HV, respectively. The Nyquist plots in Fig. 8.4b show that CPEO coatings exhibit considerably larger diameters of the capacitive loop than the low-carbon steel substrate, demonstrating that CPEO treatment enhances the corrosion resistance of Q235 low-carbon steel. The diameter of the capacitive loop of the CPEO-treated low-carbon steel increases continuously with the prolonged oxidation time, suggesting that the corrosion resistance improves with the oxidation time.

8.3.2 Growth and Properties of CPEO Coatings on Stainless Steels

Compared with carbon steel, stainless steel possesses higher Cr and Ni contents, which require different CPEO conditions. It was reported by Jin and Xue that the CPEO coatings could grow on stainless steel cathode in both glycerin or and borax + glycerin system [14, 17, 21]. Figure 8.5 indicates that an oxide coating of approximately 180 μm thick is obtained after CPEO treatment on 304 stainless steel at 260 V for 30 min [14]. A double-layered structure is formed in the CPEO oxide coating on stainless steel, with the outer layer being deficient in Cr and Ni and the dense inner layer enriched in Cr and Ni. The outer layer is composed of Fe_3O_4 columnar crystals, and the grain diameter increases with the voltage and discharge time. By contrast, the inner layer mainly consists of $FeCr_2O_4$, $NiFe_2O_4$, Fe_3O_4, and FeO phases with a dense and uniform structure with good adhesion to the steel substrate. The microhardness is approximately 600 HV for the outer layer and increases sharply to 1335 HV for the dense inner layer (see Fig. 8.5d). The maximum hardness of the CPEO coatings on stainless steels increases with the rise in voltage or oxidation time, which is significantly higher than those for CPEO coatings on carbon steels [14]. The high microhardness values in the inner layer are closely related to the presence of spinel $FeCr_2O_4$ and $NiFe_2O_4$ phases.

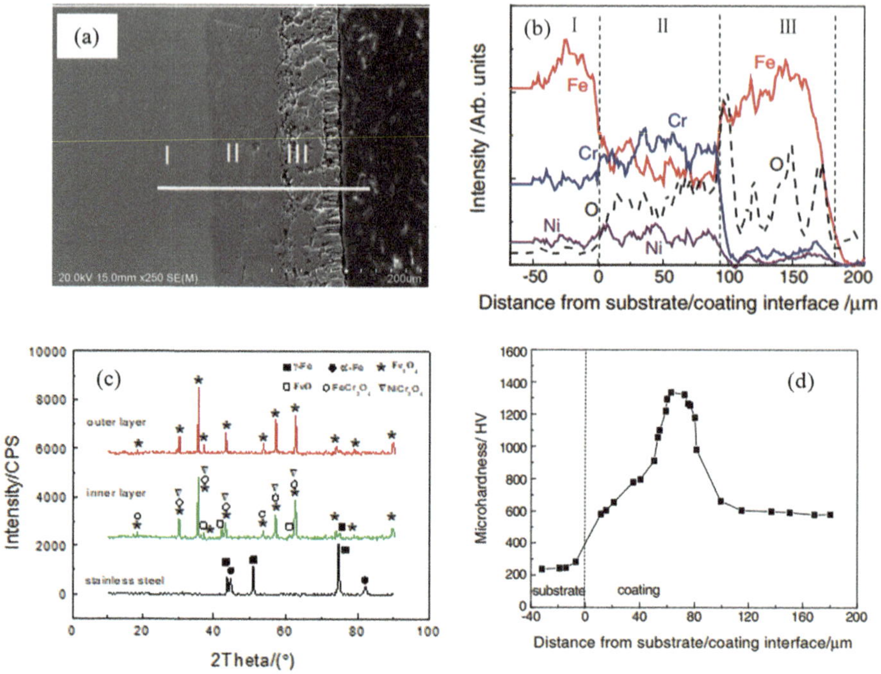

Fig. 8.5 (**a**) Cross-sectional microstructure, (**b**) composition line scans, (**c**) XRD patterns, (**d**) microhardness distribution of the CPEO coating on 304 stainless steels after 260 V/30 min discharge treatment. I steel substrate, II inner layer of coating, III outer layer of coating [14]. (Reproduced with permission from [14] © 2013 Elsevier B.V.)

Jin et al. [17] measured the near-surface temperature inside a stainless-steel cathode in borax and glycerin solution under different voltages. At 240 V, 260 V, and 280 V, the near-surface temperature during the CPEO process is 1002 °C, 1022 °C, and 1042 °C, respectively. Based on the coating thickness at different voltages and oxidation times, their oxidation rate constants are calculated as 1.34×10^{-11} m^2/s, 2.10×10^{-11} m^2/s, and 2.99×10^{-11} m^2/s, respectively. The activation energy of oxidation in the CPEO process of stainless steel is further determined as 170.5 kJ/mol.

Figure 8.6 shows the time-dependent friction coefficients of the stainless-steel substrate with and without a 260 V/5 min CPEO coating measured at room temperature (20 °C), 300 °C, and 500 °C. For the stainless-steel substrate (Fig. 8.6a), the wear experiment at 300 °C exhibits distinct fluctuations. The friction coefficient decreases gradually from 0.6 to 0.4 in the initial 120 s, then remains stable for approximately 100 s, increases gradually to 0.5, and remains stable again. In the 500 °C, the friction coefficient decreases gradually from the initial 0.65 to the lowest 0.55 at 60 s and remains stable thereafter, which demonstrates the intrinsic high-temperature wear characteristics of the stainless-steel material. The low friction coefficient of the stainless-steel substrate under high temperatures is attributed to the oxidation on the surface, which leads to the formation of a thin oxide layer with

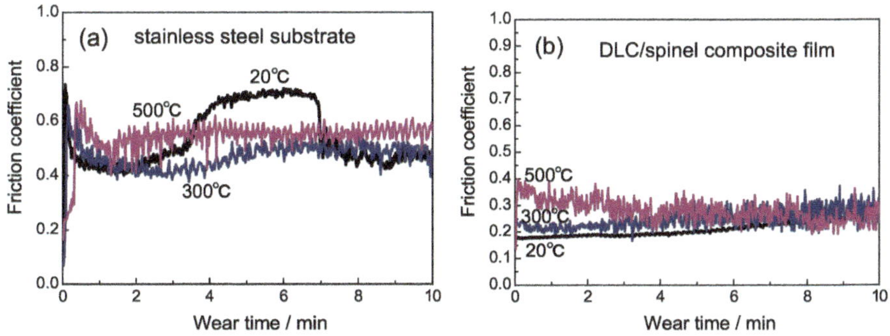

Fig. 8.6 Time-dependent friction coefficients of (**a**) bare stainless-steel substrate and (**b**) 260 V/5 min CPEO coating on a stainless-steel substrate. The experiments were performed using a ZrO_2 ball with a load of 10 N at room temperature (20 °C), 300 °C and 500 °C [21]. (Reproduced with permission from [21] © 2018 Elsevier B.V.)

higher hardness than the steel substrate. In addition, the wear debris generated by high-temperature friction is prone to reshaping at the wear defect sites and provides a lubricating effect. During the high-temperature wear experiment, the friction coefficient increases with a rise in experimental temperature.

As shown in Fig. 8.6b, the friction coefficient of the CPEO coating is approximately 0.175 at room temperature and exhibits little fluctuation with time. At 300 °C, the CPEO coating has an initial friction coefficient of 0.2, which remains stable in the early stage of the wear test and gradually increases with some fluctuation, indicating wear aggravation. When the temperature is increased to 500 °C, the friction coefficient is reduced because a small amount of plastic deformation occurs in the oxide coating at a high-temperature environment. Therefore, the friction coefficient gradually decreases from the initial 0.35 to 0.25 and remains stable thereafter. Comparing Fig. 8.6a, b, the coefficient of friction of the 260 V/5 min-CPEO coating is lower than that of the stainless-steel substrate at the corresponding temperatures.

Figure 8.7 shows that the CPEO coating on stainless steel possessed better anti-wear properties in room temperature and high-temperature environments. For the oxide coating formed at 260 V after 5–30 min of discharge, its friction coefficient is significantly lower than that of the stainless-steel substrate. With the prolongation of discharge time, the average hardness of the coating increases, and the wear rate reduces. At 30 min, the wear rate of the CPEO coating was only 1/128 that of the substrate (see Fig. 8.6a) [14]. The friction coefficient of the CPEO coating varies in the range of 0.18–0.4, which is significantly lower than that of the stainless-steel substrate and less affected by the temperature (20–500 °C). At room temperature, the wear rate of the CPEO coating is 2.39×10^{-5} mm^3/N m, which is only 1/8 that of the stainless steel substrate, and the mechanism of wear for the oxide coating is adhesive wear. However, at 500 °C, the wear rate of the CPEO coating is increased to 1.56×10^{-4} mm^3/N m, which was approximately 1/5 that of the stainless steel substrate, and the mechanism of wear involves adhesive and fatigue wear [21].

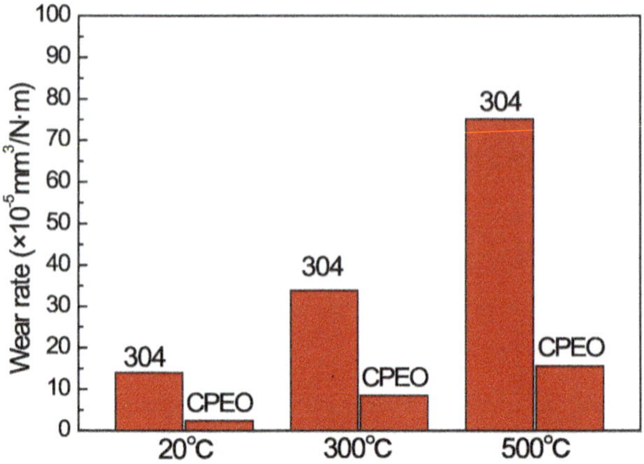

Fig. 8.7 Wear rate of stainless steel and 260 V/5 min CPEO coating at room temperature and high temperatures [14, 21]. (Reproduced with permission from [21] © 2018 Elsevier B.V.)

8.3.3 Growth of CPEO Coatings on Molybdenum

Stojadinović et al. [19] investigated the discharge behavior on the Mo cathode in borax + glycerin aqueous solution, which is similar to the electrolyte solution for the CPEO of stainless steel and electrolytic saturation processes [10, 14, 16, 22]. The optical emission spectroscopy (OES) spectrum (see Fig. 8.8a) reveals that an oxide coating had grown on the Mo surface through the CPEO process. It is observed that the multiple atomic lines of Mo (Mo I) are present in addition to the sodium (Na) and hydrogen (H) peaks. The electron concentration is calculated as 1.4×10^{21} m^{-3} based on the spectral line shape analysis of the hydrogen Balmer lines, indicating that local equilibrium has been reached in the discharge zone. By measuring the intensity of the Mo atomic lines, the highest electron temperature in the plasma channel is calculated as 15,000 K. The thickness of the CPEO coating on the Mo was 4.4 μm after discharge at 160 V for 10 min (see Fig. 8.8b). X-ray diffraction (XRD) analysis confirms that the coating is primarily composed of a MoO_3 phase.

8.4 CPEO Treatment on Valve Metals

The CPEO process was reported during electrical discharge on the stainless-steel cathode in the first time [23]. Researchers subsequently found that the CPEO process could also be applied to carbon steels and Mo [16, 19]. Recently, it was reported that in addition to non-valve metals, valve metals such as TiAl alloy can also be used for the preparation of oxide coatings by the CPEO process [18, 24]. Gui et al. [24] observed that oxide coatings with a thickness of 8–12 μm were formed on the

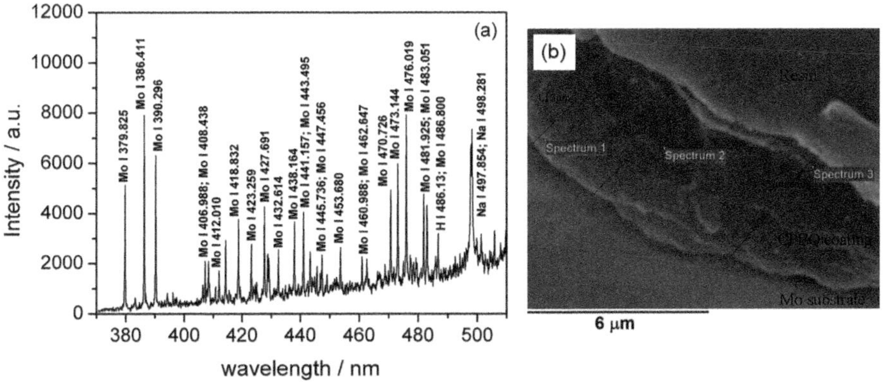

Fig. 8.8 (**a**) Optical emission spectrum and (**b**) cross-sectional image of the CPEO coating on Mo [19]. (Reproduced with permission from [19] Rights managed by AIP Publishing)

surface of high Nb content TiAl alloy (Ti-45Al-8Nb) cathode after 180 s discharge in 5–10% $NaHCO_3$ electrolyte solution in 110–130 V. The obtained CPEO coatings were composed of TiO_2, TiO, Ti_2O_3, Ti_2O_5, and Al_2O_3 phases.

PEO is actually anodic plasma electrolytic oxidation (APEO), and its coating growth results from the breakdown discharge of insulation oxide film on valve metals. CPEO is different from the PEO process in several aspects. Usually, CPEO involves higher applying voltage, higher current, more severe spark discharge phenomenon, shorter duration, and higher plasma temperature. The breakdown discharge of vapor gaseous envelope and the highly active particles in discharge plasma envelope result in the rapid growth of CPEO coatings on valve and non-valve metals.

Gao et al. [18] obtained CPEO coatings with a thickness of 10–26 μm on the surface of a γ-TiAl alloy (Ti-48Al-2Cr-2V) after 5 min discharge in aqueous glycerin solution under different voltages within 300–340 V. Figure 8.9 shows the cross-sectional microstructure of a typical CPEO coating, which is dense and adheres strongly to the TiAl substrate. Both the thickness and surface roughness of the coatings increases with the applied voltage. A higher voltage leads to a greater intensity of vapor gaseous envelope (VGE) breakdown and discharge, thus elevating the near-surface temperature within the sample and increasing the thickness of the oxide coating. The XRD patterns in Fig. 8.10a indicates that the CPEO coatings on the γ-TiAl alloy under different voltages are composed of a complex oxide (Al_2TiO_5) and anatase (A-TiO_2), rutile (R-TiO_2) and α-Al_2O_3 phases. The weak peak at $2\theta = 26°$ is related to carbon. At $2\theta = 5°–20°$, a broad diffraction peak is observed in addition to an obvious Al_2TiO_5 diffraction peak. This is attributed to the formation of a large amount of microcrystalline phases of titanium oxide and aluminum oxide, and amorphous carbon in the outer layer of the CPEO coatings. In contrast, this broad peak is not observed in the XRD patterns of PEO coatings formed on TiAl alloy anodes [25–28], suggesting the significant differences in the formation mechanisms of CPEO coatings and traditional anodic PEO coatings on TiAl alloy.

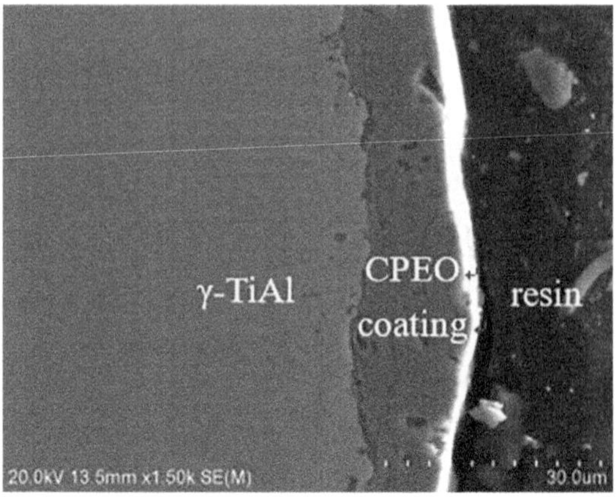

Fig. 8.9 Cross-sectional microstructure of the CPEO coating on γ-TiAl alloy [18]. (Reproduced with permission from [18] © 2022 Elsevier B.V.)

As shown in Fig. 8.10, the average microhardness of the γ-TiAl alloy substrate is 327 HV, whereas that of the CPEO coating exceeds 600 HV. It demonstrates that the CPEO treatment considerably improved the hardness of TiAl alloys. The microhardness of CPEO coatings is further enhanced with an increase in applied voltage; however, the extent of the increase is limited due to the identical composition of CPEO coatings formed under different conditions. Gao et al. [18] found that the CPEO surface treatment increased the contact angle of γ-TiAl to water from 83.8° to 110.5°. The Al_2TiO_5, TiO_2, and Al_2O_3 ceramic phases in the CPEO coatings confer hydrophobicity to the coatings, which leads to the transformation from hydrophilicity to hydrophobicity. In addition, the transient high temperature caused by discharge could decompose glycerin, leading to the deposition of carbon from the degraded glycerin onto the coating surface and further enhancing the hydrophobicity of the coatings. The surface energies of the CPEO coatings obtained at different voltages are calculated based on the contact angles with water and n-hexadecane. As shown in Fig. 8.10c, CPEO-treated oxides show lower surface energies (26–28 mJ/m^2) than γ-TiAl alloy substrate (32.56 mJ/m^2), and the surface energy value decreases with an increase in voltage.

During the microarc oxidation (MAO) process (i.e., anodic PEO coating growth), local breakdown and discharge occur in the insulation coating on the surface of valve metals, generating a large number of moving sparks. The local temperature of the spark discharge zone may reach several thousand kelvins, but the near-surface temperature within the metal remains rather low, approximately <120 °C in thicker PEO coatings [29]. However, a plasma envelope is formed by vapor-gaseous envelope (VGE) breakdown on the cathode surface during the CPEO process. Based on the OES spectra obtained from the CPEO discharge process around TiAl alloy cathode, the electron concentrations of the plasma discharge envelope are calculated in the range of 4.8×10^{22}–5.6×10^{22} m^{-3}, which demonstrates the achievement of local thermal equilibrium in plasma envelope around the cathode. The average electron temperature calculated based on the characteristic peak intensities in the OES spectra is approximately 5500 K [18]. The CPEO and anodic PEO processes have similar electron temperatures in the discharge zone. However, the near-surface temperature reaches 700–1100 °C for the CPEO process, which is significantly

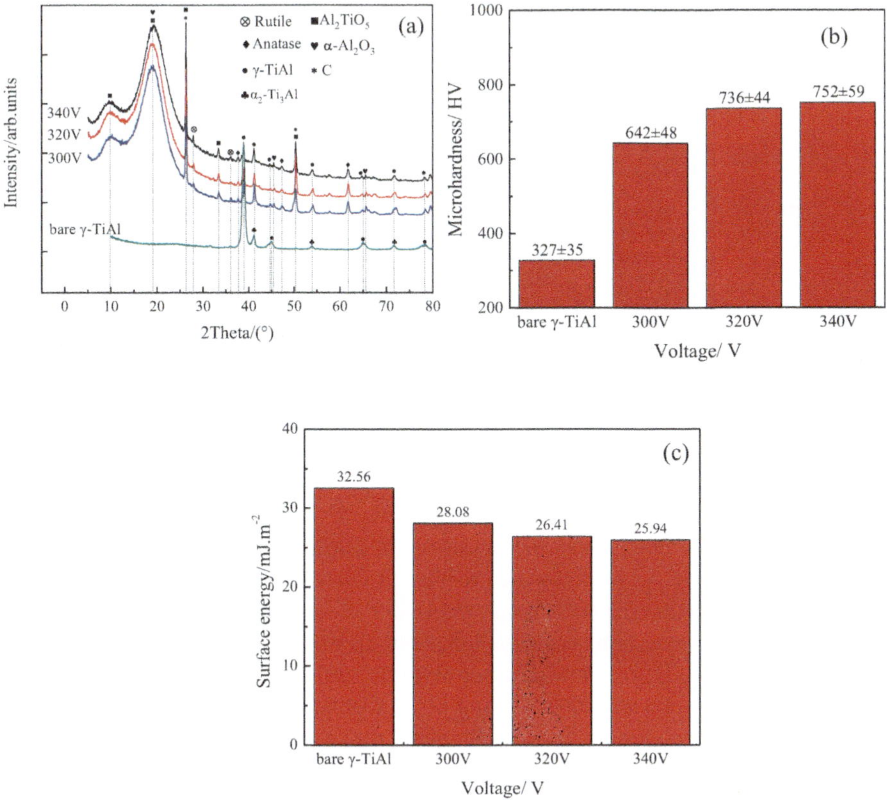

Fig. 8.10 (**a**) XRD patterns, (**b**) microhardness, (**c**) surface free energy of γ-TiAl alloy with and without CPEO coatings at different voltages [18]. (Reproduced with permission from [18] © 2022 Elsevier B.V.)

higher than that for MAO. Therefore, the growth rate of oxide coating is much higher in CPEO process than in the traditional MAO or PEO technologies, and the CPEO coatings could grow rapidly on both valve metals (e.g., Ti, TiAl) and non-valve metals (e.g., steel).

8.5 DLC-Containing CPEO Coatings and Suspended Particles in Solution

High concentrations of organic electrolytes, such as glycerin and urea, are usually added to CPEO electrolyte solutions [14, 17]. These electrolytes release large numbers of active carbon ions during high-temperature degradation in the plasma discharge envelope [20]. The ions are deposited into the interior and the surface of coatings, which causes the unique characteristics of CPEO coatings. Wu et al. [16, 29] CPEO coatings with a thickness of 35–65 μm on the T8 high-carbon steel in an

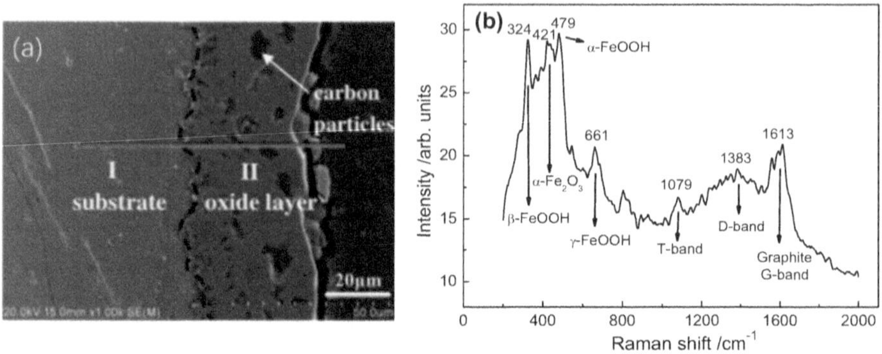

Fig. 8.11 (**a**) Cross-sectional microstructure and (**b**) Raman spectrum of the CPEO coating on T8 high-carbon steel [16]. (Reproduced with permission from [16] Rights managed by AIP Publishing)

80% glycerin aqueous solution after 5–8 min discharge at 400 V. The grains grew rapidly on the surface with an average size of 30 μm after 3 min treatment. With the prolongation of discharge time, the grains on the surface were gradually covered by more oxides. Figure 8.11 shows the cross-sectional morphology and Raman spectrum of the CPEO coating with 5 min treatment. Black areas in the cross-sectional microstructures are carbon particles generated from glycerin degradation. During the CPEO process, carbon atoms enter into the lattice spacings of iron and form some dispersed carbon particles, leading to the formation of FeO, Fe_2O_3, Fe_3O_4, and Fe_3C phases in the coating.

As shown in the Raman spectrum in Fig. 8.11b, the T peak (1079 cm^{-1}) is ascribed to the sp^3 carbon bonds in the oxide coating, implying that the CPEO coating contains diamond-like carbon (DLC) components [29]. The FeOOH peak in the Raman spectrum is attributed to the reaction of $4FeO + 2H_2O + O_2 \rightarrow 4FeO(OH)$ induced by laser irradiation during Raman spectroscopy [30]. The DLC content is high on the surface of the coating and gradually decreases toward the interior of coating. Jin et al. [21] reported that the outer layer of the CPEO coating on the surface of stainless steel had a high carbon content, with a high sp^3 carbon bond content up to 43%. Given that DLC possesses a high hardness and a low coefficient of friction, the formation of DLC components in CPEO composite coatings can further enhance the wear resistance and reduce the coefficient of friction, which is favorable for expanding the application scope of CPEO coatings.

Organic compounds such as glycerin and urea have been added to the electrolyte solution for the CPEO process. They are decomposed into small molecules in the plasma discharge envelope, meanwhile, a certain portion is directly degraded to carbon particles. These carbon particles enter into the electrolyte solution leads to the formation of a suspension solution. Wu et al. [71] treated T8 carbon steel by CPEO process in aqueous glycerin solution and observed the carbon particle distribution in solution through a laser particle size analyzer. They found that the carbon particles were uniformly distributed and exhibited a tendency for agglomeration in solution. Black powder with a particle size of approximately 100 nm was obtained

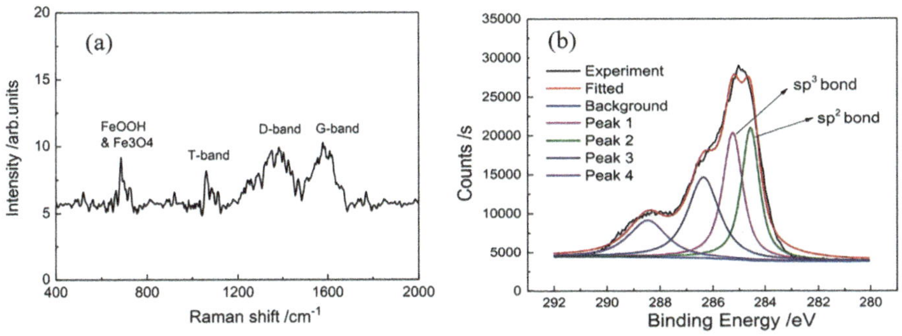

Fig. 8.12 (**a**) Raman spectrum and (**b**) XPS C1s spectrum of the particles in solution after CPEO treatment [31]. (Reproduced with permission from [31] © 2017 Elsevier B.V.)

after centrifugation treatment of suspended particle solution. It was found that these powders were mainly composed of carbon particles, meanwhile, a small amount of iron oxides detached from the sample surface. Figure 8.12 shows that the black powder contains 20–30% sp^3 carbon bonds, indicating that the high-temperature and high-pressure process in the plasma discharge envelope around cathode could synthesize the DLC components, thereby form a DLC-containing oxide coating and DLC powder in solution [32, 33]. However, further research is required to elucidate the formation mechanism of DLC component during the CPEO process.

The results described above suggest that organic molecules are directly dissociated into a large amount of active carbon, hydrogen, and oxygen ions during the CPEO process. Additionally, the high near-surface temperature (600–1100 °C) within iron-based materials can release massive iron atoms, which react with active oxygen ions to produce oxides. A certain portion of carbon ions aggregate to form carbon particles and become suspended particles in solution, while the other portion could infiltrate the lattice spacings of iron, gradually agglomerate upon the oxidation of the iron atoms, and ultimately be embedded in the oxide coating.

8.6 Basic Principles of CPEO Process

8.6.1 Optical Emission Spectroscopy and Calculation of Plasma Parameters

The plasma parameters during the plasma electrolysis can be monitored using an optical emission spectroscope (OES). The types of atoms and ions involving participate in the chemical reactions may be distinguished in terms of spectral lines of OES. Figure 8.13 displays the OES spectra of the CPEO coatings on TiAl alloy, carbon steel, and stainless steel [16–18], and their calculated electron density and electron temperature in the plasma envelope during CPEO process are shown in

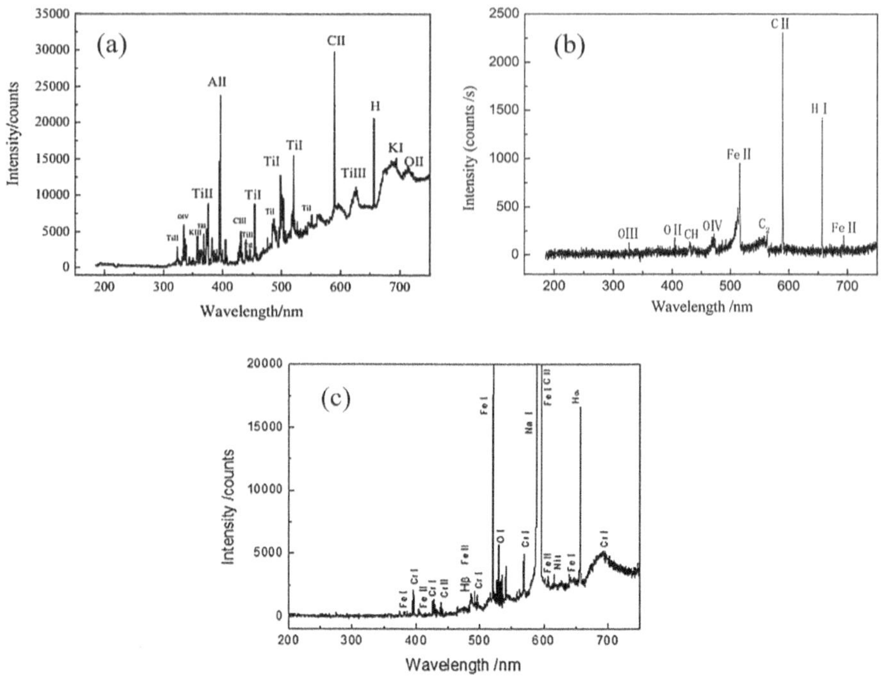

Fig. 8.13 OES spectra of the CPEO coatings on (**a**) TiAl alloy, (**b**) T8 carbon steel, and (**c**) stainless steel [16–18]. (Reproduced (**a**) with permission from [18] © 2022 Elsevier B.V., (**b**) from [16] Rights managed by AIP Publishing and (**c**) from [17] © 2019 Elsevier B.V.)

Fig. 8.14. It is found that the carbon (C), iron (Fe), oxygen (O), and hydrogen (H) from the cathode metal and electrolyte solution participate in the PEO process. Carbon exhibits a strong peak in OES, indicating that glycerin is decomposed, enters the plasma envelope, and participates in the discharge process. Fe I and Fe II, shown in Fig. 8.13, originate from the low-carbon steel substrate, and C II, H_α, and O II originate from the glycerin solution. The H_α and Fe I are neutral atoms, while C II, Fe II, and OII correspond to singly ionized atoms. Figure 8.14b, c reveals the dependence of electron concentrations and electron temperatures on discharge time, which are calculated from the ratio of Fe spectral lines (358.16 nm/526.95 nm) [20]. The electron concentration in the plasma discharge envelope is maintained at around 9×10^{22} m^{-3}, which is significantly higher than 4.1×10^{21} m^{-3}. That implies that the plasma region reaches local thermodynamic equilibrium during the CPEO discharge process, and the electron temperature is equivalent to the ion temperature. As shown in Fig. 8.14, the electron temperature range during CPEO discharge is 5000–8000 K. Therefore, the high-temperature plasma heats the steel substrate surface and promoted the formation of oxide coatings.

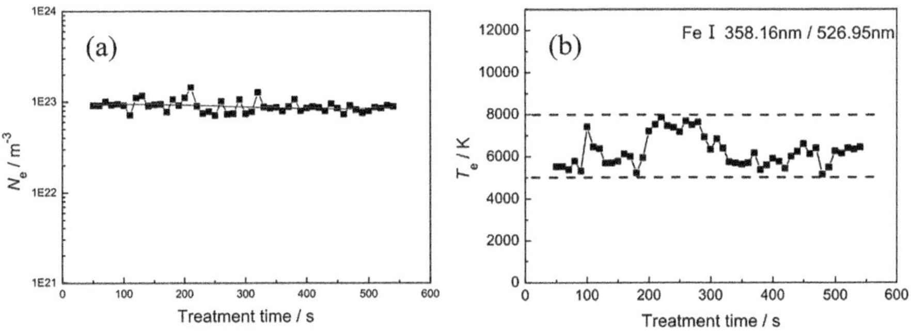

Fig. 8.14 The dependence of plasma parameters on discharge time on Q235 low carbon steel during the CPEO process. (**a**) Electron concentration (N_e), (**b**) electron temperature (T_e) [20]. (Reproduced with permission from [20] © 2019 Elsevier B.V.)

8.6.2 Electrolysis Mechanism During the CPEO Process

Figure 8.15 depicts a schematic diagram of the reaction in the plasma envelope around a Q235 steel cathode in a glycerin system during the CPEO process. At the initial stage of the reaction, the cathode surface is covered by water vapor, glycerin vapor, and hydrogen gas from the water electrolysis. A sufficiently high voltage causes breakdown and discharge in the vapor–gaseous envelope (VGE). The degradation of glycerin produces a large amount of hydrocarbon and hydroxyl active groups. These active groups possess weak bonding force in a high-energy environment and are thus rapidly decomposed into active carbon and oxygen ions. During the CPEO process, the plasma sheath is located from the sample surface to the gas–liquid interface. This facilitates the continuous heating of the steel substrate by the electric arc; thus, the near-surface temperatures can reach 600–1100 °C in the carbon steel. Consequently, the activated iron atoms produce massive iron ions, which bind with active oxygen ions to form iron oxide compounds. The rapid oxide formation increases the growth rate of CPEO coatings.

Therefore, upon the breakdown of the vapor–gaseous envelope on the surface of the metallic cathode at the threshold voltage, the plasma discharge envelope separates the electrolyte solution from the metallic cathode surface, thus preventing direct contact between the cathode and electrolyte solution. This unique environment facilitates cathodic metal oxidation, which has been proven as an innovative approach to enhance plasma electrolysis.

8.7 Conclusions and Outlook

The CPEO process utilizes the high temperature produced by plasma discharge at the vapor–gaseous envelope for rapidly oxidizing steel and titanium based cathode materials, which presents a groundbreaking method to achieve plasma electrolysis. The CPEO process is a highly promising technology for producing oxide coating

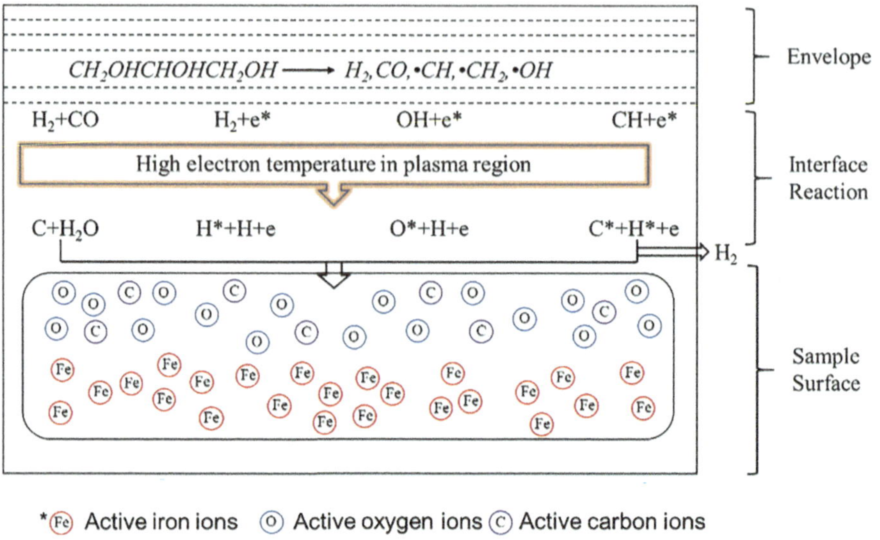

Fig. 8.15 Schematic diagram of reactions in plasma envelope around Q235 steel cathode in a glycerin solution [20]. (Reproduced with permission from [20] © 2019 Elsevier B.V.)

with good wear and corrosion resistance. It broadens the traditional plasma electrolysis theory and allows the rapid formation of oxide coating on a wider range of materials. However, compared with anodic PEO process the development of CPEO process is restricted by insufficient theoretical research and limited exploratory efforts regarding its applications. Future research may focus on the following aspects:

1. Some factors can affect the CPEO process, such as substrate material, electrical parameters, and solution systems. Existing research has largely been focused on the general exploration about parameter ranges, however, the growth patterns of CPEO coatings and related theoretical models are less reported. The feasibility of CPEO treatment for low-melting-point metals such as aluminum and magnesium remains unexplored. Moreover, the differences in process parameters and coating formation mechanisms between CPEO and anodic PEO processes remain unclear. In future research, researchers may consider adjusting process parameters, incorporating functional elements, and optimizing reaction systems to improve the properties of CPEO coatings. More researches are also required to deepen the understanding of CPEO formation theory and optimize the CPEO process.
2. Monitoring the growth process of CPEO coatings relies on observing and diagnosing the formation and breakdown of the vapor–gaseous envelope and plasma discharge. The relationships between the discharge process and morphological changes in CPEO coatings remain unclear and require further exploration.

3. The degradation pathways of organic compounds within the plasma envelope and the synthesis mechanisms and DLC component control during CPEO process should also be investigated.
4. The CPEO process is closely related to plasma electrolytic carburizing, nitriding, and boronizing processes. Therefore, it is necessary to investigate the different process conditions of these technologies. Preparing composite hardened coatings by combining the CPEO process with plasma electrolytic saturation also represents a promising area for future research.
5. At present, the CPEO process has not been commercially applied. More efforts should be considered for the application of the CPEO process in the future.

References

1. Yerokhin, A. L., Nie, X., Leyland, A., Matthews, A., & Dowey, S. J. (1999). *Surface and Coating Technology, 122*, 73–93.
2. Nie, X., Tsotsos, C., Wilson, A., Yerokhin, A. L., Leyland, A., & Matthews, A. (2001). *Surface and Coating Technology, 139*, 135–142.
3. Wang, B., Xue, W., Wu, J., Jin, X., Hua, M., & Wu, Z. (2013). *Journal of Alloys and Compounds, 578*, 162–169.
4. Wu, J., Xue, W., Wang, B., Jin, X., Du, J., & Li, Y. (2014). *Surface and Coating Technology, 245*, 9–15.
5. Cheng, Y., Feng, T., & Cheng, Y. (2022). *Journal of the Electrochemical Society, 169*, 071505.
6. Wang, S. Q., Wang, Y. M., Zou, Y. C., Chen, G. L., Wang, Z., Ouyang, J. H., Jia, D. C., & Zhou, Y. (2021). *China Surface Engineering, 50*(6), 1–22.
7. Yang, X. Z., He, Y. D., Wang, D. R., & Gao, W. (2002). *Chinese Science Bulletin, 47*, 525–529.
8. Taheri, P., Dehghanian, C., Aliofkhazraei, M., & Rouhaghdam, A. S. (2007). *Plasma Processes and Polymers, 4*, S721–S727.
9. Yerokhin, A. L., Leyland, A., Tsotsos, C., Wilson, A. D., Nie, X., & Matthews, A. (2001). *Surface and Coating Technology, 142–144*, 1129–1136.
10. Wang, B., Jin, X., Xue, W., Wu, Z., Du, J., & Wu, J. (2013). *Surface and Coating Technology, 232*, 142–149.
11. Wang, B., Xue, W. B., Jin, X. Y., & Du, J. C. (2019). *Surface Technology, 48*(7), 37–48.
12. Belkin, P. N., Kusmanov, S. A., & Parenov, E. V. (2020). *Applied Surface Science Advances, 1*, 100016.
13. Aliofkhazraei, M., & Rouhaghdam, A. S. (2011). *Surface and Coating Technology, 205*, S57–S62.
14. Jin, X., Wang, B., Xue, W., Du, J., Wu, X., & Wu, J. (2013). *Surface and Coating Technology, 236*, 22–28.
15. Wu, J., Lu, P., Dong, L., Zhao, M., Li, D., & Xue, W. (2020). *Heat Treatment and Surface Engineering, 2*(1), 1–8.
16. Wu, J., Xue, W., Jin, X., Wang, B., Du, J., & Wu, Z. (2013). *Applied Physics Letters, 103*, 031905.
17. Jin, X., Chen, L., Yu, J., Liu, R., Xue, W., & Shang, H. (2019). *Surface and Coating Technology, 363*, 314–321.
18. Gao, C., Zhu, M., Wang, B., Liao, Y., Li, H., Jin, X., Xu, C., Du, J., & Xue, W. (2022). *Surface and Coating Technology, 444*, 128666.
19. Stojadinović, S., Tadić, N., Šišović, N. M., & Vasilić, R. (2015). *Journal of Applied Physics, 117*, 233304.

20. Yu, J., Zhang, Y., Jin, X., Chen, L., Du, J., & Xue, W. (2019). *Surface and Coating Technology, 363*, 411–418.
21. Jin, X., Zhang, Y., Chen, L., Yu, J., & Xue, W. (2018). *Surface and Coating Technology, 338*, 38–44.
22. Wang, B., Wu, J., Zhang, Y., Wu, Z., Li, Y., & Xue, W. (2015). *Surface and Coating Technology, 269*, 302–307.
23. Jin, X., Wang, B., Xue, W., Du, J., Wu, X., & Wu, J. (2012). *20th Congress of the International Federation for Heat Treatment and Surface Engineering, Oct. 23–25, Beijing* (pp. 491–497).
24. Gui, W., Hao, G., Liang, Y., Li, F., Liu, X., & Lin, J. (2016). *Applied Surface Science, 389*, 1161–1168.
25. Xu, C., Zhu, M., Guan, H., Gao, C., Jin, X., Du, J., & Xue, W. (2022). *Corrosion Science, 209*, 110711.
26. Li, X., Cheng, G., Xue, W., Zheng, R. T., & Cheng, Y. J. (2008). *Materials Chemistry and Physics, 107*, 148–152.
27. Li, X., Wu, X., Xue, W., Cheng, G., Zheng, R., & Cheng, Y. (2007). *Surface and Coating Technology, 201*, 5556–5559.
28. Wang, Y. H., Ouyang, J. H., Liu, Z. G., Wang, Y. M., & Wang, Y. J. (2014). *Applied Surface Science, 307*, 62–68.
29. Wu, J., Dong, L., Deng, J., Hou, D., Li, G., Li, D., & Xue, W. (2018). *Surface and Coating Technology, 338*, 63–68.
30. Chattopadhyay, A., Bandyopadhyay, N., Das, A. K., & Panigrahi, M. K. (2005). *Scripta Materialia, 52*, 211–215.
31. Wu, J., Fan, L., Dong, L., Deng, J., Li, D., Zhang, Y., & Xue, W. (2017). *Materials Chemistry and Physics, 199*, 289–294.
32. Kong, X., Wang, S., Zhao, H., & He, Y. (2010). *Thin Solid Films, 518*, 4211–4214.
33. Yan, Z., Deng, L., & Chen, L. (2013). *Electrochimica Acta, 105*, 612–617.

Part III
In-Situ Particles Incorporation

Chapter 9
Functionalisation of Plasma Electrolytic Oxidation Coatings via Particles Addition: Types, Approaches, and Mechanisms

Beatriz Mingo, Lizhuo Zhang, and Xiaopeng Lu

9.1 Introduction

Plasma electrolytic oxidation (PEO) is an advanced, environmentally friendly surface treatment technology used to enhance the performance of valve metals, primarily aluminium (Al), magnesium (Mg), and titanium (Ti). The process creates highly stable, crystalline ceramic coatings with exceptional corrosion and wear resistance. PEO is an electrochemical technique based on applying DC unipolar, DC bipolar, or AC polarisation to the metal substrate, above the dielectric breakdown voltage of the oxide that is being form. This leads to the formation of short-lived microdischages that trigger chemical, electrochemical and plasma assisted reactions responsible for the formation of the high temperature crystalline phases [1].

The choice of electrolyte plays a crucial role in optimising the coating for specific applications. The electrolyte typically contains an alkaline base, such as potassium or sodium hydroxide, along with passivating agents like silicates or phosphates, which enhance corrosion resistance, wear protection, and thermal stability. The use of additives such as fluorides and other salts like aluminates and borates is a common strategy to tailor the coating's structure and improve specific properties such as hardness, corrosion resistance, or wear protection [2, 3]. Other additives such as complexing agents, like EDTA (ethylenediaminetetraacetic acid), can bind metal ions, preventing their early precipitation, leading to more uniform and effective coatings [4]. Furthermore, PEO allows for the incorporation of suspended particles

B. Mingo (✉) · L. Zhang
Department of Materials, Henry Royce Institute, The University of Manchester, Manchester, UK
e-mail: beatriz.mingo@manchester.ac.uk

X. Lu
Corrosion and Protection Center, Northeastern University, Shenyang, China

V. S. Saji, M. Zheludkevich (eds.), *Plasma Electrolytic Oxidation – Fundamentals, Advances and Applications*,
https://doi.org/10.1007/978-3-031-92429-3_9

(nano and microparticles) from the electrolyte into the coating, further enhancing its functional properties and making the process highly versatile [5].

Over the last decade, there has been significant interest in functionalising the surface of these types of coatings to either introduce new functionalities or further enhance their inherent corrosion resistance and mechanical properties. The most common approach to functionalise PEO coatings is the application of a sealing post-treatment, where the inherent open porosity of the coating acts as a reservoir for active species [6]. This simultaneously seals the coating's porosity, further enhancing its barrier properties. While this method offers advantages, such as ease of application, there are several challenges that limit its effectiveness. These include the early release of the active component into the environment, poor adhesion between the coating and the incorporated functional particles, and most notably, this approach only functionalises the most external layer of the coating, which limits its long-term effectiveness.

An alternative functionalisation approach relies on the in situ incorporation of nano- and microparticles suspended in the electrolyte during the coating process. This method allows for the direct integration of functional particles into the growing oxide layer. By embedding the particles throughout the thickness of the coating, this approach provides more uniform and long-lasting functionality compared to surface post-treatments, which are limited to the outer layer of the coating system. This approach has been used to improve specific functionality in a wide variety of applications, including enhanced barrier properties, corrosion and wear resistance, as well as other functional properties such as catalytic, photocatalytic, antibacterial properties, and biocompatibility. The possibility of tailoring the composition of PEO coatings via electrolyte design based on the incorporation of particles opens a range of opportunities for advanced material applications across diverse industries.

9.2 Characteristics and Types of Particles Used for Functionalisation

The functionalisation via particle addition requires the incorporation of selected particles into the electrolyte. Several factors must be considered when selecting these particles. They should be chemically stable in alkaline electrolytes, as PEO processes typically operate in a pH range of 9–13, depending on the substrate; they should have a relatively high melting or degradation temperature as they will be exposed to extreme temperature conditions during the coating processing. Additionally, since these particles are often insoluble in water, they tend to form suspensions. Achieving a stable dispersion to avoid agglomeration or sedimentation is crucial for ensuring their uniform incorporation into the coating structure.

Achieving a stable dispersion to avoid agglomeration or sedimentation is crucial for ensuring their uniform incorporation into the coating structure. The main factors that affect the stability of a suspension are particle size and distribution and surface charge (zeta potential). Nano-sized (typically less than 100 nm) and sub-micron particles (ranging from 100 nm up to 1 μm) are more prone to agglomeration due to their higher surface energy, which can disrupt the suspension stability. On the other hand, micro-particles (typically in the 1–10 μm range) tend to be more stable in suspension but may not be as easily incorporated into the microstructure of the coating. A stable dispersion is also influenced by the surface charge of the particles, which is measured by the zeta potential. A higher absolute zeta potential typically indicates stronger repulsion between particles, preventing agglomeration and ensuring a more uniform distribution within the electrolyte. In addition, the stability of the electrolyte also depends on the particle concentration. While it is easier to achieve a better dispersion in lower concertation electrolytes, an optimal concentration must be maintained to ensure sufficient particle incorporation without leading to excessive sedimentation.

The most common strategies used to improve the dispersion of particle containing electrolytes during PEO, involve the use of mechanical agitation, gas bubbling and/or ultrasonication. Surfactant additives such as sodium dodecyl sulphate [7], sodium oleate [8] and triethanolamine [9] are also used to increase the stability of the suspension.

Asgari et al. [10] evaluated the influence of stirring rate on the incorporation of Al_2O_3 particles into PEO coatings formed on a commercial magnesium alloy and revealed that the amount of incorporated particles directly depends on the intensity of the stirring (Fig. 9.1). This effect was even more pronounced when working with electrolytes with a higher concentration of particles in suspension, resulting in an overall increase in coating thickness. At lower stirring rates (50 rpm), the stirring force is insufficient to promote significant absorption of Al_2O_3 particles into the PEO coating, resulting in relatively low incorporation. However, as the electrolyte stirring rate increases to 100 rpm, the level of nanoparticle absorption reaches a maximum; beyond this point, a decrease in uptake occurs. This reduction in absorption at higher stirring rates is likely due to increased turbulence and collisions, which displace the particles from the anode surface with a "washing" effect. Additionally, increasing the Al_2O_3 concentration in the electrolyte up to 30 g/L leads

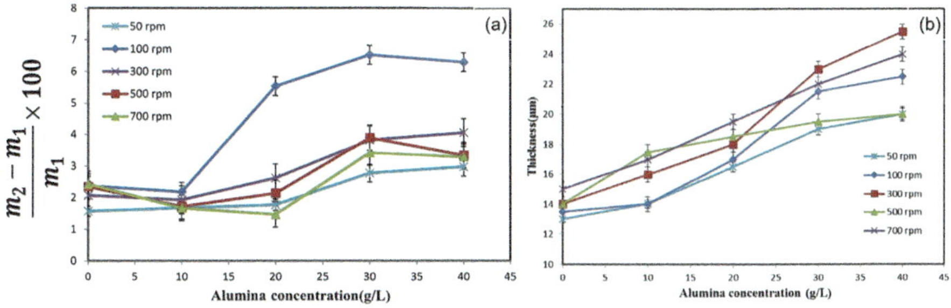

Fig. 9.1 Influence of stirring rate on (**a**) Al_2O_3 particle adsorption and (**b**) coating thickness of PEO coatings on AZ31B magnesium alloy. (Reproduced with permission from Ref. [10] © 2016 Elsevier B.V.)

to greater absorption within the coating, reaching a saturation point beyond which no further absorption occurs.

Three different types of particles have been used to functionalise PEO coatings, bulk particles, 2D structured and 3D structured particles, which are commonly used as containers for active species, e.g. corrosion inhibitors.

9.2.1 Bulk Particles

9.2.1.1 Oxides

Oxides are possibly the most commonly evaluated type of particles to date. Their high thermal stability, hardness, and electrical insulation make these composite systems quite popular options for improving both corrosion and wear resistance in PEO coatings. These include Al_2O_3, TiO_2, ZrO_2, CeO_2, and SiO_2, amongst many others. Their positive influence in the corrosion resistance is related to their ability to fill pores and defects forming a dense, protective layer that hinders the diffusion of aggressive species, improving further the protective barrier properties of the systems. This type of passive corrosion protection was achieved in Mg [11–13], Al [14–16], and Ti [17, 18] based materials. In some cases, they can also provide specific corrosion inhibition, for instance, CeO_2 is known to exhibit corrosion inhibition and self-healing properties by releasing Ce^{3+} ions when exposed to corrosive environments [11].

The increase in wear resistance is associated with the incorporation of particles formed by crystalline phases with high hardness. This results in an overall increase in the coating's hardness, not only due to the presence of these hard particles but also because they effectively fill pores and voids that could otherwise compromise the mechanical integrity of the coating. By reducing these structural defects, the particles help to distribute applied forces more evenly, thereby enhancing the coating's resistance to wear and abrasion. Additionally, the interaction between the hard particles and the PEO matrix contributes to the formation of a more compact and cohesive layer, further improving the coating's durability under mechanical stress. The most common oxides used to improve the wear resistance are α-Al_2O_3 [16, 19, 20], and ZrO_2 [19, 21, 22], although a wide variety of compositions have also shown promising performance such as TiO_2 [23], Ta_2O_3 [24], CeO_2 [25], and SiO_2 [22].

Additionally, the incorporation of various oxide particles into PEO coatings can significantly enhance their multifunctional properties beyond wear resistance. For instance, the addition of Ta_2O_5 [4] to PEO coatings on magnesium alloys has been shown to improve biocompatibility by promoting the formation of a stable and inert surface layer that minimises cytotoxic responses and promoting cell adhesion. This is achieved through the formation of a denser and thicker PEO film, which effectively reduces the corrosion rate of magnesium alloys in physiological environments, thereby making them more suitable for biomedical implant applications.

ZnO is another oxide with strong antibacterial properties, and it has been incorporated into Ti-based PEO coatings [26]. ZnO nanoparticles generate reactive oxygen species (ROS), which are toxic to bacterial cells [27]. Additionally, ZnO can disrupt bacterial cell membranes through the release of Zn^{2+} ions, leading to increased membrane permeability and eventual cell death [27]. This dual-action mechanism makes ZnO-enhanced PEO coatings highly effective in preventing bacterial colonisation and biofilm formation, which are critical concerns in medical device applications. CuO and Ag_2O nanoparticles further enhance the antimicrobial efficacy of PEO coatings. CuO nanoparticles contribute to the generation of ROS and the release of Cu^{2+} ions, which interfere with microbial metabolism and enzyme function [28]. Similarly, Ag_2O provides sustained antibacterial effects through the gradual release of Ag^+ ions, which bind to thiol groups in bacterial enzymes, disrupting their vital functions [29]. The incorporation of CuO and Ag_2O has been achieved in Al, Mg and Ti alloys [30–32] and they not only improve the antimicrobial properties but also reinforce the barrier properties of the PEO matrix, resulting in more resistant coatings. Moreover, the combination of Cu_2O and ZnO nanoparticles has been shown to impart antifouling properties to PEO coatings. This combination disrupts the adhesion and proliferation of marine organisms on coated surfaces by releasing metal ions that inhibit microbial attachment and growth, thereby maintaining surface cleanliness and integrity in marine environments [33].

In addition to antimicrobial and antifouling functionalities, these oxide particles can also contribute to photocatalytic activities within PEO coatings. For example, TiO_2 is well known for its ability to catalyse reactions under UV light, making it useful in self-cleaning and air purification applications [34]. TiO_2 coatings are commonly developed by direct PEO of titanium or titanium alloys [35]; however, the catalytic activity can be improved further by incorporating nanoparticles such as ZrO_2 [36]. Incorporating ZrO_2 into TiO_2 coatings creates mixed anatase, rutile, and ZrO_2 phases, facilitating electron transfer and reducing electron-hole recombination. Additionally, the porous microstructure and enhanced surface acidity provide more active sites, thereby boosting photocatalytic efficiency [36].

This approach also allows the functionalisation of not only titanium-based substrates but also non-titanium materials such as magnesium and aluminium alloys through the in situ incorporation of catalytically active nanoparticles. For instance, the incorporation of TiO_2 particles into PEO coatings formed on Mg results in the development of photocatalytic effects and the initiation of photocurrent upon UV illumination [23, 37]. ZnO [38] V_2O_5 [39] have also shown promising catalytic properties on PEO coatings formed on magnesium and aluminium alloys, respectively.

The strategic selection and incorporation of these oxide particles into PEO coatings not only tailor the surface properties to meet specific application requirements but also enable the development of advanced multifunctional coatings. By combining the unique chemical and physical properties of each oxide, PEO coatings can be designed to offer a combination of enhanced biocompatibility, antimicrobial activity, antifouling capabilities, and photocatalytic efficiency, thereby broadening the scope of their applicability in diverse fields such as biomedical engineering, environmental remediation, and marine technology.

9.2.1.2 Carbides and Nitrides

Carbides exhibit exceptional hardness, which is key to enhancing the mechanical resistance of PEO coatings. Among them, silicon carbide (SiC) is arguably the most widely used, thanks to its availability, affordability, and proven effectiveness in demanding environments. SiC particles significantly increase the hardness, high-temperature oxidation resistance and creep resistance of coatings. When added to PEO coatings [40–43], SiC particles fill microcracks and pores, creating a denser, more compact structure. They reinforce the coating mechanically, enhancing hardness and wear resistance, reducing surface damage under stress. This also reduces pathways for oxygen and corrosive agents to penetrate, improving oxidation and corrosion resistance. During high-temperature exposure, SiC is partially oxidised to form SiO_2, which offers oxidation resistance by sealing surface defects, limiting oxygen diffusion, and reducing the growth of cracks at high temperatures. Other carbides, such as WC [44], B_4C [45] have also shown positive results in resulting in denser coatings with improved corrosion resistance.

Carbides are particularly effective in reducing friction and wear, especially in high-stress environments. SiC [43], TiC [46] have shown positive results in improving the tribological performance of PEO coatings. During the PEO process, TiC particles penetrate and accumulate in pores and microcracks created by plasma discharges. This infiltration reinforced the coatings in the inner layer, increasing their hardness and wear resistance. The strengthened inner layer helped prevent crack propagation from the outer surface to the substrate, enhancing the ability of PEO coatings to withstand mechanical stress and friction.

While PEO coatings are usually characterised by a low thermal conductivity, the in situ incorporation of particles can provide further functionality by increasing their thermal conductivity, making them highly suitable for applications where heat dissipation is crucial, such as in engine components and heat exchangers. They also generally present improved dielectric properties, expanding their applicability to other fields such as electronics. Several attempts have been perform to incorporate AlN [47, 48], TiN [49, 50], Si_2N_4 [51] into PEO coatings, and in addition to improving both the corrosion and wear resistance, some authors have also reported improved thermal properties [48].

9.2.1.3 Metallic Particles

Metallic particles are commonly added to coatings for biomedical applications, primarily to enhance antibacterial properties. Cu and Ag are known for their excellent antibacterial properties, which are retained after incorporation into PEO coatings on Mg for biodegradable implants [52], on Ti for fixed orthopaedic implants and on Al for antifouling applications [53]. This antibacterial effect is related to the ability of these metals to disrupt bacterial cell membranes due to their phytotoxicity that inhibits microbial growth, providing long-term protection in biological environments. Ag nanoparticles have also been used to improve the photocatalytic properties of TiO_2 PEO coatings formed on Ti while increasing the degradation efficiency [54].

The in situ incorporation of metallic particles such as Co and Fe has proven to be an effective strategy for imparting magnetic and ferromagnetic properties to ceramic-based PEO coatings [55, 56]. Alternatively, Co and Fe metallic phases have been incorporated into PEO coatings on titanium by using iron salts instead of pure metallic sources. This approach results in the incorporation of these elements in both reduced and oxidised states [57].

9.2.1.4 Polymeric Particles

The use of polymeric particles in PEO coatings is often limited by their tendency to decompose or degrade at the high temperatures generated during the process, presenting challenges in creating stable organic–inorganic composite coatings in a single step. Polytetrafluoroethylene (PTFE) is a popular polymer due to its exceptional chemical inertness, hydrophobicity, and a low friction coefficient [58] and it is possibly the most commonly polymer-based particle used for the in situ incorporation of PEO coatings to improve their tribological properties. Wang et al. [59] demonstrated that by adjusting electrolyte temperature and process parameters, it is possible to incorporate polymeric particles into PEO coatings effectively. This approach influences the transient nature of PEO discharges and controlled deposition conditions to form a nanocomposite outer layer with minimal particle degradation.

Zhang et al. [60] fabricated PTFE/PEO composite coatings on aluminium alloys using an electrolyte containing PTFE nanoparticles and styrene-acrylic emulsion. This method ensured uniform embedding of PTFE, resulting in significantly reduced porosity and enhanced surface smoothness, with the coatings exhibiting superior corrosion resistance compared to conventional PEO coatings. Similarly, Rudnev et al. [61] incorporated PTFE into PEO coatings on aluminium and titanium alloys using a siloxane-acrylate emulsion-based electrolyte, achieving hybrid coatings with PTFE-rich surface layers that improved hydrophobicity and wear resistance.

9.2.2 1D and 2D Structured Particles

9.2.2.1 Carbon-Based Particles

The incorporation of carbon-based 2D particles, such as graphene, graphene nanoplatelets (GNPs), graphene oxide (GO), or carbon nanotubes, offers significant advantages in PEO coatings, particularly in enhancing mechanical strength, thermal stability, electrical conductivity and even lubrication, reducing friction and wear in dynamic systems. For instance, adding graphite to PEO coatings increases coating density, decreases porosity, and improves wear and corrosion resistance due to its exceptional lubrication properties and thermal resilience. Kaplan et al. [62] found that graphite significantly reduced friction during dry sliding tests by acting as a natural lubricant, facilitating layer sliding and reducing resistance between surfaces. Similarly, Lv et al. [63] demonstrated that fine graphite grains strengthened aluminium PEO coatings and enhanced corrosion resistance in salty environments. Wang et al. [64] reported that graphite reduced pore size and smoothed surfaces on aluminium alloys, enhancing wear resistance and coating durability. On magnesium alloys, Rudnev et al. [61] showed that graphite created a smooth, wear-resistant layer, further improving corrosion protection.

Carbon nanotubes (CNTs), especially multi-wall carbon nanotubes (MWCNTs), are increasingly used to enhance PEO coatings, improving wear resistance, corrosion protection, and mechanical properties [65]. Their hollow cylindrical structure, formed from rolled graphene sheets, provides exceptional strength, conductivity, and chemical stability, making them valuable for automotive, aerospace, and medical applications [66]. Adding CNTs to PEO coatings reduces porosity and improves coating compactness by filling discharge channels and cracks formed during oxidation. This results in denser, more uniform coatings. Yu et al. [65] demonstrated that CNTs embedded in PEO coatings on high-silicon aluminium alloys reduced microporosity, enhanced hardness, and improved thermal stability. Yürektürk et al. [67] observed that CNTs in coatings on aluminium lowered friction and wear by creating a smoother, self-lubricating surface, which significantly increased the coating's durability under sliding conditions. Beyond corrosion protection, CNTs also improve thermal conductivity and mechanical strength. Yu et al. [65] and Lee et al. [68] found that CNT-doped coatings displayed better heat dissipation, which is crucial for applications where thermal regulation is essential. Additionally, Sabouri et al. [69] reported that CNTs enhanced the overall surface hardness of PEO coatings, contributing to better wear resistance and reducing material degradation over time. However, they also reported that excessive amounts could reduce inner layer compactness, slightly decreasing corrosion performance. However, optimal concentrations improved the coating's barrier properties against corrosion.

Graphene is widely recognised for its exceptional mechanical strength, thermal conductivity, and excellent barrier properties [70]. In coatings, graphene effectively reduces porosity, enhances compactness, and improves both corrosion and wear resistance [71]. These characteristics make it an ideal additive for enhancing the performance of PEO coatings [72].

Gao et al. [73] demonstrated that adding graphene to PEO coatings on 6063 aluminium alloy significantly improved surface properties. The addition reduced the size and number of pores while enhancing coating smoothness and thickness. However, similarly to CNTs [69], excess graphene concentrations caused particle aggregation, leading to increased porosity and reduced performance. Similarly, Nasiri Vatan and Adabi [74] incorporated graphene into PEO coatings on AZ31 magnesium alloy, observing substantial improvements in wear resistance and friction performance. The graphene-filled coatings showed fewer pores and greater tribological stability, attributed to graphene's ability to fill pores and discharge channels and enhance coating uniformity. This integration resulted in a denser structure and better load-bearing capacity under sliding conditions.

These studies highlight the potential of carbon-based particles to enhance the protective and functional properties of PEO coatings, particularly in improving wear and corrosion resistance. However, further research is needed to optimise other properties, such as change in electrical conductivity, as this aspect remains underexplored compared to the focus on corrosion and wear. Addressing this gap could lead to more versatile and multifunctional PEO coatings for advanced applications.

9.2.2.2 Other 2D Materials

In addition to graphene-based materials, other 2D materials such as hexagonal boron nitride (hBN) and MoS_2 have been explored for their excellent lubrication properties. These materials are well-known for their ability to reduce friction in coatings, making them ideal candidates for applications requiring enhanced tribological performance [75].

Hexagonal boron nitride (hBN) is chemically inert and thermally stable, providing both low friction and high wear resistance due to its solid lubrication properties. It has been demonstrated that hBN can reduce both the friction coefficient and the wear rates of PEO-coated Ti6Al4V while simultaneously improving the corrosion resistance by sealing defects in the coating [76].

Molybdenum disulfide (MoS_2) is another widely used 2D material in lubricating coatings. It has been shown to have a positive impact on the wear resistance and a short-term positive effect on the corrosion behaviour of PEO coatings formed on Mg [77], Al [78], and Ti alloys [75] as the hardness of the coating increases and the porosity reduces. However, the corrosion performance of the coating can be compromised if higher concentrations of MoS_2 are used [77].

9.2.3 Hollow and Structured Particles as Reservoirs for Active Species

9.2.3.1 Silica and Aluminosilicate-Based Particles

Mesoporous silica nanoparticles (MSNs) are highly effective as nanocontainers due to their unique porous structure, large surface area, and high thermal and chemical stability. Their pore size (2–10 nm) and tunable morphology allow for efficient encapsulation and controlled release of corrosion inhibitors or active agents, which has the potential to providing self-healing properties to coatings [79]. In PEO coatings, MSNs integrate into the oxide layer, forming a dense, sponge-like structure that enhances wear and corrosion resistance. Chen et al. [80] reported that silica-based PEO coatings on aluminium alloys significantly reduced corrosion current density and increased passivation regions, providing superior protective performance. This improvement is attributed to the MSNs filling micro-pores and forming an amorphous silica layer that blocks aggressive ions.

Aluminosilicates, characterised by their diverse structural frameworks, such as layered clays, zeolites, and tubular structures, have emerged as promising materials for corrosion protection due to their inherent chemical stability and ion exchange capabilities. These structures enable the effective encapsulation and controlled release of corrosion inhibitors, enhancing the protective performance of coatings. Zeolites improve PEO coatings by enhancing corrosion resistance and introducing self-healing properties through their unique ion exchange and barrier mechanisms [81]. As crystalline aluminosilicates with a porous structure, zeolites can store corrosion inhibitors, such as Ce^{3+} ions, and release them in situ triggered by the presence of other ions in solution. This allows zeolites to act as containers, delivering inhibitors directly to damaged areas and preventing further degradation [82, 83]. Zeolite particles enhance PEO coatings by providing both barrier protection and active ion exchange capability. During the PEO process, zeolite fills micropores and cracks, increasing coating density and compactness. Additionally, zeolite can release corrosion inhibitors by ion exchange processes, which react with hydroxides and oxygen to form protective oxides and hydroxides, actively repairing coating defects and stabilising the oxide layer in corrosive environments [82, 83].

Halloysite nanotubes (HNTs) are naturally occurring aluminosilicates with a hollow tubular structure that resembles rolled kaolinite sheets. These nanotubes typically have an outer diameter of 50–100 nm, an inner lumen of 10–30 nm, and lengths ranging from 500 nm to several micrometres [84]. HNTs are characterised by their high mechanical strength, thermal stability, and chemical inertness. At neutral pH the inner lumen is positively charged while the outer layer is negatively charged, therefore, they can immobilise both anions and cations in the inner or outer part, respectively [85], making them excellent carriers for corrosion inhibitors and other active agents [6]. HNTs have been incorporated in situ into PEO coatings on several occasions [86–89]. Similarly to silica-based nanocontainers, the incorporation of HNTs tends to enhance corrosion resistance by partially sealing the coating's porosity and defects. Coatings containing corrosion inhibitor-loaded HNTs generally exhibit slightly better corrosion performance due to their specific inhibition activity. The inhibitor release mechanism of HNTs is primarily pH-sensitive [89]. In corrosive environments, a localised rise in pH triggers the release of inhibitors stored in the nanotubes. For instance, Wu et al. [89] demonstrated that 2-aminobenzimidazole (2-ABi)-loaded HNTs in PEO coatings on AZ31 Mg alloys released inhibitors upon corrosion initiation, forming a protective passive layer that sealed cracks and prevented further degradation. Similarly, Ma et al. [86] found that 2-mercaptobenzothiazole (MBT)-loaded HNTs in PEO coatings released inhibitors under high-pH conditions, thereby enhancing the coatings' self-healing properties.

9.2.3.2 Layered Double Hydroxides (LDHs)

LDHs are gaining significant traction in nanotechnology and related fields due to their unique layered structure and versatile functionality. Structurally, LDHs consist of positively charged metal hydroxide layers interspersed with charge-balancing anions and water molecules, forming a flexible framework capable of accommodating various guest species [90]. Recently, LDHs have been utilised in corrosion protection as nanocontainers for inhibitors, enabling the controlled and responsive release of corrosion-inhibiting agents. In the context of PEO coatings, several attempts have been performed to incorporate them in situ. Although complete non-reactive incorporation has not yet been achieved, their partially reactive integration within the coating can not only improve the protective barrier properties of PEO coatings [91] but also contribute to the incorporation of additional functionalities, such as photocatalytic properties [92, 93]. Li et al. [91] incorporated Al-Zn LDH containing fumarate inhibitor into PEO coatings formed on Mg and observed an increase in thickness and compactness as a consequence of their reactive incorporation, resulting in better corrosion resistance. Similarly, Mojsilović [92, 93] incorporated Zn-Al LDH into PEO coatings formed on Al and also noted an increase in thickness and compactness. Furthermore, the incorporation of LDHs led to the formation of Zn-rich crystalline phases, which enhanced the overall corrosion resistance and slightly improved the catalytic properties of the coatings.

9.3 Approaches to Particle Addition: Influencing Parameters

9.3.1 Particle Size

Particle size is a critical parameter influencing the success of in situ particle incorporation in PEO coatings. It not only affects the suspension stability but also the capacity of particles to be incorporated into the oxide layer and the incorporation state (inert versus reactive). Understanding these influences is essential for optimising coating properties and achieving desired functionalities.

Smaller particles exhibit higher mobility and diffusion rates due to lower mass and smaller hydrodynamic diameters. This increased mobility enhances their ability to migrate towards the electrode under the influence of the electric field during the PEO process, facilitating higher incorporation rates. Additionally, smaller particles typically possess lower melting points, which makes them more susceptible to their reactive incorporation. Larger particles, on the contrary, have lower mobility and diffusion capabilities, making it more challenging for them to reach and embed into the oxide layer. This can result in lower incorporation efficiencies and necessitate longer processing times or higher energy inputs to achieve the desired embedding. Therefore, smaller particles are more likely to distribute uniformly throughout the oxide layer, contributing to consistent enhancement of coating properties such as hardness and wear resistance.

Lu et al. [13] evaluated the influence of nanosized (12 nm) and microsized (1–5 μm) SiO_2 particles on the in situ incorporation into PEO coatings formed on AM50 magnesium alloy using a phosphate-based electrolyte operating in bipolar mode. They observed that particle size had a direct influence on the current/voltage time response, with nanosized particles exhibiting a faster decrease in current and achieving higher final voltages compared to microsized particles (Fig. 9.2). This resulted in the formation of thicker and denser coatings. Additionally, coatings with nanoparticle incorporation required more time to reach the breakdown potential. The authors reported that the incorporation mechanism is also directly influenced by particle size. Smaller nanoparticles can penetrate deeply into PEO coatings through discharge channels, facilitating their reactive incorporation, especially for particles with low melting points [94]. In contrast, larger particles are primarily incorporated at the surface since they cannot be accommodated within the smaller channels. In such cases, particles are more likely to achieve inert incorporation. This results and incorporation mechanism are in agreement with the behaviour observed by other authors more recently when evaluating the incorporation of α-Al_2O_3 with a particle size of 30 and 200 nm into PEO coatings formed on AZ91 alloy [95].

Yeung and co-workers [96] investigated the in situ incorporation of nanosized and microsized hydroxyapatite (HA) particles into PEO coatings formed on a Ti6Al4V alloy using a phosphate-based electrolyte under pulsed unipolar and pulsed bipolar modes. Notably, in contrast to previously reported behaviours for Mg alloys [13, 95], they observed a reduction in the final voltage as the particle size decreased (Fig. 9.3). This decrease was attributed to the distinct dielectric properties of HA particles, which, when integrated into the TiO_2 matrix of the growing PEO coating, alter the electrical discharge characteristics. The incorporation of smaller HA nanoparticles likely facilitates a more uniform and dense oxide layer formation, enhancing the coating's protective properties while modifying the overall voltage response during the PEO process. These findings suggest that the influence of

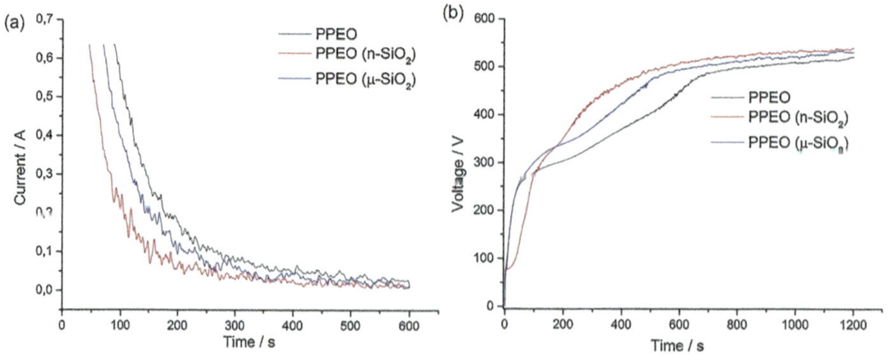

Fig. 9.2 Current (**a**) and voltage (**b**) vs. time plots during PEO processing on Mg in electrolytes with nano- and microsized particles. (Reproduced with permission from [13] © 2015 Elsevier Ltd)

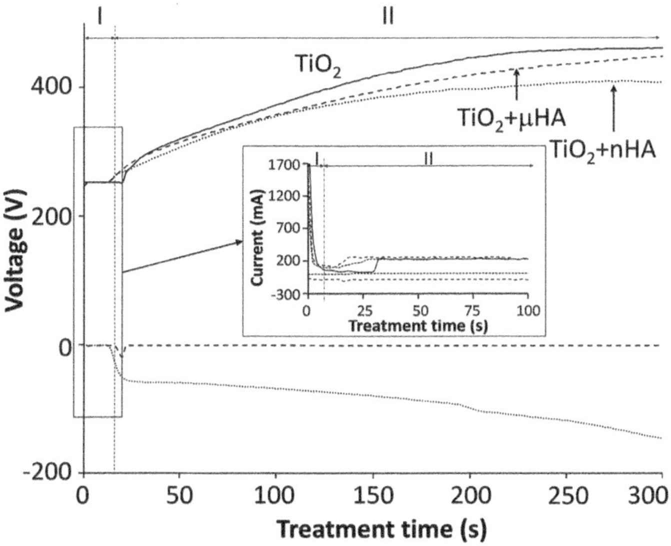

Fig. 9.3 Voltage vs. time plots during PEO processing of Ti–6Al–4V alloy without and with the addition of nano and microsized HA particles. (Reproduced with permission from [96] © Authors, published by the Royal Society of Chemistry)

particle size on the PEO electrical response is highly dependent on the specific combination of particle type, substrate, and electrolyte system.

9.3.2 Particle Concentration in the Electrolyte

Particle concentration in the electrolyte is an important parameter for the in situ particle incorporation into PEO coatings. Adequate concentrations enhance incorporation efficiency and contribute to denser, more uniform coatings with improved mechanical and protective properties. However, excessively high concentrations can lead to particle agglomeration, sedimentation, and unstable discharge behaviour, compromising coating quality. Therefore, optimising particle concentration involves balancing the benefits of increased incorporation against the challenges of maintaining suspension stability and controlling discharge characteristics.

Commonly, the particle concentration in the electrolyte ranges between 1 and 20 g/L. Within that range, higher concentrations generally lead to the formation of thinner coatings [13, 46, 95, 97–99], although in some case the opposite behaviour was observed [100]. The incorporation of more particles alters the nature of microdischarges during the PEO process, which in turn affects the growth dynamics of the oxide layer. The incorporation of particles has different effects on the electrical response during the PEO process, and this greatly depends on the specific combination of particle type, electrolyte, and substrate materials used. While several authors have reported an increase in the final voltage [13, 83, 101], others have reported a decrease [99, 102–104] while others have not found a significant change [98, 105]. This is related to several factors including the possible changes in pH and conductivity of the electrolyte [93, 106], which are key factors in the passivation response and final voltages, as well as the possible changes in the dielectric properties of the forming oxides.

Particles tend to accumulate predominantly in the outer regions of the coating when their concentration in the electrolyte is increased, sealing the coating porosity [51, 83, 107]. This selective incorporation limits the distribution of particles in the inner layers, a phenomenon commonly observed in AC and pulsed bipolar PEO. The restricted inward mass transfer of particles is attributed to size constraints that prevent larger particles from accessing narrower discharge channels during coating formation.

Mohedano et al. [97] investigated the incorporation of Al_2O_3 (300 nm), ZrO_2 (300–700 nm), and CeO_2 (560 nm) into PEO coatings formed on AM50 magnesium alloy using a phosphate-based electrolyte with increasing particle concentrations (2, 5, and 10 g/L) to enhance corrosion resistance. They observed that as particle concentration increased, the coating thickness decreased and there was an early drop in current density. This reduction in thickness was related to a faster increase in the coating's impedance during the forming oxide layer process. Additionally, the sealing effect became more pronounced with higher particle concentrations, resulting in reduced surface porosity and roughness. The study also highlighted that the high temperatures generated by microdischarges (~1400 °C) during the PEO process promoted the formation of specific phases within the coatings. Specifically, tetragonal zirconia and $MgAl_2O_4$ were formed in coatings with incorporated ZrO_2 and Al_2O_3 particles, respectively. These phase transformations enhanced the structural and protective performance of the coatings, demonstrating that both particle size and concentration play crucial roles in optimising the in situ incorporation process and achieving desired coating functionalities.

Al Abri et al. [83] evaluated the influence of zeolite (<1 μm) and Ce-modified zeolite particle concentrations (0, 2, 6, and 10 g/L) on the in situ incorporation into PEO coatings formed on 1050 aluminium alloy using an aluminate-based electrolyte operating in bipolar mode. They reported an increase in both anodic and cathodic voltages with rising particle concentrations, a trend that was more pronounced for the Ce-containing zeolites (Fig. 9.4a), which was also observed in other works [82]. This increase was attributed to the adsorption of charged zeolite nanoparticles on the surface and their subsequent incorporation into the oxide layer, thereby promoting coating growth and densification. As expected, the total amount of incorporated particles increased with their concentration in the solution. Regarding the incorporation mechanism, they observed inert (non-reactive) incorporation predominantly at the coating surface through adsorption and entrapment into the molten oxide during coating growth (Fig. 9.4b). In contrast, within the inner layers, particles were incorporated reactively via field-assisted migration through flaws and defects in the coating (Fig. 9.4c).

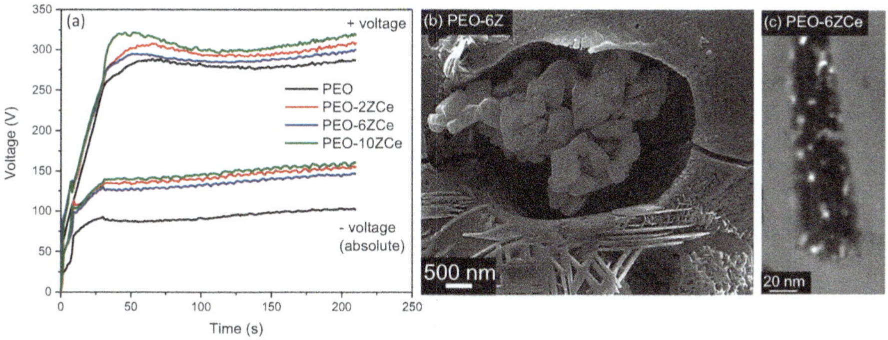

Fig. 9.4 (**a**) Voltage/time for PEO process of A1050 alloy with the addition of Ce-zeolite, (**b**) SEM micrograph of the outer layer of PEO coating formed with 6 g/L of zeolite particles, and (**c**) cross-sectional TEM micrograph of the inner region of the coating showing the reactive incorporation. (Reproduced with permission from [83] © 2024 The Authors. Published by Elsevier B.V.)

Tang [46] optimised the incorporation of TiC particles (20–50 nm) into PEO coatings formed on TC4 titanium alloy to enhance corrosion and wear resistance. To achieve this, they evaluated the influence of particle concentrations (2, 4, and 6 g/L) in a phosphate-based electrolyte. They observed that, at a fixed current density, the thickness of the coating decreased as particle concentration increased, which was related to a reduction in microdischarge intensity. Additionally, they also observed a direct relationship between the number of incorporated particles and the electrolyte particle concentration. Interestingly, the distribution of particles within the coating varied depending on the particle concentration in the electrolyte. Low particle concentrations (2 g/L) resulted in uniform particle incorporation throughout the coatings, whereas higher concentrations (4 and 6 g/L) led to an enrichment of particles in the inner layers of the coating, which is a phenomenon rarely observed. The authors linked this enrichment to the reactive incorporation of TiC particles during the formation process of the inner barrier layer.

9.3.3 Electrical Processing Parameters

PEO process is directly influenced by the electrical parameters applied during treatment, which play a pivotal role in determining the efficiency, distribution, and integration of nanoparticles within the resultant oxide coatings. Key parameters such as applied voltage, current density, pulse frequency, and duty cycle are critical in controlling the behaviour of microdischarges, thereby influencing the mechanisms by which particles are incorporated into the coating.

Generally, higher applied voltages and current densities lead to more energetic microdischarges, resulting in accelerated oxidation kinetics and increased localised temperatures. These intense discharges generate larger melting pools on the coating surface, providing more opportunities for particles to adhere and become embedded within the oxide layer [46, 108]. Consequently, an increase in the size of these melting pools facilitates the incorporation of a greater number of particles. Additionally, nanoparticles typically carry surface charges and, under alternating current (AC) or pulsed bipolar direct current (DC) conditions, migrate towards the electrode under the influence of the electric field [109]. Stronger electric fields enhance this migration, thereby promoting more efficient particle incorporation from the electrolyte into the coating. Pulse frequency and duty cycle also play significant roles in particle incorporation [108, 110]. Lower pulse frequencies combined with higher duty cycles generate more intense and sustained microdischarges, similar to the effects of increased voltage and current density leading to a higher rate of particle incorporation, which is related to the extended duration of each pulse, allowing more time for particles to interact with the molten oxide and become integrated into the coating.

While these intense thermodynamic conditions allow for the incorporation of higher concentrations of bulk particles from the electrolyte, this incorporation can occur through inert, partially reactive, or reactive mechanisms that often do not significantly affect the coating's passive functionalities. However, these same conditions pose challenges when dealing with more complex 2D and 3D structured nanoparticles. The aggressive thermodynamic conditions generated by high-energy discharges can lead to thermal oxidation or degradation of these structures, compromising their structural integrity and functional performance. Therefore, the successful incorporation of such nanoparticles requires meticulous optimisation of electrical parameters to control the intensity and duration of microdischarges, ensuring effective particle integration without damaging the nanoparticles.

However, it is important to highlight that in most reported studies, the incorporation of nanoparticles with complex structures predominantly occurs through reactive or partially reactive mechanisms, resulting in a reduced number of these nanoparticles within the PEO coatings. This limited incorporation, coupled with the degradation of active species (e.g. encapsulated corrosion inhibitors or drugs), significantly questions the effectiveness of using nanoparticles as carriers for active species in a single-step functionalisation process of PEO coatings. For instance, several attempts have been made to incorporate in situ LDH particles with different compositions into PEO coatings formed on Mg [91] and Al alloys [92, 93]. In all cases, the incorporation led to some microstructural changes associated mainly with changes in electrolyte conductivity (Fig. 9.5), which slightly modified the nature of the microdischarges and, therefore, the coating growth mechanism. In neither case were characteristic XRD peaks of LDH identified in the coatings, suggesting either a limited amount of incorporation or their reactive incorporation. It is worth noting that, despite this, some improvement in corrosion [91] and photocatalytic activity [92, 93] was observed. To the best of the authors' knowledge, no studies in the literature have attempted to optimise electrical processing conditions as a strategy for promoting the non-reactive incorporation of LDH particles. Only a few works have explored this approach with other types of 2D and 3D structured nanoparticles [76, 87].

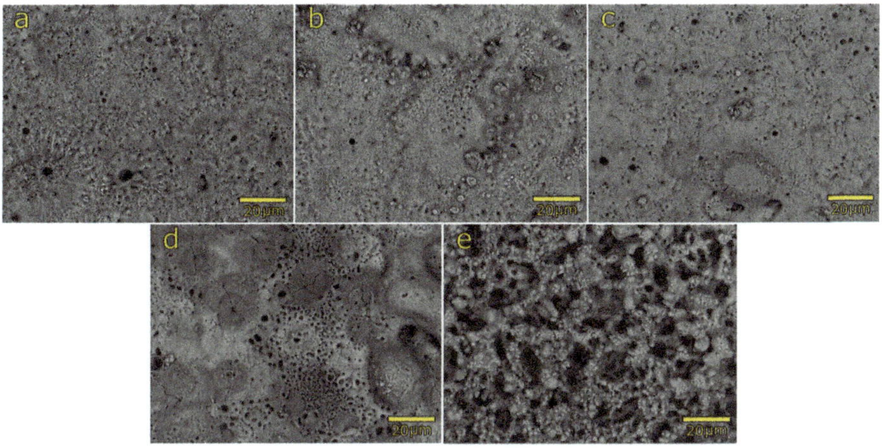

Fig. 9.5 SEM micrographs of PEO coatings formed on AA2024 Al alloy using (**a**) $K_2TiO(C_2O_4)_2$ electrolyte and modified with (**b**) MBT; (**c**) LDH[CO_3]; (**d**) LDH[CO_3] + MBT; (**e**) SE + LDH[MBT]. (Reproduced with permission from [92] © 2024 Elsevier B.V.)

Mingo et al. [87] evaluated the influence of processing frequency (100–5000 Hz) on the incorporation of HNTs as potential carriers for corrosion inhibitors in PEO coatings formed on AM50 Mg alloy using a silicate-based electrolyte operating in pulsed unipolar mode. They observed that lower frequencies (i.e. longer pulse durations) resulted in more intense and sustained microdischarges with higher temperatures, which generated larger melting pools. These conditions promoted the thermal degradation and decomposition of HNTs, causing them to lose their original tubular structure. Conversely, higher frequencies (e.g. 5000 Hz) involved shorter pulse durations, producing less intense microdischarges and smaller melting pools. This milder thermal environment preserved the tubular morphology of some HNTs, facilitating their non-reactive incorporation into the coating. As a result, coatings formed at higher frequencies retained more intact HNTs, potentially enhancing their role as effective carriers for active species. It is worth mentioning that the total number of incorporated particles was not sufficient to play a significant role in the overall corrosion performance. Wu [89] also evaluated the incorporation of HNTs loaded with an organic inhibitor (2-aminobenzimidazole) into PEO coatings formed on AZ31 Mg alloy using a silicate-based electrolyte operating in bipolar mode. They observed some partially reactive HNT incorporation into the outermost layer of the coating but provided no evidence of their incorporation into the inner layers. The authors noted a slight increase in corrosion resistance with the incorporation of loaded HNTs, but this is most likely related to changes in the coating microstructure as a consequence of altered electrolyte conductivity after adding the encapsulated corrosion inhibitor to the electrolyte.

Zhao and co-workers [76] evaluated the influence of duty cycles (10%, 20%, 30%) on the incorporation of h-BN particles into PEO coatings formed on Ti6Al4V using a silicate-phosphate electrolyte for anti-friction applications. Increasing the duty cycles enhances the incorporation of h-BN nanoparticles into the coating, as higher duty cycles prolong the duration of each discharge pulse, allowing more h-BN particles to migrate, become entrapped within the molten oxides, and sinter within the coating matrix. The authors claimed that the incorporation is primarily reactive. The high-energy plasma discharges facilitate chemical reactions between h-BN and components of the coating matrix (such as SiO_2), leading to the formation of bonded glass phases that enhance coating density and corrosion resistance. However, they also identified small XRD peaks at higher duty cycles (30%) corresponding to h-BN, which suggest a minor non-reactive incorporation.

Saji [111] evaluated the in situ incorporation of both h-BN nanosheets and organo-modified nanoclays into PEO coatings on AZ31B magnesium alloy using a silicate-fluoride alkaline electrolyte. Although they did not attempt to optimise the electrical parameters, they interestingly observed that the presence of low melting point nanoclays, incorporated reactively, facilitates the effective incorporation of h-BN nanosheets into the PEO layer. It is worth noting that h-BN was not analytically detected by XRD, leaving it unclear whether successful non-reactive incorporation was achieved.

The optimisation of electrical processing parameters has been demonstrated as a successful strategy for promoting the incorporation of bulk particles into PEO coatings, achieving inert, partially reactive, and reactive incorporations. Notably, in cases of reactive incorporation, the functionality of the incorporated particles within the coating is not necessarily compromised. However, limited research exists on the optimisation of electrical parameters for the incorporation of more complex 2D and 3D nanoparticles. These nanoparticles, which often require precise structural integrity to act as carriers for other substances, present unique challenges under typical PEO conditions. Current studies suggest that the optimisation strategies used for complex particles are insufficient to achieve a consistently non-reactive incorporation, indicating a significant area for further exploration and innovation.

9.4 Mechanisms of Particle Incorporation

The nature of the particle, including its size, structure, melting point, and chemical stability, significantly determines the incorporation mode into PEO coatings, which can be inert (non-reactive), partially reactive, or fully reactive. Bulk particles, such as large-sized oxides, are typically inertly incorporated due to their high melting point and chemical stability. In contrast, nano-sized particles of the same material exhibit lower melting points and reduced chemical stability, making them more susceptible to reactive incorporation into the coating. This often leads to the formation of new phases, such as glassy or crystalline structures, enhancing coating properties like density and corrosion resistance [93, 112]. Under high-energy discharge conditions, particles may not only partially melt but can also experience a range of additional reactions. These may include thermal decomposition, oxidation, or other chemical transformations, depending on the particle's composition and the local plasma environment. For instance, some particles may break down into simpler compounds, form oxide layers, or even react with the surrounding electrolyte to create new phases within the growing oxide coating [111].

Oxides are generally incorporated in a non-reactive manner since they are chemically and thermally stable, limiting further chemical transformations. However, carbides and nitrides, although commonly introduced in a largely inert form, may still undergo partial surface reactions if the discharge energies and local temperatures are sufficient. In some instances, surface oxidation can occur, creating mixed oxide-carbide (Fig. 9.6) or oxide-nitride interfaces [40, 49]. Nevertheless, their incorporation often remains predominantly non-reactive, preserving much of the particle's original chemistry and structure.

Metallic particles are commonly incorporated in a non-reactive way [5, 113], but if they are sufficiently small or exposed to prolonged discharge events, they may be oxidised during the deposition process [52]. Depending on particle size and local conditions, this can result in an outer oxide shell surrounding a metallic core [55, 56], or complete oxidation of the particle [105]. Polymeric particles, such as PTFE, can also be incorporated into PEO coatings. However, due to their generally low melting points, they are typically incorporated reactively, as the high temperatures and reactive species in the plasma degrade or modify their surface properties [114]. Depending on the particles properties and processing conditions, the partially reactive incorporation is also feasible, resulting in some inert incorporation alongside the formation of degradation secondary products [115, 116].

The structural complexity of particles also plays a vital role. Simpler bulk shapes like spherical particles may integrate relatively easily, more complex 2D and 3D structured nanoparticles are frequently subject to reactive or partially reactive incorporation. Their larger surface area, lower thermal stability, and layered or porous structure make them more susceptible to melting, partial chemical conversion, or bonding with the oxide matrix. This complexity poses challenges for achieving strictly non-reactive incorporation, especially when the objective is to use these particles as carriers for inhibitors, drugs, or other functional agents that require preserved structural and chemical integrity within the coating.

The general consensus identifies particle incorporation as a multi-step process: (1) particles migrate from the suspension to the electrode surface, often assisted by the electric field, (2) they adsorb onto the coating substrate, and (3) they become physically entrapped within molten oxides. However, other theories highlight that particles may also penetrate through defects, open pores and discharge channels, adding complexity to this phenomenon [5].

The surface charge of particles influences their migration and eventual incorporation during the PEO process, driven by the electric field. Positively charged particles migrate towards the substrate during the negative pulse, while negatively charged particles move during the positive pulse. This electrophoretic migration is critical for ensuring particle delivery to the coating surface, especially under high-voltage conditions where the electric field strength significantly accelerates charged particles. The migration of particles towards the electrode can also assisted by stirring or bubbling [10, 94]. When particles reach the coating surface, their

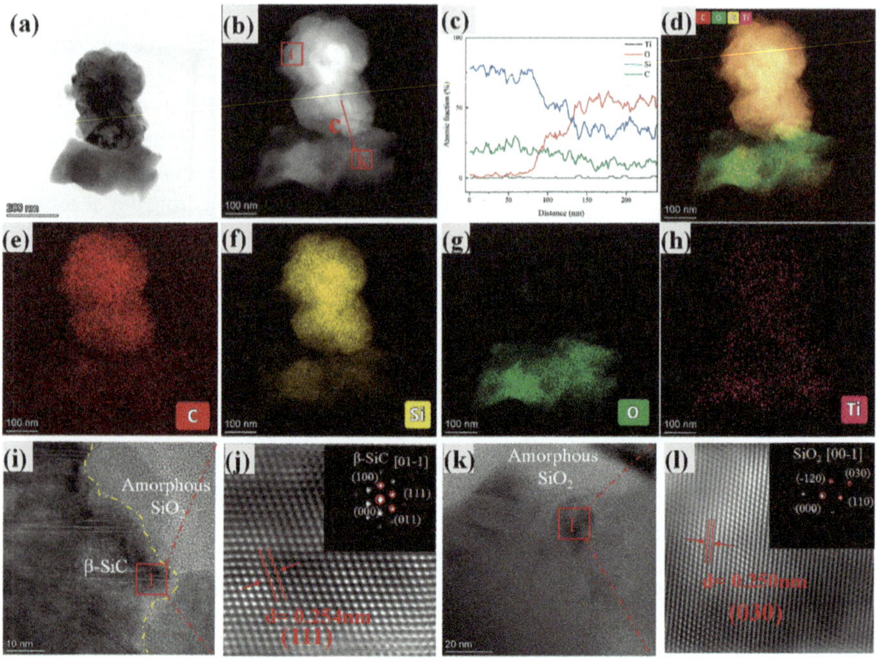

Fig. 9.6 TEM characterisation of the interface between a partially oxidised SiC particle and SiO_2 embedded in a PEO coating formed on a Ti–6Al–4V alloy using a silicate–phosphate electrolyte containing SiC particles. (**a**) TEM bright-field micrograph; (**b**) HAADF-STEM micrograph; (**c**) Linear elemental distribution along line c in (**b**); (**d–h**) Elemental maps (**d**: all elements, **e**: C, **f**: Si, **g**: O, **h**: Ti); (**i**) HRTEM micrograph of nanoparticles in region I of (**b**); (**j**) Fourier transform and SAED pattern from the red boxed region of (**i**); (**k**) HRTEM image of nanoparticles in region k of (**a**); (**l**) Fourier transform and SAED pattern from the red boxed region of (**k**). (Reproduced with permission from [40] © 2024 Elsevier Ltd and Techna Group S.r.l.)

incorporation is influenced by local discharge conditions, the coating matrix's thermal and chemical environment, and the particle's intrinsic properties. One common mechanism is physical entrapment, where molten oxides at the discharge sites encapsulate the particles, effectively embedding them preferentially into the outer layer of the coating (Fig. 9.7) [15, 16, 117]. Surface pores on the coatings are widely regarded as primary pathways for particle uptake, as these pores are often observed to be filled with particles after PEO treatment. This typically results in non-reactive incorporation, even for complex 3D structured nanoparticles (Figs. 9.4 and 9.5b) [83].

However, some authors propose that particles can also penetrate into the coating through short-circuit paths, defects and discharge channels in the porous outer layer, reaching deeper interfaces between the inner and outer layers [97]. Such deeper penetration generally exposes particles to more extreme conditions, increasing the likelihood of reactive incorporation (Fig. 9.4c).

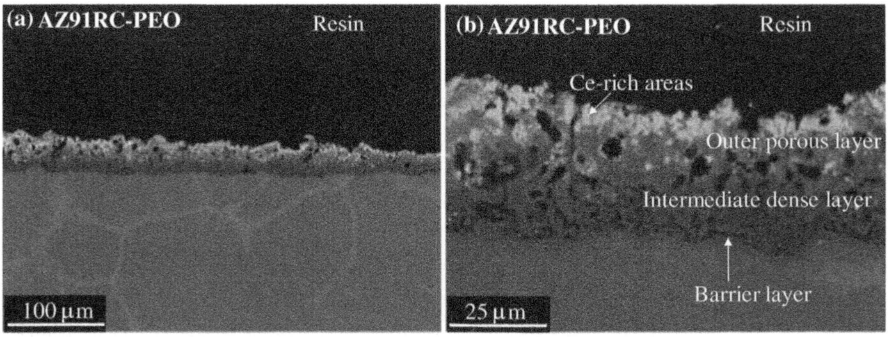

Fig. 9.7 (**a**, **b**) SEM micrographs of PEO coatings formed on AZ91RC Mg alloy in a phosphate-based electrolyte containing CeO_2 particles, showing their preferential incorporation in the outer layer. (Reproduced with permission from [117] © 2016 Elsevier B.V.)

Interestingly, Polunin [118] observed that the incorporation of particles considerably alters the PEO process, primarily by modifying local discharge conditions rather than contributing a large mass fraction to the coating. Although the incorporated particles (TiC) themselves remain mostly inert (due to their high melting point and chemical stability), their inclusion lowers the breakdown voltage required for microarc discharges. As a result, more discharges occur, or they last longer, facilitating additional molten metal ejection and increasing the overall oxide layer thickness. They proposed a mechanism in which TiC particles enter the coating through micropores, discharge channels, and vapour–gas bubbles (VGBs) (Fig. 9.8). Negatively charged TiC particles become embedded in the pore walls and microchannel interfaces under the electric field, sometimes acquiring a positive charge via triboelectric effects. These charged particles can then locally trigger sparking, promoting electron avalanche formation that leads to microarc breakdown. With more frequent or sustained discharges, temperatures rise locally, favouring phase transformations—such as increased crystallinity, conversion of metastable oxides to stable phases, and reduced residual stresses. Consequently, while TiC particles themselves remain largely unreacted, their ability to enhance and extend microarc discharge activity significantly affects coating formation, improving coating density, stabilising phases (e.g. mullite), and thickening the oxide layer.

In short, while the general framework of migration, adsorption, and entrapment offers a baseline understanding of how particles incorporate into PEO coatings, the actual mechanisms are considerably more complex. Factors such as particle size, structure, melting point, and surface charge, combined with the evolving local discharge conditions, can result in a range of incorporation modes, from inert to fully reactive. Furthermore, the potential for chemical transformations, limited penetration depths, and alternative pathways beyond the coating surface complicate our ability to reliably achieve a desired incorporation mode. This complexity highlights the need for additional research, aiming to refine our control over these variables and ultimately guide the engineering of PEO coatings with tailored properties and enhanced functionalities.

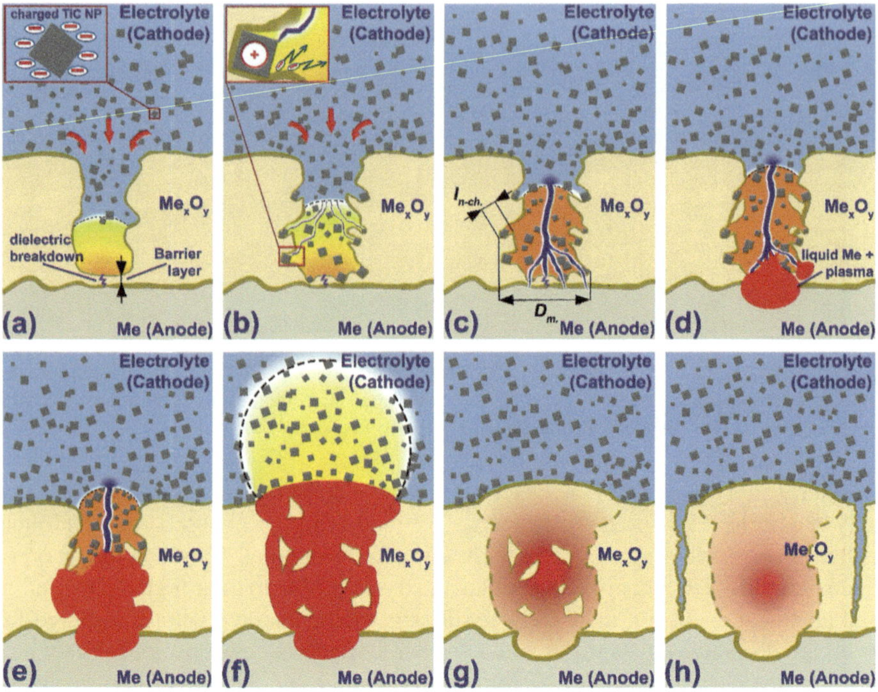

Fig. 9.8 Simplified cycle of oxide layer formation with TiC NPs: (**a**) barrier layer breakdown and formation of a vapour–gas bubble (VGB), (**b**) embedding of negatively charged TiC NPs into microchannel walls and local sparking, (**c**) transformation of the VGB into a plasma bubble and microarc discharge branching on TiC sites, (**d**) melting of the barrier layer and substrate, (**e**) expansion of the microchannel due to detachment of oxide chunks, (**f**) outflow of molten material, quenching of the microarc discharge, and bubble transition, (**g**) cooling and solidification, and (**h**) formation of new adjacent microchannels due to stress-induced cracking. (Reproduced with permission from [118] © 2021 Elsevier B.V.)

9.5 Conclusions and Outlook

The incorporation of bulk particles into PEO coatings has proven a successful approach to introduce additional functionalities—including enhanced corrosion resistance, improved wear behaviour, photocatalytic activity, thermal conductivity, and antibacterial properties—without significantly compromising their effectiveness. This achievement stems from the fact that, for bulk particles, the incorporation mode (whether inert, partially reactive, or fully reactive) does not drastically diminish their inherent functionality.

However, achieving similar outcomes with more complex 2D and 3D structured nanoparticles that serve as carriers for inhibitors, drugs, or other active agents remains a significant challenge. The inherently reactive conditions within the PEO environment often lead to partial melting, oxidation, or other chemical transformations of these nanostructures. As a result, their three-dimensional integrity and, consequently, their capacity to function as carriers can be compromised. This limitation highlights a critical gap in the current understanding and control of the incorporation process.

Moving forward, the primary goal is to refine the PEO process to achieve a stable, non-reactive incorporation of these advanced nanoparticles. This may involve carefully tailoring the electrical parameters, electrolyte chemistry, and particle surface characteristics to minimise undesirable reactions and preserve nanoparticle morphology. By combining systematic experimentation, advanced characterisation techniques, and modelling approaches, researchers can work towards unlocking the full potential of complex nanoparticle incorporation, ultimately enabling the development of multifunctional PEO coatings with increasingly sophisticated and responsive properties.

Acknowledgements The authors are grateful for the funding provided by UK EPSRC (EP/V026097/1) and from the Royal Academy of Engineering under the Research Fellowship scheme (RF\202021\20\123).

References

1. Yerokhin, A. L., Nie, X., Leyland, A., Matthews, A., & Dowey, S. J. (1999). *Surface and Coating Technology, 122*, 73–93.
2. Molaei, M., Babaei, K., Fattah-alhosseini, A. (2021). *Journal of Magnesium and Alloys, 9*, 1164–1186.
3. Babaei, K., Fattah-Alhosseini, A., & Molaei, M. (2020). *Surfaces and Interfaces, 21*, 100677.
4. An, L., Ma, Y., Liu, Y., Sun, L., Wang, S., & Wang, Z. (2018). *Surface and Coating Technology, 354*, 226–235.
5. Lu, X., Mohedano, M., Blawert, C., Matykina, E., Arrabal, R., Kainer, K. U., & Zheludkevich, M. L. (2016). *Surface and Coating Technology, 307*, 1165–1182.
6. Mingo, B., Guo, Y., Leiva-Garcia, R., Connolly, B. J., Matthews, A., & Yerokhin, A. (2020). *ACS Applied Materials & Interfaces, 12*, 30833–30846.
7. Fathyunes, L., Hosseini, M., & Khalil-Allafi, J. (2024). *Materials Today Communications, 40*, 109443.
8. Vasilyeva, M. S., Rudnev, V. S., Korotenko, I. A., & Nedozorov, P. M. (2012). *Protection of Metals and Physical Chemistry of Surfaces, 48*, 106–115.
9. Asgari, M., Aliofkhazraei, M., Barati Darband, G., & Sabour Rouhaghdam, A. (2020). *Surface and Coating Technology, 383*, 125252.
10. Asgari, M., Aliofkhazraei, M., Darband, G. B., & Rouhaghdam, A. S. (2017). *Surface and Coating Technology, 309*, 124–135.
11. Mohedano, M., Blawert, C., & Zheludkevich, M. (2015). *Materials and Design, 86*, 735–744.
12. Wang, K., Wu, S., Huang, Y., Wen, R., Chen, Y., Zhou, D., Wu, L., Song, J., Dong, Z., & Jiang, B. (2022). *Journal of Materials Research and Technology, 21*, 1344–1352.
13. Lu, X., Blawert, C., Huang, Y., Ovri, H., Zheludkevich, M. L., & Kainer, K. U. (2016). *Electrochimica Acta, 187*, 20–33.
14. Ma, G., Li, Z., Zhao, X., Wang, Z., Kou, R., Sun, S., Sun, Y., Wang, S., & Yang, Y. (2024). *Journal of Alloys and Compounds, 1008*, 176673.
15. Lopez-Martinez, E., Mohedano, M., Matykina, E., & Arrabal, R. (2024). *Journal of Materials Research and Technology, 31*, 1786–1796.

16. Arrabal, R., Mohedano, M., Matykina, E., Pardo, A., Mingo, B., & Merino, M. C. (2015). *Surface and Coating Technology, 269*, 64–73.
17. Molaei, M., Nouri, M., Babaei, K., & Fattah-Alhosseini, A. (2021). *Surfaces and Interfaces, 22*, 100888.
18. Fattah-Alhosseini, A., Molaei, M., & Babaei, K. (2020). *Surfaces and Interfaces, 21*, 100659.
19. Muhaffel, F., Kaba, M., Cempura, G., Derin, B., Kruk, A., Atar, E., & Cimenoglu, H. (2019). *Surface and Coating Technology, 377*, 124900.
20. Wang, P., Pu, J., Zhou, J. H., Cao, W. J., Xiao, Y. T., Gong, Z. Y., & Hu, J. (2018). *International Journal of Electrochemical Science, 13*, 8995–9006.
21. Li, H., Sun, Y., & Zhang, J. (2015). *Applied Surface Science, 342*, 183–190.
22. Gnedenkov, S., Sinebryukhov, S., Mashtalyar, D., Imshinetskiy, I., Samokhin, A., & Tsvetkov, Y. V. (2015). *Journal of Nanomaterials, 2015*, 154298.
23. Mashtalyar, D. V., Imshinetskiy, I. M., Nadaraia, K. V., Gnedenkov, A. S., Suchkov, S. N., Opra, D. P., Pustovalov, E. V., Yu Ustinov, A., Sinebryukhov, S. L., & Gnedenkov, S. V. (2023). *Journal of Magnesium and Alloys, 11*, 735–752.
24. Madhankumar, A., Thangavel, E., Ramakrishna, S., Obot, I. B., Jung, H. C., Shin, K. S., Gasem, Z. M., Kim, H., & Kim, D.-E. (2014). *RSC Advances, 4*, 24272–24285.
25. Atapour, M., Blawert, C., & Zheludkevich, M. L. (2019). *Surface and Coating Technology, 357*, 626–637.
26. Roknian, M., Fattah-alhosseini, A., Gashti, S. O., & Keshavarz, M. K. (2018). *Journal of Alloys and Compounds, 740*, 330–345.
27. Sirelkhatim, A., Mahmud, S., Seeni, A., Kaus, N. H. M., Ann, L. C., Bakhori, S. K. M., Hasan, H., & Mohamad, D. (2015). *Nano Micro Letters, 7*, 219–242.
28. Ma, X., Zhou, S., Xu, X., & Du, Q. (2022). *Frontiers in Surgery, 9*, 905892.
29. Gudkov, S. V., Serov, D. A., Astashev, M. E., Semenova, A. A., & Lisitsyn, A. B. (2022). *Pharmaceuticals, 15*, 968.
30. Lv, P., Xu, X., Zhou, W., Dong, Q., Guan, L., Li, Z., Sun, S., & Li, L. (2024). *Ceramics International, 50*, 55449–55460.
31. Chen, J., Zhang, Y., Ibrahim, M., Etim, I. P., Tan, L., & Yang, K. (2019). *Colloids and Surfaces. B, Biointerfaces, 179*, 77–86.
32. Lv, Y., Wu, Y., Lu, X., Yu, Y., Fu, S., Yang, L., Dong, Z., & Zhang, X. (2019). *Ceramics International, 45*, 22357–22367.
33. Zhao, D., Lu, Y., Wang, Z., Zeng, X., Liu, S., & Wang, T. (2016). *International Journal of Refractory Metals and Hard Materials, 54*, 417–421.
34. Chakravorty, A., & Roy, S. (2024). *Sustainable Chemistry for the Environment, 8*, 100155.
35. Coto, M., Troughton, S. C., Knight, P., Joshi, R., Francis, R., Kumar, R. V., & Clyne, T. W. (2021). *Surface and Coating Technology, 411*, 127000.
36. Luo, Q., Cai, Q.-z., Li, X.-w., Pan, Z.-h., Li, Y.-j., Chen, X.-d., & Yan, Q.-s. (2013). *Transactions of the Nonferrous Metals Society of China, 23*, 2950.
37. Li, W., Tang, M., Zhu, L., & Liu, H. (2012). *Applied Surface Science, 258*, 10017–10021.
38. Stojadinović, S., Perković, M., & Radić, N. (2024). *Solid State Sciences, 156*, 107661.
39. Kaseem, M., Hussain, T., Rehman, Z. U., Banisalman, M. J., & Ko, Y. G. (2020). *Surface and Coating Technology, 399*, 126200.
40. Wang, J., Zhang, L., Cheng, J., Liu, J., Qi, D., Zou, Y., Wei, D., Cheng, S., & Wang, Y. (2024). *Ceramics International, 50*, 46288–46299.
41. Qi, D., Zhang, L., Zou, Y., Fu, Y., Wang, J., Wang, Z., Zhang, K., Wang, Y., Guo, Y., Cheng, S., & Lin, L. (2023). *Materials Characterization, 206*, 113442.
42. Pezzato, L., Lorenzetti, L., Tonelli, L., Bragaggia, G., Dabalà, M., Martini, C., & Brunelli, K. (2021). *Surface and Coating Technology, 428*, 127901.
43. Yu, L., Cao, J., & Cheng, Y. (2015). *Surface and Coating Technology, 276*, 266–278.
44. Zhang, Y., Xu, Y., Miao, C., Tu, X., Yu, J., & Li, J. (2018). *International Journal of Electrochemical Science, 13*, 7923–7929.
45. Molaeipour, P., Allahkaram, S. R., & Akbarzadeh, S. (2022). *Surface and Coating Technology, 430*, 127987.

46. Tang, Y., Yang, C.-P., Sun, Q.-Q., Wu, L.-K., & Cao, F.-H. (2024). *Corrosion Communications, 14*, 1–10.
47. Yaakop, N. (2019). *Plasma electrolytic oxidation of aluminium for power electronics applications*. The University of Manchester (UK).
48. Yang, W., Wu, S., Xu, D., Gao, W., Yao, Y., Guo, Q., & Chen, J. (2020). *Ceramics International, 46*, 17112–17116.
49. Mashtalyar, D. V., Sinebryukhov, S. L., Imshinetskiy, I. M., Gnedenkov, A. S., Nadaraia, K. V., Ustinov, A. Y., & Gnedenkov, S. V. (2020). *Applied Surface Science, 503*, 144062.
50. Zhang, Y., Chen, F., Zhang, Y., Liu, M., Pang, Y., & Du, C. (2022). *Surface and Coating Technology, 444*, 128662.
51. Lu, X., Blawert, C., Scharnagl, N., & Kainer, K. U. (2013). *Journal of Magnesium and Alloys, 1*, 267–274.
52. Han, Q., Li, Y., Lu, X., Mei, D., Chen, Q., Zhang, T., & Wang, F. (2021). *Materials Letters, 293*, 129731.
53. Cerchier, P., Pezzato, L., Gennari, C., Moschin, E., Moro, I., & Dabalà, M. (2020). *Surface and Coating Technology, 393*, 125774.
54. Radić, N., Ilić, M., Stojadinović, S., Milić, J., Avdalović, J., & Šaponjić, Z. (2024). *Journal of Physics and Chemistry of Solids, 188*, 111918.
55. Gnedenkov, S., Sinebryukhov, S., Tkachenko, I., Mashtalyar, D., Ustinov, A. Y., Samokhin, A., & Tsvetkov, Y. V. (2012). *Inorganic Materials: Applied Research, 3*, 151–156.
56. Rudnev, V. S., Adigamova, M. V., Lukiyanchuk, I. V., Ustinov, A. Y., Tkachenko, I. A., Kharitonskii, P. V., Frolov, A. M., & Morozova, V. P. (2012). *Protection of Metals and Physical Chemistry of Surfaces, 48*, 543–552.
57. Adigamova, M. V., Lukiyanchuk, I. V., Morozova, V. P., Tkachenko, I. A., & Kilin, K. N. (2022). *Surface and Coating Technology, 446*, 128790.
58. Tanaka, K., & Kawakami, S. (1982). *Wear, 79*, 221–234.
59. Wang, S., Wen, L., Wang, Y., Cheng, Y., Cheng, Y., Zou, Y., Zhu, Y., Chen, G., Ouyang, J., Jia, D., & Zhou, Y. (2022). *Applied Surface Science, 592*, 153043.
60. Zhang, R., Zhao, J., & Liang, J. (2016). *Surface and Coating Technology, 299*, 90–95.
61. Rudnev, V., Vaganov-Vil'kins, A., Nedozorov, P., Yarovaya, T., & Chigrinova, N. (2012). *Russian Journal of Applied Chemistry, 85*, 1147–1152.
62. Kaplan, E., Sukuroglu, E. E., & Cuvalci, O. (2021). *Journal of Adhesion Science and Technology, 35*, 1305–1319.
63. Lv, G.-H., Chen, H., Gu, W.-C., Feng, W.-R., Li, L., Niu, E.-W., Zhang, X.-H., & Yang, S.-Z. (2009). *Current Applied Physics, 9*, 324–328.
64. Wang, P., Pu, J., Xiao, Y. T., Hu, W. J., Wu, T., Cao, W. J., Gong, Z. Y., & Huang, M. (2019). *Surface Review and Letters, 26*, 1850178.
65. Yu, L., Jia, P., Song, Y., Zhao, B., Pan, Y., Wang, J., Cui, H., Feng, R., Li, H., Cui, X., Gao, Z., Fang, X., & Zhang, L. (2022). *Journal of Materials Research and Technology, 18*, 3541–3552.
66. Rogala-Wielgus, D., & Zieliński, A. (2024). *Carbon Letters, 34*, 565–601.
67. Yürektürk, Y., Muhaffel, F., & Baydoğan, M. (2015). *Surface and Coating Technology, 269*, 83–90.
68. Lee, K. M., Ko, Y. G., & Shin, D. H. (2011). *Materials Letters, 65*, 2269–2273.
69. Sabouri, M., & Mousavi Khoei, S. M. (2018). *Surface and Coating Technology, 334*, 543–555.
70. Huang, X., Yin, Z., Wu, S., Qi, X., He, Q., Zhang, Q., Yan, Q., Boey, F., & Zhang, H. (2011). *Small, 7*, 1876–1902.
71. Kirkland, N. T., Schiller, T., Medhekar, N., & Birbilis, N. (2012). *Corrosion Science, 56*, 1–4.
72. Qiu, Z., Wang, R., Wu, J., Zhang, Y., Qu, Y., & Wu, X. (2015). *RSC Advances, 5*, 44149–44159.
73. Gao, H., Wang, C., Jiang, B., & Song, R. (2024). *International Journal of Materials Research, 115*, 59–68.
74. Nasiri Vatan, H., & Adabi, M. (2018). *Transactions of the Indian Institute of Metals, 71*, 1643–1652.
75. Mu, M., Liang, J., Zhou, X., & Xiao, Q. (2013). *Surface and Coating Technology, 214*, 124–130.

76. Zhao, X., Chen, Y., Ji, R., Xu, M., Ye, Z., Shao, W., Wang, S., Zou, Y., Wang, Y., Ouyang, J., Jia, D., & Zhou, Y. (2024). *Surface and Coating Technology, 494*, 131471.
77. Lou, B.-S., Lee, J.-W., Tseng, C.-M., Lin, Y.-Y., & Yen, C.-A. (2018). *Surface and Coating Technology, 350*, 813–822.
78. Ghorbanian, B., Tajally, M., Khoie, S. M. M., & Tavakoli, H. (2020). *Surface Innovations, 8*, 252–262.
79. Mehmood, A., Ghafar, H., Yaqoob, S., Gohar, U. F., & Ahmad, B. (2017). *Journal of Developing Drugs, 6*, 1000174.
80. Chen, Q., Lei, M., Chen, Y., Deng, Y., & Chen, M.-a. (2023). *Ceramics International, 49*, 32679–32693.
81. Sherry, H. S., & Walton, H. F. (1967). *The Journal of Physical Chemistry, 71*, 1457–1465.
82. Chen, Q., Lu, X., Serdechnova, M., Wang, C., Lamaka, S., Blawert, C., Zheludkevich, M. L., & Wang, F. (2022). *Corrosion Science, 209*, 110785.
83. Al Abri, S., Rogov, A., Aliasghari, S., Bendo, A., Matthews, A., Yerokhin, A., & Mingo, B. (2024). *Journal of Materials Research and Technology, 30*, 2365–2376.
84. Zahidah, K. A., Kakooei, S., Ismail, M. C., & Bothi Raja, P. (2017). *Progress in Organic Coating, 111*, 175–185.
85. Guimaraes, L., Enyashin, A. N., Seifert, G., & Duarte, H. A. (2010). *The Journal of Physical Chemistry C, 114*, 11358–11363.
86. Ma, Q., Liu, L., Wang, P., Wang, S., Wang, Y., Li, W., Wang, Z., & Zhao, Y. (2024). *Materials Today Communications, 38*, 108541.
87. Mingo, B., Guo, Y., Němcová, A., Gholinia, A., Mohedano, M., Sun, M., Matthews, A., & Yerokhin, A. (2019). *Electrochimica Acta, 299*, 772–788.
88. Sun, M., Yerokhin, A., Bychkova, M. Y., Shtansky, D. V., Levashov, E. A., & Matthews, A. (2016). *Corrosion Science, 111*, 753–769.
89. Wu, W. X., Wang, W. P., & Lin, H. C. (2021). *Surface and Coating Technology, 416*, 127116.
90. Cao, Y., Zheng, D., Zhang, F., Pan, J., & Lin, C. (2022). *Journal of Materials Science and Technology, 102*, 232–263.
91. Li, Y., Lu, X., Serdechnova, M., Blawert, C., Zheludkevich, M. L., Qian, K., Zhang, T., & Wang, F. (2023). *Journal of Magnesium and Alloys, 11*, 1236–1246.
92. Mojsilović, K., Serdechnova, M., Blawert, C., Zheludkevich, M. L., Stojadinović, S., & Vasilić, R. (2024). *Applied Surface Science, 654*, 159450.
93. Mojsilović, K., Serdechnova, M., Blawert, C., Zheludkevich, M. L., Stojadinović, S., & Vasilić, R. (2024). *Applied Surface Science, 677*, 161065.
94. Lu, X., Blawert, C., Zheludkevich, M. L., & Kainer, K. U. (2015). *Corrosion Science, 101*, 201–207.
95. Li, Y., Li, C., Wang, Q., Zhao, X., Zhang, T., & Wang, F. (2024). *Materials Today Communications, 38*, 108234.
96. Yeung, W. K., Sukhorukova, I. V., Shtansky, D. V., Levashov, E. A., Zhitnyak, I. Y., Gloushankova, N. A., Kiryukhantsev-Korneev, P. V., Petrzhik, M. I., Matthews, A., & Yerokhin, A. (2016). *RSC Advances, 6*, 12688–12698.
97. Mohedano, M., Arrabal, R., Mingo, B., Pardo, A., & Matykina, E. (2018). *Surface and Coating Technology, 334*, 328–335.
98. Hakimizad, A., Raeissi, K., Golozar, M. A., Lu, X., Blawert, C., & Zheludkevich, M. L. (2017). *Surface and Coating Technology, 324*, 208–221.
99. Bordbar-Khiabani, A., Yarmand, B., & Mozafari, M. (2019). *Surface and Coating Technology, 360*, 153–171.
100. Zehra, T., Kaseem, M., Hossain, S., & Ko, Y.-G. (2021). *Metals, 11*, 1182.
101. Wang, X., Ju, P., Lu, X., Chen, Y., & Wang, F. (2022). *Ceramics International, 48*, 3615–3627.
102. Li, X., & Luan, B. L. (2012). *Materials Letters, 86*, 88–91.
103. Han, J., Yu, Y., Yang, J., Xiaopeng, L., Blawert, C., & Zheludkevich, M. L. (2022). *Surface and Coating Technology, 433*, 128112.
104. Zhu, M., Song, Y., Xu, J., Dong, K., & Han, E.-H. (2023). *Surface and Coating Technology, 457*, 129334.

105. Zhang, X., Zhang, Y., Lv, Y., Dong, Z., Yang, L., Zhang, E., Hashimoto, T., & Zhou, X. (2023). *Journal of Magnesium and Alloys, 11*, 2182–2195.
106. Sedelnikova, M. B., Komarova, E. G., Sharkeev, Y. P., Ugodchikova, A. V., Mushtovatova, L. S., Karpova, M. R., Sheikin, V. V., Litvinova, L. S., & Khlusov, I. A. (2019). *Surface and Coating Technology, 369*, 52–68.
107. Lu, X., Sah, S. P., Scharnagl, N., Störmer, M., Starykevich, M., Mohedano, M., Blawert, C., Zheludkevich, M. L., & Kainer, K. U. (2015). *Surface and Coating Technology, 269*, 155–169.
108. Lu, X., Blawert, C., Mohedano, M., Scharnagl, N., Zheludkevich, M. L., & Kainer, K. U. (2016). *Surface and Coating Technology, 289*, 179–185.
109. Kaseem, M., Min, J. H., & Ko, Y. G. (2017). *Journal of Alloys and Compounds, 723*, 448–455.
110. Liu, X. H., Shi, C. X., Zhang, B., Liu, L., Dong, S., & Dong, J. (2023). *Surface Technology, 52*, 41–50.
111. Saji, V. S., Ogunlakin, N., & Madhan Kumar, A. (2024). *Applied Surface Science, 669*, 160476.
112. Ji, R., Zou, Y., Wang, S., Chen, G., Wang, Y., Ouyang, J., Jia, D., & Zhou, Y. (2023). *Surface and Coating Technology, 459*, 129382.
113. Necula, B. S., Apachitei, I., Tichelaar, F. D., Fratila-Apachitei, L. E., & Duszczyk, J. (2011). *Acta Biomaterialia, 7*, 2751–2757.
114. Chen, Y., Lu, X., Blawert, C., Zheludkevich, M. L., Zhang, T., & Wang, F. (2018). *Surface and Coating Technology, 337*, 379–388.
115. Rudnev, V. S., Vaganov-Vil'kins, A. A., Yarovaya, T. P., & Pavlov, A. D. (2016). *Surface and Coating Technology, 307*, 1249–1254.
116. Wang, S., Wang, Y., Zou, Y., Wu, Y., Chen, G., Ouyang, J., Jia, D., & Zhou, Y. (2020). *Chemical Engineering Journal, 402*, 126116.
117. Mingo, B., Arrabal, R., Mohedano, M., Pardo, A., & Matykina, E. (2017). *Surface and Coating Technology, 309*, 1023–1032.
118. Polunin, A. V., Cheretaeva, A. O., Borgardt, E. D., Rastegaev, I. A., Krishtal, M. M., Katsman, A. V., & Yasnikov, I. S. (2021). *Surface and Coating Technology, 423*, 127603.

Chapter 10
Plasma Electrolytic Oxidation Coatings with Enhanced Corrosion Resistance via Particle Incorporation

T. S. N. Sankara Narayanan

10.1 Introduction

Plasma electrolytic oxidation (PEO) (also referred to as microarc oxidation) is a well-established surface modification process, leading to the formation of an oxide coating on titanium (Ti), aluminium (Al), magnesium (Mg), tantalum (Ta) and zirconium (Zr). Anodic dissolution of metal and development of a barrier oxide film accompanied by O_2 evolution occur during the initial stage of PEO. Dielectric breakdown occurs at weaker points of the barrier oxide film. The microarc discharges produced during the PEO process increase the localized temperature and pressure and trigger plasma chemical reactions, thus promoting the formation of an oxide coating. PEO coatings are porous with a typical pancake-like structure emanated by gas evolution through the molten oxide, commonly referred to as discharge channels. The rapid solidification of the molten oxide by the cooler electrolyte generates thermal stresses, leading to the formation of microcracks in the oxide layer [1, 2]. The architecture of PEO coatings consists of three different layers: (1) a thin and dense inner barrier layer, (2) an intermediate pore band, and (3) a porous outer layer [3].

10.2 Incorporation of Particles in PEO Coatings

Incorporation of particles in PEO coatings involves two stages, namely uptake and incorporation. During the initial periods of PEO, electrophoretic movement due to the negative ζ-potential of the particles is the main driving force, leading to

T. S. N. Sankara Narayanan (✉)
Department of Analytical Chemistry, University of Madras, Chennai, India

V. S. Saji, M. Zheludkevich (eds.), *Plasma Electrolytic Oxidation – Fundamentals, Advances and Applications*,
https://doi.org/10.1007/978-3-031-92429-3_10

adsorption of the particles on the anode [4, 5]. Fortunately, most of the electrolytes used for PEO are alkaline in which most of the particles assume a negative ζ-potential. The convection of the particles is further assisted by the mechanical agitation of the electrolyte. Additives such as surfactants help to induce a surface charge on the particles and increase stability of the suspension. Once the breakdown potential is reached, uptake of nanoparticles (NPs) preferentially occur through the surface and open pores, generated during the PEO process while uptake of the micron-size particles occurs by sticking to the melted oxide coating. The incorporation stage is governed by how the particles interact with the microarc discharges produced during the PEO process. The NPs are preferentially incorporated in the open pores while micron-sized particles adhere to the molten mass of the oxide coating. As evidenced by the nature of uptake and incorporation, it would be easy to incorporate NPs in the PEO coating rather than the micron-sized particles [4, 6, 7]. A schematic representation of microarc discharge during PEO using an electrolyte solution dispersed with particles is shown in Fig. 10.1. The growth path of PEO composite coatings through which the suspended particles in the electrolytes are incorporated in the oxide coating is discussed by Tsai and Chou [5]. The mechanism of incorporation of ZnO NPs in the PEO coating deposited on AZ91 Mg alloy is schematically represented in Fig. 10.2.

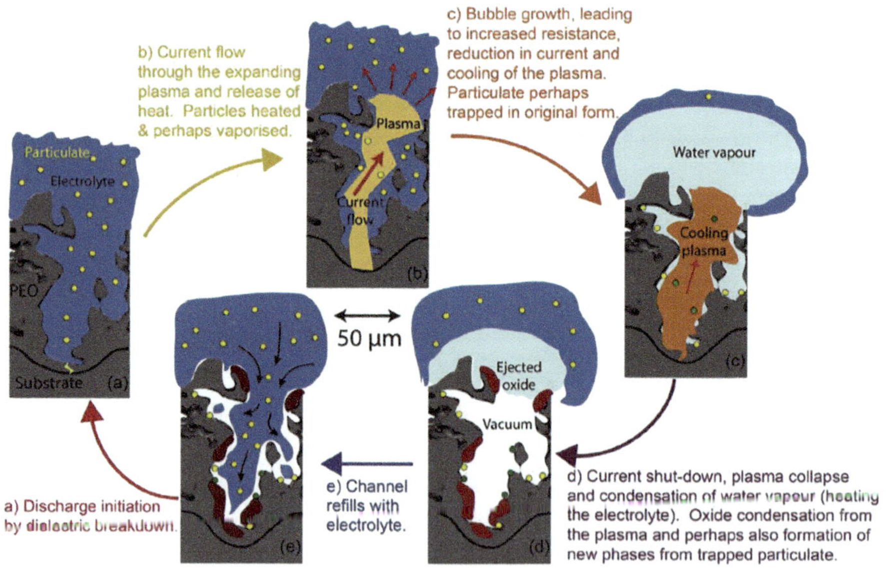

Fig. 10.1 Schematic representation of a single microarc discharge with the electrolyte containing suspended particles. (Reproduced with permission from Ref. [4] © 2020 Elsevier B.V.)

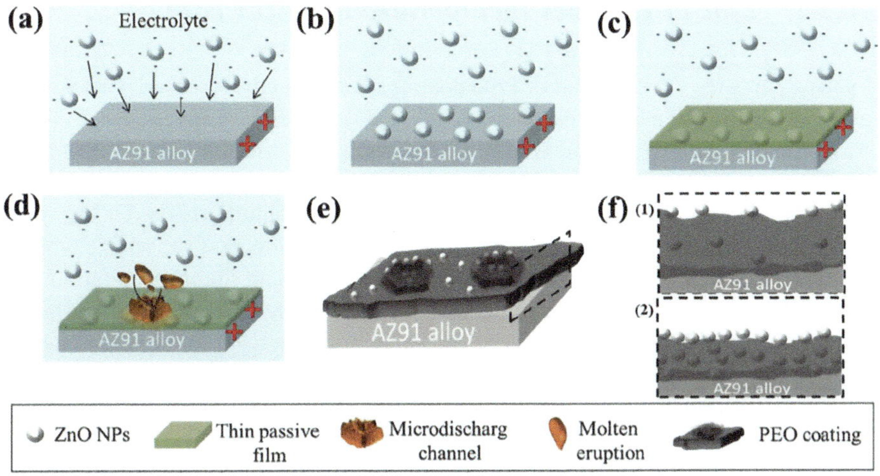

Fig. 10.2 Mechanism of incorporation of ZnO NPs in the PEO coating formed on AZ91 Mg alloy: (**a**) electrophoretic movement of negatively charged ZnO NPs towards the AZ91 Mg alloy anode; (**b**) adsorption of the ZnO NPs on the anode; (**c**) formation of a thin passive oxide film and trapping of the ZnO NPs in it; (**d**) microarc discharge, local melting and molten oxide eruption on the AZ91 Mg alloy; (**e**) sintering and incorporation of ZnO NPs in the oxide coating; and (**f**) cross-sectional view of the PEO–ZnO composite coatings formed using (1) low concentration and (2) high concentration of ZnO NPs in the electrolyte. (Reproduced with permission from Ref. [3] © 2019 Elsevier B.V.)

10.2.1 Mode of Incorporation of Particles in PEO Coatings

The modes of incorporation of particles in the PEO coating is classified as inert (non-reactive) and reactive or partially reactive. Inert incorporation is defined by the incorporation of particles with no reaction with the coating forming elements or with no new phase formation. Reactive incorporation involves melting of the particles following the high energy microarc discharges generated during PEO and its reaction with other phases in the oxide coating. If the discharge energy is sufficiently high, then it could promote liquid phase sintering. The two important factors that determine the reactivity and mode of incorporation of particles in the PEO coating are the size and melting point. As short-lived microarc discharges could not completely melt micron-sized particles, the mode of incorporation of micron-sized particles occurs by embedding of them in the PEO coating. However, due to their lower melting temperature, the possible mode of incorporation of NPs by reactive incorporation is much higher [6, 7].

10.2.2 Type of Particles Incorporated in PEO Coatings

Type of particles incorporated in PEO coatings include: carbon based particles such as graphite, graphene (G), graphene oxide (GO), reduced graphene oxide (rGO), carbon nanotube (CNT), diamond; oxides such as cerium oxide (CeO_2), silica (SiO_2), zirconium oxide (ZrO_2), zinc oxide (ZnO), titania (TiO_2), tungsten oxide (WO_3), alumina (Al_2O_3), lanthanum oxide (La_2O_3), chromium oxide (Cr_2O_3), tantalum oxide (Ta_2O_5), hafnium oxide (HfO_2), yttrium oxide (Y_2O_3), yttria-stabilized zirconia; carbides viz., tungsten carbide (WC), silicon carbide (SiC), titanium carbide (TiC); nitrides, namely, titanium nitride (TiN), hexagonal boron nitride (h-BN), silicon nitride (Si_3N_4), aluminium nitride (AlN); sulphides, viz., molybdenum disulphide (MoS_2), tungsten disulphide (WS_2); metallic particles such as silver (Ag), copper (Cu), cobalt (Co), chromium (Cr), zinc (Zn); bioactive particles such as hydroxyapatite (HA), bioactive glass (BAG), β-tricalcium phosphate (β-TCP); other types of particles such as silica sol, alumina sol, clay, polytetrafluoroethylene (PTFE), halloysite nanotubes, layered double hydroxides (LDH), glass particles viz., borosilicate (BS) glass, soda-lime (SL) glass, liquid crystal display (LCD) glass; and metal-organic frameworks (MOFs). In addition to single type of particles, mixtures of particles, such as, SiC + borosilicate glass, HA + GO and $SiO_2 + TiO_2 + ZrO_2$ have also been explored. The various types of particles incorporated to improve the corrosion resistance of PEO composite coatings formed on Mg, Al and Ti and their alloys are compiled in Table 10.1.

10.3 Corrosion Resistance of Particle-Incorporated PEO Composite Coatings

Several researchers have reviewed the principles of PEO, effect of process parameters on the characteristic properties and corrosion resistance of the PEO coatings formed on Mg, Al and Ti and their alloys [59–65]. Similarly, many researchers have reviewed the corrosion behaviour of PEO composite coatings incorporated with (1) carbon-based additives on Al and its alloys [9], (2) ZrO_2 NPs on Mg and its alloys [22], (3) ZrO_2 NPs on Ti and its alloys [66], (4) graphene and its derivatives on Ti and its alloys [10], (5) graphene and graphene oxide on Mg alloys [67], (6) graphene family particles on Ti alloys [11], (7) HA particles on Mg alloys [68], and (8) different types of particles on Mg and its alloys [69–72]. These reviews provide a detailed outline of how the incorporation of various types of particles influences the corrosion resistance of PEO coatings formed on Mg, Al and Ti and their alloys.

The thickness of the oxide coating, size, density and distribution of the pores, type of phases formed and their chemical stability determine the corrosion protective ability of PEO coatings. The higher the thickness of the oxide coating, the better is the corrosion resistance. However, irrespective of the thickness of the oxide coating, if the porosity is higher, then the corrosion resistance will be decreased. As

Table 10.1 Types of particles incorporated to improve the corrosion resistance of PEO composite coatings formed on Mg, Al and Ti and their alloys

Type of particle incorporated	References
Graphite	[8, 9]
Graphene	[9–11]
Graphene oxide	[2, 10–13]
Reduced graphene oxide	[10, 11]
Carbon nanotube	[9, 14]
Diamond	[9, 15]
Alumina	[16]
Titania	[17, 18]
Cerium oxide	[19, 20]
Zirconium oxide	[21–23]
Silica	[24, 25]
Tungsten oxide	[26]
Zinc oxide	[27, 28]
Lanthanum oxide	[29]
Chromium oxide	[30]
Tantalum oxide	[31]
Hafnium oxide	[32]
Yttria-stabilized zirconia	[33]
Silicon carbide	[34, 35]
Tungsten carbide	[1]
Titanium carbide	[36]
Silicon nitride	[37, 38]
Hexagonal boron nitride	[39, 40]
Titanium nitride	[41]
Molybdenum disulphide	[42, 43]
Tungsten disulphide	[44]
Silver	[45]
Copper	[46]
Hydroxyapatite	[47]
Bioactive glass	[48, 49]
β-Tricalcium phosphate	[50]
Clay	[6, 7, 51]
Polytetrafluoroethylene	[52]
Halloysite nanotubes	[53]
Layered double hydroxide	[54]
Glass particles	[34, 55]
Metal-organic frameworks	[56]
Sepiolite	[57, 58]

particle incorporation preferentially occurs at the pores and cracks of the PEO coating, they could prevent the permeation of the corrosive medium through the pores. In this perspective, particle incorporation has been suggested as one of the strategies to improve the corrosion-protective ability of PEO coatings [59]. Nevertheless, several other factors such as the type of particles and their concentration, conductivity of the particles, mode of incorporation, change in phase content, densification of the oxide coating following particle incorporation, dependence on the type of anions, process parameters, and, surface modification of particles on the level of particle incorporation, could determine the corrosion resistance offered by the PEO composite coatings. The present chapter elaborates how these factors could influence the corrosion resistance of PEO composite coatings.

Roknian et al. [27] have studied the formation of PEO-ZnO composite coatings on Ti using an alkaline phosphate electrolyte dispersed with 5, 10 and 15 g/L of ZnO NPs (25 nm) at 10 A/cm^2 for 7 min and evaluated their corrosion resistance in Ringer's solution at 37 °C. The average pore size of the PEO coating was decreased from 1.72 to 0.98 μm following the incorporation of ZnO NPs. In spite of a similarity in the thickness and phase contents (anatase and rutile phase), the PEO–ZnO composite coatings exhibited a better corrosion resistance than the particle-free PEO coating. The corrosion current density (i_{corr}) was decreased from 9.78×10^{-8} to 0.77×10^{-8} A/cm^2 when the concentration of ZnO NPs was increased from 0 to 15 g/L. The polarization resistance (R_p) of the outer porous layer was increased from 9.6×10^2 to 1.9×10^4 kΩ cm^2, while the R_p of the inner barrier layer was increased from 5.4×10^5 to 1.5×10^7 kΩ cm^2 when the ZnO NPs concentration was increased from 0 to 15 g/L. The enhancement in corrosion protection was attributed due to the incorporation of ZnO NPs, which filled the pores, reduced the defects and modified the microstructure of the PEO coating.

The corrosion resistance of PEO–ZnO composite coatings formed on AZ91 Mg alloy using an alkaline phosphate electrolyte containing 1.5, 3.0 and 4.5 g/L of ZnO NPs (25–65 nm) at 100 mA/cm^2 with a duty cycle of 50% and frequency of 1000 Hz for 30 min in simulated body fluid (SBF) solution at 37 °C was evaluated by Bordbar-Khiabani et al. [3]. When compared to the particle-free PEO coating, addition of 4.5 g/L of ZnO NPs in the electrolyte has decreased the average thickness from 15.3 ± 2.7 to 12.2 ± 2.3 μm, the average roughness (R_a) from 0.84 ± 0.42 to 0.44 ± 0.15 μm and the porosity from 36.4% to 14.6%. The i_{corr} of the PEO-ZnO composite coating formed using 4.5 g/L of ZnO NPs was 0.063×10^{-6} A/cm^2 which was much lower than the i_{corr} of 6.17×10^{-6} A/cm^2 obtained for the particle-free PEO coating (Fig. 10.3). In spite of a decrease in thickness, the decrease in R_a and decrease in porosity has offered an improvement in corrosion resistance of the PEO–ZnO composite coatings.

A similar trend of an increase in corrosion resistance following particle incorporation has been observed for (1) PEO–WC composite coatings formed on AZ31B Mg alloy [1]; (2) PEO–PTFE composite coatings formed on AZ90 Mg alloy [52]; and (3) PEO–MoS_2 composite coatings formed on Ti-6Al-4V alloy [43]. The inferences made from these studies clearly point out that the decrease in porosity and increase in compactness of the inner barrier layer are mainly responsible for the observed improvement in corrosion resistance of PEO composite coatings.

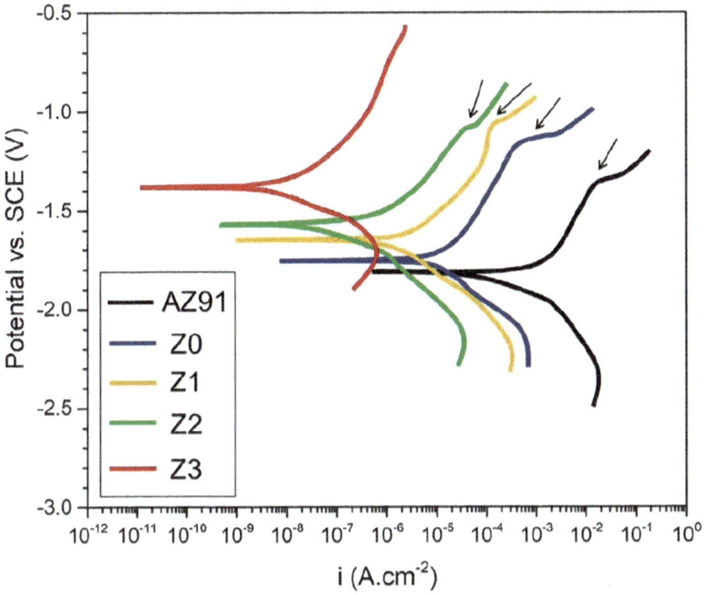

Fig. 10.3 Potentiodynamic polarization curves of uncoated (AZ91), PEO coated (Z0) and PEO–ZnO composite coated AZ91 Mg alloy prepared using 1.5 g/L (Z1), 3.0 g/L (Z2) and 4.5 g/L (Z3) of ZnO NPs in simulated body fluid (SBF) solution at 37 °C. (Reproduced with permission from Ref. [3] © 2019 Elsevier B.V.)

10.3.1 Dependence on the Type of Particle Incorporated

It is evident that particle incorporation could seal the pores and offer an improvement in corrosion resistance for the PEO composite coatings. However, it is essential to verify whether this is true irrespective of the type of particle incorporated and whether all of them could fill the pores and offer a better corrosion resistance. If a mixture of two or three different particles was dispersed in an electrolyte, then will there be any specificity for particle incorporation in terms of the chemical nature of particles such as oxides, nitrides and carbides and their conductivity? The following paragraphs address these aspects.

Shokouhfar and Allahkaram [73] have studied the effect of incorporation 7.5 g/L of three different types of NPs, viz., SiC (40 nm), Al_2O_3 (25 nm) and TiO_2 (25 nm) in the PEO coating formed on Ti using an alkaline aluminate based electrolyte at 400 V for 3 min and the corrosion resistance of the coatings in 3.5% NaCl. All the three types of NPs were incorporated in the PEO coating while filling up of the pores has reduced the R_a of all the three types of PEO composite coatings (19–26 nm) when compared to the particle-free PEO coating (44 nm). $TiAl_2O_5$ and γ-Al_2O_3 were identified as the main phases in all the coatings. Irrespective of the type of NPs incorporated in the PEO coating, all three PEO composite coatings offered better corrosion resistance when compared to their particle-free counterpart. Based on the extent of corrosion protection, the coatings were ranked as follows: Particle-free PEO coating < PEO–SiC composite coating < PEO–Al_2O_3 composite coating < PEO–TiO_2 composite coating. Based on their inferences, Shokouhfar and Allahkaram [73] have suggested that the ability of the NPs to fill the pores of the PEO coating plays a significant role in determining the extent of corrosion protection than their types and compositions.

Askarnia et al. [74] have prepared PEO composite coatings on AZ31 Mg alloy incorporated with 1 g/L of ZrO_2, 1 g/L of Al_2O_3 and a mixture of 1 g/L of ZrO_2 and 1 g/L of Al_2O_3 particles using an alkaline silicate electrolyte at 350 V at 500 Hz for 10 min and evaluated their corrosion resistance in simulated body fluid (SBF) after 7 days of immersion. The ZrO_2, Al_2O_3 as well as the mixture of ZrO_2 and Al_2O_3 particles were inertly incorporated. Incorporation of individual ZrO_2, Al_2O_3 as well as the mixture of ZrO_2 and Al_2O_3 particles has reduced the surface pores and the extent of reduction in porosity was relatively higher for the mixture of ZrO_2 and Al_2O_3 particles. MgO and Mg_2SiO_4 were identified as the main phases of the oxide coating. When a mixture of ZrO_2 and Al_2O_3 particles was used, simultaneous incorporation of both types of particles occurred without any specificity. The outer layer of the PEO composite coating becomes much denser with relatively lower amount of pores following the incorporation of a mixture of ZrO_2 and Al_2O_3 particles. Incorporation of ZrO_2, Al_2O_3 as well as a mixture of ZrO_2 and Al_2O_3 particles in the PEO coating has increased the R_a and decreased the contact angle. After 7 days of immersion in SBF, the i_{corr} of the particle-free PEO coating, PEO–ZrO_2 composite coating, PEO–Al_2O_3 composite coating, PEO-ZrO_2-Al_2O_3 composite coating was 0.91×10^{-3} A/cm^2, 0.16×10^{-3} A/cm^2, 0.13×10^{-3} A/cm^2 and 0.08×10^{-3} A/cm^2, respectively with a corresponding R_p of 19.95 Ω cm^2, 84.21 Ω cm^2, 95.63 Ω cm^2 and 167.62 Ω cm^2, respectively. The decrease in surface porosity, which was the main path for the permeation of the corrosive medium, has been correlated to the observed improvement in corrosion resistance.

Ma et al. [75] have prepared PEO composite coatings on Mg–Li alloy using an alkaline silicate-fluoride-based electrolyte dispersed with 3 g/L of B_4C particles (<1 μm), 1 g/L of C particles (3–5 μm) as well as a mixture of 3 g/L of B_4C particles and 1 g/L of C particles at 5 A/cm^2 with a duty cycle of 50% at 500 Hz for 10 min and evaluated their corrosion resistance in 3.5% NaCl. B_4C particles are semiconducting, while C particles are conducting. Addition of B_4C particles alone has decreased the conductivity of the electrolyte, while addition of C particles alone slightly increased the conductivity. Addition of a mixture of B_4C and C particles has slightly decreased the conductivity of the electrolyte. The B_4C, C as well as the mixture of B_4C and C particles were inertly incorporated. The surface porosity was decreased from 8.5% for the particle-free PEO coating to 4.5% following the incorporation of B_4C particles in the pores. Incorporation of C particles has led no obvious decrease in surface porosity. Incorporation of a mixture of B_4C and C particles has decreased the surface porosity to 5.9%. The R_a was increased following the incorporation of individual as well as the mixture of particles. The coating thickness was increased from 22.7 ± 3.7 μm for the particle-free PEO coating to 31.4 ± 0.9 μm for the PEO-B_4C composite coating and 28.7 ± 1.1 μm for the PEO-B_4C-C composite coating. The thickness of the PEO-C composite coating was 23.9 ± 2.3 μm, similar to the particle-free PEO coating. MgO and Mg_2SiO_4 were identified as the main phases of the oxide coating. The i_{corr} of the particle-free PEO coating, PEO-B_4C composite coating, PEO-C composite coating and PEO-B_4C-C composite coating in 3.5% NaCl was 7.485×10^{-6} A/cm^2, 5.758×10^{-6} A/cm^2, 7.146×10^{-6} A/cm^2 and 4.845×10^{-6} A/cm^2, respectively. Incorporation of B_4C particles has reduced the surface porosity, promoted the compactness and increased thickness of the PEO coating, thus improving the corrosion resistance. Incorporation of C particles showed no obvious decrease in surface porosity as well as thickness and compactness of the PEO coating, thus limiting the improvement in corrosion resistance. Incorporation of mixture of B_4C and C particles has led to a reasonable decrease in surface porosity with a moderate increase in thickness and compactness of the PEO coating, thus offering moderate corrosion resistance.

The formation of PEO composite coatings on Ti-6Al-4V alloy using alkaline phosphate-tungstate-fluoride based electrolyte containing copper sulphate dispersed with varying concentrations of mixture of Si_3N_4 NPs (~50 nm) and TaC microparticles (1–2 μm) (1 g/L of Si_3N_4 + 0.5 g/L of TaC; 2 g/L of Si_3N_4 + 1 g/L of TaC and 4 g/L of Si_3N_4 + 2 g/L of TaC) at 450 V with a duty cycle of 3% at 750 Hz for 10 min and their corrosion behaviour in artificial seawater was studied by Gao et al. [76]. The thickness of the particle-free PEO coating was ~13.5 μm. Addition of a mixture of 1 g/L of Si_3N_4 + 0.5 g/L of TaC and 4 g/L of Si_3N_4 + 2 g/L of TaC has increased the thickness to 20.2 μm and 22 μm, respectively. The surface porosity of the particle-free PEO coating was ~19%. Addition of a mixture of 1 g/L of Si_3N_4 + 0.5 g/L of TaC and 4 g/L of Si_3N_4 + 2 g/L of TaC has decreased the surface porosity to 11.85% and 12.35%, respectively. Anatase-type TiO_2 and rutile-type TiO_2 were identified as the main phases of the particle-free PEO coating. The PEO composite coatings incorporated with a mixture of Si_3N_4 TaC particles also showed the presence of these phases besides the particles. Nevertheless, the volume fraction of these phases was altered. In addition, additional phases were also formed. An increase in concentration of the mixture of Si_3N_4 TaC particles has increased the volume fraction of the rutile-type TiO_2 and TaC phases with a concurrent decrease in the anatase-type TiO_2 phase. In addition, the $Al_4Ti_2SiO_{12}$ phase was also formed following the reaction between Al_2TiO_5 and Si_3N_4 due to the localized high temperature and high pressure generated during the PEO process. The i_{corr} of the particle-free PEO coating, PEO-1 g/L Si_3N_4 + 0.5 g/L TaC composite coating, PEO-2 g/L Si_3N_4 + 1 g/L TaC composite coating and PEO-4 g/L Si_3N_4 + 2 g/L TaC composite coating in artificial seawater was 2.67×10^{-4} A/cm^2, 4.23×10^{-8} A/cm^2, 6.43×10^{-8} A/cm^2 and 4.15×10^{-8} A/cm^2, respectively. The PEO-1 g/L Si_3N_4 + 0.5 g/L TaC composite coating with a surface porosity of 11.85% and a thickness of 20.2 μm offered a better corrosion resistance than those offered by PEO-4 g/L Si_3N_4 + 2 g/L TaC composite coating with a surface porosity of 12.35% and a thickness of 22 μm, suggesting the importance of surface porosity than the thickness in determining the corrosion resistance of PEO composite coatings.

Ahangari et al. [77] have studied the formation of PEO composite coatings on Ta incorporated with a mixture of 1 g/L of SiO_2 (20–30 nm), 1.5 g/L of TiO_2 (10–25 nm) and 2.5 g/L of ZrO_2 (20 nm) NPs using an alkaline silicate electrolyte at 500 V and 20% and 40% duty cycles for 10 min and evaluated their corrosion resistance in 3.5% NaCl. Addition of a mixture of SiO_2, TiO_2 and ZrO_2 NPs has increased the conductivity of the electrolyte and promoted the number and intensity of microarc discharges. The average thickness of PEO composite coatings formed at 20% and 40% duty cycles was 2.28 ± 0.17 μm and 1.65 ± 0.15 μm, respectively. The higher thickness of the PEO composite coating formed at 20% duty cycle was due to the formation of higher number of micro-discharges. The PEO composite coating formed at 40% duty cycle has experienced less melting due to limited occurrence of microdischarges when compared to those prepared at 20% duty cycle. As a result, the PEO composite coating formed at 40% duty cycle was uniform and dense with fewer microcracks, consisting of higher extent of coating forming elements. The i_{corr} and R_p of the PEO composite coating formed at 40% duty cycle were 2.93×10^{-6} A/cm^2 and 9.40×10^3 Ω cm^2, respectively, suggesting a better corrosion resistance than the PEO composite coating formed at 20% duty cycle. The formation of a uniform and dense oxide coating with fewer microcracks, consisting of higher extent of coating forming elements has been correlated for observed improvement in corrosion resistance in spite of a lower thickness for the PEO composite coating formed at 40% duty cycle. Addition of a mixture of SiO_2, TiO_2 and ZrO_2 NPs in the alkaline silicate electrolyte has imparted a negative charge on these particles, which helped their electrophoretic movement towards the anode during PEO and enabled simultaneous incorporation of all of them in the PEO composite coating formed on Ta, with no specificity for any of the particle.

10.3.2 Influence of Particle Concentration in the Electrolyte

The formation of PEO–CNTs composite coatings on AZ80 Mg alloy using an alkaline silicate-silicofluoride based electrolyte incorporated with varying concentrations (0, 0.5, 1, 2 and 4 g/L) of carbon nanotube (CNTs) (size: 5–15 nm) at 100 mA/cm^2 for 5 min and their corrosion behaviour in 3.5% NaCl was studied by Kara and Zengin [78]. The conductivity of the electrolyte was increased from 17.74 to 19.96 ms/cm when the concentration of CNTs was increased from 0 to 4 g/L, which in turn increased the reaction kinetics, growth rate and thickness of the oxide coating from ~4.7 to ~10.1 μm. However, concurrent with the increase in concentration of CNTs, the pore size was increased from 0.55 to 3.55 μm, along with an increase in the number of pores. In addition, the R_a of the PEO coating was increased from 0.2 ± 0.02 to 2.7 ± 0.29 μm with an increase in concentration of CNTs from 0 to 4 g/L. MgO and Mg_2SiO_4 were identified as the main phases of the PEO coatings. Addition of 0.5 g/L of CNTs showed a onefold decrease in i_{corr} from 2.5×10^{-6} to 2.4×10^{-7} A/cm^2 due to the formation of a homogeneous and dense coating with a lower roughness. However, a further increase in concentration of CNTs up to 4 g/L has decreased the corrosion resistance due to the decrease in quality of the inner and outer layers with an increase in pore size and roughness.

PEO composite coatings incorporated with different concentrations (0, 1, 2 and 3 g/L) of GO were prepared on AZ31 Mg alloy using an alkaline phosphate electrolyte in two steps; 100 mA/cm^2 for 1 min; and 400 V for 9 min [12]. The zeta potential of GO (−22.4 mV) has enabled its electrophoretic movement towards the anode and its subsequent incorporation in the PEO coating. The thickness of the PEO coating was decreased slightly from 40.7 to 36.7 μm with an increase in concentration of GO from 0 to 3 g/L while the porosity of the PEO coatings was decreased from 10.91% to 6.43%. MgO was identified as the main phase in all the PEO coatings. Addition of 2 g/L of GO has decreased the i_{corr} of the PEO composite coating in 3.5% NaCl from 0.124×10^{-6} to 0.033×10^{-6} A/cm^2. However, a further increase in concentration of GO to 3 g/L has increased the i_{corr}, which was mainly attributed due to the aggregation of GO in the electrolyte. EIS studies indicated that the resistance of the outer porous layer (R_1) and the inner barrier layer (R_2) were increased with the addition of 2 g/L of GO. The R_1 was increased from 6.89 to 17.37 kΩ cm^2, while the R_2 was increased from 12.72 to 27.44 kΩ cm^2. The higher resistance of R_2 substantiated the formation of a compact inner barrier layer. The improved corrosion resistance of PEO-GO composite coating prepared using 2 g/L of GO has been correlated to the formation of a uniform coating consisting of a compact inner layer with reduced porosity. The mechanism of corrosion protection offered by the PEO–GO composite coating is depicted in Fig. 10.4.

Peitao et al. [8] have prepared PEO-graphite composite coatings on AZ31 Mg alloy using an alkaline silicate-based electrolyte containing 5, 10 and 25 g/L of

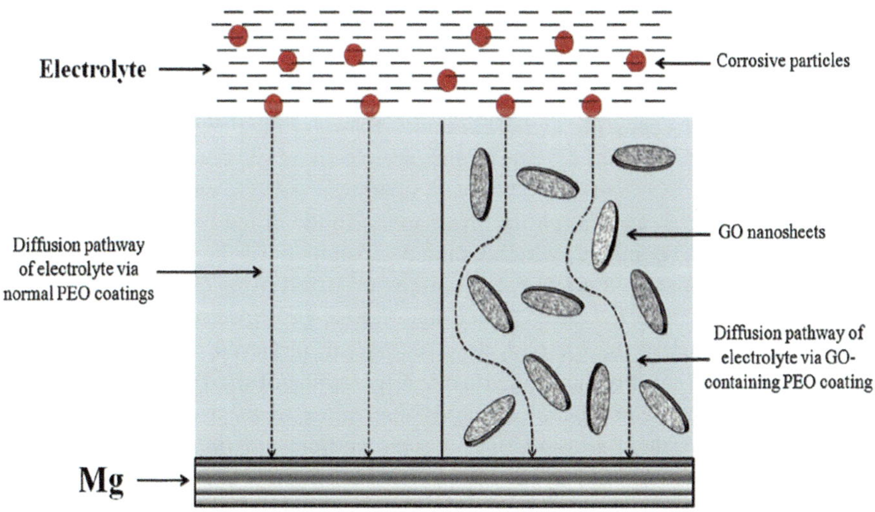

Fig. 10.4 Schematic representation of the mechanism of corrosion attack of PEO coating with and without incorporation of GO nanosheets. (Reproduced with permission from Ref. [12] © 2017 Elsevier B.V.)

graphite particles (mesh size: 15,000) and 0.125 g/L of sodium dodecyl sulfate under pulsed current mode at 11.25 A/dm^2 with a duty cycle of 50% at 600 Hz for 900 s and evaluated their corrosion resistance in 3.5% NaCl. Addition of graphite particles has increased the electrical conductivity of the electrolyte, which thereby promoted the formation of microarc discharges and gas medium layer, resulting in an increase in the thickness of the PEO-graphite composite coatings. MgO and Mg_2SiO_4 were identified as the major phases in all the PEO coatings. The i_{corr} of PEO-graphite composite coatings prepared using 5 and 10 g/L were of the order of 0.748 × 10^{-6} A/cm^2, whereas a further increased concentration of the graphite particles to 25 g/L has drastically increased the corrosion rate, as evidenced by the high i_{corr} of 6.23 × 10^{-6} A/cm^2. EIS study indicated that the resistance of the outer porous layer was increased from 84.1 to 238 Ω cm^2 with the addition of 5 g/L of graphite particles. However, addition of 10 and 25 g/L has decreased the outer layer resistance to 165 Ω cm^2 and 60 Ω cm^2, respectively. Similarly, the resistance of the inner barrier layer was increased from 62.4 to 60,500 Ω cm^2 with the addition of 5 g/L of graphite particles. However, addition of 10 and 25 g/L of graphite particles has decreased the inner barrier layer resistance to 10,800 and 127 Ω cm^2, respectively. The decrease in corrosion resistance of the PEO–graphite composite coatings prepared using 10 and 25 g/L of graphite particles was due to the change in discharge characteristics that has led to an increase pore size and generation of cracks, through which aggressive chloride ions have permeated and attacked the Mg alloy.

The effect of adding varying concentrations (0, 3, 6 and 9 g/L) of diamond particles (1.5 μm) on the formation of PEO-diamond composite coatings on 6061 Al alloy using a sodium metasilicate and boric acid based electrolyte under pulsed current mode at 10 A/dm^2 with a duty cycle of t_{ON}: 20% and t_{OFF}: 80% at 1000 Hz for 30 min and their corrosion behaviour in 3.5% NaCl was studied by Tran et al. [15]. The level of incorporation of the diamond particles in the PEO coating was increased with an increase in their concentration in the electrolyte. The addition of 3, 6 and 9 g/L of diamond particles in the electrolyte has increased thickness from 54 μm for the particle-free PEO coating to 98, 121 and 129 μm for the PEO-diamond composite coatings and the R_a from 2.3 μm for the particle-free PEO coating to 4.5, 5.7 and 10.8 μm for the PEO-diamond composite coatings. γ-Al_2O_3 was identified as the main phase. Besides, an amorphous phase was also identified following the formation of molten glassy matter, which enabled entrapment of the diamond particles. The i_{corr} of PEO-diamond composite coatings prepared using 3 and 6 g/L of diamond particles were 1.46 × 10^{-9} A/cm^2 and 9.90 × 10^{-10} A/cm^2, respectively that were much lower than the i_{corr} of 1.44 × 10^{-7} A/cm^2 observed for the particle-free PEO coating. The improvement in corrosion resistance of the PEO-diamond composite coatings has been attributed mainly due to the increase in thickness of the coating. Although incorporation of diamond particles in the PEO coating has increased the porosity and roughness, the cracks were considerably reduced. However, when the concentration of the diamond particles was increased to 9 g/L, the i_{corr} was increased to 1.20 × 10^{-9} A/cm^2. In spite of an increase in thickness, the higher roughness and large increase in porosity were suggested as the main factors affecting the corrosion resistance.

Han et al. [29] have prepared PEO-La_2O_3 composite coatings on pure Mg using an alkaline silicate-fluoride-based electrolyte containing various amounts (0.5, 1.0, 1.5 and 2.0 g/L) of La_2O_3 NPs (200 nm) under pulsed current mode at 39 mA/cm^2 staged to a voltage of 400 V for 10 min and then increased to 450 V for another 10 min with a duty cycle of 8% at 600 Hz for 10 min and evaluated their corrosion resistance in simulated body fluid (SBF) at 37 ± 0.5 °C. The zeta potential of La_2O_3 NPs in the alkaline electrolyte (−14.8 mV) enabled an electrophoretic movement towards the Mg anode and facilitated their incorporation in the PEO coating. The level of incorporation of La_2O_3 NPs was increased with an increase in their concentration in the electrolyte. Nevertheless, their distribution becomes uneven at 1.5 and 2.0 g/L. The mode of incorporation was inert. The addition of La_2O_3 NPs in the electrolyte has increased the electrical resistivity against the current flow and decreased the growth rate, which was evidenced by the decrease in coating thickness from 24 μm for particle-free PEO coating to 22 μm and 16 μm for coatings prepared using 0.5 g/L and 2.0 g/L of La_2O_3 NPs, respectively. Since the electrical discharges were impeded, open pores and spilled gas were considerably reduced that has resulted in the formation of dense PEO coatings in presence of the La_2O_3 NPs in spite of a decrease in their thickness. The porosity of the particle-free PEO coating was 14%. The addition of 0.5 and 1 g/L of La_2O_3 NPs has reduced the porosity to 8% and 5.7%, respectively. Nevertheless, the porosity of the PEO-La_2O_3 composite coatings was increased when the concentration of La_2O_3 NPs was increased to 1.5 and 2.0 g/L. The i_{corr} of particle-free PEO coating in SBF at 37 ± 0.5 °C was 3.36×10^{-5} A/cm^2. PEO-La_2O_3 composite coatings prepared using 0.5 and 1.0 g/L of La_2O_3 NPs exhibited a lower i_{corr} of 2.12×10^{-6} A/cm^2 and 6.80×10^{-7} A/cm^2, respectively. However, a further increase in concentration of La_2O_3 NPs to 1.5 and 2.0 g/L has increased the i_{corr} to 1.01×10^{-6} A/cm^2 and 2.43×10^{-6} A/cm^2, respectively. EIS study indicated that the |Z| at 0.01 Hz of particle-free PEO coating was 400 Ω cm^2 whereas the |Z| for PEO-La_2O_3 composite coating prepared using 1.0 g/L of La_2O_3 NPs was 10^4 Ω cm^2 after 6 h of immersion in SBF at 37 ± 0.5 °C. After 24 h of immersion in SBF, the |Z| of particle-free PEO coating at 0.01 Hz has considerably reduced, whereas the PEO-La_2O_3 composite coating prepared using 1.0 g/L of La_2O_3 NPs has maintained a |Z| of 10^4 Ω cm^2.

A similar trend of an increase in corrosion resistance following particle incorporation up to a threshold level followed by a decrease in corrosion resistance at higher particle concentration has been observed for (1) PEO-GO composite coatings on AZ31 Mg alloy [13]; (2) PEO-GO composite coatings formed on AZ91 Mg alloy [2]; (3) PEO-ZnO composite coatings formed on AZ131 Mg alloy [28]; (4) PEO-halloysite nanotubes composite coatings formed on Mg-Mn-Ce Mg alloy [53]; (5) PEO-TiO_2 composite coatings formed on Ti [17]; and (6) PEO-SiO_2 composite coatings formed on 7075 Al alloy [24].

Hence, it is clear that although incorporation of lower amounts of particles such as CNTs, GO and graphite NPs in PEO composite coatings could improve the corrosion resistance of the respective coatings, the trend is reversed at higher particle concentrations. The increased conductivity of the electrolyte, reactivity, pore size and number of pores, roughness and aggregation of NPs have contributed to the decrease in corrosion resistance.

10.3.3 Sub-Micron/Micron vs. Nanoparticles in PEO Coatings

Asgari et al. [79] have compared the incorporation of Al_2O_3 NPs (~40 nm) and sub-micron (~300 nm) particles in PEO coatings formed on AZ31B Mg alloy using an alkaline phosphate electrolyte containing 30 g/L of Al_2O_3 particles at 100 mA/cm^2 for 5 min. Methanol (5 g/L), ethanol (5 g/L), formamide (5 g/L), triethanolamine (5 g/L), sodium dodecyl sulphate (5 g/L), glycerin (5 g/L) and urea (0–20 g/L) were used as additives to disperse the Al_2O_3 particles in the electrolyte. When compared to the submicron-sized particles, Al_2O_3 NPs were preferentially incorporated in the PEO coating due to the formation of a thinner hydrate layer formed over them, which increased the possibility of them to come closer towards the anode for incorporation. Among the additives, due to its ability to act as a surfactant and improve the stability of the suspension by preventing agglomeration, urea was shown to be effective in increasing the level of incorporation of Al_2O_3 NPs in the PEO coating. Addition of urea has increased the conductivity of the electrolyte and increased the surface porosity, which was filled by the Al_2O_3 NPs. An increase in the concentration of urea from 0 to 20 g/L has led to a decrease in thickness as well as porosity. The increase in the level of incorporation of Al_2O_3 NPs and the decrease in porosity have enabled an increase in corrosion resistance of the PEO composite coating incorporated with Al_2O_3 NPs.

Fattah-alhosseini et al. [80] have reviewed the effects adding nano- and micro-particles on the corrosion resistance of PEO composite coatings formed on titanium and its alloys. Based on the comparison of the i_{corr} values of PEO composite coatings, they have suggested that the extent of decrease in i_{corr} is much higher for PEO composite coatings incorporated with NPs than those incorporated with microparticles.

10.3.4 Role of Electrolyte Anions

Zehra et al. [26] have studied the influence of the type of anions (AlO_2^-, SiO_3^{2-} and PO_4^{3-}) on the incorporation of WO_3 NPs in PEO coatings formed on AZ31 Mg alloy and the corrosion resistance of the resultant PEO composite coatings in 3.5% NaCl. The PEO coatings were prepared using alkaline aluminate, alkaline silicate and alkaline phosphate-based electrolytes containing 2 g/L WO_3 NPs (~290 nm) at 100 mA/cm^2 and at 60 Hz for 300 s, termed as Al-film, Si-film and P-film, respectively. Irrespective of the type of anion, all PEO coatings exhibited a typical porous morphology with cracks in the coated layer. The i_{corr} of the Si-film, Al-film and P-film were 5.60×10^{-10} A/cm^2, 2.17×10^{-8} A/cm^2 and 7.70×10^{-8} A/cm^2 with a corresponding R_p of $1.32 \times 10^8\ \Omega\ cm^2$, $2.34 \times 10^6\ \Omega\ cm^2$ and $8.12 \times 10^5\ \Omega\ cm^2$, respectively. EIS studies indicated that the resistances of the outer porous layer of the Si-film, Al-film and P-film were $1.75 \times 10^7\ \Omega\ cm^2$, $5.11 \times 10^6\ \Omega\ cm^2$ and $2.21 \times 10^5\ \Omega\ cm^2$ while the resistances of their inner barrier layer were $4.65 \times 10^8\ \Omega\ cm^2$, $1.37 \times 10^8\ \Omega\ cm^2$ and $3.13 \times 10^8\ \Omega\ cm^2$, respectively. The difference in the corrosion resistance of the Si-, Al- and P-films was attributed due to the difference in discharge behaviour during the formation of these films by the PEO process, which also determines the level of incorporation and location of WO_3 NPs (Fig. 10.5).

The formation of Si-film involved three types of discharges, namely A, B and C (Fig. 10.5a). Among them, type A and type C discharges were responsible for the pores and cracks in the top layer, while the type B discharges that occur due to the dielectric breakdown were the strongest ones. Such types of discharges have promoted a higher level of incorporation of WO_3 NPs not only in the outer layer but also in the inner layer of the Si-film. As a result, the WO_3 NPs have properly sealed the micropores and deposited in the cracks, thus displaying a pore-filled morphology capable of proving a better corrosion resistance. The formation of Al-film involved two types of discharges, namely, A and C that were responsible for the pores and cracks in the top layer that provided sufficient space for the incorporation of WO_3 NPs (Fig. 10.5b). The formation of P-film involved three types of discharges, namely, B, D and E (Fig. 10.5c). Among them, type B discharges were the

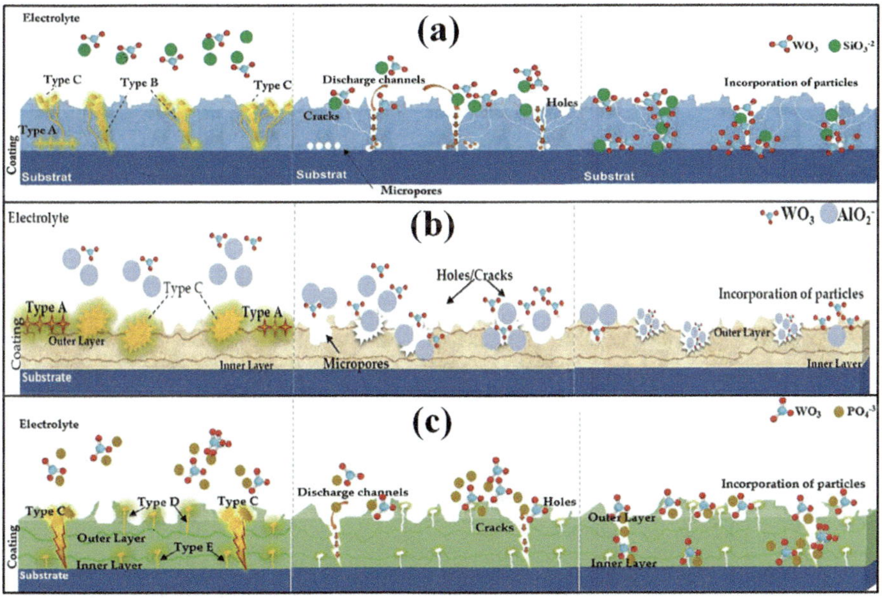

Fig. 10.5 Schematic illustration of the discharge types and incorporation of SiO_3^{2-}, AlO_2^-, PO_4^{3-} anions along with WO_3 NPs into (**a**) Si-film; (**b**) Al-film; and (**c**) P-film. (Reproduced with permission from Ref. [26] © 2022 Elsevier B.V.)

strongest ones that covered both the outer and inner layers of the P-film. Type D discharges have led to the development of large pores while type E discharges have penetrated the outer layer without very much affecting the inner layer. As a result, the amount of WO_3 NPs incorporated in the P-film was limited when compared to that of the Si- and Al-films.

Hence, it is clear that the type of anions in the electrolyte plays a crucial role in determining the discharge characteristics, which in turn influences the development of a porous structure, size and number of pores and the level of incorporation of NPs in the PEO composite coatings and their corrosion resistance.

10.3.5 Influence of Conductivity of the Incorporated Particles

Mashtalyar et al. [41] have prepared PEO-TiN composite coatings incorporated with varying concentrations (1, 2, 3 and 4 g/L) of TiN NPs (20 nm) on MA8 Mg alloy using a silicate-fluoride-based electrolyte containing 0.5 g/L of sodium dodecylsulfate. The PEO coatings were formed under bipolar mode at 300 Hz with a duty cycle of 50% in two stages. The first stage involved an anodic current density of 0.1 A/cm^2 and a cathodic component of −30 V applied for 200 s, while the second stage involved an anodic component of 200 V and a cathodic component of −10 V applied for 600 s. The surface porosity of the PEO-TiN composite coatings was reduced from 5.0% to 3.9% while the average pore size was decreased from 1.5 to 1.1 μm when the concentration of TiN NPs was increased from 0 to 4 g/L (Fig. 10.6). In spite of a similar thickness (~20 ± 1 μm), the R_a of the PEO-TiN composite coatings was increased from 1.21 to 2.13 μm when the concentration of TiN NPs was increased from 0 to 4 g/L. Magnesium orthosilicate and magnesium oxide were identified as the main phases and the incorporation of TiN NPs has not altered the phase content of the PEO-TiN composite coatings. The corrosion resistance of PEO-TiN composite coatings in 3% NaCl was relatively lower than the particle-free PEO coating. The i_{corr} was decreased from 1.2×10^{-7} to 7.9×10^{-7} A/cm^2 with a corresponding decrease in R_p from 7.1×10^5 to 6.0×10^4 Ω cm^2. EIS study indicated a decrease in the resistance of the outer porous layer from 1.91×10^4 to 1.56×10^3 Ω cm^2 and the inner barrier layer from 1.20×10^5 to 2.13×10^4 Ω cm^2. In spite of the formation of a dense coating with a homogeneous distribution of TiN NPs that decreased the apparent porosity, the corrosion resistance of the PEO-TiN composite coatings was decreased. Agglomeration of TiN NPs has been shown to increase the R_a (arrow marks in Fig. 10.6). The deterioration of the protective properties of the PEO-TiN composite coatings was mainly attributed due to the high conductivity of the TiN NPs.

Hence, it is clear that the conductivity of the particle is an important factor in determining the corrosion resistance of PEO composite coatings. Particles with higher conductivity, such as TiN could decrease the corrosion resistance of PEO-TiN composite coatings irrespective of a homogeneous distribution of the TiN NPs in the PEO coating, decrease in surface porosity and average pore size.

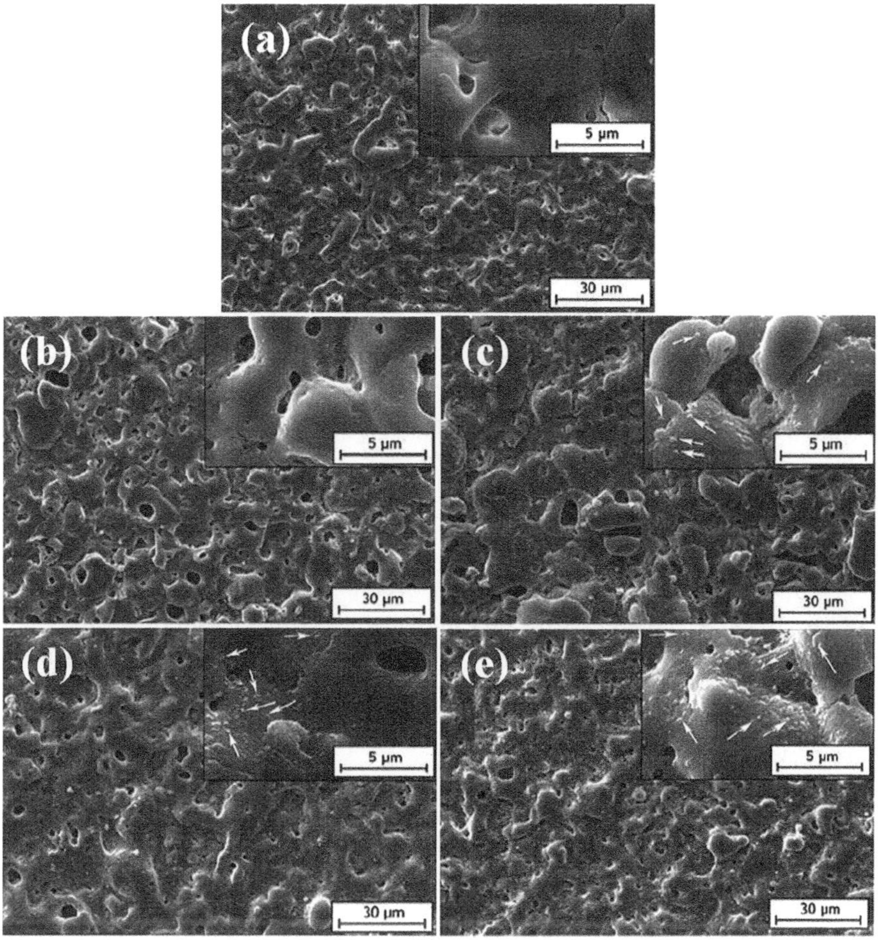

Fig. 10.6 Surface morphology of PEO-TiN composite coatings prepared using 0 to 4 g/L of TiN NPs: (**a**) 0 g/L; (**b**) 1 g/L; (**c**) 2 g/L; (**d**) 3 g/L; and (**e**) 4 g/L. (Reproduced with permission from Ref. [41] © 2017 Elsevier B.V.)

10.4 Influence of PEO Process Parameters on the Corrosion Resistance of PEO Composite Coatings

The process parameters employed for preparing the PEO coating such as applied current density, current waveform and treatment time largely influence the microarc discharge characteristics and the porosity of the PEO coating. Particle incorporation has the ability to seal the pores of the PEO coatings and improve corrosion resistance. Nevertheless, it is important to understand how the process parameters influence the characteristics of the PEO coatings so that particle incorporation can be manipulated accordingly to achieve better corrosion resistance for the PEO composite coatings.

10.4.1 Effect of Current Mode

Aljohani et al. [35] have prepared PEO-SiC composite coatings on AA 2014 alloy using an alkaline silicate electrolyte dispersed with 1 g/L of SiC particles under both unipolar and bipolar pulsed current modes at varying frequencies (1000 and 2500 Hz) and duty cycles (30% and 50%) with a positive component of current density of +0.30 mA/cm^2 and a negative component of current density −0.1 mA/cm^2 for 440 s and evaluated their corrosion resistance in 0.5 M NaCl. The PEO-SiC composite coating prepared under unipolar current mode with a duty cycle of 30% exhibited relatively larger pores with a lower pore density. Conversely, the PEO-SiC composite coating prepared under bipolar current mode at a frequency of 1000 Hz and at a duty cycle of 50% has reduced the density and size of the pores when compared to those produced under unipolar current mode. The PEO-SiC composite coating prepared under bipolar current mode at 2500 Hz and at a duty cycle of 50% exhibited a more uniform distribution of smaller pores with a relatively higher pore density when compared with the coatings processed at 1000 Hz. The PEO-SiC composite coatings prepared under both unipolar and bipolar current modes offered excellent corrosion resistance. The extent of corrosion protection provided by the composite coatings was correlated to their porosity. The PEO-SiC composite coating prepared under bipolar current mode at a frequency of 2500 Hz and at a duty cycle of 50% exhibited the best corrosion protection due to its lower porosity when compared to the coating produced under bipolar current mode at 1000 Hz and the one formed unipolar current mode with a duty cycle of 30%.

The effect of different waveforms on the formation of PEO-TiO_2 composite coatings on 7075 Al alloy using an alkaline silicate electrolyte containing 3 g/L TiO_2 NPs (30 nm) (rutile phase) and their corrosion behaviour in 3.5% NaCl was studied by Hakimizad et al. [81]. The PEO-TiO_2 composite coatings were prepared at 5.6 A/dm^2 and at 2 kHz for 1 h. For coatings prepared using the bipolar mode, the cathodic duty cycles were varied at 20% and 40%. The surface morphology of the particle-free PEO and PEO-TiO_2 composite coatings showed a strong dependence on the applied waveform. The morphology of the oxide coating was changed from a pancake-like structure to crater-like structure when the applied waveform was changed from unipolar to bipolar mode. Irrespective of the applied waveform, the TiO_2 NPs were distributed uniformly in the PEO coating. The PEO-TiO_2 composite coating prepared using unipolar waveform exhibited a lower thickness, higher R_a with a higher number of cracks and pores when compared to the particle-free PEO coating, which has limited its long-term corrosion resistance. Conversely, the lower porosity and higher thickness of the PEO-TiO_2 composite coating prepared using bipolar waveform has offered better long-term corrosion protection in spite of its higher R_a. Among them, the coatings prepared using a cathodic duty cycle of 40% were very effective due to the highest intrinsic resistance of the inner barrier layer and its ability in plugging the micropores and providing a repair mechanism.

10.4.2 Effect of Treatment Time

Nasiri Vatan and Adabi [82] have studied the effect of treatment time on the formation of PEO coatings formed on AZ31B Mg alloy using an alkaline phosphate electrolyte dispersed with 5 g/L of WC NPs (size: <80 nm) at 27.5 A/cm^2 for 5, 10, 15 and 20 min. The mean thickness, average pore diameter and surface porosity of both particle-free PEO and PEO-WC composite coatings were increased following an increase in treatment time from 5 to 20 min. For a given treatment time, the mean thickness, average pore diameter and surface porosity were relatively lower for the PEO-WC composite coating. Both the particle-free PEO and PEO-WC composite coating were composed of MgO and $Mg_3(PO_4)_2$ phases. An increase in treatment time from 5 to 20 min has decreased the i_{corr} of both the particle-free PEO and PEO-WC composite coating and for a given time, the i_{corr} of the PEO-WC composite coating was much lower than that of the particle-free PEO coating. In spite of an increase in average pore diameter and surface porosity, the increase in thickness with time from 5 to 20 min has offered an improvement in corrosion resistance. For a given time, the better corrosion resistance displayed by the PEO-WC composite coating was attributed to the formation of a homogeneous and dense oxide coating in the presence of WC NPs, which served as a barrier against the permeation of the highly aggressive chloride ions through the coating.

It is evident that PEO composite coatings prepared under bipolar current mode at a relatively higher frequency and duty cycle offered higher corrosion protection than those prepared under bipolar current mode at a relatively lower frequency and duty cycle due to their lower porosity. Among the PEO composite coatings formed using bipolar waveform, those prepared using a higher cathodic duty cycle offer better long-term corrosion protection due to their ability in plugging the micropores. The increase in thickness of the PEO composite coatings with treatment time could outweigh the deleterious influence of increase in porosity with time. Fortunately, particle incorporation filled the pores and enabled better corrosion resistance for the PEO composite coatings.

10.5 Reactive vs. Inert Incorporation of Particles on the Corrosion Resistance of PEO Composite Coatings

Incorporation of particles without a reaction with the electrolyte or components of the coating forming elements or any new phase formation is defined as inert incorporation. Particles such as TiO_2, Al_2O_3 and SiC follow inert incorporation. Reactive incorporation of particles involves melting of the particles by the high energy microarc discharges generated during the PEO process followed by its reaction with other phases in the oxide coating. If the discharge energy is sufficiently high, then it could promote liquid phase sintering. Particles such as clay and glass follow reactive incorporation. Incorporation of CeO_2 occurs partially by inert and partially by reactive incorporation.

Lu et al. [7] have prepared PEO composite coatings on AM50 Mg alloy incorporated with 5 g/L of four different micro-sized particles, namely, clay (~12 μm), SiO_2 (1–5 μm), Si_3N_4 (1–5 μm) and SiC (1–5 μm), using an alkaline phosphate-based electrolyte at 450 V for 10 min and evaluated their corrosion resistance in 3.5% NaCl. The addition of these particles in the electrolyte has reduced the growth rate and has resulted in a decrease in thickness from 45 ± 5 μm (particle-free coating; PPEO) to 33 ± 4 μm for PPEO (SiO_2), 35 ± 8 μm for PPEO (Si_3N_4) and 33 ± 5 μm for PPEO (SiC). Incorporation of these particles has increased the R_a of the PEO coatings from 2.8 ± 0.3 μm for PPEO to 3.0 ± 0.3 μm for PPEO (clay), 3.1 ± 0.2 μm for PPEO (SiO_2), 3.3 ± 0.1 μm for PPEO (Si_3N_4) and 3.6 ± 0.4 μm for PPEO (SiC). X-ray diffraction patterns identify the presence of SiO_2, Si_3N_4 and SiC particles while the presence of clay particles could not be identified due to the difference in the mechanism of incorporation. The clay particles with a low melting point (<1200 °C) demonstrated reactive incorporation whereas SiO_2 (1600 °C), Si_3N_4 (1900 °C) and SiC (2730 °C) particles with high melting points have undergone inert incorporation, suggesting the influence of melting point of the particles on the mode of incorporation. The i_{corr} of PPEO (SiO_2), PPEO (Si_3N_4) and PPEO (SiC) were of the order of 2.0×10^{-7} A/cm^2, which was slightly higher than that of the PPEO having an i_{corr} of 1.2×10^{-7} A/cm^2. However, PPEO (clay) offered better corrosion resistance, as evidenced by a lower i_{corr} of 8.7×10^{-9} A/cm^2. EIS study indicated that the corrosion resistance of all the PEO coatings was decreased with an increase in immersion time from 0 to 72 h. The |Z| at 0.01 Hz provided a useful measure of the corrosion protective ability of the PEO coatings. For any given time (0–72 h), the PPEO (clay) coating offered a better corrosion resistance when compared to that of the other coatings. The PPEO (SiO_2), PPEO (Si_3N_4) and PPEO (SiC) displayed a higher |Z| than PPEO coating during the initial periods of immersion. However, the trend was reversed with an increase in immersion time up to 72 h. The improved corrosion resistance of PPEO (clay) coating was due to the reactive incorporation of the clay particles that facilitated the formation of a compact and dense layer.

A comparison of incorporation of SiC, borosilicate glass and a mixture of SiC + borosilicate glass particles in the PEO coatings formed AZ91D Mg alloy using an alkaline silicate-phosphate based electrolyte containing 10 g/L of SiC particles and 80 g/L of borosilicate glass particles at 0.5 A/cm^2 for 1 and 3 min and evaluation of their corrosion resistance in 0.1 M Na_2SO_4 and 0.05 M NaCl was made by Pezzato et al. [34]. The addition of SiC particle has increased the thickness, while the addition of borosilicate glass particles has exerted no significant influence on thickness of the oxide coating. SiC particles followed inert incorporation as the melting point of SiC (~2730 °C) was higher than the temperature experienced at the microarc discharge during PEO and they were identified at the pores and outside the pores. Conversely, the borosilicate glass particles followed reactive incorporation due to its low melting point (~1650 °C). During PEO, the rapid melting and re-solidification of the borosilicate glass particles generate an amorphous phase that partially seals the pores of the oxide coating. The thickness of the PEO coating has increased following the incorporation of particles and the extent of improvement in thickness was much higher when a mixture of SiC + borosilicate glass particles were incorporated in the oxide coating. Incorporation of all the particles has led to an increase in R_a while the extent of increase in R_a was much higher for the PEO-SiC composite coating. MgO, Mg_2SiO_4 and $Mg_3(PO_4)_2$ were identified as the major phases in all the coatings. The PEO composite coatings incorporated with a mixture of SiC + borosilicate glass particles showed a higher resistance of 4199 Ω cm^2, followed by the coating incorporated with borosilicate glass particles alone with a resistance of 2397 Ω cm^2 and the coating incorporated with SiC particles alone with a resistance of 2122 Ω cm^2.

Pezzato et al. [55] have studied the formation of PEO-glass composite coatings by incorporating glass particles from three different sources, namely liquid crystal display (LCD) glass, borosilicate (BS) glass, and soda-lime (SL) glass on 6061 AA alloy using an alkaline silicate-based electrolyte dispersed with 40 g/L of glass particles (1–10 μm) under direct and pulsed current modes at 0.65 A/cm^2 with a duty cycle of 50% at 200 Hz for 3 min. Since the melting point of the glass additives (between 1000 and 1500 °C) is significantly lower than the temperature experienced during PEO (between 1800 and 2370 °C), the glass particles are incorporated by melting and re-solidification forming an amorphous phase. The ability of the glass particles to form an amorphous phase and to seal the pores increased the corrosion resistance of the PEO-glass composite coatings. Among the different types of glass particles, LCD glass particles were found to be effective in improving the corrosion resistance due to its higher chemical stability due to the absence of alkaline elements.

The mode of incorporation of nano- and micron-sized SiO_2 particles in the PEO coatings formed on AM50 Mg alloy using an alkaline phosphate-based electrolyte under pulsed DC power source with a pulse ratio of t_{on}:t_{off} = 0.4 ms:3.6 ms at 450 V for 10 min and the corrosion resistance of PEO-SiO_2 composite coatings in 2.5% NaCl was studied by Lu et al. [69]. The PEO coatings prepared without addition and addition of nano- and micron-sized SiO_2 particles were termed as PPEO, PPEO (n-SiO_2) and PPEO (m-SiO_2). Addition of both the nano- and micron-sized SiO_2 particles in the electrolyte has increased the final voltage. The PEO coating prepared using nano-sized SiO_2 particles required more time to reach the breakdown potential. The intensity of the microdischarges during PEO was much higher for the nano-sized SiO_2 particles than those observed for the micron-sized SiO_2 particles. The mean thickness of the PPEO, PPEO (n-SiO_2) and PPEO (m-SiO_2) coatings were 45 μm, 25 μm and 33 μm, respectively. The PPEO (n-SiO_2) coating exhibited a higher number of pores, but their size was smaller when compared to the PPEO (m-SiO_2) and PPEO and a large amount of pores were partially filled. Addition of micron-sized SiO_2 particles has not changed the R_a of the PEO coating, which was of the order of 3.1 μm similar to that of the PPEO coating. Conversely, the addition of nano-sized SiO_2 particles has enabled the formation of a PPEO (n-SiO_2) coating with a relatively smoother surface (R_a: 2.6 μm). MgO was identified as the main phase of PPEO coating. Identification of diffraction peaks pertaining to SiO_2 in the PPEO (m-SiO_2) coating suggested an inert incorporation as the short-lived microdischarges could not melt the micron-sized SiO_2 particles. Conversely, the nano-sized SiO_2 particles with a lower melting point and the ability to serve as a sintering additive have enabled the melting of the SiO_2 nanoparticles, leading to a reactive incorporation of them in PPEO (n-SiO_2) coating. The nano-sized SiO_2 particles showed a higher level of incorporation of 9 at. % as opposed to 5 at. % observed for the micron-sized SiO_2 particles. Uptake of the nano-sized SiO_2 particles occurred mainly via discharge channels and open pores, while the micron-sized SiO_2 particles mainly entered via the surface, validating the dependence on the mode of incorporation of SiO_2 particles based on their size. The i_{corr} of PPEO (n-SiO_2) and PPEO (m-SiO_2) coatings were 2.4×10^{-7} and 1.9×10^{-7} A/cm^2, which was relatively higher than that of the PPEO coating with an i_{corr} of 1.2×10^{-7} A/cm^2. The decrease in thickness of the coating following the incorporation of nano- and micron-sized SiO_2 particles was one of the reasons for the lower corrosion protective ability of PPEO (n-SiO_2) and PPEO (m-SiO_2) coatings when compared to that of PPEO coating. The incorporation of nano-sized SiO_2 particles has reduced the surface porosity. Nevertheless, a substantial reduction in the porosity at the cross-section has not occurred. In terms of phase composition, reactive incorporation of nano-sized SiO_2 particles has led to the formation of nanocrystalline MgO and amorphous material whereas inert incorporation of micron-sized SiO_2 particles has not altered the phase content. Although reactive incorporation of nano-sized SiO_2 particles could fill the surface pores and facilitate the formation of nanocrystalline MgO and amorphous material, its inability to strengthen the barrier layer has limited its corrosion protection ability.

The formation of PEO-CeO_2 composite coatings on AM50 Mg alloy using an alkaline silicate electrolyte containing 3, 6, 10 and 13 g/L of CeO_2 microparticles (5 μm) under AC pulsed current reverse mode with a positive-to-negative pulse ratio of 420/−60 V at 500 Hz applied for 600 s and evaluation of their corrosion behaviour in 0.5% NaCl was studied by Mohedano et al. [19]. The presence of CeO_2 microparticles was identified both at the surface as well as throughout the thickness of the PEO-CeO_2 composite coating, suggesting the uptake of these particles through electrophoretic movement towards the anode during the PEO process. In addition, the local heating effect has enabled liquid phase formation and subsequent solidification, leading to reactive incorporation of the CeO_2 microparticles. The reactive incorporation mechanism was substantiated by the (1) formation of cellular microstructures with the cell bands enriched in Ce, (2) presence of Ce element detected by EDS, and (3) identification of an amorphous material containing Ce by XRD. The PEO-CeO_2 composite coating prepared using 3 g/L of CeO_2 microparticles exhibited a higher corrosion resistance (1617 ± 30 kΩ cm^2) than the particle-free PEO coating (1030 ± 40 kΩ cm^2). The observed improvement in corrosion resistance has been correlated to the accumulation of CeO_2 particles closer to the inner layer and the formation of corrosion product that offered an additional protection besides those provided by the original inner layer. The absence of severe corrosion at the coating/substrate interface was believed due to the conversion of cerium-containing degradation products to Ce^{3+} ions, which could have enabled a "self-healing" effect in oxide layers.

Mohedano et al. [83] have studied the formation of PEO composite coatings with the incorporation varying concentrations (2, 5 and 10 g/L) of three different types of particles, namely, α-Al_2O_3 (~300 nm) monoclinic ZrO_2 (300–700 nm) and CeO_2 (~560 nm) on AM50 Mg alloy using an alkaline phosphate-fluoride based electrolyte under alternating current mode with a square wave electrical signal of +420 V/−60 V at a duty cycle of 50% and at a frequency of 50 Hz for 900 s. All of these particles followed inert incorporation as the melting point of α-Al_2O_3 (2072 °C) monoclinic ZrO_2 (2700 °C) and CeO_2 (2600 °C) was much higher than the temperature experienced at the microarc discharge during PEO. However, some reactive incorporation was also observed for α-Al_2O_3 and monoclinic ZrO_2 particles as evidenced by the formation of tetragonal ZrO_2 and $MgAl_2O_4$ phases. All the three PEO composite coatings offered a higher corrosion resistance than their particle-free counterpart. The improvement in corrosion resistance has been attributed to the decrease in size and population density of the surface pores following the preferential incorporation of these particles on the outer layer of the oxide coating. Among the three composite coatings, a better long-term corrosion resistance was observed for PEO-CeO_2 coating due to the conversion of cerium containing degradation products to Ce^{3+} ions, which could have enabled a "self-healing" effect in oxide layers.

Rehman et al. [84] have studied the formation of PEO-CeO_2 composite coatings on AZ91D Mg alloy using an alkaline silicate-fluoride-based electrolyte containing 6 g/L of CeO_2 NPs for 5, 15, 30 and 45 min and evaluated their corrosion resistance in 3.5% NaCl. The CeO_2 particles were preferentially incorporated in the pores, cracks and the radial zones around the pores by inert incorporation. In addition, the formation of cerium based-complexes confirmed reactive incorporation of the CeO_2 NPs. The decrease in intensity of the CeO_2 peaks with an increase in treatment time substantiated the involvement of CeO_2 particles in chemical reaction during the PEO process. The PEO-CeO_2 composite coating prepared for 30 min exhibited the lowest i_{corr} of 4.8×10^{-7} A/cm^2 due to its ability to form a compact Ce-rich coating along with preferential incorporation of CeO_2 NPs by inert incorporation and cerium-based complexes by reactive incorporation that acted as sealing agents to block the coating defects.

Incorporation of monoclinic ZrO_2 NPs (45 nm) in the PEO coating formed on Ti using an alkaline phosphate electrolyte at 150 mA/cm^2 under pulsed unipolar mode at 95% duty cycle and at 1000 Hz for 15 min followed both inert as well as partially reactive incorporation [85] as evidenced by the formation of $ZrTiO_4$ and incorporation of m-ZrO_2. An increase in alkalinity of the electrolyte by increasing the KOH content from 0 to 6 g/L has resulted in an increase in thickness, surface porosity and roughness of the PEO-ZrO_2 composite coating. In spite a higher thickness, the higher porosity limit the extent of corrosion protection offered by PEO-ZrO_2 composite coating prepared using 6 g/L of KOH. With a comparable thickness and relatively lesser surface porosity and roughness, the PEO-ZrO_2 composite coating prepared using 4 g/L of KOH offered a better corrosion resistance.

Incorporation of 1–4 g/L of Si_3N_4 NPs (50 nm) in the PEO coating formed on AZ31 Mg alloy using an alkaline phosphate-aluminate based electrolyte at 83.3 mA/cm^2 at a duty cycle of 50% and at a frequency of 1000 Hz for 10 min followed reactive incorporation [38]. As the discharge energy during PEO was higher than the melting temperature of Si_3N_4 (1900 °C) the Si_3N_4 NPs have reacted with the molten MgO, leading to the formation of Mg_2SiO_4 phase. PEO coating formed using the particle-free electrolyte showed the formation of $MgAl_2O_4$ as the major phase along with a small fraction of MgO phase. Irrespective of the reactive incorporation and phase contents, PEO-Si_3N_4 composite coating formed using 2 g/L of Si_3N_4 NPs exhibited the lowest i_{corr} of 1.95×10^{-6} A/cm^2 with an R_p of 13.38 Ω cm^2 when compared to its counterparts. The observed improvement in corrosion resistance has been attributed to its lower porosity (6.58%), relatively higher coating thickness (17.5 μm) and the absence of any major cracks in the oxide layer.

Incorporation of 5 g/L of HfO_2 particles in the PEO coating formed on Ti-6Al-4V alloy using alkaline silicate-phosphate based electrolyte at 10 A/dm^2 at a duty cycle of 50% and at a frequency of 500 Hz for 30 min was studied by Sun et al. [32]. Addition of HfO_2 particles in the electrolyte has increased the oxidation voltage, facilitated the electrophoretic movement of SiO_3^{2-} ions towards the anode and promoted sintering of the SiO_3^{2-} ions, as evidenced by an increase in precipitation of SiO_2 in the oxide layer. As the discharge energy during PEO was sufficient to melt the HfO_2 particles (2758 °C), one would expect that the HfO_2 particles will be metallurgically combined with the PEO coating instead of being mechanically mixed in it in the form of particles. Nevertheless, identification of HfO_2 as one of the phases suggested the involvement of melting and sintering caused by the microarc discharges that increased the densification and thickness of the coating. The i_{corr} of the PEO-HfO_2 composite coating was 2.42×10^{-7} A/cm^2 was lower than that of the i_{corr} of 1.21×10^{-6} A/cm^2 for the particle-free PEO coating. EIS study revealed the resistances of outer porous layer and inner barrier layer of the particle-free PEO coating were 302.2 Ω cm^2 and 6468 Ω cm^2, respectively, whereas incorporation of HfO_2 particles has led to an increase in the resistances of the outer porous layer and inner barrier layer to 442 Ω cm^2 and 13,730 Ω cm^2, respectively.

10.6 Change in Phase Content Following Particle Incorporation and Its Influence on the Corrosion Resistance of the PEO Coating

During PEO, anodic oxidation leads to the formation of the corresponding oxides, i.e. MgO on Mg and Mg alloys and Al_2O_3 on Al and its alloys. The high energy microarc discharges might also promote reaction between the corresponding oxides and the electrolyte such as the aluminate electrolyte, leading to the formation of new phases such as magnesium aluminate ($MgAl_2O_4$) along with the conventionally formed magnesium oxide (MgO) on Mg and Mg alloys. Particle addition in the electrolyte might alter the conductivity/resistivity of the electrolyte and the discharge characteristics during the PEO process, which would modify the volume fraction of these phases. In addition, the conditions employed for the PEO process could change the discharge characteristics, facilitating phase transformation, dissolution of existing phases and formation of new phases, which would alter the corrosion resistance of the PEO coatings. As the chemical stability of the various phases formed on the PEO coating could influence the corrosion resistance, it is important to understand how particle incorporation changes the phase contents of the PEO composite coatings.

Lou et al. [42] have studied the effect of the addition of varying concentrations (0, 2.5, 5.0, 7.5 and 10 g/L) of MoS_2 NPs on the formation of PEO coatings on AZ31 Mg alloy using an alkaline phosphate-aluminate based electrolyte under pulsed current mode at 83.3 mA/cm^2 with a maximum voltage of 475 V and a duty cycle of 50% for 10 min. Incorporation of MoS_2 NPs has increased the R_a from 1.04 ± 0.02 to 4.88 ± 0.64 μm and the surface porosity from 4.31% to 7.35% with an increase in concentration of the MoS_2 NPs from 0 to 10 g/L. Although the thickness of all the PEO-MoS_2 composite coatings was higher than the particle-free PEO coating, a decrease in thickness was observed with an increase in concentration of the MoS_2 NPs from 2.5 to 10 g/L. Magnesium aluminate ($MgAl_2O_4$) has been identified as the predominant phase with a very smaller amount of MgO phase for particle-free PEO coating. Conversely, for the PEO-MoS_2 composite coatings, the volume fraction of the $MgAl_2O_4$ phase is considerably decreased, whereas MgO becomes the predominant phase. After 0.5 h of immersion in 3.5% NaCl, the i_{corr} of PEO-MoS_2 composite coatings prepared using 2.5, 5.0 and 7.5 g/L of MoS_2 NPs showed a moderate improvement in corrosion resistance when compared to that of the particle-free PEO coating whereas the corrosion resistance of composite coating prepared using 10 g/L of MoS_2 NPs was drastically affected. After 6 h of immersion in 3.5% NaCl, the corrosion resistance of all the PEO-MoS_2 composite coatings was decreased. The corrosion performance was very poor for PEO-MoS_2 composite coatings prepared using 10 g/L of MoS_2 NPs with an i_{corr} of 282.3 μA/cm^2 as opposed to an i_{corr} of 1.35 μA/cm^2 for the particle-free PEO coating. The lower surface porosity, presence of a dense and continuous inner barrier layer and enrichment of the $MgAl_2O_4$ phase provided a better long-term corrosion resistance for the particle-free PEO coating. The presence of open voids, channels, cracks, and grain boundaries along with a decrease in the volume fraction of the $MgAl_2O_4$ phase are considered responsible for the poor long-term corrosion resistance of the PEO-MoS_2 composite coatings.

Sedelnikova et al. [50] have studied the incorporation of β-tricalcium phosphate (β-TCP) in the PEO coating formed on Mg-0.8Ca alloy using an alkaline phosphate-fluoride-based electrolyte containing 10–30 g/L of β-TCP particles under potentiostatic control at 350, 400, 450 and 500 V with a pulse frequency of 50 Hz and pulse duration of 100 μs for 5 and 10 min and evaluated their corrosion resistance in 0.9% NaCl. The thickness, porosity and roughness of the PEO-β-TCP composite coatings were increased with an increase in voltage from 350 to 500 V and treatment time from 5 to 10 min. PEO-β-TCP composite coatings formed at 350 and 400 V consisted of β-TCP, α-tricalcium phosphate (α-TCP) and magnesium oxide phases, while those formed at 450 and 500 V showed the formation of hydroxyapatite (HA) phase besides the above phases (Fig. 10.7). In addition, the presence of an

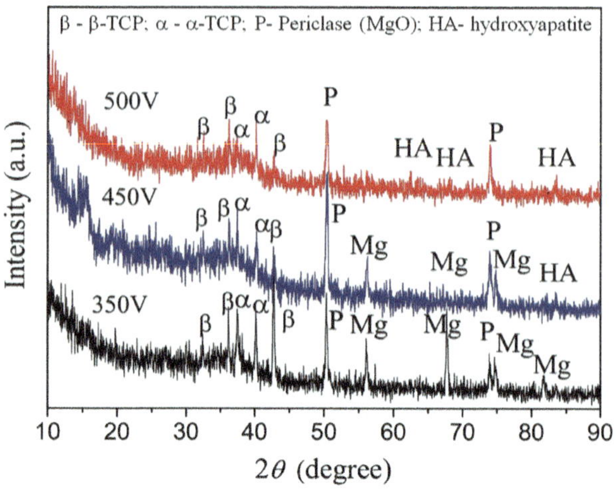

Fig. 10.7 XRD patterns of the PEO-β-TCP composite coatings formed on Mg-0.8Ca alloy at different applied voltages for 5 min. (Reproduced from Ref. [50] © 2018 Authors)

amorphous phase was also identified in all the PEO-β-TCP composite coatings. During PEO at 350 and 400 V, the temperature experienced at the microarc discharges facilitated the transformation of β-TCP to α-TCP phase. Partial dissolution of α-TCP phase has enabled the formation of the HA phase at 450 and 500 V. The PEO-β-TCP composite coating formed at 450 V for 10 min exhibited an i_{corr} of 7.2×10^{-7} A/cm^2, R_p of 4.6×10^4 Ω cm^2 and a |Z| of 6.1×10^4 Ω cm^2 at 0.01 Hz. The higher corrosion resistance of this PEO composite coating has been attributed to the formation of the HA phase, its relatively higher thickness and lower porosity.

10.7 Densification of the PEO Coating Following Particle Incorporation and Its Effect on the Corrosion Resistance of PEO Composite Coatings

The formation of PEO composite coatings incorporated with multi-walled carbon nanotubes (MWCNTs) on commercially pure Al using an alkaline phosphate aluminate-based electrolyte containing 2 g/L of MWCNTs (diameter: 25 nm; length: 1.5 μm; surface area: 300 m^2/g) at 520 V for 300 s and their long-term corrosion behaviour in 3.5% NaCl for 864 h was studied by Sabouri and Khoei [14]. The zeta potential (−26.4 mV) supported the movement of the MWCNTs towards the anode and their incorporation in the PEO coating, validated by the presence of MWCNTs within the pores and bundles of MWCNTs at the microcracks. The overall thickness of the PEO coating was decreased from ~29 to ~25 μm while the size and number of pores were increased following the incorporation of MWCNTs. Nevertheless, the thickness of the inner barrier layer of the PEO coating was increased from 782 ± 147 to 1.2 μm ± 256 nm following the incorporation of MWCNTs. During PEO, the localized increase in temperature and/or pressure in the discharge channels promoted eruption of the molten mass that has resulted in depletion of oxides along with the generation of voids and discontinuities at middle and interfacial regions. This has caused a decrease in the densification of the PEO-MWCNTs composite coating. Both α-Al_2O_3 to γ-Al_2O_3 were identified as the main phases. However, the decrease in intensity ratio of α-Al_2O_3 (113)/γ-Al_2O_3 (400) planes from ~4.2 to ~2.9 following the incorporation of MWCNTs suggested that the γ-Al_2O_3 phase was rich in the PEO-MWCNTs composite coating. For PEO coating, the resistance of the outer porous layer was decreased from 802.3 to 27.9 kΩ cm^2, while the resistance of the inner barrier layer was increased from 100 to 7800 kΩ cm^2 with an increase in immersion time from 0.5 to 864 h. For PEO-MWCNTs composite coating, the resistance of the outer porous layer was decreased from 14,579 to 5.9 kΩ cm^2, while the resistance of the inner barrier layer was decreased from 44,395 to 502.6 kΩ cm^2 under a similar time period. During the initial periods of immersion in 3.5% NaCl, the resistances of both the outer and inner layers of the PEO-MWCNTs composite coating were higher than the PEO coating. However, after 864 h of immersion, the resistances of both the outer and inner layers of the PEO-MWCNTs composite coating were considerably decreased when compared to that of the PEO coating. The inability of the PEO-MWCNTs composite coating to provide long-term corrosion protection was attributed due to the increase in size and number of pores in the outer layer and hindrance provided by the MWCNTs on sintering and densification of the coating.

10.8 Surface Modification of Particles on the Level of Incorporation and Its Effect on the Corrosion Resistance of PEO Composite Coatings

Uptake of particles in PEO coatings is governed by the electrophoretic movement. As most of electrolytes used for preparing PEO coatings are alkaline in nature, the particles assume a negative surface charge, which enables their easy movement towards the anode during the PEO process. Additives such as surfactants are added to the electrolyte to impart a surface charge on the particles. Agglomeration of particles in the electrolyte is a big challenge. Besides, it would drastically affect the quality and corrosion resistance of the PEO composite coatings. Surface modification is a viable option that helps to impart a desired charge on the particles, which would help their electrophoretic movement as well as their agglomeration. Modification of the surface of bioactive glass (BG) NPs by polydopamine (PDA) is a good example.

Aliramezani et al. [48] have shown the influence of surface modification of bioactive glass (BG) NPs (46 ± 9 nm) by polydopamine (PDA) coating on the incorporation of BG/PDA NPs in the PEO coating formed on AZ31 Mg alloy using an alkaline phosphate-fluoride based electrolyte containing 5 g/L of the BG and BG/PDA NPs under unipolar (UP) and bipolar (BP) waveforms at a frequency of 1 kHz and at a duty cycle of 20% for 15 min. The corrosion resistance of the particle-free PEO coating as well as the PEO-BG and PEO-BG/PDA composite coatings was evaluated in phosphate buffer saline (PBS) (pH: 7.40 ± 0.10) at 37 ± 1 °C. Sol-gel method was employed to synthesize the BG NPs, which was subsequently modified by oxidation and polymerization of dopamine hydrochloride. Interaction between amine groups of dopamine hydrochloride and hydroxyl groups of the BG NPs has enabled the formation of the PDA coating through hydrogen bonding, which was evidenced by the change in colour of the BG NPs from white to gray. Addition of BG NPs has increased the conductivity of the electrolyte from 23.6 to 24.1 mS/cm whereas the conductivity was increased to 24.9 mS/cm with the addition of BG/PDA NPs. Although the high negative zeta potential of BG NPs push them towards the anode, the tendency for agglomeration has limited their level of incorporation in the PEO coating under both UP and BP waveforms. The PDA coating helped to increase the repulsion among the BG NPs, thus increased the stability of the BG/PDA NPs in the electrolyte. As a result, the BG/PDA NPs had a higher chance of incorporation in the PEO coating in the open pores. The level of incorporation of BG NPs was higher under UP waveform than those prepared under BP waveform. Conversely, for BG/PDA NPs, a higher level of incorporation was observed under BP waveform (Fig. 10.8).

As the interaction of BG NPs with PDA has occurred through diverse mechanisms, both positive and negative surface charges were imparted on the BG/PDA NPs. A positive charge was imparted on the BG NPs through hydrogen bonds between the silanol groups (Si–OH) as well as through the catechol group of PDA. A negative charge was also imparted on the BG NPs through a coordination bond with the hydroxyl groups of PDA. Hence, the chance of incorporation of BG/PDA NPs becomes higher during both the anodic and cathodic pulses of the BP waveform. After 1 day of immersion in PBS at 37 ± 1 °C, PEO-BG composite coating prepared under UP waveform offered a higher corrosion resistance than those prepared under BP waveform. Conversely, PEO-PG/PDA composite coating prepared under BP waveform offered a higher corrosion resistance than those prepared under UP waveform. For a given waveform, UP or BP, the resistance of the PEO-BG composite coatings was much higher than that of the PEO-BG/PDA composite coating, due to the higher wettability of the later coating. However, after 28 days of immersion in PBS, PEO-PG/PDA composite coating could sustain better corrosion protection.

10.9 Corrosion Resistance of PEO-Composite Coatings Prepared in a Single Stage and Two Stages

Numerous modifications are constantly being made in the PEO process in preparing PEO and PEO composite coatings. Accordingly, particle incorporation in the PEO coatings can be made in a single step by dispersing them in the electrolyte. Alternatively, a particle-free porous PEO coating can be prepared in the first step

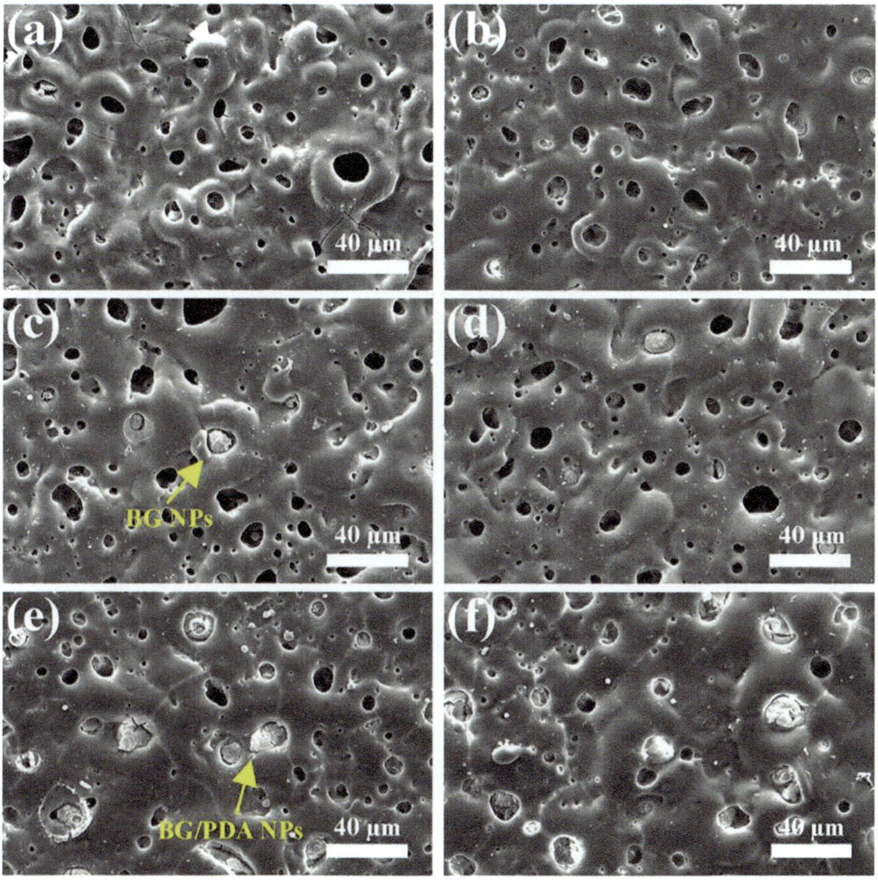

Fig. 10.8 Surface morphologies of (**a**, **b**) particle-free PEO coating; (**c**, **d**) PEO-BG composite coating; and (**e**, **f**) PEO-BG/PDA composite coatings formed on AZ31 Mg alloy under unipolar (UP) and bipolar (BP) waveforms at a frequency of 1 kHz and at a duty cycle of 20% for 15 min (**a**) UP-particle free; (**b**) BP-particle free; (**c**) UP-BG composite; (**d**) BP-BG composite; (**e**) UP-BG/PDA composite; and (**f**) BP-BG/PDA composite coatings. (Reproduced with permission from Ref. [48] © 2024 Elsevier B.V.)

and the particles can be incorporated in the pores and cracks of the PEO coating in the next steps by dispersing the particles in a suitable electrolyte. The formation of PEO-HA composite coating is a good example, as explained below:

Anawati et al. [47] have prepared PEO-HA composite coatings on AZ31 Mg alloy through a one-step and two-step fabrication by PEO and evaluated their corrosion resistance in 0.9% NaCl at 37 °C. In the one-step fabrication method, the PEO-HA composite coating was prepared using an alkaline phosphate electrolyte containing 10 g/L of HA NPs (20 nm) at 300 A/m^2 for 10 and 20 min. The two-step fabrication method involved the use of two different electrolytes wherein the coating was prepared in two different steps. In the first step, the PEO coating was formed using a particle-free alkaline phosphate electrolyte at 300 A/m^2 for 10 min, which was followed by treatment of the PEO-coated Mg alloy in an alkaline suspension containing 10 g/L of HA NPs (20 nm) in the second step. Morphological features revealed that the PEO-HA composite coating fabricated in a single-step consists of more pores and cracks than those fabricated in two-steps. The thickness of the PEO-HA composite coating fabricated in a single step was increased from 22 ± 4 and 33 ± 2 μm with an increase in treatment time from 10 to 20 min. The thickness of the PEO-HA composite coating fabricated in two steps was of the order of 29 ± 2 μm irrespective of the treatment time. Magnesium phosphate was identified as the major phase along with HA for coatings prepared in a single step as well as in two steps. However, the intensity of the diffraction peaks pertaining to the HA phase was relatively lower for the PEO-HA composite coating fabricated in a single step than those observed for the oxide coating fabricated in two steps. EDS analysis further substantiated this inference in terms of a higher atomic percentage of Ca (10.66 at. %) for the PEO-HA composite coating fabricated in two steps as opposed to those fabricated in a single step with a relatively lower Ca content (3.7 at.%).

The porosity of PEO-HA composite coatings fabricated in a single step for 10 min and 20 min was 9.2% and 14.7%, respectively. The porosity of PEO-HA composite coatings fabricated in two steps for 10 min and 20 min was 7.2% and 3.9%, respectively. The relatively higher intensity of the diffraction peaks pertaining to HA, relatively higher Ca content and relatively lower porosity point out that most of the HA NPs were accumulated inside the pores of the PEO-HA composite coatings fabricated in two steps whereas the HA NPs were spread throughout the entire coating in the PEO-HA composite coatings fabricated in a single-stage. The i_{corr} of PEO-HA composite coatings fabricated in a single step for 10 min and 20 min were 3.07×10^{-7} A/cm^2 and 1.99×10^{-7} A/cm^2, respectively. The i_{corr} of PEO-HA composite coatings fabricated in two steps for 10 min and 20 min were 1.04×10^{-6} A/cm^2 and 4.99×10^{-7} A/cm^2, respectively. The |Z| of PEO-HA composite coatings fabricated in a single step for 10 min and 20 min were 1.15×10^5 Ω cm^2 and 1.99×10^5 Ω cm^2, respectively. The |Z| of PEO-HA composite coatings fabricated in two steps for 10 min and 20 min were 1.48×10^4 Ω cm^2 and 6.92×10^4 Ω cm^2, respectively. The better corrosion resistance of PEO-HA composite coatings fabricated in a single step than those fabricated in two steps has been attributed to the uniform distribution of the HA NPs throughout the coating rather than accumulation of them mainly at the pores. The higher thickness of the PEO-HA composite coatings fabricated in a single step for 20 min is yet another reason for the better corrosion resistance.

10.10 Conclusions and Outlook

The thickness of the oxide coating, size, density and distribution of the pores and cracks, type of phases formed and their chemical stability, determine the corrosion protective ability of PEO coatings. All these factors are also applicable for PEO composite coatings. Among these parameters, the size, density and distribution of the pores and cracks are very critical. As particle incorporation preferentially occurs at the pores and cracks of the PEO coating, it has been attempted as one of the strategies to improve the corrosion resistance of PEO coatings. In principle, this is true as particle incorporation could seal the pores and cracks of the PEO coating and prevent the permeation of the corrosive medium through them to the base metal. However, several other factors, such as the type of particles and their concentration, conductivity of the particle, mode of incorporation, change in phase content, densification of the oxide coating following particle incorporation, dependence on the type of anions, process parameters, and, surface modification of particles on the level of particle incorporation, determines the corrosion resistance of the PEO composite coatings. Hence, it is suggested that the following points should also be given due consideration when particle incorporation is attempted to increase the corrosion resistance of PEO coatings.

- The ability of the particles to fill the pores and reduce the defects of the PEO coating plays a significant role in determining the extent of corrosion protection rather than the type of particles and their compositions.
- The decrease in porosity and increase in compactness of the inner barrier layer are critical to achieve a better corrosion resistance of PEO composite coatings.
- Uniform distribution of the particles throughout the oxide coating rather than accumulation of them mainly at the pores and cracks should be the focus to attain long-term corrosion resistance of PEO composite coatings.
- PEO composite coatings incorporated with NPs offer better corrosion resistance than those incorporated with microparticles.
- Agglomeration of particles beyond a threshold concentration is likely to decrease the quality of the inner and outer layers by increasing the pore size and the R_a, which in turn would deleteriously influence the corrosion resistance of the PEO composite coatings.
- Besides the pore-filling effect of the particles, the formation of a uniform and dense oxide coating consisting of higher extent of coating forming elements with better chemical stability contributes to the improvement in corrosion resistance of the PEO composite coatings.
- Inert incorporation of particles facilitates filling up of the pores and cracks and improves the corrosion resistance of the PEO composite coatings.
- Reactive incorporation of particles facilitates the formation of an amorphous phase that seals the pores and cracks and promotes the formation of a compact and dense oxide coating, which thereby increase the corrosion resistance of the PEO composite coatings.
- An increase in the volume fraction of certain phases, phase transformation as well as formation of new phases with better chemical stability could improve the corrosion resistance of PEO composite coatings.
- The type of anions in the electrolyte plays a critical role in determining the discharge characteristics, which in turn influences the development of a porous structure, the size and number of pores, and, the level of incorporation of particles in the PEO composite coatings and their corrosion resistance.

- An increase in conductivity of the electrolyte following particle addition could increase the reactivity, growth rate and thickness of the oxide coating. However, this condition could also increase the pore size, number of pores and generation of more cracks that deleteriously influence the corrosion resistance of the PEO composite coatings.
- Process parameters such as current density, current waveform and treatment time exert a strong influence on the corrosion resistance of the PEO composite coatings. An increase in current density would increase the reactivity, growth rate and thickness of the oxide coating. Nevertheless, the beneficial effect of a higher thickness on corrosion resistance is outweighed by the deleterious influence of an increase in porosity. The PEO composite coatings prepared under bipolar current mode at a relatively higher frequency and duty cycle exhibit better corrosion resistance due to the decrease in porosity and defects. An increase in treatment time could not only increase the thickness but also the defects in the PEO coating and in terms of corrosion resistance, both of them could outweigh each other.
- Conductivity of the incorporated particle is an important factor in determining the corrosion resistance of PEO composite coatings. Particles with higher conductivity could decrease the corrosion resistance of PEO composite coatings irrespective of a homogeneous distribution of the particles in the PEO coating, decrease in surface porosity and average pore size.
- Surface modification could impart a desirable charge on particles, facilitate their electrophoretic movement towards the anode and promote their incorporation in the PEO coating and the corrosion resistance of the PEO composite coatings.

References

1. Nasiri Vatan, H., Ebrahimi-Kahrizsangi, R., & Asgarani, M. K. (2016). *International Journal of Electrochemical Science, 11*(2), 929.
2. Askarnia, R., Fardi, S. R., Sobhani, M., Staji, H., & Aghamohammadi, H. (2022). *Journal of Alloys and Compounds, 892*, 162106.
3. Bordbar-Khiabani, A., Yarmand, B., & Mozafari, M. (2019). *Surface and Coating Technology, 360*, 153.
4. O'Hara, M., Troughton, S. C., Francis, R., & Clyne, T. W. (2020). *Surface and Coating Technology, 385*, 125354.
5. Tsai, D. S., & Chou, C. C. (2021). *Coatings, 11*(3), 270.
6. Lu, X., Sah, S. P., Scharnagl, N., Störmer, M., Starykevich, M., Mohedano, M., et al. (2015) *Surface and Coating Technology, 269*, 155.
7. Lu, X., Blawert, C., Kainer, K. U., Zhang, T., Wang, F., & Zheludkevich, M. L. (2018). *Surface and Coating Technology, 352*, 1.
8. Peitao, G., Mingyang, T., & Chaoyang, Z. (2019). *Surface and Coating Technology, 359*, 197.
9. Babaei, K., Fattah-Alhosseini, A., & Molaei, M. (2020). *Surfaces and Interfaces, 21*, 100677.
10. Fattah-alhosseini, A., Molaei, M., Nouri, M., & Babaei, K. (2021). *Applied Surface Science Advances, 6*, 100140.
11. Pourshadloo, M., Rezaei, H. A., Saeidnia, M., Alkokab, H., & Bathaei, M. S. (2023). *Surface Innovations, 11*(1–3), 5.
12. Zhao, J., Xie, X., & Zhang, C. (2017). *Corrosion Science, 114*, 146.

13. Zhang, Y., Chen, F., Zhang, Y., & Du, C. (2020). *Tribology International, 146*, 106135.
14. Sabouri, M., & Khoei, S. M. (2018). *Surface and Coating Technology, 334*, 543.
15. Tran, Q. P., Chin, T. S., Kuo, Y. C., Jin, C. X., Trung, T., Van Tuan, C., & Dang, D. Q. (2018). *Journal of Alloys and Compounds, 751*, 289.
16. Ghafaripoor, M., Raeissi, K., Santamaria, M., & Hakimizad, A. (2018). *Surface and Coating Technology, 349*, 470.
17. Molaei, M., Fattah-alhosseini, A., Nouri, M., Mahmoodi, P., & Nourian, A. (2022). *Ceramics International, 48*(14), 21005.
18. Mozafarnia, H., Fattah-alhosseini, A., Chaharmahali, R., Nouri, M., Keshavarz, M. K., & Kaseem, M. (2022). *Coatings, 12*(12), 1967.
19. Mohedano, M., Blawert, C., & Zheludkevich, M. L. (2015). *Materials and Design, 86*, 735.
20. Ceriani, F., Casanova, L., Massimini, L., Brenna, A., & Ormellese, M. (2023). *Coatings, 13*(10), 1718.
21. Li, J., Zhang, Z., Guo, Z., Yang, Z., Qian, W., Chen, Y., et al. (2023). *Journal of Materials Research and Technology, 24*, 2403.
22. Chaharmahali, R., Fattah-alhosseini, A., Nouri, M., & Babaei, K. (2021). *Applied Surface Science Advances, 6*, 100131.
23. Andrade, C. S., Borges, M. H. R., Silva, J. P., Malheiros, S., Sacramento, C., Ruiz, K. G., et al. (2025). *Colloids and Surfaces. B, Biointerfaces, 245*, 114237.
24. Nadimi, M., Dehghanian, C., & Etemadmoghadam, A. (2022). *Materials Today Communications, 31*, 103587.
25. Rahmadani, S., & Anawati, A. (2024). *Applied Surface Science Advances, 23*, 100631.
26. Zehra, T., Patil, S. A., Shrestha, N. K., Fattah-alhosseini, A., & Kaseem, M. (2022). *Journal of Alloys and Compounds, 916*, 165445.
27. Roknian, M., Fattah-alhosseini, A., Gashti, S. O., & Keshavarz, M. K. (2018). *Journal of Alloys and Compounds, 740*, 330.
28. Keyvani, A., Zamani, M., Bahamirian, M., Nikoomanzari, E., Fattah-alhosseini, A., & Sina, H. (2021). *Surfaces and Interfaces, 22*, 100728.
29. Han, J., Yu, Y., Yang, J., Xiaopeng, L., Blawert, C., & Zheludkevich, M. L. (2022). *Surface and Coating Technology, 433*, 128112.
30. Wang, X., Ju, P., Lu, X., Chen, Y., & Wang, F. (2022). *Ceramics International, 48*(3), 3615.
31. Mashtalyar, D. V., Imshinetskiy, I. M., Kashepa, V. V., Nadaraia, K. V., Piatkova, M. A., Pleshkova, A. I., et al. (2024). *Journal of Magnesium and Alloys, 12*(6), 2360.
32. Sun, M., Jiang, M., Huang, H., Yang, B., Lin, Y., & Wang, P. (2023). *Coatings, 13*(5), 967.
33. Hosseini, M., Khalil-Allafi, J., Etminanfar, M., Safavi, M. S., Bloise, N., & Ghalandarzadeh, A. (2023). *Materials Chemistry and Physics, 293*, 126899.
34. Pezzato, L., Lorenzetti, L., Tonelli, L., Bragaggia, G., Dabalà, M., Martini, C., & Brunelli, K. (2021). *Surface and Coating Technology, 428*, 127901.
35. Aljohani, T. A., Alawad, M. O., Elkatatny, S., Alateyah, A. I., Rubayan, M. T. B., Alhajji, M. A., et al. (2022). *Materials, 15*(10), 3724.
36. Tang, Y., Yang, C. P., Sun, Q. Q., Wu, L. K., & Cao, F. H. (2024). *Corrosion Communications, 14*, 1.
37. Lu, X., Blawert, C., Scharnagl, N., & Kainer, K. U. (2013). *Journal of Magnesium and Alloys, 1*(4), 267.
38. Lou, B. S., Lin, Y. Y., Tseng, C. M., Lu, Y. C., Duh, J. G., & Lee, J. W. (2017). *Surface and Coating Technology, 332*, 358.
39. Zhu, X., Fu, J., Ma, D., Ma, C., Fu, Y., & Zhang, Z. (2022). *Scientific Reports, 12*(1), 995.
40. Zhao, X., Chen, Y., Ji, R., Xu, M., Ye, Z., Shao, W., et al. (2024). *Surface and Coating Technology, 494*, 131471.
41. Mashtalyar, D. V., Gnedenkov, S. V., Sinebryukhov, S. L., & Imshinetskiy, I. M. (2017). *Journal of Materials Science and Technology, 33*(5), 461.
42. Lou, B. S., Lee, J. W., Tseng, C. M., Lin, Y. Y., & Yen, C. A. (2018). *Surface and Coating Technology, 350*, 813.
43. Nisar, S. S., & Choe, H. C. (2024). *Journal of Materials Research and Technology, 33*, 2035.

44. Jing-guo, F., Sheng-lin, M., Xin-he, Z., Chang-qi, X., Zhi-jun, Y., Dong, C., & Chun-sheng, M. (2020). *Materials Research Express, 6*(12), 1265c8.
45. Han, Q., Li, Y., Lu, X., Mei, D., Chen, Q., Zhang, T., & Wang, F. (2021). *Materials Letters, 293*, 129731.
46. Cerchier, P., Pezzato, L., Gennari, C., Moschin, E., Moro, I., & Dabalà, M. (2020). *Surface and Coating Technology, 393*, 125774.
47. Anawati, A., Purnamasari, Y., Khairunisa, A. P., Wulandari, M. P., & Gumelar, M. D. (2024). *Materials Research Express, 11*(8), 085403.
48. Aliramezani, R., Raeissi, K., Labbaf, S., & Hakimizad, A. (2024). *Journal of Alloys and Compounds, 1005*, 176145.
49. Uzair, S. A., Hussain, F., & Rizwan, M. (2024). *Ceramics, 7*(4), 1459.
50. Sedelnikova, M. B., Komarova, E. G., Sharkeev, Y. P., Tolkacheva, T. V., Sheikin, V. V., Egorkin, V. S., et al. (2018). *Metals, 8*(4), 238.
51. Rapheal, G., Kumar, S., Scharnagl, N., & Blawert, C. (2016). *Surface and Coating Technology, 289*, 150.
52. Mortezanejad, E., Atapour, M., Salimijazi, H., Alhaji, A., & Hakimizad, A. (2021). *Journal of Materials Engineering and Performance, 30*(6), 4030.
53. Imshinetskiy, I., Kashepa, V., Nadaraia, K., Mashtalyar, D., Suchkov, S., Zadorozhny, P., et al. (2022). *International Journal of Molecular Sciences, 24*(1), 305.
54. Zhang, G., Jiang, E., Wu, L., Ma, W., Yang, H., Tang, A., & Pan, F. (2021). *Scientific Reports, 11*(1), 2774.
55. Pezzato, L., Kostelac, L., Tonelli, L., Elsayed, H., Kajánek, D., Bernardo, E., et al. (2025). *Metals and Materials International, 31*, 636–653.
56. Chen, Y., Wu, L., Wu, M., Yao, W., Wu, J., Zhou, Y., et al. (2024). *Surface and Coating Technology, 493*, 131246.
57. Schneider, M., Kremmer, K., Fürbeth, W., Lederer, S., & Sottor, R. (2023). *Surface and Coating Technology, 459*, 129369.
58. Sottor, R., Gruen, R., Kremmer, K., Lederer, S., Schneider, M., & Fuerbeth, W. (2023). *Corrosion and Materials Degradation, 4*(3), 488.
59. Sankara Narayanan, T. S. N., Park, I. S., & Lee, M. H. (2014). *Progress in Materials Science, 60*, 1.
60. Darband, G. B., Aliofkhazraei, M., Hamghalam, P., & Valizade, N. (2017). *Journal of Magnesium and Alloys, 5*(1), 74.
61. Simchen, F., Sieber, M., Kopp, A., & Lampke, T. (2020). *Coatings, 10*(7), 628.
62. Kaseem, M., Fatimah, S., Nashrah, N., & Ko, Y. G. (2021). *Progress in Materials Science, 117*, 100735.
63. Sikdar, S., Menezes, P. V., Maccione, R., Jacob, T., & Menezes, P. L. (2021). *Nanomaterials, 11*(6), 1375.
64. Sankara Narayanan, T. S. N., & Ravichandran, K. (2023). Chapter 14: Electrochemical anodic oxidation/micro-arc oxidation-based composite coatings. In V. S. Saji (Ed.), *Advances in corrosion control of magnesium and its alloys* (pp. 215–242). CRC Press.
65. Cardoso, G. C., Grandini, C. R., & Rau, J. V. (2024). *Journal of Materials Research and Technology, 31*, 311.
66. Molaei, M., Nouri, M., Babaei, K., & Fattah-alhosseini, A. (2021). *Surfaces and Interfaces, 22*, 100888.
67. Fattah-alhosseini, A., & Chaharmahali, R. (2021). *FlatChem, 27*, 100241.
68. Fattah-alhosseini, A., Chaharmahali, R., Alizad, S., & Kaseem, M. (2023). *Journal of Magnesium and Alloys, 11*(9), 2999.
69. Lu, X., Blawert, C., Huang, Y., Ovri, H., Zheludkevich, M. L., & Kainer, K. U. (2016). *Electrochimica Acta, 187*, 20.
70. Fattah-alhosseini, A., Molaei, M., & Babaei, K. (2020). *Surfaces and Interfaces, 21*, 100659.
71. Almajidi, Y. Q., Ali, E., Jameel, M. F., Saleh, L. H., Aggarwal, S., Zearah, S. A., et al. (2023). *Lubricants, 11*(12), 519.

72. Zhao, C., Wang, X., Yu, B., Cai, M., Yu, Q., & Zhou, F. (2023). *Coatings, 13*(7), 1189.
73. Shokouhfar, M., & Allahkaram, S. R. (2017). *Surface and Coating Technology, 309*, 767.
74. Askarnia, R., Sobhani, M., Zare, M., Aghamohammadi, H., & Staji, H. (2023). *Journal of the Mechanical Behavior of Biomedical Materials, 141*, 105784.
75. Ma, X., Jin, S., Wu, R., Zhang, S., Hou, L., Krit, B., et al. (2022). *Surface and Coating Technology, 438*, 128399.
76. Gao, W., Wang, L., Jin, Y., Yao, Y., Ding, Z., Yang, W., & Liu, J. (2022). *Coatings, 12*(9), 1247.
77. Ahangari, A., Mokhtarzade, A., & Mahmoodi, M. (2023). *Materials Chemistry and Physics, 308*, 128210.
78. Kara, R., & Zengin, H. (2022). *Acta Metallurgica Sinica (English Letters), 35*(7), 1195.
79. Asgari, M., Aliofkhazraei, M., Darband, G. B., & Rouhaghdam, A. S. (2020). *Surface and Coating Technology, 383*, 125252.
80. Fattah-alhosseini, A., Chaharmahali, R., & Babaei, K. (2020). *Journal of Magnesium and Alloys, 8*(3), 799.
81. Hakimizad, A., Raeissi, K., Golozar, M. A., Lu, X., Blawert, C., & Zheludkevich, M. L. (2017). *Surface and Coating Technology, 324*, 208.
82. Nasiri Vatan, H., & Adabi, M. (2017). *Transactions of the IMF, 95*(6), 308–315.
83. Mohedano, M., Arrabal, R., Mingo, B., Pardo, A., & Matykina, E. (2018). *Surface and Coating Technology, 334*, 328.
84. Rehman, Z. U., Uzair, M., Lim, H. T., & Koo, B. H. (2017). *Journal of Alloys and Compounds, 726*, 284.
85. Gowtham, S., Hariprasad, S., Arunnellaiappan, T., & Rameshbabu, N. (2017). *Surface and Coating Technology, 313*, 263.

Chapter 11
Biomedical Plasma Electrolytic Oxidation Composite Coatings via Particles Addition

Xinyuan Wang, Di Mei, Shijie Zhu, Liguo Wang, and Shaokang Guan

11.1 Introduction

With the development of modern materials science, life sciences, and medical sciences, implant materials have been widely applied in clinical fields. Implant materials refer to materials that are completely introduced into the human body through clinical intervention or replace human tissues and remain in place after surgery [1]. Traditional implant materials can be categorized into metals (pure titanium, stainless steel, cobalt-based alloys, titanium alloys, etc.) [2–5], polymers (polyurethane, silicone rubber, Poly(vinyl chloride), polyethylene, polypropylene, polyetheretherketone, polylactic acid) [6–11], ceramics (bioglass ceramics, hydroxyapatite (HA), ceramics, carbon, alumina (Al_2O_3), zirconia, etc.) [12–16], natural materials (collagen, hyaluronic acid, chitosan, Cellulose) [17, 18], and composite materials made from two or more of the aforementioned categories [19–22]. Metal materials, as medical implant materials, have the characteristics of high strength, good toughness, and excellent processing performance [23].

Magnesium alloys have gained increasing attention in biomedical applications, particularly for vascular stents and bone fixation implants [24]. Unlike conventional engineering applications, the main advantage of magnesium alloys in biomedical applications is their biodegradability, which allows the implant to degrade and be absorbed by the body after fulfilling its function, thus avoiding secondary surgery to remove the implant [25]. In addition, the appropriate mechanical properties of magnesium alloys are also an attractive feature, as they can reduce the stress shielding effect when used as orthopedic implants [26]. In terms of biodegradable implants,

X. Wang · D. Mei (✉) · S. Zhu · L. Wang · S. Guan
School of Materials Science and Engineering & Henan Province Key Laboratory of Advanced Light Alloys, Zhengzhou University, Zhengzhou, People's Republic of China
e-mail: dimei@zzu.edu.cn

V. S. Saji, M. Zheludkevich (eds.), *Plasma Electrolytic Oxidation – Fundamentals, Advances and Applications*,
https://doi.org/10.1007/978-3-031-92429-3_11

the most essential characteristic is a controllable degradation rate that matches the healing rate of the affected tissue [27]. However, the relatively high corrosion rate of magnesium and its alloys may lead to premature failure of the implant during service, and the large amount of hydrogen gas produced may also exacerbate the inflammatory response in surrounding tissues. These factors limit further development of their use in biodegradable implants [28].

Titanium and its alloys possess excellent properties such as high strength, low density (high specific strength), high corrosion resistance, and enhanced biocompatibility, making them promising materials for medical applications [29]. Compared to magnesium alloys, titanium alloys exhibit superior mechanical properties and better corrosion resistance, making them ideal nondegradable materials for medical use. They are employed in the repair of bones and the replacement of specific components, such as dental and orthopedic implants [30]. However, their relatively poor wear resistance may lead to the generation of debris through friction, potentially causing implant loosening and inflammation [31]. Additionally, titanium alloys have limited osteointegration and osteoinductive capabilities, preventing direct bonding with bone tissue. Surface morphology, composition, hydrophilicity, and roughness are key factors influencing implant-tissue interactions and bone integration [32].

PEO is one of the commonly employed coating technologies. It is a process derived from anodic oxidation that forms a porous ceramic coating. This process can enhance the wear resistance, corrosion resistance, and other properties of magnesium, aluminum, and titanium alloys [33, 34]. This chapter focuses on improving coating performance and adding new functionalities by incorporating particles into the electrolyte. It reviews the effects of different particles on the PEO process and the coating performance of medical metals. Finally, the application and future research directions of biomedical-grade particle incorporation into electrochemical conversion coating technologies are discussed.

11.2 Biomedical PEO Composite Coatings via Particles Addition

Biomaterial-based PEO Coatings typically prepared in environmentally friendly alkaline electrolytes, which contain components such as silicates, phosphates, fluorides, or aluminate salts, are processed under high-pressure conditions. The final performance of the PEO coating is determined by both its microstructure, which is influenced by process parameters, and its chemical composition, which is determined by the electrolyte components [35]. Therefore, by optimizing process parameters and electrolyte composition, the performance of alloy materials used in biomedical applications can be effectively improved [36].

Introducing particles into the PEO electrolyte and incorporating them into the PEO coating during the coating formation process has become a widely used surface modification technique in recent years [37]. The incorporation of particles not only improves the porosity characteristics of the PEO coating but also imparts various functionalities to the coating due to the different physical and chemical properties of the particles [38]. The nature of the particles, the electrical parameters during the PEO process, and the composition of the electrolyte all influence the incorporation method and efficiency of the particles. Generally, the incorporation of particles can be categorized into two types: inert incorporation and (partially) reactive incorporation. Inert incorporation refers to the process where particles enter the coating without inducing new phase changes, while reactive incorporation indicates that particles undergo melting and reaction during the incorporation process. Therefore, high melting points and larger particles tend to be incorporated through inert methods, while the particles with low melting points and smaller sizes are more likely to undergo reactive incorporation [39].

Currently, substantial research has been conducted on particle-incorporated PEO composite coatings on biomedical metallic materials, resulting in significant progress. Common types of added particles include oxide particles, nonmetallic particles, and metallic particles. The introduction of these particles further expands the potential applications of PEO coatings in the biomedical field [37]. Thus, this technology not only enhances their corrosion resistance and wear resistance but also provides medical functions such as antibacterial properties and bioactivity.

11.3 Magnesium and Its Alloys

Bone fixation implants and vascular stents are two critical applications of magnesium alloys in the biomedical field [24]. One of the main advantages of PEO coating is its ability to form a strong interfacial bond with the substrate through a metallurgical bonding process, significantly enhancing the adhesion strength between the coating and the substrate. The ceramic coating formed on the surface of the magnesium alloy exhibits excellent wear resistance. It can effectively induce the formation of HA, which provides a distinct advantage for applications such as bone fixation implants that are subjected to friction and wear [40]. Although the degradation rate of PEO-coated bone fixation implants may not align perfectly with the bone healing process, the similarity of the PEO coating to natural bone tissue ensures that it does not cause unacceptable adverse effects on the biocompatibility of the implant [41]. However, for vascular stent applications, due to the high surface roughness of PEO coatings and the relatively thick coating structure, the effectiveness of PEO coatings in vascular stents is somewhat restricted [42]. Conventional PEO coatings often exhibit numerous cracks and pores on their surface, which reduces the corrosion resistance of the coating, making it challenging to meet the long-term in vivo performance requirements for implants [43]. By adding particles to the electrolyte solution to optimize the performance of PEO coatings and impart enhanced functional properties to the coating, it is possible to improve their application effectiveness in the biomedical field [44].

11.3.1 Oxide Particles

Improving the corrosion resistance and wear resistance of these biomedical metallic materials can extend the lifespan of implants in the human body, preventing the rapid loss of mechanical properties and minimizing biocompatibility issues such as inflammation [45]. Oxides typically have high melting points, so most oxide particles tend to be incorporated in an inert form during the PEO process [46]. Lu et al. added nano- and micro-sized SiO_2 particles into the PEO electrolyte. They found that the micro-sized SiO_2 particles were incorporated in an inert form, sealing some pores. In contrast, nano-sized SiO_2 particles reacted and participated in the formation process of the PEO coating, forming nano-crystalline MgO and amorphous materials. The addition of SiO_2 caused the coating to thin and slightly decreased its corrosion resistance but significantly improved its wear resistance [47]. Zhao et al. introduced graphene oxide (GO) into the electrolyte during PEO treatment on Mg alloy, which effectively reduced the number of micropores in the coating thus improving its corrosion resistance [48].

For magnesium alloys, uneven corrosion can lead to a rapid decline in the mechanical properties of the alloy, as pitting on the surface of the sample results in irregular shapes and stress concentration, which may become the source of cracks leading to alloy fracture. Mohedano et al. prepared PEO coatings on AM 50 magnesium alloy using electrolytes containing CeO_2, α-Al_2O_3, and monoclinic ZrO_2 particles. By studying the incorporation of these three particles, it was found that the size and density of surface pores decreased with increasing particle concentration (from 2 to 10 g/L), leading to smoother coatings. Particle-containing electrolytes also resulted in thinner coatings. It was observed that lower concentrations of CeO_2 in the electrolyte could lead to the blocking of pores and cracks in the coating, and due to the accumulation of corrosion products on the surface, the CeO_2-containing PEO coating exhibited better long-term corrosion resistance. However, when the particle concentration increased to 10 g/L, more uneven microstructures appeared in the coating, with more unsealed pores and cracks, affecting the short-term corrosion behavior [49]. Lee et al. prepared PEO passivation layers containing ZrO_2 nanoparticles (NPs), which were mainly distributed in the pores of the magnesium oxide layer. The improved pitting corrosion resistance of the alloy [50] is beneficial for maintaining the mechanical integrity of implants.

Chen et al. found that the addition of TiO_2 NPs changed the microstructure of the PEO coating, blocked some of the pores, and thereby enhanced the corrosion resistance of the coating [51]. It also increased the hardness and hydrophilicity of the PEO coating [52]. The concentration of TiO_2 NPs could influence the coating structure and change the defect density.

Bordbar-Khiabani et al. prepared a MgO/ZnO composite coating on the surface of AZ91 magnesium alloy using the PEO method. ZnO particles were successfully incorporated into the coating, blocking some of the pores and improving the density of the coating (as shown in Fig. 11.1a, b), which made it more difficult for corrosion ions to penetrate [54]. Although the coating thickness decreased, the corrosion

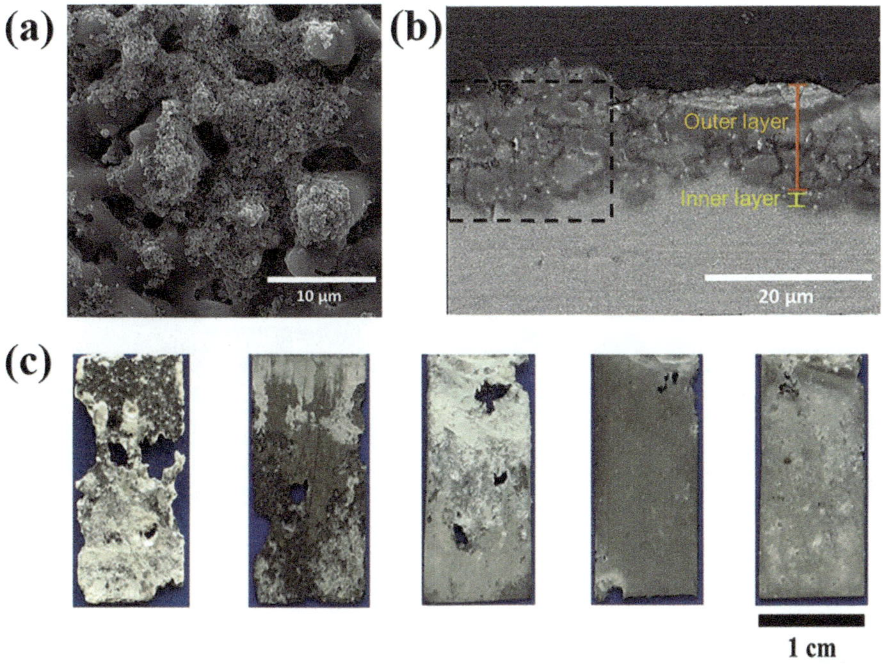

Fig. 11.1 Plasma electrolytic oxidation coating with 4.5 g/L ZnO NPs: (**a**) Surface morphology SEM image, (**b**) Cross-sectional SEM image, (**c**) Optical images of different samples after 14 days of immersion in SBF solution. (Reproduced with permission from [53] © 2019 Elsevier B.V.)

resistance of the samples was still improved, with the polarization resistance increasing from 9.26 to 683.2 kΩ cm^2. Furthermore, a 14-day immersion test in simulated body fluid (SBF) showed that as the concentration of ZnO NPs increased, Ca^{2+}, PO_2^{3-}, and HPO_4^{2-} ions were more attracted to the surface (as shown in Fig. 11.1c), promoting the growth of calcium phosphate layer on the sample [53]. Bahrampour et al. incorporated Mn_3O_4 NPs into the electrolyte to prepare a PEO composite coating with increased thickness and reduced porosity. Electrochemical and immersion corrosion experiments were conducted under normal and inflammatory conditions. The results indicated that inflammation increased the corrosion rate of the samples, while the PEO composite coating containing Mn_3O_4 exhibited higher impedance resistance and lower corrosion current density. The enhanced performance was attributed to the catalytic activity of Mn_3O_4 in removing H_2O_2 under simulated inflammatory conditions, as well as the increased thickness and reduced porosity of the coating compared to the essential PEO coating, which together hindered the penetration of corrosive agents into the substrate [55].

In addition to enhancing the corrosion resistance of PEO coatings by adding particles, particles can also be introduced to functionalize the coating biologically. Madhankumar et al. successfully prepared a PEO coating containing Ta_2O_5 NPs on AZ31 magnesium alloy using a phosphate electrolyte. The addition of the particles

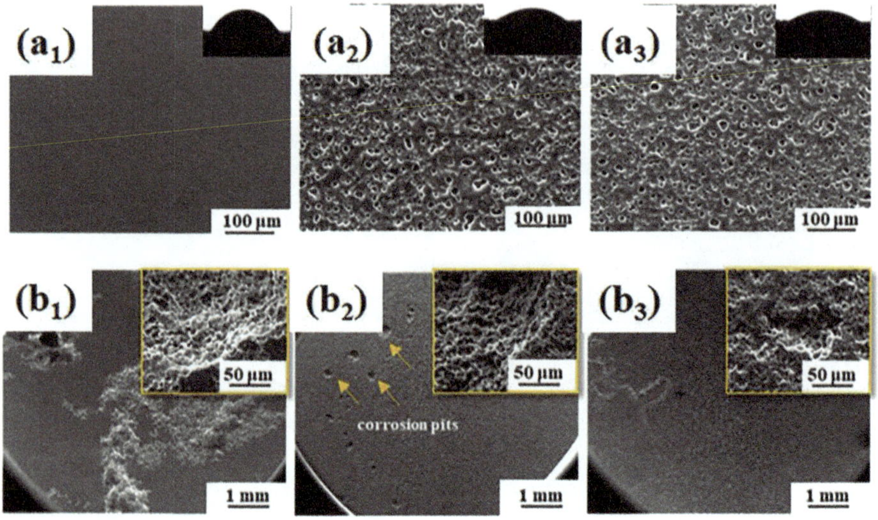

Fig. 11.2 Surface morphology and contact angle of the samples: (**a1**) Mg substrate, (**a2**) MAO samples, and (**a3**) Ta-MAO; corrosion morphology of the samples after immersion in HBSS for 7 days, showing the removal of corrosion products: (**b1**) Mg substrate, (**b2**) MAO samples, and (**b3**) Ta-MAO. (Reproduced with permission from [57] © 2024 Elsevier B.V.)

made the coating thicker, more uniform, and free from cracks and micropores. In vitro cell culture experiments with MG-63 cells demonstrated that the PEO coating containing Ta_2O_5 exhibited better cell adhesion behavior than bare magnesium and the PEO coating alone [56]. Liang et al. developed a PEO coating containing Ta_2O_5 that showed improved hydrophilicity and resistance to pitting corrosion (as shown in Fig. 11.2a). After removing the corrosion products (as shown in Fig. 11.2b), it can be observed that the number of pitting corrosion pits is significantly reduced [57]. The coating enhanced tribological performance and corrosion resistance without compromising biocompatibility. Zhang et al. prepared a series of composite coatings with varying amounts of CuO particles (3, 5, and 7 wt%) and found that the appropriate concentration of CuO particles promoted cell proliferation and adhesion while improving the corrosion resistance of the coating. The slow release of Cu^{2+} ions in the solution gradually increased the antibacterial effect, ultimately reaching a 100% inhibition rate after 72 h [58].

11.3.2 Nonmetallic Particles

Nonmetallic particles come in various types, including organic and inorganic particles as well as nano-containers, and they are widely used in various applications. To enhance the overall performance of the PEO coating on magnesium alloy substrates, researchers often add nano-containers with bioactivity or other functions into the electrolyte. Li et al. introduced layered double hydroxide (LDH) nanocontainers loaded with fumaric acid corrosion inhibitors into the electrolyte, significantly improving the thickness and compactness of the coating [59]. Sun et al. loaded benzotriazole (BTA) into halloysite nanotubes (HNTs) and added them to the electrolyte to prepare a PEO coating on the AM50 magnesium alloy surface. The coating, loaded with benzotriazole, effectively altered the corrosion behavior of the substrate, exhibited self-healing properties, and significantly suppressed pitting corrosion [60]. Guo et al. added polytetrafluoroethylene (PTFE) particles into the PEO coating, resulting in a more uniform and compact coating structure, which significantly improved the corrosion, tribological performance, and hydrophobicity of the coating [61]. Zhang et al. introduced clay particles into the electrolyte, and the study showed that this addition significantly affected the formation process of the PEO coating [62, 63].

Additionally, PEO technology can also improve the drug delivery capabilities of magnesium alloy implants. Through PEO treatment, a coating containing osteoporosis drugs can be prepared on the surface of AZ31 magnesium alloy. Studies have shown that negatively charged drug particles effectively block the coating pores, and a monolayer of bisphosphonates or alendronate forms on the coating surface, thereby enhancing the hydrophilicity and corrosion resistance of the coating [64].

HA, ($Ca_{10}(PO_4)_6(OH)_2$), as the primary inorganic component of natural bone, can promote the proliferation and differentiation of bone cells, accelerating new bone formation when HA coatings are formed on the surface of biomedical implants. HA exhibits excellent biocompatibility and bioactivity; however, its mechanical properties are relatively poor. By altering the electrolyte composition in PEO, HA particles can be in situ incorporated into the PEO coating, potentially achieving a higher bonding strength and better biocompatibility. There has been substantial research on this approach [66]. After PEO, HA particles may enter the PEO film in either an inert or reactive form. Inertly added HA particles tend to concentrate in the outer porous layer, and their accumulation in coating defects is much greater than in other areas. This results in near-ideal closure of the micropores, enhancing the corrosion resistance of the coating [67]. Furthermore, as shown in Fig. 11.3a, immersion tests in SBF indicate that the HA on the coating surface enhances the formation ability of apatite [68]. Many factors, including the alloy type, electrolyte parameters, particle size, and concentration, influence the incorporation effect of HA in the coating. The introduction of particles into the coating may disrupt its integrity, but an appropriate concentration of HA particles can reduce the porosity of the coating, improve its thickness, and enhance its corrosion resistance [69, 70]. With increasing voltage, both the HA content and corrosion resistance of the PEO film increase. Although the HA content increases slightly with prolonged preparation time, cracks appear on the coating, leading to a decline in its corrosion resistance [68]. Current density and duty cycle also influence the formation of composite coatings [71]. By selecting appropriate process parameters, coatings with a good combination of corrosion resistance and phase composition can be obtained. Under certain conditions, HA may reactively bond with the coating, and the strong micro-arc discharge generated at high voltage can cause HA to decompose into $Ca_3(PO_4)_2$. At the same time, the

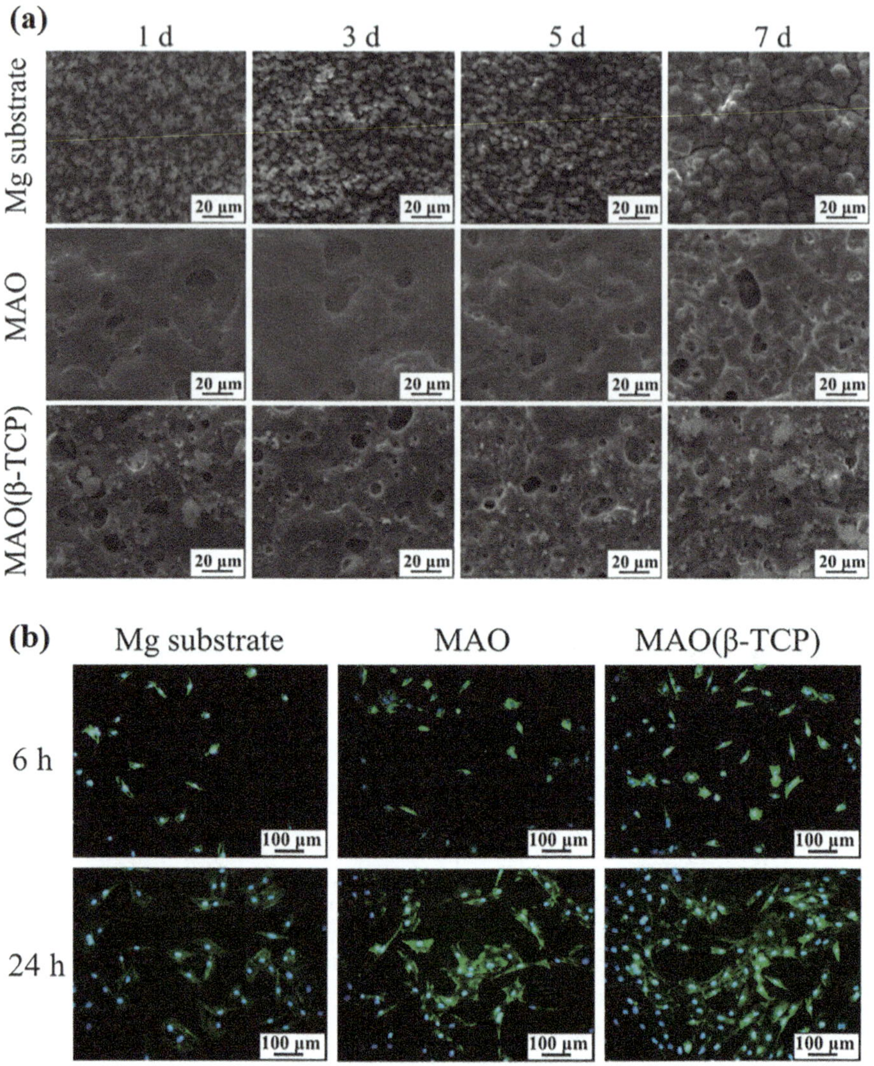

Fig. 11.3 Morphology of different samples after immersion for 1 day, 3 days, 5 days, and 7 days: (**a**) Morphology of corrosion products and MC3T3-E1 cell culture in extracts from different samples for 6 and 24 h, (**b**) Fluorescence images of cell nuclei and cytoskeleton staining. (Reproduced with permission from [65] © 2023 Elsevier Ltd)

high temperature promotes the synthesis of $Mg_3(PO_4)_2$. The sintered products and HA close some of the pores in the PEO film, thereby achieving higher bonding strength and better corrosion resistance [72], and enhancing corrosion resistance during immersion [73]. Cellular experiments have demonstrated that the introduction of phosphate particles significantly enhances osteoblast proliferation and adhesion. As shown in Fig. 11.3b, osteoblasts spread over a larger area on the coating surface, with filopodia extending in multiple directions [65]. Yazici et al. attempted to incorporate silver-doped HA (Ag-HA) NPs into the PEO coating. With an increase in the particle content in the electrolyte, the corrosion resistance of the PEO coating decreased, but its antibacterial properties were enhanced [74].

Unlike HA, inorganic particles with a higher melting point are typically incorporated into PEO coatings in an inert manner to improve the microstructure of the coating. Certain inorganic particles, due to their unique properties such as high hardness, high wear resistance, good chemical inertness, and excellent high-temperature performance, have been widely attempted to be added to PEO coatings. For example, the incorporation of Si_3N_4 particles partially fills the pores on the coating surface but does not significantly improve the compactness of the coating, and increasing particle size negatively affects the thickness of the coating [75]. Additionally, the way particles are incorporated into the coating and the microstructure of the coating significantly affect its performance. Research indicates that SiC NPs can be incorporated in higher quantities into the coating and evenly distributed along the coating depth [76]. This distribution reduces the number and size of micropores on the coating surface, significantly improving the corrosion and wear resistance of the coating [77]. Similarly, TiN NPs can be uniformly distributed in the PEO coating, filling the pores of the coating. Due to their excellent electrical conductivity, TiN particles can effectively reduce the corrosion current density and increase the polarization resistance, thereby enhancing the protective performance of the coating [78].

Graphite and its allotropes, due to their high specific surface area, excellent mechanical properties, superior chemical and thermal stability, and noncytotoxicity toward osteoblasts and mesenchymal matrix cells, have become ideal additive particles in PEO coatings. These materials promote osteogenic differentiation and enhance the expression of osteogenic-related genes [79]. The incorporation of graphene and GO can improve the conductivity of the electrolyte, reduce or prevent pore formation, and thus promote the formation of denser, thicker PEO coatings. These particles, by filling the pores of the coating, provide a more substantial barrier capability for the substrate, effectively preventing the penetration of aggressive anions [80]. The introduction of these particles not only alters the wear mechanism of the coating but also significantly reduces the wear volume [81]. Due to the high chemical stability and melting point of graphene and its oxides, their incorporation into PEO coatings occurs in an inert manner, meaning the particles are incorporated into the coating without any chemical reaction or new phase formation [48].

11.3.3 Metallic Particles

Previous attempts to incorporate metal particles into PEO electrolytes for surface modification of magnesium alloys have been limited. However, some researchers believe that PEO composite coatings containing metal particles have potential in biomedical applications [82]. As shown in Fig. 11.4, Husak et al. attempted to incorporate AgNPs into PEO coatings and found that Ag particles were uniformly incorporated into the coating with inert behavior, altering the surface morphology of the coating (as shown in Fig. 11.4a–d). During the immersion experiments, Ag^+ ions were slowly released, enhancing the antibacterial properties of the PEO coating (as shown in Fig. 11.4e), without exhibiting cytotoxicity [83]. Similarly, research by Yu

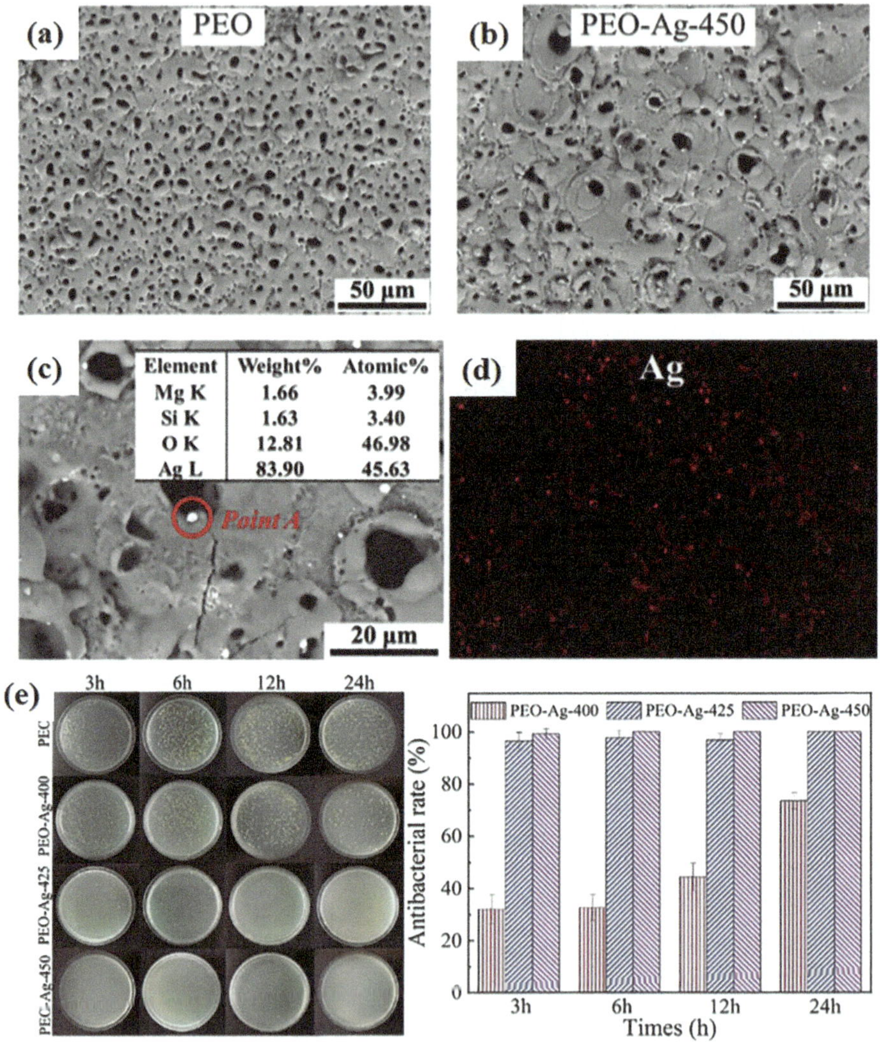

Fig. 11.4 (**a**) Surface morphology of the PEO coating, (**b**) PEO coating doped with AgNPs, (**c**) Point analysis, and (**d**) EDS mapping of silver (Ag) on the surface of PEO-Ag-450. (**e**) Photograph of *Staphylococcus aureus* (*S. aureus*) colony growth on nutrient agar plates and antimicrobial activity of the coating at different coculture times. (Reproduced with permission from [83] © 2021 Elsevier B.V.)

et al. revealed that PEO coatings containing 1 g/L of Ag particles exhibited an antibacterial rate of 94.9% against Escherichia coli. Furthermore, the coating demonstrated excellent biocompatibility, being nontoxic to cells and promoting osteoblast growth [84].

11.3.4 Mixed Particles

In addition to the addition of single particles, the co-incorporation of different types of particles can significantly improve the performance of PEO coatings. For example, Wen et al. prepared PEO coatings containing HA/GO composite particles on AZ31 magnesium alloy. The study demonstrated that HA/GO particles were successfully incorporated into the coating and effectively filled some of the pores (as shown in Fig. 11.5a). This composite coating exhibited good corrosion resistance in SBF and showed potential bio-mineralization capability (as shown in Fig. 11.5b), suggesting that the HA/GO composite particles could enhance both the bioactivity and corrosion resistance of the coating [85]. Chen et al. prepared PEO coatings containing hydroxyapatite/copper oxide (HA/CuO) particles on Mg-2Zn-1Gd-0.5Zr magnesium alloy for biomedical applications. The study found that the addition of CuO particles not only improved the corrosion resistance of the coating but also significantly enhanced its antibacterial properties. The Cu^{2+} ions released from the coating exhibited a 96% inhibition rate against *S. aureus*, indicating the potential antibacterial performance of the coating [86].

Mozafarnia et al. employed the PEO technique to modify the surface of AZ31 magnesium alloy and investigated the coating performance of TiO_2 NPs and HA composite particles at different concentrations in phosphate-based electrolytes. The results showed that the incorporation of particles reduced the corrosion resistance of the PEO coating. However, all particle-containing coatings exhibited better corrosion resistance compared to the bare substrate, with the coating containing 1 g/L TiO_2 particles showing the best wear resistance and corrosion resistance. As the TiO_2 particle concentration increased, the antibacterial properties of the coating were also enhanced. When the TiO_2 particle concentration reached 4 g/L, the coating demonstrated significant antibacterial effects with an antibacterial rate of 97.65% [87].

11.4 Titanium and Its Alloys

Titanium alloys offer superior corrosion resistance and higher strength than magnesium alloys, making them ideal nondegradable materials for medical applications, particularly for dental and orthopedic implants [88]. However, the relatively low wear resistance of titanium alloys can lead to loosening of the implants [30]. Furthermore, the wear debris generated can cause inflammation and be destructive

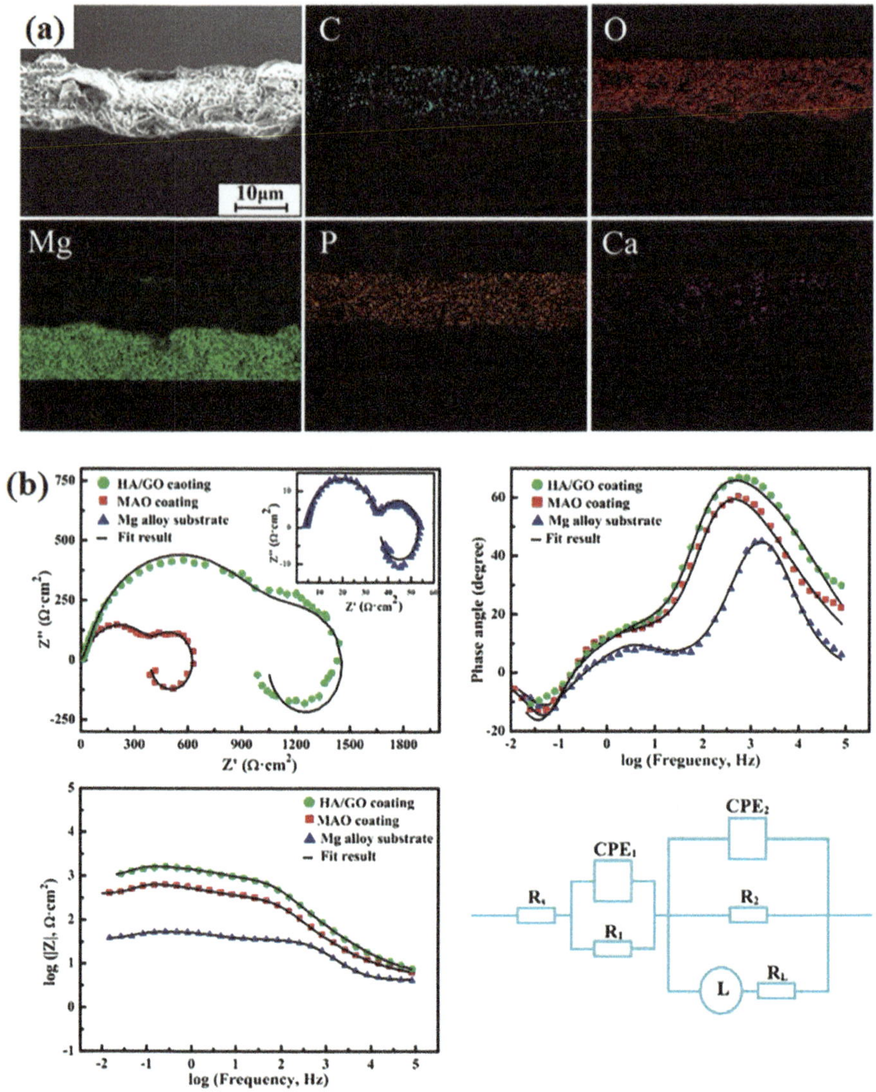

Fig. 11.5 (**a**) Cross-sectional morphology of the HA/GO coating and the corresponding distribution of typical elements. (**b**) EIS measurements (scatter plot) and fitted circuit model (solid points) of magnesium alloy substrate, MAO, and HA/GO-coated magnesium alloy in SBF solution at 37 °C. (Reproduced with permission from [85] © 2017 Elsevier B.V.)

to the bone supporting the implant [89]. One of the most significant advantages of PEO coatings for titanium alloys used in biomedical applications is the strong adhesion to the substrate, high hardness, and the ability to uniformly coat metals with complex geometries [90, 91]. The strength of the PEO coating is enhanced compared to the titanium alloy itself, and the improved friction behavior remains beneficial for biomedical use [92]. Additionally, the porous surface of the PEO coating is suitable for enhancing bone integration and osteoinductive properties, as well as carrying antibiotics to combat postsurgical infections [93–95].

11.4.1 Oxide Particles

By introducing different types of oxide NPs into the electrolyte, the composition and structure of the PEO coating can be effectively controlled, thereby enhancing the performance of titanium alloys in the biomedical field. Muhaffel et al. found that adding ZrO_2 particles to the PEO coating on commercially pure titanium could introduce ZrO_2 particles during the coating formation process, leading to the generation of a new $ZrTiO_4$ oxide phase and significantly improving the compactness of the TiO_2 layer [96]. Shokouhfar et al. further confirmed that the ZrO_2 NPs filled the micropores and cracks on the coating surface, which increased the hardness of the coating, reduced the friction coefficient and wear rate, and enhanced its corrosion resistance [97]. Additionally, microstructure observations showed that the ZrO_2 particles were uniformly dispersed within the coating, and the surface roughness increased [98]. As shown in Fig. 11.6, this rough surface facilitates the nucleation of biomimetic HA and cell attachment, further improving the bioactivity of osteoblasts on ZrO_2, which is superior to the PEO TiO_2 coatings [99]. The HA particles formed were also larger, indicating that the coating promotes the formation of HA [100]. Nikoomanzari et al. found in their study on ZrO_2-containing PEO coatings on Ti-6Al-4V titanium alloy that the coating without added particles could completely suppress the growth of pathogenic Gram-positive bacteria and *Escherichia coli*. However, the antibacterial performance of the coating decreased after adding ZrO_2 particles, possibly due to the poor photocatalytic performance of ZrO_2 and $ZrTiO_4$, along with the changes in the coating surface morphology, which led to increased bacterial attachment. As the ZrO_2 particle content increased from 0 to 5 g/L, the porosity, thickness, and surface roughness of the coating all increased. Among these, a ZrO_2 particle concentration of 3 g/L resulted in the lowest porosity and corrosion current density [101], and also promoted the deposition of HA [94].

Sarbishei et al. successfully prepared PEO coatings on titanium substrates by adding varying concentrations of Al_2O_3 NPs (0, 3, 6, 10 g/L) in silicate-based electrolytes. The study showed that Al_2O_3 NPs could effectively enter the coating structure, significantly reducing the density and size of pores. As the Al_2O_3 concentration increased, the corrosion resistance of the composite coatings was enhanced considerably. When using an electrolyte containing 10 g/L Al_2O_3, the coating exhibited the lowest porosity (11.2%) and optimal corrosion resistance [102]. In contrast, Muhaffel et al. investigated the effect of incorporating Al_2O_3 NPs into PEO coatings and found that, although the inclusion of NPs did not significantly improve the hardness of the coatings, the wear resistance was lower than that of silicate-based coatings without particles [103]. Sharifi et al. used a phosphate-based electrolyte containing high concentrations (25 g/L) of $A1_2O_3$ NPs and introduced atenolol. The

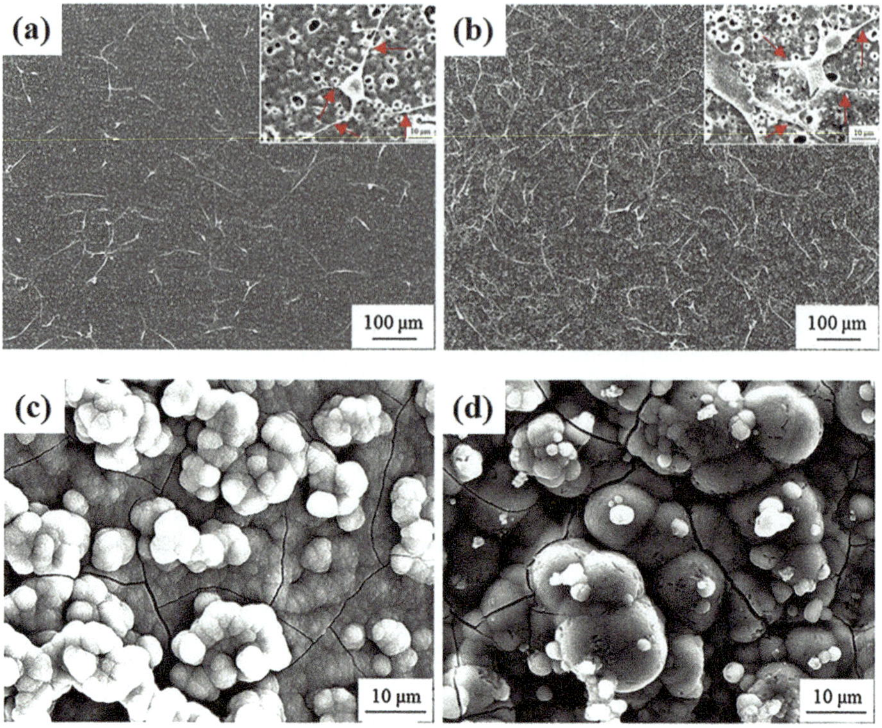

Fig. 11.6 SEM images of osteoblasts on the surface of samples cultured for 48 h and SEM images of biomimetic hydroxyapatite on the surface of samples after 14 days of immersion in SBF solution: (**a**) PEO and (**b**) PEO with added ZrO_2 NPs. Arrows indicate the filopodia of osteoblasts. Insets show high-magnification SEM images of the cells. (Reproduced with permission from [99] © 2015 Elsevier B.V.)

results indicated that the addition of atenolol facilitated the adsorption and incorporation of Al_2O_3 particles, thereby reducing the porosity of the coating. This modification enhanced the corrosion resistance of the coating and significantly improved its mechanical properties, with the wear rate decreasing by a factor of 130 [104]. In comparison, Shokouhfar et al. found that when TiO_2 NPs were incorporated into the PEO coating, the wear resistance and corrosion resistance of the coating were superior to those of coatings incorporating Al_2O_3 [97].

In addition to enhancing the inherent properties of coatings by adding oxide particles, the incorporated particles can also impart new characteristics to the coatings. Zhang et al. successfully fabricated Ag-doped TiO_2 coatings on titanium surfaces using an electrolyte containing Ag_2O NPs. The results indicated that during the incorporation process, the particles decomposed to form AgNPs, and an amorphous interfacial layer rich in Ti, O, Ca, P, and Ag was formed between the Ag NPs and the surrounding TiO_2, suggesting a reactive bond between the Ag NPs and TiO_2[105]. The addition of more Ag_2O NPs enhanced the antibacterial properties of the resulting coating, though it slightly increased cytotoxicity [106]. Roknian et al.

developed ZnO nanoparticle-modified PEO coatings on titanium metal surfaces and found that the incorporation of ZnO NPs reduced the number of pores and improved the microstructure of the coatings, thereby enhancing their corrosion resistance. Antibacterial tests showed that nano-ZnO could effectively maintain the antibacterial activity of the coating against both Gram-positive and Gram-negative bacteria, with a stronger antibacterial effect observed at higher concentrations [107]. CuO NPs can also be incorporated into PEO coatings, exhibiting a stronger antibacterial ability than nano-ZnO [108]. SiO_2 is also considered a bioactive material that can promote the adhesion, proliferation, and differentiation of osteoblast-like cells through silicon ions released from silica-based materials. After incorporating SiO_2 particles into PEO coatings, the coatings maintained good biocompatibility and reduced the friction coefficient, achieving improved tribological behavior [38].

11.4.2 Nonmetallic Particles

In the modification of titanium alloy surfaces through PEO coatings, the incorporation of nonmetallic particles into the electrolyte has become a popular and effective method to significantly enhance the wear resistance and corrosion resistance of the coatings [97]. This approach has also garnered significant attention in biomedical applications. Sedelnikova et al. modified HA particles added to the electrolyte by substituting part of the calcium and phosphorus in HA with strontium and silicon, which are natural bone-inducing elements. The results showed that the modified PEO coating exhibited higher cell proliferation activity and demonstrated significant potential for promoting bone formation and enhancing osseointegration [109].

Yigit et al. employed alternating current (AC) PEO technology to fabricate a graphene nanosheet (GNS)-reinforced PEO coating containing nano-HA on Ti6Al4V alloy. The nano-HA particles were uniformly deposited in the coating, and the addition of GNS led to a coating with smaller pore sizes and higher pore density during the PEO process. Immersion experiments in SBF demonstrated that the coating not only exhibited excellent corrosion resistance but also showed good bioactivity [110]. Wang et al. incorporated micro/nanoscale barium titanate ($BaTiO_3$) particles into the PEO coating, and the results indicated that this coating effectively inhibited bacterial proliferation and killed bacteria adhered to the coating surface. The antibacterial performance of the coating was primarily attributed to the surface contact potential difference (CPD) and the reactive oxygen species (ROS) generated by its spontaneous polarization. The CPD could suppress microbial adhesion by disrupting the electrolyte balance around the membrane and may damage the cell membrane, while ROS could alter the membrane permeability or cause leakage of cell contents, ultimately leading to bacterial death [111]. Ziółkowska et al. investigated the effect of adding copper phosphate to the coating, which, through the release of Cu ions, exhibited good antibacterial properties and promoted the attachment, proliferation, and growth of live cells, showing excellent cytocompatibility [112].

Calcium phosphate inorganic salt particles are one of the hotspots in research on titanium-based nonmetallic particles/PEO coatings for biomedical applications [115], especially for orthopedic implants [116], because calcium phosphate salts are a natural component of bone [117]. The properties of calcium phosphate salt particles and their concentration in the electrolyte have a significant impact on the performance of PEO coatings. Dzhurinskiy et al. used PEO technology to prepare modified PEO coatings on Ti-6Al-4V alloy by adding HA NPs to a phosphate electrolyte. The experiment showed that the incorporation of HA particles did not alter the porous structure of the PEO coating, which exhibited a dual-layer structure, including an inner barrier layer responsible for corrosion protection and an outer porous layer that promotes cell adhesion [118]. Compared with the bare substrate, the HA-modified PEO coating samples exhibited higher corrosion resistance [114]. Some studies have pointed out that the binding of HA particles within the coating is related to the particle size. Micron-sized HA particles are deposited on the surface of the coating, reducing pore size and surface roughness (as shown in Fig. 11.7a, b), which results in a highly irregular morphology formed by the interpenetrating network of pores and coating material [113]. In contrast, NPs are more likely to migrate to the surface and penetrate into the coating (as shown in Fig. 11.7c), partially filling the pores while retaining the primary morphological features of the porous TiO_2

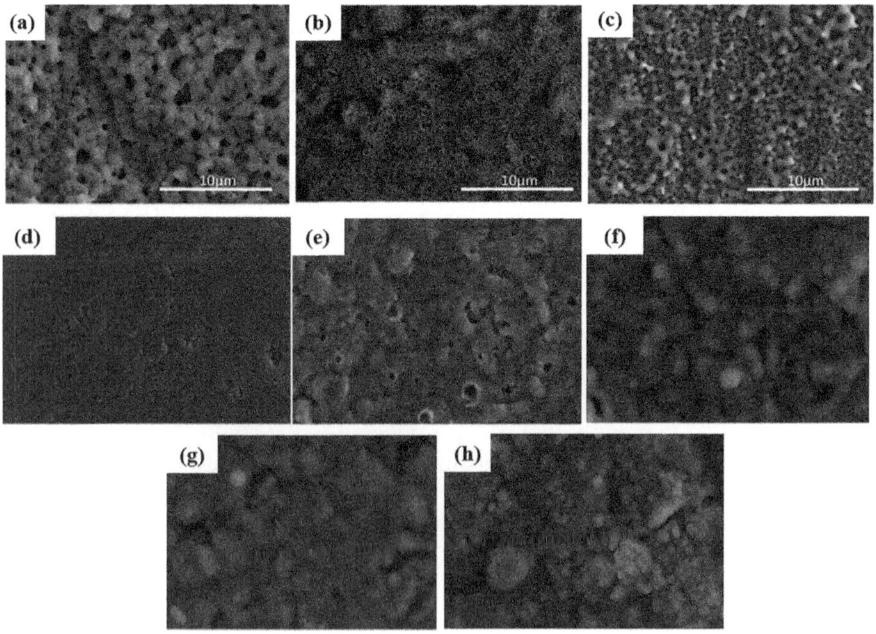

Fig. 11.7 (**a**) SEM images of TiO_2, (**b**) TiO_2 + mHA, and (**c**) TiO_2 + nHA. SEM images of pure Ti- and PEO-treated samples with different HA concentrations after 3 days of immersion in SBF: (**d**) pure Ti, (**e**) 0 g/L, (**f**) 5 g/L, (**g**) 10 g/L, and (**h**) 20 g/L. (Reproduced with permission from [113, 114] © 2016 The Royal Society of Chemistry and © 2011 Elsevier B.V.)

coating [118]. Lee et al. pointed out that the amount of HA particles in the mixed coating increases with the concentration of HA in the electrolyte solution while maintaining the microporous morphology. The coating thickness and surface roughness increase and the porosity gradually decreases [119]. Grebn et al. found that amorphous calcium phosphate particles, compared to HA particles, can achieve higher incorporation into the PEO coating and simultaneously improve the biological activity of the coating [120]. However, higher calcium phosphate content does not necessarily lead to an improvement in coating performance. A PEO coating prepared with a HA concentration of 1.5 g/L achieved a bonding strength of 2099 mN, and the maximum scratch hardness of the coating was about 5 GPa. When the HA concentration exceeds 1.5 g/L, it could hinder the breakdown of micro-arc discharges, causing the oxide film to become unstable and ultimately leading to a deterioration in the bonding strength of the coating [121]. In terms of bioactivity, no signs of apatite formation were observed on the surface of pure titanium alloy, while the ability of the coating to form apatite was positively correlated with HA concentration (as shown in Fig. 11.7d–h), with higher HA concentrations leading to coatings with better bioactivity [114]. Cell experiments showed that MC3T3-E1 osteoblasts on the coating had a high proliferation rate and expressed early differentiation markers such as alkaline phosphatase, indicating that the HA-modified PEO coating provides synergistic effects on osteoblast proliferation and differentiation [113].

11.4.3 Metallic Particles

Metal NPs, such as Ag, Cu, and Zn NPs, have attracted significant attention from scholars in the field due to their excellent stability, antimicrobial effects [58], resistance to the development of resistant bacterial strains, and relatively low toxicity to human cells [122]. Through PEO technology, ceramic composite coatings doped with metal particles can be fabricated on titanium alloy surfaces, significantly improving the wear resistance and antibacterial ability of titanium alloys and preventing complications and premature failure of implants [123, 124].

Among inorganic antimicrobial metal NPs, AgNPs have been the most widely studied due to their excellent antimicrobial activity and the lack of significant toxicity to human cells upon release of Ag^{+} ions [125]. Lan et al. employed PEO technology to fabricate composite coatings containing AgNPs on a Ti6Al4V titanium alloy substrate. The coating retained a typical porous structure with low porosity and exhibited significant improvement in both wear resistance and corrosion resistance, with a one-order magnitude decrease in corrosion current density [126]. The PEO coating doped with AgNPs demonstrated enhanced tribocorrosion resistance and excellent cellular response [92]. Studies have shown that Ag particles are well incorporated and evenly distributed within the oxide layer of the coating without altering its overall morphology [125, 127]. Furthermore, AgNPs not only embed into the dense oxide film but also appear at the pore walls and the surface of the coating, maintaining stable morphology and chemical properties [128]. Zhang et al. successfully fabricated a silver-doped TiO_2 coating on titanium surfaces using PEO technology with an electrolyte containing AgNPs, which exhibited excellent antimicrobial properties without adversely affecting cellular compatibility [105].

Ag^+ and AgNPs are widely regarded as having excellent antibacterial properties. Specifically, AgNPs can increase the permeability of bacterial membranes and generate ROS, allowing them to enter bacterial cells, and causing DNA damage and protein oxidation. Ag^+ binds to thiol groups in bacterial cell proteins, disrupting the normal structure of DNA and impairing its replication ability [129]. In the porous titanium oxide coating, AgNPs may react with water and release Ag^+, and as the content of Ag particles increases, the antimicrobial effect of the coating also strengthens [130]. However, it is important to note that excessively high Ag content in the coating may have adverse effects on the human body, especially when cells come into direct contact with the surface of the coating. This risk is more significant not merely due to the release of Ag^+ in osteogenic media [129].

By introducing particle reinforcement techniques into the PEO coating, similar methods have been applied to porous titanium alloys fabricated via additive manufacturing (AM). These materials exhibit mechanical properties comparable to natural bone, helping to reduce the stress shielding effect and extend the lifespan of implants. At the same time, the porous structure increases the material's surface area, which improves bone tissue regeneration, but it may also increase the risk of bacterial infection [90]. Untreated porous titanium alloys often induce strong pro-inflammatory responses. To address this issue, a PEO coating doped with AgNPs has been demonstrated to exhibit good antibacterial properties on the surface of porous titanium alloys. However, this coating has shown significant cytotoxicity toward adhering macrophages [132]. Building on this, Hengel et al. developed a PEO coating doped with Zn and Ag NPs by using an electrolyte containing Ca/P species as well as Ag and Zn NPs. These NPs were tightly embedded into the growing titanium oxide layer (as shown in Fig. 11.8a–c), displaying a synergistic antibacterial effect (as shown in Fig. 11.8d–f). In particular, the ratio of 75% Ag and 25% Zn resulted in biofunctionalized implants that could completely eradicate bacterial inoculum within 24 h. Moreover, the zinc-doped coating surface significantly enhanced the metabolic activity of pre-osteoblasts and had a positive effect on osteogenic differentiation [131].

Additionally, Yao et al. prepared copper-doped TiO_2 antibacterial coatings on titanium alloy substrates using the PEO method in an electrolyte containing 3 g/L Cu NPs. The Cu NPs were successfully embedded into the coating and were uniformly distributed on the coating surface and within the pores. Studies have shown that the addition of Cu NPs did not significantly alter the microstructure and phase composition of the TiO_2 coating. The Cu NPs imparted significant antibacterial properties to the coating, and their highly efficient bactericidal action may be attributed to both contact killing and the release of Cu ions into the bacterial solution [133]. In vitro cell compatibility assessments showed that coatings prepared with electrolytes containing 0.3 and 3 g/L Cu NPs exhibited no cytotoxicity to endothelial cells and promoted cell proliferation. However, high concentrations of Cu NPs

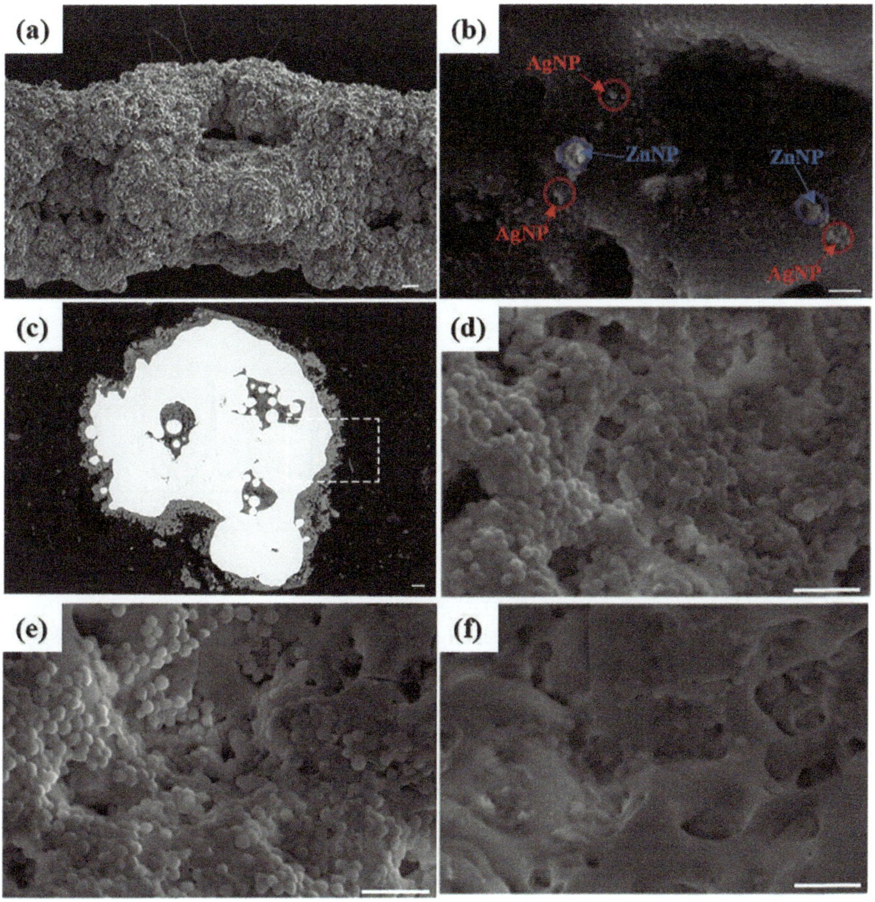

Fig. 11.8 SEM images of the surface morphology of porous titanium alloy implants after PEO treatment: (**a**) Surface images; (**b**) EDS analysis; (**c**) Cross-sectional analysis; Biofilm formation on the implants after 48 h of incubation in TSB with 1% glucose; (**d**) PEO; (**e**) ZnNP PEO; and (**f**) AgZnNP PEO. (Reproduced with permission from [131] © 2020 Acta Materialia Inc. Published by Elsevier Ltd)

were toxic to osteoblasts, while low concentrations of Cu NPs promoted osteoblast proliferation, adhesion, and extracellular matrix mineralization [134].

11.4.4 Mixed Particles

San et al. combined reduced graphene oxide (rGO) with AgNPs and incorporated them onto the surface of AM porous titanium implants using the PEO method. The addition of rGO nanosheets improved the dispersion of AgNPs in the electrolyte and

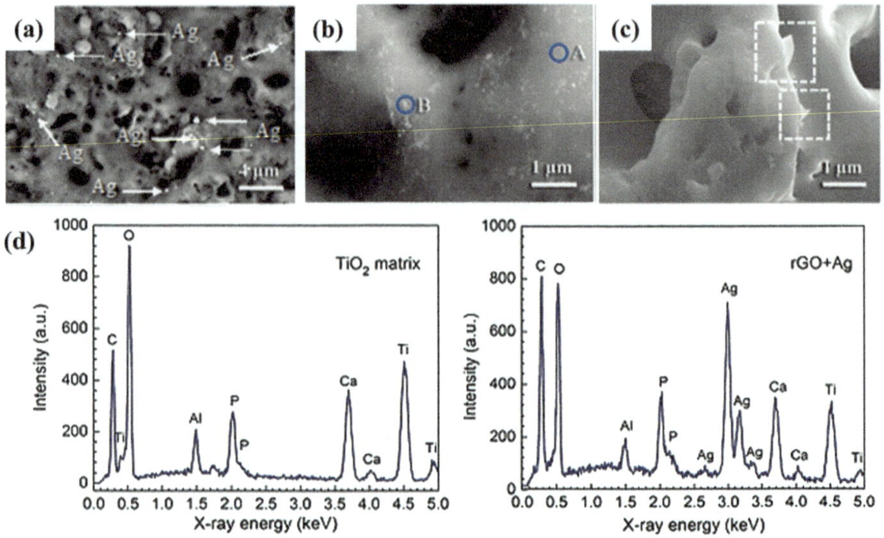

Fig. 11.9 (**a**, **b**) Backscattered electron images of Ag particle-doped PEO coating. (**c**) "Nanoknife" region on the surface of the coating. (**d**) EDS spectra of the TiO_2 matrix (labeled as A in (**b**)) and rGO + Ag NPs (labeled as B in (**b**)). (Reproduced with permission from [135] © 2022 by the authors. Licensee MDPI)

facilitated their subsequent incorporation into the PEO layer (as shown in Fig. 11.9a, b, d). Compared to PEO surfaces prepared with only AgNPs, the rGO + Ag coating exhibited enhanced in vitro antibacterial activity against methicillin-resistant *Staphylococcus aureus* (MRSA). This improvement was attributed to the generation of ROS, higher Ag^+ release levels, and the presence of the rGO "nano scalpel" structure (as shown in Fig. 11.9c). Furthermore, the particle-modified PEO coating stimulated the metabolic activity and osteogenic differentiation of MC3T3-E1 pre-osteoblasts [135].

11.5 Other Alloys

Some researchers have attempted to extend this approach to other alloys, such as NiTi shape memory alloys and aluminum alloys, with the goal of improving their antimicrobial properties [136, 137]. In the case of NiTi shape memory alloys, the Ag nanoparticle-modified PEO coating only formed a titanium oxide (TiO_2) layer on the surface, with no oxidation of Ni. The coating containing Ag NPs exhibited a stronger ability to form HA and retained the shape memory characteristics of NiTi dental alloys, preventing the release of Ni^{2+} ions and improving biocompatibility [138].

11.6 Conclusions and Outlook

This chapter introduces the development and recent advances in biomedical applications of particle-enhanced PEO coatings. The incorporation of particles is an effective way to modify and optimize the composition, structure, morphology, and properties of PEO coatings. For certain particles, the mode of incorporation can be altered and controlled through electrolyte parameters. In some cases, the inert binding of particles appears more ideal, as it imparts new functions to the PEO coating due to the specific properties of the incorporated particles.

In biological applications, the addition of particles to the PEO layer holds great potential. The developed biomedical materials and surface modification technologies must be compatible with living systems and should not cause any harm to the body. This includes considering the potential adverse effects of the materials on components of biological systems (bones, extracellular and intracellular tissues, and the ionic composition of plasma).

However, it should be noted that almost all related studies have been conducted using typical bulk samples rather than implant samples. Therefore, although some processing techniques have produced coatings with good performance, their applicability to implants with complex structures (such as screws) remains unclear. In future work, in addition to continuing the development of high-performance coatings, the practical value of these coating technologies should also be considered.

References

1. Liang, Y., Song, Y., & Wang, L. (2023). *Odontology, 111*, 813.
2. Baron, S., Desmond, D., & Ahearne, E. (2019). *Wear, 424–425*, 89.
3. Harb, S. V., Bassous, N. J., & de Souza, T. A. C. (2020). *Materials Science and Engineering C, 116*, 111149.
4. Sultana, N., Nishina, Y., & Nizami, M. Z. I. (2024). *Coatings, 14*, 248.
5. Zhang, J., Li, G., & Qu, Y. (2022). *ACS Applied Materials & Interfaces, 14*, 9807.
6. Gupta, A., Maharjan, A., & Kim, B. S. (2019). *Applied Sciences, 9*, 4694.
7. Li, G., Zhao, M., & Xu, F. (2020). *Molecules, 25*.
8. Meléndez-Ortiz, H. I., Alvarez-Lorenzo, C., & Concheiro, A. (2016). *Radiation Physics and Chemistry, 119*, 37.
9. Paxton, N. C., Allenby, M. C., & Lewis, P. M. (2019). *European Polymer Journal, 118*, 412.
10. Sarfraz, S., Mäntynen, P.-H., & Laurila, M. (2022). *Polymers, 14*, 3862.
11. Teo, A. J. T., Mishra, A., & Park, I. (2016). *ACS Biomaterials Science & Engineering, 2*, 454.
12. Soni, R. S., & Singh, V. P. (2021). *Materials Today: Proceedings, 42*, 536.
13. Baino, F., Verné, E., & Fiume, E. (2019). *International Journal of Applied Ceramic Technology, 16*, 1850.
14. Gremillard, L., Chevalier, J., & Martin, L. (2018). *Acta Biomaterialia, 68*, 286.
15. Heimke, G. u., Leyen, S., & Willmann, G. (2002). *Biomaterials, 23*, 1539.
16. Gao, X., Wang, S., & Xu, Y. (2018). *PLoS One, 13*.
17. Chen, S., Tian, H., & Mao, J. (2023). *International Journal of Biological Macromolecules, 226*, 410.

18. Li, X., Jiang, G., & Wang, G. (2024). *International Journal of Biological Macromolecules, 260*, 129600.
19. Chen, B., Wang, Y., & Tuo, X. (2020). *Journal of Applied Polymer Science, 138*, 50253.
20. Yu, S., Chi, H., & Li, P. (2024). *Additive Manufacturing, 93*, 104411.
21. Wang, X., Su, J., & Li, C. (2024). *Journal of Materials Research and Technology, 33*, 698.
22. Ma, H., Suonan, A., & Zhou, J. (2021). *Arabian Journal of Chemistry, 14*, 102977.
23. Agarwal, R., & Garcia, A. J. (2015). *Advanced Drug Delivery Reviews, 94*, 53.
24. Singh, J., Wahab Hashmi, A., & Ahmad, S. (2024). *Inorganic Chemistry Communications, 169*, 113111.
25. Guan, S., Mei, D., & Wang, J. (2023). *Journal of Magnesium and Alloys, 11*, 4011.
26. Kang, J., Dong, E., & Li, D. (2020). *Materials & Design, 191*, 108608.
27. Cheng, S., Zhang, D., & Li, M. (2021). *Bioactive Materials, 6*, 91.
28. Chen, D., Mei, D., & Chen, L. (2024). *Applied Surface Science, 672*, 160790.
29. Han, X., Ma, J., & Tian, A. (2023). *Colloids and Surfaces B: Biointerfaces, 227*, 113339.
30. Geetha, M., Singh, A. K., & Asokamani, R. (2009). *Progress in Materials Science, 54*, 397.
31. Attar, H., Prashanth, K. G., & Chaubey, A. K. (2015). *Materials Letters, 142*, 38.
32. Yoon, I.-K., Hwang, J.-Y., & Jang, W.-C. (2014). *Applied Surface Science, 301*, 401.
33. Muraev, A. A., Murzabekov, A. I., & Ivanov, S. Y. (2023). *Sovrem Tekhnologii Medicine, 15*, 18.
34. Yang, C., Chen, P., & Wu, W. (2024). *Coatings, 14*, 451.
35. Esmaeili, M., Tadayonsaidi, M., & Ghorbanian, B. (2021). *Surface Innovations, 9*, 184.
36. Xie, P., Blawert, C., & Serdechnova, M. (2024). *Journal of Magnesium and Alloys, 12*, 1386.
37. Fattah-alhosseini, A., Molaei, M., & Babaei, K. (2020). *Surfaces and Interfaces, 21*, 100659.
38. Xu, G., & Shen, X. (2019). *Surface and Coatings Technology, 364*, 180.
39. Lu, X., Mohedano, M., & Blawert, C. (2016). *Surface and Coatings Technology, 307*, 1165.
40. Fattah-alhosseini, A., Chaharmahali, R., & Babaei, K. (2022). *Journal of Magnesium and Alloys, 10*, 2354.
41. Wu, Y., Wang, Y. M., & Zhao, D. W. (2019). *Colloids and Surfaces B: Biointerfaces, 181*, 349.
42. Zhang, Z. Q., Yang, Y. X., & Li, J. A. (2021). *Bioactive Materials, 6*, 4729.
43. Chen, Z., Geng, X., & Yong, X. (2022). *Journal of Materials Engineering and Performance, 31*, 8885.
44. Mashtalyar, D. V., Imshinetskiy, I. M., & Kashepa, V. V. (2024). *Journal of Magnesium and Alloys, 12*, 2360.
45. Almajidi, Y. Q., Ali, E., & Jameel, M. F. (2023). *Lubricants, 11*, 519.
46. Lu, X., Blawert, C., & Kainer, K. U. (2016). *Electrochimica Acta, 196*, 680.
47. Lu, X., Blawert, C., & Huang, Y. (2016). *Electrochimica Acta, 187*, 20.
48. Zhao, J., Xie, X., & Zhang, C. (2017). *Corrosion Science, 114*, 146.
49. Mohedano, M., Arrabal, R., & Mingo, B. (2018). *Surface and Coatings Technology, 334*, 328.
50. Lee, K. M., Shin, K. R., & Namgung, S. (2011). *Surface and Coatings Technology, 205*, 3779.
51. Chen, W. H., Huang, S.-Y., & Chu, Y.-R. (2024). *Surface and Coatings Technology, 476*, 130238.
52. Daroonparvar, M., Yajid, M. A. M., & Yusof, N. M. (2016). *Journal of Alloys and Compounds, 688*, 841.
53. Bordbar-Khiabani, A., Yarmand, B., & Mozafari, M. (2019). *Surface and Coatings Technology, 360*, 153.
54. Keyvani, A., Zamani, M., & Bahamirian, M. (2021). *Surfaces and Interfaces, 22*, 100728.
55. Bahrampour, S., Bordbar-Khiabani, A., & Hossein Siadati, M. (2024). *Chemical Engineering Journal, 483*, 149016.
56. Madhankumar, A., Thangavel, E., & Ramakrishna, S. (2014). *RSC Advances, 4*, 24272.
57. Liang, F., Ren, Y., & Mei, D. (2024). *Journal of Alloys and Compounds, 1009*, 177073.
58. Zhang, D., Han, Q., & Yu, K. (2021). *Journal of Materials Science and Technology, 61*, 33.
59. Li, Y., Lu, X., & Serdechnova, M. (2023). *Journal of Magnesium and Alloys, 11*, 1236.
60. Sun, M., Yerokhin, A., & Bychkova, M. Y. (2016). *Corrosion Science, 111*, 753.

61. Guo, J., Wang, L., & Wang, S. C. (2009). *Journal of Materials Science, 44*, 1998.
62. Zhang, Y., Feyerabend, F., & Tang, S. (2017). *Materials Science and Engineering: C, 78*, 405.
63. Blawert, C., Sah, S. P., & Liang, J. (2012). *Surface and Coatings Technology, 213*, 48.
64. Rahmati, M., Mohammadi Zahrani, E., & Atapour, M. (2024). *Materials Chemistry and Physics, 315*, 128983.
65. Jin, X., Mei, D., & Chen, D. (2023). *Materials Today Communications, 36*, 106694.
66. Chen, Y., Dou, J., & Pang, Z. (2020). *RSC Advances, 10*, 8244.
67. Liu, C., Zhang, W., & Xu, T. (2022). *Ceramics International, 48*, 13676.
68. Lin, X., Wang, X., & Tan, L. (2014). *Ceramics International, 40*, 10043.
69. Keyvani, A., Kamkar, N., & Chaharmahali, R. (2023). *Inorganic Chemistry Communications, 158*, 111470.
70. Chaharmahali, R., Fattah-Alhosseini, A., & Esfahani, H. (2019). *Journal of Asian Ceramic Societies, 8*, 39.
71. Chaharmahali, R., Fattah-alhosseini, A., & Esfahani, H. (2021). *Journal of Ultrafine Grained and Nanostructured Materials, 54*(2), 149–162.
72. Ma, X., Zhu, S., & Wang, L. (2014). *Journal of Alloys and Compounds, 590*, 247.
73. Yang, J., Lu, X., & Blawert, C. (2017). *Surface and Coatings Technology, 319*, 359.
74. Yazici, M., Gulec, A. E., & Gurbuz, M. (2017). *Thin Solid Films, 644*, 92.
75. Lu, X., Blawert, C., & Scharnagl, N. (2013). *Journal of Magnesium and Alloys, 1*, 267.
76. Yu, L., Cao, J., & Cheng, Y. (2015). *Surface and Coatings Technology, 276*, 266.
77. Wang, S.-y., Si, N.-c., & Xia, Y.-p. (2015). *Transactions of Nonferrous Metals Society of China, 25*, 1926.
78. Mashtalyar, D. V., Gnedenkov, S. V., & Sinebryukhov, S. L. (2017). *Journal of Materials Science & Technology, 33*, 461.
79. Fattah-alhosseini, A., & Chaharmahali, R. (2021). *FlatChem, 27*, 100241.
80. Daavari, M., Atapour, M., & Mohedano, M. (2021). *Surfaces and Interfaces, 22*, 100850.
81. Pezzato, L., Angelini, V., & Brunelli, K. (2018). *Transactions of Nonferrous Metals Society of China, 28*, 259.
82. Han, Q., Li, Y., & Lu, X. (2021). *Materials Letters, 293*, 129731.
83. Husak, Y., Olszaniecki, J., & Pykacz, J. (2024). *Applied Surface Science, 654*, 159387.
84. Yu, K., Li, P., & Han, Q. (2022). *Surfaces and Interfaces, 30*, 101840.
85. Wen, C., Zhan, X., & Huang, X. (2017). *Surface and Coatings Technology, 317*, 125.
86. Chen, J., Zhang, Y., & Ibrahim, M. (2019). *Colloids and Surfaces B: Biointerfaces, 179*, 77.
87. Mozafarnia, H., Fattah-Alhosseini, A., & Chaharmahali, R. (2022). *Coatings, 12*, 1967.
88. Satyanarayana, C. P., Raju, L. S., & Dumpala, R. (2024). *Materials Today Communications, 38*, 107985.
89. Li, H.-F., Huang, J.-Y., & Lin, G.-C. (2021). *Rare Metals, 40*, 3091.
90. van Hengel, I. A. J., Riool, M., & Fratila-Apachitei, L. E. (2017). *Biomaterials, 140*, 1.
91. Kuroda, P. A. B., Rossi, M. C., & Grandini, C. R. (2023). *Journal of Materials Research and Technology, 26*, 4656.
92. Marques, I. S. V., Alfaro, M. F., & Saito, M. T. (2016). *Biointerphases, 11*, 031008.
93. Li, H., & Yang, X. (2024). *ACS Biomaterials Science & Engineering, 10*, 6017.
94. Nikoomanzari, E., Karbasi, M., & Melo, W. C. M. A. (2022). *Chemical Engineering Journal, 441*, 136003.
95. Wang, Y., Yu, H., & Chen, C. (2015). *Materials & Design, 85*, 640.
96. Muhaffel, F., & Cimenoglu, H. (2015). *Surface and Interface Analysis, 47*, 1166.
97. Shokouhfar, M., & Allahkaram, S. R. (2017). *Surface and Coatings Technology, 309*, 767.
98. Shin, K. R., Ko, Y. G., & Shin, D. H. (2012). *Journal of Alloys and Compounds, 536*, S226.
99. Shin, K. R., Kim, Y. S., & Kim, G. W. (2015). *Colloids and Surfaces B: Biointerfaces, 131*, 47.
100. Shin, K. R., Ko, Y. G., & Shin, D. H. (2010). *Materials Letters, 64*, 2714.
101. Nikoomanzari, E., Fattah-alhosseini, A., & Pajohi Alamoti, M. R. (2020). *Ceramics International, 46*, 13114.

102. Sarbishei, S., Faghihi Sani, M. A., & Mohammadi, M. R. (2016). *Ceramics International, 42*, 8789.
103. Muhaffel, F., Kaba, M., & Cempura, G. (2019). *Surface and Coatings Technology, 377*, 124900.
104. Sharifi, H., Aliofkhazraei, M., & Darband, G. B. (2016). *Surface and Coatings Technology, 304*, 438.
105. Zhang, X., Lv, Y., & Cai, G. (2021). *Progress in Natural Science: Materials International, 31*, 215.
106. Lv, Y., Wu, Y., & Lu, X. (2019). *Ceramics International, 45*, 22357.
107. Roknian, M., Fattah-alhosseini, A., & Gashti, S. O. (2018). *Journal of Alloys and Compounds, 740*, 330.
108. Zhao, D., Lu, Y., & Wang, Z. (2016). *International Journal of Refractory Metals and Hard Materials, 54*, 417.
109. Sedelnikova, M. B., Komarova, E. G., & Sharkeev, Y. P. (2019). *Bioactive Materials, 4*, 224.
110. Yigit, O., Dikici, B., & Ozdemir, N. (2021). *Surface and Coatings Technology, 415*, 127139.
111. Wang, R., Zhou, T., & Liu, J. (2021). *Materials & Design, 203*, 109584.
112. Leśniak-Ziółkowska, K., Kazek-Kęsik, A., & Rokosz, K. (2021). *Applied Surface Science, 563*, 150284.
113. Yeung, W. K., Sukhorukova, I. V., & Shtansky, D. V. (2016). *RSC Advances, 6*, 12688.
114. Bai, Y., Kim, K.-A., & Park, I. S. (2011). *Materials Science and Engineering: B, 176*, 1213.
115. Bai, Y., Park, I. S., & Park, H. H. (2010). *Journal of Biomedical Materials Research Part B: Applied Biomaterials, 95B*, 365.
116. Kim, D.-Y., Kim, M., & Kim, H.-E. (2009). *Acta Biomaterialia, 5*, 2196.
117. Wei, D., Zhou, Y., & Wang, Y. (2007). *Applied Surface Science, 253*, 5045.
118. Dzhurinskiy, D., Gao, Y., & Yeung, W. K. (2015). *Surface and Coatings Technology, 269*, 258.
119. Lee, J.-H., Kim, H.-E., & Koh, Y.-H. (2009). *Materials Letters, 63*, 1995.
120. Grebņevs, V., Leśniak-Ziółkowska, K., & Wala, M. (2022). *Applied Surface Science, 598*, 153793.
121. Adeleke, S. A., Ramesh, S., & Bushroa, A. R. (2018). *Ceramics International, 44*, 1802.
122. Fattah-alhosseini, A., Molaei, M., & Attarzadeh, N. (2020). *Ceramics International, 46*, 20587.
123. Necula, B. S., Fratila-Apachitei, L. E., & Zaat, S. A. J. (2009). *Acta Biomaterialia, 5*, 3573.
124. Necula, B. S., Apachitei, I., & Fratila-Apachitei, L. E. (2013). *Applied Surface Science, 273*, 310.
125. Thukkaram, M., Cools, P., & Nikiforov, A. (2020). *Applied Surface Science, 500*, 144235.
126. Zhou, L., Lü, G.-H., & Mao, F.-F. (2014). *Chinese Physics B, 23*, 035205.
127. Shin, K. R., Kim, Y. S., & Kim, G. W. (2015). *Applied Surface Science, 347*, 574.
128. Necula, B. S., Apachitei, I., & Tichelaar, F. D. (2011). *Acta Biomaterialia, 7*, 2751.
129. Necula, B. S., van Leeuwen, J. P. T. M., & Fratila-Apachitei, L. E. (2012). *Acta Biomaterialia, 8*, 4191.
130. Marques, I. S. V., Barão, V. A. R., & Cruz, N. C. d. (2015). *Corrosion Science, 100*, 133.
131. van Hengel, I. A. J., Putra, N. E., & Tierolf, M. W. A. M. (2020). *Acta Biomaterialia, 107*, 325.
132. Razzi, F., Fratila-Apachitei, L. E., & Fahy, N. (2020). *Biomedical Materials, 15*, 035017.
133. Yao, X., Zhang, X., & Wu, H. (2014). *Applied Surface Science, 292*, 944.
134. Zhang, X., Li, J., & Wang, X. (2018). *Materials Science and Engineering: C, 82*, 110.
135. San, H., Paresoglou, M., & Minneboo, M. (2022). *International Journal of Molecular Sciences, 23*, 9204.
136. Karabudak, F., Yeşildal, R., & Şüküroğlu, E. E. (2017). *Arabian Journal for Science and Engineering, 42*, 2329.
137. Cerchier, P., Pezzato, L., & Brunelli, K. (2017). *Materials Science and Engineering: C, 75*, 554.
138. Yeşildal, R., Karabudak, F., & Şüküroğlu, E. E. (2018). *Applied Physics A, 124*, 572.

Part IV
Applications

Chapter 12
Advancements in Anticorrosion Plasma Electrolytic Oxidation Coatings for Light-Weight Alloys: Al and Mg Alloys

Xiaopeng Lu, Qianqian Chen, Jirui Ma, Yufei Xie, Xiaobo Chen, Beatriz Mingo, and Fuhui Wang

12.1 Introduction

Light-weight materials, such as aluminum (Al) and magnesium (Mg) alloys, are increasingly used in various industries due to their high specific strength, excellent machinability, recyclability, and biocompatibility. However, their high susceptibility to corrosion remains a major challenge, necessitating the development of advanced anticorrosion coatings.

Plasma electrolytic oxidation (PEO), also known as micro-arc oxidation (MAO), is a promising technique for enhancing the surface properties of Al and Mg alloys. PEO involves the formation of oxide coatings through anodic oxidation in an electrolyte, which utilizes high voltage to create plasma discharges on the metal surface. However, the PEO process subjects the ceramic layer to intense spark discharges and thermal stress, leading to the formation of numerous pores and cracks. These defects allow the corrosive medium to penetrate and directly contact the alloy, thereby promoting corrosion.

PEO coatings prevent corrosion by separating the metal substrate from corrosive environments. The top porous layer, with large cracks and pores, offers poor protection. However, the discontinuous porous middle layer delays corrosion by hindering

X. Lu (✉) · Q. Chen · J. Ma · Y. Xie · F. Wang
Corrosion and Protection Center, Northeastern University, Shenyang, China
e-mail: luxiaopeng@mail.neu.edu.cn

X. Chen (✉)
School of Engineering, RMIT University, Melbourne, VIC, Australia
e-mail: xiaobo.chen@rmit.edu.au

B. Mingo
Department of Materials, Henry Royce Institute, The University of Manchester, Manchester, UK

V. S. Saji, M. Zheludkevich (eds.), *Plasma Electrolytic Oxidation – Fundamentals, Advances and Applications*,
https://doi.org/10.1007/978-3-031-92429-3_12

solution penetration, while the dense inner layer provides the most protection. Defects in the outer layer allow corrosive agents to breach the barrier. Enhancing densification and sealing defects, and blocking corrosive pathways are crucial for improving corrosion resistance.

12.2 Anticorrosion PEO Coatings for Al Alloys

The barrier properties of PEO-coated Al alloys are highly influenced by porosity, both apparent and cross-sectional, and thickness of the oxide layers. For Al alloys, PEO coatings primarily comprise aluminum oxide (Al_2O_3) phases, which offer excellent protective barriers. Corrosion performance of PEO coatings can be effectively enhanced by optimizing electrical parameters, adjusting electrolyte composition, and/or implementing appropriate posttreatment processes.

12.2.1 Electrical Parameters

Recent advances have focused on optimizing PEO processes for Al alloys to control cracks and pores. The simplest strategy is to adjust the electrical parameters like voltage, treatment time, current density, frequency, duty cycle, and power mode, which significantly affect the morphology, microstructure, and thickness of the PEO layer.

12.2.1.1 Processing Time Length

Effect of processing time on PEO coating is mainly related to the change of phases and microstructure, which can be elucidated by the coating formation mechanism [1]. A growth model for PEO coatings shown in Fig. 12.1 describes their evolution. In silicate-based electrolyte, starting as a uniform and thin barrier-type anodic layer in the pre-sparking stage, coatings transit to a single-layer structure and eventually to a bilayer configuration during sparking. In stage II, penetrating B-type discharges, initiated at defects in the barrier-type anodic film from stage I, are primarily responsible for forming pancakes, micro-hemisphere protrusions, and even cavities as PEO coating evolves through stages II–IV. As time increases, increased voltage resulted in larger protrusions form in the inner barrier layer due to the increased penetration of discharges, indicating a locally inhomogeneous inward growth of coating and sealed-hole pancakes become evident (Fig. 12.1e), leading to the formation of larger cavities between the pancakes and the protrusions. Furthermore, cracks often appear alongside the development of these sealed-hole pancakes.

In stage IV, the sealed-hole pancakes exhibit core areas that are less resistive than surrounding regions, making them likely sites for dielectric breakdowns. Energetic

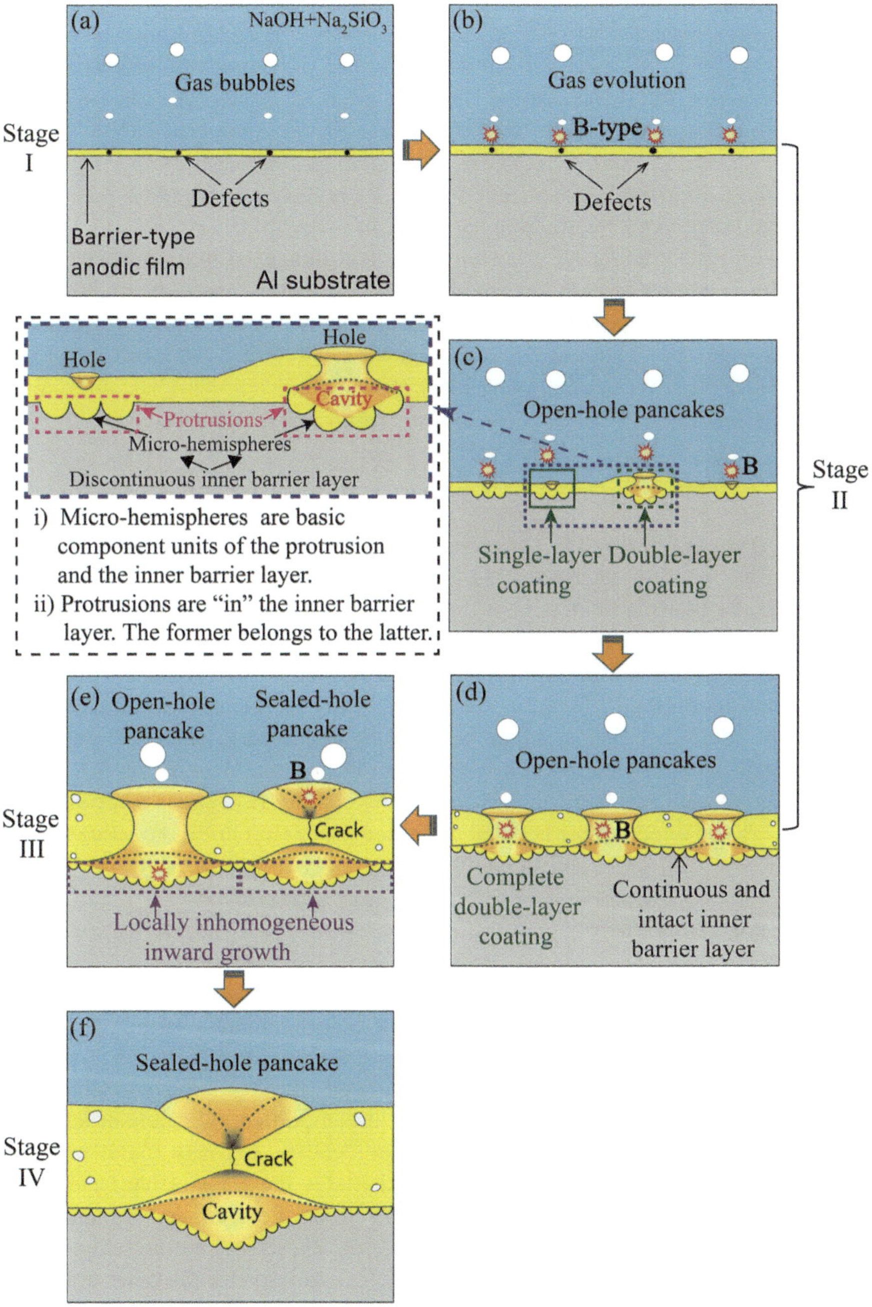

Fig. 12.1 (**a**–**f**) Schematic diagrams of the growth mechanism for PEO coatings [1]. Reproduced with permission from [1] © 2018 Elsevier B.V.

B-type discharges will penetrate these core areas and the underlying barrier layers. This process is expected to generate a new two-layered coating with larger pancakes, protrusions, and cavities in similar locations. The population of pancake features and protrusions at the coating/substrate interface decreases, consistent with the evolution of sparking micro-discharges. This decline can be attributed to two factors: the increased coating thickness, which reduces weak points and breakdown sites, and the tendency for concurrent discharge events to merge, resulting in larger pancakes and protrusions within the barrier layer.

Moreover, PEO coatings after a short treatment time (2–4 min) consisted of the γ-Al_2O_3 phase, whereas coatings formed within 10–30 min formed the mullite phase along with the γ-Al_2O_3. Therefore, it is also possible to improve the corrosion resistance of coatings by obtaining coatings with the desired microstructure and phase composition with different treatment times.

12.2.1.2 Voltage

The applied voltage during PEO has a significant impact on the corrosion resistance of the resulting coating when the PEO treatment was performed in constant voltage mode with all other parameters kept the same. Low voltages can result in thinner, more porous coatings, while excessively high voltages may cause micro-cracks or porous structures due to strong plasma discharges, which may reduce the protective performance of coatings. Moderate voltages typically facilitate stronger plasma discharges, promoting a denser coating structure that offers enhanced protection against corrosive environments. This optimization can minimize micro-defects and improve the overall mechanical integrity of the coating [2]. Thus, controlling the voltage is critical to achieving optimal corrosion protection in PEO coatings.

12.2.1.3 Current Density

The current density has a significant impact on the properties of the ceramic layer. Typically, increasing the current density during the PEO process would result in the increase of the coating thickness [3, 4]. Adjusting the current density can change the micro-discharge characteristics, such as increase the breakdown voltage and decrease the breakdown strength, which in turn modifies the microstructure by increasing the crystallinity and reducing the defects and porosity [5]. It may also occur that the increase of current density facilitates the transformation between α-Al_2O_3 and γ Al_2O_3. It is worth noting that the effect of current density on coating quality may vary in different base electrolytes [6]. Therefore, optimization of current density within a specific electrolyte is the key to improve the corrosion resistance of the coating.

12.2.1.4 Frequency and Duty Cycle

In the PEO process, frequency refers to how often the voltage pulses are applied per second [5, 7]. It affects the time between discharges and influences coating characteristics such as structure, porosity, and phase composition. The duty cycle refers to the percentage of time that the power is "on" during one complete cycle of pulsed current or voltage [5, 8–10]. Both frequency and duty cycle are crucial in determining the properties of PEO coatings. As the duty cycle and frequency increase (Fig. 12.2), the coating surface transitions from voids and pancake-like structures to a smoother surface with melted pores. These micropores in the oxide layers form due to plasma micro-discharges, which cause localized melting of the coating material and gas release within the plasma channels. At lower frequencies and duty cycle values, the energy generated by the micro-discharges is insufficient to fully melt the coating material near the plasma channels, resulting in unfilled voids and more porous coatings. Additionally, extending the PEO process to 60 min reduces porosity compared to coatings formed in 30 min. Higher energy input during the process leads to thicker coatings, which provide better corrosion protection than those produced with shorter oxidation times and lower duty cycles [8]. Optimizing both parameters is essential for achieving coatings with superior corrosion resistance and structural integrity.

12.2.1.5 Power Supply

Power supply is of particular concern due to its influence on the quality of oxide layers, and also relating to the coating process [6, 8, 9, 11, 12]. The unipolar power supply provides a positive voltage relative to the electrolyte and substrate, creating a one-way discharge between the anode (substrate) and the cathode (electrolyte). This typically results in coatings with distinct, regular discharge patterns. The coating can be less uniform if the process is not carefully controlled.

Today, pulsed bipolar current regimes have been explored to control the microstructure and reduce coating defects [6, 8, 11, 13–15]. In a bipolar power supply configuration, the electrical current alternates direction. This means the voltage alternates between positive and negative, creating a bidirectional discharge. The voltage in the discharge time of the positive pulse can be rapidly reduced to less than the breakdown voltage by introducing negative pulses, and the discharge of unextinguished sparks on the sample surface was inhibited, resulting in a relatively enhanced uniformity and improved coating quality (Fig. 12.3) [9].

The mechanism behind the advantages of a bipolar power supply over a unipolar power supply in processes was proposed. Contrary to a unipolar current operating mode, the application of bipolar current pulse regimes usually offers better control over plasma chemical processes at both coating/metal interface and coating/electrolyte interface as well as improve the coating growth rate [13]. As mentioned, the PEO coating typically consists of three distinct layers: the outer loose region characterized by pores and defects caused by the most recent discharges, the inner layer

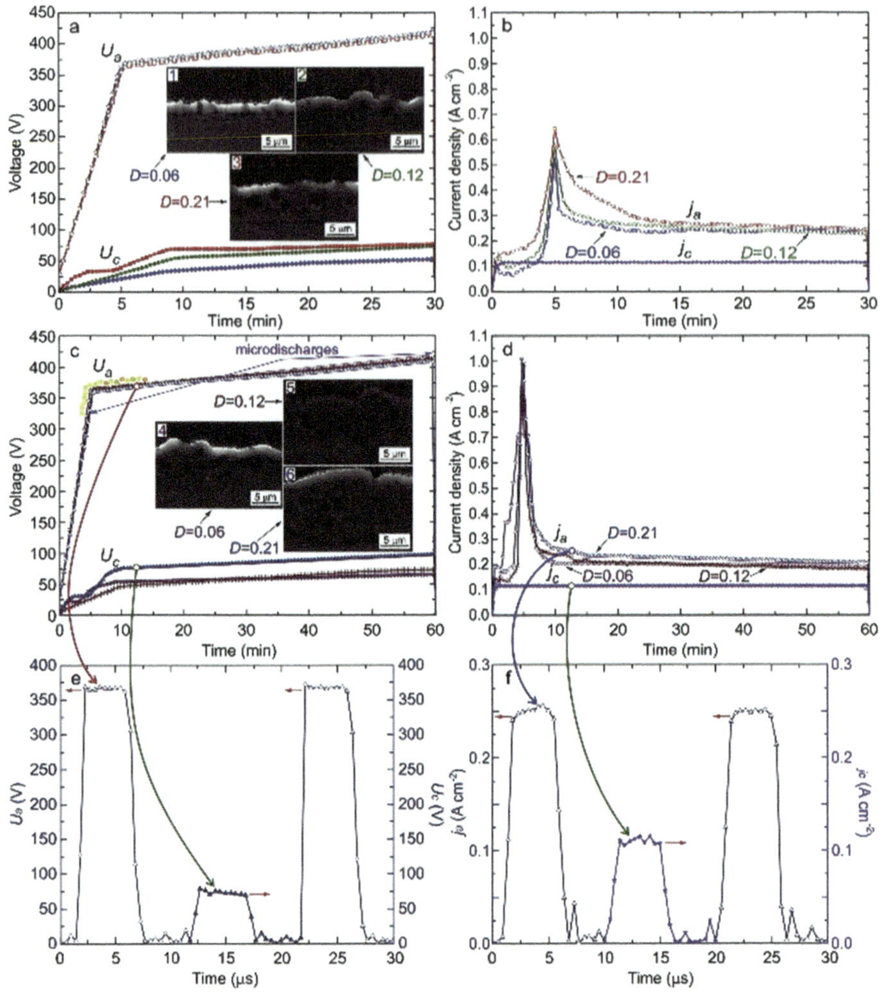

Fig. 12.2 Electrical parameters of the oxidation modes for the 30 min (**a**, **b**) and 60 min (**c**, **d**) process, and waveforms of the pulses for the 60 min process (**e**, **f**), and SEM-images of the cross-sections of the PEO-coatings produced by oxidation for 30 (1–3) and 60 (4–6) min at duty cycle (*D*) = 0.06, 0.12, and 0.21. Reproduced with permission from [8] © 2017 Elsevier B.V.

(γ-Al_2O_3 and α-Al_2O_3) subdivided into porous inner layer and dense inner layer, and the diffusion layer located at the coating/substrate interface. As the surface and cross-sectional morphology of the coating can be linked to the plasma discharging behaviors, the high temperature spikes, associated with strong B-type discharges, play a critical role in coatings microstructure, often making the coating more porous. However, during the positive pulse, oxide formation occurs, while the negative pulse serves to stabilize the process by neutralizing the surface charge. This alternating nature reduces the intensity of plasma discharges, especially the strong

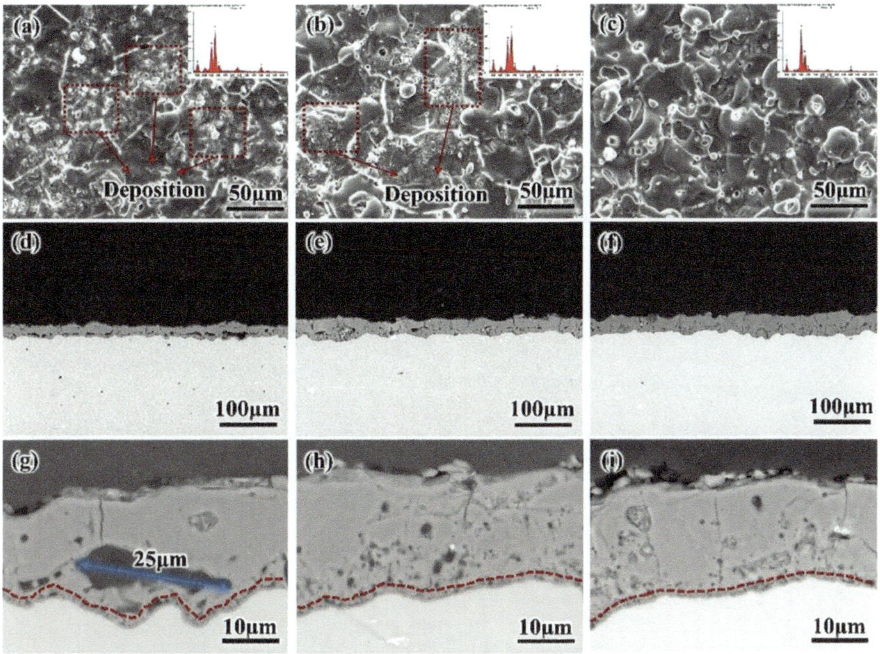

Fig. 12.3 Cross-sectional and surface morphology of PEO coatings prepared at different oxidation time with the current of 4/−4 A/dm^2: (**a**, **d**) 30 min; (**b**, **e**) 40 min; (**c**, **f**) 50 min. Reproduced with permission from [9] © 2022 Elsevier Ltd.

B-type discharges by appropriately adjusting the bipolar current and the pulse on/off intervals. The reduced or eliminated intense B-type discharges result in a decrease in the average temperature of the plasma, promoting the formation of a thick dense inner layer and a thin outer porous layer (Fig. 12.4).

In summary, the alternating nature of the bipolar power supply allows for better control over the plasma discharges, thermal management, and coating formation processes, resulting in higher-quality coatings with improved density, uniformity, and corrosion resistance.

12.2.2 Electrolyte Composition

Coating properties also depend on electrolytic bath composition. Normally the PEO process use environmentally and human-friendly electrolytes, which include silicates, phosphates, aluminates, fluorides, and borates [16]. Ions or particles present in the electrolyte can be incorporated into the coating structure and influence its properties.

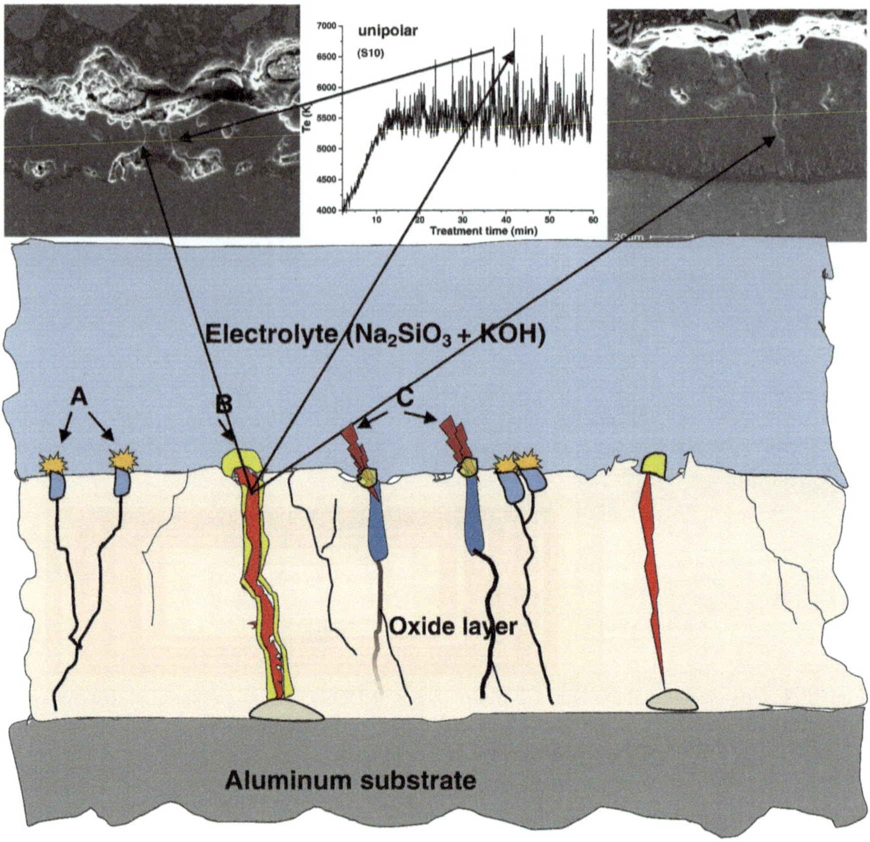

Fig. 12.4 Schematic diagram of the discharge model showing the influence of B-type discharges on plasma temperature profile and on the cross section of the oxide coating. Reproduced with permission from [13] © 2010 Elsevier B.V.

The additives can be divided into: conductive agent to improve solution conductivity, passivator (for the initial oxide coating), stabilizer ($C_3H_8O_3$, Na_2EDTA) to improve the stability of the electrolyte and the pH value, and modifier (Na_2WO_4, Na_3VO_4) to improve special performance of ceramic coatings. By using an optimal electrolyte, the densification, thickness, wear resistance, and corrosion resistance of the coatings will be improved.

12.2.2.1 Inorganic Additives

Inorganic additives can dissolve into aqueous electrolyte and participate in the oxidation process, effectively modifying surface morphology and structure of PEO layers [17–21]. For example, addition of fluorides to electrolyte can lead to a thick and uniform coating by effectively reducing surface porosity and improving coating density, which significantly enhances corrosion resistance and friction performance [16]. Similarly, incorporating sodium tungstate into electrolyte further improves corrosion performance of PEO coating. As the concentration of sodium tungstate increases, both coating thickness and α-Al_2O_3 content rise, leading to a low friction coefficient and enhanced microhardness and corrosion resistance.

When $KMnO_4$ is added, as shown in Fig. 12.5, the formation of Mn_2O_3 phase during PEO process, compared to coatings prepared in alkaline electrolytes,

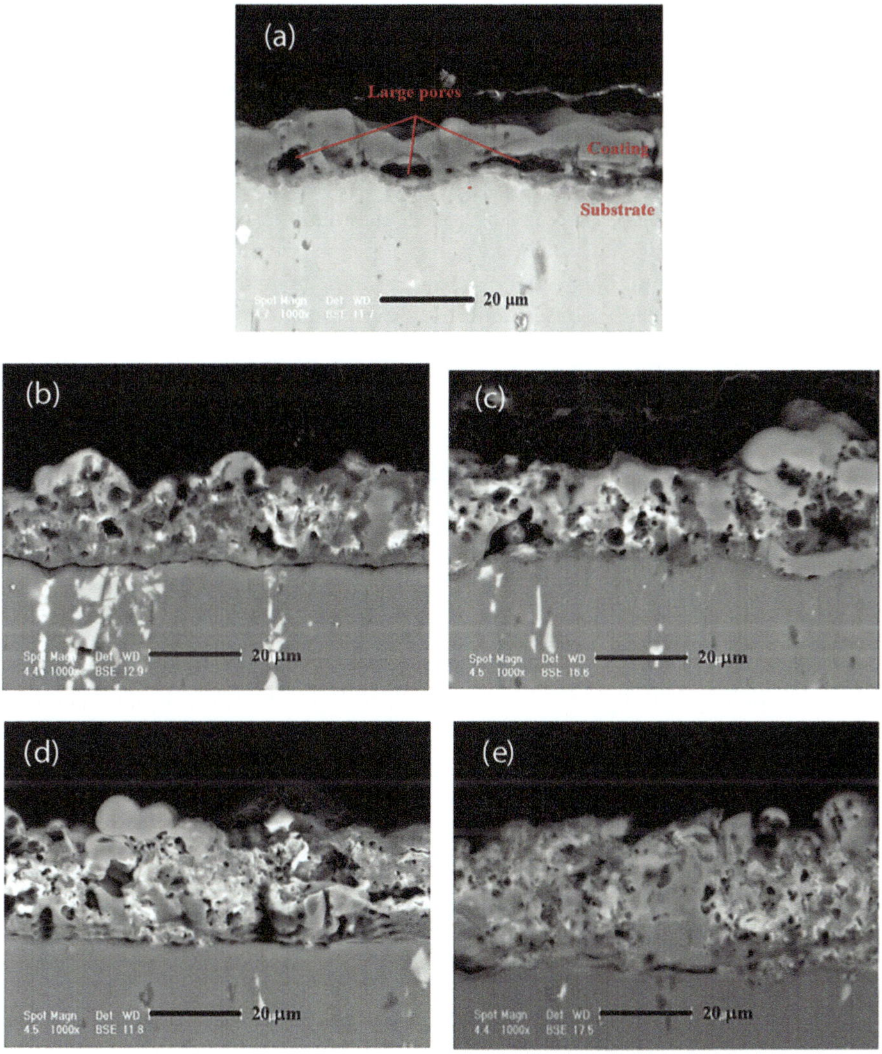

Fig. 12.5 Cross-section images of the PEO coatings obtained at the presence of: (**a**) 0, (**b**) 0.5, (**c**) 1, (**d**) 2, and (**e**) 3 g/L $KMnO_4$. Reproduced with permission from [17] © 2017 Elsevier B.V.

suppresses B-type discharges which may cause large pores at the alloy/coating interface, thereby reducing direct contact between electrolyte and substrate [17]. Studies have shown that coatings obtained with addition of $KMnO_4$ benefit from enhanced densification, which is attributed to the inhibitory effect of MnO_2 [22]. In contrast to alumina, the formed MnO_2 has a lower solubility in aqueous solution and therefore improves stability of coating. Additionally, researchers have suggested that the thermodynamically favorable reactions of MnO_2 contribute to anodic inhibition and provide cathodic protection through the formation of $Mn(OH)_2$.

Besides, in the PEO electrolyte composed of sodium hexametaphosphate and ammonium iron(III) oxalate hydrate, a compact ceramic coating with few cavities containing stable Fe and P phases can be achieved [23]. Ammonium iron(III) oxalate hydrate can increase the acidity of the electrolyte and more $H_2PO_4^-$ would be generated (pH 4–5), which is not observed in alkaline electrolytes containing silicate. As the generation of the $H_2PO_4^-$ and discharge power rises with coating thickness, Al_2O_3 reacts with $H_2PO_4^-$ to form $AlPO_4$, resulting in reduced aluminum content and increased phosphorus content in the coating. As coating develops, discharge power can exceed 500 V, leading to oxidation of $Fe(C_2O_4)_3^{3-}$ and production of Fe_2O_3 on the outer layer of the coating. Finally, the small amount of $FePO_4$ may result from the reaction between Fe_2O_3 and $H_2PO_4^-$. Finally, these PEO coatings containing stable Fe and P phases demonstrate excellent corrosion resistance.

12.2.2.2 Particles

The performance of PEO coatings on Al substrates can be enhanced by incorporating various particles, such as ZrO_2, CeO_2, Al_2O_3, and carbon-based materials [19, 24–38]. The introduction of nanoparticle to PEO electrolyte facilitates sealing of the coating through nanoparticle integration [39]. For instance, the addition of zinc oxide nanoparticles (ZnO-NPs) promotes the formation of mullite phase ($3Al_2O_3{\cdot}2SiO_2$) and inhibits the transformation of γ-Al_2O_3 to α-Al_2O_3, as shown in Fig. 12.6 [14]. Although ZnO-NPs could enter coating through discharge channels, their penetration depth is limited by particle size and migration rate. The interface layer acts as a barrier, preventing ZnO-NPs from entering the coating/substrate reaction interface, which means they do not participate in the formation of interface. Instead, ZnO-NPs can be incorporated near cavity and outer surface of the coating. These nano-composites result in a uniform and dense coating, effectively reducing pathways for corrosive agents [18, 31, 40].

It is worth noting that sometimes the incorporation of inhibitor loaded particles may decrease the corrosion resistance of the coatings remarkably compared to the standard treatment. This reduction is likely due to the introduction of additional defects within the coatings from the incorporated particles. To address this, the combination of inhibitors and nanoparticles presents a promising strategy to further enhance the corrosion resistance of coatings. The positive trend due to the inhibitor addition can be visible more effective corrosion protection than those with pure particles [41]. Al Abri et al. [42] demonstrated this approach by directly

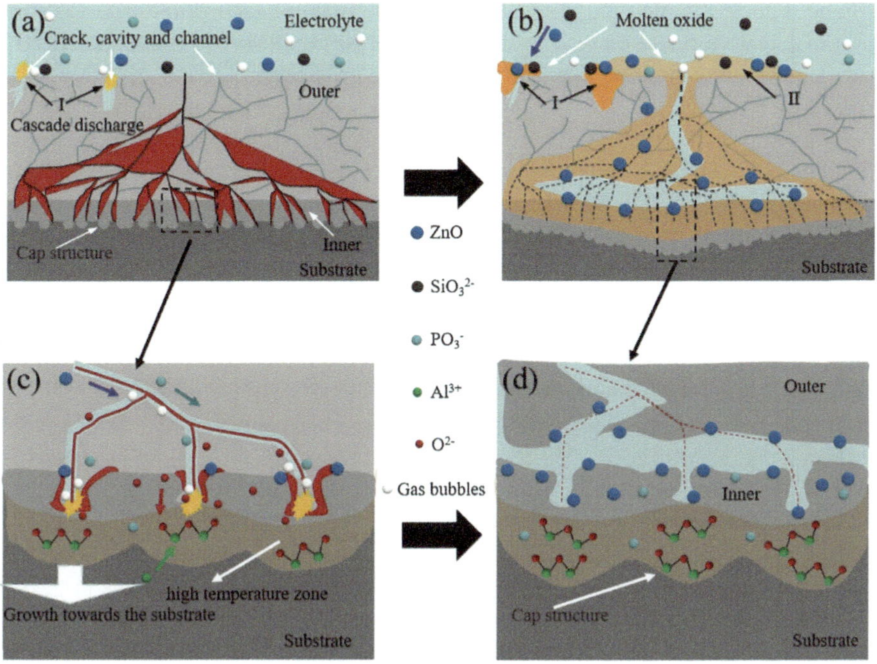

Fig. 12.6 (**a–d**) Incorporation mechanism of ZnO-NPs during PEO process. Reproduced with permission from [14] © 2021 Elsevier B.V.

incorporating cerium-loaded zeolites into PEO coatings from the electrolytic bath. These cerium zeolites served as effective corrosion inhibitors by partially sealing the pores of the oxide film, thereby blocking access to aggressive species.

In recent years, layered double hydroxides (LDHs) have attracted attention due to their unique layered structure, which facilitates the loading of corrosion-inhibiting anions [27, 43–45]. Kristina et al. [40] conducted a comprehensive investigation into the preparation and properties of PEO coatings formed in a near pH-neutral titanium oxide oxalate ($K_2TiO(C_2O_4)_2$) electrolyte with the addition of LDH nanocontainers and inhibitor on the surface of AA2024. In their study, different forms of Zn-Al LDH and/or 2-mercaptobenzothiazole (MBT) were added to the standard electrolyte. Initially, the incorporation of LDH with nitrate ions proved ineffective for initiating the PEO process. To overcome this limitation, the intercalated nitrate ions were successfully exchanged for carbonate ions through an ion-exchange process using Na_2CO_3. The modified LDH-CO_3 slurry allowed PEO processing, as did LDH-MBT powder and a combination of LDH-CO_3 slurry and MBT inhibitor. The influence of MBT alone was also explored to understand its potential effect on the porous structure or its release from LDH-MBT. The results showed that the incorporation of LDH-MBT resulted in the formation of a foamy, porous PEO structure. In contrast, the addition of LDH-CO_3 and MBT together significantly improved the corrosion performance of the coating. Notably, no distinct crystalline phase of LDH can be detected, which may be attributed to the reactive incorporation or degradation of LDH within the electrolyte. Nevertheless, in situ incorporation of LDH is still a promising method for improving the corrosion protection of PEO-coated Al alloys.

The strategy of combining particles with corrosion inhibitors has broad application prospects in the development of coatings, especially in industrial sectors facing increasingly stringent corrosion resistance requirements.

12.2.3 Others

PEO coating combined with inhibitors/nanocontainers and posttreatment to form a duplex coating system is also the effective approach to enhance the corrosion protection of Al alloys [2, 34, 37, 38, 43, 46–50]. There are three main strategies for achieving controlled release of inhibitors to ensure long-term corrosion resistance and active protection.

The first involves introducing inhibitors into both the surface and the pores of the coating, followed by sealing the pores with organic coatings [15, 38, 51]. This method allows the inhibitors to be effectively stored within the PEO structure, where they can gradually release into the substrate interface when exposed to corrosive environments. The subsequent sealing with the organic coating not only protects the PEO layer but also ensures that the inhibitors remain accessible. As a result, when corrosion threats arise, the inhibitors can migrate to the surface, forming protective layers that inhibit pitting and other forms of corrosion, thereby providing robust and long-lasting protection for the Al substrate.

Similarly, the incorporation of corrosion inhibitors or nanocontainers loaded with inhibitors into organic coatings considerably improves the corrosion resistance of PEO-coated Al alloys [52]. This combination allows for the controlled release of inhibitors, which effectively mitigate corrosion processes. When the organic coating is intact, the inhibitors remain stored within the coating matrix, providing ongoing protection. In the event of damage to the coating, these inhibitors can be released to form protective layers that inhibit corrosion, ensuring that the underlying aluminum alloy remains safeguarded against harsh environments. This approach not only improves the durability of the coating system but also extends the service life of aluminum components in corrosive conditions.

In situ synthesis of films to seal the pores of coating can also provide a sealing effect on the micropores of the PEO coating. For example, layered double hydroxides (LDHs) [2, 27, 15, 53, 54] seal the defects within the PEO film, and release inhibitor through an ion exchange mechanism and form insoluble precipitates on the metal surface to repair damaged areas, preventing corrosive ions from corrupting the film and attacking the alloys. This approach effectively protects the Al alloy substrate from corrosion. Currently, composite coatings consisting of oxide ceramics and LDHs have been successfully prepared on various Al alloys. Additionally, superhydrophilicity can be achieved for the PEO coating through posttreatment methods, such as the deposition of hydrophobic agents [47]. These composite coatings provide synergistic benefits, including improved barrier performance and reduced porosity, further enhancing their protective capabilities.

12.3 Anticorrosion PEO Coatings for Mg Alloys

12.3.1 Process of PEO Coating Formation

The growth process of the PEO coatings on Mg alloys can be divided into four stages. The initial stage involves the formation of a thin oxide layer (~1 μm thick) on the alloy surface without any visible arc phenomena. As voltage increases, the dissolution rate of oxide film rises, marking the anodic oxidation stage. Upon reaching the breakdown voltage, localized breaches in the oxide layer occur, accompanied by the emergence of bright, low-density electric arcs, signifying the spark discharge stage. These discharges progressively intensify with further voltage increase, elevating the surface temperature and inducing rapid melting and resolidification of the oxide material, thereby forming a denser and more uniform coating. This marks the onset of the micro-arc oxidation stage. In the final stage, as the coating thickness increases and the reaction continues, higher energy is required to sustain breakdown events. This can result in either the disappearance of visible discharges or their intensification, potentially leading to localized overheating and structural damage to the coating.

12.3.2 The Influence of Discharge Behavior on the Coating Structure

The PEO coatings exhibit three distinct discharge modes, each intricately linked to the coating structure and identifiable by the elemental distribution around micropores:

1. Initial Mode: This occurs when the oxide film breaks down under a strong electric field, initiating within small pores at the electrolyte-coating interface, where electrolyte elements are concentrated. This mode, marked by lower discharge intensity, promotes a reduction in coating defects.
2. Gas Discharge Mode: This stage features gas discharges that form a volcano-like topography as molten materials are expelled from micropores. Elevated temperature accelerates phase transformation but compromise corrosion resistance by allowing penetration into central pores.
3. Arc Discharge Mode: This involves arc discharges occurring between electrons within surface pores and the electrolyte. These discharges typically take place in deeper surface pores abundant with matrix elements. High-intensity discharges

in this mode can damage the coating, creating spherical topography by expelling molten material.

12.3.3 The Effect of Electrolyte Composition on Corrosion Performance

The electrolyte composition plays a crucial role in determining the corrosion resistance of PEO-coated Mg alloys. Common electrolytes such as phosphate, silicate, aluminate, and their mixtures uniquely influence the discharge characteristics, coating structure, and overall performance. In phosphate-based electrolytes, the strong adsorption of PO_4^{3-} accelerates the formation of the passive film, while the low solubility product constant of magnesium phosphate substantially improves the corrosion performance of coating. In contrast, silicate-based electrolytes, when combined with Mg^{2+}, facilitate the formation of insoluble compounds such as $MgSiO_3$ and Mg_2SiO_4. These compounds help reduce breakdown voltage and initial oxidation time, with the effect becoming more pronounced as the concentration of sodium silicate increases.

The extra addition of aluminate salts to the electrolyte solution can be particularly effective for the formation of stable $MgAl_2O_4$, which exhibits a microhardness of 1500–1800 HV, greatly enhancing the wear resistance in PEO coatings. Ghasemi et al. investigated the roles of SiO_3^{2-}, PO_4^{3-} and AlO_2^- in the formation and corrosion properties of PEO coatings [55]. Phases such as Mg_2SiO_4, $Mg_3(PO_4)_2$, and $MgAl_2O_4$ are specifically formed in Si-, P-, and Al-based coatings, respectively. It has been demonstrated that substituting SiO_3^{2-} for PO_4^{3-} or AlO_2^- results in coatings with superior corrosion properties (Fig. 12.7). This improvement is attributed to the optimal coating thickness and the reduction in open pores. These findings highlight

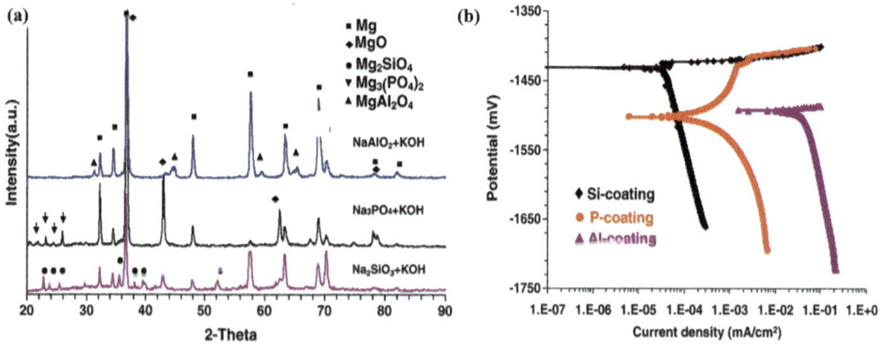

Fig. 12.7 (**a**) XRD patterns to determine the chemical composition of Si-coating, P-coating, and Al-coating. (**b**) Typical polarization plots of three types of coatings in a 3.5 wt.% NaCl solution; the limiting current density decreases in the order Al > P > Si. Reproduced with permission from [55] © 2009 Elsevier B.V.

the critical role of electrolyte selection in the development of advanced PEO coatings for improved durability and performance.

The utilization of mixture electrolytes offers significant structural and performance advantages, potentially leading to synergistic effects. Mori et al. [56] explored the influence of varying phosphate-silicate electrolyte mixture ratios on the morphology, phase composition, and corrosion resistance of PEO coatings. They found that increasing the silicate content resulted in the formation of smaller pores and micro-channels. During corrosion, the unstable amorphous phosphate dissolved into the aqueous solution, leading to the deposition of insoluble crystalline $Mg_3(PO_4)_2{\cdot}22H_2O$ (Eqs. 12.1 and 12.2), which effectively cover the pores and enhance the corrosion resistance of coatings.

$$Mg_3\left(PO_4\right)_2\left(\text{amorphous}\right) \rightarrow 3Mg^{2+} + 2PO_4^{\ 3-}\left(\text{dissolution}\right) \tag{12.1}$$

$$3Mg^{2+} + 2PO_4^{\ 3-} + 22H_2O \rightarrow Mg_3\left(PO_4\right)_2 \cdot 22H_2O \downarrow \left(\text{recrystallization}\right) \tag{12.2}$$

12.3.4 Effect of pH and Conductivity on Corrosion Performance

The pH level and conductivity of the electrolyte are key factors that significantly influence the formation of oxide films on Mg alloys. A high concentration of hydroxide ions (OH^-) facilitates gas evolution, and as the electric field intensity increases, the decomposition of these gases results in plasma formation, initiating arc discharge and spark generation. This phenomenon contributes to the enlargement of internal pores within the coating, leading to a porous, less dense structure that compromises corrosion resistance and elevates energy consumption [40, 57]. Thus, controlling the pH value is essential to minimize the formation of large arc discharges. Li et al. [58] designed neutral electrolytes based on KF and NaH_2PO_4 (Fig. 12.8). Compared to conventional electrolytes, this approach led to a significant reduction in both spark discharge intensity and spark density on the magnesium alloy surface, which in turn enhanced the compactness of the resultant coating. Additionally, the reduction in pH promoted the generation of MgF_2 thus decreasing the formation of MgO.

12.3.5 Effect of Additives on Corrosion Performance

The addition of electrolyte additives can modify the structure and overall performance of PEO coatings. These additives are typically classified into inorganic additives, organic additives, and particle additives. Inorganic additives primarily impact

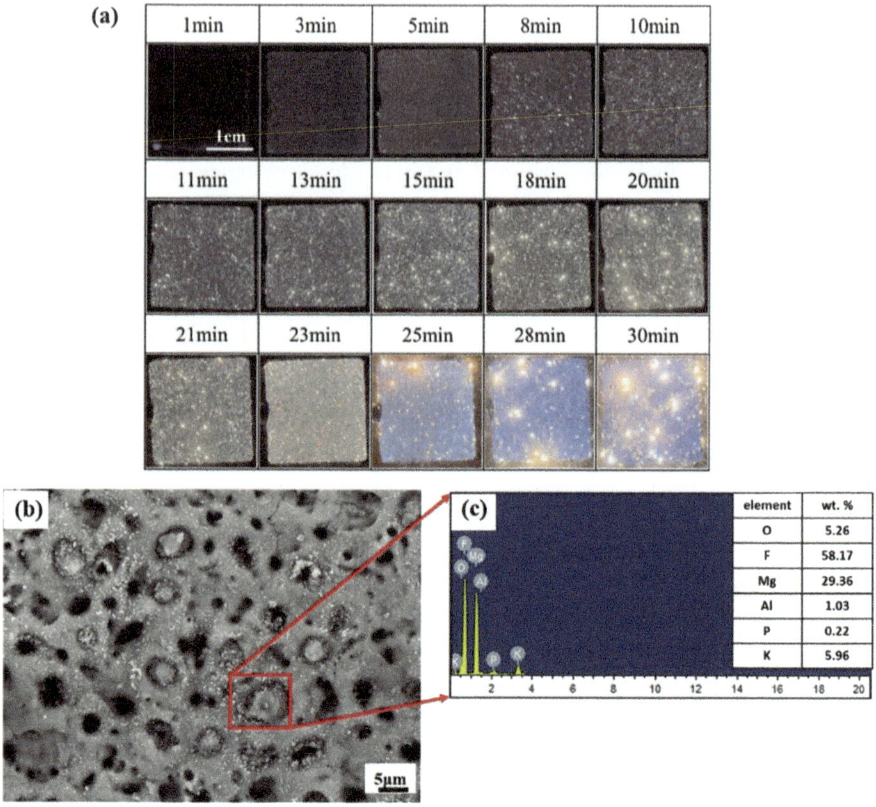

Fig. 12.8 (**a**) Macroscopic spark morphology over time during PEO discharge process. (**b**) SEM image of self-sealed pores. (**c**) Element compositions of self-sealed pores. Reproduced with permission from [58] © 2024 The Authors. Published by Elsevier B.V.

the discharge and operational voltage during the PEO process, thereby altering the microstructure and phase composition of the coating [58–69].

12.3.5.1 Inorganic Additives

Fluoride additives, particularly NaF and KF, are the most widely used inorganic additives in PEO processes [59, 61, 64, 70]. Zhu et al. [59] investigated the roles of fluoride in the growth and corrosion properties of PEO coating on Mg. Their findings revealed that the addition of NaF into the electrolyte promotes the formation of a fluoride-containing nanolayer (FNL) at the interface between the substrate and the coating (Fig. 12.9). The FNL is primarily composed of MgO nanoparticles, interspersed with insoluble MgF_2 nanoparticles. Increasing the NaF concentration results in an increase in both the thickness and the MgF_2 content of the FNL.

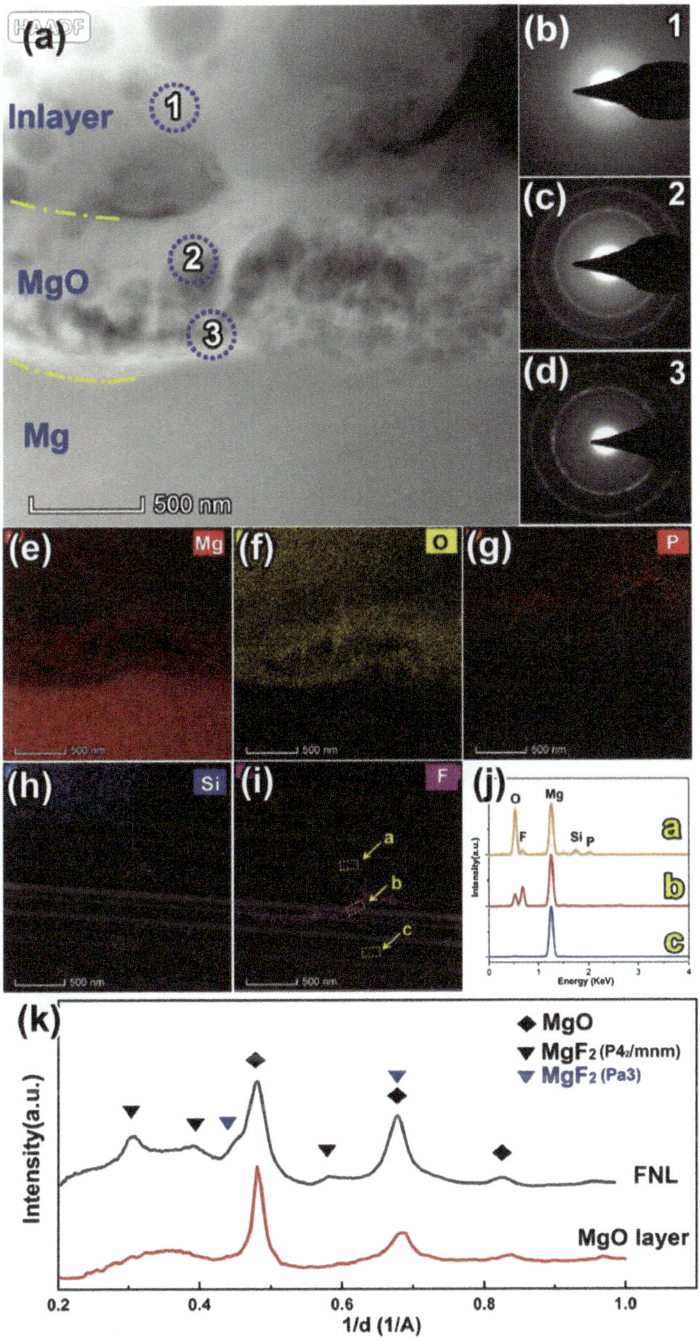

Fig. 12.9 (**a**) Cross-sectional HAADF-STEM micrograph of the PEO coating formed in the base electrolyte containing 0.8 g/L NaF with the SAED patterns. Corresponding EDS element mapping of (**e**) Mg, (**f**) O, (**g**) P, (**h**) Si, and (**i**) F; (**j**) the selected area EDS profiles. (**k**) Radial average intensity profiles of the SAED patterns. Reproduced with permission from [59] © 2021 Chongqing University. Publishing services provided by Elsevier B.V. on behalf of KeAi Communications Co. Ltd.

The addition of fluorozirconate and fluortitanate compounds, specifically K_2TiF_6 and K_2ZrF_6, has been demonstrated to enhance the properties of PEO coatings by forming protective layers on the coating surface [58, 62, 63, 65, 71–74]. Song et al. [71, 74] used K_2TiF_6 as an additive to prepare self-sealing PEO coatings on AM60 Mg alloy. K_2TiF_6 actively participates in the oxidation reactions, leading to the formation of titanium-containing oxide on the coatings surface. These additional compounds act as effective barriers against corrosive media, significantly enhancing the corrosion resistance of the coatings. Notably, these coatings maintain their structural integrity and protective properties even after undergoing a 2000-h salt spray test, demonstrating their dense structure and exceptional durability.

Furthermore, the introduction of K_2ZrF_6 into the electrolyte yields a more uniform PEO coating. Chen et al. [72] found that ZrF_6^{2-}, generated by the decomposition of K_2ZrF_6 readily combines with Mg^{2+}. The fluorine in ZrF_6^{2-} also reacts with Mg^{2+} to form MgF_2, which precipitates on the surface of the Mg substrate. Additionally, the separated Zr^{4+} oxidize to form ZrO_2 (Eqs. 12.3, 12.4, and 12.5). The presence of K_2ZrF_6 leads to the sealing of larger pores in the coating with zirconium-containing compounds, thereby increasing the coating thickness and significantly improving corrosion resistance.

$$K_2ZrF_6 \rightarrow 2K^+ + ZrF_6^{2-} \tag{12.3}$$

$$3Mg^{2+} + ZrF_6^{2-} \rightarrow Zr^{4+} + 3MgF_2 \downarrow \tag{12.4}$$

$$ZrF_6^{2-} + 4OH^- \rightarrow ZrO_2 + 6F^- + 2H_2O \tag{12.5}$$

The use of electrolytes containing inorganic salts with valence-changing metals presents a feasible and effective approach to enhancing the anticorrosion performance of PEO coatings [66, 67, 75]. Fan et al. [66] added the Na_3VO_4, $NaVO_3$, and V_2O_3 to a Si-based electrolyte to investigate the effects of inorganic salts containing valence-changing metals versus those with non-valence-changing metals (Fig. 12.10). During the PEO process, VO_3^- ions reacted to form various V-containing compounds ($Mg_2V_2O_7$, MgV_2O_4, VO_x). The results demonstrated that samples with NaVO3 additive exhibited a more intact coating with significantly reduced porosity, leading to improved corrosion resistance.

The most straightforward method for preparing composite PEO coatings is by incorporating particles or powders, which form alloy compounds with molten MgO, thereby enhancing corrosion resistance. For instance, sodium silicate particles generate SiO_2 during the discharge process, which helps to seal cracks in the coating, although this may lead to a reduction in coating uniformity. Smaller particles tend to distribute more evenly within the coating. The entry of particles into the coating is influenced by the electric field force and the pore size, with larger pores facilitating easier particle penetration. Yang et al. [76] incorporated ZnO nanoparticles into PEO-coated AZ31B Mg alloy, which enhanced corrosion resistance and extended the salt spray life to 2000 h. This improvement can be attributed to the ability of the ZnO nanoparticles to effectively seal pores and distribute uniformly throughout the coating (Fig. 12.11).

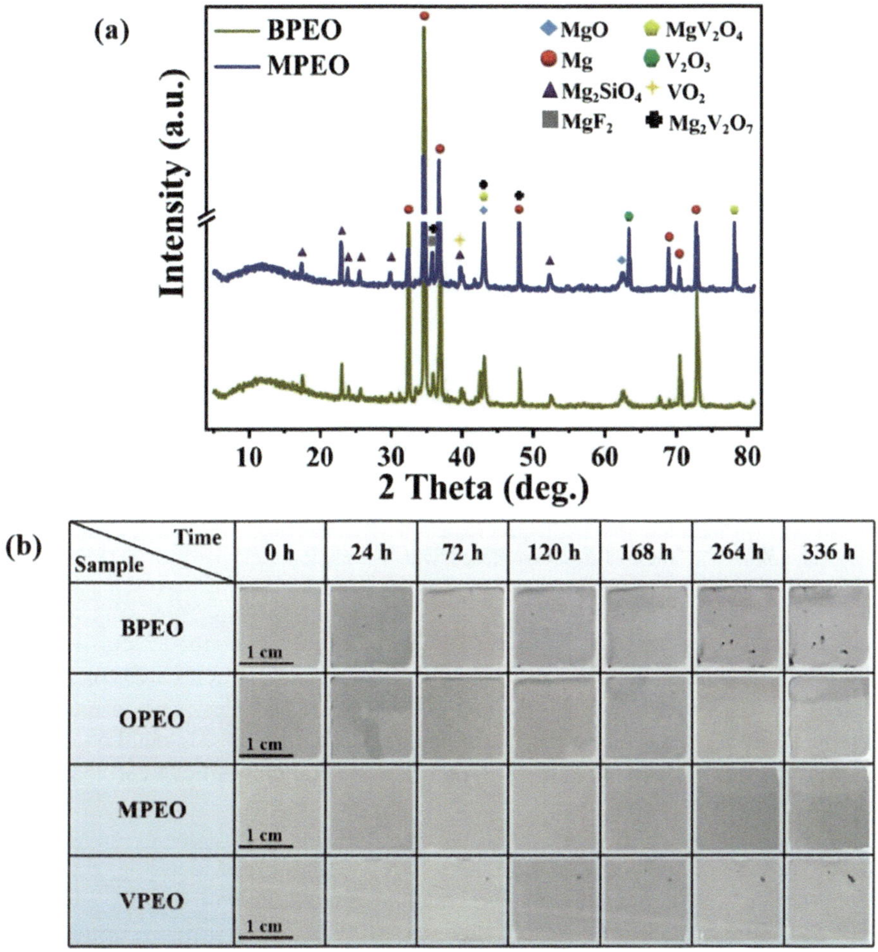

Fig. 12.10 (**a**) XRD pattern of different samples. (**b**) Surface appearance of the PEO coatings with and without additive during 14 days of salt spray test, BPEO (basic electrolyte), OPEO (the electrolyte with Na_3VO_4), MPEO (the electrolyte with $NaVO_3$), VPEO (the electrolyte with V_2O_3). Reproduced with permission from [66] © 2023 Elsevier Ltd and Techna Group S.r.l.

Particle additives can be either inert or reactive incorporated into the PEO coatings [77–79]. Inert additives do not contribute to the formation of new phases in the coating, and their size and shape remain unchanged throughout the PEO process. Conversely, reactive additives may undergo melting from discharge during the treatment, and react with other elements from both the electrolyte and substrate. This complex process is influenced by various factors, including substrate material, particle size, melting point, concentration, and zeta potential [80]. Lu et al. [81] investigated the effect of SiO_2 particle size on PEO coatings, finding that microparticles predominantly entered the coating surface, whereas nanoparticles primarily

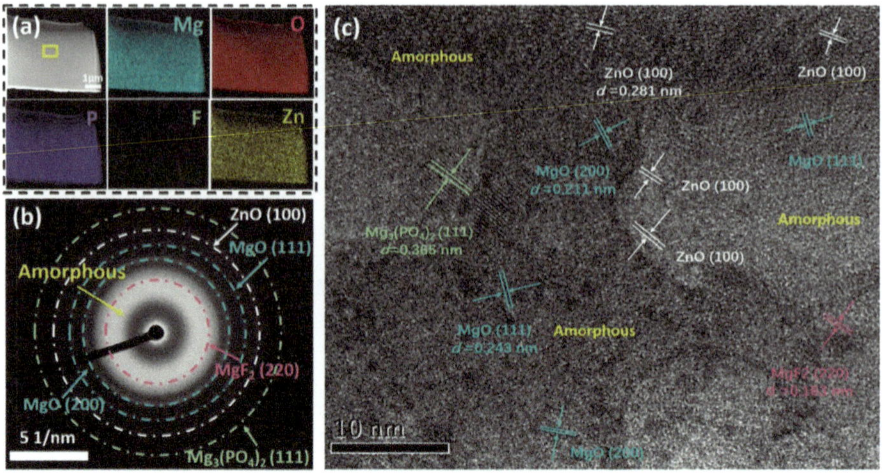

Fig. 12.11 TEM observation on the cross-section of the PEO coating doped with ZnO nanoparticles: (**a**) EDS mapping; (**b**) SAED pattern; and (**c**) HRTEM image showing the compound as nanocrystalline and amorphous. Reproduced with permission from [76] © 2023 Chongqing University. Publishing services provided by Elsevier B.V. on behalf of KeAi Communications Co. Ltd.

infiltrated discharge channels and open pores. Thus, nanoparticles are typically preferred for PEO electrolytes, including carbon-based particles (carbon nanotubes, graphene, etc.), metal-based particles (Ag, Cu, Fe, etc.) [82, 83], metal oxide particles (ZnO, TiO_2, MgO, etc.) [84–86], and ceramic particles (silicon carbide, silicon nitride, etc.) [87].

A practical challenge in the addition of particle additives is achieving a uniform distribution of particles within the PEO electrolyte. Particles with higher zeta potential are more stable, preventing sedimentation and agglomeration in the PEO solution [88, 89]. It has been observed that negatively charged particles, or those with negative zeta potentials, are often preferred in alkaline electrolytes [90]. Furthermore, the inherent properties of the particles can also affect their dispersion in the solution. For example, graphene oxide exhibits enhanced dispersion in PEO electrolytes due to the presence of oxygen functional groups on its surface. Additionally, their high specific surface area, superior permeability, and excellent chemical and thermal stability make it an attractive additive for improving coating protection [91–93]. Zhang et al. [93] studied the in situ incorporation of graphene oxide in PEO coatings, and found that while the addition of graphene oxide did not induce the formation of new phases, it increased the intensity of the main diffraction peaks as the graphene oxide concentration increased. The addition of graphene oxide significantly improved the impedance values of the resultant composite coating. However, excessive graphene oxide content led to particle agglomeration, which concentrated erosive ions on the coating surface and compromised the barrier properties of both the outer and inner PEO layers.

12.3.5.2 Organic Additives

The stability of the electrolyte or changes in its composition may significantly affect the coating performance. Incorporating appropriate organic additives can significantly enhance electrolyte stability and reduce sparking on the sample surface during the PEO process [94–102]. Alcohol-based additives, such as ethanol, ethylene glycol, and glycerol, have been found to improve the corrosion resistance of PEO coatings [94–97]. Jang et al. [96, 97] reported that the addition of glycerol during the PEO process resulted in coatings with smaller micropore sizes, higher pore density, and reduced coating thickness. The presence of glycerol promoted the formation of MgO on the coating surface while suppressing the development of the amorphous Mg_2SiO_4 phase. Additionally, the inner barrier layer of the PEO coating became thicker and denser, significantly enhancing its corrosion resistance. Asoh et al. [95] studied the effect of alcohol addition to Na_3PO_4-based electrolyte on the PEO coating. The presence of alcohol increased both the initial breakdown voltage and the oscillation voltage during the PEO process compared to the base electrolyte. This modification resulted in thicker PEO films with reduced surface roughness and enhanced corrosion resistance.

Besides, the introduction of environmentally friendly additives has become a prominent research focus for modifying PEO coatings [99]. Plant extracts offer numerous advantages, including abundant availability, renewability, biodegradability, and ecological compatibility, coupled with diverse chemical compositions, making them promising green additives [68, 98, 100]. Guo et al. [68] investigated the effects of incorporating pectin as an eco-friendly additive into the electrolyte on the microstructure and chemical composition of the coatings. The findings revealed that hydroxyl and carboxyl functional groups in pectin adsorbed onto the metal surface, forming a stable film structure in the anodic discharge region, thereby facilitating the formation and growth of the coating. Additionally, the functional groups in pectin contribute to the formation of a stable protective layer on the metal surface, significantly improving the corrosion resistance of the coating. Similarly, ionic liquids (ILs) have garnered attention as green and sustainable inhibitors due to their remarkable adsorption capacity on metal surfaces, low volatility, and exceptional thermal and chemical stability. These characteristics make ILs a novel and promising additive for enhancing the performance of PEO coatings [101–103]. Zhang et al. [101, 102] investigated the effects of incorporating varying concentrations of BmimBF4 into an alkaline silicate electrolyte on the performance of PEO coatings. The study found that the concentration of BmimBF4 played a critical role in determining coating characteristics. At higher concentrations, BmimBF4 uniformly adsorbed onto the substrate surface during the initial stage of the PEO process, resulting in uniform discharge across the substrate throughout the PEO process. This uniformity contributed to reduced surface roughness and smaller micropores of the coating. Moreover, the long hydrophobic alkyl chain in BmimBF4 enhanced the corrosion resistance of the coating by effectively adsorbing on the coating surface, forming a robust barrier against corrosive agents (Fig. 12.12).

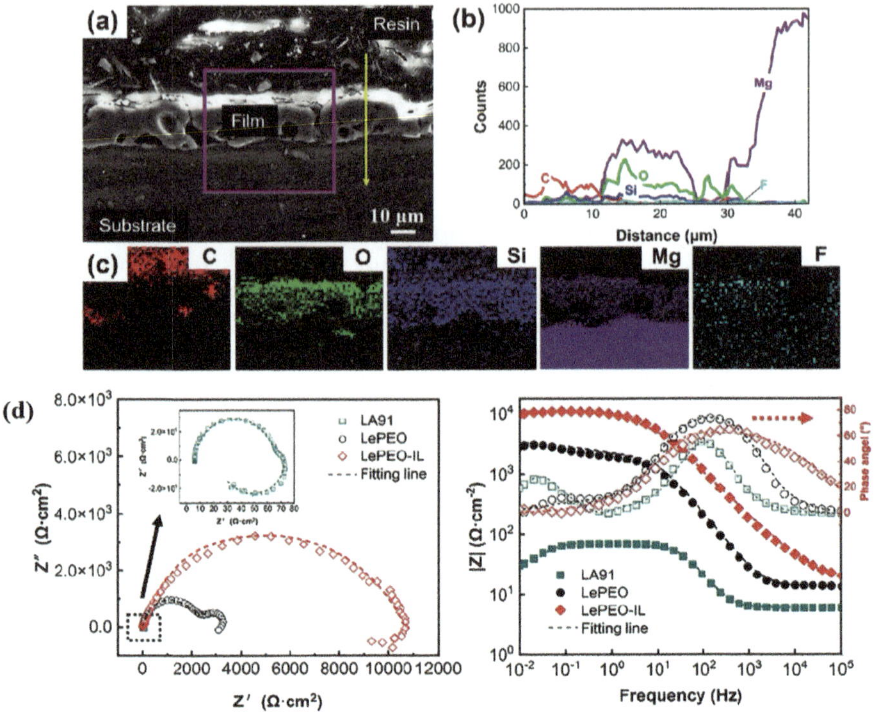

Fig. 12.12 (**a**) The cross-sectional SEM image. (**b**) Corresponding elements depth profiles. (**c**) Distribution maps and corresponding EDS depth profiles. (**d**) EIS spectra of the LA91 substrate, LePEO (PEO coating) and LePEO-IL (electrolytes with BmimBF4)-coated samples in 3.5 wt.% NaCl solution. Reproduced with permission from [102] © 2023 Chongqing University. Publishing services provided by Elsevier B.V. on behalf of KeAi Communications Co. Ltd.

12.4 Conclusions and Outlook

PEO coatings have demonstrated significant potential in enhancing the corrosion resistance of light-weight Al and Mg alloys. By employing the strategies mentioned above, the performance of PEO-coated Al and Mg alloys can be significantly enhanced, leading to longer-lasting materials that can withstand harsh environments. These improvements not only increase the lifespan of the alloys but also broaden their potential application in various industries, ultimately contributing to more sustainable and efficient material usage. Each strategy complements the others, creating a comprehensive approach to corrosion protection that addresses the inherent challenges of using lightweight Al and magnesium alloys.

The future of PEO coatings for Al and Mg alloys lies in the development of more environmentally friendly electrolytes, and smart coatings with self-healing and multifunctional capabilities. The integration of responsive materials capable of autonomously repairing damage or imparting additional functionalities, such as antibacterial and anti-icing properties, represents a promising frontier in material science. Exploring the synergistic effects of combining PEO with other advanced surface treatments. Moreover, understanding the fundamental mechanisms at the atomic and molecular levels will enable more precise control over coating properties and performance.

Acknowledgements This work was financially supported by National Natural Science Foundation of China (No. 12411530101), LiaoNing Revitalization Talents Program (XLYC2403026), Shenyang Young and Middle-aged Science and Technology Innovation Talent Support Program (RC231178) and the Fundamental Research Funds for the Central Universities (N25GFY002).

References

1. Wu, Y.-K., Yang, Z., Wang, R.-Q., Wu, G.-R., Chen, D., Wang, D.-D., Liu, X.-T., Li, D.-L., Guo, C.-H., Yu, S.-X., Shen, D.-J., & Nash, P. (2018). *Surface and Coating Technology, 351*, 136–152.
2. Mohedano, M., Serdechnova, M., Starykevich, M., Karpushenkov, S., Bouali, A. C., Ferreira, M. G. S., & Zheludkevich, M. L. (2017). *Materials & Design, 120*, 36–46.
3. Kaseem, M., Min, J. H., & Ko, Y. G. (2017). *Journal of Alloys and Compounds, 723*, 448–455.
4. Zhang, Y., Wu, Y., Chen, D., Wang, R., Li, D., Guo, C., Jiang, G., Shen, D., Yu, S., & Nash, P. (2017). *Surface and Coating Technology, 321*, 236–246.
5. Li, H., Zhang, M., Wu, X., Geng, Y., & Lei, L. (2024). *Surface and Coating Technology, 485*, 130881.
6. Cheng, Y., Cao, J., Mao, M., Xie, H., & Skeldon, P. (2016). *Surface and Coating Technology, 291*, 239–249.
7. Oh, G. H., Yoon, J. K., Huh, J. Y., & Doh, J. M. (2023). *Surface and Coating Technology, 471*, 129861.
8. Egorkin, V. S., Gnedenkov, S. V., Sinebryukhov, S. L., Vyaliy, I. E., Gnedenkov, A. S., & Chizhikov, R. G. (2018). *Surface and Coating Technology, 334*, 29–42.
9. Zhu, M., Song, Y., Dong, K., Shan, D., & Han, E.-H. (2022). *Electrochimica Acta, 411*, 140056.
10. Dehnavi, V., Luan, B. L., Shoesmith, D. W., Liu, X. Y., & Rohani, S. (2013). *Surface and Coating Technology, 226*, 100–107.
11. Oh, G. H., Yoon, J. K., Huh, J. Y., & Doh, J. M. (2024). *Corrosion Science, 233*, 112123.
12. Yerokhin, A. L., Shatrov, A., Samsonov, V., Shashkov, P., Pilkington, A., Leyland, A., & Matthews, A. (2005). *Surface and Coating Technology, 199*, 150–157.
13. Hussein, R. O., Nie, X., & Northwood, D. O. (2010). *Surface and Coating Technology, 205*, 1659–1667.
14. Yu, Z., Du, A., Wang, C., Ma, R., Fan, Y., Zhao, X., & Cao, X. (2021). *Surface and Coating Technology, 412*, 127068.
15. Fernández-López, P., Alves, S. A., Azpitarte, I., San-José, J. T., & Bayón, R. (2022). *Corrosion Science, 207*, 110548.
16. Olesiński, A., Bugla, A., Babilas, D., Stolarczyk, A., Simka, W., & Sowa, M. (2024). *Corrosion Science, 237*, 112322.
17. Aliramezani, R., Raeissi, K., Santamaria, M., & Hakimizad, A. (2017). *Surface and Coating Technology, 329*, 250–261.
18. Valentini, F., Pezzato, L., Dabalà, M., & Brunelli, K. (2023). *Journal of Materials Research and Technology, 27*, 3595–3609.
19. Chen, P., Wu, Z., Huang, Q., Ji, S., Weng, Y., Wu, Z., Ma, Z., Chen, X., Weng, M., Fu, R. K. Y., Lin, H., Tian, X., Pan, F., & Chu, P. K. (2019). *Tribology International, 138*, 157–165.

20. Rakoch, A. G., Van Tuan, T., Khabibullina, Z. V., Blawert, C., Serdechnova, M., Scharnagl, N., Zheludkevich, M. L., & Gladkova, A. A. (2022). *Surface and Coating Technology, 433*, 128075.
21. Kaseem, M., Kamil, M. P., Kwon, J. H., & Ko, Y. G. (2015). *Surface and Coating Technology, 283*, 268–273.
22. Kulinich, S. A., Farzaneh, M., & Du, X. W. (2007). *Inorganic Materials, 43*, 956–963.
23. Ji, S., Weng, Y., Wu, Z., Ma, Z., Tian, X., Fu, R. K. Y., Lin, H., Wu, G., Chu, P. K., & Pan, F. (2017). *Journal of Alloys and Compounds, 710*, 452–459.
24. Barati, N., Meletis, E. I., Golestani Fard, F., Yerokhin, A., Rastegari, S., & Faghihi-Sani, M. A. (2015). *Applied Surface Science, 356*, 927–934.
25. Tran, Q.-P., Chin, T.-S., Kuo, Y.-C., Jin, C.-X., Trung, T., Tuan, C. V., & Dang, D. Q. (2018). *Journal of Alloys and Compounds, 751*, 289–298.
26. Fatimah, S., Kamil, M. P., Kwon, J. H., Kaseem, M., & Ko, Y. G. (2017). *Journal of Alloys and Compounds, 707*, 358–364.
27. Wang, Y., Pan, Y., Xu, X., Zhang, W., Feng, R., Li, H., & Gao, Z. (2024). *Journal of Materials Research and Technology, 28*, 1044–1061.
28. Kaseem, M., Lee, Y. H., & Ko, Y. G. (2016). *Materials Letters, 182*, 260–263.
29. Chen, Q., Jiang, Z., Tang, S., Dong, W., Tong, Q., & Li, W. (2017). *Applied Surface Science, 423*, 939–950.
30. Zhang, P., Zuo, Y., & Nie, G. (2020). *Journal of Alloys and Compounds, 816*, 152520.
31. Ignjatović, S., Blawert, C., Serdechnova, M., Karpushenkov, S., Damjanović, M., Karlova, P., Wieland, D. C. F., Starykevich, M., Stojanović, S., Damjanović-Vasilić, L., & Zheludkevich, M. L. (2021). *Applied Surface Science, 544*, 148875.
32. Babaei, K., Fattah-alhosseini, A., & Molaei, M. (2020). *Surfaces and Interfaces, 18*, 100441.
33. Ghafaripoor, M., Raeissi, K., Santamaria, M., & Hakimizad, A. (2018). *Surface and Coating Technology, 349*, 470–479.
34. Maizeray, A., Marcos, G., Cappella, A., Planche, M. P., Liao, H., Henrion, G., Czerwiec, T., & Martin, J. (2024). *Surface and Coating Technology, 482*, 130713.
35. Liu, W., Liu, Y., Lin, Y., Zhang, Z., Feng, S., Talha, M., Shi, Y., & Shi, T. (2019). *Applied Surface Science, 475*, 645–659.
36. Fattah-alhosseini, A., Chaharmahali, R., & Babaei, K. (2020). *Journal of Magnesium and Alloys, 8*, 799–818.
37. Telmenbayar, L., Ramu, A. G., Erdenebat, T.-O., & Choi, D. (2022). *Materials Today Communications, 33*, 104662.
38. Zhang, S., Ma, Y., Zhao, S., Dai, W., & Liu, B. (2024). *Corrosion Science, 235*, 112203.
39. Arrabal, R., Mohedano, M., Matykina, E., Pardo, A., Mingo, B., & Merino, M. C. (2015). *Surface and Coating Technology, 269*, 64–73.
40. Mojsilović, K., Serdechnova, M., Blawert, C., Zheludkevich, M. L., Stojadinović, S., & Vasilić, R. (2024). *Applied Surface Science, 677*, 161065.
41. Mojsilović, K., Božović, N., Stojanović, S., Damjanović-Vasilić, L., Serdechnova, M., Blawert, C., Zheludkevich, M. L., Stojadinović, S., & Vasilić, R. (2021). *Surfaces and Interfaces, 26*, 101307.
42. Al Abri, S., Rogov, A., Aliasghari, S., Bendo, A., Matthews, A., Yerokhin, A., & Mingo, B. (2024). *Journal of Materials Research and Technology, 30*, 2365–2376.
43. Bouali, A. C., Straumal, E. A., Serdechnova, M., Wieland, D. C. F., Starykevich, M., Blawert, C., Hammel, J. U., Lermontov, S. A., Ferreira, M. G. S., & Zheludkevich, M. L. (2019). *Applied Surface Science, 494*, 829–840.
44. Fedel, M., Zanella, C., Ferrari, L., & Deflorian, F. (2021). *Electrochimica Acta, 381*, 138288.
45. Bouali, A. C., André, N. M., Silva Campos, M. R., Serdechnova, M., dos Santos, J. F., Amancio-Filho, S. T., & Zheludkevich, M. L. (2021). *Journal of Materials Science and Technology, 67*, 197–210.
46. Akbarzadeh, S., Coelho, L. B., Dangreau, L., Lanzutti, A., Fedrizzi, L., & Olivier, M.-G. (2023). *Corrosion Science, 222*, 111424.

47. Gnedenkov, S. V., Sinebryukhov, S. L., Egorkin, V. S., & Vyaliy, I. E. (2016). *Surface and Coating Technology, 307*, 1241–1248.
48. Huang, J., Zhao, D., Gong, Y., Yang, C., Zhu, H., Wang, M., Chen, D., & Wang, H. (2022). *Surface and Coating Technology, 441*, 128566.
49. Mohedano, M., Matykina, E., Arrabal, R., Mingo, B., & Pardo, A. (2015). *Applied Surface Science, 346*, 57–67.
50. Vaghefinazari, B., Lamaka, S. V., Gazenbiller, E., Yasakau, K., Blawert, C., Serdechnova, M., Scharnagl, N., Wieland, D. C. F., & Zheludkevich, M. L. (2024). *Corrosion Science, 226*, 111651.
51. Li, X., Di, Y., Chen, Z., & Yang, W. (2022). *Colloids and Surfaces A: Physicochemical and Engineering Aspects, 653*, 129990.
52. Udoh, I. I., Shi, H., Daniel, E. F., Li, J., Gu, S., Liu, F., & Han, E.-H. (2022). *Journal of Materials Science and Technology, 116*, 224–237.
53. Kasneryk, V., Serdechnova, M., Blawert, C., & Zheludkevich, M. L. (2023). *Applied Clay Science, 232*, 106774.
54. Fattah-alhosseini, A., Fardosi, A., Karbasi, M., & Kaseem, M. (2024). *Journal of Magnesium and Alloys, 12*, 465–489.
55. Ghasemi, A., Raja, V. S., Blawert, C., Dietzel, W., & Kainer, K. U. (2010). *Surface and Coating Technology, 204*, 1469–1478.
56. Mori, Y., Koshi, A., Liao, J., Asoh, H., & Ono, S. (2014). *Corrosion Science, 88*, 254–262.
57. Moga, S., Malinovschi, V., Marin, A., Coaca, E., Negrea, D., Craciun, V., & Lungu, M. (2023). *Surface and Coating Technology, 462*, 129464.
58. Li, C., Li, Y., Zhang, T., & Wang, F. (2024). *Journal of Materials Research and Technology, 30*, 7684–7696.
59. Lujun, Z., Hongzhan, L., Qingmei, M., Jiangbo, L., & Zhengxian, L. (2023). *Journal of Magnesium and Alloys, 11*, 2823–2832.
60. Yang, C., Sheng, L., Zhao, C., Wu, D., & Zheng, Y. (2024). *Materials Letters, 355*, 135559.
61. Cui, X.-J., Liu, C.-H., Yang, R.-S., Li, M.-T., & Lin, X.-Z. (2015). *Surface and Coating Technology, 269*, 228–237.
62. Liu, F., Shan, D., Song, Y., & Han, E.-H. (2011). *Surface and Coating Technology, 206*, 455–463.
63. Rahmati, M., Raeissi, K., Toroghinejad, M. R., Hakimizad, A., & Santamaria, M. (2020). *Surface and Coating Technology, 402*, 126296.
64. Chen, Q., Zheng, Y., Dong, S., Chen, X.-B., & Dong, J. (2021). *Surface and Coating Technology, 417*, 126883.
65. Ur Rehman, Z., Heun Koo, B., Jung, Y.-G., Lee, J. H., & Choi, D. (2020). *Materials, 13, 499.*
66. Fan, H., Ling, N., Bai, R., Zhang, J., & Wang, L. (2023). *Ceramics International, 49*, 24783–24793.
67. Shen, D., Ma, H., Guo, C., Cai, J., Li, G., He, D., & Yang, Q. (2013). *Journal of Rare Earths, 31*, 1208–1213.
68. Guo, C., Li, Y., Qi, C., Sun, H., Xue, Y., Wan, Y., & Zhang, D. (2024). *Surface and Coating Technology, 476*, 130209.
69. Pezzato, L., Brunelli, K., Napolitani, E., Magrini, M., & Dabalà, M. (2015). *Applied Surface Science, 357*, 1031–1039.
70. Němcová, A., Skeldon, P., Thompson, G. E., & Pacal, B. (2013). *Surface and Coating Technology, 232*, 827–838.
71. Dong, K., Song, Y., Shan, D., & Han, E.-H. (2015). *Corrosion Science, 100*, 275–283.
72. Chen, X. W., Zhang, M., Zhang, D. F., Cai, L. P., Song, H., & Zeng, D. Z. (2023). *Journal of Alloys and Compounds, 966*, 171474.
73. Dong, K., Song, Y., Shan, D., & Han, E.-H. (2015). *Surface and Coating Technology, 266*, 188–196.
74. Song, Y., Dong, K., Shan, D., & Han, E.-H. (2013). *Journal of Magnesium and Alloys, 1*, 82–87.

75. Kaseem, M., & Choe, H.-C. (2021). *Corrosion Science, 192*, 109764.
76. Yang, C., Huang, J., Cui, S., Fu, R., Sheng, L., Xu, D., Tian, X., Zheng, Y., Chu, P. K., & Wu, Z. (2024). *Journal of Magnesium and Alloys, 12*, 3602–3615.
77. Singh, A., Drunka, R., Smits, K., Vanags, M., Iesalnieks, M., Joksa, A., Blumbergs, I., & Steins, I. (2023). *Crystals, 13*, 508.
78. Lou, B.-S., Lin, Y.-Y., Tseng, C.-M., Lu, Y.-C., Duh, J.-G., & Lee, J.-W. (2017). *Surface and Coating Technology, 332*, 358–367.
79. Nasiri Vatan, H., Ebrahimi-Kahrizsangi, R., & Asgarani, M. K. (2016). *Tribology International, 98*, 253–260.
80. Lu, X., Mohedano, M., Blawert, C., Matykina, E., Arrabal, R., Kainer, K. U., & Zheludkevich, M. L. (2016). *Surface and Coating Technology, 307*, 1165–1182.
81. Lu, X., Blawert, C., Huang, Y., Ovri, H., Zheludkevich, M. L., & Kainer, K. U. (2016). *Electrochimica Acta, 187*, 20–33.
82. Arun, S., Ahn, S.-G., & Choe, H.-C. (2023). *Surfaces and Interfaces, 39*, 102932.
83. Han, Q., Li, Y., Lu, X., Mei, D., Chen, Q., Zhang, T., & Wang, F. (2021). *Materials Letters, 293*, 129731.
84. Han, J., Yu, Y., Yang, J., Xiaopeng, L., Blawert, C., & Zheludkevich, M. L. (2022). *Surface and Coating Technology, 433*, 128112.
85. Wang, X., Ju, P., Lu, X., Chen, Y., & Wang, F. (2022). *Ceramics International, 48*, 3615–3627.
86. Lopez-Martinez, E., Mohedano, M., Matykina, E., & Arrabal, R. (2024). *Journal of Materials Research and Technology, 31*, 1786–1796.
87. Pezzato, L., Lorenzetti, L., Tonelli, L., Bragaggia, G., Dabalà, M., Martini, C., & Brunelli, K. (2021). *Surface and Coating Technology, 428*, 127901.
88. Mehdilo, A., Irannajad, M., & Rezai, B. (2013). *Colloids and Surfaces A: Physicochemical and Engineering Aspects, 428*, 111–119.
89. Hanaor, D., Michelazzi, M., Leonelli, C., & Sorrell, C. C. (2012). *Journal of the European Ceramic Society, 32*, 235–244.
90. Lee, K. M., Lee, B. U., Yoon, S. I., Lee, E. S., Yoo, B., & Shin, D. H. (2012). *Electrochimica Acta, 67*, 6–11.
91. Fattah-alhosseini, A., & Chaharmahali, R. (2021). *FlatChem, 27*, 100241.
92. Zhao, J., Xie, X., & Zhang, C. (2017). *Corrosion Science, 114*, 146–155.
93. Zhang, Y., Chen, F., Zhang, Y., & Du, C. (2020). *Tribology International, 146*, 106135.
94. Zhu, F., Wang, J., Li, S., & Zhang, J. (2012). *Applied Surface Science, 258*, 8985–8990.
95. Asoh, H., Asakura, K., & Hashimoto, H. (2020). *RSC Advances, 10*, 9026–9036.
96. Jangde, A., Kumar, S., & Blawert, C. (2022). *Surface and Coating Technology, 450*, 128971.
97. Jangde, A., Kumar, S., & Blawert, C. (2019). *Corrosion Science, 157*, 220–246.
98. Zehra, T., Dikici, B., Dafali, A., & Kaseem, M. (2023). *Progress in Organic Coating, 182*, 107677.
99. Kaseem, M., Zehra, T., Khan, M. A., Safira, A. R., Cho, H., Lee, J., Lee, G., Yang, H. W., & Park, N. (2022). *Surfaces and Interfaces, 34*, 102403.
100. Qin, J., Chen, Y., Chen, C., Zhong, S., Yan, Z., Liu, W., Wang, Y., Lai, X., Zhao, Y., Zhao, R., & Zhang, R. (2023). *Surface and Coating Technology, 466*, 129655.
101. Chen, C., Zhang, Y., Wang, S., Zhang, Z., Wen, C., & Chen, F. (2024). *Surface and Coating Technology, 477*, 130329.
102. Zhang, Y., Chen, C., Tian, H., Wang, S., Wen, C., & Chen, F. (2024). *Journal of Magnesium and Alloys, 12*, 2380–2396.
103. Mo, M.-H., Wu, L.-K., Cao, H.-Z., Lin, J.-P., Lu, D.-H., & Zheng, G.-Q. (2016). *Surface and Coating Technology, 307*, 190–199.

Chapter 13
Advances in Anti-wear Plasma Electrolytic Oxidation Coatings for Lightweight Materials: Al and Mg Alloys

Patricia Fernández López, Sofia Afonso Alves, Ainara López Ortega, and Eva Gutierrez Berasategui

13.1 Introduction

PEO constitutes an electrochemical and plasma-assisted technology for the development of multifunctional ceramic coatings on different valve metals, outstanding aluminum (Al), magnesium (Mg), and titanium (Ti) alloys [1]. Despite the multiple applications of these light alloys, especially in the automotive and aerospace sectors where lightweighting is one of the most promising trends nowadays, their use under more demanding requirements is restricted due to their low tribological resistance. In this context, PEO technology has proven to be an efficient and sustainable alternative for the tribological improvement of light alloys. In particular, PEO coatings are generally characterized by their high tribological performance, offering excellent adhesion to the substrate [2, 3], high hardness [4, 5], and high wear resistance [6, 7].

This chapter presents the main findings in the development of PEO coatings for the improvement of the tribological properties (improvement of wear resistance or improvement of the coefficient of friction depending on the application) of Al and Mg alloys.

P. F. López (✉) · S. A. Alves · E. G. Berasategui
Plasma Coating Technologies Unit, TEKNIKER, Basque Research and Technology Alliance (BRTA), Eibar, Spain
e-mail: patricia.fernandez@tekniker.es

A. L. Ortega
Tribology Unit, TEKNIKER, Basque Research and Technology Alliance (BRTA), Eibar, Spain

V. S. Saji, M. Zheludkevich (eds.), *Plasma Electrolytic Oxidation – Fundamentals, Advances and Applications*,
https://doi.org/10.1007/978-3-031-92429-3_13

13.2 PEO Coatings with High Tribological Performance Developed on Aluminum Alloys

13.2.1 Aluminum: A Key Player in Lightweight Material Sustainable Applications

Al is pivotal to the European Union's green transition, serving critical roles in solar panels, wind turbines, batteries, heat pumps, hydrogen electrolyzers, and sustainable packaging. It also supports e-mobility and green building initiatives, with demand expected to rise by 30% by 2040 due to the expanding clean technology sector. Therefore, this is a highly versatile material that has become strategic across multiple industries, including automotive, aerospace, construction, electronics, renewable energy, and packaging [8–12]. Its popularity stems from its low density (2.7 g/cm^3)—roughly one-third that of cast iron or steel—making it ideal for lightweight applications [13]. In sectors such as transport, increased Al content in vehicles effectively contributes to lightweighting efforts, which is critical for reducing emissions and achieving decarbonization. Al's ability to form diverse alloys further enhances its adaptability, while its high ductility, ease of machining, and low cost solidify its position as the most widely used non-ferrous metal and the second most widely used metal overall, after steel. Furthermore, Al is nearly infinitely recyclable without loss of quality, contributing to its long-term sustainability [14, 15]. Although primary Al offers superior purity, its production is highly energy-intensive, contributes to the depletion of bauxite—a critical raw material—and is a significant source of greenhouse gas (GHG) emissions. In contrast, Al recycling requires only 5% of the energy used for primary production (Fig. 13.1 [16]) and prevents 9 tonnes of CO_2 emissions.

Despite the many advantages of Al alloys, these materials have significant limitations that restrict their use in certain applications. A primary challenge is their limited corrosion resistance. While a passive oxide layer naturally forms on Al surfaces,

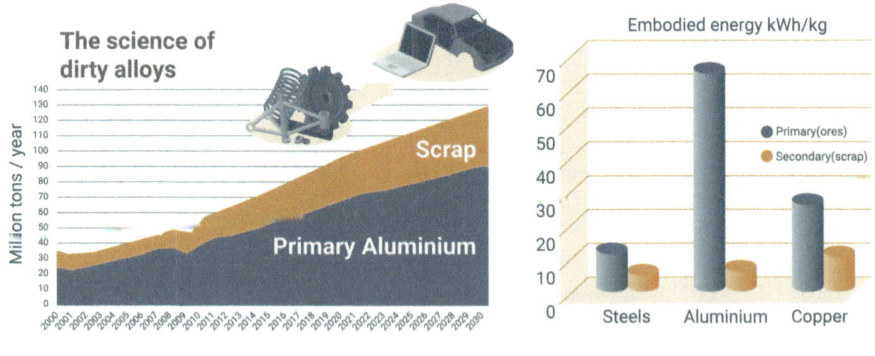

Fig. 13.1 Evolution of the Al consumed from 2000 to 2030 and comparison between the energy consumption required for the primary and secondary production. Reproduced with permission from [16] © 2022 The Authors. Published by Elsevier Ltd.

it provides inadequate protection in harsh environments, exposing the underlying metal to irreversible corrosion and ultimately reducing its service life. Furthermore, Al has poor wear resistance and low hardness, with these tribological weaknesses being even more pronounced in pure Al. Consequently, Al is often alloyed with other elements or subjected to heat treatment to enhance its mechanical properties and expand its usability. In many cases, Al alloys also require surface modifications, such as protective coatings, to further improve their performance and durability, making them more suitable for demanding applications.

13.2.2 PEO as a Breakthrough Solution to Treat Aluminum Alloys for Tribological Applications

Depending on their manufacturing method, Al alloys are broadly divided into wrought and casting alloys. Casting alloys offer significant economic benefits over wrought alloys due to their faster processing and reduced assembly costs. Indeed, around 60% of the global Al production is destined for casting applications [17]. Among the casting alloys, aluminum-silicon (Al-Si) alloys are predominant and account for more than 80% of the market [18], due to their outstanding mechanical properties, such as high strength/weight ratio, low density, good toughness, low coefficient of thermal expansion, proper electrical and thermal conductivity, low density, excellent cast ability [19–22], and low manufacturing costs [4, 23]. In addition to cast Al-Si alloys, cast Al-Cu and Al-Mg are also usual cast Al alloys [24]. Despite the benefits introduced by these materials, their final application is frequently hampered by their poor tribological and electrochemical features, which compromise their durability and safety [20, 25]. Due to the low hardness, high and unstable friction coefficient, and elevated tendency to adhesion due to relatively high ductility and reactivity, the tribological behavior of Al-Si components is seriously compromised, being generally accompanied by high levels of wear [21, 22, 26, 27]. The limited durability of cast Al alloys under abrasive wear restricts their application in mechanical engineering, particularly for machine parts such as pulleys, cylinders, pistons, guide rails, rollers, textile machinery components, and impellers exposed to abrasive environments [28].

The modification of the surface properties of cast Al-Si alloys, without detriment of bulk properties, has been increasingly explored as a very promising way to overcome most of their tribological limitations [1, 6]. Several techniques based on chemical, physical, and electrochemical approaches have been explored to overcome the current issues. Traditionally, Cr-plating and hard anodizing are favored for Al alloys due to their energy efficiency and the protective layer thicknesses of 100–200 μm with microhardness values of 800–1000 HV. However, these methods often use environmentally hazardous electrolytes (notably in chromium plating) and perform better on pure Al than on cast alloys. Environmental restrictions on chromium have accelerated the search for alternative wear-resistant coatings for Al alloys.

Anodizing has been demonstrated to be the most widely used surface treatment for primary Al materials, providing enhanced corrosion resistance, improved lubrication, adhesion, and even esthetic properties, such as color. This electrolytic process involves immersing Al parts in aggressive acidic baths and applying voltages ranging from 10–50 V (standard anodizing) to 20–120 V (hard anodizing) or current densities between 0.5 and 10 A/dm^2. Although anodizing is effective for primary Al alloys, its efficiency and quality diminish when treating secondary Al alloys, which often contain significant amounts of impurities such as Fe, Cu, and Zn. These impurities—present at levels of several percent due to recycling components with inserts, wires, and coatings—greatly reduce the quality of anodic coatings. When impurity levels exceed 3–5%, anodizing becomes impractical, rendering the process ineffective. Additionally, anodizing is unsuitable for cast aluminum-silicon (Al-Si) alloys due to differences in the conductivity and oxide growth rates of the Al and Si phases. These variations prevent the formation of thick, dense, and uniform coatings [2, 18, 21]. The presence of Si in Al alloys induces the generation of Si-enriched phases in the material, which increases as the content of this allowing element increases. The formation of Si phases in the Al matrix prevents the reaction between Al and O_2 during oxidation processes and therefore inhibits the growth of Al oxide during anodization inducing to the generation of nonhomogeneous films [4, 6]. The generation of dense, hard, and wear resistant films on cast Al-Si components by anodizing technique is therefore a challenge [3].

To address these challenges, PEO technology is emerging as a safer and more sustainable approach for enhancing the surface protection of recycled Al materials. PEO is an electrochemical process conducted in an electrolytic bath where a high voltage or current is applied, promoting the growth of Al_2O_3 at the Al-electrolyte interface through the generation of electrical plasma micro-discharges. However, unlike anodizing, PEO employs environmentally friendly, aqueous alkaline electrolytes, which facilitate the formation of ceramic-like coatings with superior hardness, thickness, and enhanced wear and corrosion resistance. This makes PEO particularly suitable for recycled Al alloys, even those with elevated impurity levels, enabling the generation of effective protective coatings that surpass the limitations of traditional anodizing techniques. The main differences between anodizing and PEO process are highlighted in Table 13.1.

Figure 13.2 illustrates a typical PEO coating produced on an aluminium alloy. These coatings typically consist of a three-layer structure: a porous outer layer, an intermediate layer, and a dense inner layer. The outermost layer, characterized by high porosity, is primarily composed of sintered Al_2O_3 and SiO_2, while the dense inner layer contains well-crystallized α-Al_2O_3 with low porosity. As shown in Fig 13.2, successive grinding significantly reduces the number of visible flaws and defects. This inner layer serves as an effective corrosion-resistant barrier and plays a key role in determining the thermo-mechanical and tribological properties of the coating. Optimizing the PEO process—selecting suitable electrolyte compositions and electrical parameters for the power supply—is crucial to achieving the desired microstructural characteristics to meet specific application requirements [7, 29].

Table 13.1 Differences between state-of-the-art anodizing and PEO techniques

Variable	Anodizing	PEO
Type of material	Primary Al alloys (low impurities)	Primary and secondary Al alloys (high impurities)
Pre-treatment	Crítical (agressive baths: ex—nitric acid, sulfuric acid, chromic acid, etc.). They are not ecological—environmentel impact, and must be treated at its EoL (added cost)	Not critical (cleaning with water and neutral soap)
Electrolyte	Acidic electrolytes (sulfuric, chromic, nitric, oxalic; tartaric; glycolic; citric; lactic; benzoic...). They are not ecological—environmentel impact, and must be treated at its EoL (added cost)	Alkaline (low concentration), non-aggressive. Environmentally friendly. Can be easily managed or recycled. Generally contain aluminates, silicates and phosphates, agents used, for example, in cleaning products.
Number of electrochemical baths in production line	Aprox. 12	Aprox. 5 (Less water/resources consumption). Less waste.
Post-treatment	Critical (agressive baths) Environmentel impact, and must be treated at its EoL (added cost)	Not critical. Re-usage of PEO electrolyte for cleaning.
Voltage range	10–50 V (anodizing); 20–120 V (hard anodizing)	400–1000 V impact on cost
Current density	0.5–10 $A{\cdot}dm^{-2}$ (DC current)	10–100 $A{\cdot}dm^{-2}$ (alternate or pulsed)

13.2.3 *Influence of PEO Process Parameters on Coating Properties*

Over the past decades, significant advances have been achieved in the application of PEO technology to Al alloys—both wrought and recycled—to develop highly protective tribological coatings with additional functionalities such as enhanced wear and tribocorrosion resistance. The microstructure and properties of PEO coatings depend significantly on factors such as substrate material and composition, electrolyte composition and additives, treatment duration, electrical parameters (frequency, duty cycle, current density [both positive and negative], and voltage [both positive and negative]), and finally, electrolyte temperature. A thorough understanding of how these parameters affect coatings' morphology, structure, phase composition, corrosion and wear/mechanical properties is essential for advancing research and industrial applications in this field [29].

Therefore, the optimization and development of PEO coatings have followed several strategies, with one of the key approaches being the adjustment of PEO process parameters. Simchen et al. [7] outlined the various variables that influence

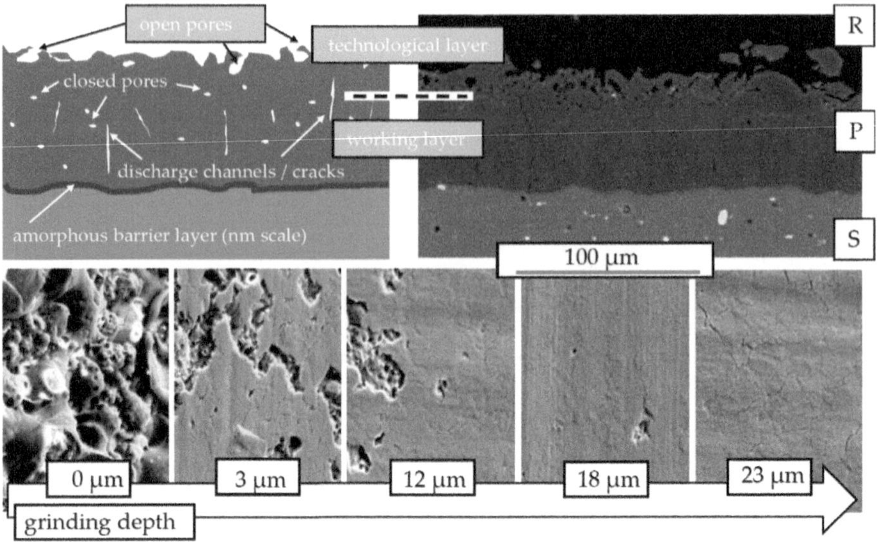

Fig. 13.2 Morphology of a PEO coating on an aluminium alloy: schematic representation and cross-sectional micrographs (R—Embedding Resin, P—PEO Layer, S—Substrate), along with top-view images at progressive grinding depths, revealing an increasingly compact layer structure. Reproduced with permission from [7] © 2020 by the authors. Licensee MDPI

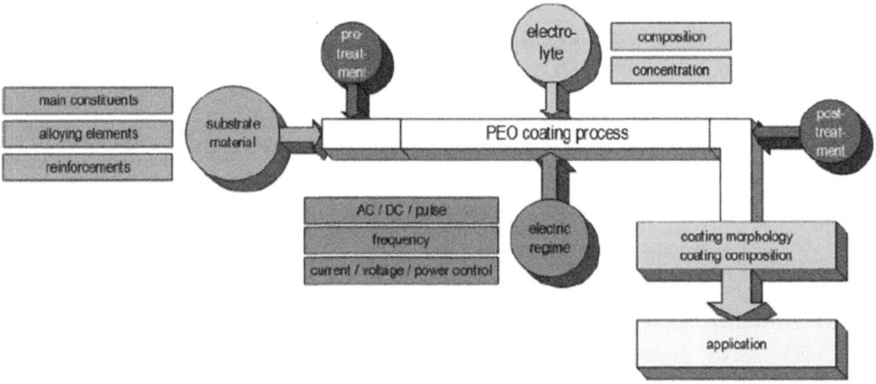

Fig. 13.3 Schematic graphic representation of the PEO process and its dependencies from the uncoated substrate over the process parameters to the application. Reproduced with permission from [7] © 2020 by the authors. Licensee MDPI

the properties of PEO coatings, emphasizing the importance of the entire process flow, from material composition to final application (Fig. 13.3). The electrolyte design is crucial for optimizing both the PEO process and coating efficiency. Typically, additives like silicates and phosphates boost process efficiency, while complexing agents and particles in the electrolyte enhance coating quality. Additionally, electrical parameters—such as current mode and density, frequency, and waveform—significantly impact the lifetime, spatial distribution, and electrical properties of plasma discharges during PEO. Careful adjustment of these factors allows for tailoring PEO coatings to meet specific performance requirements, enabling the development of multifunctional, durable coatings suited to various applications. The following section describes how these key variables may influence PEO coating characteristics.

13.2.3.1 Substrate Material Composition

The substrate material in the PEO process defines the types of oxides that can form, which directly affects the coating properties. During PEO, the metal ions participating in electrochemical reactions come from the substrate, with oxides of the substrate metal typically making up the main coating components. This process naturally incorporates alloying elements, metal precipitates, and reinforcement phases from metal matrix composites. Additionally, part of the substrate metal dissolves during PEO, actively contributing to the ceramic coating formation. As a result, the chemical composition of the alloy significantly influences the coating's composition, microstructure, and properties [30–32]. Previous investigations have demonstrated that the presence of alloying elements, such as Cu, Mg, or Zn in Al alloys inhibits the transition of γ-Al_2O_3 to α-Al_2O_3 [33, 34]. This suppression in the phase transformation is attributed to the lower bond energy between the O_2 and these alloying elements compared to the robust interaction between Al and O_2. Under the influence of the electric field, Cu^{2+}, Mg^{2+}, and Zn^{2+} ions would migrate to the outer part of the coating faster than the Al^{2+} ions. The presence of these weaker bonds in the lattice would disrupt the grain growth since the transformation from γ-Al_2O_3 to α-Al_2O_3 involves a shift from cubic to hexagonal packing, typically accompanied by grain growth and an elevation in the transition temperature [32].

In the particular case of cast Al-Si alloys, the Si will also significantly hinder the effective growth of the PEO coating [20, 32]. Fe is generally the most prevalent undesirable impurity in these alloys [35, 36], with its content being higher in recycled alloys, in which it reaches approximately 1% [37]. Due to the low solubility of Si and Fe in the Al matrix, brittle intermetallic compounds, like Al15(FeMn)3Si2 or Al15(FeMnCr)3Si2 can be formed, which usually leads to a deterioration of the mechanical properties of the alloy (Fig. 13.4) [37]. The presence of these compounds may influence the properties of the resulting PEO coatings and should be considered at the time of coating and process design for a specific application.
Plasma electrolytic oxidation (PEO) was used by Agureev et al. [39] to form coatings up to 75 μm thick on Al composites, both unalloyed and alloyed with 1–4.5% Cu, in a silicate-alkaline electrolyte under alternating current. Coatings were characterized by SEM, X-ray microanalysis, XPS, nuclear backscattering spectrometry, and XRD. After 60 min, coatings primarily contained γ-Al_2O_3 and η-Al_2O_3, with metal inclusions; by 180 min, high-temperature corundum α-Al_2O_3 formed, and inclusions disappeared. Corrosion current density was reduced in coated samples, even as Cu alloying increased corrosion in uncoated samples. As a result, PEO

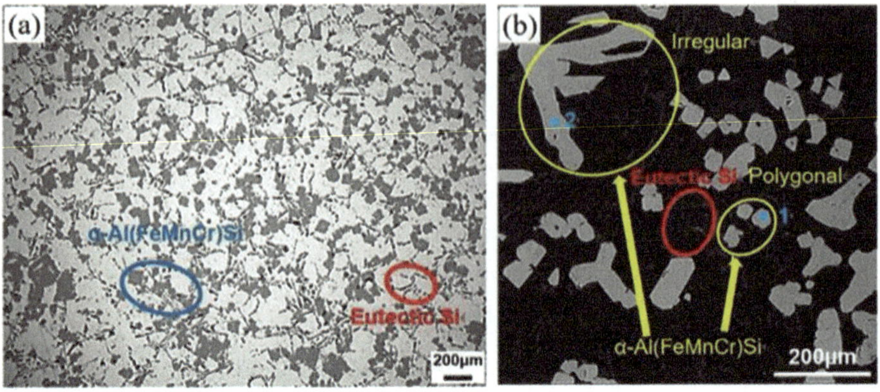

Fig. 13.4 Optical (**a**) and SEM (**b**) micrographs of an AlSi7FeMn1.5Cr0.5 alloy containing iron-rich intermetallic compounds [38]. Reproduced with permission from [38] © 2020 Elsevier B.V.

coatings hold strong potential to "mask" the substrate material, effectively protecting it against harsh conditions like corrosion and tribocorrosion.

13.2.3.2 Electrolyte Composition and Additives

The chemical composition, conductivity, and pH of the electrolyte used in the PEO process is likely one of the most decisive factors in determining the properties of the resulting PEO coatings. The elements in the electrolyte are ionized during the PEO process and, to varying degrees, incorporated into the coating, influencing critical parameters such as thickness, roughness, chemical composition, crystalline phase distribution, and hardness [40–44]. Thus, modifying or adjusting the electrolyte composition allows for the development of PEO coatings with diverse functionalities—both functional and esthetic—that can be tailored to the specific requirements of each component [45, 46]. In addition to electrolyte composition, ensuring good electrolyte stability is crucial for guaranteeing its effectiveness during processing and its proper management at the end of its service life. Improved chemical stability reduces the formation of sludge and precipitates. Furthermore, good electrolyte stability enhances both economic and environmental benefits, as it may enable the electrolyte to be reused in multiple processes.

PEO electrolytes can be classified into sodium hydroxide-based, silicate-based, phosphate-based, and aluminate-based categories. In silicate, phosphate, and aluminate based electrolytes, SiO_3^{2-}, PO_4^{3-}, and AlO_2^- ions are typically incorporated into the coatings through electrophoresis under a strong electric field, therefore influencing the mechanical, tribological, and other properties of the coatings [29]. Among these, silicate-based, with Na_2SiO_3 being the most prevalent reagent, are the most widely used electrolytes in PEO technology [47–49]. These electrolytes are among the most commonly used in PEO processes on Al alloys [2, 50] as they promote the formation of thick ceramic layers [51]. However, their use often leads to PEO films with higher SiO_2 content, which tends to reduce hardness and wear resistance [52] while increasing surface roughness [53]. Additionally, in cast Al-Si alloys, the oxidation and incorporation of Si from the substrate can exacerbate these effects. Consequently, aluminate-based electrolytes are being increasingly studied for producing PEO films with enhanced properties on Al alloys. In this context, Fernández-López et al. [4] developed high-performance tribological coatings on recycled cast Al-Si alloys using PEO technology, facilitated by a new aluminate-based electrolyte. The findings of this study indicate that the aluminate-based electrolytes increased the coating thickness and density while significantly enhancing surface hardness compared to the uncoated cast Al-Si alloy. All PEO coatings tested demonstrated excellent wear resistance and improved corrosion protection compared to the untreated cast Al-Si alloy. Additionally, the authors observed that the inclusion of K_2TiF_6 in the electrolyte resulted in PEO coatings that offered the highest protection against wear and corrosion for the treated cast Al-Si alloy. In collaboration with previous work, Alves et al. [54] utilized the newly developed aluminate-based electrolyte to enhance the tribological performance of PEO-treated cylinder liners made of cast Al with high Si content in a setup simulating the real contact between a cylinder liner and a piston ring. The novel PEO-treated cylinder liners exhibited significantly improved wear resistance when subjected to tribological actions, which was attributed to the substantially lower porosity, defects and higher hardness of the coatings (Fig. 13.5). These PEO-treated cylinder liners also demonstrated a lower coefficient of friction (COF) compared to the commercial reference throughout the entire sliding test. Moreover, tribological tests revealed that the PEO-treated Al cylinder liners exhibited mechanical wear damage levels comparable to those of steel cylinder liners.

Phosphate and aluminate electrolytes were also employed by Peng et al. [55] to produce PEO coatings on 6061 Al alloy. Friction and wear properties of the new coatings were evaluated through pin-on-disk sliding tests under dry conditions, while corrosion resistance was assessed via electrochemical and salt spray tests in acidic environments. Results indicated that both phosphate and aluminate PEO coatings predominantly comprised α-Al_2O_3 and γ-Al_2O_3. A bilayer coating was formed in the phosphate electrolyte, while a single-layer dense alumina coating with a hardness of 1300 HV was achieved in the aluminate electrolyte. The aluminate PEO coating exhibited a lower wear rate, whereas the phosphate PEO coating demonstrated enhanced corrosion resistance, attributed to the presence of amorphous phosphorus at the substrate/coating interface.

Porous structures and other defects may accelerate corrosion by allowing corrosive fluids to infiltrate through the pores, prompting various efforts to create defect-free coatings [25]. Furthermore, porous and fragile structures may reduce the performance of tribological components, as hard oxide particles can become dislodged between contacting surfaces. To address this, numerous approaches have been explored, such as adding particles or additives to the electrolytes to achieve in situ incorporation or sealing of the porous films while also imparting new functionalities to the coatings [14, 26–29]. As reviewed by Famiyeh and Huang [29], many researchers have investigated the effects of various additives—including KF,

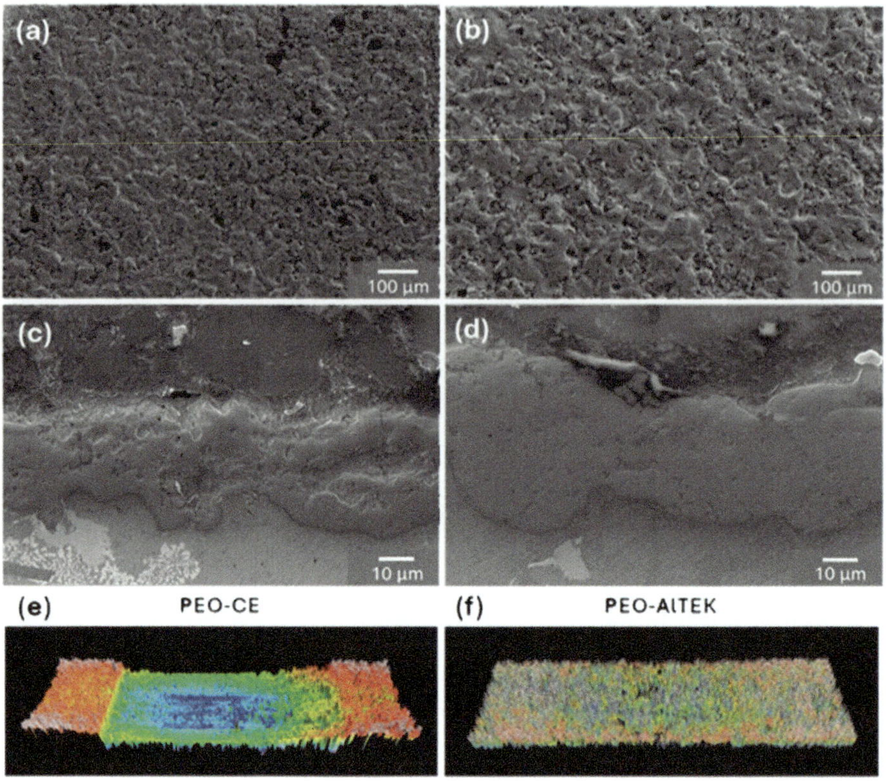

Fig. 13.5 SEM micrographs of PEO coatings developed in (**a**) silicate and (**b**) aluminate-based electrolytes. SEM cross sections of PEO coatings developed in (**c**) silicate and (**d**) aluminate-based electrolytes. Confocal images of the wear tracks extracted from PEO coatings developed in (**e**) silicate and (**f**) aluminate-based electrolytes, after tribological tests mimicking piston ring/cylinder liner real contact. Reproduced with permission from [54] © 2022 Elsevier B.V.

$NaAlO_2$, NH_4VO_3, ZrO_2, $Na_2WO_4{\cdot}2H_2O$, $(NH_4)_6Mo_7O_{24}{\cdot}4H_2O$, K_2TiF_6, and Na_3AlF_6—on the morphology, phase composition, structure, thickness, hardness, and corrosion and wear resistance of PEO coatings. Hwang et al. [56] examined the impact of adding 0.08 M Na_2WO_4 to a KOH (0.14 M)/K_2HPO_4 (0.05 M) electrolyte on PEO coatings for an Al alloy substrate. WO_3 was incorporated into the coating, enhancing adhesion strength between the substrate and coating layers. Extended treatment time resulted in black ceramic coatings with large micropores. Recently, Fernández-López et al. [57] developed an electrolyte aimed at protecting recycled Al substrates from corrosion and tribocorrosion. In this study, PEO coatings were produced using a novel aluminate-based electrolyte containing $Na_4P_2O_7$, Na_2WO_4, and K_2TiF_6. The novel electrolyte formulation included other complex chemical reagents, such as Na_2WO_4 or K_2TiF_6, as it has been previously reported that their addition to the electrolyte promotes the obtention of PEO coatings with enhanced corrosion resistance [58–60], generally related with an improved coatings' thickness and density. Also, Wang et al. [61] investigated the effect of Na_3AlF_6 on the structure and mechanical properties of PEO coatings on 6061 Al alloy. The PEO process was performed using an electrolyte of 10 g/L $Na_2SiO_3{\cdot}5H_2O$ and 1.0 g/L KOH, with a Na_3AlF_6 additive at 0.5 g/L. The presence of the additive also increased the content of γ-Al_2O_3 and α-Al_2O_3, leading to improved micro-hardness and Young's modulus of the coatings. Additionally, the hydrolysis of AlF_6^{3-} during the PEO process produced F^- ions, which were incorporated into the coatings, significantly influencing the phase composition and enhancing hardness and corrosion performance [62]. Student et al. [28] investigated the microhardness and abrasive wear resistance of cast Al-Si alloys treated with PEO in a mildly alkaline electrolyte (3 g/L KOH + 2 g/L Na_2SiO_3) and with added H_2O_2. XRD revealed the PEO layer to consist of γ-Al_2O_3, α-Al_2O_3, sillimanite ($Al_2O_3{\cdot}SiO_2$), and minor mullite ($3Al_2O_3{\cdot}2SiO_2$). The authors found that PEO-treated AK9 and AK12 silumins exhibited microhardness up to 900–1000 HV and wear resistance improvement by 14–57 times, with H_2O_2 further enhancing high-temperature oxide formation and wear resistance by 30–70%.

By introducing nanoparticles or salts containing Mo, P, Ti and Fe into the electrolyte, insoluble or poorly soluble compounds can be generated in the coatings to form a physical barrier and improve corrosion resistance [7]. Huang et al. [33] have produced an insoluble zinc phosphate layer on the surface of PEO coating to achieve super hydrophilicity and self-healing of the pores, and the coatings could withstand 11,000 h of salt spraying. Notwithstanding, the authors found out that the incorporation of the anticorrosion phases reduced the mechanical properties and wear resistance of the PEO films. On the other hand, elements such as Zr, W, and Mn have been incorporated into PEO electrolytes to form hard oxides, and therefore increase the hardness and wear resistance of the coatings [30–32]. Muhaffel et al. [12], aiming to increase the success of PEO coatings, worked on the fabrication of films on Al alloys with improved ability to withstand degradation under sliding contact conditions at high temperatures. The authors reported that the coating manufactured in a ZrO_2-added electrolyte exhibited enhanced wear resistance during dry sliding wear tests carried out at room temperature and had higher durability during the tests done at 300 °C.

Lubricating phases of MoS_2, graphene, graphite particles, and $CePO_4$ have been also added to PEO electrolytes to reduce friction coefficient and therefore provide the coating self-lubricating properties [7]. Ma et al. [34] prepared composite coatings by PEO by adding graphite particles to the electrolyte, which granted them self-lubricating properties. Also, in view of particle addition, Tran et al. [35] added to PEO electrolyte diamond powder at different amounts and fabricated films on Al 6061 with improved tribological features. From this study, it is observed that PEO coatings treated in an electrolyte enriched with diamond powder, displayed an improved hardness, corrosion, and a significantly lower coefficient of friction, and consequently, an improved wear resistance. Also, Matykina et al. [2] demonstrated that the wear of Al 6082 alloy is two times lower when covered with a PEO treatment with added α-Al_2O_3 particles, compared to electrolytic hard chrome. Furthermore, as Lou et al. [36] have found out, by incorporating Si_3N_4

nanoparticles, the coating thickness, surface roughness, hardness, and elastic modulus of the PEO film increased and also exhibited improved corrosion resistance and low coefficient of friction. Still, Ghafaripoor et al. [40] have also found a significant improvement in the corrosion performance of PEO films due to the efficient pore blocking provided by α-Al_2O_3 particles.

In conclusion, particle incorporation presents a highly effective strategy for modifying and optimizing the composition, microstructure, morphology, and functional properties of PEO coatings. The resulting coating composition depends significantly on the incorporation mode of these particles, which is largely influenced by their potential to chemically react with primary metallic oxides formed from specific substrate-electrolyte combinations. This interaction shapes the integration of particles, ultimately tailoring the coating's characteristics for enhanced performance [63].

13.2.3.3 Treatment Time

The treatment time in the PEO process plays a crucial role in determining the tribological properties of the resulting coatings. Longer treatment times typically result in thicker oxide layers, which can enhance wear resistance by providing a more durable barrier against mechanical abrasion. However, excessive treatment times may increase surface roughness and porosity due to prolonged micro-arc activity, potentially leading to higher friction coefficients and reduced sliding performance. Striking the right balance in treatment time is essential to optimize the coatings' wear resistance and frictional behavior, ensuring they meet the specific tribological demands of their application. Ayday et al. [64] investigated the effect of oxidation time on the surface free energy of PEO coatings. Their findings revealed that as the oxidation time increased, the surface free energy initially rose, reaching a peak of 37.10 mJ/cm^2 at 40 s. However, with further extension of the treatment time to 1200 s, the surface free energy decreased to 25.95 mJ/cm^2.

13.2.3.4 Electrical Parameters

The type of power supply is a critical factor in conducting plasma PEO treatments, as it directly influences the process dynamics and the properties of the resulting coatings. Different power supply modes—such as direct current (DC), alternating current (AC), and pulsed bipolar—affect key process parameters like voltage, current density, and frequency of the applied electrical field.

In DC mode, thinner coatings are typically produced due to the absence of a dense intermediate layer and the lack of the α-Al_2O_3 phase formation. The DC mode also results in a slower oxide growth rate and higher porosity, stemming from its limited control over discharge characteristics. This restricts its applicability in producing high-performance PEO coatings. AC mode, on the other hand, offers better process control by preventing electrode polarization and interrupting arcs during the discharge phase. However, the cathodic half-cycle in AC mode can partially reduce the oxide layer electrochemically, prolonging the formation of a sufficient barrier layer for discharge initiation in the early stages of the process. Discharges in the cathodic phase are generally less intense than those in the anodic phase, resulting in a lower energy input into the coating. However, these discharges can locally heat the oxide layer, potentially reducing the field strength required for breakdown in subsequent anodic cycles, thereby affecting coating uniformity and quality. Pulsed bipolar power supplies have emerged as a preferred choice for PEO treatments due to their precise control over plasma discharges. This mode reduces thermal stress and enhances coating uniformity and quality by minimizing strong plasma discharges and excessive surface roughness. Unlike DC or AC modes, bipolar pulsed power can produce thicker coatings with excellent corrosion resistance and improved tribological properties. The ability to minimize temperature spikes and control discharge dynamics during the process makes it particularly suitable for producing advanced coatings. By tailoring the power supply mode and parameters to the specific substrate material and application requirements, the PEO process can be optimized to achieve superior performance in terms of wear resistance, corrosion protection, and surface morphology [7, 29, 65–67].

Precise control over various electrical parameters is essential in the PEO process, as these parameters significantly influence the quality and functionality of the resulting coatings. Key parameters requiring careful regulation include anodic and cathodic potentials, applied current density, frequency, duty cycle, pulse width, process time, and temperature. Among these, the anodic/cathodic current densities and voltages, cathodic-anodic current ratio, frequency, and pulse width are particularly critical for achieving desired coating characteristics.

The composition, microstructure, morphology, and mechanical and tribological properties of PEO coatings are directly affected by these electrical parameters. For instance, the duty cycle—a parameter that determines the proportion of time the current is applied within one cycle—plays a vital role in coating development. Akbar et al. [68] demonstrated that an increase in the duty cycle results in a decrease in coating thickness, likely due to reduced off-times during the PEO process, which may limit oxide growth and affect the plasma discharge dynamics.

The current density applied during the PEO process is also a critical parameter that directly determines the growth rate of the coating and is largely influenced by the nature of the treated alloy. For Al alloys with low alloying element content, optimal current densities typically range between 5 and 15 $A{\cdot}dm^2$. In contrast, Al alloys with higher alloying content, particularly cast Al-Si alloys, require higher current densities, around 25 $A{\cdot}dm^2$ [4, 54, 57]. Applying current densities that are too low results in a slow coating growth rate, reducing the energy efficiency of the process. Conversely, excessive current densities can degrade coating quality, leading to lower mechanical properties, increased surface roughness, excessive gaseous by-products, and even burning of the treated sample [1, 69]. Frequency is another important parameter, as higher frequencies generally produce coatings with reduced surface roughness. This is achieved by shortening plasma discharge lifetimes, which minimizes the occurrence of large, destructive discharges [69]. Additionally, the process time has a proportional effect on the coating's roughness. Longer treatment durations increase the arc forces acting on the surface, promoting stronger discharges and resulting in more uneven surfaces [69].

Understanding and optimizing these parameters are crucial for tailoring PEO coatings to meet specific application requirements, ensuring the best combination of wear resistance, corrosion protection, and mechanical performance.

13.2.3.5 Temperature

The temperature of the electrolyte in the PEO process significantly influences the properties of the coatings. At very low temperatures, the oxidation potential of the substrate is inadequate, resulting in reduced coating thickness and hardness. Conversely, at high temperatures, the dissolution of the oxide layer occurs, leading to a substantial decrease in both thickness and hardness. While the effect of electrolyte temperature on coating properties is less pronounced than in conventional anodization, maintaining temperature control remains essential. Studies typically regulate electrolyte temperatures within the range of 20–40 °C [29].

Cheng et al. [70] investigated the PEO of pure Al in a dilute aluminate electrolyte, examining effects of controlled electrolyte temperature ranges—low (10–30 °C), medium (40–50 °C), and high (70–80 °C)—on coating morphology, microstructure, composition, and phase. SEM, EDS, and XRD analyses indicated that higher temperatures increased coating growth rate and surface roughness but reduced α-Al_2O_3 content. Electrochemical tests showed the highest corrosion resistance at low temperatures and the lowest at high temperatures due to enhanced PEO reaction kinetics at elevated temperatures. Thus, optimal temperature control is essential for achieving balanced coating properties.

13.2.4 PEO in Combination with Other Technologies

Combining PEO with other surface modification technologies can significantly enhance the performance and durability of the coatings, offering a broader range of functional benefits. While PEO is highly effective in improving wear resistance, corrosion protection, and surface hardness, its performance in tribological applications can be further optimized when integrated with other advanced surface technologies. For example, Shin et al. [71] investigated the use of silicone oil impregnation as a posttreatment for PEO of Al alloys to enhance corrosion and wear resistance. The authors formed a monolayer of polydimethylsiloxane (PDMS) on the porous oxide layer, creating a water-repellent, slippery surface with a low contact angle hysteresis of <5°. This water repellence improved corrosion resistance by over four orders of magnitude compared to PEO-treated Al alloys without silicone oil impregnation. Additionally, the silicone oil within the porous oxide layer provided lubrication, reducing friction coefficients from approximately 0.6–0.1 and enhancing wear resistance. Importantly, the PDMS monolayer can be restored through frictional heat, enabling the water-repellent surface to endure physical damage to the oxide layer. These findings offer a novel approach for posttreatment of PEO coatings on Al alloys.

Also, Fernández-López et al. [57] developed PEO coatings using a novel aluminate-based electrolyte containing $Na_4P_2O_7$, Na_2WO_4, and K_2TiF_6, which exhibited enhanced protective performance when combined with two posttreatment processes: polishing of the outermost coating layer and sealing with a sol-gel layer. The sol-gel treatment significantly improved the corrosion resistance and dielectric properties of the coatings by effectively sealing surface pores, while polishing reduced surface roughness, thereby enhancing tribocorrosion performance. These posttreatments highlight the potential of PEO coatings in applications where components are exposed to highly corrosive environments, offering improved durability and wear resistance. This combination of PEO with advanced posttreatments demonstrates a promising strategy for enhancing the functional properties of coatings in demanding industrial applications.

Several researchers have pioneered the use of PEO treatments on deposited Al coatings, unlocking potential for new and enhanced applications across various fields. For example, Rao et al. [72] investigated the PEO process to grow an alumina coating on cold-sprayed 7075 Al alloy, enhancing its surface properties. They analyzed phase composition, microstructure, mechanical properties, corrosion, and wear resistance of the alloy, both pre- and post-PEO treatment. The study found that PEO treatment using an aluminate electrolyte produces a dense, single-layer alumina coating, with a thickness of 15–20 μm and hardness of 1353 HV0.01. The alumina-coated alloy exhibited significant improvements in corrosion and wear resistance, with corrosion current density and wear rate of 6.27×10^{-7} A/cm^2 and 1.22×10^{-6} mm^3/N·m, respectively—three and two orders of magnitude lower than the untreated rolled 7075 Al alloy (T6) substrate, indicating superior protective performance of the PEO-treated surface. These results highlight the superior protective performance of the PEO-treated surface, making it an ideal solution for demanding applications where enhanced durability and resistance to corrosion and wear are essential.

Martin et al. [73] explored the limitations of PEO in enhancing corrosion and wear resistance on lightweight metals, noting that while PEO successfully forms ceramic oxide coatings on Al alloys, it proves less effective for ferrous metals and generally yields thin, porous oxide coatings on Mg alloys. To address these challenges, the study assessed a duplex treatment that combines cold spray (CS) deposition with PEO. Cold-sprayed Al coatings were applied to Mg (EV31) and steel (S235) substrates prior to PEO treatment. This duplex approach demonstrated that thick, dense, and crystalline alumina coatings could be achieved on both Mg and steel substrates. The study found that the growth kinetics of the oxide layer in this CS-PEO duplex process improved threefold compared to conventional PEO on bulk Al. Another interesting study on combining cold-spray with PEO technology has been published by Popov et al. [74]. The researchers aimed to develop a novel combined technology for restoring liners and cylinder blocks independent of the material used in their manufacture. The method centers on using cold gas dynamic spraying with Al powder to restore the geometry of the liner surface. Following this, PEO is applied to create a ceramic coating with a thickness of 80–100 microns, which is suitable for prolonged operation in internal combustion engines. This technology was tested on a GZ-195N liner, where PEO treatment confirmed a hardened coating depth of 100 microns as required. The resulting coating demonstrated superior wear resistance, with a wear factor 1.27 times lower than that of the reference sample. The flexibility of cold spray Al coating on various metals makes this combined technology viable for cylinder liner repair, enhancing both performance and wear resistance of cylinder blocks and liners. The promising results indicate the need for further industrial development and testing in an operational engine environment.

Similarly to CS-PEO duplex system, López-Ortega et al. [75] studied the enhancement of thermally sprayed aluminum (TSA) coatings, widely used for corrosion protection of steel in marine environments but prone to wear damage during transport due to Al's low wear resistance. To address this, the research applied PEO to create a TSA/PEO duplex system with improved durability. The performance of this duplex system was evaluated through sliding wear, electrochemical corrosion, and tribocorrosion tests in synthetic seawater. Characterization techniques, including SEM, confocal microscopy, and X-ray diffractometry, provided insights into the microstructure and composition of the coatings, highlighting the PEO treatment's effectiveness in reinforcing the TSA layer against wear and corrosion.

These studies collectively illustrate the significant advantages of combining PEO with other surface modification techniques, particularly cold spray, to enhance the mechanical and tribological properties of coatings. The duplex treatments offer a versatile solution to improve wear resistance, corrosion protection, and overall durability across a wide range of materials and applications.

13.3 PEO Technology for the Development of Anti-wear Coatings on Magnesium Alloys

Valued for their high strength-to-weight ratio, high specific stiffness, dimensional stability, weldability, and castability, Mg and its alloys are also the lightest engineering metals [76, 77], whose density is ¼ that of Fe and 2/3 that of Al [78]. Mg and its alloys also exhibits excellent thermal conductivity, dimensional stability and machinability, along with effective electromagnetic shielding, superior damping properties, and ease of recyclability [78], making them an attractive choice for aerospace, automotive, consumer electronics, and communication fields [79]. Moreover, due to its excellent biodegradability, Mg is considered a relevant biomaterial [80]. Nevertheless, their broad use in several applications, especially within the automotive and aerospace industries, has been hindered by their poor wear and corrosion resistance [76, 77]. Mg is chemically active (−2.36 V vs. SHE) and easily corrodes in humid terrestrial and marine environments [81, 82]. Furthermore, it has a low hardness and poor tribological properties [83, 84]. To address these inherent issues of Mg alloys, several surface technologies have been implemented, including PVD, PE-CVD, electroplating, sol-gel, or plasma spraying, among others [76, 85].

13.3.1 From Challenges to Solutions: PEO Coatings on Magnesium Substrates for Tribological Applications

PEO technology is currently gaining attention due to its ability to form coatings on Mg alloys with high hardness and high adhesion to the substrate, as well as to significantly enhance the corrosion resistance of these alloys. The current number of publications on the improvement of the tribological properties of PEO coatings on Mg alloys is scarce [86]. Nevertheless, it is well known that PEO treatment can involve a significant improvement of these properties in order to pave the way to applications that seek for a weight reduction in which their current use is limited by their high wear rate [79]. PEO coatings on Mg alloys display inferior mechanical properties compared to those developed on Al alloys, since the primary oxide phase formed on Mg substrates (periclase, i.e., cubic MgO) has a lower hardness than the Al_2O_3 phases formed on Al alloys [7]. Furthermore, PEO coatings developed on Mg alloys tend to have a porous structure, with voids and micro-cracks, constituting a limitation for their use in applications with more demanding requirements. Nonetheless, some works have demonstrated the potential of PEO coatings on Mg alloys. This is the case of Zhang et al. [87], who observed an improved wear behavior of a PEO-treated AJ62 alloy employed in engine cylinders, even better than that of a hypereutectic Al-Si alloy commonly used in this same application.

Therefore, the potential of PEO treatment for improving the tribological performance of Mg alloys seems to be a promising route to be explored. Thus, some important aspects that can affect the coating growth as well as its final properties are discussed in this chapter, i.e., the process parameters, pretreatments, or incorporation of particles.

13.3.2 Process Parameters Determining the Tribological Properties of the PEO Coatings Developed on Magnesium Alloys

The tribological properties of PEO-coated Mg substrates are determined by several process parameters, being the most influential the composition and characteristics of the Mg substrate, the application of pretreatments, the chemical composition of the electrolyte used, as well as the electrical parameters applied during the coating process, including the processing time, or particle addition into the coating [86]. In this section, the different trends to improve the tribological properties of Mg-PEO coatings are reviewed.

13.3.2.1 Properties of Magnesium Substrate

The properties of Mg alloys strongly depend on their chemical composition. Certain elements such as Mn, Al, Zn, Si, Cu, Zr, and rare-earth metals affect the final microstructure, mechanical, and corrosion properties of Mg alloys [88]. For instance, Al improves the tensile strength of Mg, as well as the ductility and the castability [89], whereas Zn is added to enhance tensile properties by solid solution [90], but it can decrease their weldability [91]. On the other hand, Mn improves its corrosion resistance [92], Ca enhances the mechanical properties and the corrosion behavior [93], Y improves elevated temperature plasticity and the creep resistance [94], and rare earth elements can refine grain size to enhance creep and corrosion resistance at high temperatures through solid solution or precipitation hardening [95].

Based on the process of operation, Mg alloys can be classified as cast and wrought alloys (forging, extrusion, or rolling operations to shaping). Typical casting Mg alloys are those containing Al, Zn, and Mn (such as AM50, AZ91, and AM60). Other Al-free alloys containing Zr, RE, Y, and Zr are the WE43 and WE54 [96, 97]. The main elements of wrought alloys are Al, Mn, and Zn [98]. The properties of both cast and wrought alloys can be improved by strengthening mechanisms, such as alloying, precipitation, or grain refinement, among others [99]. The formation of certain intermetallic phases with alloying elements lead to improved properties, even at high temperature [100–103]. The volume fraction and morphology of these particles are directly affected by the composition of the alloy and the cooling rate during casting process [104].

In general, PEO coatings developed on Mg alloys exhibit higher hardness compared to those formed on pure Mg [105, 106], as well as improved wear and corrosion resistance [107–112]. One of the main routes to improve the properties of PEO coatings developed on Mg alloys has been closely related to the substrate composition. The phase composition and surface morphology of the PEO coatings (Fig. 13.6), and with that their final properties, are strongly influenced by the substrate composition [106, 113]. The addition of alloying elements such as Al, Zr, Mn, Zn, or Ca has a positive effect on the corrosion and mechanical properties of the resulting coating [80]. For instance, Al or Zn can increase the average PEO growth rate. In Mg-Al-Zn alloys, Zn plays a catalytic role that improves the oxidation rate of Al and Mg during the PEO process [86, 105]. Khaselev et al. [114] observed that the addition of Al and Zn to an Mg-Al alloy reduced the thickness of the coating, and bigger discharge channels with rougher surfaces were formed.

Krishna et al. [115] found that the inner barrier layer in the substrate/coating interface of ternary alloys is more dense and crack-free than that of binary alloys or pure Mg, which present a higher density of defects, and that Al promotes the formation of the harder $MgAl_2O_4$ phase within the coating. In particular, it was found that Al contents between 6 wt.% and 9 wt.% led to a more pronounced increase in the

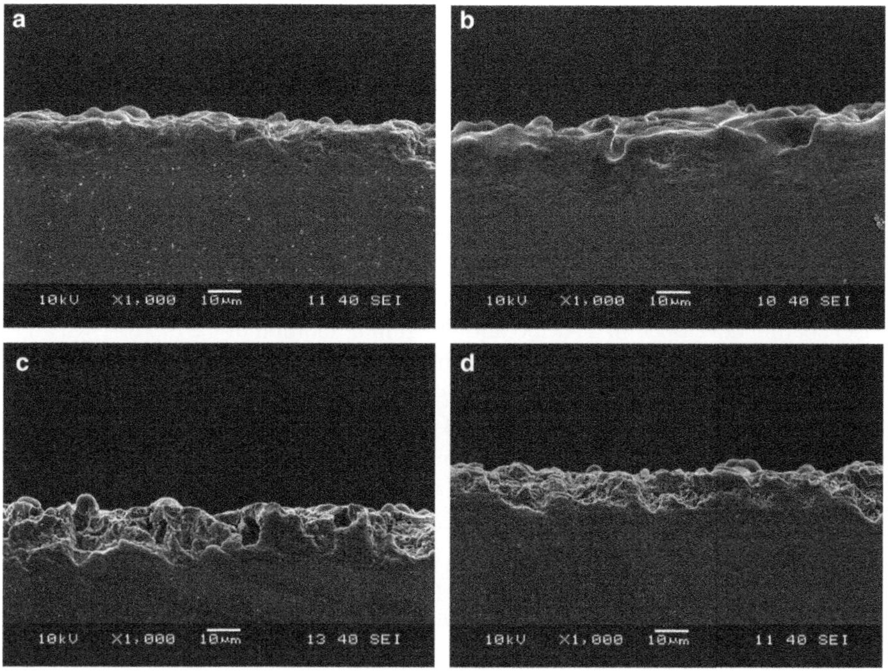

Fig. 13.6 Cross-sections of PEO coatings grown on (**a**) AZ91D, (**b**) AZ31B, (**c**) AM60B, and (**d**) AM50B Mg alloys [106]. Reproduced with permission from [106] © 2009 Elsevier B.V.

Table 13.2 Micro-hardness of the PEO coatings developed on Mg alloys with different Al concentrations [115]

Substrate	Al wt.%	PEO coating microhardness $HV_{0.015}$
Pure Mg	–	320
Mg-3Al	2.97	380
Mg-6Al	6.02	580
Mg-9Al	8.97	775
AZ31	2.95	370
AZ63	5.96	560
AZ91	9.04	750

hardness of the PEO coating (Table 13.2). Gunduz et al. [116] observed a decrease in the coatings' thickness with the addition of Al or Zn to the Mg alloy. The addition of Al into Mg resulted in a wavy substrate/coating interface that retarded the growth rate and increased the coating microhardness and presented a coarse surface with bigger discharge channels. The growth retardation was lower for the addition of Zn, which presented smaller discharge channels and negligible effect on the microhardness. Finally, Al was found to be incorporated both in the outer and inner regions of the coatings, while Zn was only detected in the inner region as bright precipitates.

Generally, coatings generated on Mg-Al-Zn alloys are denser, smoother, and present higher adhesion properties than those formed on Mg-Al-Mn (Fig. 13.6). Cakmak et al. [106] attributed this to a higher proportion of Periclase (MgO) and the formation of Spinel ($MgAl_2O_4$). They also observed a close dependency on the alloy composition to the hardness and adhesion of the coatings, which was higher for that formed on Mg-Al-Zn alloys, compared to those on Mg-Al-Mn ones. On the other hand, the addition of rare earth metals (RE) has also been found to improve the resistance of Mg alloys [117]. Muhaffel et al. [118] observed an improvement of the PEO coating developed in a AM50 alloy by minor addition of Ce (up to 1 wt.%) to the bare alloy's composition, obtaining more uniform layers with improved corrosion properties. Similarly, Song et al. [119] modified a Mg-Al-Zn (AZ91) alloy with neodymium and saw an improvement of the resulting PEO coating in terms of a denser resulting barrier layer that led to higher corrosion resistance than that of the untreated alloy. Tekin et al. [117] studied the coatings developed on RE-containing E21 (Elektron21) and WE43 Mg alloys compared to a AZ31B Mg alloy. The PEO coatings obtained for the RE-containing alloys were thicker, mainly composed of periclase (MgO) and forsterite (Mg_2SiO_4) crystalline phases, and presented a two-layered structure composed of a dense barrier layer and a porous outer top layer. This led to an improved corrosion resistance compared to the bare AZ31B Mg alloy. Similarly, Arrabal et al. [120] compared the coatings obtained on an AZ91D Mg alloy with a WE43-T6 Mg alloy containing Nd, Zr, and other RE and observed that the coatings were thicker and denser on the RE-containing alloys.

Since the properties of the coating seem to be in coherence with those of the substrate alloys, the addition of even minor amounts of alloying elements to the substrate composition can change the formation process of the coating and its final properties.

13.3.2.2 Pretreatments

Apart from the substrate chemical composition, the surface morphology of the Mg alloy (i.e., roughness or grain size) plays a major role in the formation of PEO coatings. Yoo et al. [121] evaluated the influence of the surface roughness of an AZ91Mg alloy substrate on the final properties of the resulting PEO coating. For that, they polished the substrate with silicon carbide papers of different grit number obtaining several roughness grades (Ra = 2.5 μm, Ra = 1.0 μm, and Ra = 0.5 μm), and they found that the size and density of pores generated during the PEO process decreased with decreasing surface roughness (Fig. 13.7). Similar observations were made by Sovik et al. [122]. They evaluated the effects of mechanical pretreatment by grounding an AZ80 Mg alloy with SiC emery papers of different grain sizes on the resulting PEO coating and observed an improvement in the corrosion resistance of the coating developed on substrate with lower roughness. Finally, Li et al. [123] found

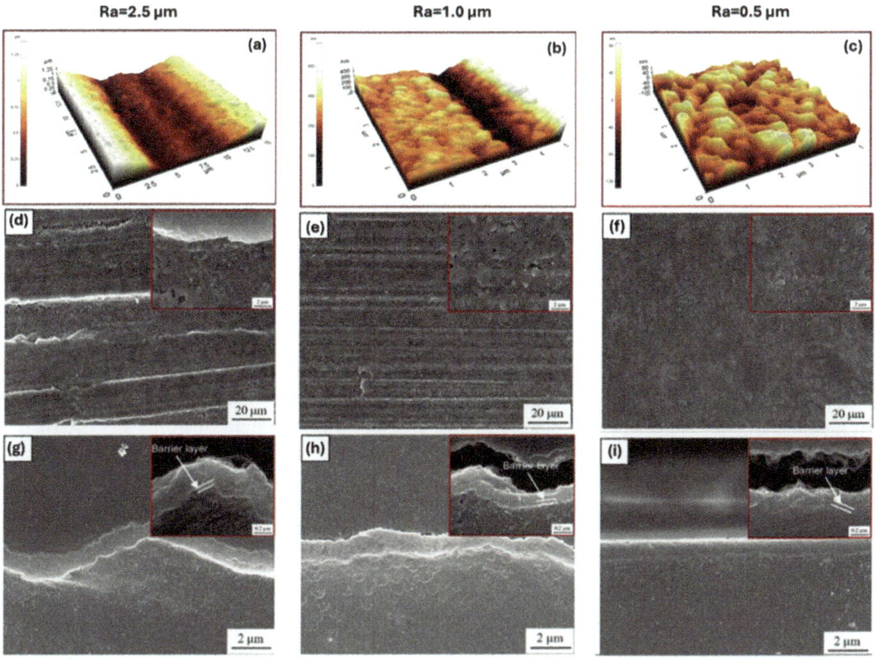

Fig. 13.7 AFM images of the Mg substrates with different roughness values: (**a**) $R_a = 2.5$ μm, (**b**) $R_a = 1.0$ μm, and (**c**) $R_a = 0.5$ μm; Surface morphologies and cross-sectional SEM images of the PEO coatings grown on the Mg substrates with (**d**, **g**) $R_a = 2.5$ μm, (**e**, **h**) $R_a = 1.0$ μm, and (**f**, **i**) $R_a = 0.5$ μm [121]. Reproduced with permission from [121] © 2010 Elsevier B.V.

that the porosity and bonding strength of the coating generated in a ZM5 Mg alloy was highly affected by the substrate roughness. The surface roughness of the resulting coating was found to depend on that of the substrate, being lower for the lower Ra values of the substrates. Furthermore, no effect of the substrate roughness was found on the final thickness and chemical composition of the coating, while the porosity and bonding strength increased with increasing substrate surface roughness. The highest porosity of the coatings led to a lower corrosion resistance at the beginning of corrosion tests, but the higher bonding strength of the coating to the substrate grooves showed an improved resistance at longer immersion times, inhibiting the extension of corrosion and slowing down the degradation of the substrate.

All in all, most of the research on pretreatments' influence on PEO coatings has focused on the improvement of corrosive resistance, and little research can be found regarding their effect on the tribological performance of the coatings. Even if thicker and denser coatings have directly been related to improved tribological performance of PEO coatings [14, 124, 125], further investigations are needed to confirm the influence of the different pretreatments on the tribological properties of Mg-PEO coatings.

13.3.2.3 Chemical Composition of the Electrolyte

The chemical composition of the electrolyte used in the PEO process determines the composition, morphology, and microstructure of the PEO coating, which inherently affects its tribological behavior. Depending on the electrolyte, amorphous and crystalline phases such as MgO, Mg_2SiO_4, $Mg(PO_4)$, or $MgAl_2O_4$ can be obtained [126]. When Mg is exposed to aqueous solutions, a double layer composed of $Mg(OH)_2$ and MgO is formed. The former is close to the substrate and the latter is on top in direct contact with the solution [113].

Regarding electrolyte composition, silicate- and phosphate-based electrolytes have been most commonly used in the development of PEO coatings on Mg alloys [127–131]. Several investigations have shown that the tribological performance of the coatings developed in phosphate-based electrolytes is inferior to those obtained in silicate-based ones at higher loads (Fig. 13.8) [132]. Durdu et al. [133] developed PEO coatings on pure commercial-grade Mg in sodium silicate- and sodium phosphate-containing electrolytes. The predominant phases formed in the silicate-containing coatings were forsterite (Mg_2SiO_4) and periclase (MgO), whereas those in the phosphate-containing electrolyte were farringtonite ($Mg_3(PO_4)_2$) and periclase (MgO). The coating developed in sodium silicate electrolyte presented higher adhesion and wear resistance than those produced in sodium phosphate ones. This was attributed to the presence of Mg_2SiO_4, which possesses higher hardness than MgO and $Mg_3(PO_4)_2$. Similar observations were made by Liang et al. [79], who also used silicate- and phosphate-based electrolytes with an AM60B Mg alloy. The layer developed in the phosphate electrolyte was less compact, presenting higher porosity and less hardness, which resulted in larger wear rates.

On the other hand, some studies have demonstrated that aluminate-based electrolytes (containing $NaAlO_2$) provide coatings with enhanced tribological properties [134]. The resulting coatings were thicker and presented higher surface roughness, as well as higher hardness that led to improved wear resistance and lower volume losses. This improvement would be mainly attributed to the higher Al content, both in the substrate or in the electrolyte, which increases the concentration of $MgAl_2O_4$ in the coating to the detriment of the predominant MgO phase. $MgAl_2O_4$ not only has a higher hardness than MgO ($HVMgAl_2O_4$ = 1500–1800, HVMgO = 900) but also improves the wear resistance of the coating [134, 135]. Similarly, Zhang et al. [87] employed aluminate-based electrolytes for the growth of PEO coatings on the engine-used AJ62 (MgAl6Mn0.34Sr2) alloy, using different $NaAlO_2$ concentrations (10 and 20 g/L). The results obtained revealed that a higher $NaAlO_2$ concentration promoted PEO coatings with higher thickness and roughness, as well as an improved tribological performance, even surpassing those of a hypereutectic Al-Si alloy used for engine applications under high load contacts (Table 13.3). Da Silva Rodrigues et al. [136] evaluated the corrosion and tribological behavior of PEO coatings formed on ZK30 Mg alloy in an electrolyte composed of sodium metasilicate, potassium hydroxide, and sodium borate. The resulting coating composed of Si- and Mg-containing oxides exhibited an improved wear resistance, showing lower coefficients of friction and smaller material loss.

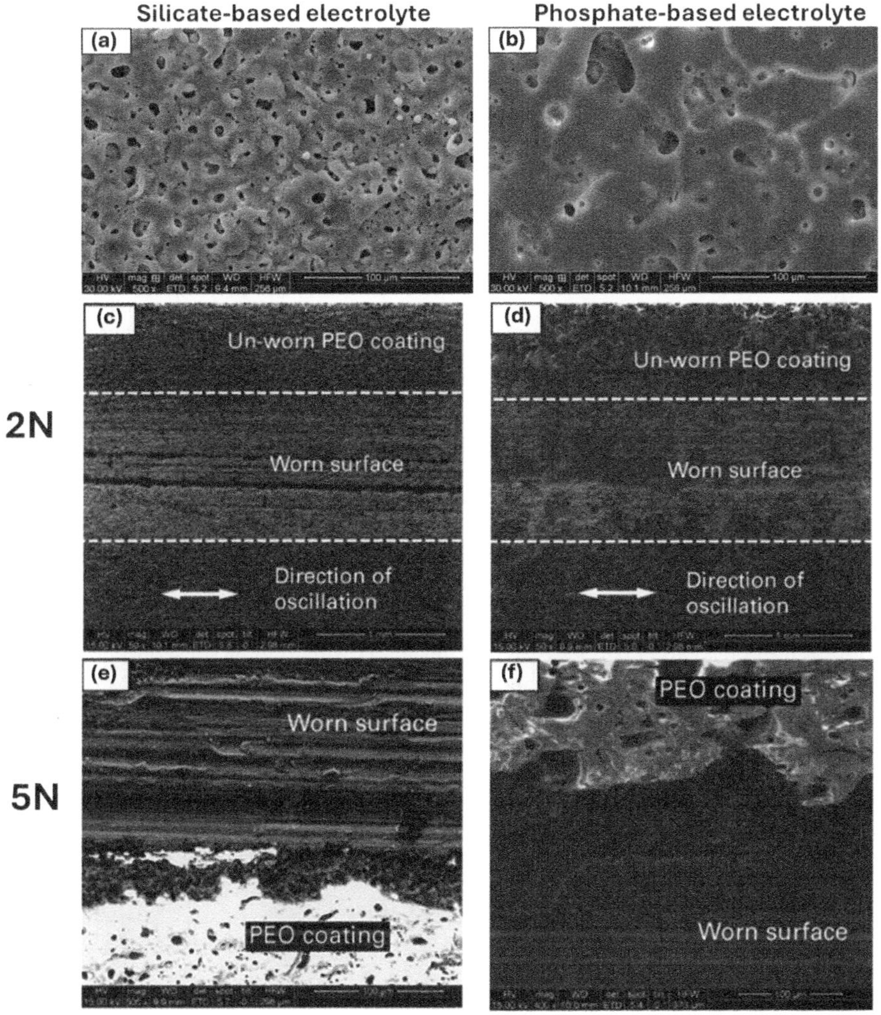

Fig. 13.8 Cross-sectional SEM images of the as-coated PEO layers developed on the Mg alloy using (**a**) a silicate-based and (**b**) a phosphate-based electrolyte; wear tracks of the silicate-based coating after tribological tests under loads of (**c**) 2N and (**e**) 5N and of the phosphate-based PEO coating under loads of (**d**) 2N and (**f**) 5N. The wear track depths after the tribological tests are also shown. Reproduced with permission from [132] © 2011 Elsevier B.V.

Finally, incorporation of different constituents to the electrolytes have also been found to be beneficial in improving the tribological performance of PEO coatings. For instance, introduction of sodium tungstate (Na_2WO_4) into silicate and KOH-containing electrolytes resulted in higher microhardness and lower wear rates [137]. This was attributed to the higher content of hard Mg_2SiO_4, which formation was promoted by the addition of Na_2WO_4. Duan et al. [138] investigated the effect of

Table 13.3 PEO process parameters (all the coatings were developed under the same current density) and wear rates for each sample after the tribological tests [87]

Sample	Electrolyte concentration NaAlO2 (g/L)	Process duration (min)	COF		Wear rate (mm^3/Nm)	
			2 N	5 N	2 N	5 N
PEO S1	10	10	0.15	0.22	6.60E-07	3.32E-06
PEO S2	10	20	0.21	0.24	3.39E-06	5.81E-06
PEO S3	20	10	0.16	0.17	1.38E-06	1.43E-06
PEO S4	20	20	0.19	0.19	4.71E-06	6.33E-06
Alusil	/	/	0.13	0.26	1.47E-07	1.51E-05
AJ62	/	/	0.50	0.51	6.79E-04	8.14E-04

different electrolyte additives such as phosphate, fluoride, and borate in the PEO coating formed on an AZ91D alloy, and found a positive influence of borate on the coating growth and of fluoride and phosphate on the formation of denser inner layers. The layers were mainly composed of Mg_2SiO_4, MgO, $MgAl_2O_4$, and a few α-Al_2O_3, regardless of the electrolyte constituents. Other investigations revealed that addition of borax ($Na_2B_4O_7$) to sodium silicate electrolytes resulted in a better performance in terms of growth of the coating [139], higher compactness, as well as in the reduction of the surface roughness, increased hardness, and significantly reduction of the wear rate of the developed coatings [140]. Furthermore, the incorporation of borax to phosphate electrolytes also resulted in increased microhardness and a lower coefficient of friction favored by the existence of the P element in the coating, improving the tribological performance of the Mg alloy [141]. Lastly, the utilization of organic additives such as glycerol, ethylene glycerol (EG), or polyethylene glycols (PEG) to silicate-based electrolytes has also been investigated. In this case, the incorporation of glycerol reduced the surface roughness and micropore size and number [142], while the addition of EG or PEG ductile segments into the coatings can be a promising route for antifriction coating development [143], though this statement must be confirmed by further studies.

13.3.2.4 Electrical Parameters Applied

PEO processes can be performed under direct (DC), alternating (AC), or pulsed current regimes [130, 144]. While DC and AC are the most well stablished electrical regimes, in the last years the application of bipolar pulses (BP) has demonstrated a wide range of benefits in the properties of PEO coatings [145, 146]. Jin et al. [147] evaluated the composition, structure, and mechanical properties of PEO coatings grown on an AZ91D Mg alloy applying DC and BP current modes. In both cases, an average current density of 3 A/dm^2 and a treatment time of 20 min were used, while the BP processes were performed under a frequency of 800 Hz with four stages. Both PEO coatings were mainly composed of MgO, yielding improved friction coefficients, higher micro-hardness, and lower wear rates compared with the untreated Mg alloy, although the enhancement effects were more substantial for the

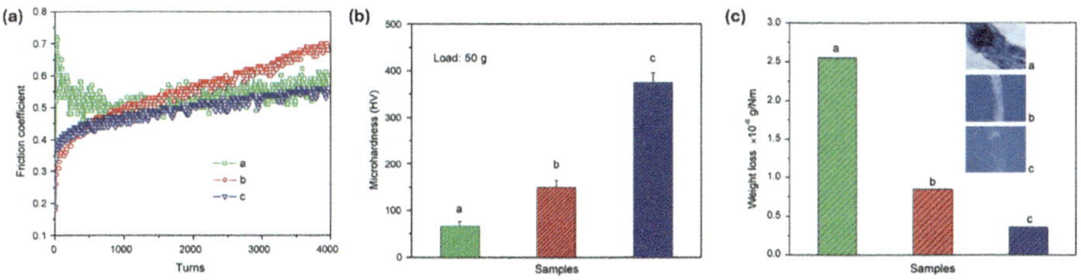

Fig. 13.9 (**a**) Evolution of the friction coefficients with the sliding turns during the wear tests, (**b**) microhardness results measured using a load of 50 g and (**c**) weight loss and wear tracks after the tribological tests for the untreated Mg alloy (green), DC PEO coating (red) and BP PEO coating (blue). Reproduced with permission from [147] © 2006 Elsevier B.V.

BP mode (Fig. 13.9). This was attributed to the higher frequency current pulses that enabled the creation of shorter and more energetic microdischarges.

Niedźwiedź et al. [148] investigated the influence of process parameters, i.e., peak current density, process time, and pulse frequency, on the surface wettability and tribological properties of the coatings produced on a AZ31B Mg alloy. They employed a trapezoidal voltage waveform (voltage limit of 400 V to the positive current cycles); peak current densities of 10, 15, and 20 A/dm^2; pulse frequencies of 50, 75, and 100 Hz; and process times of 30, 45, and 60 min. The electrolyte used was a two-component alkaline electrolyte consisting of 5 g/L of sodium metasilicate (Na_2SiO_3) and 5 g/L of sodium hydroxide (NaOH). They observed no significant influence of the process parameters on the resulting coatings, whereas the surface wettability was affected by the oxidation parameters, with higher contact angles for average values of production parameters. Since wettability influences the ability to create sliding films reducing resistance to movement, the highest contact angles led to lowest coefficient of friction. They also observed low friction coefficients for the minimum peak current density and process time values at maximum frequency, and for maximum peak current density and process time values at minimum frequencies.

13.3.3 Incorporation of Particles into the PEO Coating

One of the current trends to enhance the properties of the PEO coatings developed on Mg alloys, also reducing their high intrinsic porosity, is the incorporation of particles or nanoparticles into the coating that contribute to the sealing of these pores, leading to a more compact structure that will provide enhanced tribological features (Fig. 13.10) [63, 85, 149]. These particles will be firstly added to the electrolyte and, during the process, will be in situ incorporated into the coating formed [85]. The in situ incorporation of particles into the PEO coatings will also modify their thickness and phase composition [63].

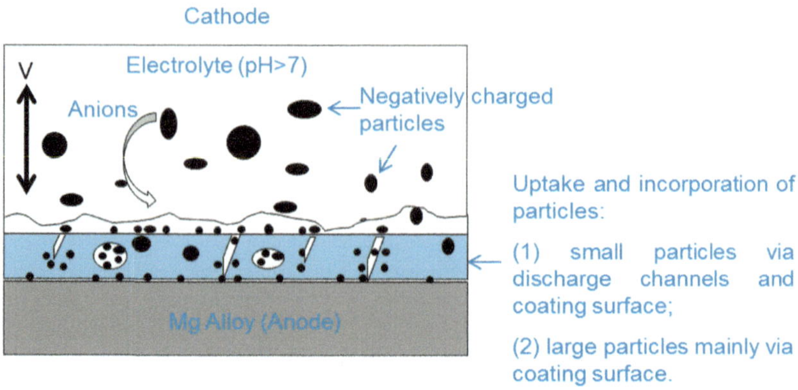

Fig. 13.10 Schematic representation of the incorporation mechanism of particles into a PEO coating on a Mg alloy. Reproduced with permission from [149] © 2015 Elsevier Ltd.

The particles present in the electrolyte can be incorporated into the PEO coating in an inert, partly reactive, or reactive form [63, 149]. In cases of inert incorporation, the shape and size of the particles are barely modified since no reaction or changes or formation of phases occurs. In contrast, during reactive or partly reactive incorporation, the particles are melted by the plasma discharges during the process, reacting with elements of the substrate or the electrolyte and forming new phases in the PEO coating [85, 149, 150]. The type of incorporation of the particles is mainly determined by their melting temperature. In this way, particles with a high melting temperature (mainly ceramic particles) are generally incorporated into the coating in an inert form, regardless of the particle size. In contrast, particles with lower melting temperatures are generally reactively or partially reactively incorporated, as the energy of the plasma micro-discharges is enough to melt them [85, 149].

Since the incorporation of particles into the electrolyte will modify its physicochemical properties, also altering the properties of the coatings obtained, it is necessary to control the physicochemical characteristics of both the particles and the electrolyte, such as pH, conductivity, particle concentration and size, or zeta potential, among others, in order to control the properties of the obtained coatings [85, 108, 151].

13.3.3.1 Oxide Particles

Cerium Dioxide (CeO_2)

It is well known that CeO_2 can reduce the porosity of oxide coatings [152], having a beneficial effect on the microstructural stability and oxidation resistance of oxide coatings [153] and direct improvement of the corrosion and wear resistance [154].

Lim et al. [155] developed CeO_2-containing PEO coatings on an AZ31 Mg through the incorporation of CeO_2 particles to a Na_2-SiO_3-based electrolyte. Based on a previous unsuccessful work in which the development of PEO coatings using cerium nitrate ($Ce(NO_3)_3$) electrolytes [156], as well as others that effectively incorporated CeO_2 to the PEO coating after a conversion coating pretreatment [157], they demonstrated the suitability of CeO_2 incorporation in an one-step direct process. The CeO_2 particles were distributed throughout the outer porous layer, where some of them were located in pores and cracks, minimizing the defects of the coating. Similar results were observed by other researchers [158–160]. Furthermore, Mohedano et al. [159] observed a direct relation between the coating performance with the concentration of CeO_2 in the coatings developed on AM50 alloy. Lower amounts of particles lead to a partial blocking of the pores and cracks, whereas higher concentrations increased the defect density, worsening the coating performance.

Silicon Dioxide (SiO_2)

Another oxide particle that has shown promising results in the improvement of the tribological performance of Mg-PEO coatings is SiO_2. The incorporation of inertly embedded SiO_2 NPs into the PEO layer provides an increase of hardness and promotes the formation of thick oxide layers [151, 161], that results in lower wear and the toleration of higher load cycles [149, 162]. Furthermore, the size of the SiO_2 NPs influences the performance of the coating. Krishtal et al. [163] observed a considerable improvement on the wear resistance with the incorporation 25 nm, while no significant difference was observed for particles of 35 nm. Finally, the addition of NPs of 50 nm showed the worst wear resistance. This was attributed to the composition of the coating showing the presence of magnesium phosphate $Mg_3(PO_4)_2$ with the addition of 25 nm SiO_2 NPs, and lower presence of forsterite (Mg_2SiO_4) for the larger NPs. Lu et al. [149] evaluated the effect of the in situ incorporation of nano- and microparticles of SiO_2 into a PEO coating growth on a cast AM50 Mg alloy. It was corroborated that the incorporation of SiO_2 particles into the PEO coating modifies both the microstructure and the properties of the coating for both nano- and microparticles. Moreover, it was observed that while the nanoparticles were incorporated by penetrating through the open channels and pores of the coating and were sintered (reactive incorporation), the microparticles were absorbed on the surface of the coating (inert incorporation). Regarding the tribological performance, it was observed that the PEO coating without particles (PPEO) did not pass the sliding wear test, showing high fluctuations in the friction coefficient (Fig. 13.11a) together with a depth wear profile (Fig. 13.11b), while the coatings with nano- and micro-SiO_2 particles did not fail during the tribotests, thus improving the tribological response of the Mg substrate.

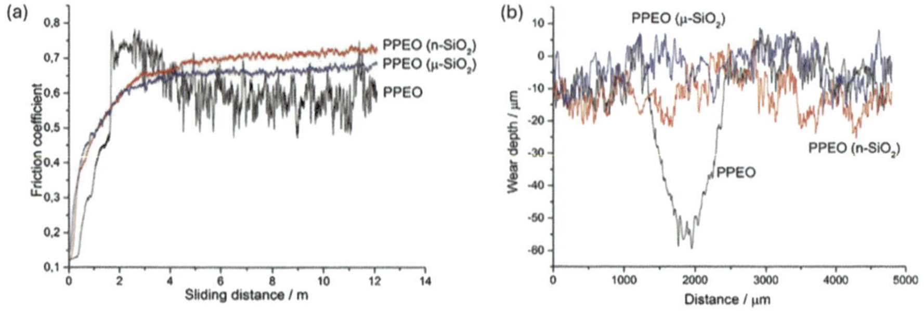

Fig. 13.11 (**a**) Evolution of the friction coefficient during the sliding wear tests of the PEO coatings applying a load of 5 N and (**b**) wear depth profile of the PEO coatings after the tribological tests. Reproduced with permission from [149] © 2015 Elsevier Ltd.

Zirconium Dioxide (ZrO_2)

The addition of ZrO_2 also promotes the formation of more compact and harder coatings, which results in the improvement of the tribological performance [112, 164]. In their review, Lu et al. [63] highlighted the improvement of the abrasive wear resistance of PEO coatings due to the high hardness and toughness of ZrO_2, making these coatings very promising Mg implants of high durability. Furthermore, the combination of different NPs such as ZrO_2 and SiO_2 has also shown a further improvement on the wear resistance of PEO coatings, due to the high hardness of ZrO_2/SiO_2 NPs [165]. Gnedenkov et al. [162] investigated the effect of ZrO_2 and SiO_2 nanoparticles to the PEO coating developed on a MA8 alloy. The incorporation of the nanoparticles led to higher thickness of the resulting coating, and the characterization performed showed a greater improvement for the coating containing ZrO_2 particles.

Aluminum Oxide (Al_2O_3)

Al_2O_3 has also been incorporated into PEO coatings with promising results. These NPs were found to fill the pores of the coating [166, 167], and once pulled out from the coating structure during abrasive wear, they played the role of lubrication reducing the wear extent (Fig. 13.12) [168]. Furthermore, the high hardness of Al_2O_3 led to a conversion from sliding friction to rolling friction, resulting in a lower volume of material loss [167, 169]. The wear mechanisms also depend on the NP concentration, switching from abrasive wear with high concentrations to adhesive wear for lower NP concentration [166]. Li et al. [170] developed PEO coatings on AM60B Mg by suspending Al_2O_3 particles in an aluminate electrolyte. They proposed a chemical reaction mechanism for the incorporation of Al_2O_3 particles into the coating, in contraposition to the trapping mechanism reported in previous research, and observed an increase of $MgAl_2O_4$/MgO formation and no presence of Al_2O_3

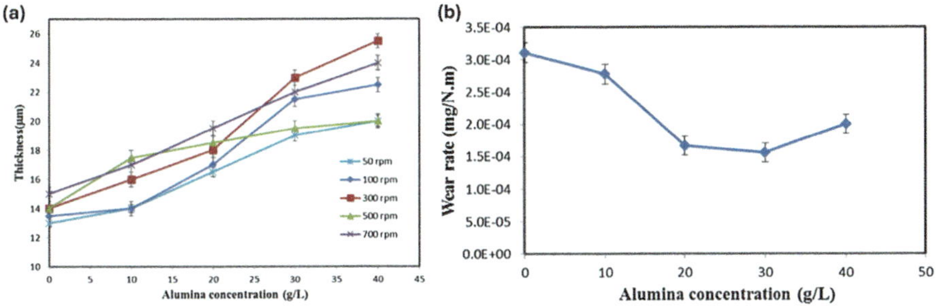

Fig. 13.12 (**a**, **b**) Influence of the alumina concentration in the electrolyte in the thickness and wear rate of the PEO coatings developed on an Mg alloy [166]. Reproduced with permission from [166] © 2016 Elsevier B.V.

nanoparticles in the resulting coatings. Wang et al. [171] developed PEO coatings on the AZ91D Mg alloy, investigating the effect of adding different concentrations (0, 5, 10, and 15 g/L) of Al_2O_3 nanoparticles to the electrolyte on the mechanical properties of the coatings. The Al_2O_3 nanoparticles were not only entrapped in the coating but also reacted chemically during the second stage of the PEO process, where the plasma discharges were significantly more intense. As a result, the $MgAl_2O_4$ phase was formed within the coating, which provides enhanced mechanical properties compared to the typical MgO phase found in PEO coatings on Mg alloys. In particular, the PEO coating containing 15 g/L of Al_2O_3 nanoparticles displayed a significantly increased hardness (291.5 HV) compared to the base metal (57.9 HV) and the PEO coating without Al_2O_3 (156.3 HV). Li et al. [172] observed a reduction in the porosity and improvement of abrasion resistance of the PEO coating on an AZ91 Mg alloy through the in situ incorporation of Al_2O_3 particles of varying size to the coating. Smaller particles were doped into the coating, while larger particles were deposited on the surface. The abrasion resistance of the coating improved with larger size Al_2O_3 particles.

Titanium Dioxide (TiO_2)

The addition of TiO_2 to Mg-PEO coatings has been found to improve the corrosion resistance [173, 174] and photocatalytic properties [175]. Moreover, the additivation of these NPs has shown an increase in microhardness and adhesive strength [176]. However, little is known about the effect of these particles on the tribological performance of the coatings. Based on the improvements in the wear resistance observed on the coatings developed for other materials, e.g., titanium [177], similar results could be expected for Mg substrates. For instance, Mozfarnia et al. [178] incorporated TiO_2 NPs (1–4 g/L) together with hydroxyapatite NPs on the coatings developed on AZ31 Mg alloy. The obtained coatings containing crystalline rutile and anatase presented more uniform morphology with less structural imperfections, which exhibited improved antibacterial ability as well as higher wear resistance and corrosion performance as a consequence of the higher hardness and lower COFs provided by TiO_2 concentrations up to 3 g/L. For the coatings with a higher amount of TiO_2 (4 g/L), the positive effect of the NPs was negligible, and the coating presented higher porosity and was more brittle, which led to higher material loss and COFs.

13.3.3.2 Non-oxide Ceramic

Titanium Nitride (TiN)

Some ceramic nanoparticles that have proven to have improved tribological properties of PEO coatings on Mg alloys are TiN. Mashtalyar et al. [164, 179] observed that when incorporated in an adequate concentration, these nanostructured materials promoted the formation of solid ceramic-like coatings with increased microhardness and lower porosity that helped tolerating more load cycles, leading to improved wear resistance. On the other hand, some studies have also shown that higher concentrations of TiN NPs led to embrittlement of the ceramic PEO coating, enhancing the wear loss extent [180, 181].

Tungsten Carbide (WC)

Incorporation of hard particles such as WC leads to a considerable increase in the hardness of Mg-PEO coatings, improving their wear resistance [182, 183]. Nasiri Vatan et al. [184] developed PEO coatings on AZ31B Mg using a WC nanopowder containing phosphate-based electrolyte. The developed coatings with embedded WC particles were found to be more compact and less thick, showing an average reduction of the 15% and 6% on the pore diameter and porosity, respectively. The presence of WC nanoparticles led to an increase in the hardness of the coating and improvement of the wear performance of the coating, showing a reduction in the COF of 45% (Fig. 13.13) and a significant reduction in the wear rate and wear track width.

Silicon Carbide (SiC)

Adding SiC to the coating can improve the wear performance of Mg-PEO coatings leading to lower and more stable COFs, by producing a rolling-effect on the surface and thus functioning as lubricants [151, 185]. The incorporation of SiC particles increases the thickness and hardness of the coating [186, 187]. Some investigations found the wear tracks to be smooth without any evidence of detachment [188, 189]. Pezzato et al. [190] improved the tribological and corrosion performance of an AZ91D Mg alloy through the incorporation of SiC and borosilicate glass particles into the developed PEO coating. They found a positive influence of the borosilicate

Sample code	Nanopowder	Time of coating deposition (min)
NP5	WC	5
NP10	WC	10
NP15	WC	15
NP20	WC	20
SS5	–	5
SS10	–	10
SS15	–	15
SS20	–	20

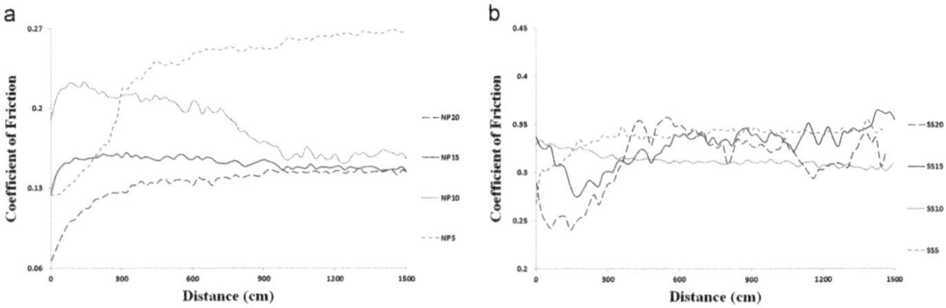

Fig. 13.13 PEO process parameters and WC nanoparticles concentrations added and evolution of the COF during the wear tests for (**a**) PEO coatings with WC nanopowders and (**b**) PEO coatings without WC nanopowders [184]. Reproduced with permission from [184] © 2016 Elsevier Ltd.

glass on the improvement of the corrosion resistance of the PEO coatings, while the SiC resulted in the opposite effect, reducing the resistance. On the other hand, SiC particles considerably improved the wear resistance due to enhanced load support. However, the incorporation of SiC increased the friction coefficient. Thus, a compromise between the concentration of both additives is necessary to improve both wear and corrosion resistance of the Mg alloy. Lu et al. [191] incorporated SiC nanoparticles to the PEO coating on an AZ31 Mg alloy and obtained thicker coatings composed of mainly Mg_2SiO_4 and MgO, and with an uniform distribution of SiC along the whole thickness. While SiC particles greatly improved the wear resistance of the coatings, they observed this positive influence of SiC only on the coatings developed under lower current densities, which presented lower porosity.

Silicon Nitride (Si_3N_4)

When added in appropriate amounts, Si_3N_4 NPs have shown the ability to improve the wear resistance of PEO coatings developed on Mg alloys [151]. Lou et al. [192] observed that Si_3N_4 NPs were reactively incorporated into the coating promoting the formation of Mg_2SiO_4 in addition to $MgAl_2O_4$ and MgO present in the coating with no Si_3N_4. The hardness of the resulting coating increases with higher amounts of Si_3N_4 NPs (1–4 g/L), as a consequence of the higher presence of Mg_2SiO_4. However, the highest concentration studied (4 g/L) showed poor adhesion and higher porosity. The study evinced the importance of correct NP concentration to obtain the best tribological performance.

Boron Nitride (BN)

Hexagonal boron nitride (h-BN) has been positioned as a promising solution to reduce wear [193–195] due to its unique lamellar crystalline structure that enables an easy sliding between its layers, providing excellent lubrication. Ji et al. [196] developed a nanocomposite PEO coating doped with carbon nanotubes (CNTs) and hexagonal h-BN on an AZ31 Mg alloy. The resulting coating presented the ability of accommodating abrasive particles into its porous structure, which coupled with the lubricating effect of CNTs and h-BN resulted in a reduction of the coefficient of friction and lower wear loss compared to the MgO coating without these particles addition.

13.3.3.3 Metallic Particles

Silver Nanoparticles (Ag NPs)

Several works demonstrate the benefits of incorporating Ag NPs into Mg-PEO coatings, such as improvement of corrosion resistance or antibacterial with low cytotoxicity [197–199]. For instance, Husak et al. [200] incorporated Ag MPs into silicate-based electrolytes to obtain antibacterial PEO coatings in pure Mg. The resulting coatings with a porous network improved the corrosion resistance of the Mg, with low cytotoxicity, antibacterial properties vs. *Staphylococcus aureus* B918. Even though no specific studies have been published yet confirming the improvement of the tribological performance of Mg-PEO coatings by Ag NP addition, results obtained for PEO coatings on Ti alloys have shown the benefits of these nanoparticles on the wear resistance of the coating, due to lubricating ability of Ag under wear solicitations that led to a reduced COF and material loss [201]. Therefore, similar results could be expected for Mg-PEO coatings, though further research is needed to confirm this statement.

13.3.3.4 Inorganic Particles

Molybdenum Disulfide (MoS_2)

Lou et al. [202] investigated the effect of MoS_2 particle addition into the PEO coating generated on an AZ31 alloy and found the resulting coatings to possess a high hardness, good adhesion, and improved wear performance. The addition of MoS_2 reduced the amount of $MgAl_2O_4$ but increased the presence of MgO. With increasing percentages of MoS_2, the particle segregation and production of discontinuous and more porous inner films led to lower mechanical properties of the oxide coating.

Clay

Clay has been found to reactively incorporate into the coatings promoting the formation of dense films and enhancing the overall wear performance. Lu et al. [151] studied the influence of different particles on the properties of PEO coating developed on an AM50 Mg alloy. They employed clay, SiO_2, Si_3Ni_4, and SiC micro-sized particles as additives. They observed that the incorporation mode of particles depended on their melting point. For those with a lower melting point, a fully reactive incorporation was achieved, whereas for those with a higher melting point, the particles were mainly inertly incorporated. The reactive incorporation of clay contributed to the formation of a dense layer that improved the wear and corrosion performance of the coating. On the other hand, SiO_2 and Si_3Ni_4 were found to improve the wear resistance of the coating.

13.3.3.5 Carbon-Based Materials

Graphene

The presence of the C element and the formation of hard phases can significantly reduce the COF by acting as solid lubricants inside the coating [76, 77, 203]. Fattahalhosseini et al. [124] include in their review a wide number of research studying the effect of graphene and graphene oxide addition to PEO coatings on Mg on their corrosion and wear behaviors. Due to their small particle size, graphene can change the morphology of the coating by filling up the porosity and becoming embedded in the coating structure [76, 204]. A synergistic effect has been found between the lubricating ability that reduces the COF, the rise in hardness, and the reduction of roughness by filling microcracks and micropores, which leads to an overall improvement in the wear performance [76, 77, 205, 206].

Carbon Nanotubes (CNTs)

Several studies have demonstrated the improvement in the mechanical properties and wear resistance of PEO coatings on Al and Ti by the incorporation of CNTs, which minimize friction and surface damage by rolling between contact surfaces [207–209]. Imshinetskiy et al. [210] incorporated CNTs into a PEO coating formed on an MA8 Mg alloy and observed an improvement in the adhesive strength, microhardness, and Young Modulus of the resulting coatings. Daavari et al. [211] developed PEO coatings containing multi-walled carbon nanotubes (MWCNTs) via a one-step process on a wrought AZ31B Mg alloy. The resulting coatings presented bigger size pores and a thick and dense barrier layer composed of Mg_2SiO_4 in addition to MgO. All in all, they observed an improvement in the wear resistance of the coatings by reducing the coefficient of friction by 15% and the wear depth by 60%.

13.4 Conclusions and Outlook

In recent years, PEO technology has been consolidated as an advanced and sustainable solution for the improvement of the tribological properties of light alloys, including Al and Mg alloys. These alloys, widely used in sectors like the automotive, space, and biomedical, face important limitations due to their low wear resistance, which restricts their use in applications with more demanding requirements. In this context, PEO coatings confer alloys high hardness, excellent adhesion to the substrate, and high wear resistance.

In the case of Al alloys, numerous advances have been made in the development of PEO coatings with enhanced tribological behavior, obtaining very competitive results even on cast Al-Si alloys, which are more difficult to modify superficially due to their chemical composition. For Mg alloys, on the other hand, there are fewer studies about the development of PEO coatings with high tribological functionality compared to, for example, research conducted on the improvement of the corrosion resistance of Mg alloys by means of PEO coatings. Nevertheless, the number of studies is increasing, and the addition of nanoparticles to PEO coatings has proven to be an effective technique for improving the wear resistance of both alloys.

Therefore, PEO coatings currently provide a promising and increasingly feasible way to improve the versatility and performance of Al and Mg alloys in applications with more demanding requirements.

References

1. Kaseem, M., Fatimah, S., Nashrah, N., & Ko, Y. G. (2021). *Progress in Materials Science, 117*, 100735.
2. Curran, J. A., Kalkanci, H., Magurova, Y., & Clyne, T. W. (2007). *Surface and Coatings Technology, 201*, 8683–8687.
3. Arunnellaiappan, T., Kishore Babu, N., Rama Krishna, L., & Rameshbabu, N. (2015). *Surface and Coatings Technology, 280*, 136–147.
4. Fernández-López, P., Alves, S. A., López-Ortega, A., San José-Lombera, J. T., & Bayón, R. (2021). *Ceramics International, 47*, 31238–31250.
5. Wu, X., Qin, W., Guo, Y., & Xie, Z. (2008). *Applied Surface Science, 254*, 6395–6399.
6. Algahtani, A., & Mahmoud, E. R. I. (2019). *Journal of Materials Research and Technology, 8*, 2699–2709.
7. Simchen, F., Sieber, M., Kopp, A., & Lampke, T. (2020). *Coatings, 10*, 1–19.
8. Tanwir Alam, M., & Husain Ansari, A. (2017). *International Journal of Advanced Technology in Engineering and Science, 5*, 278–294.
9. Heinz, A., Haszler, A., Keidel, C., Moldenhauer, S., Benedictus, R., & Miller, W. S. (2000). *Materials Science and Engineering A, 280*, 102–107.
10. Verma, R. P., & Kumar Lila, M. (2021). *Materials Today: Proceedings, 46*, 10687–10691.
11. Bai, J., Li, Y., Zhao, Y., Luo, F., Sun, T., Liang, Y., Tang, Y., & Zhang, S. (2024). *Applied Thermal Engineering, 257*, 124332.
12. Varshney, D., & Kumar, K. (2021). *Ain Shams Engineering Journal, 12*, 1143–1152.
13. Sjögren, B., Iregren, A., Montelius, J., & Yokel, R. A. (2015). Aluminum, *In* Handbook on the Toxicology of Metals (pp. 549–564) (4th ed.). Elsevier.

14. Nie, X., Leyland, A., Song, H. W., Yerokhin, A. L., Dowey, S. J., & Matthews, A. (1999). *Surface and Coatings Technology, 116–119*, 1055–1060.
15. Cullen, J. M., & Allwood, J. M. (2013). *Environmental Science & Technology, 47*, 3057–3064.
16. Raabe, D., Ponge, D., Uggowitzer, P. J., Roscher, M., Paolantonio, M., Liu, C., Antrekowitsch, H., Kozeschnik, E., Seidmann, D., Gault, B., De Geuser, F., Deschamps, A., Hutchinson, C., Liu, C., Li, Z., Prangnell, P., Robson, J., Shanthraj, P., Vakili, S., Sinclair, C., Bourgeois, L., & Pogatscher, S. (2022). *Progress in Materials Science, 128*, 100947.
17. Berlanga-Labari, C., Biezma-Moraleda, M. V., & Rivero, P. J. (2020). *Metals, 10*, 1–30.
18. Ervina Efzan, M. N., Kong, H. J., & Kok, C. K. (2014). *Advances in Materials Research, 845*, 355–359.
19. Wang, P., Li, J. P., Guo, Y. C., Yang, Z., & Wang, J. L. (2016). *Surface Engineering, 32*, 428–434.
20. Xu, F., Xia, Y., & Li, G. (2009). *Applied Surface Science, 255*, 9531–9538.
21. Babaei, K., Fattah-alhosseini, A., & Molaei, M. (2020). *Surfaces and Interfaces, 21*, 100677.
22. Li, Q., Liang, J., Liu, B., Peng, Z., & Wang, Q. (2014). *Applied Surface Science, 297*, 176–181.
23. Gulec, A. E., Gencer, Y., & Tarakci, M. (2015). *Surface and Coatings Technology, 269*, 100–107.
24. Glazoff, M. V., Khvan, A. V., Zolotorevsky, V. S., Belov, N. A., & Dinsdale, A. T. (2019). *Industrial and Perspective Casting Alloys*, Casting Aluminum Alloys (Second Edition) (pp. 405–510).
25. Yang, C., Zhu, J., Cui, S., Chen, P., Wu, Z., Ma, Z., Fu, R. K. Y., Tian, X., Chu, P. K., & Wu, Z. (2021). *Surface and Coatings Technology, 409*, 126885.
26. Muhaffel, F., Baydogan, M., & Cimenoglu, H. (2021). *Surface and Coatings Technology, 409*, 126843.
27. Javidi, M., & Fadaee, H. (2013). *Applied Surface Science, 286*, 212–219.
28. Student, M., Pohrelyuk, I., Padgurskas, J., Posuvailo, V., et al. (2023). *Coatings, 13*, 637.
29. Huang, X. (2019). *Modern Concepts in Material Science, 2*, 1–13.
30. Sieber, M., Simchen, F., Morgenstern, R., Scharf, I., & Lampke, T. (2018). *Metals, 8*, 1–17.
31. Wang, L., Wang, G., Dong, H., Ye, M., Li, X., Liu, L., Pan, J., & Ye, Z. (2022). *Surface and Coatings Technology, 435*, 128246.
32. Oh, Y. J., Il Mun, J., & Kim, J. H. (2009). *Surface and Coatings Technology, 204*, 141–148.
33. Zhu, L., Guo, Z., Zhang, Y., Li, Z., & Sui, M. (2016). *Electrochimica Acta, 208*, 296–303.
34. Cheng, Y., Mao, M., Cao, J., & Peng, Z. (2014). *Electrochimica Acta, 138*, 417–429.
35. Robles Hernandez, F. C., Herrera Ramírez, J. M., & Mackay, R. (2017). *Al-Si alloys: Autmotive, aeronautical, and aerospace applications*. Springer.
36. Shabestari, S. G. (2004). *Materials Science and Engineering A, 383*, 289–298.
37. Otani, L. B., Soyama, J., Zepon, G., Costa e Silva, A., Kiminami, C. S., Botta, W. J., & Bolfarini, C. (2017). *Journal of Phase Equilibria and Diffusion, 38*, 298–304.
38. Wang, D., Zhang, X., Nagaumi, H., Li, X., & Zhang, H. (2020). *Materials Letters, 277*, 128384.
39. Agureev, L., Savushkina, S., Ashmarin, A., Borisov, A., Apelfeld, A., Anikin, K., Tkachenko, N., Gerasimov, M., Shcherbakov, A., Ignatenko, V., & Bogdashkina, N. (2018). *Metals, 8*, 1–12.
40. Yang, W., Liu, W., Peng, Z., Liu, B., & Liang, J. (2017). *Ceramics International, 43*, 16851–16858.
41. Clyne, T. W., & Troughton, S. C. (2019). *International Materials Review, 64*, 127–162.
42. Martin, J., Leone, P., Nominé, A., Veys-Renaux, D., Henrion, G., & Belmonte, T. (2015). *Surface and Coatings Technology, 269*, 36–46.
43. Liu, Y. J., Xu, J. Y., Lin, W., Gao, C., Zhang, J. C., & Chen, X. H. (2013). *Reviews on Advanced Materials Science, 33*, 126–130.
44. Saikiran, A., Hariprasad, S., Arun, S., Rama Krishna, L., & Rameshbabu, N. (2019). *Surface and Coatings Technology, 372*, 239–251.

45. Jiang, B. L., & Wang, Y. M. (2010). Plasma electrolytic oxidation treatment of aluminium and titanium alloys. In *Surface engineering of light alloys. Aluminium, magnesium and titanium alloys* (pp. 110–154). Woodhead Publishing.
46. Toulabifard, A., Rahmati, M., Raeissi, K., Hakimizad, A., & Santamaria, M. (2020). *Coatings, 10*, 1–19.
47. Lee, J. H., Jung, K. H., & Kim, S. J. (2020). *Applied Surface Science, 516*, 146049.
48. Sharma, A., Jang, Y. J., & Jung, J. P. (2017). *Journal of Materials Engineering and Performance, 26*, 5032–5042.
49. Yang, Z., Zhang, X. Z., Wu, Y. K., Wu, G. R., Wang, D. D., Liu, X. T., Han, H. P., Su, Y., & Shen, D. J. (2019). *Ceramics International, 45*, 19388–19394.
50. Xue, W., Deng, Z., Chen, R., & Zhang, T. (2000). *Thin Solid Films, 372*, 114–117.
51. Salehi Doolabi, D., Ehteshamzadeh, M., & Mirhosseini, S. M. M. (2012). *Journal of Materials Engineering and Performance, 21*, 2195–2202.
52. Sundararajan, G., & Rama Krishna, L. (2003). *Surface and Coatings Technology, 167*, 269–277.
53. Moon, S., & Jeong, Y. (2009). *Corrosion Science, 51*, 1506–1512.
54. Alves, S. A., Fernández-Lopéz, P., Lopéz-Ortega, A., Fernández, X., Quintana, I., San-José, J. T., & Bayón, R. (2022). *Surface and Coatings Technology, 433*, 128146.
55. Peng, Z., Xu, H., Liu, S., Qi, Y., & Liang, J. (2021). *Materials*, 14, 4037.
56. Hwang, I. J., Shin, K. R., Lee, J. S., Ko, Y. G., & Shin, D. H. (2012). *Materials Transactions, 53*, 559–564.
57. Fernández-López, P., Alves, S. A., Azpitarte, I., San-José, J. T., & Bayón, R. (2022). *Corrosion Science, 207*, 110548.
58. Tang, M., Li, W., Liu, H., & Zhu, L. (2012). *Current Applied Physics, 12*, 1259–1265.
59. Nagumothu, R. B., Thangavelu, A., Nair, A. M., Sukumaran, A., & Anjilivelil, T. (2019). *Transactions of the Indian Institute of Metals, 72*, 47–53.
60. Arunnellaiappan, T., Rama Krishna, L., Anoop, S., Uma Rani, R., & Rameshbabu, N. (2016). *Surface and Coatings Technology, 307*, 735–746.
61. Wang, X., Wu, X., Wang, R., & Qiu, Z. (2013). *International Journal of Electrochemical Science, 8*, 4986–4995.
62. Kaseem, M., Yang, H. W., & Ko, Y. G. (2017). *Scientific Reports, 7*, 1–10.
63. Lu, X., Mohedano, M., Blawert, C., Matykina, E., Arrabal, R., Kainer, K. U., & Zheludkevich, M. L. (2016). *Surface and Coatings Technology, 307*, 1165–1182.
64. Ayday, A., & Durman, M. (2015). *Acta Physica Polonica A, 127*, 886–887.
65. Matykina, E., Arrabal, R., Mohedano, M., Mingo, B., Gonzalez, J., Pardo, A., & Merino, M. C. (2017). *Transactions of Nonferrous Metals Society of China, 27*, 1439–1454.
66. Khan, R. H. U., Yerokhin, A., Li, X., Dong, H., & Matthews, A. (2010). *Surface and Coatings Technology, 205*, 1679–1688.
67. Rogov, A. B., Yerokhin, A., & Matthews, A. (2018). *Journal of Physics D: Applied Physics, 51*, 405303.
68. Akbar, A., Adnan Qaiser, M., Hussain, A., Ali Mustafa, R., Xiong, D., & Adnan Akbar, E. (2017). *International Journal of Engineering Works, 4*, 114–123.
69. Molaei, M., Fattah-alhosseini, A., Nouri, M., & Nourian, A. (2022). *Ceramics International, 48*, 6322–6337.
70. Cheng, Y., Shi, X., Lv, Y., & Zhang, X. (2024). *Metals, 14*, 615.
71. Shin, Y., Bae, K., Lee, S., Kim, H., Shin, D., Kim, D., Choi, E., Moon, H. S., & Lee, J. (2023). *Nanomaterials, 13*, 2582.
72. Rao, Y., Wang, Q., Oka, D., & Ramachandran, C. S. (2020). *Surface and Coatings Technology, 383*, 125271.
73. Martin, J., Akoda, K., Ntomprougkidis, V., Ferry, O., Maizeray, A., Bastien, A., Brenot, P., Ezo'o, G., & Henrion, G. (2020). *Surface and Coatings Technology, 392*, 125756.
74. Popov, N., & Chavdarov, A. (2024). *E3S Web of Conferences, 583*, 03020.

75. López-Ortega, A., Arana, J. L., Rodríguez, E., & Bayón, R. (2018). *Corrosion Science, 143*, 258–280.
76. Chen, F., Zhang, Y., & Zhang, Y. (2017). *International Journal of Electrochemical Science, 12*, 6081–6091.
77. Nasiri Vatan, H., & Adabi, M. (2018). *Transactions of the Indian Institute of Metals, 71*, 1643–1652.
78. Gray, J. E., & Luan, B. (2002). *Journal of Alloys and Compounds, 336*, 88–113.
79. Liang, J., Hu, L., & Hao, J. (2007). *Applied Surface Science, 253*, 4490–4496.
80. Tan, J., & Ramakrishna, S. (2021). *Applied Sciences, 11*, 6861.
81. Liu, L. J., & Schlesinger, M. (2009). *Corrosion Science, 51*, 1733–1737.
82. Song, G., Atrens, A., Wu, X., & Zhang, B. (1998). *Corrosion Science, 40*, 1769–1791.
83. Wu, G., Li, L., Chen, X., Zhu, L., Wang, Y., Wen, C., & Yao, J. (2025). *Journal of Magnesium and Alloys, 13*, 760–776.
84. Yamauchi, N., Demizu, K., Ueda, N., Cuong, N. K., Sone, T., & Hirose, Y. (2005). *Surface and Coatings Technology, 193*, 277–282.
85. Fattah-alhosseini, A., Chaharmahali, R., & Babaei, K. (2020). *Journal of Magnesium and Alloys, 8*, 799–818.
86. Barati Darband, G., Aliofkhazraei, M., Hamghalam, P., & Valizade, N. (2017). *Journal of Magnesium and Alloys, 5*, 74–132.
87. Zhang, P., Nie, X., Hu, H., & Liu, Y. (2010). *Surface and Coatings Technology, 205*, 1508–1514.
88. Loukil, N. (2016). Alloying Elements of Magnesium Alloys: A Literature Review, *Magnesium Alloys Structure and Properties* (Eds. Tański, T., Jarka, P.), Intech.
89. Dargusch, M. S., Pettersen, K., Nogita, K., Nave, M. D., & Dunlop, G. L. (2006). *Materials Transactions, 47*, 977–982.
90. Shaw, C., & Jones, H. (1997). *Materials Science and Engineering A, 226–228*, 856–860.
91. Liu, Y. Q., Yu, D., Zhang, Y., Zhou, J. P., Sun, D. Q., & Li, H. M. (2023). *Journal of Materials Research and Technology, 25*, 3458–3481.
92. Li, J., Liu, C., Wang, Z., Cai, Y., & Li, J. (2023). *Journal of Materials Engineering and Performance, 33*, 10574–10588.
93. Wan, Y., Xiong, G., Luo, H., He, F., Huang, Y., & Zhou, X. (2008). *Materials and Design, 29*, 2034–2037.
94. Wu, B. L., Zhao, Y. H., Du, X. H., Zhang, Y. D., Wagner, F., & Esling, C. (2010). *Materials Science and Engineering A, 527*, 4334–4340.
95. Kong, L. (2017). *IOP Conference Series: Materials Science and Engineering, 242*, 012026.
96. Aghion, E., & Golub, G. (1996). *Acta Materialia, 1*, 1–31.
97. Kiełbus, A. (2018). Microstructure and Properties of Casting Magnesium Alloys Designed to Work in Elevated Temperature, *Magnesium Alloys - Selected Issue* (Ed. Tański, T., Borek, W., Król, M.). IntechOpen.
98. You, S., Huang, Y., Kainer, K. U., & Hort, N. (2017). *Journal of Magnesium and Alloys, 5*, 239–253.
99. Kwak, T. Y., Lim, H. K., & Kim, W. J. (2016). *Journal of Alloys and Compounds, 658*, 157–169.
100. Kumar, R., & Katyal, P. (2022). *Materials Today Proceedings, 56*, 2443–2450.
101. Jain, S., Gokhale, A., Jain, J., Singh, S. S., & Hariharan, K. (2017). *Materials Science and Engineering A, 684*, 652–659.
102. Naik, G. M., Gote, G. D., Narendranath, S., & Satheesh Kumar, S. S. (2018). *Materials Research Express, 5*, 086513.
103. Liao, H. J., Zhou, X. F., Li, H. Z., Deng, M., Liang, X. P., & Liu, R. M. (2015). *Transactions of the Nonferrous Metals Society of China, 25*, 3921–3927.
104. Khan, M. I., Mostafa, A. O., Aljarrah, M., Essadiqi, E., & Medraj, M. (2014). *Journal of Materials, 2014*, 1–18.

105. Arrabal, R., Matykina, E., Hashimoto, T., Skeldon, P., & Thompson, G. E. (2009). *Surface and Coatings Technology, 203*, 2207–2220.
106. Cakmak, E., Tekin, K. C., Malayoglu, U., & Shrestha, S. (2010). *Surface and Coatings Technology, 204*, 1305–1313.
107. Buling, A., & Zerrer, J. (2019). *Surface and Coatings Technology, 369*, 142–155.
108. Molaei, M., Babaei, K., & Fattah-alhosseini, A. (2021). *Journal of Magnesium and Alloys, 9*, 1164–1186.
109. Hussein, R. O., & Northwood, D. O. (2014). *WIT Transactions on the Built Environment, 137*, 531–544.
110. White, L., Koo, Y., Neralla, S., Sankar, J., & Yun, Y. (2016). *Materials Science and Engineering B, 208*, 39–46.
111. Kennedy, E., Shivappa, Sriraman, & Mohan. (2020). *Materials Today Proceedings, 27*, 2236–2238.
112. Almajidi, Y. Q., Ali, E., Jameel, M. F., Saleh, L. H., Aggarwal, S., Zearah, S. A., Alamula, A. F., Alsaalamy, A., Sharifianjazi, F., & Bathaei, M. S. (2023). *Lubricants, 11*, 1–35.
113. Hussein, R. O., Northwood, D. O., & Nie, X. (2013). *Surface and Coatings Technology, 237*, 357–368.
114. Khaselev, O., Weiss, D., & Yahalom, J. (2001). *Corrosion Science, 43*, 1295–1307.
115. Krishna, L. R., Poshal, G., Jyothirmayi, A., & Sundararajan, G. (2015). *Materials and Design, 77*, 6–14.
116. Gunduz, K. O., Oter, Z. C., Tarakci, M., & Gencer, Y. (2017). *Surface and Coatings Technology, 323*, 72–81.
117. Tekin, K. C., Malayoğlu, U., & Shrestha, S. (2013). *Surface and Coatings Technology, 236*, 540–549.
118. Muhaffel, F., Mert, F., Cimenoglu, H., Höche, D., Zheludkevich, M. L., & Blawert, C. (2015). *Surface and Coatings Technology, 269*, 200–211.
119. Song, Y. L., Liu, Y. H., Yu, S. R., Zhu, X. Y., & Wang, Q. (2008). *Applied Surface Science, 254*, 3014–3020.
120. Arrabal, R., Matykina, E., Viejo, F., Skeldon, P., & Thompson, G. E. (2008). *Corrosion Science, 50*, 1744–1752.
121. Yoo, B., Shin, K. R., Hwang, D. Y., Lee, D. H., & Shin, D. H. (2010). *Applied Surface Science, 256*, 6667–6672.
122. Sovík, J., Kajánek, D., Pastorek, F., Štrbák, M., Florková, Z., Jambor, M., & Hadzima, B. (2023). *Materials, 16*, 5650.
123. Li, J., Bian, Y., Tu, X., Li, W., & Song, D. (2022). *Journal of Electroanalytical Chemistry, 910*, 116206.
124. Fattah-alhosseini, A., & Chaharmahali, R. (2021). *FlatChem, 27*, 100241.
125. Egorkin, V. S., Gnedenkov, S. V., Sinebryukhov, S. L., Vyaliy, I. E., Gnedenkov, A. S., & Chizhikov, R. G. (2018). *Surface and Coatings Technology, 334*, 29–42.
126. Hussein, R. O., Northwood, D. O., & Nie, X. (2014). *Materials Sciences and Applications, 05*, 124–139.
127. Blawert, C., Heitmann, V., Dietzel, W., Nykyforchyn, H. M., & Klapkiv, M. D. (2007). *Surface and Coatings Technology, 201*, 8709–8714.
128. An, L., Ma, Y., Sun, L., Wang, Z., & Wang, S. (2020). *Journal of Magnesium and Alloys, 8*, 523–536.
129. Lu, X., Sah, S. P., Scharnagl, N., Störmer, M., Starykevich, M., Mohedano, M., Blawert, C., Zheludkevich, M. L., & Kainer, K. U. (2015). *Surface and Coatings Technology, 269*, 155–169.
130. Gao, Y., Yerokhin, A., & Matthews, A. (2014). *Applied Surface Science, 316*, 558–567.
131. Bala Srinivasan, P., Liang, J., Blawert, C., & Dietzel, W. (2010). *Applied Surface Science, 256*, 3265–3273.
132. Rapheal, G., Kumar, S., Blawert, C., & Dahotre, N. B. (2011). *Wear, 271*, 1987–1997.
133. Durdu, S., & Usta, M. (2012). *Applied Surface Science, 261*, 774–782.

134. Li, X., Liu, X., & Luan, B. L. (2011). *Applied Surface Science, 257*, 9135–9141.
135. Mu, W., & Han, Y. (2010). *Xiyou Jinshu Cailiao Yu Gongcheng/Rare Metal Materials and Engineering, 39*, 1129–1134.
136. da Silva Rodrigues, J., Marasca Antonini, L., da Cunha Bastos, A. A., Zhou, J., & de Fraga Malfatti, C. (2021). *Surface and Coatings Technology, 410*, 126983.
137. Ding, J., Liang, J., Hu, L. T., Hao, J. C., & Xue, Q. J. (2007). *Transactions of Nonferrous Metals Society of China, 17*, 244–249.
138. Duan, H., Yan, C., & Wang, F. (2007). *Electrochimica Acta, 52*, 3785–3793.
139. Lu, S., Wang, Z. X., Chen, J., & Zhou, X. S. (2011). *Transactions of the Nonferrous Metals Society of China, 21*, 929–935.
140. Shen, M. J., Wang, X. J., & Zhang, M. F. (2012). *Applied Surface Science, 259*, 362–366.
141. Peibo, S., Xiaohong, W., & Zhaohua, J. (2008). *Materials Letters, 62*, 3124–3126.
142. Wu, D., Liu, X., Lu, K., Zhang, Y., & Wang, H. (2009). *Applied Surface Science, 255*, 7115–7120.
143. Zhu, F., Wang, J., Li, S., & Zhang, J. (2012). *Applied Surface Science, 258*, 8985–8990.
144. Tsai, D. S., & Chou, C. C. (2018). *Metals, 8*, 1–22.
145. Jaspard-Mécuson, F., Czerwiec, T., Henrion, G., Belmonte, T., Dujardin, L., Viola, A., & Beauvir, J. (2007). *Surface and Coatings Technology, 201*, 8677–8682.
146. Cheng, Y., Wang, T., Li, S., Cheng, Y., Cao, J., & Xie, H. (2017). *Electrochimica Acta, 225*, 47–68.
147. Jin, F., Chu, P. K., Xu, G., Zhao, J., Tang, D., & Tong, H. (2006). *Materials Science and Engineering A, 435–436*, 123–126.
148. Niedźwiedź, M., Bara, M., Korzekwa, J., Kaptacz, S., Sowa, M., Olesiński, A., & Simka, W. (2024). *Coatings, 14*, 1446.
149. Lu, X., Blawert, C., Huang, Y., Ovri, H., Zheludkevich, M. L., & Kainer, K. U. (2016). *Electrochimica Acta, 187*, 20–33.
150. Hakimizad, A., Raeissi, K., Golozar, M. A., Lu, X., Blawert, C., & Zheludkevich, M. L. (2017). *Surface and Coatings Technology, 324*, 208–221.
151. Lu, X., Blawert, C., Kainer, K. U., Zhang, T., Wang, F., & Zheludkevich, M. L. (2018). *Surface and Coatings Technology, 352*, 1–14.
152. Salman, S. A., Ichino, R., & Okido, M. (2009). *Transactions of Nonferrous Metals Society of China, 19*, 883–886.
153. Lin, P. Y., Zhou, H., Sun, N., Li, W. P., Wang, C. T., Wang, M. X., Guo, Q. C., & Li, W. (2010). *Corrosion Science, 52*, 416–421.
154. Aliofkhazraei, M., Gharabagh, R. S., Teimouri, M., Ahmadzadeh, M., Darband, G. B., & Hasannejad, H. (2016). *Journal of Alloys and Compounds, 685*, 376–383.
155. Lim, T. S., Ryu, H. S., & Hong, S. H. (2012). *Corrosion Science, 62*, 104–111.
156. Li, J., Tian, Y., Gui, Z., & Huang, Z. (2008). *Rare Metals, 27*, 50–54.
157. Cai, J., Cao, F., Chang, L., Zheng, J., Zhang, J., & Cao, C. (2011). *Applied Surface Science, 257*, 3804–3811.
158. Stojadinović, S., & Radić, N. (2024). *Metals, 14*, 366.
159. Mohedano, M., Blawert, C., & Zheludkevich, M. L. (2015). *Materials and Design, 86*, 735–744.
160. Zhang, G., Jiang, E., Wu, L., Ma, W., Yang, H., Tang, A., & Pan, F. (2021). *Scientific Reports, 11*, 1–14.
161. Lu, X., Chen, Y., Blawert, C., Li, Y., Zhang, T., Wang, F., Kainer, K. U., & Zheludkevich, M. (2018). *Coatings, 8*, 306.
162. Gnedenkov, S. V., Sinebryukhov, S. L., Mashtalyar, D. V., Imshinetskiy, I. M., Samokhin, A. V., & Tsvetkov, Y. V. (2015). *Journal of Nanomaterials, 2015*, 54298.
163. Krishtal, M. M., Ivashin, P. V., Polunin, A. V., & Borgardt, E. D. (2019). *Materials Letters, 241*, 119–122.
164. Mashtalyar, D. V., Imshinetsky, I. M., & Sinebryukhov, S. L. (2017). *MATEC Web of Conferences, 129*, 10–13.

165. Imshinetsky, I. M., Gnedenkov, S. V., Sinebryukhov, S. L., Mashtalyar, D. V., Samokhin, A. V., & Tsvetkov, Y. V. (2018). *Defect and Diffusion Forum, 386 DDF*, 321–325.
166. Asgari, M., Aliofkhazraei, M., Darband, G. B., & Rouhaghdam, A. S. (2017). *Surface and Coatings Technology, 309*, 124–135.
167. Asgari, M., Aliofkhazraei, M., Barati Darband, G., & Sabour Rouhaghdam, A. (2020). *Surface and Coatings Technology, 383*, 125252.
168. Gheytani, M., Bagheri, H. R., Masiha, H. R., Aliofkhazraei, M., Sabour Rouhaghdam, A., & Shahrabi, T. (2014). *Surface Engineering, 30*, 244–255.
169. Long, Y., Wu, L., Zhang, Z., Atrens, A., Pan, F. S., Tang, A., & Zhang, G. (2018). *International Journal of Electrochemical Science, 13*, 7157–7174.
170. Li, X., & Luan, B. L. (2012). *Materials Letters, 86*, 88–91.
171. Wang, Y., Wei, D., Yu, J., & Di, S. (2014). *Journal of Materials Science and Technology, 30*, 984–990.
172. Li, Y., Li, C., Wang, Q., Zhao, X., Zhang, T., & Wang, F. (2024). *Materials Today Communications, 38*, 108234.
173. Liang, J., Hu, L., & Hao, J. (2007). *Electrochimica Acta, 52*, 4836–4840.
174. White, L., Koo, Y., Yun, Y., & Sankar, J. (2013). *Journal of Nanomaterials, 2013*, 319437.
175. Mashtalyar, D. V., Imshinetskiy, I. M., Nadaraia, K. V., Gnedenkov, A. S., Suchkov, S. N., Opra, D. P., Pustovalov, E. V., Yu Ustinov, A., Sinebryukhov, S. L., & Gnedenkov, S. V. (2023). *Journal of Magnesium and Alloys, 11*, 735–752.
176. Imshinetsky, I. M., Nadaraia, K. V., Sokolov, A. A., Suchkov, S. N., Pleshkova, A. I., Mashtalyar, D. V., Sinebryukhov, S. L., & Gnedenkov, S. V. (2020). *Solid State Phenomena, 312*, 372–376.
177. Molaei, M., Fattah-Alhosseini, A., Nouri, M., & Kaseem, M. (2023). *Metals, 13*, 821.
178. Mozafarnia, H., Fattah-alhosseini, A., Chaharmahali, R., Nouri, M., Keshavarz, M. K., & Kaseem, M. (2022). *Coatings, 12*, 1967.
179. Mashtalyar, D. V., Sinebryukhov, S. L., Imshinetskiy, I. M., Gnedenkov, A. S., Nadaraia, K. V., Ustinov, A. Y., & Gnedenkov, S. V. (2020). *Applied Surface Science, 503*, 144062.
180. Imshinetsky, I. M., Mashtalyar, D. V., Sunebryukhov, S. L., & Gnedenkov, S. V. (2017). *AIP Conference Proceedings, 1874*, 040012.
181. Mashtalyar, D. V., Gnedenkov, S. V., Sinebryukhov, S. L., Imshinetskiy, I. M., & Puz, A. V. (2017). *Journal of Materials Science and Technology, 33*, 461–468.
182. Nasiri Vatan, H., & Adabi, M. (2017). *Transactions of the Institute of Metal Finishing, 95*, 308–315.
183. Nasiri Vatan, H., Ebrahimi-Kahrizsangi, R., & Kasiri Asgarani, M. (2016). *International Journal of Electrochemical Science, 11*, 929–943.
184. Nasirivatan, H., Kahrizsangi, R. E., & Asgarani, M. K. (2016). *Tribology International, 98*, 253–260.
185. Nasiri Vatan, H., Ebrahimi-Kahrizsangi, R., & Kasiri-Asgarani, M. (2016). *Journal of Alloys and Compounds, 683*, 241–255.
186. Vatan, H. N., Kahrizsangi, R. E., & Asgarani, M. K. (2016). *Protection of Metals and Physical Chemistry of Surfaces, 52*, 859–868.
187. Yang, Y., & Wu, H. (2010). *Journal of Materials Science and Technology, 26*, 865–871.
188. Ebrahimi-kahrizsangi, R., & Vatan, H. V. (2016). *Proceedings of the 10th World Congress on New Technologies, 2*, 16–20.
189. Nasiri Vatan, H., Ebrahimi-Kahrizsangi, R., & Kasiri-Asgarani, M. (2016). *International Journal of Electrochemical Science, 11*, 5631–5654.
190. Pezzato, L., Lorenzetti, L., Tonelli, L., Bragaggia, G., Dabalà, M., Martini, C., & Brunelli, K. (2021). *Surface and Coatings Technology, 428*, 127901.
191. Yu, L., Cao, J., & Cheng, Y. (2015). *Surface and Coatings Technology, 276*, 266–278.
192. Lou, B. S., Lin, Y. Y., Tseng, C. M., Lu, Y. C., Duh, J. G., & Lee, J. W. (2017). *Surface and Coatings Technology, 332*, 358–367.
193. Chen, B., Zhang, M., Li, X., Dong, Z., Jia, Y., & Li, C. (2020). *Progress in Organic Coatings, 147*, 105767.

194. Gao, Y., Xiao, S., Wu, H., Wu, C., Chen, G., Yin, Y., & Chu, P. K. (2023). *Ceramics International, 49*, 37475–37485.
195. Ao, N., Liu, D., Wang, S., Zhao, Q., Zhang, X., & Zhang, M. (2016). *Journal of Materials Science and Technology, 32*, 1071–1076.
196. Ji, R., Wang, S., Zhao, X., Zou, Y., Zhang, T., Qian, X., Chen, G., Wang, Y., Ouyang, J., Jia, D., & Zhou, Y. (2024). *Surface and Coatings Technology, 494*, 131438.
197. Zhang, X., Zhang, Y., Lv, Y., Dong, Z., Yang, L., Zhang, E., Hashimoto, T., & Zhou, X. (2023). *Journal of Magnesium and Alloys, 11*, 2182–2195.
198. Wang, X., Yan, H., Hang, R., Shi, H., Wang, L., Ma, J., Liu, X., & Yao, X. (2021). *Journal of Materials Research and Technology, 11*, 2354–2364.
199. Han, Q., Li, Y., Lu, X., Mei, D., Chen, Q., Zhang, T., & Wang, F. (2021). *Materials Letters, 293*, 129731.
200. Husak, Y., Olszaniecki, J., Pykacz, J., Ossowska, A., Blacha-Grzechnik, A., Waloszczyk, N., Babilas, D., Korniienko, V., Varava, Y., Diedkova, K., Kyrylenko, S., Hodzic, A., Krichbaum, M., Lu, X., Dryhval, B., Pogorielov, M., & Simka, W. (2024). *Applied Surface Science, 654*, 159387.
201. López-Ortega, A., Sáenz de Viteri, V., Alves, S. A., Mendoza, G., Fuentes, E., Mitran, V., Cimpean, A., Dan, I., Vela, A., & Bayón, R. (2022). *Biomaterials Advances, 138*, 212875.
202. Lou, B. S., Lee, J. W., Tseng, C. M., Lin, Y. Y., & Yen, C. A. (2018). *Surface and Coatings Technology, 350*, 813–822.
203. Tonelli, L., Pezzato, L., Dolcet, P., Dabalà, M., & Martini, C. (2018). *Wear, 404–405*, 122–132.
204. Han, B., Yang, Y., Li, J., Deng, H., & Yang, C. (2018). *International Journal of Electrochemical Science, 13*, 9166–9182.
205. Zhang, Y., Chen, F., Zhang, Y., & Du, C. (2020). *Tribology International, 146*, 106135.
206. Zhang, Y., Chen, F., Zhang, Y., Liu, Z., Wang, X., & Du, C. (2019). *Surface and Coatings Technology, 364*, 144–156.
207. Keshri, A. K., & Agarwal, A. (2011). *Journal of Thermal Spray Technology, 20*, 1217–1230.
208. Abbasi, S., Mahboob, A., Bakhtiari Zamani, H., Bilesan, M. R., Repo, E., & Hakimi, A. (2022). *Journal of Nanomaterials, 2022*, 675038.
209. Arslan, E., Totik, Y., Demirci, E. E., & Efeoglu, I. (2013). *Surface and Coatings Technology, 214*, 1–7.
210. Imshinetskiy, I., Kashepa, V., Nadaraia, K., Mashtalyar, D., Sinebryukhov, S., & Gnedenkov, S. (2023). *MATEC Web of Conferences, 376*, 01010.
211. Daavari, M., Atapour, M., Mohedano, M., Arrabal, R., Matykina, E., & Taherizadeh, A. (2021). *Surfaces and Interfaces, 22*, 100850.

Chapter 14
Advances in Anticorrosion and Anti-wear Plasma Electrolytic Oxidation Coatings for Titanium and Its Alloys

Aydin Bordbar-Khiabani

14.1 Introduction

14.1.1 The Engineering Importance of Titanium and Its Alloys

Titanium (Ti) and its alloys are widely used in engineering due to their unique combination of high strength-to-weight ratio, excellent corrosion resistance, and biocompatibility [1, 2]. These properties make titanium essential in industries requiring reliable performance under extreme conditions [2]. With a low density of approximately 4.5 g·cm^{-3} and high mechanical strength, titanium is particularly suited for lightweight structural applications in aerospace, automotive, and sports equipment [3]. Its ability to form a stable and protective titanium oxide (TiO_2) layer ensures exceptional corrosion resistance in aggressive environments, including marine, chemical, and biomedical applications [4]. This natural passivation significantly extends the material's service life in demanding conditions [5].

Titanium's versatility spans multiple industries. A key advantage of titanium is its outstanding biocompatibility, making it the preferred material for implants, dental devices, and prosthetics [6]. Its resistance to bodily fluids and nontoxic nature ensure long-term compatibility with human tissues [7]. In high-temperature applications, such as jet engine components and industrial heat exchangers, titanium maintains stability where many conventional materials degrade [3]. Additionally, its excellent fatigue and corrosion resistance make it highly reliable in cyclic loading conditions,

A. Bordbar-Khiabani (✉)
Materials Performance and Interactions, VTT Technical Research Centre of Finland Ltd., Espoo, Finland
e-mail: aydin.bordbar-khiabani@vtt.fi

V. S. Saji, M. Zheludkevich (eds.), *Plasma Electrolytic Oxidation – Fundamentals, Advances and Applications*,
https://doi.org/10.1007/978-3-031-92429-3_14

which are common in aerospace and biomedical applications [8]. In aerospace, it is used for airframe structures, turbine blades, and fasteners, contributing to improved fuel efficiency and performance. Its corrosion resistance makes it highly durable in seawater, supporting applications in ship propellers, offshore structures, and desalination systems [1–5]. In chemical processing, titanium is essential for equipment handling aggressive chemicals, such as reactors and heat exchangers. The energy sector also benefits from its corrosion-resistant properties, particularly in nuclear and thermal power plants, where it extends the lifespan of condenser tubes [8]. Additionally, titanium is incorporated into high-performance vehicles and sporting goods to achieve strength and durability while reducing weight [9].

14.1.2 Addressing Corrosion and Wear Challenges

Titanium and its alloys are widely recognized for their excellent corrosion resistance and mechanical properties, yet they face challenges in environments where both wear and corrosion act simultaneously [10]. The naturally formed oxide layer provides strong protection against many corrosive media, including seawater and acidic solutions [1–5]. However, under aggressive conditions, such as high chloride concentrations, oxidative environments, elevated temperatures, or fluctuating pH levels, this passive film can degrade [10, 11]. This degradation increases susceptibility to localized corrosion phenomena, including pitting and crevice corrosion, which can compromise structural integrity [12]. In applications where mechanical stresses and corrosive environments coexist, such as marine structures and biomedical implants, additional concerns arise, including corrosion fatigue and stress corrosion cracking, which can significantly reduce material durability [8, 13].

Wear is another critical issue in titanium applications, particularly in tribological conditions involving sliding, abrasion, or impact [14]. Despite its high strength, titanium exhibits relatively poor wear resistance due to its low hardness and strong tendency to gall, where material transfer occurs between sliding surfaces [14, 15]. This limitation is particularly problematic in aerospace components, biomedical implants, and industrial machinery, where continuous surface interactions lead to progressive material loss [16]. The combination of wear and corrosion, known as tribocorrosion, further accelerates degradation by simultaneously weakening the protective oxide layer and increasing mechanical wear, ultimately reducing the service life of components and raising maintenance costs [17, 18].

A major challenge in mitigating these effects is balancing wear resistance with corrosion protection [18]. Conventional surface treatments, such as nitriding and carburizing, can significantly improve wear resistance but often compromise the integrity of the passive oxide layer, reducing corrosion protection [19]. This trade-off is particularly critical in biomedical applications, where surface modifications must maintain biocompatibility while enhancing durability in body fluids [20]. Similarly, in marine and chemical processing environments, any surface treatment designed to improve wear performance must withstand highly corrosive conditions without sacrificing structural stability [21, 22].

14.1.3 Overview of Plasma Electrolytic Oxidation

Plasma electrolytic oxidation (PEO), also known as micro-arc oxidation (MAO), anodic spark deposition (ASD), micro-plasma oxidation (MPO), and electrolytic plasma oxidation (EPO), is an advanced surface modification technique designed to enhance the properties of both valuable metals and their alloys, such as aluminum [23, 24], titanium [25], and magnesium [26, 27], as well as nonvaluable metals and alloys, including tantalum [28], niobium [29], zirconium [30], zinc [31, 32], copper [33], silicon [34, 35], nickel-titanium [36], steels [37], brass [38], and metal matrix composites [39]. By combining electrochemical and plasma-assisted processes, PEO facilitates the formation of ceramic-like oxide coatings that significantly improve corrosion resistance, wear resistance, and thermal stability [40]. This makes it a valuable approach for applications where conventional surface treatments fall short [41].

The PEO process operates by applying high-voltage electrical discharges in an electrolytic solution. This electrochemical surface modification technique involves connecting the workpiece to the anode and a suitable cathode—often made of materials like stainless steel—with both electrodes immersed in the electrolyte. The high voltage induces micro-discharges on the metal surface, leading to localized plasma reactions that transform the surface into a dense, wear-resistant oxide layer. During oxidation, the behavior of the treated samples can be monitored by a camera, as depicted schematically in Fig. 14.1a (top) [40, 41]. Plasma micro-discharges generated during the process induce localized melting and rapid solidification (Fig. 14.1a, bottom), resulting in a coating with a characteristic microstructure, typically consisting of a porous outer layer and a compact inner barrier layer (Fig. 14.1b, top) [42]. Titanium and its alloys, in particular, exhibit excellent compatibility with the process due to their ability to naturally form a stable and protective oxide layer, which is further enhanced through PEO treatment [43].

PEO coatings on titanium provide substantial improvements in corrosion resistance by forming a dense inner barrier layer that effectively shields the substrate from aggressive environments, including chloride-rich and acidic solutions [44]. Additionally, the ceramic-like nature of the coating increases hardness and reduces friction, improving wear resistance under tribological conditions [44]. These coatings also maintain their protective properties at elevated temperatures, making them suitable for aerospace and high-temperature industrial applications [43, 45]. The versatility of PEO allows for tailoring coating properties by adjusting process parameters such as electrolyte composition, current density, and pulse waveform, enabling optimization for specific applications [40].

PEO coatings on titanium have gained significant attention for their potential in biomedical implants, aerospace components, and industrial tooling [43]. Their ability to enhance the durability of titanium surfaces while also contributing to multifunctionality—such as biocompatibility and photocatalytic activity—makes them particularly valuable in advanced engineering applications [43–45]. While several comprehensive review papers on PEO coatings have been published since 2020 [43,

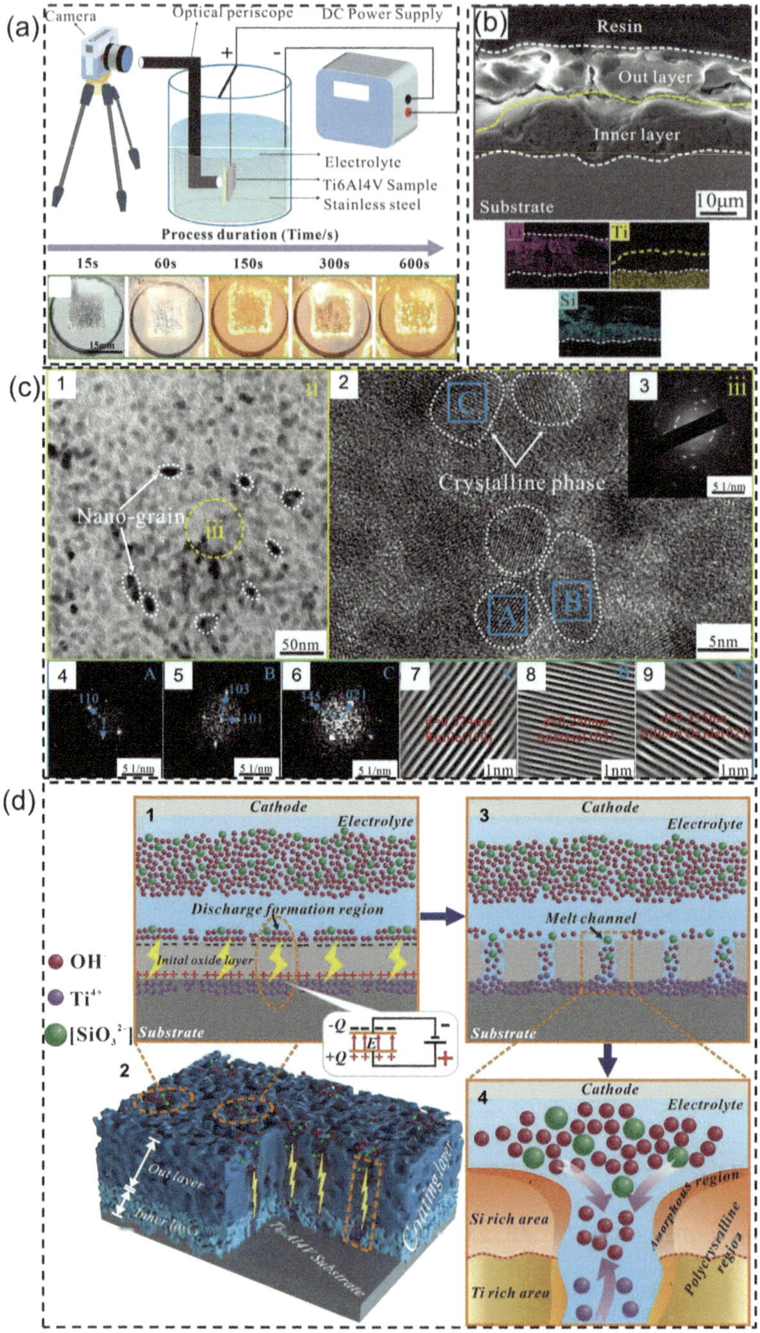
(a)
Camera
Optical periscope
DC Power Supply
Electrolyte
Ti6Al4V Sample
Stainless steel
Process duration (Time/s)
15s
60s
150s
300s
600s
(b)
Resin
Out layer
Inner layer
Substrate
10μm
Ti
Si
(c)
Nano-grain
Crystalline phase
50nm
5nm
(d)
Cathode
Electrolyte
Discharge formation region
Inital oxide layer
Substrate
Melt channel
OH⁻
Ti⁴⁺
[SiO₃²⁻]
Out layer
Coating layer
Si rich area
Ti rich area
Amorphous region
Polycrystalline region

Fig. 14.1 (**a**) Schematic representation of the PEO process setup for Ti-6Al-4V samples, illustrating process duration and the evolution of micro-discharges over time. (**b**) Cross-sectional SEM image of the coated sample, showing the outer and inner layers of the PEO coating along with elemental mapping. (**c**) Transmission electron microscopy (TEM) analysis of the PEO coating, revealing nano-grains, crystalline phases, and corresponding selected area electron diffraction (SAED) patterns indicating different crystallographic orientations. (**d**) Proposed mechanism of coating formation during PEO, depicting discharge formation, melt channel evolution, and the development of outer and inner coating layers with Si-rich and Ti-rich regions. Reproduced with permission from Ref. [42] © 2024 Elsevier B.V.

46–48], this chapter aims to build upon that foundation by focusing on recent advancements and technological refinements in enhancing the anticorrosion and anti-wear properties of PEO coatings on titanium and its alloys.

14.2 Anticorrosion Mechanisms of PEO Coatings on Titanium

14.2.1 Corrosion Resistance Fundamentals

The typical morphology of PEO coatings features micro- and nano-pores resulting from the rapid solidification of molten substrate material expelled through micro-discharge channels during processing (Fig. 14.1c1). The corrosion resistance of PEO coatings on titanium is primarily derived from the formation of a dense and adherent oxide layer during the PEO process [49]. This oxide layer serves as a protective barrier, preventing corrosive species such as water, oxygen, and chloride ions from reaching the underlying titanium substrate [49]. The ceramic-like nature of the PEO coating, characterized by high chemical stability and low solubility in aggressive environments, significantly enhances its ability to withstand corrosive conditions [50]. The primary component of the oxide layer is TiO_2, which exists in multiple crystalline forms, including anatase and rutile (Fig. 14.1c2–9) [50]. The corrosion resistance is often associated with the compact structure of the rutile phase, which has lower permeability to aggressive ions and provides better long-term stability compared to anatase [49, 50].

A key factor influencing the corrosion performance of PEO coatings is their microstructural integrity [43]. These coatings typically exhibit a multilayered structure, consisting of an outer porous layer and an inner dense layer directly in contact with the substrate (Fig. 14.1b, top) [42]. While the porous outer layer can allow the ingress of corrosive agents, the inner compact layer serves as the primary barrier that prevents electrolyte penetration [49]. Additionally, the microstructure is influenced by the incorporation of electrolyte-derived elements into the oxide layer during the PEO process [51]. Applying a high voltage causes negatively charged ions (anions) in the electrolyte to migrate toward the positively charged anode (the metal substrate). These anions are drawn into the molten channels created by micro-discharges on the substrate's surface, where they participate in plasma-assisted reactions, leading to the incorporation of electrolyte-derived elements into the growing oxide layer and contributing to the formation of a protective coating (Fig. 14.1d1–4) [42]. These incorporated species, such as silicates, phosphates, or vanadates, can improve the chemical stability of the coating and further enhance its corrosion resistance by forming secondary protective phases [49]; for instance, silicon incorporation from a silicate electrolyte, as evidenced by EDS elemental mapping in Fig. 14.1b (bottom), contributes to the formation of a more robust oxide layer [42].

The electrical properties of the oxide layer also contribute to its corrosion resistance [52]. The high electrical resistivity of the PEO coating minimizes electrochemical activity at the surface, reducing the potential for anodic dissolution of the substrate [40, 43, 44]. This dielectric behavior is particularly beneficial in aggressive environments with high chloride concentrations or acidic conditions, where conventional passive films on titanium may break down [49]. Additionally, some PEO coatings exhibit self-healing properties, where minor defects in the oxide layer can be repaired through in situ oxide reformation, thereby improving long-term durability.

The corrosion performance of PEO coatings is strongly influenced by process parameters, including voltage, current density, electrolyte composition, and oxidation time [43]. These parameters dictate critical coating characteristics, such as thickness, porosity, and phase composition, all of which determine the coating's ability to provide effective corrosion protection [53]. Tailoring the electrolyte composition can significantly enhance coating performance by reducing porosity, promoting the formation of more stable oxide phases, and increasing thickness [54]. The incorporation of silicates, phosphates, and nanoparticles into the electrolyte has been shown to improve coating uniformity and limit defect formation, leading to superior barrier properties [54, 55]. Extending the oxidation time in the PEO process can also influence the coating's microstructure by increasing the intensity and duration of micro-discharges (Fig. 14.1a, bottom), which may lead to the formation of larger pores and a rougher surface [42].

To further enhance the corrosion resistance of PEO coatings, various pretreatment and posttreatment strategies can be employed [46–48]. Pretreatment processes, such as surface cleaning, anodizing, and additive manufacturing (AM), influence the coating's adhesion and microstructural uniformity, ultimately affecting its protective performance [47, 48]. Posttreatment methods, including sealing treatments, secondary topcoats, and chemical or thermal modifications, are essential for optimizing the coating's protective properties [43, 46–48]. These techniques help to minimize porosity, increase the coating's thickness, and enhance its chemical stability in highly aggressive environments [56].

14.2.2 Advances in Substrate Engineering

Substrate engineering plays an important role in optimizing the performance of PEO coatings for titanium and its alloys, as it directly influences the coating's microstructure and, consequently, its corrosion resistance [57]. Recent advances in alloying, surface texturing, laser pretreatments, and additive manufacturing have demonstrated significant potential in improving the corrosion performance of PEO-coated titanium through tailored substrate modifications [43, 46–48].

The development of new titanium alloy systems, such as Ti-Nb-Zr-X alloys (where X may be Ta, Mo, or other metals [58, 59]), and high-entropy alloys (HEAs) like Ti-Nb-Ta-Zr-Hf [60], has gained significant attention for enhancing PEO coating performance. These alloying approaches introduce elements such as niobium, zirconium, and hafnium into the substrate, which promote the formation of stable oxide phases, including niobium pentoxide (Nb_2O_5), zirconium oxide (ZrO_2), and hafnium oxide (HfO_2), during the PEO process [58–60]. These oxides, when coexisting with TiO_2, contribute to a dense and protective coating structure, reducing porosity and enhancing resistance to corrosive environments [58]. For instance, studies have shown that the incorporation of Nb_2O_5 in PEO coatings reduces the susceptibility of titanium alloys to pitting corrosion in chloride-containing solutions by forming a barrier layer with low ionic conductivity [61]. Similarly, the addition of Zr enhances the growth of ZrO_2 in the coating, which is chemically inert and stable, thereby improving the long-term corrosion resistance in harsh environments [58]. The synergy of these alloying elements with titanium ensures the development of coatings with optimized microstructures that outperform coatings formed on conventional pure titanium or Ti-6Al-4V substrates [62].

Surface texture modifications prior to PEO treatment also play a substantial role in determining the final properties of the coatings [63]. Pretreatments such as acid etching, sandblasting, and polishing alter the roughness and surface energy of the substrate, which influences the coating's adherence and growth during PEO. For example, substrates with increased surface roughness from sandblasting promote enhanced coating thickness and uniformity, as the textured surface provides a greater contact area for the discharge events during the PEO process [64]. This thicker and denser coating structure contributes to improved corrosion resistance by minimizing pathways for electrolyte penetration [49]. On the other hand, excessively rough surfaces can lead to localized current densities during PEO, potentially resulting in coating defects. Thus, optimizing the degree of surface roughness is essential to achieve a balance between coating thickness, porosity, and corrosion performance.

Laser irradiation has been explored as an auxiliary energy source to enhance the functional properties of PEO coatings on titanium by influencing both the oxidation process and microstructural evolution [65]. Studies have shown that applying laser-assisted irradiation during PEO promotes the phase transformation of TiO_2 from anatase to rutile, reduces oxygen vacancy defects, and contributes to the formation of a distinctive porous "ant nest" microstructure (Fig. 14.2a). The introduction of

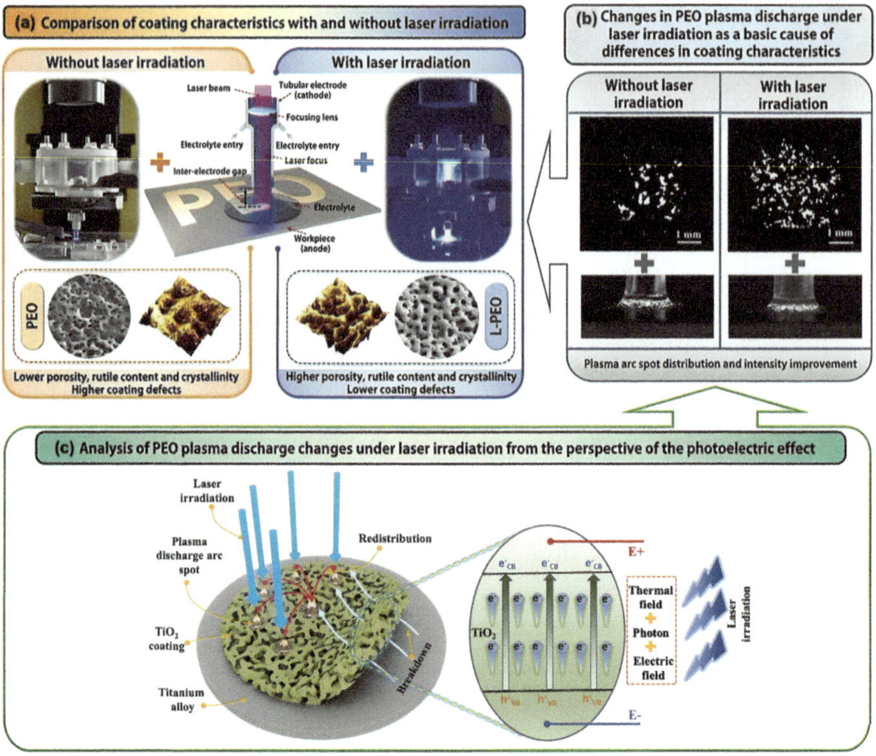

Fig. 14.2 (**a**) Comparison of coating characteristics with and without laser irradiation, showing differences in porosity, crystallinity, and coating defects between PEO and L-PEO coatings. (**b**) Changes in PEO plasma discharge under laser irradiation, illustrating improved plasma arc spot distribution and intensity, which influence coating characteristics. (**c**) Analysis of PEO plasma discharge modifications under laser irradiation, highlighting the role of the photoelectric effect in redistributing charge carriers within the TiO_2 coating and enhancing plasma discharge behavior. Reproduced with permission from Ref. [65] © 2024 Elsevier B.V.

laser irradiation affects the plasma discharge characteristics by restricting electric field diffusion at the anode surface, thereby lowering the termination voltage while increasing both discharge intensity and uniformity. Experimental observations indicate that the maximum plasma discharge percentage rises significantly under laser irradiation, leading to a more homogeneous and refined coating morphology (Fig. 14.2b). Additionally, electrochemical impedance spectroscopy (EIS) and instantaneous photoresponse measurements confirm that the laser-modified PEO coatings exhibit an improved and sustained photoelectric response. The enhanced plasma discharge behavior is attributed to the photogenerated electrons from the semiconductor TiO_2 layer, which act as initial carriers in an electron avalanche, facilitating plasma generation at the anode surface. This process fundamentally improves the coating characteristics, yielding denser and more corrosion-resistant PEO layers with refined microstructural features (Fig. 14.2c) [65].

Laser-based pretreatments, such as femtosecond laser surface modification, have emerged as a precise and effective approach for substrate engineering [66]. In the case of Ti-6Al-4V alloys, femtosecond laser treatment generates a controlled micro-textured surface with periodic patterns, which enhances the nucleation and growth of PEO coatings [67]. The resulting coatings exhibit a more compact structure with reduced porosity and improved thickness distribution. For example, femtosecond laser-treated Ti-6Al-4V substrates have been shown to form PEO coatings with superior corrosion resistance in 0.9% NaCl solution, attributed to the uniform microstructure and reduced coating defects [67]. Additionally, the laser-induced surface features can influence the phase composition of the coating, encouraging the formation of corrosion-resistant phases such as anatase TiO_2, which further strengthens the coating's protective capabilities [67].

Surface patterning of titanium via AM has been shown to enhance its corrosion resistance by allowing precise control over grain orientation, which influences the material's susceptibility to corrosive environments [68]. The application of AM techniques, such as selective laser melting (SLM) and electron beam melting (EBM), has introduced another dimension to substrate engineering for PEO coatings [69, 70]. AM enables the fabrication of complex titanium components with tailored microstructures that differ from their conventionally manufactured counterparts. These differences, including increased porosity, grain refinement, and residual stresses, affect the PEO coating process and its resulting properties [70]. For instance, PEO coatings formed on AM Ti-6Al-4V substrates produced by SLM tend to have higher porosity compared to coatings on conventionally fabricated alloys [71]. However, AM substrates with refined microstructures often allow for a more uniform oxide layer growth, leading to improved corrosion resistance. EBM, with its inherent ability to reduce residual stresses in printed components, has also been shown to positively influence the adhesion and thickness of PEO coatings, resulting in better protection against corrosion in saline environments [72]. Comparative studies between AM and bulk Ti substrates reveal that the AM substrates, despite some initial challenges such as porosity, can achieve comparable or superior corrosion resistance when optimized with PEO coatings.

14.2.3 Role of Electrical Parameters

The influence of electrical parameters on the corrosion resistance of PEO coatings arises from their ability to control plasma discharge behavior, which dictates the coating's microstructure, phase composition, and overall barrier properties [52]. A study explored the corrosion resistance of PEO coatings on Ti-6Al-4V alloys in silicate electrolytes, focusing on the impact of cathode current and soft sparking mechanisms. Findings showed that higher cathode current reduced TiO_2 formation, resulting in denser coatings but thinner oxide layers. Corrosion resistance initially increased with rising cathode current but declined at higher values due to changes in amorphous phase distribution. Thicker coatings alone did not guarantee better protection; instead, coatings with higher density and lower amorphous content exhibited superior electrochemical stability. The study highlighted that achieving optimal corrosion resistance in PEO-treated titanium alloys requires balancing coating density, thickness, and amorphous content, as excessive cathode current can negatively impact protective properties despite increasing density [73].

The influence of pulse modes and waveform shapes on the microstructure and corrosion resistance of PEO coatings has been closely examined, particularly regarding the role of soft-spark discharges [74]. Studies have shown that soft sparks primarily occur near the outer porous layer, penetrating weak regions in the amorphous interfacial layer between the coating and the substrate. This process facilitates the formation of filamentary discharges, leading to increased rutile phase content and bubble-like pores. As illustrated in Fig. 14.3a, a treatment duration of 1200 s results in a compact barrier layer (≤0.5 μm) with a porous outer layer (~15 μm). Extending the oxidation time to 12,000 s (Fig. 14.3c) enhances the soft-spark layer thickness, with molten oxides filling micro-sized pores, increasing coating density. Scanning electron microscopy (SEM) and energy dispersive spectroscopy (EDS) analysis reveal that Si incorporation is primarily detected in the initial layer, suggesting co-deposition of TiO_2 and SiO_2 during microarc discharge. Further TEM and SAED patterns (Fig. 14.3b, d) show that prolonged soft-sparking reduces the amorphous interfacial layer thickness from ~20 nm at 1200 s to ~5 nm at 12000 s, while rutile-phase nanocrystals (~10 nm) become more prominent. Potentiodynamic polarization (PDP) curves (Fig. 14.3e) indicate that although extended processing increases coating thickness, corrosion resistance does not improve significantly due to porosity and filamentary channels introduced by prolonged discharge activity, highlighting the importance of optimizing pulse parameters for a balance between coating density and electrochemical stability [74].

Voltage regimes also determine the nature and size of plasma discharges, impacting the coating's structure and corrosion performance. A voltage range of 400–600 V is typically used for titanium alloys, with higher voltages promoting phase transformations and thicker coatings [75]. For example, when a voltage of 550 V was applied during PEO of a Ti-6Al-4V alloy, the resulting oxide layer contained both anatase and rutile phases, with the latter improving corrosion resistance in saline conditions. However, voltages exceeding this range can lead to uncontrolled discharge events that cause microcracks and reduced coating effectiveness. Precise voltage control ensures a balance between coating growth and structural integrity, minimizing the risk of electrolyte penetration [76].

Frequency adjustments in pulsed PEO processes directly impact the discharge dynamics and oxide layer properties. Higher pulse frequencies, such as those exceeding 500 Hz, reduce the energy per discharge and result in finer microstructures [77]. For example, PEO coatings on Ti-6Al-4V produced at 1 kHz exhibited significantly lower porosity compared to those produced at 200, 400, 600, and 800 Hz. The finer structure improved the coating's resistance to localized corrosion, as demonstrated by lower i_{corr} values in potentiodynamic polarization tests. This indicates that higher pulse frequencies promote the formation of compact and dense oxide layers, effectively preventing the ingress of corrosive species [77].

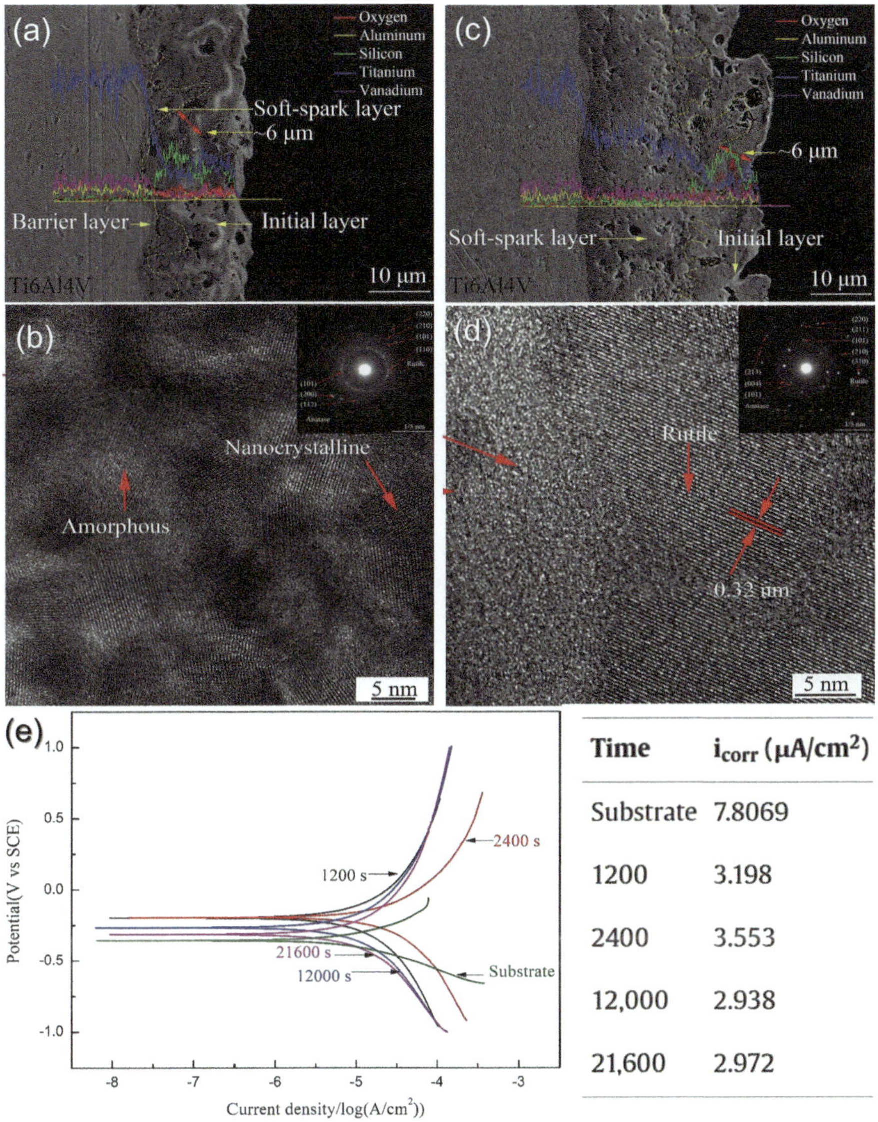

Time	i_{corr} (μA/cm^2)
Substrate	7.8069
1200	3.198
2400	3.553
12,000	2.938
21,600	2.972

Fig. 14.3 (**a**) SEM cross-sectional images of coatings obtained with soft-sparking mode at 1200 s and (**c**) 12,000 s. (**b**, **d**) TEM high-magnification images and SAED patterns of coatings formed at 1200 s and 12,000 s, respectively. (**e**) PDP curves of samples with extracted corrosion current density (i_{corr}) values. Reproduced with permission from Ref. [74] © 2023 Elsevier B.V.

The duty cycle, defined as the percentage of time the current or voltage is applied during a cycle, also has a profound effect on coating properties. A lower duty cycle reduces the heat accumulation at the substrate surface, preventing excessive thermal stresses that can cause microcracks [78]. For example, a duty cycle of 30% applied during the PEO of a Ti-29Nb-13Ta-4.6Zr alloy produced a uniform oxide layer with minimal defects. This coating exhibited enhanced corrosion resistance in acidic environments due to the reduced porosity and the presence of corrosion-resistant oxides such as ZrO_2 and TiO_2 [79].

Polarity adjustments, particularly in bipolar waveforms, influence the incorporation of electrolyte-derived species into the coating. For instance, applying cathodic pulses during the PEO of titanium in phosphate-based electrolytes promotes the incorporation of phosphate ions into the oxide layer. This results in the formation of phosphate-rich phases that act as additional barriers to corrosive ions. Coatings produced using this approach demonstrated lower passive current densities and improved pitting resistance compared to those formed under unipolar conditions [80].

14.2.4 Advances in Electrolyte Formulations

The formulation of electrolytes in PEO has been extensively refined to enhance the corrosion resistance of titanium-based coatings. Among the most promising recent approaches, the addition of soluble salts and insoluble particles into the electrolyte has been widely investigated. These strategies significantly influence the microstructure and chemical composition of the oxide layer by modifying its porosity, thickness, and phase distribution, all of which are critical factors for improving corrosion resistance [43, 46–48].

The introduction of soluble salts into the electrolyte results in the dissociation of ions, with cations preferentially adsorbing onto micro-discharge channels due to their opposite charge [47, 49]. This mechanism facilitates the in situ formation of stable oxides during the growth of TiO_2, which can either enhance the chemical stability of the coating or modify its microstructural characteristics to reduce corrosion susceptibility [81]. For example, the addition of sodium aluminate ($NaAlO_2$) in an alkaline electrolyte during PEO leads to the incorporation of aluminum oxide (Al_2O_3) within the oxide layer [80]. The presence of Al_2O_3 strengthens the passive layer by increasing its chemical stability, particularly in chloride-rich environments, which significantly reduces pitting corrosion susceptibility [80]. Another example is the introduction of sodium molybdate (Na_2MoO_4), which promotes the formation of chemically stable molybdenum oxide (MoO_x) phases, primarily MoO_2 (Mo^{4+}) and MoO_3 (Mo^{6+}). These oxides enhance the protective nature of the coating by acting as diffusion barriers, reducing ionic conductivity, and improving long-term corrosion resistance. While MoO_2 contributes to enhanced electrochemical stability, MoO_3 can provide oxidation resistance but may be susceptible to dissolution in chloride-rich environments such as 0.9 wt% NaCl solution, potentially influencing coating durability over time [82]. The addition of potassium fluorozirconate (K_2ZrF_6) to a silicate-based electrolyte significantly enhances the PEO process by increasing ion concentration and electrolyte conductivity, thereby intensifying plasma discharges and facilitating coating growth [83]. Under a strong electric field, negatively charged ZrF_6^{2-}, SiO_3^{2-}, and OH^- ions migrate toward the anode, contributing to the deposition of molten oxides and the thickening of the ceramic layer. The interaction between ZrF_6^{2-} and hydroxide ions leads to the formation of $Zr(OH)_4$ colloids, which undergo rapid dehydration and sintering at high plasma temperatures, forming stable ZrO_2 phases that reinforce the coating structure (Fig. 14.4a). PDP curves (Fig. 14.4d) show that coatings containing 6 g·L^{-1} K_2ZrF_6 exhibit the

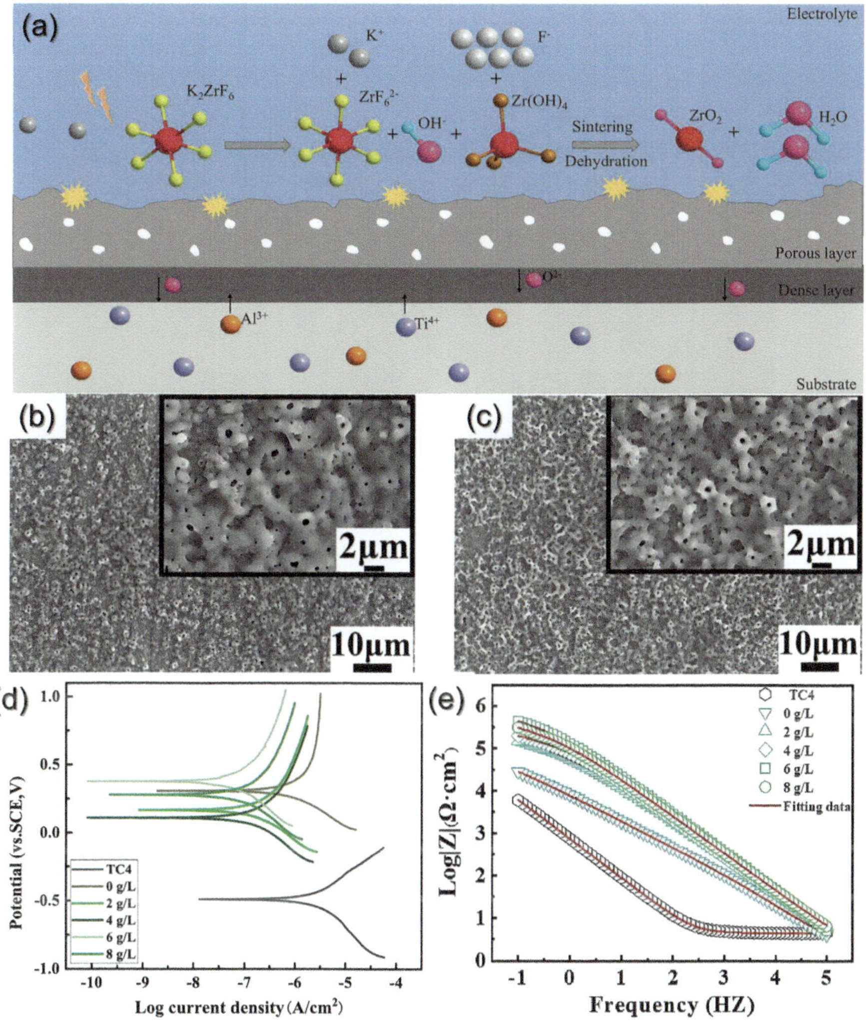

Fig. 14.4 (**a**) Schematic illustration of the evolution of PEO ceramic coatings on a TC4 substrate, depicting the incorporation mechanism of ZrO_2 into the coatings during PEO treatment. SEM surface morphologies of PEO coatings formed on TC4 with varying K_2ZrF_6 concentrations: (**b**) 0 g·L^{-1} and (**c**) 6 g·L^{-1}. (**d**) PDP curves and (**e**) Bode modulus plots in 3.5 wt% NaCl solution. Reproduced with permission from Ref. [83] © 2023 Elsevier B.V.

highest E_{corr} (0.377 V), lowest i_{corr}, and highest polarization resistance, indicating improved corrosion protection. This enhancement is further supported by Bode modulus plots (Fig. 14.4e), where coatings with 6 g·L^{-1} K_2ZrF_6 show a higher $|Z|_{0.1}$Hz value, reflecting superior electrochemical stability due to a denser microstructure (Fig. 14.4c). However, in pure PEO (0 g·L^{-1}), excessive internal stress leads to crack formation and increased porosity (Fig. 14.4b), reducing corrosion resistance by promoting the infiltration of aggressive ions into the substrate [83].

The incorporation of calcium, phosphorus, and copper elements into PEO coatings enhances microstructure, wettability, and corrosion resistance [84]. Surface analysis (Fig. 14.5a) shows that the co-incorporated sample (T2) exhibits reduced porosity, leading to improved discharge stability compared to the pure PEO coating (T0). Wettability measurements (Fig. 14.5b) indicate that ternary ion incorporation increases hydrophobicity, further contributing to corrosion resistance. PDP curves

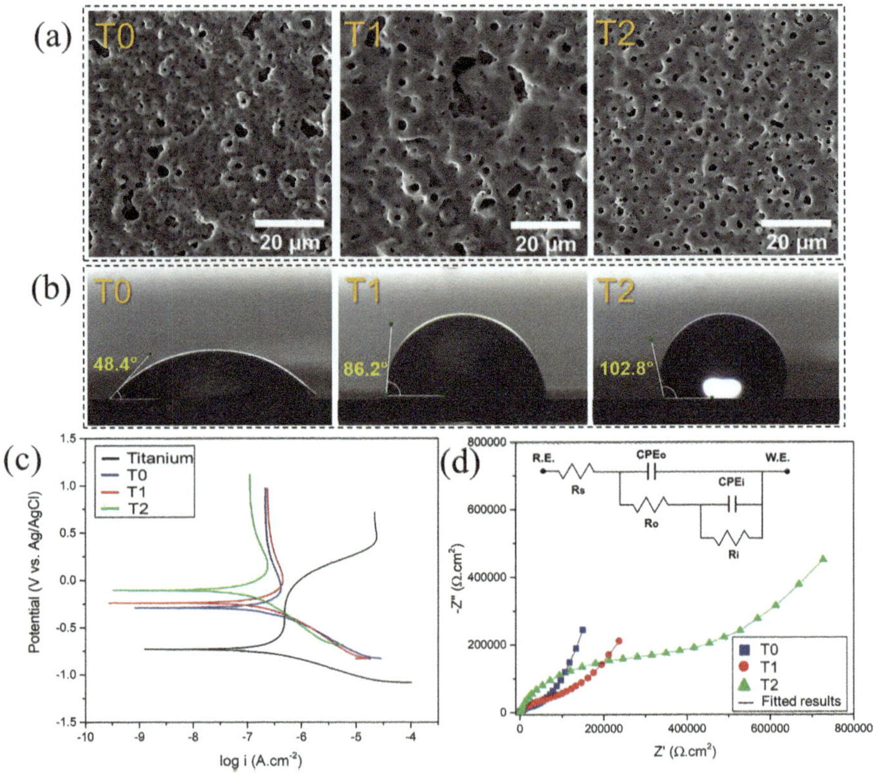

Fig. 14.5 (**a**) SEM images of TiO_2 coatings prepared in different electrolyte compositions: T0 (base electrolyte with Na_3PO_4 and KOH), T1 (addition of 1 g·L^{-1} $C_4H_6CaO_4$), and T2 (further addition of 2 g·L^{-1} $C_4H_6CuO_4$), showing variations in surface morphology. (**b**) Water contact angle measurements of T0, T1, and T2. (**c**) PDP curves, and (**d**) Nyquist plots with the corresponding equivalent circuit in artificial saliva solution at 37 °C. Reproduced with permission from Ref. [84] © 2022 Elsevier B.V.

(Fig. 14.5c) reveal that T2 has the lowest i_{corr} (0.095 μA·cm^{-2}) and nobler corrosion potential (E_{corr}, −107 mV), with copper playing a dominant protective role. Nyquist plots (Fig. 14.5d) confirm enhanced dielectric properties, while EIS results show an increase in $R_{i/o}$ and a decrease in $CPE_{i/o}$, indicating a more stable layer with minimized defects, reducing corrosive ion diffusion and improving coating durability [84].

Another widely adopted approach involves the addition of insoluble particles such as metal oxides and carbon-based materials to the electrolyte. These particles, when positively charged, are attracted to the negatively charged micro-discharge channels, where they are deposited within the oxide layer, effectively reducing porosity and improving barrier properties [85]. One of the most studied examples is the addition of ZrO_2 nanoparticles into the electrolyte. These particles become embedded within the TiO_2 matrix, decreasing the overall porosity of the coating and increasing its density [86]. The resulting composite layer exhibits superior corrosion resistance by limiting electrolyte penetration pathways. Similarly, the incorporation of hematite (Fe_2O_3) nanoparticles has been shown to enhance the corrosion resistance of PEO coatings by reducing surface porosity and promoting the transformation of anatase to rutile. Additionally, the addition of Fe_2O_3 nanoparticles to the electrolyte resulted in increased coating thickness while reducing surface roughness. PDP results indicated that the polarization resistance of the coatings increased from 435.88 to 1735.25 kΩ·cm^2 with the addition of 3 g·L^{-1} Fe_2O_3 to the base electrolyte [87].

Graphene oxide (GO) and reduced graphene oxide (rGO) have also been explored as potential reinforcements for PEO coatings due to their exceptional barrier properties [88–91]. When dispersed in the electrolyte, these carbon-based particles integrate into the oxide layer, forming a more compact and defect-free structure [92]. The incorporation of rGO is particularly effective in reducing porosity and increasing the thickness of the PEO coating by influencing the voltage-time (V-T) response (Fig. 14.6a–d). The rGO/TiO_2 coatings improve corrosion resistance by increasing oxide layer thickness, reducing porosity, extending diffusion pathways (L2 > L1), and increasing the water contact angle, thereby minimizing corrosive agent penetration and substrate degradation (Fig. 14.6e) [93].

The presence of insoluble particles in the electrolyte also influences the distribution of discharge channels, modifying the microstructural development of the oxide layer [92]. For example, studies on the incorporation of boron carbide (B_4C) nanoparticles in the electrolyte show that the particles preferentially accumulate in the outer porous regions, effectively sealing discharge pores and reducing ionic conductivity. This sealing effect significantly improves the corrosion resistance of PEO coatings in saline media [94]. A similar effect is observed with diamond nanoparticles, where their deposition in microcracks and pores creates a more compact layer, preventing aggressive ions in phosphate buffer solution (PBS) from penetrating the PEO coatings on Ti-6Al-4V alloy [95].

Moreover, by co-adding these insoluble particles, researchers have developed coatings with superior corrosion resistance compared to conventional PEO

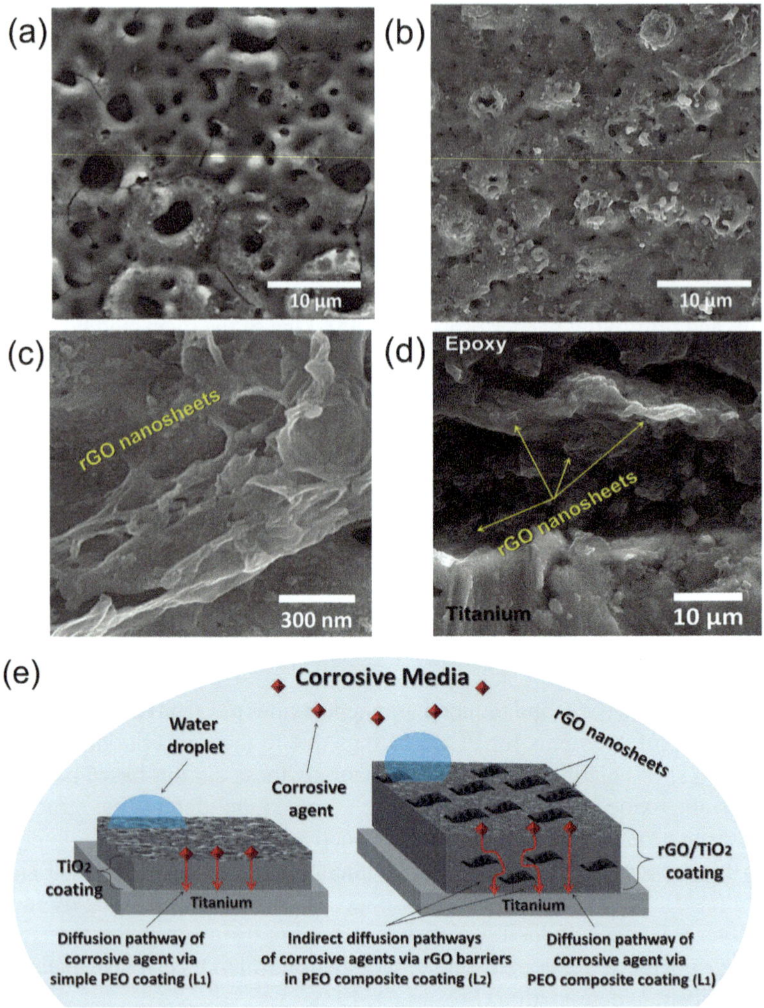

Fig. 14.6 SEM surface morphologies of (**a**) TiO_2 and (**b**) rGO/TiO_2 coatings. (**c**) High-magnification SEM image highlighting the presence of rGO nanosheets within the composite coating. (**d**) Cross-sectional SEM image showing rGO nanosheets embedded within the coating structure. (**e**) Schematic illustration of the corrosion protection mechanism, comparing the diffusion pathway of corrosive agents in a simple PEO coating versus the rGO/TiO_2 composite coating, where rGO nanosheets act as barriers, reducing direct diffusion and enhancing corrosion resistance. Reproduced with permission from Ref. [93] © 2019 Elsevier B.V.

coatings. The simultaneous use of germanium dioxide (GeO_2) and tantalum pentoxide (Ta_2O_5) in sodium polyphosphate results in hybrid coatings that benefit from both in situ phosphate incorporation and particle reinforcement through electrophoretic movement during TiO_2 growth in PEO. This synergistic approach not only enhances coating thickness and reduces porosity but also optimizes the chemical composition to maximize electrochemical stability, improve osteoblastic differentiation due to Ta_2O_5, and enhance antibacterial capability due to GeO_2, thereby providing long-term corrosion protection [96].

14.2.5 Enhancing Resistance via Posttreatment

Various posttreatment techniques have been explored to enhance the corrosion resistance of PEO-coated titanium by modifying the microstructure of the oxide layer [97]. These treatments primarily focus on sealing the inherent porosity and microcracks present in PEO coatings, which are the main pathways for electrolyte penetration [98–100]. Additionally, some posttreatment methods increase the coating thickness, creating a more effective diffusion barrier against corrosive ions [97–100]. Certain treatments also modify the surface chemistry, leading to the development of hydrophobic top layers that further reduce corrosive interactions. One of the most effective approaches to achieving these improvements involves duplex or hybrid PEO coatings, where secondary surface engineering techniques such as electrophoretic deposition (EPD), physical vapor deposition (PVD), sol-gel methods, and polymeric top coatings are applied to further enhance corrosion resistance [97–100].

The deposition of additional ceramic phases onto PEO coatings through techniques such as EPD has been widely investigated as a method to improve barrier properties [99–101]. For example, hydroxyapatite (HAp) and magnesium oxide (MgO) particles have been electrophoretically deposited onto PEO-coated titanium to seal surface defects and enhance corrosion protection in Ringer's solution [101, 102]. The HA and MgO top layer on TiO_2 PEO coatings not only blocks electrolyte penetration but also enhances bioactivity for biomedical applications and improves chemical stability in aggressive environments [101, 102]. The EPD of rGO on PEO-coated Ti-6Al-4V significantly enhances corrosion resistance, as evidenced by the TiO_2/rGO coating reducing i_{corr} to 2.19 $\mu A \cdot cm^{-2}$ compared to 9.85 $\mu A \cdot cm^{-2}$ for pure TiO_2 in palmitic acid-incorporated physiological solutions, which simulate the elevated PA levels found in elderly patients' blood post-implantation [103]. PVD has also been utilized to apply dense ceramic layers onto PEO coatings, particularly for increasing corrosion resistance in chloride-rich environments [104]. The deposition of TiN films onto PEO-coated CP-Ti via gas nitriding in a tube furnace with flowing N_2 gas has significantly reduced passive current densities. The polarization resistance of TiO_2 and TiO_2/TiN coatings was measured at 632.2 and 1451.9 $k\Omega \cdot cm^2$, respectively, demonstrating their effectiveness in limiting ionic diffusion through the coating [104].

In addition to ceramic deposition, polymeric and polymer-ceramic hybrid coatings have emerged as effective posttreatment strategies for PEO-coated titanium. A study investigated the impact of a polymethyl methacrylate (PMMA) sealing layer on PEO-coated dental implants exposed to pure and fluoride-containing saliva [105]. Camera image of the dental implant, micro-discharge events after 300 s of

PEO, and SEM image of the dental implant are shown in Fig. 14.7a–c, respectively. SEM images of PEO-coated implants (Fig. 14.7d) revealed widespread pores and cracks, while the surface SEM image of the duplex-coated implant (Fig. 14.7e) confirmed the successful deposition of PMMA, effectively sealing defects and creating a smoother, more compact structure [105]. Electrochemical analysis using PDP and

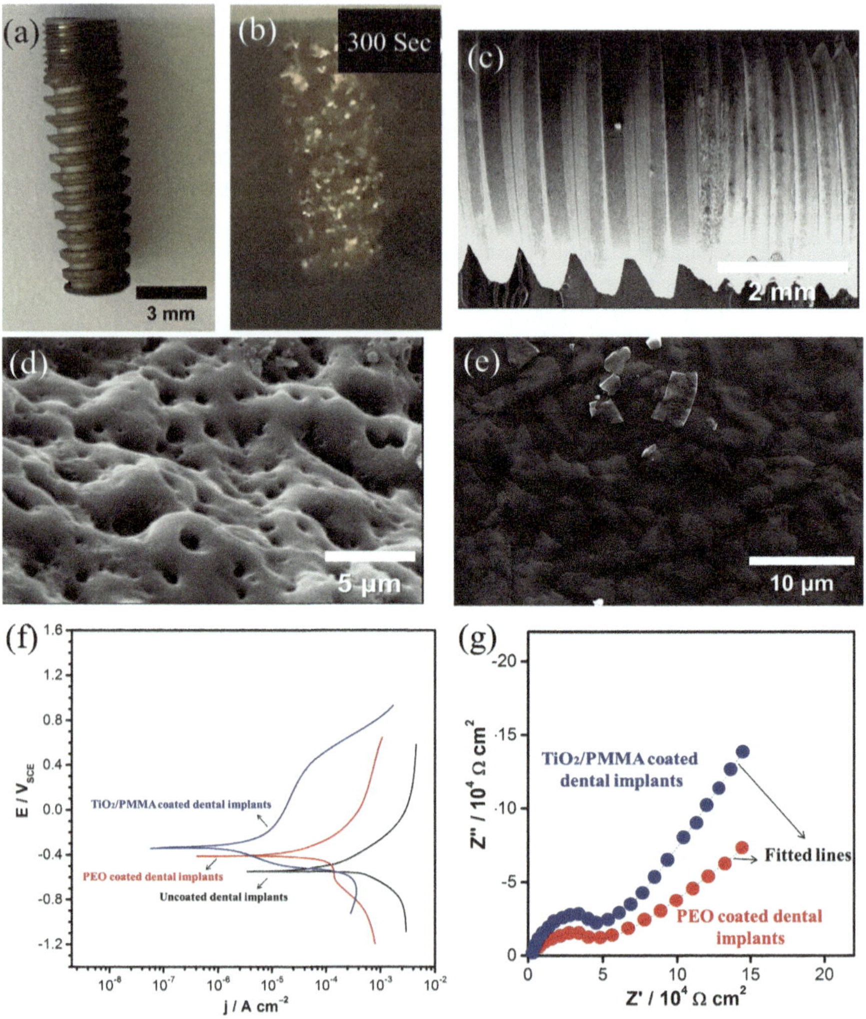

Fig. 14.7 (**a**) Optical microscopy image of a titanium dental implant, (**b**) micro-discharge evolution on the implant during the first 300 s of PEO, (**c**) SEM micrographs of PEO-coated dental implants, (**d**) detailed view of PEO coatings, (**e**) PMMA layer on the PEO-coated implant, (**f**) PDP curves, and (**g**) Nyquist plots in fluoride-containing saliva solutions at 36.5 °C. Reproduced from Ref. [105] under Creative Commons Attribution 4.0 International License (CC BY 4.0), published by IOP Publishing

Nyquist diagrams (Fig. 14.7f, g) demonstrated that the TiO_2/PMMA composite coating provided superior corrosion resistance, ranking TiO_2/PMMA > PEO > bare implant substrate in both saliva solutions. The PMMA layer acted as a protective barrier, reducing corrosive ion penetration, especially in acidic fluoride media, thereby enhancing the long-term stability of titanium dental implants under oral conditions [105].

The use of silane-based coatings, particularly those incorporating functional nanomaterials, has demonstrated significant improvements in corrosion resistance. A notable example is the application of GO-incorporated silane layers on PEO-coated CP-Ti using the sol-gel method [106, 107]. In simulated body fluid (SBF) environments, this composite coating effectively blocked ion diffusion through the PEO layer, leading to lower corrosion rates [106]. Moreover, when tested under simulated inflammatory conditions using an H_2O_2 and HCl-modified SBF, the GO-silane coating exhibited superior protection by reducing the dissolution of the underlying oxide layer, which is typically accelerated under oxidative stress [107]. The hybrid nature of this coating combines the chemical barrier properties of silane with the impermeability and electrical stability of GO, making it particularly effective for biomedical applications [106, 107].

Dip coating is another widely used posttreatment method for depositing polymeric and hybrid layers onto PEO-treated titanium [108]. The dip-coated polyethylene glycol (PEG) layer infused with betamethasone (BET) has been demonstrated to enhance the stability of PEO coatings in aggressive environments by forming a responsive protective barrier. This BET-loaded PEG layer enables sustained drug release under acidic inflammatory conditions, while also reinforcing the corrosion resistance of the PEO coating in such environments [108]. In a pH 5.0 solution, i_{corr} measurements indicated values of 1.75 $\mu A \cdot cm^{-2}$ (PEO/PEG-BET), 8.87 $\mu A \cdot cm^{-2}$ (PEO), and 49.17 $\mu A \cdot cm^{-2}$ (bare titanium). These findings underscore the effectiveness of the PEG-BET layer in sealing pores within PEO coatings, thereby minimizing corrosive ion infiltration and significantly enhancing corrosion protection in inflammatory conditions [108].

By tailoring the microstructure and chemical composition of PEO coatings through posttreatment techniques, researchers have achieved coatings with significantly enhanced corrosion resistance. The combination of ceramic deposition, polymeric sealing, and hydrothermal modifications creates multifunctional coatings that not only reduce porosity but also increase coating thickness, improve adhesion, and introduce additional chemical stability. The selection of an appropriate posttreatment strategy depends on the intended application, with ceramic deposition being preferred for high-temperature or mechanically demanding environments and polymeric top layers offering superior chemical resistance in biomedical and marine applications.

14.3 Anti-wear Mechanisms of PEO Coatings on Titanium

14.3.1 Key Tribological Properties of PEO Coatings

The tribological properties of PEO coatings are crucial for enhancing the wear resistance of titanium and its alloys, especially in demanding applications where friction-induced damage is a concern [109]. These properties are primarily governed by the microstructure, phase composition, hardness, surface roughness, and thickness of the coatings, all of which are influenced by the PEO processing conditions and substrate characteristics [110].

PEO coatings on titanium typically exhibit a ceramic-like structure composed of TiO_2 phases, such as anatase and rutile, which are known for their high hardness and low coefficient of friction (COF) [110]. The presence of these phases significantly reduces material loss under sliding or abrasive conditions. The rutile phase, in particular, contributes to superior wear resistance due to its higher density and stability compared to anatase [111]. The distribution and volume fraction of these phases within the coating are affected by electrical parameters, electrolyte composition, and substrate properties during the PEO process [43]. Coatings enriched with harder phases, such as rutile, or reinforced with secondary oxides, such as Al_2O_3, ZrO_2, or Nb_2O_5, formed due to the presence of alloying elements in the substrate, demonstrate exceptional resistance to abrasive and adhesive wear [112, 113].

The hardness of PEO coatings is another critical factor determining their tribological performance. Coatings with higher hardness values effectively resist penetration and deformation under applied loads, which is particularly beneficial for applications involving high contact pressures [114]. The hardness of PEO coatings is largely influenced by their microstructure, including the density and connectivity of ceramic phases. Homogeneous coatings with dense ceramic layers exhibit higher hardness and better wear resistance, while coatings with porous structures or microcracks are more susceptible to wear-induced damage [115]. Advances in PEO processing, such as the optimization of duty cycles and electrolyte formulations, have been shown to enhance coating hardness by promoting the formation of dense and uniform oxide layers.

Surface roughness, a characteristic feature of PEO coatings, also plays a pivotal role in determining their tribological behavior [116]. The rough surface morphology, resulting from localized micro-discharges during PEO, can act as a double-edged sword [116]. On one hand, the surface roughness provides reservoirs for lubricant retention, which can reduce friction in lubricated sliding conditions. On the other hand, excessive roughness can lead to increased friction and wear in dry sliding environments, as the asperities on the surface can concentrate stresses and exacerbate material removal [117]. Therefore, posttreatment processes, such as polishing or sealing, are often employed to reduce surface roughness and improve the tribological performance of PEO coatings [118].

The thickness of PEO coatings is another important parameter influencing their wear resistance [118]. Thicker coatings provide a more substantial barrier against material loss, as the load is distributed over a larger volume of protective material [117, 118]. However, excessively thick coatings can sometimes exhibit increased porosity or weak interfacial adhesion, which may compromise their wear resistance [119]. Striking the right balance in coating thickness is critical to achieving optimal tribological properties [119]. Recent advancements in multistage PEO processes have enabled the development of coatings with tailored thickness profiles, enhancing their resistance to wear without sacrificing adhesion or structural integrity [118].

The wear mechanisms of PEO coatings are largely dependent on the operating conditions, such as the type of counterface material, applied load, and sliding velocity. Under mild wear conditions, PEO coatings primarily experience abrasive wear, where the hard ceramic phases effectively resist scratching and material removal. In contrast, under severe conditions, delamination or microcracking may occur if the coating's integrity is compromised.

14.3.2 Advances in Substrate Engineering

Advances in substrate engineering have influenced the tribological performance of PEO coatings on titanium and its alloys by modifying the substrate properties and microstructure to enhance wear resistance [120]. Recent research highlights several approaches to improving the wear performance of PEO coatings through alloying, surface modifications, laser treatments, and additive manufacturing, each offering unique contributions to the tribological behavior of the coatings [120].

The development of advanced titanium alloys, such as Ti30Zr5Al3V, has opened new avenues for enhancing the wear resistance of PEO coatings [112]. These alloying strategies incorporate elements that form hard, wear-resistant oxides during the PEO process. For instance, the oxidation of Zr and Al produces ZrO_2 and Al_2O_3, respectively, which significantly enhance hardness and stability, effectively minimizing wear under sliding and abrasive conditions [112]. Additionally, the inclusion of Nb and Sn facilitates the formation of Nb_2O_5 and tin oxides (SnO and SnO_2), which act as self-lubricating phases within the coating, further reducing friction and wear [121]. The coexistence of these oxides with TiO_2 results in a composite structure that exhibits superior hardness and lower wear rates [112, 121]. Experimental studies confirm that the addition of such alloying elements in titanium alloys not only enhances coating hardness but also reduces microcracking and delamination during wear. This improved structural integrity significantly enhances the durability of coated surfaces in tribological applications, making them more effective than PEO coatings on CP-Ti.

Surface pretreatments prior to PEO, such as PVD of β-Ti39Nb on Ti-6Al-4V substrates, have demonstrated notable enhancements in wear resistance [113]. Coatings produced by performing PEO on substrates pretreated with PVD exhibit improved hardness and tribological properties compared to coatings formed on bulk β-Ti39Nb. This improvement is attributed to the synergistic effect of the PVD layer and the subsequent PEO treatment, which enhances the adhesion and uniformity of the coating [113]. The intermediate β-Ti39Nb layer provides a stable and wear-resistant interface that supports the formation of a denser oxide layer during PEO, resulting in coatings with better load-bearing capacity and reduced material loss under wear conditions. Such combined surface treatments have proven effective in addressing the limitations of traditional PEO processes on unmodified titanium substrates [113].

Laser surface pretreatment has emerged as another promising method for enhancing the tribological performance of PEO coatings. To enhance fretting wear resistance, PEO coatings were fabricated on TC6 alloy pretreated with laser remelting (LRM). Characterization confirmed that pretreatment increased the rutile TiO_2 phase from 68% to 89%, improving coating uniformity and reducing porosity by 2.8%. Fretting wear tests in SBF demonstrated superior wear resistance, with the LRM-PEO coating reducing wear depth to 5 μm, significantly improving fretting wear resistance compared to PEO (12 μm) and TC6 alloy (14 μm). The improved performance was primarily attributed to slight delamination and fatigue wear [122]. A study combined laser texturing (LT) and MAO (PEO) to create an anti-wear composite coating with 600 μm microgrooves on TC4 alloy (Fig. 14.8a) [123]. The LT/MAO coating contained anatase and rutile phases, with higher rutile content improving wear resistance. Friction tests revealed that LT lowered the friction coefficient to 0.66, 21.4% lower than untextured TC4 (0.84), while MAO further reduced it to 0.32. The LT/MAO coating had the lowest friction coefficient (0.23), enhancing durability (Fig. 14.8b). Long-term wear tests (30 min) showed MAO coatings deteriorated over time, exposing the substrate. However, LT/MAO coatings maintained lower friction, demonstrating extended anti-wear life (Fig. 14.8c). The micro-textured oxide coating reduced contact area, distributing loads effectively. High-hardness SiO_2 and TiO_2 micro-convex structures absorbed shear stress, reducing friction. Over time, abrasive debris filled surface gaps, smoothing wear tracks and transforming sliding friction into rolling friction, further improving lubrication (Fig. 14.8d). This composite coating significantly prolongs substrate protection and enhances tribological performance [123].

A recent study examined the effect of substrate plastic deformation on the tribological performance of PEO coatings on CP-Ti [124]. Hydrostatic extrusion (HE) was used to refine the microstructure, producing ultrafine-grained (UFG) titanium, which was compared with coarse-grained (CG) CP-Ti. Both were subjected to the PEO process. The results showed that HE-treated CP-Ti facilitated the formation of thicker (8 μm) and more wear-resistant PEO coatings than Ti-CG (5 μm). Wear track analysis revealed that the Ti-UFG sample experienced only 2 μm wear depth, while Ti-CG exhibited 9 μm, extending beyond the coating into the substrate. The COF of the PEO-coated Ti-UFG remained stable at 0.22 ± 0.02 under 2 N load, whereas Ti-CG's COF increased beyond 18 m, leading to coating delamination at ~24 m [124].

SLM of titanium has revolutionized the design and performance of titanium components, including their wear-resistant coatings. AM substrates differ significantly from conventionally processed titanium due to their unique microstructures,

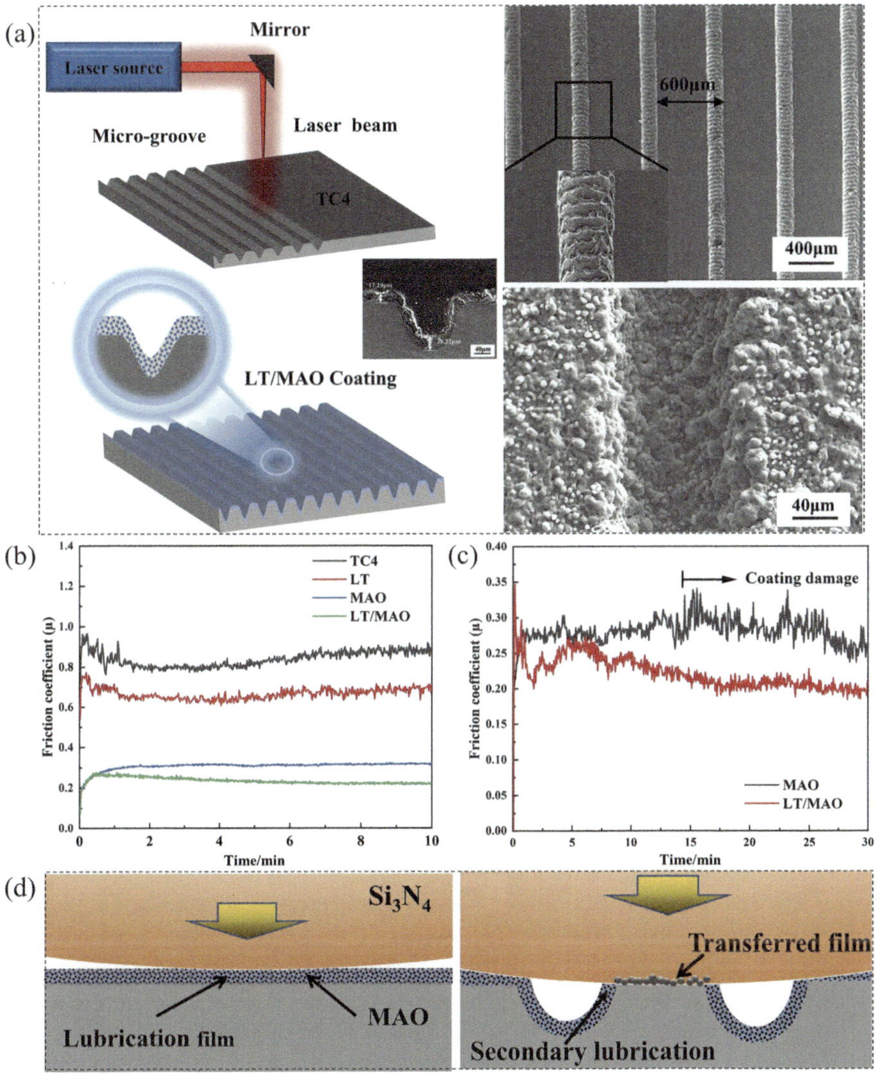

Fig. 14.8 (**a**) Schematic illustration of laser texturing (LT) on a TC4 substrate followed by MAO coating, along with SEM images of the laser-induced micro-grooves and coating morphology. (**b**) Friction coefficient variations over time for TC4, LT, MAO, and LT/MAO samples, highlighting differences in tribological performance. (**c**) Friction coefficient evolution of MAO and LT/MAO samples, indicating coating damage over time. (**d**) Schematic representation of the lubrication mechanism, showing the formation of a lubrication film on the MAO coating and secondary lubrication effects due to transferred film formation in LT/MAO coatings. Reproduced with permission from Ref. [123] © 2023 Elsevier B.V.

which feature refined grains, directional solidification patterns, and controlled porosity [125]. These characteristics directly impact the microstructure of PEO coatings, thereby influencing their tribological properties. For instance, PEO coatings on SLM-fabricated Ti_2AlNb substrates demonstrate superior wear resistance compared to those on CP-Ti counterparts. This enhancement is attributed to the refined microstructure of the AM substrate and the presence of hard phases such as Al_2O_3, Nb_2O_5, and aluminum titanate (Al_2TiO_5), which promote the formation of a more uniform and adherent oxide layer during the PEO process [125]. Moreover, the ability of AM to produce complex geometries enables the custom integration of wear-resistant coatings into intricate components, further broadening the potential applications of PEO-coated titanium in advanced engineering fields.

14.3.3 Role of Electrical Parameters

The intensity and distribution of plasma discharges, governed by the applied voltage or current density, also significantly influence the wear resistance of PEO coatings. A study developed PEO coatings on Ti-6Al-4V alloys using a calcium- and phosphate-based electrolyte at 200 V (Ti-PEO200V) and 300 V (Ti-PEO300V) with a pulsed electrical regime [126]. Coating thickness and roughness increased with higher voltage, while Ca and P incorporation also intensified. X-Ray diffraction (XRD) analysis confirmed coatings consisted mainly of rutile and anatase phases. Both coatings exhibited similar tribological behavior, but Ti-PEO300V had higher wear rates (4.5×10^{-4} μm·s^{-1}) compared to Ti-PEO200V (2.5×10^{-4} μm·s^{-1}). Despite its greater thickness, Ti-PEO300V's surface roughness and microcracks weakened its wear resistance, making it more susceptible to degradation [126].

Frequency adjustments in pulsed PEO systems have been employed to control discharge size and distribution, with direct implications for tribological behavior. High-frequency pulsing, for example, reduces the energy per discharge, creating a finer microstructure and more uniform coating surface [125]. For example, the Vickers hardness (HV0.5) of as-printed Ti_2AlNb was ~321.2 HV, increasing to 402.6 HV after heat treatment due to the formation of B2 and orthorhombic phases. PEO coatings exhibited superior hardness, ranging from 513.2 HV to 821.3 HV as pulse frequency increased (250–1000 Hz). Friction coefficient analysis showed high values (~0.7) for Ti_2AlNb, while PEO coatings reduced friction from 0.6 (250 Hz) to 0.56 (750 Hz). However, at 1000 Hz, friction slightly increased due to poor coating adhesion, causing direct contact between the Si_3N_4 ball and substrate. SEM-EDS analysis confirmed coating detachment, revealing Ti, Al, and Nb in the wear track [125].

A thick composite PEO coating was developed on Ti-10V-2Fe-3Al alloys using an environmentally friendly malic acid electrolyte containing PTFE nanoparticles, with varying pulse frequencies from 0.5 to 20.0 Hz [127]. The transformation of TiO_2 from anatase to rutile was influenced by electrochemical dynamic equilibrium, significantly affecting the coating's wear resistance and tribological properties. The wear tests indicated that coatings formed at 1.0 Hz and 2.0 Hz exhibited the best wear resistance, with narrower wear tracks (0.251 mm and 0.383 mm) and smoother surfaces. At higher frequencies (20.0 Hz), wear track width and depth peaked at 0.958 mm and 25.03 μm, with increased friction coefficient (~0.4) due to coating detachment. The role of PTFE nanoparticles was crucial. More PTFE remained on wear tracks at 1.0 and 2.0 Hz, entrapped in pores and cracks, reducing wear. However, high temperatures during severe friction softened PTFE, increasing visco-elastic deformation and reducing load-carrying capacity, leading to higher wear rates. Friction coefficient analysis showed 1.0 Hz coatings maintained the lowest COF, while higher frequencies (5.0 Hz and 10.0 Hz) exhibited increasing COF trends. At 20.0 Hz, fewer PTFE nanoparticles resulted in fluctuating COF and higher wear. The lowest wear rate was recorded at 1.0 Hz, confirming the optimal frequency for wear resistance [127].

14.3.4 Advances in Electrolyte Formulations

The formulation of electrolytes used in PEO has been extensively modified to enhance the wear resistance of coatings on metallic alloys [128]. As discussed earlier, two of the most effective strategies involve the incorporation of soluble salts and the suspension of insoluble reinforcement particles in the electrolyte. While soluble salts can influence the phase composition and hardness of the coating, the addition of hard ceramic or lubricating particles is particularly effective in reducing wear by modifying the coating's microstructure, hardness, and tribological properties [129].

The addition of soluble salts influences the in situ growth of secondary oxide phases within the TiO_2 matrix, modifying the mechanical properties of the PEO coating. A recent study optimized electrical parameters and electrolyte composition in PEO to promote the in situ growth of molybdenum diselenide ($MoSe_2$) by adding sodium selenite (Na_2SeO_3) and Na_2MoO_4 [130]. This modification led to the formation of a lubricating $MoSe_2$ layer interlocked with a robust TiO_2 ceramic coating, significantly improving adhesion and microporous structure (Fig. 14.9a). Wear tests revealed sliding widths of 624.51 μm (substrate), 583.87 μm (TC6/PEO), and 511.39 μm (PEO/$MoSe_2$), confirming superior wear resistance in the PEO/$MoSe_2$ coating (Fig. 14.9b). During friction, $MoSe_2$ gradually diffuses from micro-pits, forming a low-shear lubrication layer that spreads over the wear surface, reducing friction, and distributing stress. Additionally, abrasive debris fills micro-cracks, reinforcing the interface and preventing further material loss (Fig. 14.9c). The synergistic effect of $MoSe_2$'s lubricating properties and TiO_2's hardness creates a dual-function structure, reducing wear in both dry and liquid lubrication conditions. This system, benefiting from the PEO microporous structure's hydrophilic and oleophilic nature, achieves a remarkably low COF and near-negligible wear, providing insights for advanced self-lubricating wear-resistant coatings (Fig. 14.9d) [130].

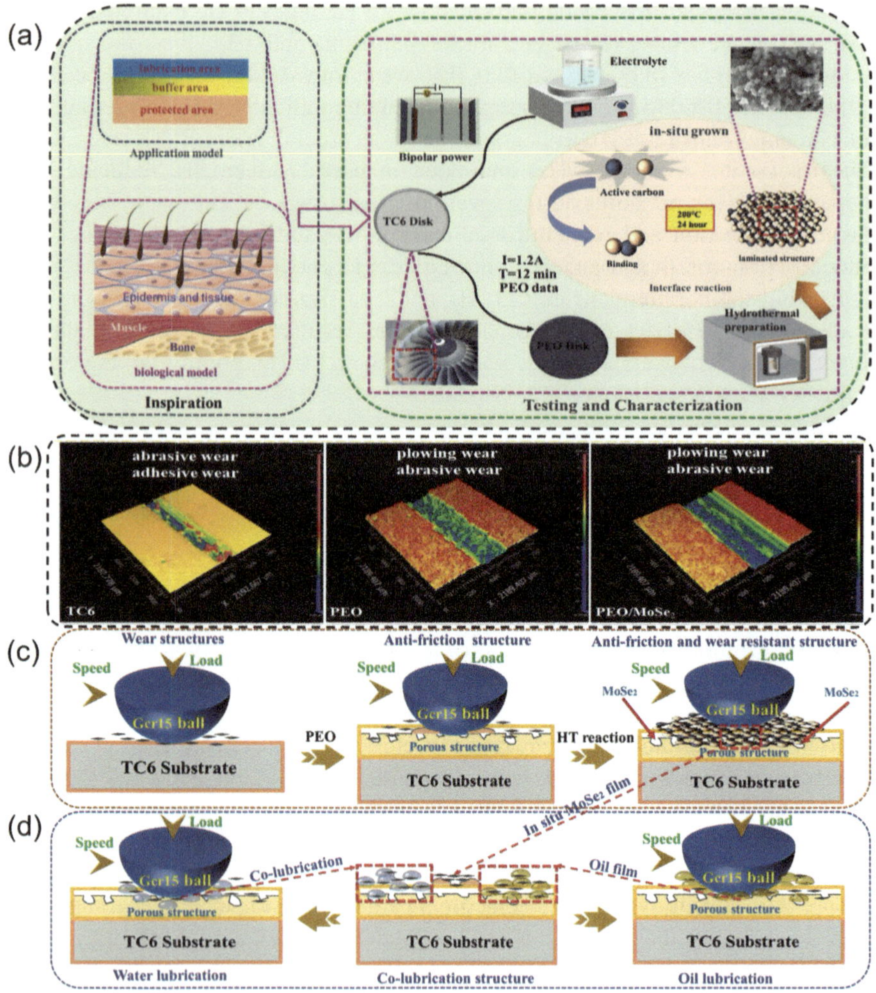

Fig. 14.9 (**a**) Schematic representation of the inspiration, testing, and characterization of PEO coatings on a TC6 substrate, illustrating the biological model concept and the fabrication process involving in situ grown active carbon and hydrothermal preparation. (**b**) 3D surface wear morphology of TC6, PEO, and PEO/$MoSe_2$ coatings, showing differences in wear mechanisms such as adhesive, abrasive, and plowing wear. (**c**) Illustration of the wear structures and antifriction effects of PEO and PEO/$MoSe_2$ coatings under load, highlighting the impact of $MoSe_2$ in enhancing wear resistance. (**d**) Schematic depiction of lubrication mechanisms, demonstrating water and oil lubrication effects and the formation of a co-lubrication structure facilitated by the in situ $MoSe_2$ film. Reproduced with permission from Ref. [130] © 2024 Elsevier B.V.

While soluble salts provide some enhancement in wear resistance through the formation of protective oxides, the addition of hard or lubricating insoluble particles into the electrolyte is particularly effective in reducing material loss under severe tribological conditions. ZrO_2 nanoparticles, when introduced into the electrolyte, become embedded in the oxide layer, forming a composite structure that significantly increases coating hardness [131]. The Al_2O_3 nanoparticles enhance resistance to abrasive wear by impeding the penetration of hard counterface asperities, thereby limiting the depth of wear tracks. Microhardness analysis also demonstrated a progressive increase in coating hardness from 452.3 HV0.5 to 656.7 HV0.5 with rising Al_2O_3 nanoparticle concentration. This enhancement is primarily due to the formation of the Al_2TiO_5 phase and a notable decrease in coating porosity [132]. A similar effect is observed with co-doping of ZrO_2 and $CePO_4$ particles, which not only increase hardness but also reduce the COF due to their ability to form a smooth tribofilm under sliding conditions. The optimized coating exhibits exceptional durability, achieving a hardness of 626 HV, an ultra-low friction coefficient of 0.12, and a minimal wear rate of $3.1 \times 10^{-5} \pm 0.15$ $mm^3 \cdot N^{-1} \cdot m^{-1}$ [133]. Coatings synthesized in calcium phosphate-based electrolytes with varying TiO_2 nanoparticle concentrations demonstrated that higher nanoparticle levels led to improved uniformity, smaller pores, and enhanced hardness. Wear resistance values improved from 7.23×10^{-6} $mm^3 \cdot N^{-1} \cdot m^{-1}$ (initial sample) to 3.98×10^{-6} $mm^3 \cdot N^{-1} \cdot m^{-1}$ (optimal sample), with wear track width decreasing to 0.78 mm, confirming superior mechanical stability and stress distribution. However, excessive TiO_2 increased wear track width to 1.03 mm, leading to microcrack formation and reduced stability [134].

To enhance the wear resistance and corrosion behavior of TC4 alloy, PEO was performed in an electrolyte containing TiC particles [135]. Increasing PEO current density improved coating thickness, enhanced TiC incorporation, and significantly boosted wear resistance, while higher TiC concentrations reduced coating thickness. Under optimized conditions (2 $g \cdot L^{-1}$ TiC, 5 $A \cdot dm^{-2}$ current density), the PEO coating exhibited high wear and corrosion resistance. Wear tests and 3D surface imaging revealed severe abrasion and ploughing wear on 1-PEO-4 and 3-PEO-4 coatings, exposing the substrate. However, 5-PEO-6, 5-PEO-4, and 5-PEO-2 coatings maintained structural integrity, showing fluctuations in friction coefficients. Wear rate analysis ranked coatings as 5-PEO-6 (11.6×10^{-4} $mm^3 \cdot N^{-1} \cdot m^{-1}$) > 5-PEO-4 ($10.6 \times 10^{-4}$ $mm^3 \cdot N^{-1} \cdot m^{-1}$) > 5-PEO-2 ($10.5 \times 10^{-4}$ $mm^3 \cdot N^{-1} \cdot m^{-1}$). The results confirm that TiC incorporation significantly enhances wear resistance, particularly at higher PEO current densities [135].

Hexagonal boron nitride (hBN) particles have been successfully incorporated into PEO coatings to enhance the tribological properties of titanium alloys [136, 137]. Acting as a solid lubricant, hBN offers a unique advantage due to its hexagonal structure, which enables easy shear between layers, thereby reducing friction and improving the sliding wear resistance of the coating [137]. Recent studies indicate that the formation of a dense and stable plasma at the coating/electrolyte interface facilitates the deposition and sintering of hBN particles, effectively sealing surface imperfections and achieving a coating density of 86.7%. The incorporation of hBN into TiO_2 nanocomposite coatings significantly reduces the friction coefficient from 0.54 (titanium alloy) to 0.28, maintaining stability even after 2000 sliding cycles. Additionally, the wear rate of the nanocomposite coating decreases by 70.8% (4.3×10^{-4} $mm^3 \cdot N^{-1} \cdot m^{-1}$) compared to the substrate, primarily due to the self-lubricating properties of hBN, which effectively lower frictional shear stress and enhance wear resistance [136]. Further studies also confirm that PEO coatings incorporating BN can achieve up to a 50% reduction in the COF compared to conventional TiO_2 coatings when 1.5 $g{\cdot}L^{-1}$ of BN is added to the electrolyte, demonstrating its effectiveness in minimizing adhesive wear [137].

Carbon-based reinforcements, such as GO and rGO, have also been extensively investigated for their ability to enhance wear resistance [138]. These particles, when introduced into the electrolyte, integrate into the growing oxide layer, forming a composite coating with significantly reduced porosity and improved load distribution under wear conditions [138]. The layered structure of graphene reduces friction by acting as a boundary lubricant, minimizing the severity of adhesive wear [139]. Additionally, rGO-reinforced coatings exhibit increased microhardness and better resistance to surface damage due to the high mechanical strength of the embedded graphene layers [139]. Studies have shown that the addition of GO to the silicate-hypophosphite electrolyte enhances PEO coating properties, increasing hardness from 331 HV to 366 HV as GO concentration rises from 0.1 to 0.5 $kg{\cdot}m^{-3}$. Simultaneously, friction coefficient decreases from 0.73 to 0.69, and wear rate drops significantly from 8.04 to 5.2 $mm^3{\cdot}N^{-1}{\cdot}m^{-1}$, demonstrating improved mechanical strength and wear resistance [140].

Controlling the type and concentration of particles in the electrolyte allows for the customization of PEO coatings' mechanical and tribological properties to suit specific applications. The integration of hard ceramic phases with self-lubricating phases within a single coating has proven to be a highly effective method for enhancing wear resistance and optimizing overall durability.

14.3.5 Posttreatment Strategies

Posttreatment strategies have been increasingly utilized to improve the wear resistance of PEO coatings on titanium by addressing surface roughness, modifying microstructure, and introducing additional protective layers that enhance tribological performance. The deposition of ceramic-based hard coatings onto PEO-treated titanium using PVD methods has proven highly effective in minimizing wear damage. A notable example is the application of Nb/NbN multilayer coatings on PEO-treated additively manufactured porous Ti-6Al-4V alloy [141]. These multilayers provided significant improvements in wear resistance by modifying surface morphology and sealing micropores formed during PEO. The wear mechanisms changed noticeably after PVD treatment, shifting from a combination of adhesive and abrasive wear in untreated samples to predominantly abrasive and delamination wear in coated samples (Fig. 14.10) [141]. The PVD coatings, particularly with optimized thickness, effectively concealed most of the PEO-induced porosity, leading to a smoother surface with fewer wear initiation sites. Among different PVD-layered structures, the PVD-2 layers coating exhibited the highest wear resistance, reducing wear rates by 90% compared to the untreated sample. The enhanced wear performance was attributed to the synergy between the soft and hard Nb/NbN

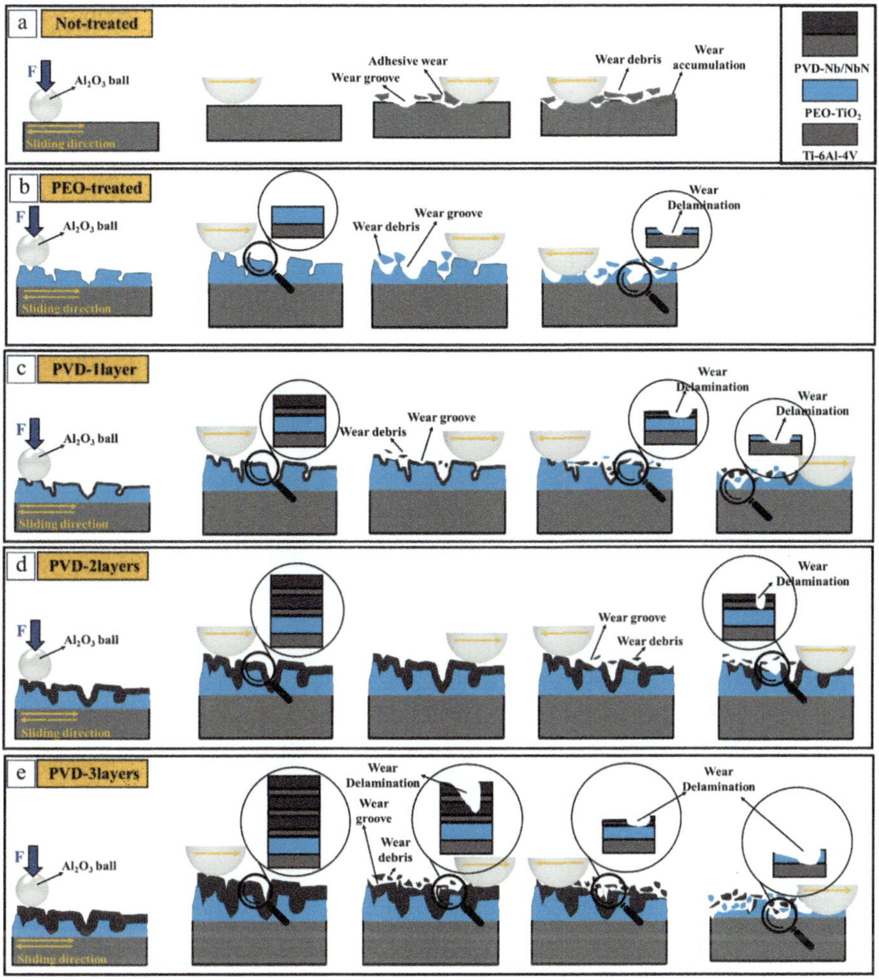

Fig. 14.10 Wear mechanisms observed in different sample treatments. The Not-treated sample exhibited severe surface degradation due to a combination of abrasive and adhesive wear (**a**). In the PEO-treated sample, adhesive, abrasive, and delamination wear were evident due to localized coating detachment (**b**). The PEO/PVD series primarily experienced abrasive wear, but with significantly reduced surface damage. The PVD-1 layer sample showed coating detachment due to high surface roughness, leading to debris-induced abrasive wear (**c**). In the PVD-3 layers sample, residual stress caused extensive coating detachment, generating hard debris that aggravated wear (**e**). The PVD-2 layers sample displayed minimal grooves and no visible debris, indicating superior wear resistance (**d**). The multilayer PVD coatings contributed to enhanced abrasion and deformation resistance by combining soft and hard layers, improving adhesion and preventing crack propagation. Reproduced with permission from Ref. [141] © 2024 Elsevier B.V.

layers, where the hard layers provided high load-bearing capacity, while the soft layers acted as energy absorbers, mitigating crack propagation and plastic deformation [141]. Insufficient thickness in the PVD-1 layer coating, however, resulted in higher surface roughness and localized detachment, leading to wear debris generation and intensified abrasive wear. On the other hand, excessive thickness in the PVD-3 layers coating caused stress accumulation and coating detachment, increasing abrasive wear through the presence of loose debris on the worn surface [141].

In addition to niobium-based coatings, other ceramic coatings such as titanium nitride (TiN), chromium nitride (CrN), and zirconium nitride (ZrN) have been widely explored as posttreatments for PEO-coated titanium [142]. The tribological properties of TC4 titanium alloy were enhanced by PEO, followed by CrN coating deposition via multi-arc ion plating. This PEO/CrN composite film significantly improved wear resistance. The friction coefficient of untreated TC4 alloy was 0.6, which decreased to 0.5 after PEO treatment and further dropped to 0.2 with the CrN coating. The surface roughness of the TC4/PEO/CrN composite was measured at 1.4 μm, while the porosity, initially 4.5%, was reduced to 0.66% after sealing. The PEO layer, consisting of anatase and rutile phases, formed a porous structure (~10 μm thick). The CrN layer introduced CrN, Cr, and Cr_2N, and phases, further improving wear resistance. After PEO/CrN treatment, wear scar width and depth decreased significantly, reducing wear volume by two orders of magnitude. Despite abrasive wear being the primary mechanism, the composite coating demonstrated superior performance in reducing friction and wear [142].

An Al_2O_3 ceramic coating was formed on CP-Ti using hot-dip aluminizing and PEO. To enhance coating properties, diffusion annealing was applied, creating a 125 μm composite layer. This treatment increased microhardness to 1230.1 HV and elastic modulus to 241.3 GPa, while reducing compressive residual stress from −850 MPa to −1530 MPa. Tribological performance improved significantly, with the COF decreasing by ~29% and relative wear resistance increasing ~8.4 times compared to PEO without diffusion annealing [143].

Another effective approach to improving wear performance is the application of polymer-based or hybrid polymer-ceramic top coatings. These coatings aim to reduce friction and provide a self-lubricating effect, lowering wear rates under dry and lubricated conditions. The deposition of polytetrafluoroethylene (PTFE)-based coatings on PEO-treated Ti-6.5Al-2Zr-1Mo-1V has been shown to significantly decrease the COF, leading to improved mechanical wear resistance. The PTFE layer formed a stable boundary lubrication film, reducing direct contact between asperities and minimizing adhesive wear [144]. Similarly, incorporating graphene into an epoxy layer on PEO-treated TC4 titanium alloy significantly reduced material loss by 70% during tribological testing. This improvement resulted from the combined effects of surface smoothing and friction reduction, effectively lowering wear rates. The graphene network functioned as a lubricant, preventing severe abrasion and maintaining coating integrity over prolonged sliding cycles [145].

A study explored carbon nanotubes (CNT)-assisted PEO with hBN nanoparticles to develop a double-layer coating on Ti-6Al-4V alloy, achieving exceptional tribological performance [146]. CNTs functioned as both lubricating agents and structural enhancers, facilitating the transition from a nanocomposite to a double-layer coating. During the later PEO stage, the CNT-doped electrolyte triggered a sudden rise in current density, leading to intensified discharge, which aided in hBN deposition and sintering. Consequently, the double-layer coating exhibited higher adhesion, increased hardness, and enhanced hBN incorporation compared to coatings without CNTs. The tribological tests showed that the PEO-hBN/CNTs-20 coating achieved a COF of 0.11, significantly lower than PEO (0.91) and PEO-hBN (0.43). The self-lubricating effect resulted from hBN nanoparticle release under shear, forming a lubrication film with CNTs. The COF initially increased as contact points formed but later stabilized as hBN and CNTs filled surface irregularities, reducing shear stress. Microhardness tests confirmed that PEO-hBN/CNTs-20 had the highest hardness due to reduced defects and increased coating thickness. Despite its minimal COF, this coating exhibited the largest wear loss (0.32 mg). The combined hBN and CNT lubrication effect significantly improved wear resistance, making this self-lubricating ceramic coating ideal for antifriction applications on light alloys [146].

The marine environment demands high-performance surface protection for titanium alloys, making advanced coatings essential [147]. This study developed PEO/ta-C composite coatings on TC4 titanium alloy using PEO and filtered cathodic vacuum arc. Compared to monolayer PEO or ta-C films, the composite coating maintained porous characteristics while significantly improving bonding strength. Wear tests revealed wear widths of 522.20 μm (TC4), 477.36 μm (PEO), 571.04 μm (ta-C), and 356.84 μm (PEO/ta-C). The PEO/ta-C coating exhibited the narrowest and shallowest wear scar, indicating strong adhesive wear resistance. In contrast, the ta-C film alone showed the widest wear scar due to poor substrate adhesion. EDS analysis found Cr and Fe in wear scars of all coatings except PEO/ta-C, confirming its excellent lubrication and anti-wear properties. With a friction coefficient of 0.13, PEO/ta-C provided superior tribological performance, protecting TC4 titanium alloy effectively [147].

By integrating PEO with other surface modification techniques, researchers have developed coatings with significantly improved tribological performance. Laser shock peening (LSP) created a fine-grained transition layer on TC4 titanium alloy, followed by PEO for composite coating formation. LSP refined surface grains, introduced high-density dislocations, and formed nanocrystal clusters. The LSP-PEO coating exhibited smaller pores, greater thickness, higher microhardness, and superior adhesion, corrosion, and wear resistance [148]. High porosity from micro-arc discharges weakens PEO coatings. To address this, a study introduced anodic oxidation (AO) posttreatment to enhance hardness, smoothness, and density, forming a composite coating with increased thickness and reduced roughness. AO primarily reacted within PEO discharge channels, creating an in situ amorphous oxide layer. The composite coating exhibited superior wear resistance due to sealed discharge channels. This method optimizes PEO coatings and holds potential for aerospace applications, offering a high-performance surface treatment approach for Ti-6Al-4V titanium alloy [149].

14.4 Tribocorrosion Performance of PEO Coatings

14.4.1 Synergy Between Corrosion and Wear Resistance

Tribocorrosion of PEO coatings occurs due to the simultaneous action of mechanical wear and electrochemical corrosion, which can accelerate material degradation beyond what would be expected from either process alone. The synergy between wear and corrosion is critical in applications where titanium components are exposed to sliding or fretting in corrosive environments, such as biomedical implants, marine structures, and aerospace components [150].

PEO coatings provide a protective ceramic-like oxide layer that improves both wear and corrosion resistance. However, under tribocorrosion conditions, this protective layer can degrade due to cyclic mechanical stresses, leading to localized coating failure [151]. Wear-induced damage, such as microcracking and delamination, exposes the underlying titanium substrate to direct interaction with the corrosive medium, accelerating corrosion rates. Conversely, corrosion can weaken the mechanical properties of the coating, making it more susceptible to material removal under wear conditions [152].

The primary tribocorrosion mechanisms in PEO coatings on titanium involve abrasion-assisted corrosion, where wear strips away the protective oxide layer, exposing fresh metal to electrochemical attack [111]. Additionally, corrosion-assisted wear occurs when chemical degradation compromises the coating's mechanical integrity, increasing its susceptibility to wear-induced material loss. The extent of tribocorrosion damage depends on factors such as applied load, sliding speed, environmental conditions, and coating properties [153]. Optimizing the composition, thickness, and surface characteristics of PEO coatings is essential to mitigating tribocorrosion effects, ensuring long-term performance in demanding environments [154].

14.4.2 Testing and Evaluation of Tribocorrosion Behavior

Tribocorrosion testing is essential for evaluating the combined effects of mechanical wear and electrochemical corrosion on PEO coatings applied to titanium and its alloys. These studies provide critical insights into how PEO coatings respond under simultaneous sliding and corrosive environments, which is particularly relevant for biomedical implants, aerospace components, and marine applications [111, 150–154]. The methodologies typically involve in situ electrochemical monitoring during wear testing, allowing researchers to correlate wear track evolution with electrochemical stability. Recent investigations have focused on the effects of in situ oxide transformations, third-body interactions, and self-lubricating mechanisms on tribocorrosion resistance.

Research investigated the tribocorrosion performance of PEO- and thermal oxidation (TO)-treated Ti-B_4C titanium matrix composites in PBS under applied loads of 0.5 N and 10 N [111]. Untreated samples exhibited a sharp open circuit potential (OCP) drop upon sliding due to passive film removal, followed by partial OCP recovery under 0.5 N, indicating oxide debris formation offering temporary protection. However, under 10 N, OCP remained significantly lower, showing severe wear damage. PEO-treated samples under 0.5 N showed stable OCP, while under 10 N, an initial sharp OCP drop was followed by a slower decline before stabilizing. Fluctuations in OCP and COF were linked to wear debris formation and disintegration. TO-treated samples showed a sharp OCP drop, a transition phase, and then stabilization, with higher OCP retention than PEO-treated ones. Post-sliding OCP recovery in TO-treated in situ composites indicated superior corrosion resistance, attributed to the oxygen diffusion zone, making them better suited for biomedical applications than PEO-treated composites [111].

A study examined the tribocorrosion behavior of CuxO/TiO_2 coatings on Ti-6Al-4V, produced via PEO with 3, 7, and 10 $g{\cdot}L^{-1}$ Na_2Cu-EDTA [151]. The $[Cu\text{-}EDTA]^{2-}$ complex facilitated Cu^{2+} migration into the anodic region through electromigration, promoting Cu_2O formation over CuO (Fig. 14.11a) and reducing surface porosity (Fig. 14.11b, c). OCP analysis in Fig. 14.11d showed that PEO-Cu7 $g{\cdot}L^{-1}$ and PEO-Cu 10 $g{\cdot}L^{-1}$ maintained a stable electrochemical state, whereas traditional PEO coatings exhibited severe fluctuations due to coating removal during sliding. The COF of PEO-Cu 7 $g{\cdot}L^{-1}$ decreased from 0.57 to 0.39, demonstrating superior lubrication. Compared to traditional PEO coatings, PEO-Cu 7 $g{\cdot}L^{-1}$ reduced wear track width by 19.7% and mass loss by 53.5%. This improvement was attributed to the formation of CuO and HAp on the worn surface, which acted as a lubricating layer, reducing friction and preventing excessive material loss [151].

Researchers also examined the tribocorrosion behavior of PEO coatings with varying CNT concentrations [153]. OCP analysis showed that uncoated samples rapidly lost their passive film, dropping to −0.8 V vs. Ag/AgCl under sliding, while PEO-coated samples maintained a more positive OCP. The optimal CNT concentration (0.15 $g{\cdot}L^{-1}$) resulted in higher OCP stability (0.57 V vs. Ag/AgCl) and lowest COF (0.34), demonstrating enhanced corrosion resistance and lubrication. However, at 0.2 $g{\cdot}L^{-1}$ CNTs, OCP declined to 0.42 V vs. Ag/AgCl, and COF increased to 0.40, indicating reduced performance. The wear rate followed the COF trend, with 0.15 $g{\cdot}L^{-1}$ CNTs providing the best tribocorrosion resistance. The results confirmed that CNT incorporation improved PEO coatings, offering superior corrosion protection, lower friction, and enhanced durability for Ti-6Al-4V alloy in corrosive environments [153].

Researchers also investigated the tribocorrosion behavior of PEO-coated Ti6-Al-4V at −3 °C and +20 °C electrolyte temperatures [154]. The oxide layers contained Ca and P, with the −3 °C PEO layer having fewer open porosities due to self-sealing from erupted material deposition. Corrosion tests revealed an 80 mV positive shift in E_{corr} and over 50% reduction in passive current density for auto-sealed layers, demonstrating enhanced electrochemical stability. Unlike untreated

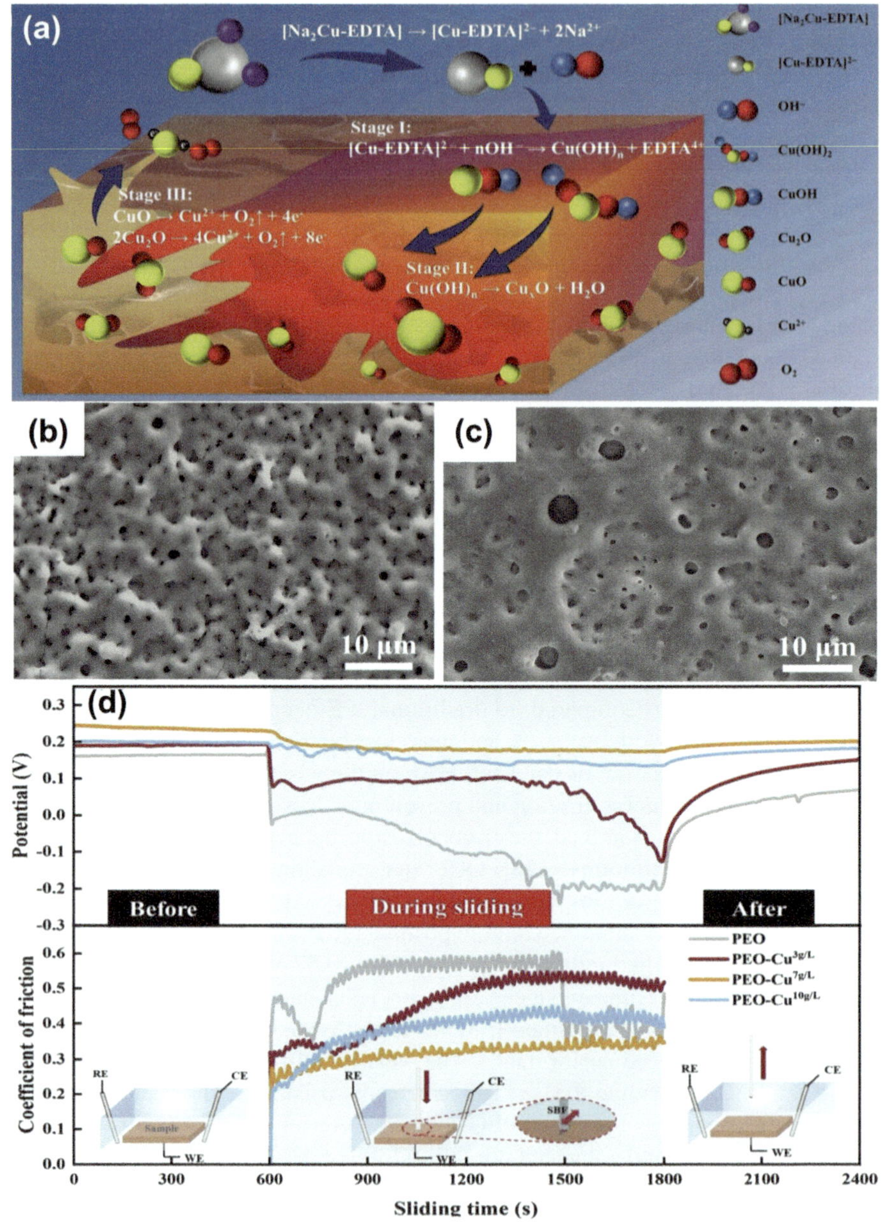

Fig. 14.11 (**a**) Schematic illustration of the formation mechanism of Cu-based compounds in PEO coatings, depicting different reaction stages involved in Cu_xO incorporation. SEM images of the surface morphology of PEO coatings: (**b**) without Cu additives and (**c**) with Cu additives, showing differences in porosity and microstructure. (**d**) Electrochemical and tribological performance of PEO and PEO-Cu coatings, displaying potential variation and COF before, during, and after sliding tests. Insets illustrate the experimental setup and interaction between the coated surface and SBF during wear testing. Reproduced with permission from Ref. [151] © 2022 Elsevier B.V.

Ti6-Al-4V, which exhibited a sudden OCP drop, PEO layers remained stable under sliding. During tribocorrosion tests, 20 °C PEO layers maintained constant OCP for 15 min, followed by a 200 mV drop. Conversely, −3 °C PEO layers showed no significant OCP reduction for 30 min, followed by a 40 mV decrease, remaining stable thereafter. The dominant wear mechanism was mild abrasive wear, characterized by parallel scratches on −3 °C PEO layers, confirming superior tribocorrosion resistance [154].

Another study examined the effects of a zinc-doped polypyrrole (PPy) top coating on PEO-treated CP-Ti, demonstrating significant tribocorrosion resistance [152]. Tribocorrosion in physiochemical body environments, exacerbated by microbial activity, accelerates titanium dental implant degradation, leading to surface damage, increased ion release, and biofilm formation, ultimately compromising implant longevity and biocompatibility (Fig. 14.12a). Wear track analysis revealed that untreated CP-Ti exhibited wide wear scars with plate-like debris, transitioning from adhesive to abrasive wear over time. PEO-coated samples suffered from complete oxide removal, leading to three-body wear interactions and greater material loss. During 8600 s of tribocorrosion testing, OCP measurements showed that PEO + PPy/Zn coatings maintained a stable potential (~−0.2 V), unlike untreated CP-Ti (−1.1 V) and PEO (−0.8 V), which experienced significant drops due to passive layer degradation (Fig. 14.12b). The PPy film provided superior passivation, minimizing wear loss under anodic conditions, while cathodic polarization suppressed repassivation, increasing wear damage. SEM micrographs confirmed PEO + PPy coatings exhibited narrower wear tracks, retaining polymeric remnants post-testing. The PPy layer's adhesion, hardness, and self-lubrication properties enhanced mechanical stability, reducing wear severity and preventing excessive material loss, proving its effectiveness in tribocorrosion protection (Fig. 14.12c, d) [152].

14.5 Conclusions and Outlook

PEO has become a highly effective method for improving the corrosion and wear resistance of titanium and its alloys. By forming a ceramic-like oxide layer with a unique multilayered microstructure, PEO coatings provide significant improvements in material durability, particularly in aggressive environments where conventional treatments fall short. The synergy between wear and corrosion resistance in PEO coatings is dictated by their structural characteristics, including phase composition, coating thickness, porosity, and secondary phase inclusions derived from electrolyte modifications. Recent advancements in PEO processing parameters, electrolyte formulations, and posttreatment strategies have demonstrated significant progress in tailoring these coatings for specific applications, such as biomedical implants, aerospace components, and industrial machinery.

For corrosion protection, PEO coatings rely on their dense inner layer to act as an electrochemical barrier, minimizing ionic transport and shielding the titanium

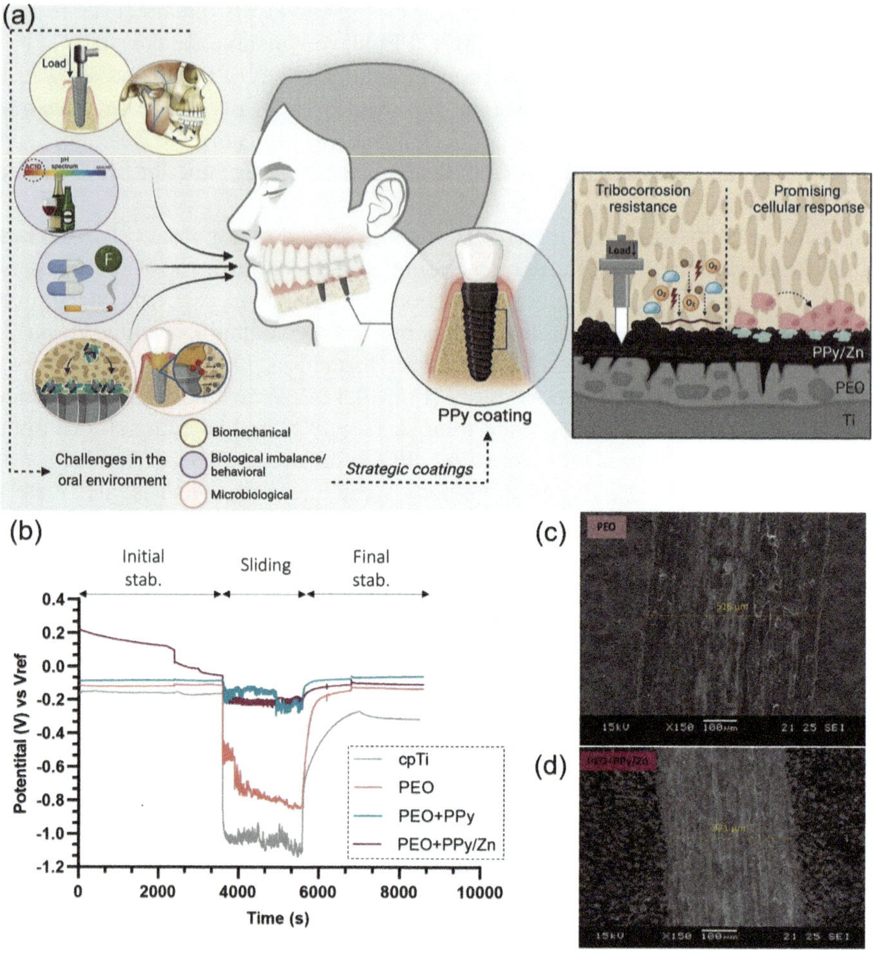

Fig. 14.12 (**a**) Schematic representation of challenges in the oral environment affecting dental implants, including biomechanical, biological, and microbiological factors, along with the strategic role of PEO and PPy/Zn coatings in enhancing tribocorrosion resistance and cellular response. (**b**) Tribocorrosion behavior of CP-Ti, PEO, PEO + PPy, and PEO + PPy/Zn coatings, showing potential variation during initial stabilization, sliding, and final stabilization. (**c**, **d**) SEM images of the wear track on PEO and PEO + PPy/Zn coatings, highlighting differences in surface damage and wear resistance. Reproduced with permission from Ref. [152] © 2023 Elsevier B.V.

substrate from aggressive environments. The incorporation of various elements into the oxide layer has further strengthened this barrier, reducing susceptibility to localized corrosion. Posttreatment strategies, including polymeric and ceramic sealing methods, have proven effective in mitigating the inherent porosity of PEO coatings, thereby extending their corrosion resistance in highly aggressive conditions such as marine, biomedical, and industrial processing environments.

In terms of tribological performance, PEO coatings significantly enhance the wear resistance of titanium by forming a hard, ceramic-based protective layer. However, challenges such as high surface roughness, brittle failure, and the presence of microcracks remain concerns. To address these issues, recent research has focused on integrating self-lubricating phases into the PEO layer. Additionally, duplex treatments have demonstrated superior wear resistance by modifying surface morphology, reducing porosity, and improving adhesion between the coating and the substrate.

Despite these advancements, several research gaps remain, offering promising directions for future studies:

1. Current PEO coatings primarily focus on either wear or corrosion resistance, but further development is needed to engineer coatings that provide superior protection against both degradation mechanisms simultaneously. Future research should explore multilayered and hybrid coatings that integrate hard phases for wear protection with chemically stable oxides for corrosion resistance, ensuring a balanced performance across different applications.
2. One of the inherent challenges of PEO coatings is their porous surface morphology, which can act as a pathway for corrosive species and serve as wear initiation sites. Future work should focus on refining PEO processing conditions—such as voltage waveform optimization, pulsed DC techniques, and controlled discharge behavior—to minimize porosity without compromising coating adhesion and thickness.
3. The next generation of PEO coatings should incorporate self-healing functionalities through electrolyte modifications. The integration of microcapsules containing corrosion inhibitors or self-healing agents into the coating could provide long-term durability in harsh environments. Additionally, the development of "smart" coatings capable of responding to environmental stimuli, such as pH shifts or mechanical stress, would further enhance protection.
4. With the increasing use of AM for titanium-based components, there is a need to optimize PEO treatments specifically for AM-produced substrates. The unique microstructure of AM titanium, characterized by residual stresses and anisotropic properties, may influence the growth and adhesion of PEO coatings. Research should focus on understanding how AM parameters affect the PEO process and how coatings can be tailored for AM-specific applications.
5. While traditional posttreatments such as polymeric sealing and PVD coatings have improved PEO performance, alternative techniques such as atomic layer deposition (ALD) and laser surface modification should be further investigated. These methods offer precise control over coating composition and microstructure, enabling the development of ultra-thin, defect-free protective layers that maintain the mechanical integrity of PEO coatings.
6. A critical gap in PEO research is the limited understanding of long-term tribocorrosion performance in real-world applications. Advanced computational models that integrate electrochemical and mechanical degradation mechanisms should be developed to predict the lifespan of PEO coatings under combined

wear and corrosion conditions. Such models will be invaluable for optimizing coating designs based on specific industrial requirements.

7. Future advancements should focus on reducing the environmental impact of PEO processes by developing eco-friendly electrolytes and energy-efficient processing methods. The replacement of hazardous chemicals with biodegradable alternatives and the reduction of energy consumption during PEO processing will be key factors in making the technology more sustainable for widespread industrial adoption.
8. The use of artificial intelligence (AI) and machine learning can revolutionize PEO coating development by predicting optimal process parameters, identifying correlations between processing conditions and coating properties, and accelerating material design. AI-driven models can analyze large datasets from PEO experiments to optimize electrolyte composition, voltage regimes, and pulse patterns, reducing trial-and-error approaches and enabling more precise control over coating microstructure and performance. Additionally, ML algorithms can be employed for real-time monitoring of PEO processes, detecting anomalies, and predicting coating lifespan under varying wear and corrosion conditions.

These research gaps must be addressed to further optimize the next generation of PEO coatings for greater durability, multifunctionality, and adaptability in various engineering applications. Continuous advancements in PEO technology will enhance the performance of titanium-based materials while contributing to the broader goal of developing sustainable and high-performance surface engineering solutions.

References

1. Pesode, P., & Barve, S. (2023). *Journal of Engineering and Applied Science, 70*, 25.
2. Najafizadeh, M., Yazdi, S., Bozorg, M., Ghasempour-Mouziraji, M., Hosseinzadeh, M., Zarrabian, M., & Cavaliere, P. (2024). *Journal of Alloys and Compounds Communications, 3*, 100019.
3. Zhao, Q., Sun, Q., Xin, S., Chen, Y., Wu, C., Wang, H., Xu, J., Wan, M., Zeng, W., & Zhao, Y. (2022). *Materials Science and Engineering A, 845*, 143260.
4. Shokri, A., & Fard, M. S. (2022). *Chemosphere, 307*, 135640.
5. Yuan, S., Zou, J., Lin, N., Zhang, H., Li, D., & Wu, Y. (2022). *Ceramics International, 48*, 31896–31901.
6. Sidhu, S. S., Singh, H., & Gepreel, M. A. (2021). *Materials Science and Engineering: C, 121*, 111661.
7. Simon, A. P., da Silva, B. L., Belusso, E. F., Mazzucatto, T. S., Santos, J. S., Trivinho-Strixino, F., & de Souza Sikora, M. (2023). *Surface and Coatings Technology, 474*, 130077.
8. Campanelli, L. C. (2021). *Journal of Materials Research, 36*, 151–165.
9. Anil Kumar, V., Gupta, R. K., Prasad, M. J., & Narayana Murty, S. V. (2021). *Journal of Materials Research, 36*, 689–716.
10. Yang, J., Song, Y., Dong, K., & Han, E. H. (2023). *Corrosion Reviews, 41*, 5–20.
11. Bordbar-Khiabani, A., & Gasik, M. (2023). *Scientific Reports, 13*, 2312.

12. Haruna, T., Imamura, M., & Hirohata, Y. (2024). *Journal of the Electrochemical Society, 171*, 011502.
13. Zheng, X., Xu, C., Zhang, B., Cai, Y., & Zhou, X. (2024). *Materials Today Communications, 39*, 108732.
14. Sivaprakasam, P., Hailu, T., & Elias, G. (2023). *Results in Materials, 19*, 100422.
15. Chen, Y., Zhang, H., Wang, B., Huang, J., Zhou, M., Wang, L., Xi, Y., Jia, H., Xu, S., Liu, H., & Wen, L. (2024). *Coatings, 14*, 786.
16. Zhao, S., Meng, F., Fan, B., Dong, Y., Wang, J., & Qi, X. (2023). *Wear, 512*, 204532.
17. Li, Y., Zhou, Z., & He, Y. (2023). *Materials, 17*, 65.
18. Dong, K., Song, Y., Bian, G., Cai, Y., & Han, E. H. (2024). *Tribology International, 192*, 109292.
19. Xie, B., & Gao, K. (2023). *Coatings, 13*, 1486.
20. Han, X., Ma, J., Tian, A., Wang, Y., Li, Y., Dong, B., Tong, X., & Ma, X. (2023). *Colloids and Surfaces. B, Biointerfaces, 227*, 113339.
21. Huang, R., Suo, W., Wang, Y., Pan, Y., Ren, G., Hu, H., & Huang, L. (2024). *Applied Surface Science, 671*, 160766.
22. Gao, K., Zhang, Y., Yi, J., Dong, F., & Chen, P. (2024). *Coatings, 14*, 148.
23. Fernández-López, P., Alves, S. A., San-Jose, J. T., Gutierrez-Berasategui, E., & Bayón, R. (2024). *Coatings, 14*, 217.
24. Rahimi, S., Bordbar-Khiabani, A., Yarmand, B., & Kolahi, A. (2018). *Materials Today Proceedings, 5*, 15667–15676.
25. Molaei, M., Fattah-alhosseini, A., Nouri, M., Mahmoodi, P., Navard, S. H., & Nourian, A. (2022). *Surfaces and Interfaces, 30*, 101967.
26. Bordbar-Khiabani, A., Yarmand, B., & Mozafari, M. (2019). *Surface and Coating Technology, 360*, 153–171.
27. Bahrampour, S., Bordbar-Khiabani, A., Siadati, M. H., & Gasik, M. (2024). *Ceramics International, 50*, 29703–29710.
28. Fattah-alhosseini, A., Chaharmahali, R., Dikici, B., & Kaseem, M. (2025). *International Journal of Refractory Metals and Hard Materials, 128*, 107059.
29. Babaei, K., Fattah-alhosseini, A., & Chaharmahali, R. (2020). *Surfaces and Interfaces, 21*, 100719.
30. Karimi, M., Eskandari, M. J., & Araghchi, M. (2024). *Results in Surfaces and Interfaces, 14*, 100205.
31. Bordbar-Khiabani, A., Ebrahimi, S., & Yarmand, B. (2020). *Corrosion Science, 173*, 108781.
32. Ebrahimi, S., Bordbar-Khiabani, A., & Yarmand, B. (2020). *Journal of Alloys and Compounds, 845*, 156219.
33. Fattah-alhosseini, A., Molaei, M., & Kaseem, M. (2024). *Surfaces and Interfaces, 46*, 104179.
34. Bordbar-Khiabani, A., Ebrahimi, S., & Yarmand, B. (2018). *Applied Surface Science, 462*, 913–922.
35. Ebrahimi, S., Bordbar-Khiabani, A., & Yarmand, B. (2019). *Materials Letters, 239*, 151–154.
36. Meng, J., Li, H., Dong, X., Yu, H., Zhou, H., Zhan, S., & Tang, Y. (2023). *Journal of Materials Science, 58*, 13816–13830.
37. Sun, S., Cheng, B., Liu, A., Liu, Z., Xu, G., Liu, P., & Lu, H. (2024). *Tribology International, 200*, 110114.
38. Guan, Q., Meng, J., Wang, S., Dong, X., Zhou, H., Yu, H., & Li, H. (2022). *Surface Topography: Metrology and Properties, 10*, 045030.
39. Saji, V. S. (2024). *Chemical Engineering Journal, 498*, 155066.
40. Kaseem, M., Fatimah, S., Nashrah, N., & Ko, Y. G. (2021). *Progress in Materials Science, 117*, 100735.
41. Sikdar, S., Menezes, P. V., Maccione, R., Jacob, T., & Menezes, P. L. (2021). *Nanomaterials, 11*, 1375.
42. Liu, Z., Shi, W., Xiang, S., Liu, F., & Ramamurty, U. (2024). *Surface and Coating Technology, 487*, 130985.

43. Aliofkhazraei, M., Macdonald, D. D., Matykina, E., Parfenov, E. V., Egorkin, V. S., Curran, J. A., Troughton, S. C., Sinebryukhov, S. L., Gnedenkov, S. V., Lampke, T., & Simchen, F. (2021). *Applied Surface Science Advances, 5*, 100121.
44. Abbasi, S., Mahboob, A., Bakhtiari Zamani, H., Bilesan, M. R., Repo, E., & Hakimi, A. (2022). *Journal of Nanomaterials, 2022*, 5675038.
45. Madhuri, D., Ghosh, R., Hasan, M. A., Dey, A., Pillai, A. M., Anantharaju, K. S., & Rajendra, A. (2021). *Journal of Materials Engineering and Performance, 30*, 4072–4082.
46. Makurat-Kasprolewicz, B., & Ossowska, A. (2023). *Materials Today Communications, 34*, 105425.
47. Fattah-alhosseini, A., Molaei, M., & Babaei, K. (2020). *Surfaces and Interfaces, 21*, 100659.
48. Li, G., Ma, F., Liu, P., Qi, S., Li, W., Zhang, K., & Chen, X. (2023). *Journal of Alloys and Compounds, 948*, 169773.
49. Hu, Y., Wang, Z., Ai, J., Bu, S., & Liu, H. (2021). *Coatings, 11*, 230.
50. Ming, X., Wu, Y., Zhang, Z., & Li, Y. (2023). *Coatings, 13*, 2064.
51. Xi, F., Zhang, X., Jiang, X., Kang, Y., Wen, X., & Liu, Y. (2023). *Surface and Coating Technology, 467*, 129685.
52. Abbas, A., Kung, H. P., & Lin, H. C. (1950). *Micromachines, 14*, 2023.
53. Kuroda, P. A., Rossi, M. C., Grandini, C. R., & Afonso, C. R. (2023). *Journal of Materials Research and Technology, 26*, 4656–4669.
54. Zhai, D., Qiu, T., Shen, J., & Feng, K. (2022). *International Journal of Minerals, Metallurgy, and Materials, 29*, 1991–1999.
55. Molaei, M., Nouri, M., Babaei, K., & Fattah-Alhosseini, A. (2021). *Surfaces and Interfaces, 22*, 100888.
56. Fattah-alhosseini, A., Chaharmahali, R., & Kaseem, M. (2024). *Hybrid Advances, 5*, 100151.
57. Ao, N., Liu, D., Zhang, X., & He, G. (2020). *Journal of Alloys and Compounds, 823*, 153823.
58. López-Ortega, A., de Viteri, S., Alves, S. A., Mendoza, G., Fuentes, E., Mitran, V., Cimpean, A., Dan, I., Vela, A., & Bayon, R. (2022). *Biomaterials Advances, 138*, 212875.
59. Costa, N. A., Correa, D. R., Lisboa-Filho, P. N., Sousa, T. S., Grandini, C. R., & Rocha, L. A. (2021). *Surface and Coating Technology, 408*, 126856.
60. Zhang, G., Khanlari, K., Huang, S., Li, X., Zhao, D., Wu, H., Cao, Y., Liu, B., & Huang, Q. (2023). *Surface and Coating Technology, 456*, 129254.
61. Farrakhov, R., Melnichuk, O., Parfenov, E., Mukaeva, V., Raab, A., Sheremetyev, V., Zhukova, Y., & Prokoshkin, S. (2021). *Coatings, 11*, 401.
62. Li, A., Wang, Q., Chen, R., Ding, X., Su, Y., & Fu, H. (2024). *Materials Today Communications, 42*, 111111.
63. Zhang, X., Cai, G., Lv, Y., Wu, Y., & Dong, Z. (2020). *Surface and Coating Technology, 400*, 126202.
64. Xi, F., Huang, Y., Zhao, Y., Liu, Y., Dai, W., & Tian, Y. (2022). *Materials, 15*, 4256.
65. Yin, Y., Li, L., Wu, G., Wang, Y., Yang, Z., Wen, C., & Yao, J. (2024). *Journal of Alloys and Compounds, 1002*, 175501.
66. Kolarovszki, B., Ficsor, S., Frank, D., Katona, K., Soos, B., & Turzo, K. (2024). *Lasers in Medical Science, 39*, 162.
67. Nisar, S. S., Arun, S., & Choe, H. C. (2023). *Surface and Coating Technology, 464*, 129553.
68. Bordbar-Khiabani, A., & Gasik, M. (2023). *Journal of Materials Research and Technology, 26*, 356–370.
69. Mora-Sanchez, H., Ramos, C., Mohedano, M., Torres, B., Arrabal, R., & Matykina, E. (2024). *Transactions of the Nonferrous Metals Society of China, 34*, 1150–1166.
70. Zakaria, A., Hamdi, M., Todoh, M., Yusof, F., & Arai, K. (2024). *Surface and Coating Technology, 488*, 131059.
71. Wang, X. M., & Zhang, F. Q. (2022). *Surface and Coating Technology, 444*, 128696.
72. Grigoriev, S., Peretyagin, N., Apelfeld, A., Smirnov, A., Yanushevich, O., Krikheli, N., Kramar, O., Kramar, S., & Peretyagin, P. (2022). *Materials, 15*, 4535.
73. Qiu, T., Tan, L., Zhai, D., Ni, P., & Shen, J. (2023). *Metallurgical and Materials Transactions A: Physical Metallurgy and Materials Science, 54*, 333–345.

74. Zhai, D., Tang, Q., Ni, P., Li, X., & Shen, J. (2023). *Surface and Coating Technology, 473*, 130030.
75. Chu, Y., Liu, P., Chen, Y., & Li, X. (2020). *Materials Research, 23*, e20200002.
76. Muhaffel, F., Jarzębska, A., Trelka, A., Balin, K., Bieda, M., & Cimenoglu, H. (2024). *Surface and Coating Technology, 476*, 130224.
77. Sobolev, A., Kossenko, A., & Borodianskiy, K. (2019). *Materials, 12*, 3983.
78. Chen, Y., Zhou, P., & Yuan, M. (2023). *Journal of Physics Conference Series, 2419*, 012025.
79. Songur, F., Dikici, B., Niinomi, M., & Arslan, E. (2019). *Surface and Coating Technology, 374*, 345–354.
80. Wu, T., Blawert, C., Serdechnova, M., Karlova, P., Dovzhenko, G., Wieland, D. F., Stojadinovic, S., Vasilic, R., Wang, L., Wang, C., & Mojsilovic, K. (2022). *Applied Surface Science, 595*, 153523.
81. Manojkumar, P., Premchand, C., Lokeshkumar, E., Subrahmanyam, C., Viswanathan, A., Krishna, L. R., & Rameshbabu, N. (2022). *Journal of Alloys and Compounds, 919*, 165781.
82. Kaseem, M., & Choe, H. C. (2021). *Corrosion Science, 192*, 109764.
83. Li, H., Ma, G., & Wang, Z. (2023). *Surface and Coating Technology, 469*, 129814.
84. Jamali, R., Bordbar-Khiabani, A., Yarmand, B., Mozafari, M., & Kolahi, A. (2022). *Materials Chemistry and Physics, 276*, 125436.
85. Bahrampour, S., Bordbar-Khiabani, A., Siadati, M. H., Gasik, M., & Mozafari, M. (2024). *Chemical Engineering Journal, 483*, 149016.
86. Nikoomanzari, E., Fattah-alhosseini, A., Karbasi, M., & Nourian, A. (2022). *Surfaces and Interfaces, 32*, 102128.
87. Hoseini, A., Yarmand, B., & Kolahi, A. (2021). *Surface and Coating Technology, 409*, 126938.
88. Pourshadloo, M., Rezaei, H. A., Saeidnia, M., Alkokab, H., & Bathaei, M. S. (2022). *Surface Innovations, 11*, 5–14.
89. Ebrahimi, S., Bordbar-Khiabani, A., Yarmand, B., & Asghari, M. A. (2019). *Ceramics International, 45*, 1746–1754.
90. Hoseini, A., & Yarmand, B. (2023). *Surfaces and Interfaces, 42*, 103424.
91. Hoseini, A., & Yarmand, B. (2024). *Journal of Alloys and Compounds, 984*, 173976.
92. Ogur, E., Alves, A. C., & Toptan, F. (2024). *Materials Today Communications, 41*, 110343.
93. Bordbar-Khiabani, A., Ebrahimi, S., & Yarmand, B. (2019). *Applied Surface Science, 486*, 153–165.
94. Molaeipour, P., Allahkaram, S. R., & Akbarzadeh, S. (2022). *Surface and Coating Technology, 430*, 127987.
95. Li, W., Gao, J., Ma, Y., Zheng, K., Zhi, J., Xin, Y., Xie, S., & Yu, S. (2023). *Surface and Coating Technology, 458*, 129340.
96. Jiang, W., Sun, Y., Wang, F., Zhang, X., Hua, Y., Wang, X., Wei, J., & Zhang, Y. (2023). *Journal of Materials Science, 58*, 18169–18186.
97. Casanova, L., Belotti, N., Pedeferri, M., & Ormellese, M. (2021). *Materials and Corrosion, 72*, 1894–1898.
98. Ghanbari, A., Bordbar-Khiabani, A., Warchomicka, F., Sommitsch, C., Yarmand, B., & Zamanian, A. (2023). *Surfaces and Interfaces, 36*, 102495.
99. Bordbar-Khiabani, A., Rahimi, S., Yarmand, B., & Mozafari, M. (2018). *Materials Today: Proceedings, 5*, 15603–15612.
100. Bordbar-Khiabani, A., Yarmand, B., & Mozafari, M. (2020). *Materials Letters, 258*, 126779.
101. Muntean, R., Brîndușoiu, M., Buzdugan, D., Nemeș, N. S., Kellenberger, A., & Uțu, I. D. (2023). *Materials, 16*, 1410.
102. Fan, X., Du, J., Li, Y., Duan, K., & Liu, G. (2023). *Journal of Orthopaedic Surgery and Research, 18*, 901.
103. Pourshadloo, M., Jameel, M. F., Romero-Parra, R. M., Yeslam, H. E., Shafik, S. S., Kareem, A. K., Zabibah, R. S., Sharifianjazi, F., & Bathaei, M. S. (2023). *Ceramics International, 49*, 33598–33606.
104. Bakhtiari-Zamani, H., Saebnoori, E., Bakhsheshi-Rad, H. R., & Berto, F. (2022). *Materials, 15*, 8300.

105. Mohammed, N. B., Daily, Z. A., Alsharbaty, M. H., Abullais, S. S., Arora, S., Lafta, H. A., Jalil, A. T., Almulla, A. F., Ramírez-Coronel, A. A., Aravindhan, S., & Bathaei, M. S. (2022). *Materials Research Express, 9*, 125401.
106. Hamrahi, B., Yarmand, B., & Massoudi, A. (2021). *Surface and Coating Technology, 422*, 127558.
107. Hamrahi, B., Yarmand, B., & Massoudi, A. (2023). *Materials Letters, 333*, 133587.
108. Al-Hawary, S. I., Habash, R. T., Abosaooda, M., Hjazi, A., Saleh, E. A., Hassan, Z. F., & Bathaei, M. S. (2024). *Heliyon, 10*, e25605.
109. Abakay, E., Armağan, M., Yıldıran Avcu, Y., Guney, M., Yousif, B. F., & Avcu, E. (2024). *Frontiers in Materials, 11*, 1452288.
110. Garcia-Cabezón, C., Rodríguez-Méndez, M. L., Borrás, V. A., Bayón, R., Salvo-Comino, C., Garcia-Hernandez, C., & Martin-Pedrosa, F. (2021). *Friction, 9*, 1303–1318.
111. Sousa, L., Costa, N. A., Rossi, A., Simões, S., Toptan, F., & Alves, A. C. (2024). *Surface and Coating Technology, 485*, 130854.
112. Zhong, H., Liu, L., Yang, L. Q., Ma, M. Z., Hu, E. Z., & Hu, K. H. (2023). *Surface and Coating Technology, 474*, 130095.
113. Gabor, R., Cvrček, L., Mašek, K., Hlinka, J., Motyka, O., Walter, J., Martynková, G. S., Mikesková, G., & Seidlerová, J. (2024). *International Journal of Refractory Metals and Hard Materials, 118*, 106502.
114. Rodriguez, L., Paris, J. Y., Denape, J., & Delbé, K. (2023). *Coatings, 13*, 373.
115. Al-Asadi, M. M., & Al-Tameemi, H. A. (2022). *Tribology International, 176*, 107919.
116. Student, M. M., Pohrelyuk, I. M., Hvozdetskyi, V. M., Zadorozhna, K. R., & Veselivs'ka, H. H. (2023). *Materials Science, 58*, 488–493.
117. Student, M. M., & Pohrelyuk, I. M. (2021). *Materials Science, 57*, 377–386.
118. Buling, A., & Zerrer, J. (2023). *Wear, 523*, 204825.
119. Gao, P., Guo, Q., Xing, Y., & Guo, Y. (2023). *Coatings, 13*, 325.
120. Zeng, Q., Wang, Z., He, W., Pang, Z., Ning, Z., Chen, R., Zheng, C., Yan, C., & Guo, L. (2023). *Advanced Engineering Materials, 25*, 2300228.
121. Rossi, M., Kuroda, P., Danelon, M. R., Rossino, L. S., & Afonso, C. R. (2024). *Materials Letters, 370*, 136773.
122. Sun, D., Wu, X., Xie, F., He, P., Li, Z., & He, J. (2023). *Ceramics International, 49*, 4979–4986.
123. Wu, G., Yin, Y., Zhang, S., Wang, Y., Xiang, Y., Li, L., & Yao, J. (2023). *Surface and Coating Technology, 453*, 129114.
124. Maj, Ł., Muhaffel, F., Jarzębska, A., Trelka, A., Trembecka-Wójciga, K., Kawałko, J., Kulczyk, M., Bieda, M., & Çimenoğlu, H. (2025). *Journal of Alloys and Compounds, 1010*, 178246.
125. Zhou, Y. H., Chen, P. H., Huang, D. N., Wu, Z. Z., Yang, T., Kai, J. J., & Yan, M. (2022). *Surface and Coating Technology, 445*, 128719.
126. Santos, P. B., Baldin, E. K., Krieger, D. A., de Castro, V. V., Aguzzoli, C., Fonseca, J. C., Rodrigues, M., Lopes, M. A., & de Fraga Malfatti, C. (2021). *Surface and Coatings Technology, 417*, 127179.
127. Xu, D. A., Chen, W. E., Liang, W. U., Lei, L. I., Wu, Y., Di, X., Wu, J., Ci, W., Ta, A., & Pa, F. (2021). *Chinese Journal of Aeronautics, 34*, 228–242.
128. Almajidi, Y. Q., Ali, E., Jameel, M. F., Saleh, L. H., Aggarwal, S., Zearah, S. A., Alamula, A. F., Alsaalamy, A., Sharifianjazi, F., & Bathaei, M. S. (2023). *Lubricants, 11*, 519.
129. Guo, H., Liu, Z., Wang, Y., & Li, J. (2021). *Ceramics International, 47*, 7344–7352.
130. Liu, A., Gao, S., Du, S., Lu, H., & Guo, J. (2024). *Tribology International, 193*, 109409.
131. Zhou, K., Xie, F., Wu, X., & Wang, S. (2021). *Surface and Coating Technology, 421*, 127429.
132. Shi, L., Jiang, C., Zhao, R., Si, T., Li, Y., Qian, W., Gao, G., & Chen, Y. (2024). *Ceramics International, 50*, 18484–18496.
133. Yang, C., Cui, S., Wu, Z., Zhu, J., Huang, J., Ma, Z., Fu, R. K., Tian, X., Chu, P. K., & Wu, Z. (2021). *Tribology International, 160*, 107018.

134. Rakhadilov, B., Zhassulan, A., Ormanbekov, K., Shynarbek, A., Baizhan, D., & Aldabergenova, T. (2024). *Crystals, 14*, 945.
135. Tang, Y., Yang, C. P., Sun, Q. Q., Wu, L. K., & Cao, F. H. (2024). *Corrosion Communications, 14*, 1–10.
136. Zhao, X., Chen, Y., Ji, R., Xu, M., Ye, Z., Shao, W., Wang, S., Zou, Y., Wang, Y., Ouyang, J., & Jia, D. (2024). *Surface and Coating Technology, 494*, 131471.
137. Chen, X. W., Ren, P., Zhang, D., Hu, J., Wu, C., & Liao, D. (2022). *Surface Innovations, 11*, 49–59.
138. Fattah-alhosseini, A., Molaei, M., Nouri, M., & Babaei, K. (2021). *Applied Surface Science Advances, 6*, 100140.
139. Gao, Q., Liu, S., Hou, K., Li, Z., & Wang, J. (2022). *Lubricants, 10*, 273.
140. Grigoriev, S., Peretyagin, N., Apelfeld, A., Smirnov, A., Morozov, A., Torskaya, E., Volosova, M., Yanushevich, O., Yarygin, N., Krikheli, N., & Peretyagin, P. (2023). *Materials, 16*, 3928.
141. Ebrahimi, M., Kermanpur, A., Kharaziha, M., & Bagherifard, S. (2024). *Surface and Coating Technology, 494*, 131400.
142. Chen, J., Guo, Q., Yang, Z., Li, J., Guo, Y., Yang, W., & Xu, D. (2024). *Journal of Materials Engineering and Performance, 33*, 1391–1400.
143. Yürektürk, Y. (2022). *Surface and Coating Technology, 433*, 128170.
144. Wang, S., Wang, Y., Zou, Y., Wu, Y., Chen, G., Ouyang, J., Jia, D., & Zhou, Y. (2020). *Chemical Engineering Journal, 402*, 126116.
145. Chen, X. W., Tang, S., Xie, W. L., Zhang, M., Song, H., Ran, Q. Z., Zhang, D. F., & Zeng, D. Z. (2024). *Materialwissenschaft und Werkstofftechnik, 55*, 1037–1044.
146. Ji, R., Wang, S., Zou, Y., Chen, G., Wang, Y., Ye, Z., Ouyang, J., Jia, D., & Zhou, Y. (2024). *Tribology International, 198*, 109885.
147. Lan, N., Yang, W., Gao, W., Guo, P., Zhao, C., & Chen, J. (2021). *Diamond and Related Materials, 117*, 108483.
148. Luo, J., Tang, W., & Cui, S. (2023). *Journal of Materials Research and Technology, 26*, 5851–5859.
149. Fan, Z., Lu, H., Liu, P., Pan, H., Ding, Y., Xu, G., & Tu, N. (2024). *Ceramics International, 50*, 44993–45005.
150. Sourani, F., Raeissi, K., Enayati, M. H., Kharaziha, M., Hakimizad, A., Blugan, G., & Salimijazi, H. R. (2022). *Journal of Alloys and Compounds, 920*, 165856.
151. He, B., Xin, C., Chen, Y., Xu, Y., Zhao, Q., Hou, Z., Tang, Y., Liu, H., Su, X., & Zhao, Y. (2022). *Applied Surface Science, 600*, 154096.
152. Borges, M. H., Kanniyappan, H., Sun, Y., Nagay, B. E., Da Cruz, N. C., Mathew, T. M., & Barão, V. A. (2023). *Surface and Coating Technology, 475*, 130165.
153. Guo, Y., Xu, L., Luan, J., Wan, Y., & Li, R. (2022). *Surfaces and Interfaces, 28*, 101626.
154. Fazel, M., Shamanian, M., & Salimijazi, H. R. (2020). *Biotribology, 23*, 100131.

Chapter 15
Advances in Anticorrosion and Anti-Wear Plasma Electrolytic Oxidation Coatings for Steel

Lujun Zhu and Manling Sui

15.1 Introduction

Steel is one of the most widely used engineering materials due to its desirable properties, including high toughness, high tensile strength, good machinability, and low cost. However, like light alloys, steel still faces challenges with corrosion and wear resistance. For example, carbon steels are commonly used in industries such as construction, pipelines, and kitchen appliances. However, when carbon steel is used in sectors like automotive and shipbuilding, it is susceptible to corrosion, especially in aqueous environments.

Various methods, including gas flame spraying, arc discharge plasma, and vacuum deposition, have been used to create ceramic coatings on steel. However, these methods typically require high temperatures to ensure strong adhesion of the coating, which can weaken the mechanical properties of the steel. Additionally, these methods may not be suitable for treating parts with complex shapes [1].

Plasma electrolytic oxidation (PEO) has been successfully applied to valve metals like aluminum, magnesium, titanium, and zirconium, significantly improving their wear and corrosion resistance [2]. However, applying PEO to non-valve metals, such as carbon steel, iron, copper, and nickel, is challenging. In the early stage of the PEO process, a barrier layer needed to initiate the anodic oxidation process

L. Zhu (✉)
School of Physics and Information Technology, Shaanxi Normal University, Xi'an, China
e-mail: zhulujun@snnu.edu.cn

M. Sui
State Key Laboratory of Materials Low-Carbon Recycling, Beijing Key Laboratory of Microstructure and Property of Advanced Materials, College of Materials Science and Engineering, Beijing University of Technology, Beijing, China
e-mail: mlsui@bjut.edu.cn

V. S. Saji, M. Zheludkevich (eds.), *Plasma Electrolytic Oxidation – Fundamentals, Advances and Applications*,
https://doi.org/10.1007/978-3-031-92429-3_15

does not form on the metal surface [3]. One of the reasons for this is the loose structure of Fe_2O_3 [4], a key product in the oxidation process.

In recent years, PEO technology has been further developed to improve the corrosion and wear resistance of steel surfaces. Steel parts treated with PEO coatings, as well as parts made from other alloys, exhibit in situ growth of porous, highly crystalline, and thick oxide coatings (typically 1–100 μm). These coatings provide excellent adhesion to the substrate, high surface microhardness, good shock tolerance at high temperatures, and significant resistance to both corrosion and wear. Additionally, they show noticeable insulation resistance under working conditions.

This chapter will focus on the important research progress of PEO technology for enhancing the surfaces of stainless steel and other metals. To enable the PEO process, an electrically insulated oxide film and an arc trigger are necessary. The oxide film accumulates charge, while the arc trigger facilitates self-oxidation of the substrate. However, steel's self-oxidation product, Fe_3O_4, is highly conductive, making it difficult to form PEO coatings directly on steel surfaces. Unlike valve metals, where anodic discharges can form easily, steel requires different strategies to overcome this challenge. In the following sections, we will discuss three main strategies that researchers have developed to address this issue.

15.2 PEO Coatings on Aluminized Steel

15.2.1 Mechanism

Since steel has low passivation in PEO electrolytes, the anodic voltage does not increase with treatment time, preventing anodic discharge from occurring. To enable the use of advanced surface modification technologies for steel, researchers have explored various indirect methods. One common approach is to apply an aluminum layer to the steel surface before performing PEO process. For example, hot-dipping or arc spraying aluminum onto steel surfaces has been used prior to the PEO process [5]. The device used for hot-dipping is typically a self-made crucible resistance furnace, as shown in Fig. 15.1.

The aluminum layer on the steel surface is oxidized into Al_2O_3 during the PEO process. Several conventional techniques, such as arc spraying [7], hot-dip aluminum (HDA) [8–11], plasma sputtering of aluminum [12, 13], and electro-spark deposition [14], can be used to fabricate aluminum layers on steel. The adhesion between the aluminized layer and the steel is a crucial factor in the process design. Figure 15.2a, b shows the PEO coating prepared by Z. Wu et al. [7] on the surface of hot-dip aluminized steel. The aluminized layer consists of a 30–50 μm thick aluminum layer and a 60 μm-thick FeAl layer. The PEO coating was produced using a bipolar pulse power in an alkali-silicate solution (16 g/L Na_2SiO_3 and 2 g/L NaOH). Figure 15.2c illustrates the variation in the average anodic voltages of HDA and pure aluminum during the PEO treatment. The average voltage for pure aluminum

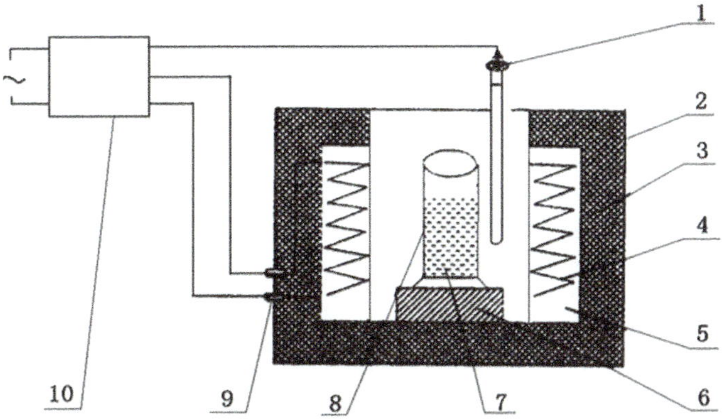

Fig. 15.1 Schematic drawing of the aluminum hot-dipping equipment. 1–Thermocouple; 2–furnace shell; 3–asbestos insert; 4–resistance wire; 5–gray porcelain tube; 6–firebrick; 7–molten aluminum; 8–aluminum melting pot; 9–quartz tube; 10–temperature controller for resistance furnace. Reproduced with permission from [6] © 2011 Elsevier B.V.

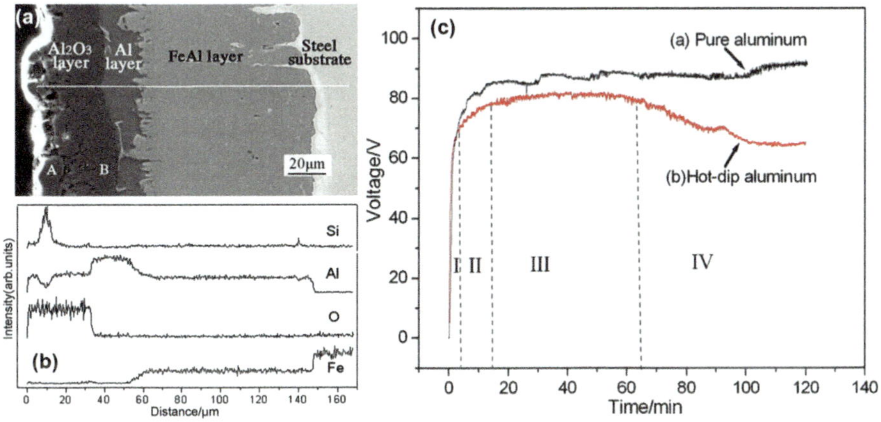

Fig. 15.2 (**a**) Cross-section structure and (**b**) element distribution of the PEO coating on aluminized steel. (**c**) The average anodic voltage vs. time response. Reproduced with permission from [7] © 2007 Elsevier B.V.

is approximately 90 V, while for hot-dip aluminum, it ranges from 70 to 80 V. Although the average voltage is relatively low, the peak voltage of each pulse can reach between 200 and 500 V, which is high enough to generate discharge sparks at the sample surface during the treatment process. The anodic voltage plots for hot-dip aluminum can be divided into three or four stages. Stages I–III correspond to the PEO stages for aluminum alloys. In contrast to the pure aluminum sample, the anodic voltage of the hot-dip aluminum sample drops in the final stage, and only a few discharges are observed at the surface.

15.2.2 Phase Component and Microstructure

The phase compositions of PEO coatings on aluminized steel substrates primarily consist of α-Al_2O_3 and γ-Al_2O_3. The mullite phase has been detected in the XRD patterns of PEO coatings prepared in silicate-based solutions, as shown in Fig. 15.3 [15]. TEM studies [6, 16, 17] were conducted to analyze the phase composition of PEO coatings on aluminized steel. Their results revealed the presence of additional alumina phases, including θ-Al_2O_3, κ-Al_2O_3, β-Al_2O_3, and an amorphous phase. In fact, the amorphous Al_2O_3 phase has been shown to play a role in the coating's growth [2, 18].

The surface morphology of PEO coatings on aluminized steel is quite similar to that of coatings on aluminum alloys. Figure 15.4 shows the surface topography of PEO coatings on aluminized Q235 steel after varying treatment times. The micrographs reveal that pores are distributed across the surface of the ceramic coatings. As treatment time increases, the pore density decreases, while the diameter of the pores increases noticeably [19]. These pores exhibit typical crater-like structures. Some studies have suggested that the instantaneous temperature within the discharge channels can reach several thousand degrees Celsius, leading to the formation of molten oxides. These molten oxides react transiently with the electrolyte, solidifying quickly, with some floccules deposited around the discharge zones [20]. Z. Wu et al. [7] reported that the diameters of some pores could reach 30 μm after 60 min of PEO processing, which is larger than the typical pore sizes observed in PEO coatings on bulk Al or Mg alloys.

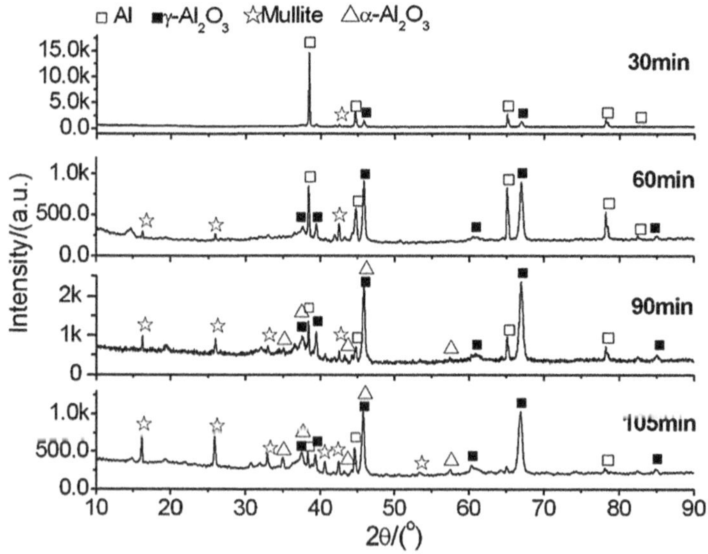

Fig. 15.3 XRD analysis of PEO coating on aluminized steel at different PEO treatment times. Reproduced with permission from [7] © 2007 Elsevier B.V.

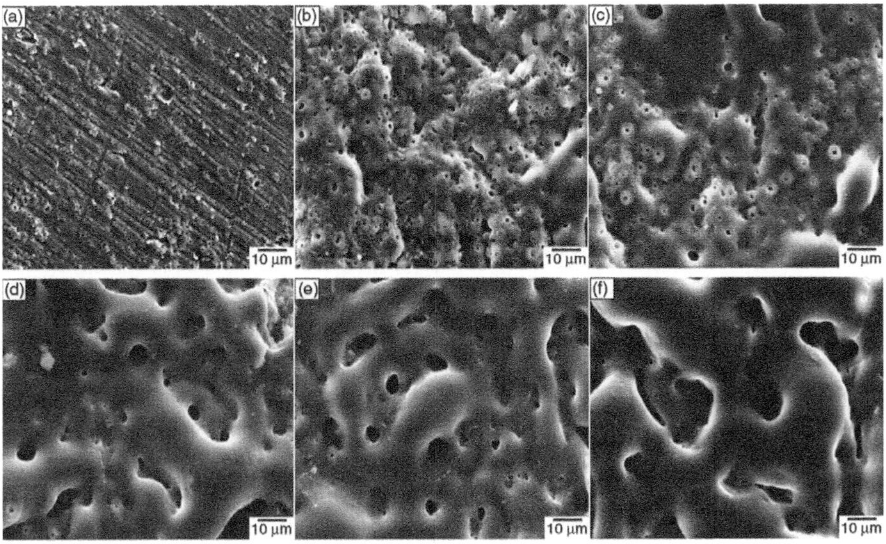

Fig. 15.4 SEM surface view micrographs of the coatings processed for (**a**) 45 s, (**b**) 10 min, (**c**) 15 min, (**d**) 20 min, (**e**) 25 min, and (**f**) 35 min. Reproduced with permission from [19] © 2005 Elsevier B.V.

The addition of additives to the electrolyte is also an important factor affecting the coating morphology. Y.L. Wang et al. [21] studied the effects of Na_2WO_4 and Na_2SiO_3 additives in electrolytes on microstructure and properties of PEO coatings on Q235 carbon steel. All coatings exhibit typical features, such as dark circular pores and molten regions distributed across the coating's surface. When additives are introduced into the electrolytes, the pore density increases and the pore diameter decreases. This demonstrates that a proper concentration of Na_2WO_4 or Na_2SiO_3 in the electrolyte can result in a more uniform and finer coating surface.

15.3 Single PEO Process in Aluminate or Silicate Electrolytes for Steel

15.3.1 Mechanism

The use of PEO for anodizing non-valve metals such as carbon steel, iron, copper, nickel, and other practically important metals is hindered by the fact that, at the initial stage of oxidation, the barrier layer essential for transitioning the process into the spark regime does not form on the metal surface. In addition to the method of pre-coating the steel surface with aluminum, as mentioned in the previous section, a second approach involves the formation of a barrier-like layer on the iron surface. This is achieved not by directly oxidizing the iron substrate but through the deposition of an Al_2O_3 or SiO_2 layer formed via micro-arc decomposition of aluminate or silicate electrolytes. These electrolytes include sodium aluminate and sodium hydroxide [22, 23], sodium silicate and sodium carbonate [24], silicate electrolytes with hydrophobic agents [25], sodium aluminate and sodium dihydrogen phosphate or sodium silicate and sodium hypophosphite [26], or sodium aluminate and sodium phosphate [27].

Typically, aluminate electrolytes decompose near the electrode surface, where the resulting species, such as alumina, form an Al_2O_3 film at the initial stages of the PEO process. This Al_2O_3 film acts as a barrier layer, allowing the process to transition from anodizing to the sparking regime. Similarly, the formation of an SiO_2 layer is crucial to initiate the PEO process, as this layer supports the appearance of discharge channels, enabling the PEO reactions to proceed [28]. Li et al. [29] suggested that the formation of stable plasma discharges on steel is linked to the development of a thin aluminum phosphate ($AlPO_4$ and $AlPO_4{\cdot}2H_2O$) or alumina aluminum phosphate (AAP) layer at the initial PEO stage. Figure 15.5 illustrates the role of the $AlPO_4$ layer in stimulating plasma discharge generation and initiating the PEO coating on steels. Figure 15.5a shows the presence of primary anions in the electrolyte before the PEO process. After applying the anodic current, numerous oxygen bubbles appear, as depicted in Fig. 15.5b, causing local acidification at the substrate surface. This results in a thin electrolyte layer forming on the surface, as shown in Fig. 15.5c. Anion species like H_2PO_4 may form within the thin electrolyte layer due to equilibrium reactions between different phosphate ions. Once a uniform $AlPO_4$ layer forms on the surface, dielectric breakdown occurs, and plasma discharges are generated, as illustrated in Fig. 15.5d.

15.3.2 Phase Component and Microstructure

PEO coatings on steel, obtained using aluminate or silicate electrolytes, are primarily composed of deposits derived from the electrolyte. For example, when treated in an aluminate system, the main components of the coating are α-Al_2O_3 and γ-Al_2O_3. V. Malinovschi et al. [23] prepared PEO coatings on stainless steel using an electrolyte solution of sodium aluminate ($NaAlO_2$—7 g/L) and sodium hydroxide (NaOH—2 g/L). To study the influence of the immersion depth of the sample in the electrolyte upon the deposited layer structure, the other set of PEO experiments was performed with the same oxidation process duration and working voltage (5 min, 350 V) at immersion depths (h_c) of 3 cm, 4 cm, 5 cm, 6 cm, and 7 cm. The resulting coatings primarily consisted of α-Al_2O_3, γ-Al_2O_3, and small amounts of the substrate oxide α-Fe_2O_3. As shown in Fig. 15.6, after a 1-min treatment, the oxide layer was composed solely of γ-Al_2O_3. However, for treatment durations of 3 min or longer, the α-Al_2O_3 phase also appeared, with its concentration increasing as the treatment time was extended. A qualitative phase analysis of samples treated at different depths under the same conditions (U = 350 V; τ = 5 min) revealed the presence of the same polycrystalline phases. However, when using an electrolyte

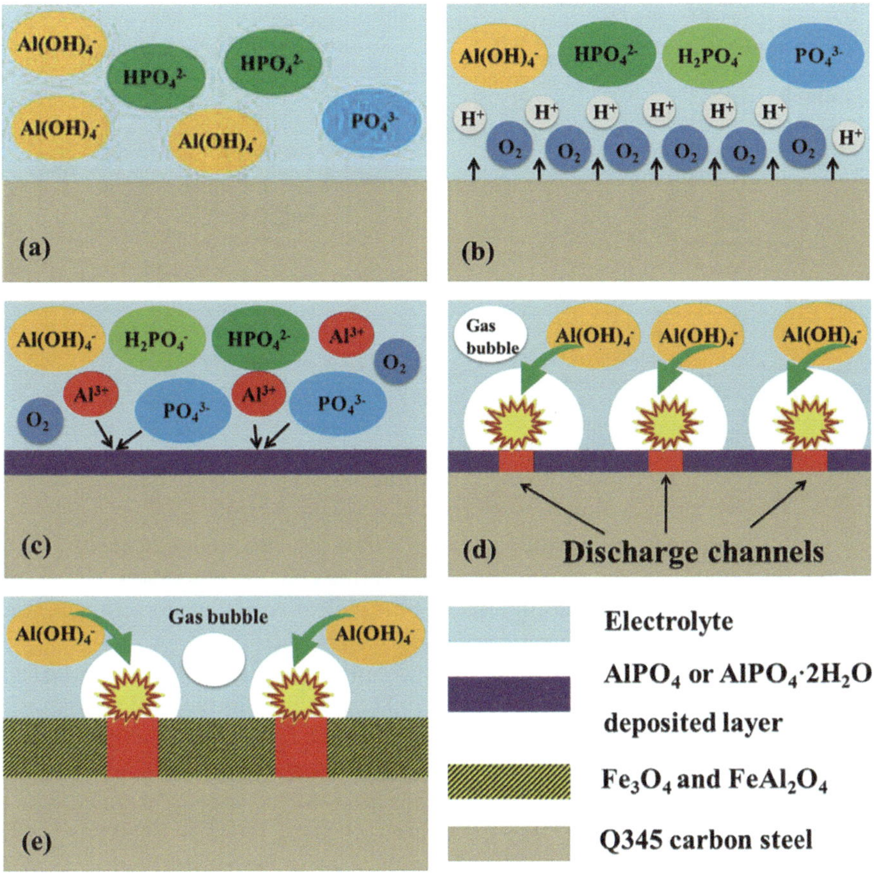

Fig. 15.5 Schematic presentation of the PEO process on steel in 8 g/L $NaAlO_2$ + 2 g/L NaH_2PO_4: (**a**) the sample before applying the anodic current, (**b**) a short time after applying anodic current, (**c**) the depositing aluminum phosphates or AAP layer on the steel surface, (**d**) the beginning of plasma discharges through the breakdown of the dielectric layer, and (**e**) steady-state regime of discharges at the final stage of the PEO process. Reproduced with permission from [29] © 2018 Elsevier Ltd.

solution of sodium silicate and sodium carbonate for PEO treatment on stainless steel, the coatings consisted of a mixture of the amorphous phase of SiO_2 and polycrystalline iron oxides, Fe_2O_3 and Fe_3O_4, as shown in Fig. 15.7 [24].

In many cases, the surface morphology of PEO coatings on steel treated in aluminate or silicate electrolytes differs significantly from that of coatings typically formed on valve metals. L. Pezzato et al. [30] prepared PEO coatings on steel using a NaOH-Na_2SiO_3-$NaAlO_2$ electrolyte. Figure 15.8a shows BSE-SEM images of the cross-section (left) and surface (right) of a PEO coating obtained using an electrolyte composed of 10 g/L NaOH + 10 g/L Na_2SiO_3 + 10 g/L $NaAlO_2$. The resulting oxide layer is cracked and nonhomogeneous, with a thickness of approximately

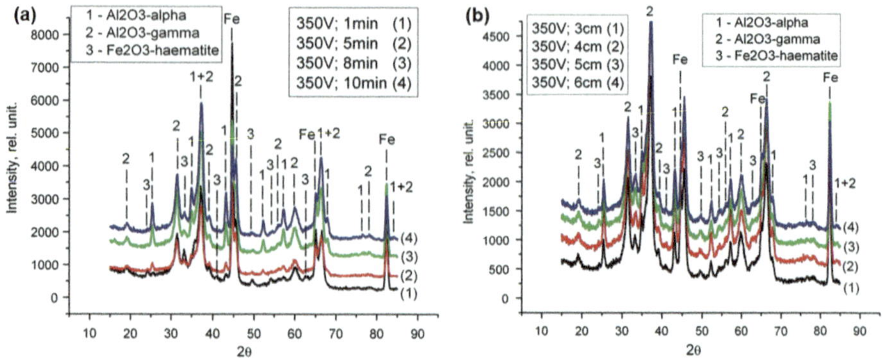

Fig. 15.6 The qualitative phase analysis of the PEO samples obtained in aluminate electrolytes: (**a**) with different periods of oxidation at a depth $h_c \cong 7$ cm, (**b**) with different sample depth diving at oxidation time of 5 min. The depth h_c is the distance from the center of the sample to the free surface of the electrolyte solution. Reproduced with permission from [23] © 2014 Elsevier B.V.

11 μm. The surface also exhibits visible cracking and inhomogeneity. Semi-quantitative EDS analysis revealed that the coating consists of oxides of Fe, Al, Si, and Cr. This variation in surface morphology is indicative of the interaction between different components in the electrolyte and the substrate.

In contrast, when a different electrolyte mixture—10 g/L NaOH + 100 g/L Na_2SiO_3 + 10 g/L $NaAlO_2$—was used, a PEO coating was formed with two distinct surface zones. One zone was darker, exhibiting numerous cracks and being rich in Si oxides, while the other, lighter zone contained several pores and was rich in iron oxides, as shown in Fig. 15.8b. Figure 15.8c shows a PEO coating obtained with an electrolyte of 10 g/L NaOH + 10 g/L Na_2SiO_3 + 100 g/L $NaAlO_2$. The PEO process resulted in a coating approximately 15 μm thick, with two distinct zones: Zone 1, primarily composed of Fe oxides, and Zone 2, mainly consisting of Al oxides. At the interface between the substrate and the coating, a thinner, irregular layer (Zone 3), <1 μm thick, was observed. This zone was enriched in iron oxides and exhibited a higher oxygen content compared to Zone 1. The formation of distinct zones in the coating structure suggests a complex interaction between the electrolyte and substrate, leading to different oxide phases at various depths. The complete microstructure of the coating consists of four distinct layers, two of which are very thin and not visible. The surface of the coating shows two main zones: a darker zone (Zone 4), rich in micro-cracks, and a lighter, smoother, and more homogeneous zone (Zone 5). EDS analysis revealed that Zone 4 is predominantly composed of aluminum oxide, while Zone 5 is primarily iron oxide. The white zones scattered across the lighter regions are enriched in iron. This multilayered structure highlights the complexity of the PEO process, where different oxide phases form at varying stages, contributing to the coating's overall properties.

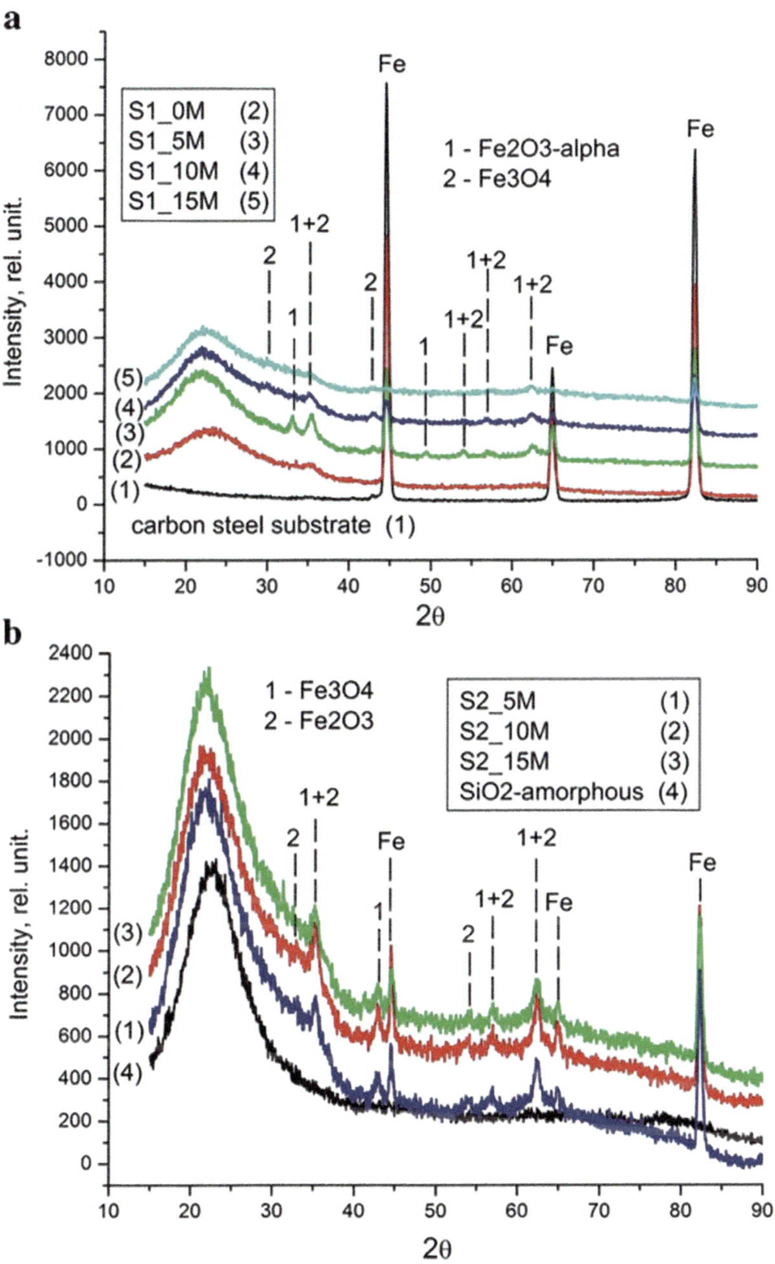

Fig. 15.7 (**a**) XRD patterns of steel substrate and oxide coatings formed during 0 min, 5 min, 10 min, and 15 min of PEO samples obtained in silicate electrolytes. (**b**) XRD patterns of SiO_2 amorphous powder and oxide coatings formed during 5 min, 10 min, and 15 min of the PEO process. Reproduced with permission from [24] © 2016 Elsevier B.V.

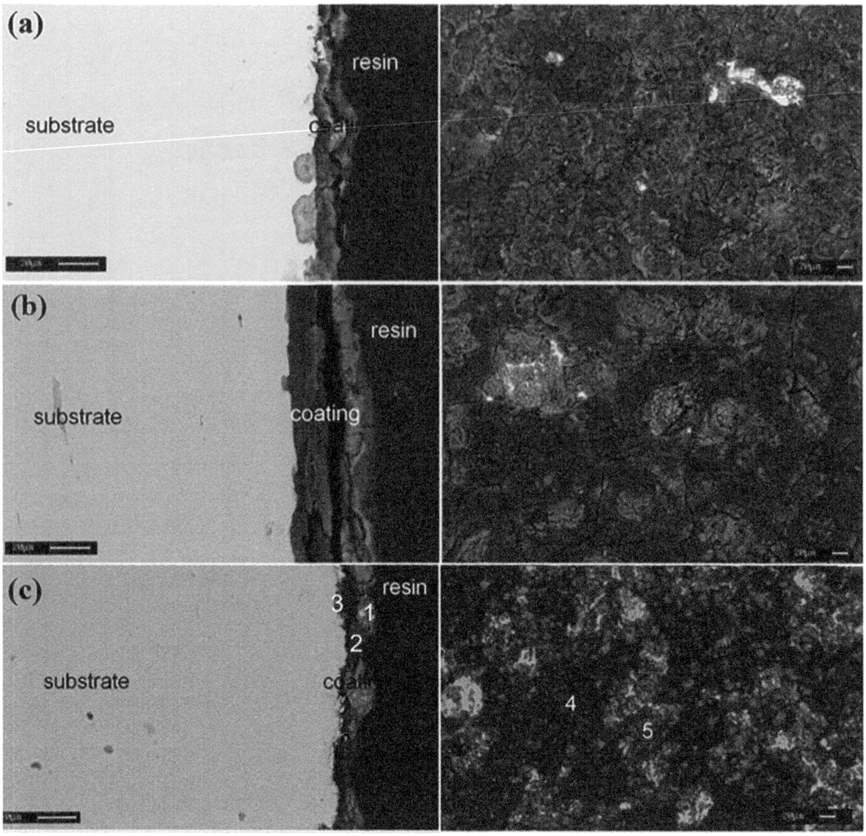

Fig. 15.8 BSE-SEM images of the cross section (left) and the surface (right) of the PEO coating obtained with electrolyte of (**a**) 10 g/L NaOH + 10 g/L Na_2SiO_3 + 10 g/L $NaAlO_2$, (**b**) 10 g/L NaOH + 100 g/L Na_2SiO_3 + 10 g/L $NaAlO_2$, and (**c**) 10 g/L NaOH + 10 g/L Na_2SiO_3 + 100 g/L $NaAlO_2$. Reproduced with permission from [30] © 2016 Elsevier B.V.

15.4 Cathodic PEO for Steel

15.4.1 Mechanism

In traditional PEO, the workpiece as anode is immersed in the inorganic electrolyte solution at ambient temperature. By increasing the voltage between workpiece and counter electrode, the anodic passive film is broken down, leading to a plasma discharge and in situ ceramic film growth. However, the PEO process cannot take effect on steel materials directly, due to the loose characteristic of Fe_2O_3 phase. For this, researchers [23, 27] have tried to introduce Al element into the electrolyte solution, forming Al_2O_3 phase to improve the insulativity of PEO coating. The mechanism of this method is essentially the same with the PEO process on aluminum or aluminum alloys, and the electrolyte composition is complex [4, 31].

Unlike the PEO process, the Cathodic PEO (CPEO) process sets the workpiece/sample as cathode. When the high voltage is applied to the steel sample, the aqueous solution around the sample as a cathode is heated to form a continuous gas envelope. After the applied voltage is over a critical value, this gas envelope is broken down to result in the plasma discharge around the sample. The gas discharge envelope contains lots of active oxygen species. The surface temperature of the sample is over several hundred degrees centigrade. Then, the steel surface is quickly oxidized to form an oxide coating [31]. Figure 15.9 shows a schematic illustration of the experimental setup of CPEO process and a profile of the applied voltage [32].

15.4.2 Phase Component and Microstructure

Due to the differences in growth mechanisms, the composition of CPEO coatings is significantly different from that formed during anodizing. Wu et al. [4] prepared CPEO coatings for steel using a glycerol-based electrolyte. The oxide layer primarily consists of the FeO phase, with some carbon particles embedded within it. XRD patterns shown in Fig. 15.10a were obtained from both the bare steel substrate and the CPEO-treated steel after 5 and 8 min of discharge. The diffraction peaks at 36.0°, 41.9°, 60.8°, 72.7°, and 76.6° correspond to the (111), (200), (220), (311), and (222) planes of the cubic FeO structure. These results indicate that the main phase of the CPEO coating is FeO. Additionally, small amounts of Fe_3O_4, Fe_2O_3, and some carbide phases were detected. Increased discharge time leads to thicker oxide layers and weaker α-Fe peaks from the steel substrate, suggesting an ongoing transformation of the substrate surface.

Figure 15.10b displays the Raman spectrum of the CPEO-treated steel, with laser excitation at 633 nm. This spectrum differs significantly from that of the

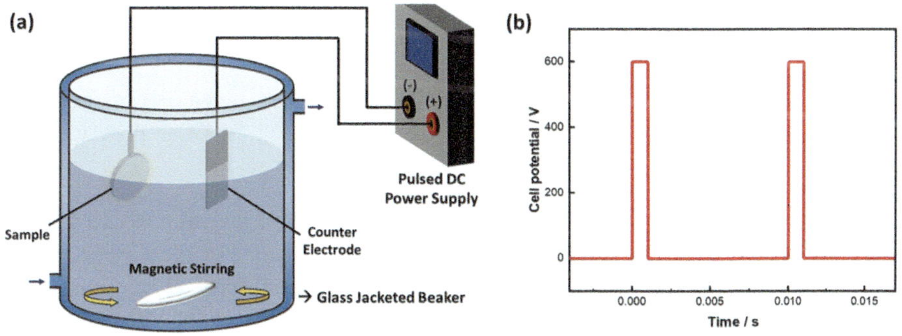

Fig. 15.9 (**a**) Schematic illustration of CPEO setup, and (**b**) applied voltage profile during CPEO process. Reproduced with permission from [32] © 2023 Elsevier B.V.

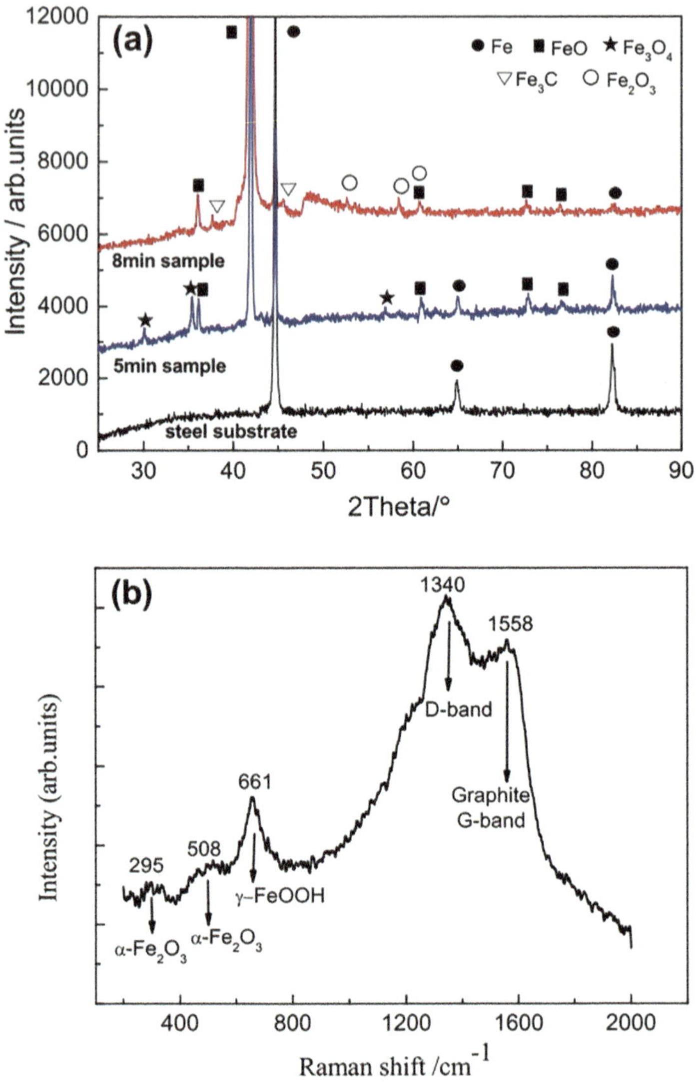

Fig. 15.10 (**a**) XRD patterns of the bare steel substrate and the CPEO-treated steel after 5 min and 8 min discharge, (**b**) Raman spectrum of the oxide layer on steel with laser excitation at 633 nm. Reproduced with permission from [4] © 2023 Elsevier B.V.

untreated steel. Raman scattering depends on changes in molecular polarizability, which is why no Raman signal is observed for pure metal. For the CPEO-treated steel, the Raman spectrum provides a detailed examination of the surface composition. The spectra for the 5-min and 8-min samples are almost identical. The peak at 661 cm^{-1} corresponds to FeO(OH) [33], which results from the oxidation of FeO in air through the reaction:

$$4FeO + 2H_2O + O_2 \rightarrow 4FeO(OH) \quad (15.1)$$

Additionally, peaks at 295 cm^{-1} and 508 cm^{-1} correspond to the Fe_2O_3 phase [33], while no Fe_3O_4 phase is detected in the Raman spectrum. This suggests that the Fe_3O_4 phase is located in the inner layer beneath other oxides. Furthermore, the D peak at 1340 cm^{-1} and the G peak at 1558 cm^{-1} indicate the presence of residual carbon particles on the steel surface after CPEO treatment [34]. These carbon particles could serve as lubricants during friction and wear processes. The presence of these carbon particles may enhance the coating's wear resistance.

Figure 15.11a–c shows the typical surface morphologies of CPEO coatings on AISI 304 stainless steel prepared at 240 V, 260 V, and 280 V for 5 min [35]. It is observed that the surface of the CPEO coatings is very rough with a lot of fine particles and some hill-shaped embossments. These particles get less with the increase of applied voltage. The grain boundary on the surface of coatings can be observed and the average grain size ranges are 0–5 μm, 3–8 μm, and 6–14 μm for the CPEO coatings under 240 V, 260 V, and 280 V, respectively. Several pores, identified as discharge channels, are visible through the oxide layer. Longer CPEO treatment times result in smoother surfaces, larger grains, and fewer pores. Yu et al. [36] proposed that the smooth surface morphology may be a result of the sintering-like effect caused by the longer plasma discharge duration at the surface. This sintering effect leads to a more compact and uniform surface structure. The cross-sectional images of the CPEO coatings in Fig. 15.11e–g exhibit a loose layer on the top and a compact layer at the bottom. The loose layer is characterized by a weaker structure with numerous cracks and pores, while the compact layer below is denser with a large number of dots. The thickness of the double-layer structure CPEO coatings is 32 μm, 42 μm, and 54 μm, respectively, which has an ascending order with the increase of oxidation voltage. The EDS line composition profiles in Fig. 15.11e–g show that the Cr and Ni elements aggregate in the compact layer; meanwhile, the Fe element aggregates in the loose layer. The O element can be found both in the compact layer and the loose layer. A clear boundary between the Fe and Cr contents suggests that Fe has a higher tendency to diffuse outward compared to Cr [35]. This observation implies a diffusion-driven segregation of elements during the CPEO process, contributing to the formation of a dual-layer structure.

15.5 Wear Properties of PEO Coating for Steel

Material wear processes include one or a combination of wear mechanisms such as abrasion, fretting, adhesion, fatigue, oxidation, and other tribochemical reactions [37]. To mitigate material damage or loss due to wear, wear-resistant materials have been developed or designed. Since wear is a surface phenomenon, the wear behavior of materials is closely related to their surface properties and structures. One of the most effective strategies to reduce wear is surface engineering [38].

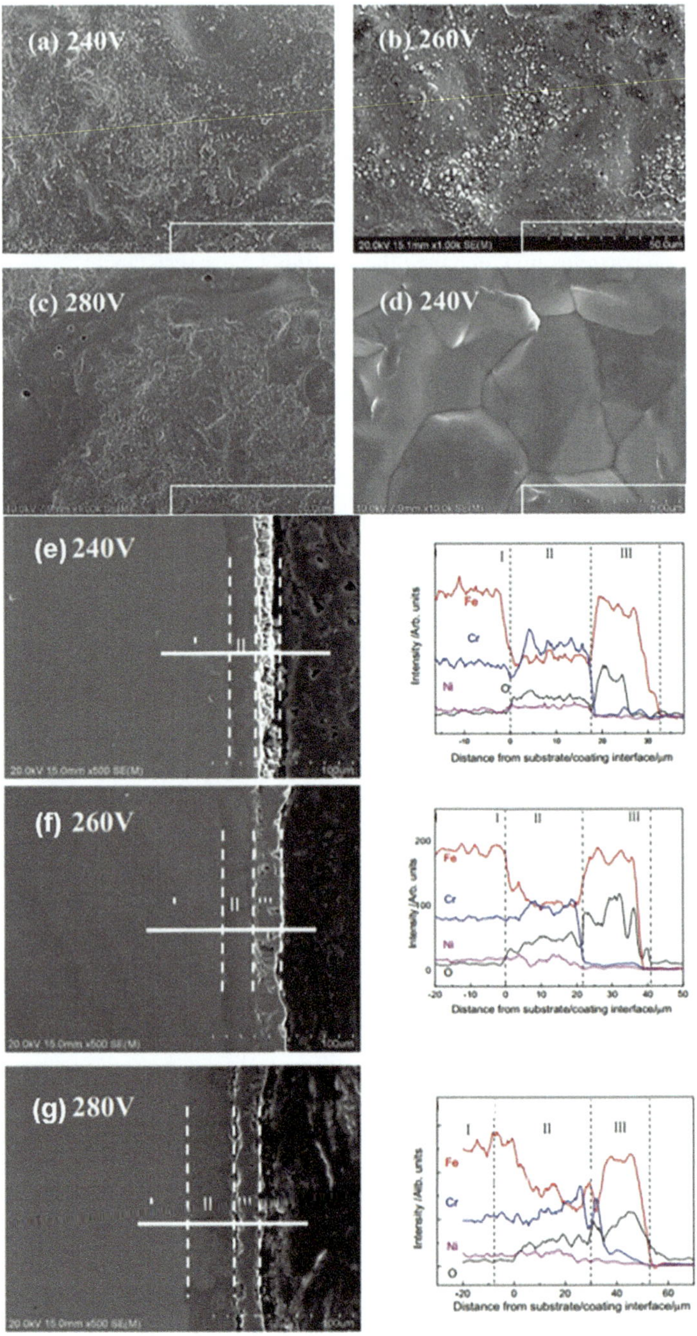

Fig. 15.11 Surface SEM morphologies of CPEO films on 304 stainless steel at different voltages: (**a**) 240V, (**b**) 260V, (**c**) 280V; (**d**) is the magnified picture of (**a**). Cross-section microstructure (**e–g**) and corresponding composition profiles of the CPEO films treated under 240 V, 260 V, and 280 V, respectively. I 304 stainless steel substrate, II compact layer, III loose layer. Reproduced with permission from [35] © 2019 Elsevier B.V.

In general, PEO-coated steel substrates exhibit lower friction coefficients compared to bare steel. Several factors contribute to the lower friction coefficients of PEO coatings, including the formation of a worm-like structure [26], an increase in surface microhardness [16, 39], and the presence of phosphorus (P) in the chemical composition of the coatings [26]. The combined effect of these factors enhances the tribological performance of PEO coatings.

The wear resistance of the samples was measured by abrasion weight loss, with less abrasion weight loss indicating better wear resistance. L. Lu et al. [6] compared the wear resistance of PEO-coated steel samples with heat-treated 45# steel samples. Figure 15.12 shows that the wear resistance of the PEO samples was

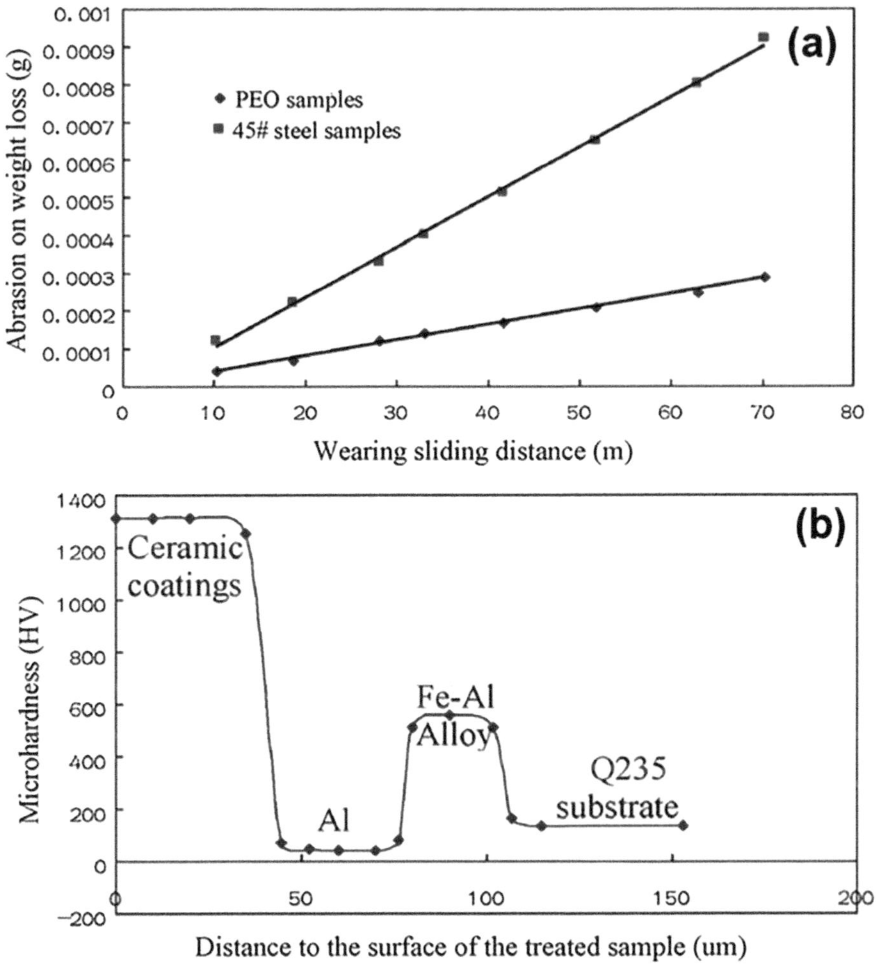

Fig. 15.12 (**a**) Abrasion weight-loss curves of the coatings and the heat-treated 45# steel. (**b**) Hardness distribution along the cross-section of the HDA-PEO-coated sample. Reproduced with permission from [6] © 2011 Elsevier B.V.

approximately three times higher than that of the heat-treated steel. The high hardness of the coatings contributed to their superior wear resistance. The wear resistance improvement is attributed to both the hard phases in the coating and the dense microstructure. Their study found that the PEO coating consisted of θ-Al_2O_3, κ-Al_2O_3, and β-Al_2O_3 phases. The hard phase composition and dense microstructure were identified as key factors contributing to the high wear resistance of PEO coatings on steel.

S. Durdu et al. [40] prepared a titanium alloy (Ti6Al4V) deposit layer on steel substrates via the electro-spark deposition (ESD) method, followed by the PEO process. The results showed that the average microhardness of the ESD-PEO coating was significantly higher than that of the ESD-only coating, even though the coating produced by ESD was denser than the one produced by PEO. This suggests that the microhardness of the coating is not only influenced by the coating density but also by the phase composition of the coating. They attributed the increased microhardness to the content of α-Al_2O_3 (corundum) and γ-Al_2O_3 phases in the PEO coating. The corundum phase, with a theoretical hardness value of 2500 HV [41], is a stable phase of Al_2O_3. Therefore, the presence of the corundum phase significantly enhances the average hardness of the duplex coating.

Wang Y et al. [26] prepared the PEO coating on Q234 carbon steel in carbonate electrolyte and silicate electrolyte, respectively. The results indicated that the coating obtained in carbonate electrolyte was Fe_3O_4, while the coating achieved from silicate electrolyte was proven to be amorphous. Both kinds of coatings showed coarse and porous surfaces. The Fe_3O_4 coatings obtained in carbonate electrolyte showed a high bonding strength to the substrate up to 20 ± 2 MPa and the value was 15 ± 2 MPa for the amorphous coatings obtained in carbonate electrolyte. The micro hardness of the amorphous coating and the Fe_3O_4 coating was 1001 HV and 1413 HV, respectively, which was more than two and three times that of the Q235 alloy substrate (415 HV). The friction coefficient exhibited by amorphous coating and Fe_3O_4 coating was 0.13 and 0.11, respectively, both lower than the uncoated Q235 substrate which ranged from 0.17 to 0.35.

Wu et al. [4] prepared CPEO coatings for steel in glycerol electrolyte. It is found that the microhardness gradually increases from the steel substrate to the surface. The maximum hardness of 8 min sample is about 650 HV, which is much higher than that of T8 carbon steel substrate (110 HV). Beyond the distance plotted for each oxidation time, the microhardness will reach the sample surface hardness, and it will not degrade beyond a certain distance since there is no loose layer near the sample surface. The high hardness of the oxide layer will be conducive to the enhancement of wear resistance for carbon steel. Figure 15.13 shows SEM micrographs of the wear tracks of the steel substrate and CPEO samples against ZrO_2 ceramic ball under 10 N load. They revealed that the wear mechanism has transformed to adhesive wear and fatigue wear. Results showed the oxide layer of 8 min sample is more compact, and its wear track presents more smooth and shallow as compared with 5 min sample. Therefore, the CPEO treatment has significantly improved the wear resistance of carbon steel.

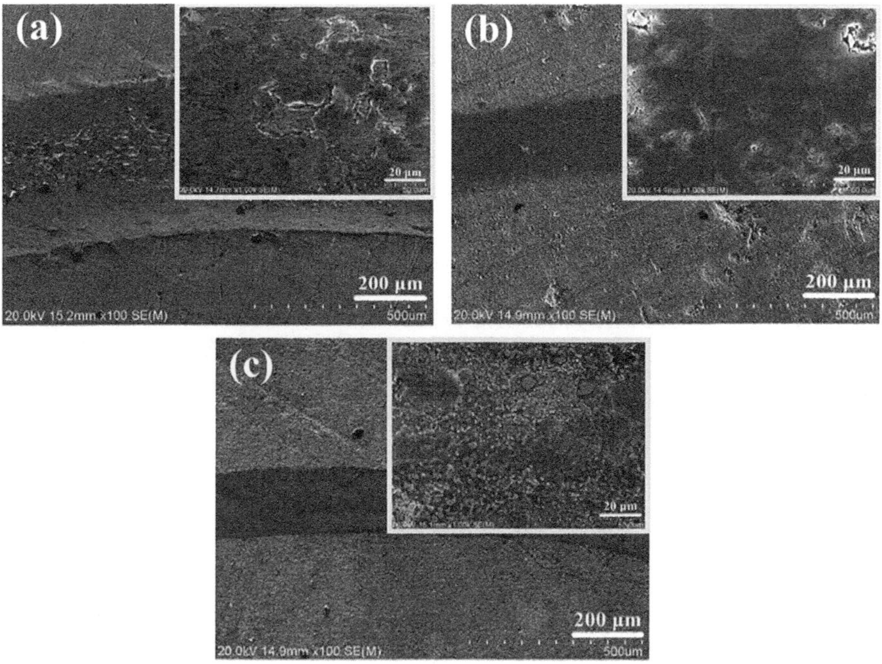

Fig. 15.13 SEM micrographs of the wear tracks of steel substrate (**a**), 5 min CPEO-treated sample (**b**), and 8 min CPEO-treated sample (**c**). Insets are their enlarged images. Reproduced with permission from [4] © 2018 Elsevier B.V.

Figure 15.14 demonstrates a comparative wear study on the PEO coatings produced on AerMet100 steel and the bare substrate. The two different PEO coatings showed greater wear performance in terms of wear depth and width, in which both declined significantly compared with the bare substrate. The PEO coatings on no aluminized steel and aluminized steel showed 30% and 14% decreases in their wear depth and width, respectively. Noticeably, substrate suffered from serious adhesive spalling and plowing during the wear test. This happened not only due to the lower shear resistance of the substrate, but also a thinner oxide layer influenced the wear behavior. Therefore, the bare substrate is not strong enough to tolerate the friction process and the wear test. Generally, the greater wear performance of the PEO coating was because of the micro-convex body on the surface of the oxide film, making the plastic deformation and fracture harder under the action of shear stress. However, slight adhesive wear could be noticed, while furrow on the surface of HPEO coating became lower. These phenomena could be attributed to the greater hardness of the PEO coating and the stronger plastic resistance. Interestingly, the debris left after the wear test filled the pores and some remained on the surface. Filling pores with left debris protected the surface from harder direct contact between the friction pair and the coating and could be used as a lubricant during the wear test. Besides, a smooth wear mark was found for IPEO coating with shallow furrows and lower exfoliated film [42].

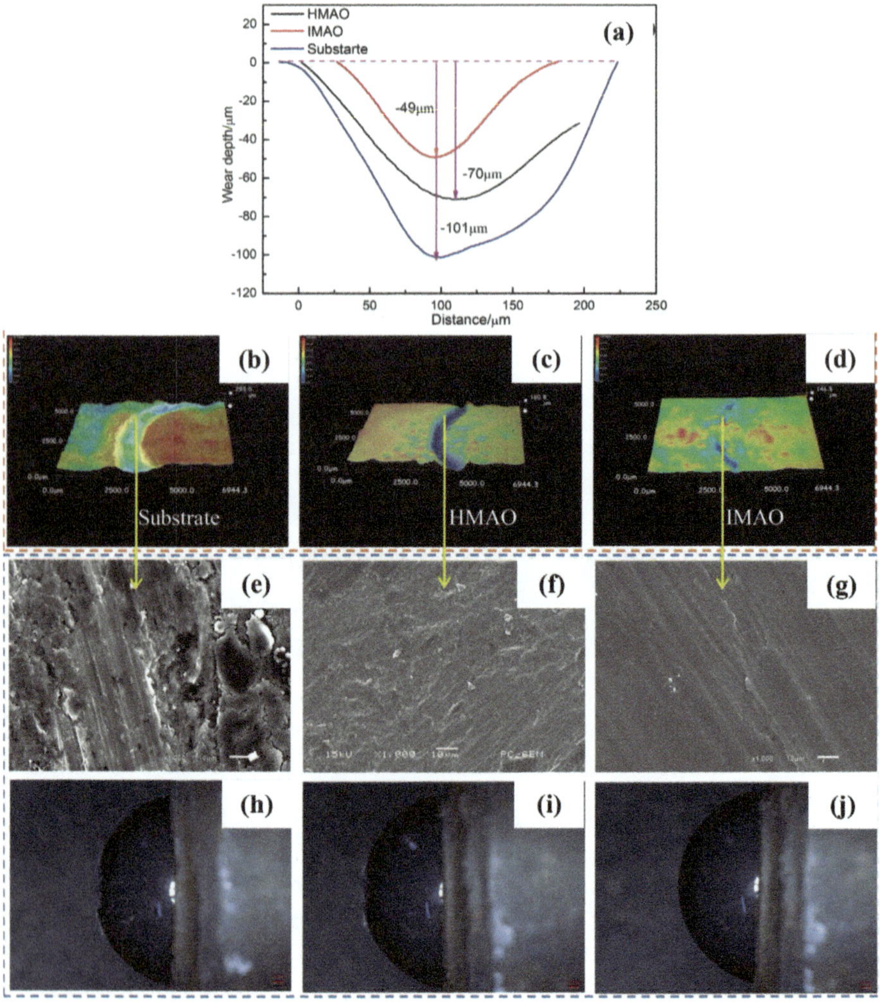

Fig. 15.14 Wear test results: (**a**) wear depth and width, (**b–d**) three-dimensional wear morphology, (**e–g**) high power wear morphology, and (**h–j**) surface morphology of friction pair. Reproduced with permission from [42] © 2021 Elsevier B.V.

15.6 Corrosion Properties of PEO Coating for Steel

Generally, corrosion assessment of PEO-coated steel samples was carried out using potentiodynamic polarization (PDP) and electrochemical impedance spectroscopy (EIS) techniques. It was reported that PEO-coated steels have demonstrated superior protective behavior in comparison to bare steels.

Jun Heo et al. [32] prepared the PEO coating on stainless steel by cathodic PEO. Two distinct layers, a porous-structured outer layer (~7 μm-in-thickness) comprising Fe oxides and a compact inner layer (~12 μm-in-thickness) comprising Cr oxides, were created on the stainless steel surface after the CPEO process. The PDP test showed the improved corrosion resistance of the CPEO-treated samples with a lower corrosion current. Accordingly, the corrosion rate showed a decrease of 64.93% and 68.84% for 20 s and 30 s samples, respectively. The EIS analysis, the outer oxide layer showed high CPE values, indicating a higher capacitive behavior with negative charges accumulated on the porous surface that may be the cause of the negatively formed corrosion potential. In contrast, the inner oxide layer showed high resistance against corrosion, acting as a dominant contributor to the low corrosion rate. The indentation test verified the strong bonding of the oxide layer to the substrate. This outstanding protective oxide layer can be fabricated in an extremely short processing time of <30 s.

Figure 15.15 shows typical Nyquist and Bode plots of PEO coatings on steel. A representative capacitive semicircle was drawn from the Nyquist plot using one time constant, which was composed of a parallel arrangement of a capacitor (or constant phase element) and resistor. The CPEO-processed samples exhibit depressed semicircles. This phenomenon can be interpreted as the overlap of several capacitive loops with different frequency ranges [36]. J. Heo et al. confirmed that by increasing the duration of the CPEO process, the corrosion resistance improvement ratio increased with an increase in the thickness of the protective oxide layer [32].

The content of α-Al_2O_3 in PEO coating plays the key roles on the corrosion resistance. α-Al_2O_3 with a hexagonal structure not only has characteristics of high crystallinity and hardness but also has excellent wear resistance and insulation. Additionally, the thermal expansion coefficient of α-Al_2O_3 is close to Fe, making an ideal ceramic coating for automotive rivets [43]. Pezzato et al. [30] successfully prepared hard ceramic PEO coatings on 39NiCrMo3 steel using electrolytes containing $NaAlO_2$ (100 g/L), Na_2SiO_3 (10 g/L), and NaOH (10 g/L). The PEO coatings were mainly composed of Fe_2O_3 together with a small amount of Al_2O_3. The self-corrosion potential of the coatings increased from −0.9 V to −0.3 V, and the corrosion current density decreased significantly, proving that the Al_2O_3-containing Fe_2O_3 film remarkably improves the corrosion resistance of steel. Zhang et al. [44] used aluminum nitrate to prepare an α-Al_2O_3 film on the surface of Fe-Cr-Al alloy, a dense and uniform microstructure was observed and its corrosion resistance reached up to 40 MΩ. Li et al. [45] focused on the electrolytes including Na_2SiO_3 and NaH_2PO_4, and studied the effects of Na_2CO_3 and $Na_2B_4O_7$ on the PEO process of the 10B21 low-carbon steel. The results showed that the addition of Na_2CO_3 and $Na_2B_4O_7 \cdot 10H_2O$ in the electrolyte effectively regulated the ionization and chemical reaction of the PEO process and inhibited the oxidation and phosphatization reactions of the matrix. At the same time, the α-Al_2O_3 formation was significantly accelerated, and there was no gap between the films and the matrix, leading to good bonding. After 7 days of neutral salt spray corrosion, the corrosion rate of the coating decreased to 53 $g \cdot m^{-2}\,h^{-1}$, and the service life of the steel components increased.

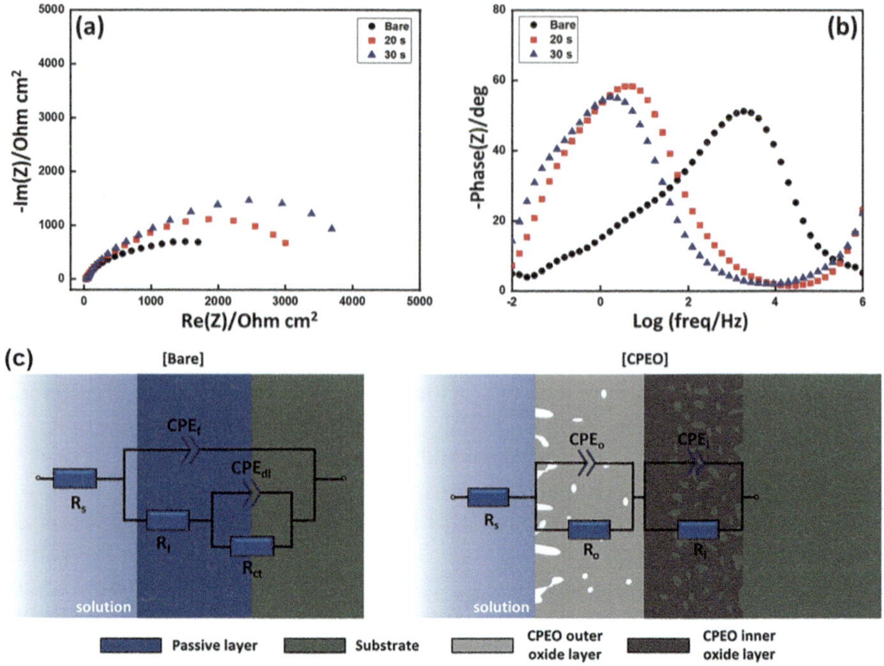

Fig. 15.15 (**a**) Nyquist impedance diagram, (**b**) Bode phase diagram of bare 316 L SS and CPEO films with various processing durations, and (**c**) physical models of bare 316 L SS and CPEO-processed SS samples with ECs. Reproduced with permission from [32] © 2023 Elsevier B.V.

Based on the electrolyte including $NaAlO_2$, NaH_2PO_4, Na_2CO_3, and $Na_2B_4O_7$, Zhao et al. [46] studied the effects of different concentrations of the α-Al_2O_3 additive (0, 1, 2, 3, and 4 g/L) in the electrolytes on the PEO layers of 10B21 and investigated the mechanisms for the effect of the added α-Al_2O_3. The results showed that the added α-Al_2O_3 particles diffused and were adsorbed into the pores in the PEO layer, and as the conversion rate from Al_2O_3 to the a phase in the layer increased, the pores in the films became smaller. Since the surface of the PEO films prepared by adding α-Al_2O_3 had an increased α-Al_2O_3 content, excellent electrochemical corrosion performance in a 3.5% NaCl solution was obtained, and the electrochemical self-corrosion current density was 0.2 mA/cm^2. After a neutral salt spray test was carried out for 7 days, the corrosion rate was 20 $g{\cdot}m^{-2}{\cdot}d^{-1}$, and the corrosion resistance was greatly improved.

Ribeiro et al. [47] evaluated the effect of several KOH and $Ta(OH)_5$ concentrations in PEO coating for stainless steel 304 samples. Subsequent analysis revealed that PEO-treated specimens displayed a heterogeneous coating comprising Ta-rich agglomerates, iron oxide plates, and an exposed substrate. Chemical analyses confirmed successful Ta enrichment within the iron oxide coatings, resulting in the phase composition of the samples changing from a biphasic ($\alpha + \gamma$) structure to a single γ-iron phase. EIS results revealed the presence of a double layer associated with Ta- and Fe-rich regions within the oxide layer, effectively enhancing the corrosion of PEO coatings.

Yang W et al. [48] carried out PEO coating on low carbon steel sheets in the electrolyte of 28 g/L Na_2SiO_3 and 1.2 g/L NaH_2PO_2, and then coated the sample with polyurethane. Results show that the PEO process produces an oxide layer with a porous and rough surface structure on the low-carbon steel substrate. The porous and rough PEO layer is beneficial for enhancing the adhesion strength and thickness of the organic top coatings. The organic-coated sample with PEO pretreatment exhibits improved corrosion resistance and longer service life in corrosive environment compared to that without PEO pretreatment, as shown in Fig. 15.16.

Pezzato L et al. [49] employed the base electrolyte of 25 g/L Na_2SiO_3 and 2.5 g/L NaOH produced the copper-containing PEO coating on zinc-aluminized steels with the addition of 15 g/L metallic copper particles. The result showed that the presence of copper particles can clearly be trapped in the pores of PEO coating. A general increase in the porosity of the coating in the PEO + Cu sample in comparison to the PEO sample can also be observed from the analysis of the sample's surface [49]. A higher magnification micrograph of the cross section of the sample PEO + Cu can be found reported in Fig. 15.17. It can be noted that the copper particles are present both in the external part of the coating and in the inner part. The addition of copper into the PEO coating not only enhanced the anodic regions but also the anodic intensity (*j* values). Optical images were taken at the end of the immersion test, confirming the adverse impact of the copper added to the PEO coating on corrosion

Fig. 15.16 Optical photos of the (**a**) single organic coating and (**b**) PEO/organic duplex coating after immersing for different times. Reproduced with permission from [48] © 2015 Elsevier B.V.

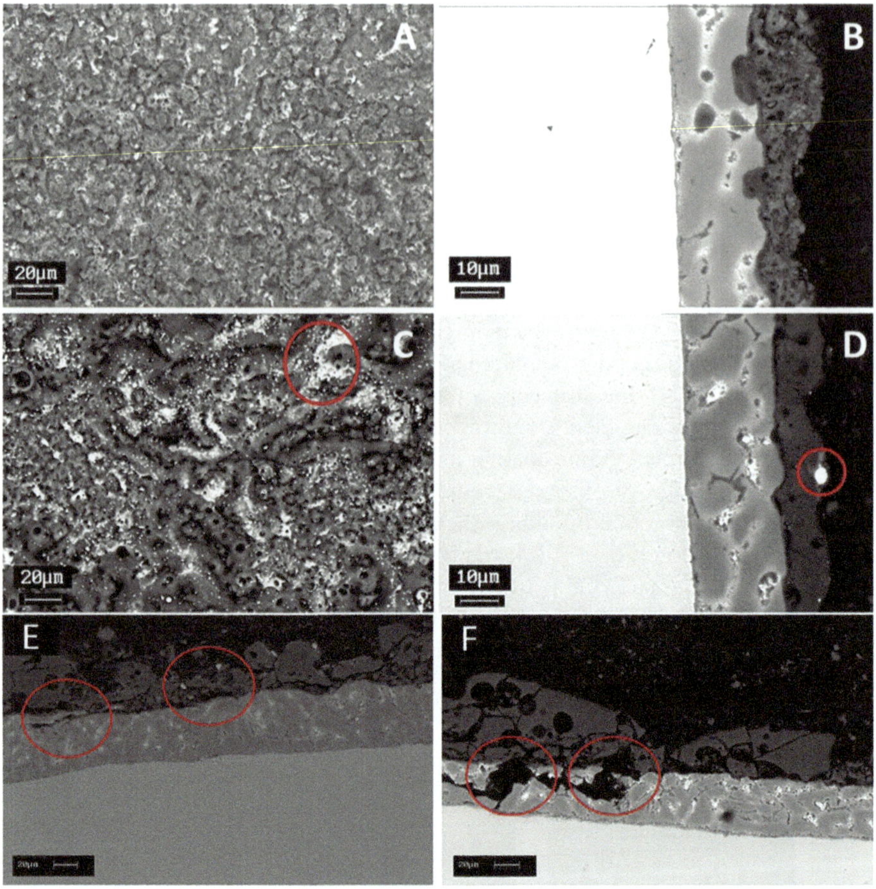

Fig. 15.17 SEM images of the surface of the PEO-treated sample (**a**) and of the PEO + Cu sample (**c**) and SEM images of the cross-section of the PEO-treated sample (**b**) and the PEO + Cu sample (**d**). Cross-sectional SEM micrographs of the PEO-treated sample (**e**) and the PEO + Cu sample (**f**) after 28 days of immersion. Reproduced with permission from [49] © 2024 The Author(s). Published by Elsevier B.V.

protection performance. The order of the samples, in terms of the corrosion protection properties, is PEO > substrate > PEO + Cu.

15.7 Conclusions and Outlook

This chapter introduces the recent advances in the PEO treatment of steel substrates. The preparation of PEO coatings on steel surfaces can be classified into three types based on their growth mechanisms. The first method involves applying an aluminum layer to the steel surface using other techniques before the PEO process, resulting in a coating similar to those formed on aluminum alloys. The second method uses appropriate concentrations of silicate and aluminate electrolytes to form a coating on the steel surface, primarily composed of oxides of electrolyte elements, such as SiO_2 or Al_2O_3. The third method is cathodic PEO technology, which grows a protective coating mainly composed of compact iron oxides on the steel surface. All three methods can effectively enhance the wear and corrosion resistance properties of steel by improving surface microhardness, coating compactness, and coating thickness.

Research on PEO coatings is becoming more in-depth, with future studies expected to focus on the following key areas: (1) enhancement of coating performance: researchers are exploring new electrolyte systems, such as those incorporating nanoparticles, ionic liquids, organic compounds, and composite electrolytes, to improve the microstructure and functionality of the coatings; (2) optimization of treatment parameters: the quality and performance of PEO coatings are significantly influenced by treatment parameters, such as current density, voltage, and processing time; and (3) development of multifunctional coatings: there is a growing effort to develop coatings with additional properties, such as antibacterial, anti-fouling, and thermal insulation capabilities, to meet broader application demands.

In conclusion, PEO coating technology on steel surfaces has made significant progress, with notable improvements in corrosion resistance, wear resistance, and thermal stability.

References

1. Gu, W.-C., Lv, G.-H., & Chen, H. (2007). *Journal of Alloys and Compounds, 429*, 238–244.
2. Zhu, L., Guo, Z., Zhang, Y., Li, Z., & Sui, M. (2016). *Electrochimica Acta, 208*, 296–303.
3. Meletis, E., Nie, X., Wang, F., & Jiang, J. (2002). *Surface and Coating Technology, 150*(2–3), 246–256.
4. Wu, J., Dong, L., Deng, J., Hou, D., Li, G., Li, D., & Xue, W. (2018). *Surface and Coating Technology, 338*, 63–68.
5. Wei, T., Jun, F. Y., & Tian, J. (2005). *Journal of Alloys and Compounds, 389*, 169–176.
6. Lu, L., Zhang, J., Shen, D., Wu, L., Jiang, G., & Li, L. (2012). *Journal of Alloys and Compounds, 512*, 57–62.
7. Wu, Z., Li, X. Y., Xu, G., & Xu, F. (2007). *Applied Surface Science, 253*, 8398–8403.
8. Xie, S. Y., Wu, C. Z., & Kou, B. D. (2003). *Light Alloy Fabrication Technology, 31*, 35 (in Chinese).
9. Niu, Z., Li, Z. L., Zhang, J., & Wang, A. (2010). *Advances in Materials Research, 139–141*, 394–397.
10. Zhao, Z., & Liu, J. W. (2010). *Advances in Materials Research, 139–141*, 406–409.
11. Barekatain, H., & Khoei, S. M. M. (2020). *Surface and Coatings Technology, 384*, 125339.
12. Chen, Z., Li, G., Wu, Z., & Xia, Y. (2011). *Materials Science and Engineering A, 528*, 1409–1414.
13. Efremov, A. P. (1989). *Protection of Metals, 25*(2), 176–180.
14. Durdu, S., Korkmaz, K., Aktuğ, S., & Çakır, A. (2017). *Surface and Coating Technology, 326*, 111–120.
15. Yilmaz, M., Çakir, A., Ribalko, A., & Korkmaz, K. (2015). *Acta Physica Polonica A, 127*(4), 1277–1281.

16. López-Ortega, A., Arana, J. L., Rodríguez, E., & Bayón, R. (2018). *Corrosion Science, 143*, 258–280.
17. Lihong, L., Shen, D., Zhi, J., Sui, J., & Li, L. (2011). *Applied Surface Science, 257*, 4144–4150.
18. Zhu, L., Ke, X., Li, J., Zhang, Y., Zhang, Z., & Sui, M. (2021). *Surfaces and Interfaces, 25*, 101186.
19. Gu, W., Shen, D., Wang, Y., Chen, G., Feng, W., Zhang, G., & Yang, S. (2006). *Applied Surface Science, 252*, 2927–2932.
20. Xue, W., Deng, Z., Chen, R., & Zhang, T. (2000). *Thin Solid Films, 372*, 114–117.
21. Wang, Y., Jiang, Z., & Yao, Z. (2009). *Journal of Alloys and Compounds, 481*, 725–729.
22. Karpushenkov, S. A., Shchukin, G. L., Belanovich, A. L., Savenko, V. P., & Kulak, A. I. (2009). *Journal of Applied Electrochemistry, 40*, 365–374.
23. Malinovschi, V., Marin, A., Moga, S., & Negrea, D. (2014). *Surface and Coating Technology, 253*, 194–198.
24. Malinovschi, V., Marin, A., Mihalache, M., & Iosub, I. (2016). *Surface and Coating Technology, 296*, 96–103.
25. Boinovich, L. B., Gnedenkov, S. V., Alpysbaeva, D. A., Egorkin, V. S., Emelyanenko, A. M., Sinebryukhov, S. L., & Zaretskaya, A. K. (2012). *Corrosion Science, 55*, 238–245.
26. Wang, Y., & Jiang, Z. (2009). *Applied Surface Science, 255*(12), 6240–6243.
27. Wang, Y., Jiang, Z., Yao, Z., & Tang, H. (2010). *Surface and Coating Technology, 204*(11), 1685–1688.
28. Fattah-alhosseini, A., Molaei, M., & Kaseem, M. (2024). *Surfaces and Interfaces, 46*, 104179.
29. Li, Z., Cheng, Y., Kang, S.-H., Tu, W., & Cheng, Y. (2018). *Electrochimica Acta, 284*, 681–695.
30. Pezzato, L., Brunelli, K., Dolcet, P., & Dabalà, M. (2016). *Surface and Coating Technology, 307*, 73–80.
31. Jin, X., Wang, B., Xue, W., Du, J., Wu, X., & Wu, J. (2013). *Surface and Coating Technology, 236*, 22–28.
32. Heo, J., Lee, J., Kim, S., Alfantazi, A., & Cho, S. O. (2023). *Surface and Coating Technology, 462*, 129448.
33. Chattopadhyay, A., Bandyopadhyay, N., Das, A. K., & Panigrahi, M. K. (2005). *Scripta Materialia, 52*, 211–215.
34. Wu, J., Fan, L., Dong, L., Deng, J., Li, D., Zhang, Y., & Xue, W. (2017). *Materials Chemistry and Physics, 199*, 289–294.
35. Jin, X., Chen, L., Yu, J., Liu, R., Xue, W., & Shang, H. (2019). *Surface and Coatings Technology, 363*, 314–321.
36. Yu, J., Zhang, Y., Jin, X., Chen, L., Du, J., & Xue, W. (2019). *Surface and Coating Technology, 363*, 411–418.
37. Baydoun, S., & Fouvry, P. A. (2020). *Tribology International, 151*, 106490.
38. Zhai, W., Zhou, L. B., Fan, X., Kang, G., Liu, Y., & Zhou, K. (2021). *Advancement of Science, 8*, 2003739.
39. Yang, W., Liu, W., Peng, Z., Liu, B., & Liang, J. (2017). *Ceramics International, 43*, 16851–16858.
40. Durdu, S., Aktuğ, S. L., & Korkmaz, K. (2013). *Surface and Coating Technology, 236*, 303–308.
41. Gent, M., Menendez, M., Toraño, J., & Torno, S. (2012). *Powder Technology, 224*, 217–222.
42. Attarzadeh, N., Molaei, M., Babaei, K., & Fattah-alhosseini, A. (2021). *Surfaces and Interfaces, 23*, 100997.
43. Li, J., Hu, R., Liu, H., Zhou, M., Gao, Z., & Luo, X. (2021). *Ceramics International, 47*(23), 32915–32926.
44. Zhang, Y., Chen, C., Chen, W., Cheng, H., Wang, Y., & Wang, L. (2017). *Journal of Industrial and Engineering Chemistry, 56*(51), 15058–15064.
45. Li, Y., Chen, M., Li, W., Wang, Q., Wang, Y., & You, C. (2019). *Surface and Coating Technology, 358*, 637–645.

46. Zhao, Z., Chen, M., You, C., Li, W., Tie, D., & Liu, H. (2020). *Journal of Materials Research and Technology, 9*(3), 3875–3884.
47. de Proença, J. P., Ribeiro, R. P., Rangel, E. C., da Cruz, N. C., Pinto, B. D. O., Grandini, C. R., & Correa, D. R. N. (2023). *Crystals, 13*(10), 1480.
48. Yang, W., Li, Q., Xiao, Q., & Liang, J. (2015). *Progress in Organic Coating, 89*, 260–266.
49. Pezzato, L., Akbarzadeh, S., Settimi, A. G., Moschin, E., Moro, I., Olivier, M. G., Brunelli, K., & Dabalà, M. (2024). *Surface and Coating Technology, 482*, 130631.

Chapter 16
Active Corrosion Protection Coatings Based on Plasma Electrolytic Oxidation: Multilayer and Self-Healing Coatings

Liang Wu, Zuojun Jiao, Wenhui Yao, and Fusheng Pan

16.1 Introduction

Plasma electrolytic oxidation (PEO), also known as micro-arc oxidation (MAO), is an electrochemical method that use valve metals or alloys (Mg, Al, Ti, etc.) as anodes and inert materials such as stainless steel as cathodes, placed in an electrolyte containing anions such as O^{2-}, SiO_3^{2-}, PO_4^{3-}, and AlO_2^-. When subjected to a strong electric field, the anode metal dissolves and undergoes electrochemical or plasma chemical reactions with oxygen and anions in the electrolyte, forming an oxide film onto the metal surface [1–3]. PEO coating can usually be divided into inner and outer layers. The inner layer has a dense structure and few defects, while the outer layer is more porous and looser. It is usually considered that the denser inner layer contributes the major corrosion resistance due to the excellent physical barrier effect [4]. Besides, PEO coating has good wear resistance, biocompatibility, and significantly higher corrosion resistance than conventional anodic oxidation. Although pores and cracks of PEO coatings accelerate the erosion of aggressive ions and increase the corrosion tendency, they can enhance the adhesion of organic small molecules, cells, and inorganic polymer molecules due to the coarse surface. Nowadays, PEO coating has become a mature precursor for Mg alloy composite coatings [5].

In recent years, extensive research has been carried out on corrosion-resistant coatings. These include sol-gel coating, Ca-P coating, layer-by-layer assembly coating (LBL), LDH coating, polymer coating, etc. However, the abovementioned coatings have a common problem, which is that they may crack, swell, and peel off

L. Wu (✉) · Z. Jiao · W. Yao · F. Pan
National Engineering Research Center for Magnesium Alloys, College of Materials Science and Engineering, Chongqing University, Chongqing, China
e-mail: wuliang@cqu.edu.cn

V. S. Saji, M. Zheludkevich (eds.), *Plasma Electrolytic Oxidation – Fundamentals, Advances and Applications*,
https://doi.org/10.1007/978-3-031-92429-3_16

during long-term service, leading to premature failure of the coatings [6]. LDH coatings are popular in sealing the defects of PEO coatings, which can be grown in situ on the outer layer of PEO through hydrothermal, water bath, and other methods, significantly improving the adhesion of the composite coating [7, 8]. It is even unnecessary to add external cations into the electrolyte [7]. However, a single LDH coating is still difficult to achieve effective long-term protection. Organic and polymer resin coatings can effectively hinder the invasion of corrosive media (water, oxygen, and ions) into metal substrates, but they face complex service environments and mechanical damage [9, 10]. Environmental factors such as temperature changes, high humidity, salt spray, ultraviolet radiation, and mechanical damage, including macroscopic damage caused by local scratches and stress concentration, deteriorate their protective effect [11]. If the composite coating is endowed with self-healing function, its long-term stability during service will be greatly improved. It is worth mentioning that PEO coating itself possesses a certain self-healing function. For Mg alloys, local enrichment of Mg^{2+} ions occurs when the PEO coating dissolves, and the insoluble salts produced by its reaction with anions in the solution precipitate at the scratching site of PEO, forming self-healing function. However, this self-healing effect is quite limited.

The multilayered active protective (self-healing) coating based on PEO-coated Mg alloys further enhances the long-term stability of the coating while improving its corrosion resistance. According to different protective mechanisms, self-healing coatings comprised two types: intrinsic self-healing (Fig. 16.1a) and extrinsic self-healing (Fig. 16.1b). Intrinsic self-healing indicates the use of the interaction among coating material molecules to re-bond with or without certain conditions (light, heat, electromagnetic, and humidity) after fracture occurs, thereby restoring or approaching the original state of the coating. As shown in Fig. 16.1a, the intermolecular interactions commonly involved in intrinsic self-healing coatings include strong covalent bonding interactions (e.g., disulfide bond, imine bond, and coordination bond) and relatively weak noncovalent bonding interactions (e.g., hydrogen bond, hydrophobic interaction, and electrostatic interaction). These coatings based on the structural characteristics of coating molecules do not require filler assistance and are relatively fast and effective in repair, but it has drawbacks that are difficult to design. Unlike intrinsic self-healing, extrinsic self-healing relies on external fillers to achieve self-healing performance. The repair efficiency is affected by the compatibility between the containers and substrate, the dispersion uniformity of the container, and the content of corrosion inhibitors in the container. Commonly used containers can be classified into the inorganic (e.g., SiO_2, TiO_2, CeO_2, halloysite nanotubes [HNTs], and LDHs), the organic (e.g., nanofiber, micro-gel, chitosan [CS], and other polymer microcapsules), and carbon nanomaterials (e.g., carbon nanotubes, carbon fiber, and graphene oxide) (Fig. 16.1b). Self-healing functional agents are usually loaded onto containers such as micro/nano capsules, micro/nano tubes, glass fibers, and nanoparticles. When the carrier in the coating is damaged, functional fillers will overflow and undergo chemical reactions such as cross-linking and solidification with the matrix molecules, thereby repairing the damaged area.

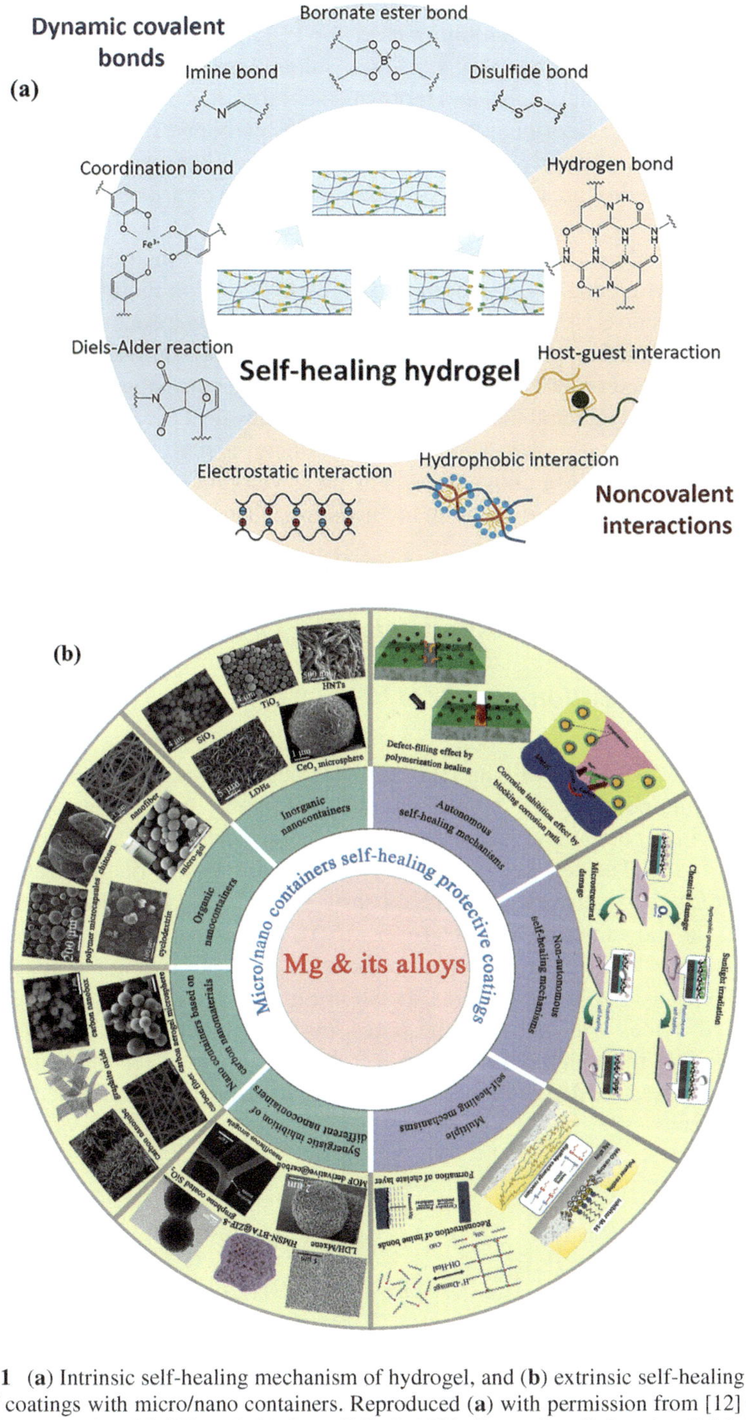

Fig. 16.1 (**a**) Intrinsic self-healing mechanism of hydrogel, and (**b**) extrinsic self-healing mechanism of coatings with micro/nano containers. Reproduced (**a**) with permission from [12] © 2018 Liu and Hsu, under CC BY, and (**b**) from [13] © 2023 Chongqing University. Publishing by Elsevier B.V. on behalf of KeAi

Corrosion inhibitors used for self-healing coatings mainly include organic and inorganic types. Inorganic inhibitors used for PEO self-healing function include metal cations such as cerium salts, lithium salts, aluminum salts, rare metal elements such as lanthanum salts, and inorganic acid ions that can form insoluble precipitates with Mg^{2+}: fluoride, phosphate, vanadate, carbonate, silicate, etc. Organic corrosion inhibitors mainly include organic molecules that can chelate with metal substrates, including oleate, 8-hydroxyquinoline (8-HQ), various amino acids, and polymers.

The commonly used post-coating approaches for PEO include saline, sol-gel, LDH, and polymer coatings. After modifying with corrosion inhibitors, abovementioned coatings possess active protective abilities. Recently, researchers have found that metal organic frameworks (MOFs) and slippery liquid-infused porous surface (SLIPS) developed on PEO coatings could alleviate the corrosion of alloys ascribed to their dense structure and identified chemical and physical properties. This chapter primarily discusses the corrosion inhibitors incorporated self-healing coatings based on PEO coated Mg alloys, as well as the development of multilayer coatings and their characterization/evaluation methods for self-healing.

16.2 Corrosion Inhibitors Used in PEO Coatings

16.2.1 Metal Cations

Metal cations, like Ce^{3+}, Nd^{3+}, and rare earth elements (La^{3+}), are commonly applied in self-healing coatings. Ce^{3+} ion is the most commonly used inorganic corrosion inhibitor with a good active protection enhancement effect on PEO pretreated valve metals [14, 15]. The self-healing effect of metal cations was demonstrated as the creation of hydroxide/oxide, which happens because of local alkalization at corrosion sites (as shown in Eqs. (16.1), (16.2), (16.3), and (16.4)). Taking the most commonly used Ce^{3+} ions as an example, in the atmosphere's existing oxygen, Ce (III) is possible to oxidize to Ce (IV) (as shown in Eq. 16.5). The newly formed cerium oxides/hydroxides will gather in the pores, thereby hindering the invasion of the aggressive ions through the passive layer at the interface between the metal and the coating. As a result, the coating exhibits improved corrosion resistance and self-healing effect.

$$Mg \rightarrow Mg^{2+} + 2e^- \tag{16.1}$$

$$2H_2O + 2e^- \rightarrow H_2 + 2OH^- \tag{16.2}$$

$$Ce^{3+} + 3OH^- \rightarrow Ce(OH)_3 \tag{16.3}$$

$$Ce^{3+} + O_2 + 6H_2O \rightarrow 4CeO_2 + 12H^+ \tag{16.4}$$

$$4Ce^{3+} + O_2 + 14H_2O \rightarrow 4Ce(OH)_4 + 12H^+ \tag{16.5}$$

Olmo et al. [16] coated silicon based mixed sol-gel coating on anodized AA2024-T3 alloy samples. $LiNO_3$ and $Ce(NO_3)_3{\cdot}6H_2O$ were doped into the coating as corrosion inhibitors. Experiments have shown that, after 21 days of immersion in 0.5 M NaCl solution, a large number of Ce^{3+} ions were released at the scratch site to form hydroxides and a few oxides at different pH environments (Fig. 16.2), and their dense structure effectively protected the alloy substrate. Qiao et al. [17] investigated the influence of the concentration of La^{3+} cations added into 3-methacryloxypropyl trimethoxysilane (MPS) coating formed on AZ31 Mg alloy on their corrosion resistance. They found that a low concentration of the La^{3+} cations improved the anticorrosive effect. However, the corrosion resistance decreased as the concentration of La^{3+} cations increased due to the formation of cracks and pores during the dissolution of $La(NO)_3{\cdot}6H_2O$. Consequently, the compatibility of corrosion inhibitor and coating molecules is of great significance in coating design.

Metal ions are typically incorporated into coatings in the form of metal salts or nanoparticles of metal oxides, which are then loaded onto the coatings through in situ growth or vacuum impregnation [18]. Relevant research has been presented in Table 16.1 as follows. Noteworthy, self-healing coatings using cationic corrosion inhibitors usually have the characteristics of simple preparation process and

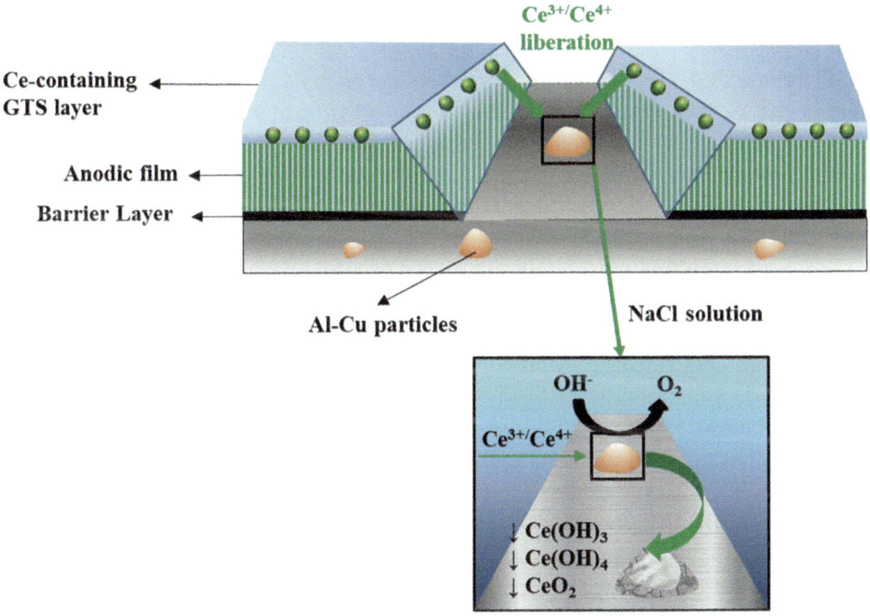

Fig. 16.2 Schematic diagram of active protection of AA2024-T3 Al alloy sample surface with corrosion-resistant coating achieved by Ce^{3+} ions. Reproduced with permission from [16] © 2021 Olmo, Tiringer, Milošev, Visser, Arrabal, Matykina, and Mol. Published by Elsevier B.V.

Table 16.1 Metal cations as corrosion inhibitors applied on PEO coatings

Mg substrate	Corrosion inhibitor	Applied coatings	Anticorrosion mechanism	Reference
AA2024-T3	$Li(NO_3)_3$, $Ce(NO_3)_3$	Sol-gel	Dense precipitate of $Ce(OH)_3$ and Ce_2O_3 at scratch	[16]
Pure Mg	CeO_2, KF	PEO/Ca-P	Long-term corrosion resistance	[19]
As-cast 7085 Al	$Ce(NO_3)_3$/ CeO_2	PEO	Promote the Al^{3+} precipitating under acidic environment	[20]
AZ31	$Ce(DEHP)_3$	EP/Saline	The synergistic effect of Ce^{3+} and PO_4^{3-} ions	[21]
AZ31	$V_2O_7^{4-}$, La	PEO/LDH	Ion exchange ability of $V_2O_7^{4-}$ and the rapid deposition of La	[22]
Pure Mg	MnOOH	PEO	Spontaneous transformation to generate LDH	[23]
AZ31	PVA, $Nd(NO_3)_3$	PEO	Complexation of PVA and Mg substrate and the coverage of insoluble Nd^{3+} salts	[10]

environmental friendliness, but their self-healing effect is usually affected by low loading and rapid release.

16.2.2 *Anions and Ligands*

Acid anions and organic ligands capable of forming insoluble precipitates with metal cations, including organic ligands that can coordinate with host metal atoms, are considered to have corrosion inhibition effects. This is because they can precipitate or adsorb onto the metal surface, forming a protective film. Although this film may not be perfectly dense, it can still provide a certain level of protection against corrosive ions.

Inorganic acid anions such as PO_4^{3-}, $V_2O_7^{4-}$, SiO_3^{2-}, and CO_3^{2-} can react with degraded Mg^{2+} ions to form insoluble precipitation at damaged sites of coatings on Mg alloys. The most popular method of introducing anions into coatings is the ion exchange process during the preparation of LDH. Besides, some anions can get encapsulated into composite coatings in the form of salts. In certain circumstances, such as human body fluid environment, PO_4^{3-} and CO_3^{2-} exist naturally as buffer ions, which can also be taken to realize self-healing performance. It is worth noting that although acid anions and metal ions can form stable precipitates through chemical reactions, the film-forming capability and compactness of these precipitates are typically low. This limitation leads to significant constraints on the self-healing effect.

Negatively charged organic ligands mainly take effect by their negative groups (e.g., hydroxyl, carboxyl, etc.), which are able to chelate to the metal substrate and slow down degradation of metal substrate. Besides, some of the ligands can combine with the corrosive impurities, blocking their deposition onto substrate, and thereby actively protect the alloys. Commonly used organic ligands include oleate, 8-hydroxyquinoline (8-HQ), dopamine, ethylene diamine tetraacetic acid (EDTA), sodium dodecyl sulfate (SDS), etc. Phytic acid (PA) is also a usual assistance agent in self-healing coating, which on the one hand adsorbs onto the metal surface and on the other hand grabs soluble metal ions to form precipitation on exposed substrate. As shown in Fig. 16.3, when coatings underwent degradation in solution, PA loaded in polycaprolactone (PCL) microcapsules came out and gradually chelated with metal cations. Mg^{2+} ions were seemed to be easier to chelate with PA and firstly grabbed, and then the Ca^{2+} ions. Simultaneously, the chelation deposited onto damaged sites under the influence of electrostatic interaction [24]. Zhang et al. simultaneously utilized the active protective properties of MoO_4^{2-} and 8-HQ. MoO_4^{2-} was firstly introduced into the PEO coating, and then 8-HQ was loaded onto the coating to form a multilayer active protective coating [25].

Tri(bis(2-ethylhexyl)phosphate) ($Ce(DEHP)_3$) simultaneously utilizes the active protective effect of Ce^{3+} ions and PO_4^{3-} ions. Ce^{3+} ions and PO_4^{3-} ions can, respectively, combine with OH^- and Mg^{2+} to form precipitates and repair the coating under local alkaline conditions at the corrosion site [21]. The synergistic corrosion enhancement effect of Ce^{3+} ions and PO_4^{3-} can also be found in the study of Jayaraj et al. [15].

Oleic acid is an unsaturated fatty acid commonly seen in vegetable oils and animal fats, whose sodium salt (sodium oleate, SO), is known as a corrosion inhibitor. Chirkunov et al. [26] impregnated a PEO-coated AZ31 Mg alloy with sodium oleate

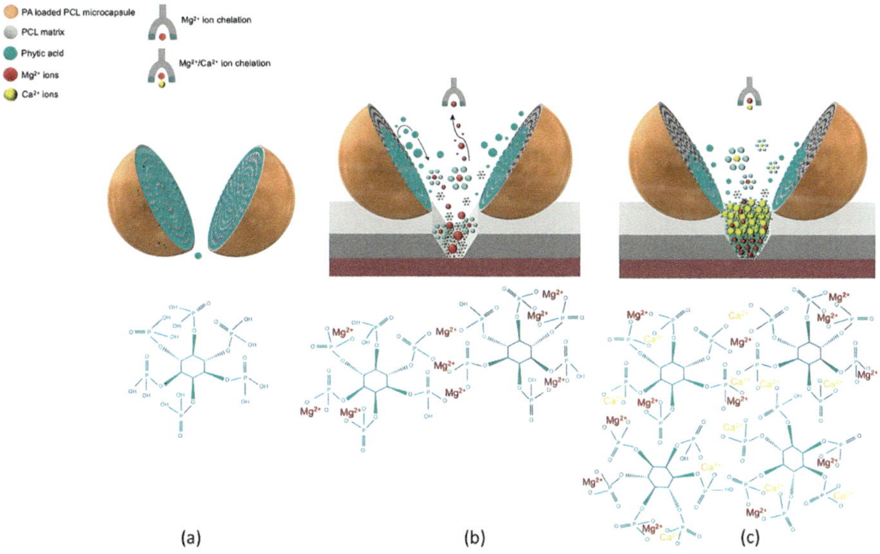

Fig. 16.3 **(a–c)** Self-healing mechanism of PA capsulated into PCL nanocontainers after mechanical scratching on ZM21 Mg alloy. Reproduced with permission from [24] © 2023 Chongqing University. Publishing services provided by Elsevier B.V. on behalf of KeAi Communications Co. Ltd.

(SOl), sodium oleyl sarcosinate (SOS), sodium linolenic acid (SLin), sodium 3-amino-1,2,4-triazole-5-carboxylate (SATC), 8-hydroxyquinoline (8-HQ), and sodium fluoride (IFKhAN-25F). The results have shown that SOl has the best corrosion inhibition ability. Gnedenkov et al. [27] fabricated a mixed coating of NaOl and PCL on PEO coating by impregnation method. Electrochemical tests indicated that the mixed coating possesses excellent long-term protection and active protection characteristics. In addition, sodium oleate can also be used as a corrosion inhibitor for steel, aluminum, and their alloys [28].

Anions and ligands possessed better active protective effects compared with metal cations, which are related to the better adhesion strength and denser defense layer. But a major hidden danger is that the ion concentration in the coating rapidly releases and decays, resulting in the rapid loss of the coating's protective effect. Besides, the distribution and uniformity of the containers and corrosion inhibitors play important roles in the self-healing effect. Table 16.2 lists some recent research on negatively charged radicals or anions as self-healing agents.

16.2.3 Metal-Organic Frameworks (MOFs)

Metal-organic frameworks (MOFs), consisting of metal clusters and organic ligands, are known for their porous structure, high specific surface area, and thermal and chemical stability. MOFs are recently applied to protective coatings due to the progress in water-stable MOFs. However, the poor adhesion of MOFs usually bothers its application. Fortunately, it seems that the porous surface of PEO can evidently enhance the adhesion ability of MOFs, even if the pores are sealed by the MOF particles. Additionally, a lot of research has proved that MOFs can endow coatings with self-healing properties, which are achieved by forming precipitation with metal ions and adsorption onto bare metals. Consequently, PEO/MOF composite coatings have recently become popular in self-healing coatings.

Commonly used MOF materials in coatings include MOF-5, ZIF-8, HKUST-1, etc., as listed in Table 16.3. Chen et al. [44] recently prepared Ce-MOF, HKUST-1, Mg-MOF-74, MOF-5, and Zn-MOF on PEO/AZ31 alloy by hydrothermal methods. All five MOFs enhanced the coatings' corrosion resistance, and Ce-based MOF exhibited the highest improvement effect.

16.3 Single PEO Self-Healing Coating

Usually, PEO coatings can achieve self-healing effect by loading corrosion inhibitors into their inherent pores, and the pores are simultaneously sealed to form even and dense coatings. The methods of achieving self-healing function directly through loading corrosion inhibitors into PEO micropores are mainly divided into two categories: one is to directly introduce corrosion inhibitors during the preparation of

Table 16.2 Anions and ligands corrosion inhibitors applied on PEO coatings

Mg substrate	Corrosion inhibitor	Applied coatings	Anticorrosion mechanism	Reference
AZ31	CO_3^{2-}	PEO/LDH	CO_3^{2-} ions promote the precipitation of Mg^{2+}, thereby generate $Mg(OH)_2$ in alkaline environment	[29]
AZ31B	SiO_3^{2-}	PEO/LDH	Enhanced effect of promoting Mg^{2+} ions precipitation compared with CO_3^{2-}	[30]
AZ31	ALD-Al_2O_3	PEO	ALD-Al_2O_3 particles spontaneously generate LDH	[31]
AZ31	$V_2O_7^{4-}$, La	PEO/LDH	Ion exchange ability of $V_2O_7^{4-}$ and the rapid deposition of La	[22]
AZ31	SOl; SOS; SLin; SATC; 8-HQ; IFKhAN-25F	PEO	Delay the occurrence time of first corrosion damages	[26]
Mg-Mn alloy	NaOl	PEO	Long-term corrosion resistance	[27]
AZ91	Sodium salts of glycolic, 4-aminosalicylic, 2,6-pyridinedicarboxylic acids	PEO/ sol-gel	Inhibit impurity deposition and promote corrosion inhibitor adsorption	[32]
AM60B	C60@SDS	Sol-gel	Accelerate the formation of dense oxide films in Mg alloys, and hinder galvanic corrosion	[33]
AZ31	MSN@SB	PEO/EP	Chelation between SB and Mg substrate	[11]
AZ91	PDA	PEO	The catechol group combines with Mg^{2+} ions and deposits at the scratch site	[34]
AZ31	PMTMS	PEO	Water swelling and hydrolysis condensation of PMTMS	[35]
AZ31	8-HQ@GO	PEO/LDH	Chelation of 8-HQ and Mg^{2+} ions and substrate	[36]
AZ31	M-16	PEO/PU	The chelation of M-16 and Mg substrate and the shape memory function of the outer layer PU	[37]

(continued)

Table 16.2 (continued)

Mg substrate	Corrosion inhibitor	Applied coatings	Anticorrosion mechanism	Reference
AZ31	ASP	PEO	Forming insoluble precipitation at scratching site	[38]
AZ31B	CCNT-IL	PEO/PP/CCNT-IL	The CCNT-IL deposits on exposed Mg substrate by forming CCNT-IL/Mg^{2+} chelate	[39]
AZ91	MoO_4^{2-}, PFDS, PFPE	PEO/LDH	PFPE has good fluidity, and coating scratches can heal on their own after a certain period of time. Insoluble MoO_4^{2-} salt at scratching site	[40]
AZ31B	MoO_4^{2-}, 8-HQ	PEO/WPU	Chelation of 8-HQ and Mg^{2+} ions and substrate, insoluble MoO_4^{2-} salt at scratching site	[25]

Table 16.3 MOFs as corrosion inhibitors applied on PEO coatings

Mg substrate	Corrosion inhibitor	Applied coatings	Anticorrosion mechanism	Reference
AZ31	ZIF-67	PEO/LDH	Complexation between 2-MI and Mg substrate	[41]
AZ31	Ce-MOF, HKUST-1, Mg-MOF-74, MOF-5, Zn-MOF	PEO/MOF	Metal ions and organic ligands of MOFs react with metal cations and form insoluble precipitates	[42]
AZ31	ZIF-8	PEO/ZIF-8	Superhydrophobicity and barrier effect	[43]

PEO, that is, in situ growth; another approach is to load corrosion inhibitors onto the prepared PEO through a posttreatment.

16.3.1 *In Situ Growth Methods*

In situ growth methods, which introduces corrosion inhibitors during the preparation process of PEO coatings, are simpler and more effective compared with post-treatment methods, and the produced coatings typically exhibit good adhesion to the substrate. Usually, corrosion inhibitors, as such or encapsulated in nanocontainers, are introduced to the PEO electrolyte. Chen et al. [45] in situ-fabricated $CePO_4$-contained PEO coating on LY12 Al alloy. The sodium hexametaphosphate and cerium acetate were added into the electrolyte of PEO. Although they focused on the self-lubrication ability of $CePO_4$, it was unignorable that the formed PEO coating loaded with $CePO_4$ might possess self-healing performance. Heretofore, the active protective PEO coatings are mostly achieved by posttreatment. It is necessary to develop efficient in situ methods for self-healing PEO coatings.

16.3.2 Post-Processing Methods

Post-processing methods are applied on well-prepared PEO coatings, where the inherent pores of PEO coatings can be used as the containers to load inhibitors. Otherwise, chelation agents or other containers can be used to load inhibitors. The most popular posttreatment method is negative pressure (vacuum) impregnation, during which the negative pressure offers self-healing agents sufficient time and power to enter the pores. Additionally, optimizing the immersion time and temperature may be beneficial to improving the corrosion resistance of obtained coatings [44]. The hydrothermal method is commonly used in the preparation of posttreatment self-healing coatings.

16.3.2.1 Filling the Pores of PEO with Only Inhibitors

Zhang et al. [23] prepared a MnOOH reinforcement layer on pure Mg pretreated with PEO using a simple impregnation method. Gong et al. [20] put PEO coated Al alloy in $Ce(NO_3)_3$ solution, which was composed of $Ce(NO_3)_3$ and H_2O_2 dissolved in deionized water under 20 °C with constant stirring. In their self-healing coatings, the corrosion inhibitor existed as CeO_2 particles. They found that the corrosion resistance of the coatings did not keep rising with the prolonged immersion time. They believed that after 60 min, the dissolution of CeO_2 particles had dominated the process.

16.3.2.2 Filling the Pores of PEO with Assistance of Chelators or Containers

Yang et al. [46] prepared a coating rich in Ce^{3+} ions on a PEO pretreated AZ31B Mg alloy using a simple hydrothermal method with assistance of PA. Similar research was carried out on PEO-coated 6063 Al alloy, and PA was applied to assist the loading of MoS_2, which obviously improved the active protection effect and wear resistance [47]. Chen et al. [44] successfully sealed the defects with abovementioned five kinds of MOFs to form an even and uniform surface with better corrosion resistance and self-healing effect.

As shown in Fig. 16.4, Samadianfard et al. [33] stabilized SDS surfactant on fullerene C60 nanoparticles, which were subsequently dispersed in the sol-gel coating for actively protecting AM60B Mg alloy. After adding $F\text{-}NH_2$ and F-SDS functional nanofillers, the micro defects in the sol-gel coating disappeared. This is

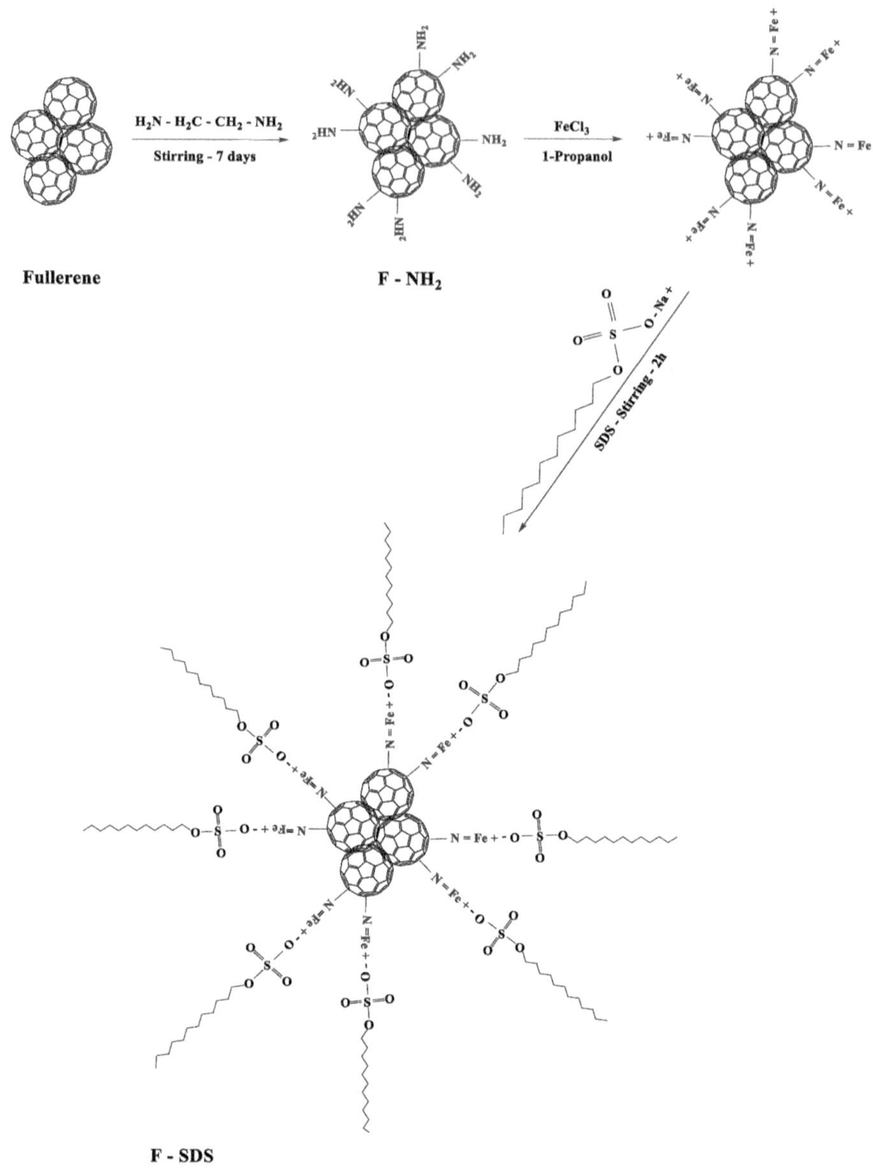

Fig. 16.4 F-SDS synthesis route. Reproduced with permission from [33] © 2021 Elsevier B.V.

ascribed to the localized alkaline atmosphere caused by corrosion, which destroyed the Fe-N double bonds in F-SDS. The SDS molecules released from coatings accelerate the formation of dense oxide films on bare alloys and inhibit the galvanic effect of Mg alloys through complexation with intermetallic compounds.

16.3.3 Self-Healing Mechanism

Yang et al. [46] fabricated a PEO/PA/Ce coating, which was uniform and dense ascribed to the fore-deposited PA layer. Ce^{3+} ions are released continuously due to the degradation of coatings with the increase of immersion time. The released Ce^{3+} ions reacted with the OH^- produced by hydrogen evolution of Mg substrate, forming precipitates covering the cracks, and the self-healing effect was accomplished. Chen et al. [48] prepared PEO-zeolite-Ce coatings in situ by adding Ce^{3+} loaded zeolite into electrolyte. Ce^{3+} was successfully loaded by ion exchange reaction with Na^+ in zeolite. At the same time, when immersed and degraded, the Ce^{3+} loaded in zeolite can come out by ion exchange, form precipitation under a localized base environment. However, as Chen et al. said, when large-sized damage appeared on the coatings, the self-healing effect would be limited and needed assistance with other active protective methods. In addition, Gong et al. [20] used the micropores of PEO as containers to load CeO_2 particles, and the coating exhibited good self-healing performance under acidic conditions. They believed that CeO_2 reacted with H^+ and H_2O_2 in acidic solutions to release O_2 and Ce^{2+}, which would promote the formation of $Al(OH)_3$ in cracks. Another study has used CeO_2 nanoparticles to modify silicone-based mixed coatings, thereby achieving self-healing of localized corrosion in coatings [49]. A self-healing coating designed by Jia et al. [50] that would respond to local pH changes was composed of an in situ-grown PEO coating, a storage medium shell (chitosan, CS), and a healing agent (nano cerium) composite. The porous PEO coating served as both an anchor and a barrier, utilizing the $Ce\text{-}NH_2$ complexation reaction to immobilize Ce onto CS molecular chains and form $(CSCe)_n$. The results demonstrated that pH triggered the formation of Ce_2O_3/$Ce(OH)_3$ precipitates, as well as the pH buffering activity and flow swelling ability of chitosan macromolecules, which helped to suppress anodic activity and dynamically heal damage/defects.

In the study of Zhang et al., MnOOH reinforcement layer could spontaneously generate LDH nanosheets during the soaking process to repair the coatings' damaged parts. As depicted in Fig. 16.5, during 28 days immersion in 0.9 wt. % NaCl solution at 4 °C, bare PEO coatings gradually dissolved with the defects enlarged. In comparison, the coatings containing MnOOH grew flaky LDHs between the pores and cracks, which alleviated the further attack of corrosive ions. On a large scale, MnOOH enforcement layer evidently inhibited the production of microcracks in the coating. The presence of Cl^- at the cracks in the coating further confirmed the formation of LDHs and the active protective ability of the coating [23]. Because the newly formed LDHs from MnOOH are able to catch the Cl^- by anion exchangeability.

In the work of Huangfu et al. [39], carboxyl carbon nanotubes were firstly loaded onto PEO coatings by vacuum soaking, and then 1-(2-aminoethyl)-3-methylimidazolium chloride ([MamAE]Cl) was grafted onto carboxyl carbon nanotubes through acylation reaction. The ion exchange reaction between Cl^- and β-aminopropionic acid was successfully utilized to achieve the loading of

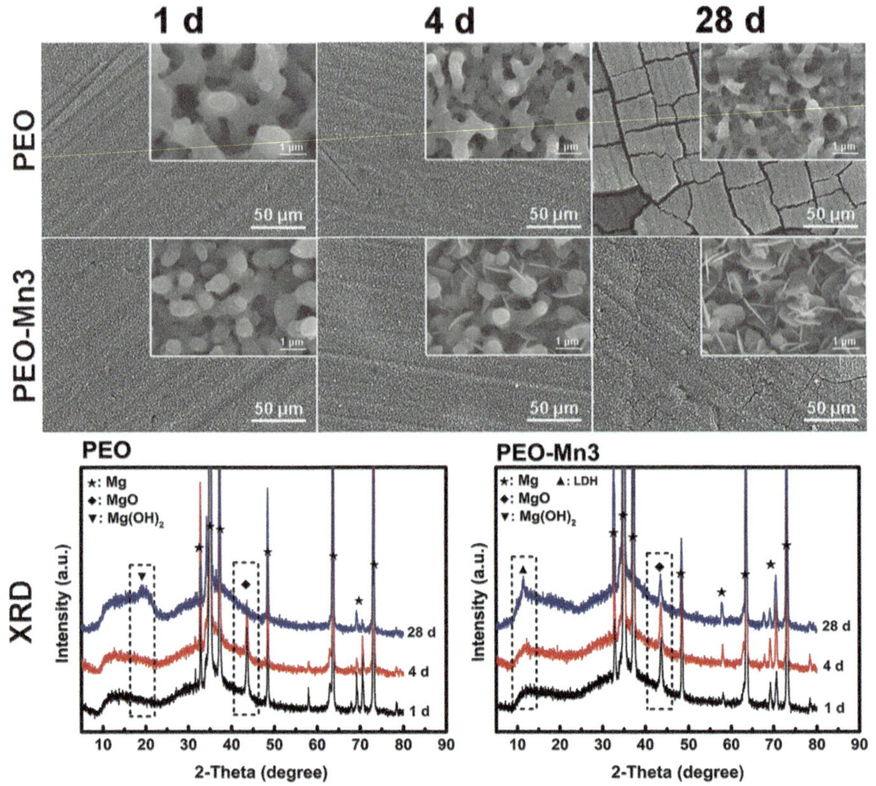

Fig. 16.5 SEM surface images and XRD patterns of the PEO and PEO–Mn3 samples immersed in 0.9 wt. % NaCl solution at 4 °C for 1, 4, and 28 days. Reproduced with permission from [23] © 2021 Elsevier B.V.

β-aminopropionic acid. At the damaged location of coating, the grafted carboxyl carbon nanotubes will chelate with the exposed metal, thereby covering the metal surface and forming a self-healing function.

16.4 Multilayer Self-Healing Coating Based on PEO

As we have mentioned before, although in situ preparation and posttreatment methods endow the coatings with active protection performance and possess the advantage of simpleness, their self-healing effects are limited when serious corrosion happens. Furthermore, the still existing defects lead to premature failure of alloys. Multilayer or composite coatings are necessary for further enhancing the anticorrosive ability of alloys.

16.4.1 PEO/LDH

The PEO/LDH composite coatings are popular in anticorrosive coatings of Mg alloys, both in industrial and medical application, because of the ion exchange ability, physical barrier effect, and excellent adhesion with PEO coatings [7]. The anion intercalation of LDH can serve as a trap to capture Cl^- in corrosive media, and inhibit its further attack on metal substrate. Furthermore, the metal ions used to form LDH are changeable, which can endow the LDH coatings with various performances. For example, Ga^{3+} ion, with a certain antibacterial ability, can form Mg-Al-Ga/LDH and endow the coating with its performance [31].

Based on the changeable metal cations, Ce^{3+} and La^{3+} ions can serve as trivalent cations and corrosion inhibitors to form an LDH coating with self-healing ability. Besides, the anion intercalated into the intermediate layer of LDH is changeable, which allows acid radical ions to enter. When corrosion happens, the acid radical ions will be released and combine with metal ions to precipitate and be deposited on the damaged sites by the influence of micro-galvanic. Commonly used acid radical ions with self-healing ability include NO_3^-, $V_2O_7^{4-}$, MoO_4^-, SiO_3^{2-}, and CO_3^{2-}, etc. Among the anions, NO_3^- and CO_3^{2-} are the most common. CO_3^{2-} anions are usually introduced by simple hydrothermal reaction from CO_2 in the air or the solved CO_2 in the solution. The NO_3^- anions are often from the additional nitrates for the formation of LDHs. Noteworthy, it has been exhibited that the stability of different anions intercalated LDH follows the order: $CO_3^{2-} > SO_4^{2-} > OH^- > F^- > Cl^- > Br^- > NO_3^-$, and the stable structure is not conducive to anion exchange processes [51]. Consequently, NO_3^- intercalated LDHs are excellent precursors to prepare LDHs with other intercalated anions.

Chen et al. [22] first prepared GO doped Mg-Al-La LDHs/GO (L/G-N) on AZ31 Mg alloy pretreated with PEO and generated Mg-Al-La-$V_2O_7^{4-}$-LDHs/GO through ion exchange reaction. The composite coating has a synergistic barrier effect of GO and LDH and has excellent corrosion resistance. When the coating is damaged, La^{3+} ions can play an active protective role, and the K_{sp} of La^{3+} hydroxide is low, quickly forming precipitates to repair the damaged area. In addition, $V_2O_7^{4-}$ ions can capture and exchange with corrosive Cl^- to achieve active corrosion protection.

The study by Li et al. [29] indicated that Mg-Al LDH itself possesses a certain self-healing property. They believed that LDH naturally contained intercalated CO_3^{2-} ions. During the dissolution and diffusion process of the coating, they reacted with Mg^{2+} ions generated by Mg dissolution to form $MgCO_3$, which was then converted into $Mg(OH)_2$ in an alkaline solution and deposited at the site of coating rupture to form a self-healing effect. Similarly, some studies have utilized SiO_3^{2-} ions for the simple substitution of CO_3^{2-} in LDH coatings, further enhancing the basic repair performance of LDH [30], while others have simultaneously utilized the self-healing function of LDH and rGO [52], but this repair effect is often limited.

Although LDH coatings are able to achieve a certain self-healing effect, it is limited. Fortunately, LDH can work as containers to load other corrosion inhibitors, which endow the coatings with a better protective effect. Chen et al. [36] first synthesized 8-hydroxyquinoline@oxidized graphene composite materials (8-HQ@GO) through ring-opening reaction, which were incorporated into LDH coatings (PEO/LDHs/HG) through hydrothermal reactions. When the coating was damaged, 8-HQ was released and formed insoluble chelates $Mg(HQ)_2$ and $Mg(OH)_2$ onto the Mg substrate, covering the scratch area to achieve self-healing performance (Fig. 16.6). In addition, the ability of GO and 8-HQ to bind through ring opening reactions ensured the long-term stability of corrosion inhibitors.

In another work, Chen et al. [41] grew Co-Fe LDH coating on Mg alloy via in situ hydrothermal method and loaded ZIF-67 onto the LDH coatings to produce a composite coating by impregnation. The study showed that the dimethyl-imidazole (2-MI) produced by the degradation of the composite coating complexed with the Mg^{2+} ions and deposited at the damaged portions of the coating, producing a good self-healing effect.

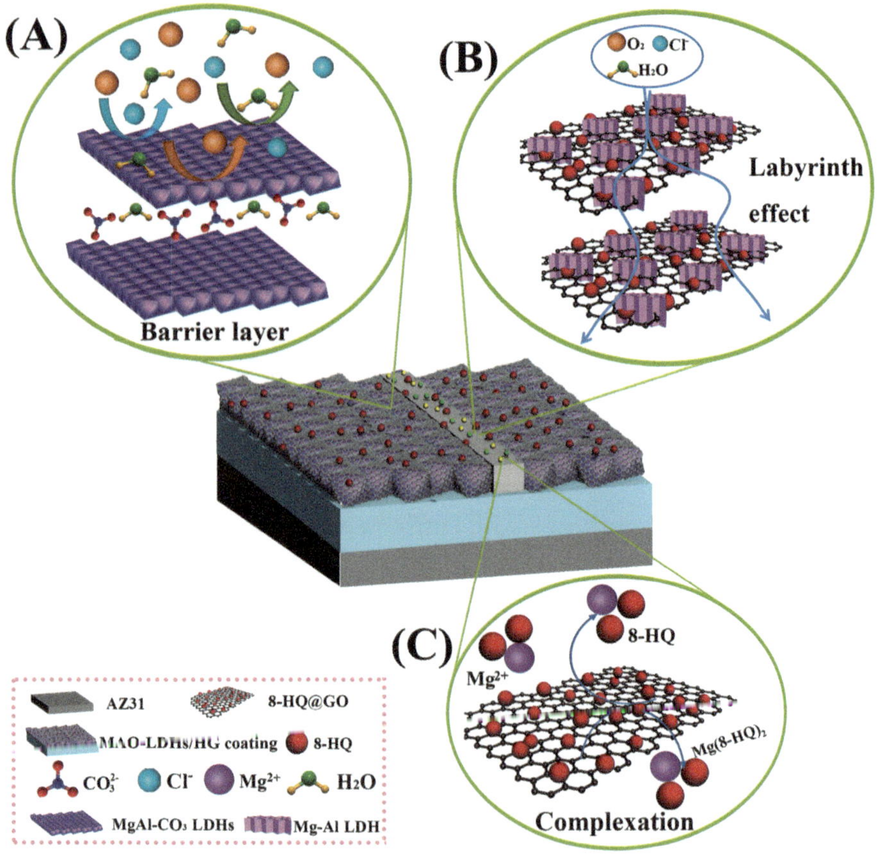

Fig. 16.6 (**a–c**) Schematic diagram of anticorrosive mechanism of the PEO-LDHs/HG coating. Reproduced with permission from [36] © 2021 Elsevier B.V.

16.4.2 PEO/MOF

As demonstrated before, MOFs have recently been investigated as effective protective membranes in anticorrosion of alloys. MOFs themselves have various functions when forming coatings on alloys, such as hydrophobicity (Cu-MOFs, Zn-MOFs). Besides, MOFs are environmentally friendly and biocompatible, which can be applied in both industrial and medical fields. When applied onto PEO-based alloys, it can seal the defects to produce dense and uniform composite coatings.

Telmenbayar and his colleagues grew in situ Zn-MOFs firstly on the PEO coating to construct slippery surface. ZIF-8 played a role as intermediate layer and provided the coatings with superhydrophobic ability [43]. However, the self-healing effect was not referred to in their work. Chen et al. [44] fabricated various MOF coatings on PEO-pretreated AZ31 alloy. All of the coatings had a better protective effect than conventional PEO because of the physical barrier effect and the corrosive ions trapping by MOFs. Furthermore, they found that the metal cations and negative ligands from the dissolution of MOFs offered the coatings with self-healing properties. Among the five kinds of prepared MOF coatings, the Ce-MOF and MOF-5 coatings exhibited higher resistance than others. Ce-MOF achieves the highest corrosion resistance by dual-reaction of Ce^{3+} and ligands. Zn-MOF and MOF-5, with Zn as metallic ions, were both observed with uniform distribution of P element, which indicated the valid protective effect of PEO in Zn-based MOFs.

Besides the sorts of inhibitors, Yang et al. [53] presented that the sizes and shapes of MOF particles had an influence on the corrosion resistance of coatings. They indicated that ~100 nm ZIF-7 particle sizes showed the lowest corrosion resistance, and ~1500 nm showed the highest, ascribing to the inducing corrosion and poor adhesion on PEO of large-sized MOFs. Additionally, they found that 0.5–1.7 wt.% ZIF-7 concentration allowed easier homogenous dispersion of particles together with lower diffusion paths for corrosive ions attacking. Higher concentration may deteriorate the integrity and barrier effect of the coating.

16.4.3 PEO/Organic Coating

16.4.3.1 Intrinsic Self-Healing Resin

Intrinsically self-healable resin materials are also applied in the fabrication of self-healing coatings based on PEO. Unlike the abovementioned repair methods, the self-healing of these coatings relies on the interaction between resin molecules, which restores the coating's own state through the reconstruction of dynamic bonds. For example, the commonly used polyurethanes (PU) are a kind of polymer with urethane repeated units, which can achieve self-healing ability with or without outer stimuli.

It can be found that PU and polymethyltrimethoxysilane (PMTMS) have been applied to encapsulate composite coatings; ascribed to their shape memory effect, coatings exhibit a certain self-healing effect [35, 37]. Liu et al. [37] developed a new Mg alloy corrosion inhibitor M-16 (Fig. 16.7a), which was encapsulated in the micropores of PEO coating under negative pressure. Subsequently, a layer of disulfide bond modified self-healing PU (Fig. 16.7b) was coated to further encapsulate the sealed PEO, resulting in a multilayered coating with dual self-healing effects: shape memory function of the dynamic disulfide bond in modified PU coating and the repairing function of M-16. Similar to most organic anions or ligands, negatively charged Br element in M-16 will adsorb onto the uncoated substrate of Mg alloy and inhibit further erosion of aggressive ions. In their study, the concentrated distribution of the characteristic element Br at the scratch site of the coating was observed, which proved the adsorption and self-healing effect of M-16 molecule on the Mg alloy substrate.

16.4.3.2 Organic Corrosion Inhibitors

Organic corrosion inhibitors mainly achieve damage repair by chelating with the exposed metal substrate at the damaged location of the coating. This is because organic inhibitors typically have larger molecular weights and volumes, which naturally achieve a spatial hindrance effect. In its application, it can passively prolong

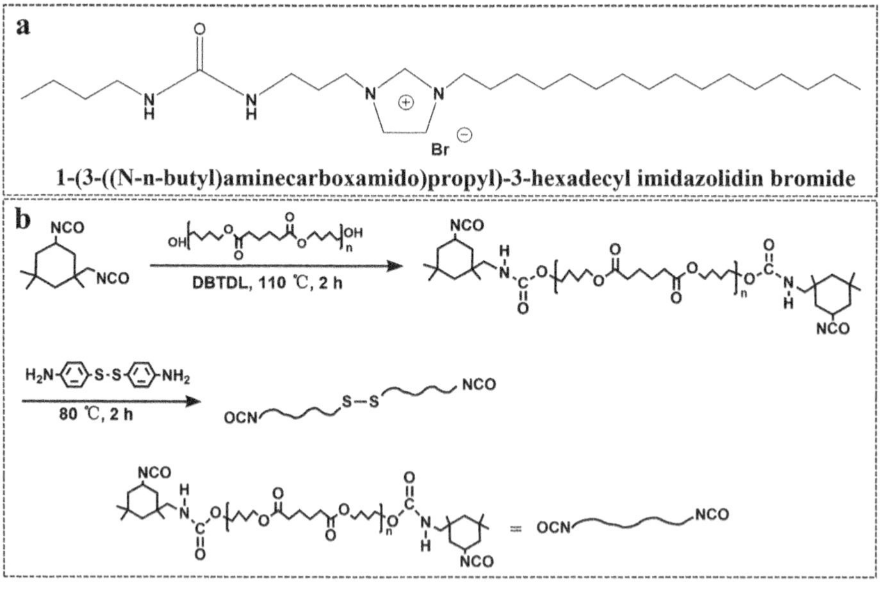

Fig. 16.7 (**a**) Chemical structure of inhibitor M-16, (**b**) Synthetic routes for the self-healing polymer samples. Reproduced with permission from [37] © 2021 Elsevier B.V.

the corrosion path of the corrosive ions. In addition, the ligands can adsorb onto the bare substrate and produce a dense passive film, achieving an active protection effect. Besides, organic inhibitors can also combine with impure ions in solution to form complexes that cover the bare substrate, achieving active protective effects.

Chen et al. [32] prepared PEO/sol-gel multilayered coating on the surface of AZ91 Mg alloy, doped and compared the corrosion protection effect of glycolic acid, 4-aminosalicylic acid, and 2,6-pyridinedicarboxylic acid sodium salt. Three types of organic inhibitors were added into the micropores of PEO, and the composite coating achieved active protective performance by inhibiting the precipitation of impurities and promoting the deposition of corrosion inhibitors at the exposed areas of the alloy. Wang et al. [54] prepared poly-p-xylene outer layer on PEO coated AZ31 alloy by vapor deposition, whose wear resistance evidently increased. The poly-p-xylene released from the damaged PEO pores transformed onto bare substrate to imply a prolonged corrosion resistance.

Farshid et al. [34] applied a layer of dopamine (PDA) on AZ91 Mg alloy using electrodeposition technique. The electrochemical anode branch of potentiodynamic polarization curve of the PEO/PDA dual coating prepared has a wide passivation zone (Fig. 16.8a), indicating its good corrosion resistance and active protective performance. Its self-healing performance comes from the combination between $Mg(OH)_2$ and catechol groups, which spontaneously heals corrosion defects, forming zipper-like and pore-blocking effects (Fig. 16.8b).

Cui et al. [35] sealed alkali-treated PEO coatings with poly-methyltrimethoxysilane (PMTMS) and prepared PEO/PMTMS hybrid coatings. The hydroxylated PEO coating obtained after alkaline pretreatment was beneficial for the adsorption of PMTMS based on PEO. Although PMTMS achieved active protection through water expansion and hydrolysis condensation, the coating inevitably produced bubbles and eventually peeled off because of the volume expansion of corrosion products.

Chen et al. [38] prepared Mg-Al LDH based on PEO pretreated AZ31 Mg alloy using a one-step hydrothermal method and added aspartic acid (ASP) as a corrosion inhibitor. The composite film achieved self-healing function by forming insoluble precipitates at the scratch site through ASP.

Recently, many studies have utilized the dual active protective effects of both organic and inorganic inhibitors to realize active protection performance of PEO coatings. As exhibited in Fig. 16.9, Cai et al. [10] prepared a polyvinyl alcohol (PVA)–$Nd(NO_3)_3$ hybrid coating for corrosion protection of PEO pretreated AZ31 Mg alloy. The adsorption effect of PVA on Nd^{3+} ions promoted the enrichment of Nd^{3+} on the coating surface. At the alkaline corrosion circumstance, the enriched Nd^{3+} was converted into its carbonate, which deposited and sealed the damaged sites of the coating and achieved active protection. In addition, the chelation effect of PVA on bare metal substrate hinders the further attack of corrosive ions.

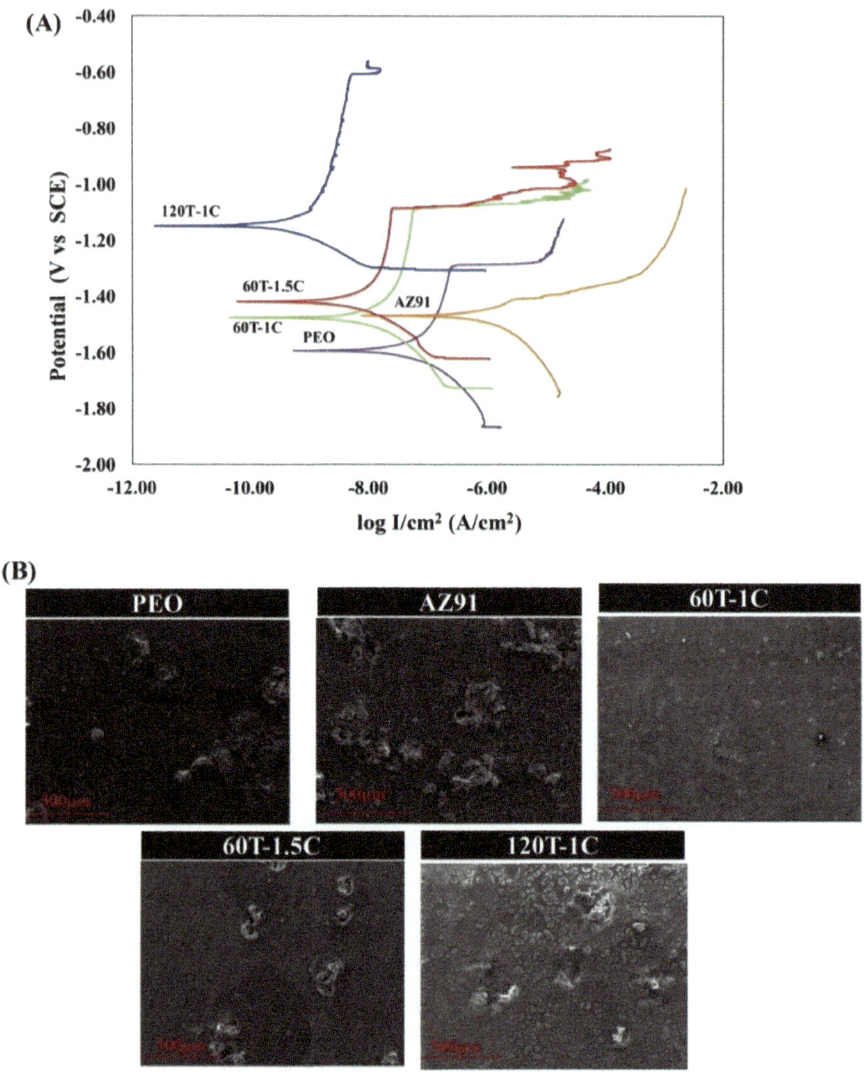

Fig. 16.8 (**a**) PDP curves of duplex PEO/PDA coatings, (**b**) SEM surface images of samples after polarization test. Reproduced with permission from [34] © 2022 Chongqing University. Publishing services provided by Elsevier B.V. on behalf of KeAi Communications Co. Ltd.

16.4.3.3 Epoxy Resin

Besides the intrinsic self-healing resin and organic corrosion inhibitors, organic coatings such as epoxy resin are usually used as outer layers, which play roles in physical barrier and encapsulation of self-healing agents.

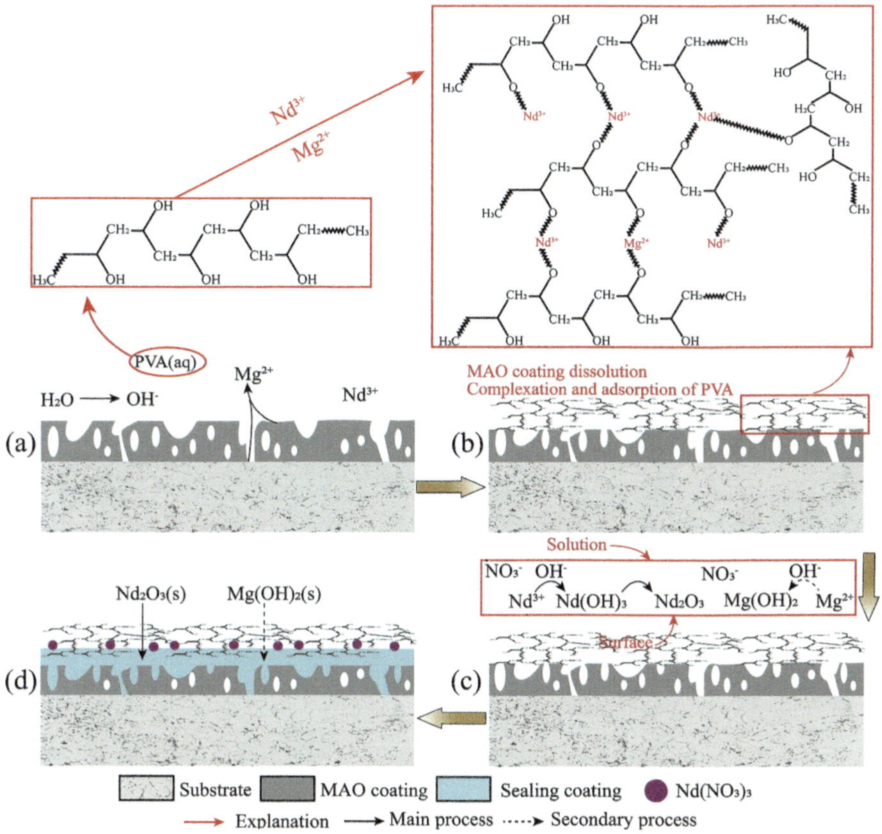

Fig. 16.9 (**a–d**) Schematic diagram of self-healing during sealing and corrosion process of PVA/$Nd(NO_3)_3$ mixed coating. Reproduced with permission from [10] © 2022 Elsevier Inc

Zhang et al. [11] encapsulated the PEO pretreated AZ31 using epoxy resin (EP) and dispersed mesoporous silica (MSN) containers loaded with sodium benzoate (SB) as corrosion inhibitor to the PEO (Fig. 16.10a, b) and EP layers (Fig. 16.10c), respectively. The effects of different loading positions on the anticorrosive and self-healable performance of the multilayered coating were studied. When the coating was damaged, the SB carried in MSN reacts with Mg^{2+} ions and deposits at the damaged area to form self-healing function. Additionally, research has shown that MSN@SB loaded in the PEO layer, due to its proximity to the alloy substrate, exhibits lower admittance values and better corrosion resistance. The MSN@SB loaded in the EP layer is further away from the substrate, whose contribution to self-healing performance is limited.

Li et al. [55] prepared PVA-TA mixed solution, which was loaded on PEO-coated Mg-Li alloys by negative pressure impregnation. EP was coated to form the outer defense layer. The composite coating exhibited a synergistic active protection effect. PVA exhibited water-swelling properties, which allowed it to become filling

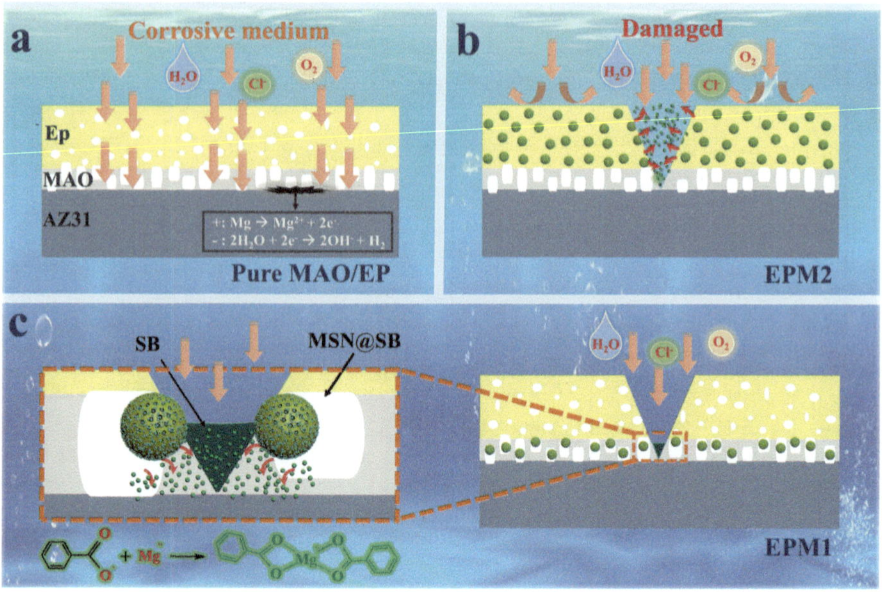

Fig. 16.10 Active protection mechanisms of the MSN@SB in different positions in the smart composite coatings. (**a**) pure PEO/EP, (**b**) EPM2, and (**c**) EPM1 coatings. Reproduced with permission from [11] © 2023 Chongqing University. Publishing services provided by Elsevier B.V. on behalf of KeAi Communications Co. Ltd.

materials at outer layer damaged site. In addition, the hydrogen bonds between PVA and TA accelerate the self-healing process. When destroyed into inner layer, the released Mg^{2+} owing to corrosion can chelate with TA to form TA-Mg, which subsequently adhered onto the damage site and prevented further corrosion.

Liu et al. [56] compared the Na_3PO_4, MBT, and their synergistic self-healing effect on PEO/EP and polyamide composite coating. It was obvious that the simultaneous adding of two inhibitors further enhanced the self-healable effect. Noteworthy, the single adding of Na_3PO_4 in coatings exhibited a current density decrease of 42% and the single adding of MBT exhibited a decrease of 8%. Although both of them are cathodic inhibitors, the inhibition efficiency has a large difference, which may be affected by the content and connection with substrate. Additionally, the serious fluctuation of the corrosion current of a single MBT-added coating indicated the lack of long-term self-healing stability of MBT.

16.4.4 PEO/SLIPSs

A slippery liquid-infused porous surface (SLIPS) is another posttreatment approach to realize self-healing for PEO coatings. SLIPSs are known for their extremely low adhesion and excellent self-healing effect. Different from the air pockets of conventional superhydrophobic coatings, the lubricants in SLIPSs are more stable and smooth [57]. Jiang et al. [40] applied hydrothermal method to generate MoO_4^{2-} modified LDH film on PEO-treated AZ91 Mg alloy. The superhydrophobic treatment of AZ91 surface was carried out under hydrothermal conditions using 1H, 1H, 2H, 2H perfluorodecyltrimethoxysilane (PFDS). In addition, after modification with perfluoropolyether (PFPE), the coating scratches could heal their own defects due to the good fluidity of PFPE. However, the rough surface of LDH limits the lateral migration of PFPE, so the healing period usually takes a long time, and its fluidity needs to be considered. After the coating on the surface layer fails, the MoO_4^{2-} ions carried by LDH can still form insoluble precipitation with metal ions to achieve self-healing effect. Tian et al. [58] prepared similar composite coatings; what the difference is that the self-healing agent in their work was KrytoxTM GPL 100. This coating method that relies on fluid lubricants to achieve self-healing is simple and efficient, but it is obvious that its coating application environment is greatly limited (Fig. 16.11).

16.4.5 Other Coatings

Hiromoto [42] prepared hydroxyapatite (HAp) and octacalcium phosphate (OCP) coatings on the surface of AZ31 Mg alloy and studied the active protective performance of the coatings via scratch tests. Research has shown that the pH value at the scratch site rapidly increases, contributing to the promoted deposition of Mg phosphate and Mg hydroxide and protecting the scratch site. Meanwhile, the rapid

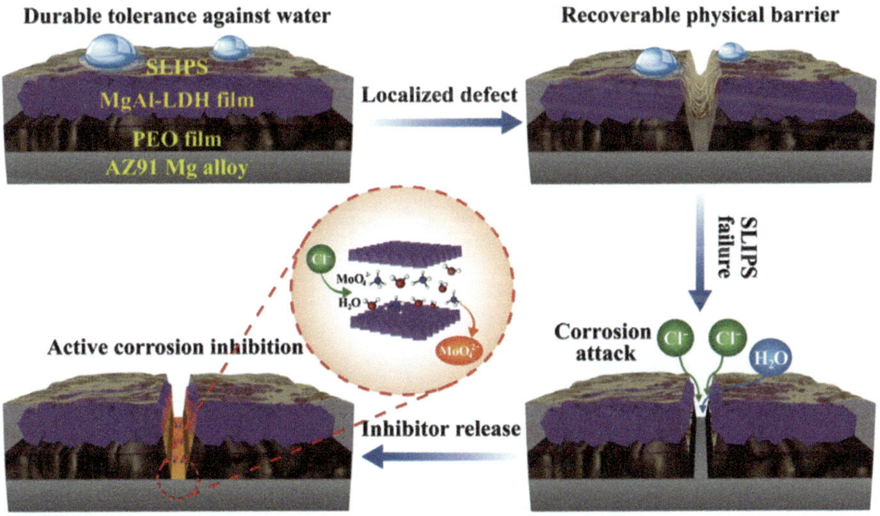

Fig. 16.11 Schematic diagram of intelligent active protection mechanism of PEO-LDH-SILIPS coating. Reproduced with permission from [57] © 2019 Elsevier B.V.

anodic reaction of Mg alloy in the scratches accelerates the cathodic precipitation of corrosion products.

Atomic layer deposition (ALD) is a high-precision thin film deposition technology, which deposits material in the form of a single-layer atomic film layer by layer on the substrate surface based on chemical phases. By introducing two or more precursor chemicals (each containing a chemical element) onto the substrate surface, a single-layer thin film material is formed. ALD technology can achieve precise control of film thickness by controlling the number of deposition cycles, and chemical adsorption ensures its good adhesion. Jiao et al. [31] used ALD technology to perform 1000 cycles of deposition on the PEO coating of AZ31, forming a dense Al_2O_3 film for encapsulation (Fig. 16.12). Research has shown that composite coatings exhibit layered structures similar to LDH at the crack locations after long-term immersion, and self-healing behavior has been observed. Its active protection is achieved through the release of Al^{3+} ions in the Al_2O_3 film and redeposition under local alkaline conditions.

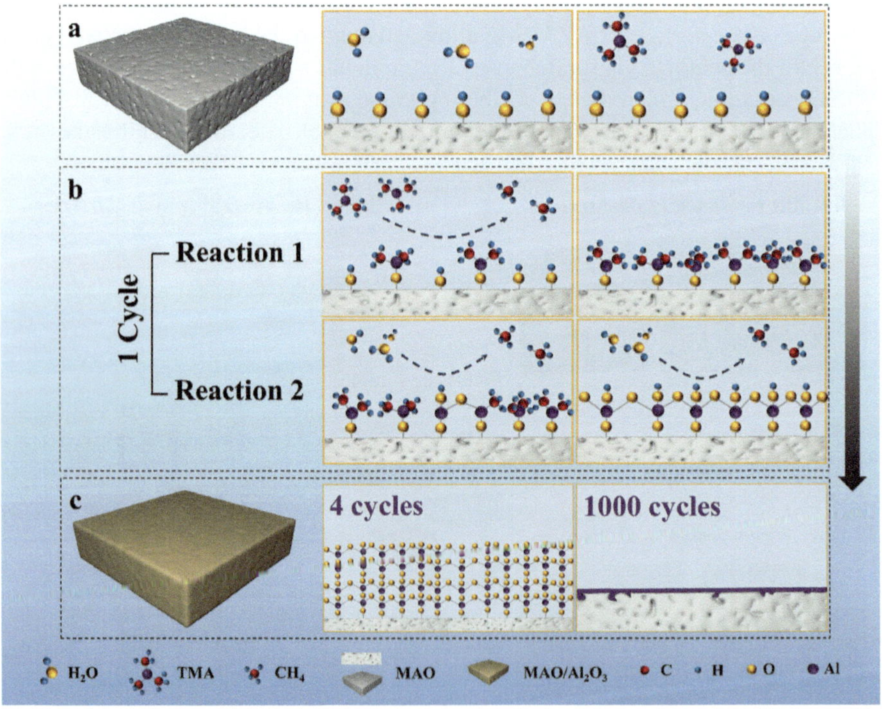

Fig. 16.12 (**a–c**) Schematic illustration of ALD technique applied on PEO coated AZ31 alloy. Reproduced with permission from [31] © 2023 Elsevier Ltd and Techna Group S.r.l.

16.5 Self-Healing Performance Testing Methods

16.5.1 Electrochemical Tests

16.5.1.1 OCP, EIS, and PDP Tests

OCP, EIS, and PDP testing are commonly used for corrosion resistance testing and for self-healing testing. EIS can determine the active protective performance of different coatings by comparing the |Z| values of coating systems before and after artificial scratching. Toorani et al. [59] designed a PEO/Silane/epoxy coating system, incorporating 8-HQ, sodium diethyldithiocarbamate (DDTC), 2-mercaptobenzimidazole (MBO), and indole-3-carbaldehyde (I3C) as corrosion inhibitors, whose performance was tested and compared. By analyzing the |Z| value corresponding to 0.01 Hz in EIS testing, the greater the |Z| value, the more difficult it is for the corrosive agent to penetrate, indicating that the coating exhibits a strong corrosion inhibition effect. As can be seen in Fig. 16.13, the |Z| value at 0.01 Hz varied in the order: Triplex-Ce-HQ > Triplex-Ce-I3C > Triplex-Ce > Triplex. In

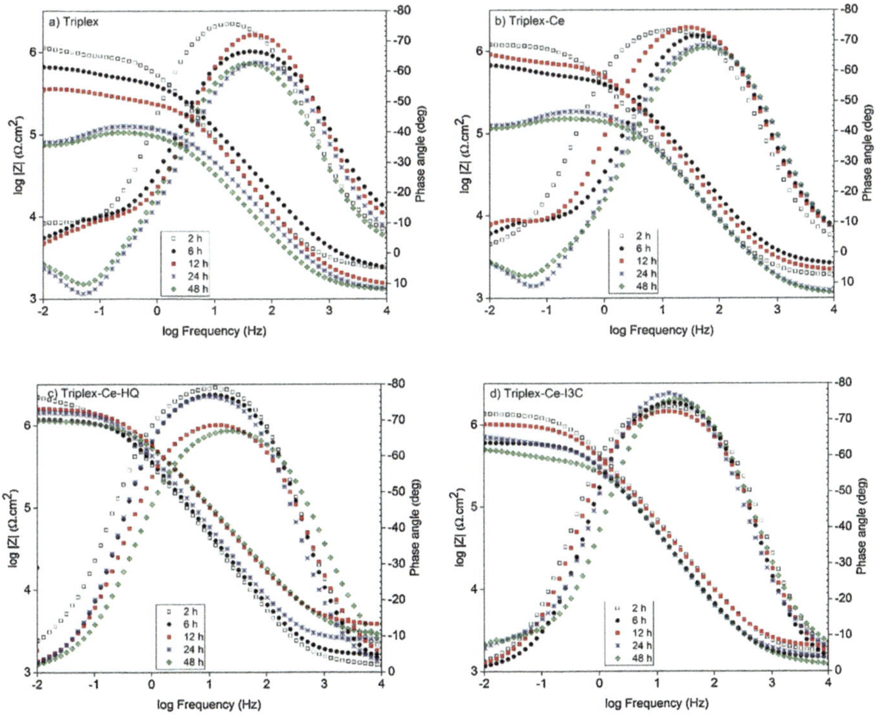

Fig. 16.13 Bode diagrams of triplex coating samples with artificial scratching (**a**) Triplex, (**b**) Triplex-Ce, (**c**) Triplex-Ce-HQ, (**d**) Triplex-Ce-I3C, during 48 h of immersion in 0.5 wt.% NaCl solution. Reproduced with permission from [59] © 2020 Elsevier Ltd.

addition, the peak width at half-height (PWH) can reflect the corrosion resistance of samples. It is evident that the PWH of I3C doped triplex coating has not significantly decreased, indicating the better anticorrosion effect of triplex-Ce-I3C coating.

In PDP testing, observing the changes in an anodic half reaction curve (Breakdown potential, E_b) can indirectly reflect the changes in the coating. Zeng et al. [60] indicated that compared with the porous PEO coating, the polarization curve of MoO_4^{2-}-LDH/PEO coating showed multiple passivation phenomena (Fig. 16.14a), proving that MoO_4^{2-}-LDH endow the coating with self-healing ability. The self-healable effect of the coating is closely related to the ion exchange of MoO_4^{2-} ions and the hollow structure of LDH. As shown in Fig. 16.14b, MoO_4^{2-} ions in this coating can be replaced with cathodic depolarizing agents (Cl^- ions) to enhance corrosion resistance and realize self-healing effects.

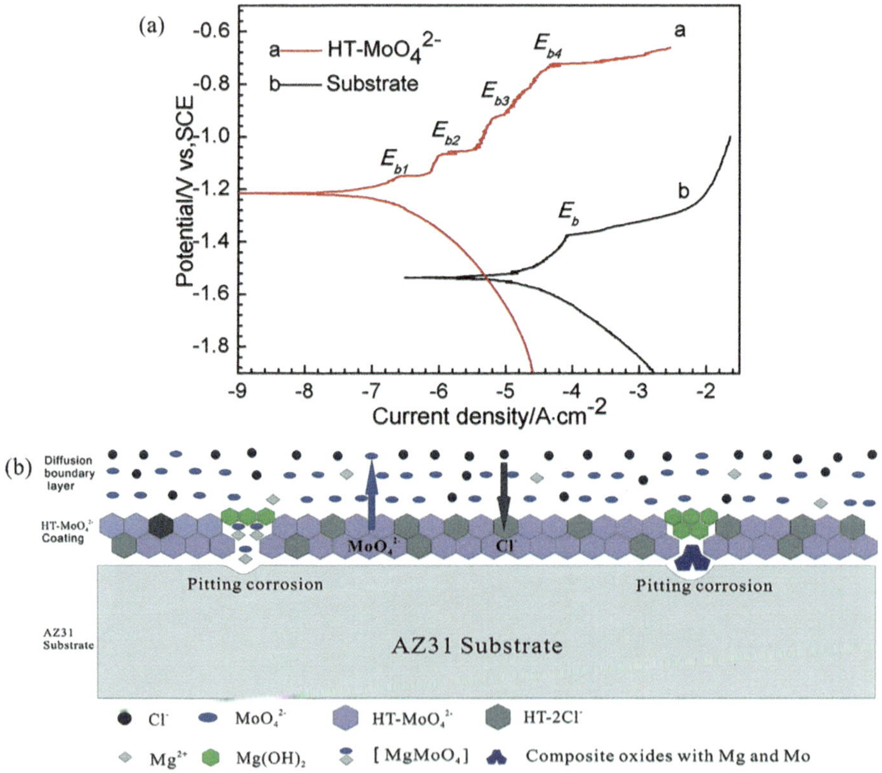

Fig. 16.14 (**a**) PDP curves of the bare substrate and HT-MoO_4^{2-} coated sample in 3.5 wt. % NaCl solution; (**b**) Schematic diagram of self-healing process of MoO_4^- intercalated LDH coating during corrosion in 3.5 wt. % NaCl solution. Reproduced with permission from [60] © The Royal Society of Chemistry 2014

16.5.1.2 Local Electrochemical Impedance Spectroscopy (LEIS)

Local electrochemical impedance spectroscopy (LEIS) is a novel electrochemical testing technology that can measure the local impedance of organic coating systems and establish a direct relationship between changes in local impedance, changes in coating capacitance, and the degree of coating delamination. It compensates for the limitation of traditional EIS, which can only reflect the overall average information of the measured area. Therefore, it is of great importance for evaluating the environmental failure of organic coatings. Zhang et al. [11] tested the LEIS data at scratch positions by 3 h and 10 h of immersion to investigate the damaged areas during the corrosion process and presented them in the form of admittance graphs in Fig. 16.15. Admittance is the reciprocal of impedance, and the higher its value, the thinner the coating and the poorer its corrosion resistance. The comparison results showed that the active protective performance of the composite film was better, and the method of loading the inhibitor into the PEO pores resulted in a better self-healable effect of the coating.

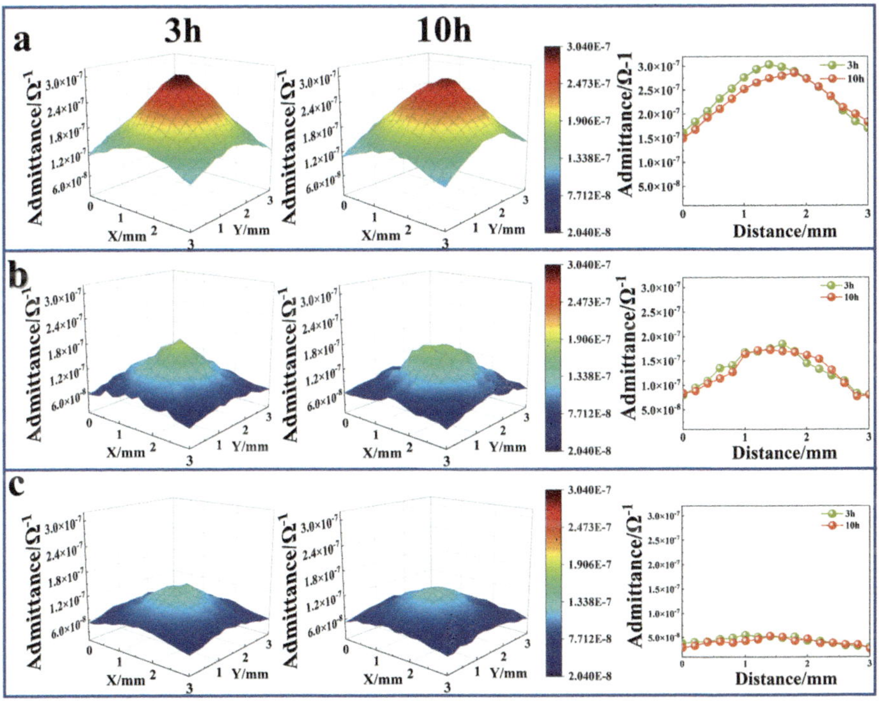

Fig. 16.15 LEIS maps of (**a**) pure PEO/EP, (**b**) EPM2, and (**c**) EPM1 coatings after 3 h and 10 h immersion in 3.5 wt. % NaCl solution and the impedance values at $x = 1.4$ mm. Reproduced with permission from [11] © 2023 Chongqing University. Publishing services provided by Elsevier B.V. on behalf of KeAi Communications Co. Ltd.

16.5.1.3 Scanning Vibrating Electrode Technique (SVET)

Scanning vibrating electrode technique (SVET) is able to evaluate the localized corrosion current density (i_{corr}) of samples [61], which can directly reflect the change of corrosion resistance. Obviously, the self-healable effect can be characterized by testing the i_{corr} of damaged sites after different healing times.

Liu et al. [62] fabricated ionic liquid-graphene oxide (IR-GO) through a grafting reaction between GO and imidazole ionic liquid (IL), which was subsequently coated on carbon steel with EP encapsulation. SVET was applied to characterize the localized i_{corr} of damaged sites of coatings. The coating with IR-GO exhibited the lowest i_{corr} even after 30 h immersion in 3.5 wt. % NaCl solution. It can be concluded that the coating with IR-GO maintained a protective effect due to the physical barrier and the active protection effect of imidazole IL.

Wu et al. [63] prepared MgAlLa-LDH sealing PEO coating, which subsequently loaded mxene@4-aminophenol composite (M@A). SVET technique was applied to observe the potential fluctuation of the damaged part on coating after different healing times. Generally, a higher potential change reflects a more serious corrosion activity. The result indicated that the damaged site of composite coating loaded with M@A possessed the lowest potential fluctuation after immersion, indicating the best active protective effect.

It is typically believed that corrosion current density directly reflects the rate of alloy corrosion. Akbarzadeh et al. [64] initially prepared PEO coatings on the surface of AA2024 aluminum alloy and sealed the pores with sol-gel coatings. In order to achieve self-healing properties of the coatings, benzotriazole (BTA) and 8-HQ were introduced between the PEO and sol-gel coatings. After artificial scratching of the coatings, the corrosion current density of the scratched area immersed for 12 h was tested using the SVET technique. As shown in Fig. 16.16, the tested coatings included PEO coating, PEO/sol-gel (P-S) coating, PEO/BTA/sol-gel (P-H5-S) coating, PEO/8-HQ/sol-gel (P-H5-S) coating, and PEO/BTA&8-HQ/sol-gel (P-HB5-S) coating. The results showed that the single PEO and PEO/sol-gel coatings lacked inhibitors and thus did not exhibit self-healing effects, displaying increased anodic activity. With the addition of BTA, the anodic activity was inhibited within 6 h, but continued to increase at 12 h, suggesting that BTA only acted on the surface of PEO. Coatings with added 8-HQ demonstrated good corrosion inhibition effects within 12 h, indicating that 8-HQ could easily enter the pores of PEO to inhibit corrosion. The combination of BTA and 8-HQ completely suppressed surface anodic activity within 12 h, indicating the successful achievement of self-healing effects by the coatings.

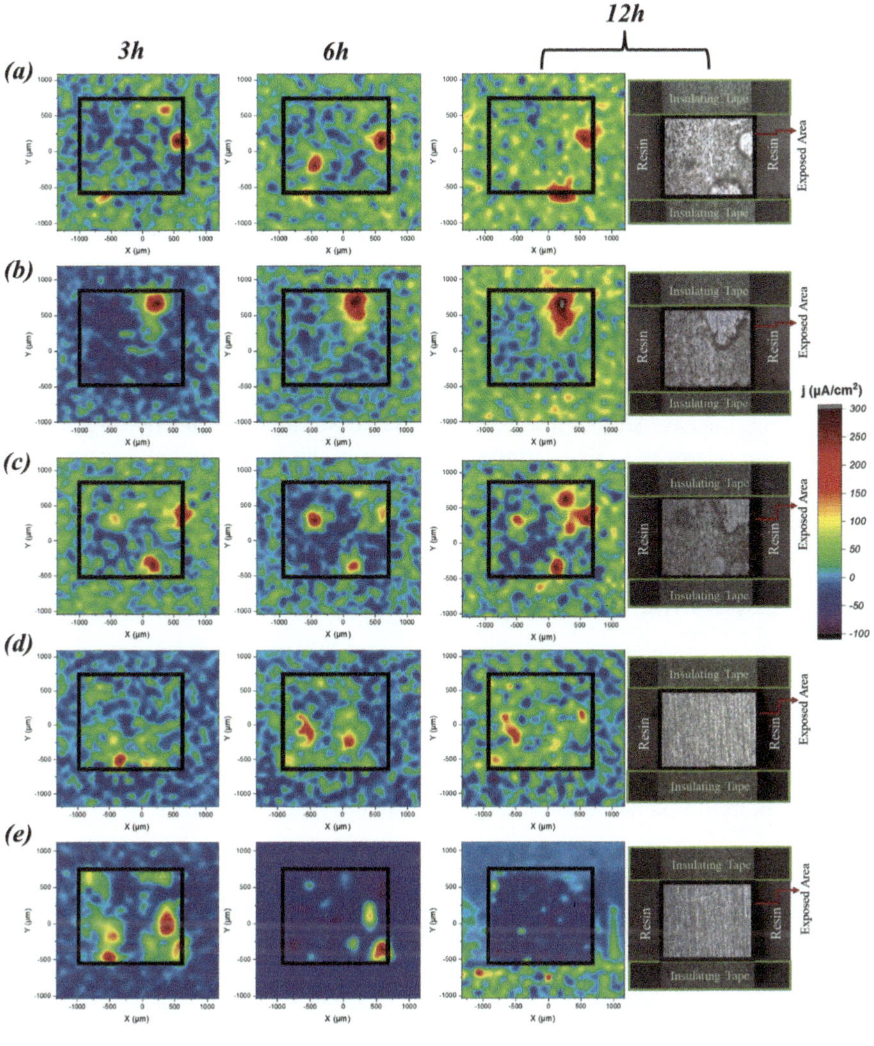

Fig. 16.16 SVET maps of the PEO (**a**), P-S (**b**), P-B5-S (**c**), P-H5-S (**d**), and P-BH5-S (**e**) samples after 3 h, 6 h, and 12 h immersion in 15 mM NaCl solution as well as the optical image at the end of the test. Reproduced with permission from [64] © 2023 Elsevier Ltd.

16.5.2 *Internal Reflection Infrared Imaging (IR)*

Internal reflection infrared imaging can exhibit the healing process at molecular level, which is a powerful technique to characterize the molecular activity in organic coatings. Liu et al. [65] prepared a new supramolecular PU coating, which possessed self-healable effect relying on dynamic interactions of hydrogen bonds. IR technique was applied on the damaged and undamaged regions to test the self-healing effect (Fig. 16.17). The results indicated that the peak intensity of carbamate carbonyl and urea carbonyl in damaged sites improved. After healing process, the intensity of carbamate carbonyl and urea carbonyl was more approximate to original state or undamaged sites, which revealed the reforming of hydrogen bonds and the excellent self-healable effect of the coating.

16.5.3 Hydrogen Evolution

The negative differential effect of Mg alloys presents as hydrogen evolution in solution, which is a positive correlation with corrosion rate. Therefore, hydrogen evolution test is able to exhibit the change of corrosion rate during the corrosion and self-healing process. Although some research [31, 35] indicates the hydrogen evolution rate can reflect the self-healing behavior of oxide film or coatings, it is lateral and one-sided, which needs to be assisted with other methods.

16.5.4 Scanning Electron Microscope (SEM)

SEM and the affiliated energy dispersive spectrometer (EDS) are commonly used in the analysis of micro morphology and compositions of coatings. Although the in situ observation is difficult to achieve, the tests of damaged or artificial scratching site after different immersion time can be easily carried out. For the observation of coatings, both the surface and cross section should be concerned.

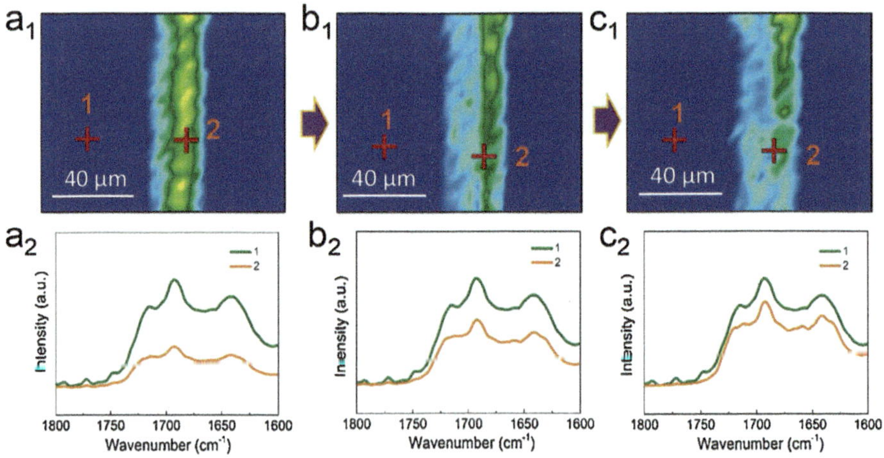

Fig. 16.17 IR spectra and intensity of scratching sites of PU coating prepared in Liu's study after various healing times: ($\mathbf{a}_1$) 0 h, ($\mathbf{b}_1$) 1 h, and ($\mathbf{c}_1$) 2 h. Reproduced with permission from [65] © 2021 Elsevier Ltd.

Wang et al. [66] prepared PEO/$CaCO_3$ coatings by ultrasound treatment and characterized the morphology and composition before and after immersion via SEM and EDS. They found that the surface of PEO/$CaCO_3$ coatings was covered by newly formed calcite, which compared to bare PEO and PEO/($CaHPO_4 \cdot H_2O$ DCPD) coatings evidently sealed the cracks and pores and therefore exhibited self-healing performance.

In the study of Chen et al. [22], in order to further explore the self-healing mechanism of composite coatings, MgAlLa-$V_2O_7^{4-}$ LDHs/GO, SEM and EDS were respectively applied to observe the morphology and composition of scratching sites, in which it could be found that the O, La, and V signals appeared around the scratches. It can be summarized that precipitation film comprising of Mg, O, La, and V are formed and covered the damaged site, improving the anticorrosive effect of coatings and exhibiting self-healable effect.

SEM/EDS can also be applied in the characterization self-healable ability of organic coatings, for example, epoxy resins. Sun et al. [67] fabricated autonomous corrosion detecting and self-healing EP coating on AZ31 alloy, and the dual-function was achieved by introducing (phenolphthalein, PhPh)/EP contained micro-capsules into EP coatings. The self-healable effect was proved by observing the scratching area of coatings under SEM. As shown in Fig. 16.18b, c, the uncured EP encapsulated in microcapsules flowed out and cured at scratch to form a new film, thereby achieving the active protection.

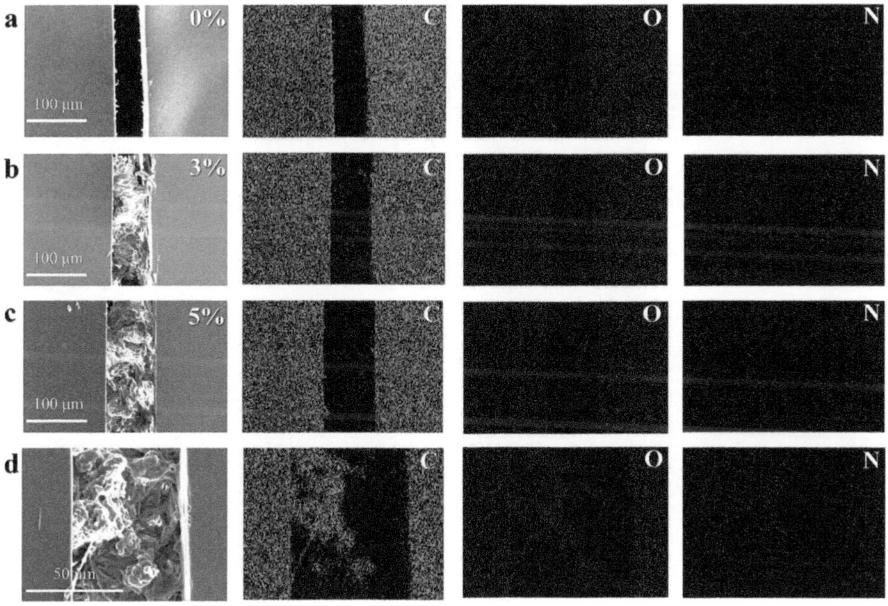

Fig. 16.18 SEM morphology and EDS mapping of (**a**) pure EP, (**b**) EP with 3% microcapsules, (**c**) EP with 5% microcapsules after artificial scratched, and (**d**) samples cured for 24 h. Reproduced with permission from [67] © 2023 Elsevier B.V.

16.5.5 Electron Probe Microanalysis (EPMA)

EPMA is commonly applied to scan elemental composition and distribution. Chen et al. [48] used EPMA to prove the self-healable effect of their in situ prepared PEO/Ce coating. EPMA was used to characterize the elements distribution of PEO/Ce coatings after immersion in NaCl solution for 48 h and 312 h. By comparing the content change of element between above two immersed samples both from surface and cross section shown in Fig. 16.19, the Ce element content in coating has increased after immersion, especially in corrosion pits, which indicated that the Ce ions are transformed from the surface to the corrosion cracks. Combining with other data obtained in their work, it can be inferred that Ce ions play the role of self-healing agents in this coating.

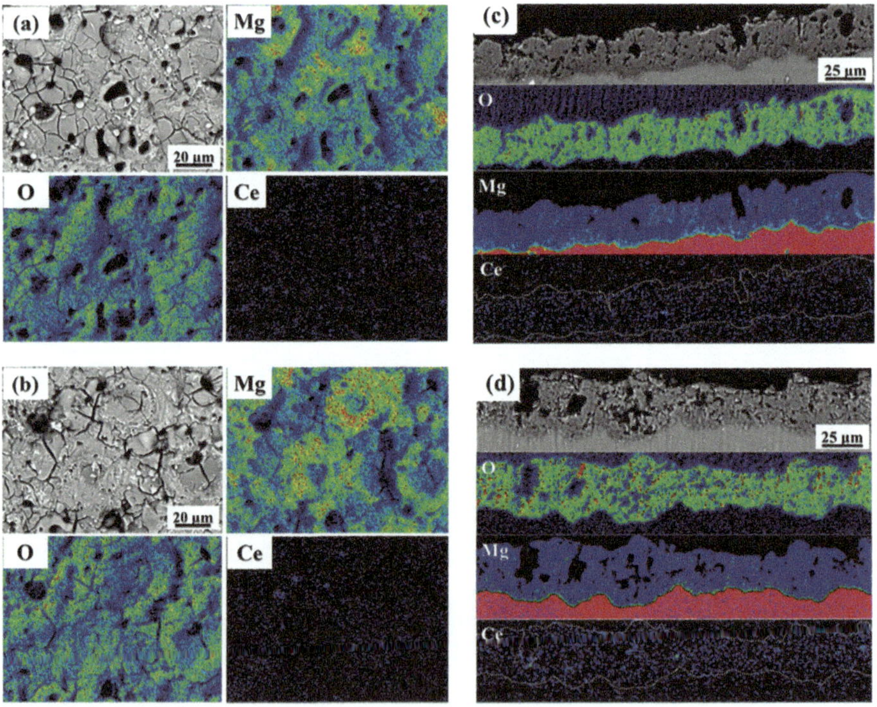

Fig. 16.19 EPMA maps of Mg, O, and Ce distribution on the surface section of PEO-Zeolite-Ce coatings after (**a**) 48 h and (**b**) 312 h immersion in NaCl solution; EPMA maps of Mg, O, and Ce distribution on the cross section of PEO-Zeolite-Ce coatings after (**c**) 48 h and (**d**) 312 h immersion in NaCl solution. Reproduced with permission from [48] © 2022 Elsevier Ltd.

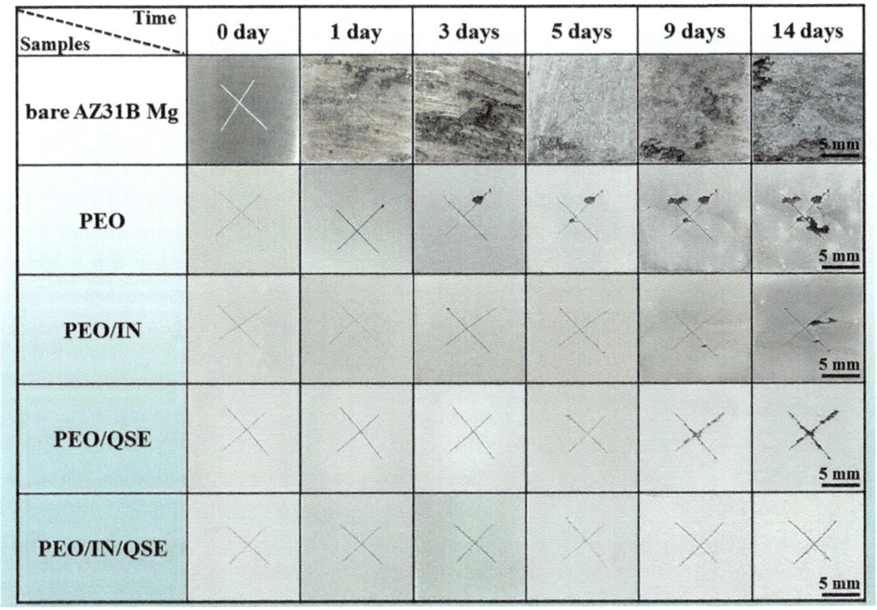

Fig. 16.20 Optical images of artificial scratching region of various samples during the immersion in 3.5 wt.% NaCl solution. Reproduced with permission from [68] © 2022 Elsevier B.V.

16.5.6 Optical Microscope

Li et al. [68] fabricated a multilayered coating based on PEO with imidazolium-based ionic liquid as inhibitor and saline-based superhydrophobic coatings as outer layer (PEO/IN/QSE). The optical microscope was applied to observe the macro morphology of artificial scratching site during 14 days corrosion process (Fig. 16.20). The result indicated that the scratching site of PEO/IN/QSE exhibited no obvious corrosion expansion, which was related to the dual effect of superhydrophobic and self-healable.

16.6 Conclusions and Outlook

PEO coatings are always known as valid methods for protecting valve metals from premature corrosion, but the inherent pores and cracks influence their application. Multilayer coatings and active protective coatings based on PEO coatings not only compensate the shortcomings of sole PEO coating but endow the coatings with self-healing ability. To achieve self-healing ability, corrosion inhibitors including metal cations, negative charged anions and ligands, and MOFs are doped onto the PEO coatings as supporting agents of extrinsic self-healing via either in situ methods or post-processing methods. Simultaneously, coatings like LDHs, MOFs, SLIPSs, and organic coatings are further applied to sealing the defects to prevent the premature failure of bare PEO coatings. Although some progress has been made in the enhancement of corrosion resistance, there is still a long way to transform research into applications. Further research is needed to optimize and simplify the process, thereby decreasing the time and money costs. Besides, computational material science and machine learning should be applied to develop active protective coatings based on PEO. As the conventional testing methods of self-healing are mostly time-consuming, it will be highly effective and saving if the results are simulated in advance. Last but not least, long-term environmental sustainability requires selecting environmentally friendly corrosion inhibitors and self-healing coatings to prevent pollution and achieve sustainable development of the economy and society.

References

1. Yang, C., Sheng, L., Zhao, C., Chen, P., Ouyang, W., Xu, D., Zheng, Y., & Chu, P. K. (2024). *Journal of Materials Research and Technology, 33*, 2275–2291.
2. Barati Darband, G., Aliofkhazraei, M., Hamghalam, P., & Valizade, N. (2017). *Journal of Magnesium and Alloys, 5*, 74–132.
3. Yao, W., Wu, L., Wang, J., Jiang, B., Zhang, D., Serdechnova, M., Shulha, T., Blawert, C., Zheludkevich, M. L., & Pan, F. (2022). *Journal of Materials Science and Technology, 118*, 158–180.
4. Jiao, Z.-J., Li, C.-Y., Du, Y.-K., Cui, L.-Y., Chen, X.-B., Xi, Y.-M., & Zeng, R.-C. (2023). *Surface and Coating Technology, 472*, 129922.
5. Wei, P., Chen, L., Li, X., Gu, H., & Chen, D. (2024). *Journal of Adhesion Science and Technology, 38*, 991–1013.
6. Fayyad, E. M., Almaadeed, M. A., Jones, A., & Abdullah, A. M. (2014). *International Journal of Electrochemical Science, 9*, 4989–5011.
7. Guo, L., Wu, W., Zhou, Y., Zhang, F., Zeng, R., & Zeng, J. (2018). *Journal of Materials Science and Technology, 34*, 1455–1466.
8. Wu, L., Yang, D., Zhang, G., Zhang, Z., Zhang, S., Tang, A., & Pan, F. (2018). *Applied Surface Science, 431*, 177–186.
9. Liu, R., Yao, Q., Liu, L., Meng, F., & Wang, F. (2022). *Progress in Organic Coating, 166*, 106774.
10. Cai, L., Song, X., Liu, C.-B., Cui, L.-Y., Li, S.-Q., Zhang, F., Bobby Kannan, M., Chen, D.-C., & Zeng, R.-C. (2023). *Journal of Colloid and Interface Science, 630*, 833–845.
11. Zhang, A.-M., Liu, C., Sui, P.-S., Sun, C., Cui, L.-Y., Kannan, M. B., & Zeng, R.-C. (2023). *Journal of Magnesium and Alloys, 11*, 4562–4574.
12. Liu, Y., & Hsu, S.-H. (2018). *Frontiers in Chemistry, 6*, 449.
13. Chen, Y., Wu, L., Yao, W., Wu, J., Serdechnova, M., Blawert, C., Zheludkevich, M. L., Yuan, Y., Xie, Z., & Pan, F. (2023). *Journal of Magnesium and Alloys, 11*, 2230–2259
14. Liu, Y., Yong, Q., Wu, L., Yu, G., & Xie, Z.-H. (2024). *Surface and Coating Technology, 478*, 130461.
15. Jayaraj, J., Kumar, S. A., Srinivasan, A., Raghu, K. G., Arunchandran, C., & Rajinikanth, V. (2024). *Applied Surface Science, 644*, 158797.
16. del Olmo, R., Tiringer, U., Milošev, I., Visser, P., Arrabal, R., Matykina, E., & Mol, J. M. C. (2021). *Surface and Coating Technology, 419*, 127251.
17. Qiao, Y., Li, W., Wang, G., & Zhang, X. (2015). *Journal of Rare Earths, 33*, 647–654.

18. Jian, S.-Y., Yang, C.-Y., & Chang, J.-K. (2020). *Applied Surface Science, 510*, 145385.
19. Zheng, Z., Zhao, M.-C., Tan, L., Zhao, Y.-C., Xie, B., Yin, D., Yang, K., & Atrens, A. (2020). *Surface and Coating Technology, 386*, 125456.
20. Gong, Y., Geng, J., Huang, J., Chen, Z., Wang, M., Chen, D., & Wang, H. (2021). *Surface and Coating Technology, 417*, 127208.
21. Calado, L. M., Taryba, M. G., Morozov, Y., Carmezim, M. J., & Montemor, M. F. (2020). *Corrosion Science, 170*, 108648.
22. Chen, Y., Wu, L., Yao, W., Wu, J., Yuan, Y., Xie, Z., Jiang, B., & Pan, F. (2022). *Colloids and Surfaces, A: Physicochemical and Engineering Aspects, 655*, 130339.
23. Zhang, D., Peng, F., Qiu, J., Tan, J., Zhang, X., Chen, S., Qian, S., & Liu, X. (2021). *Corrosion Science, 192*, 109840.
24. Grewal, N. S., Batra, U., Kumar, K., & Mahapatro, A. (2023). *Journal of Magnesium and Alloys, 11*, 1440–1460.
25. Zhang, S., Li, G., Xu, L., Liao, B., Wan, S., & Guo, X. (2024). *Progress in Organic Coating, 192*, 108511.
26. Chirkunov, A., Rakoch, A. G., Monakhova, E. V., Gladkova, A. A., Khabibullina, Z. V., Ogorodnikova, V. A., Serdechnova, M., Blawert, C., Kuznetsov, Y. I., & Zheludkevich, M. L. (2019). *International Journal of Corrosion and Scale Inhibition, 8*, 1170–1188.
27. Gnedenkov, A. S., Sinebryukhov, S. L., Filonina, V. S., Ustinov, A. Y., & Gnedenkov, S. V. (2023). *Polymer, 15*, 3035.
28. Wang, Y., Zuo, Y., & Tang, Y. (2018). *Construction and Building Materials, 167*, 197–204.
29. Li, C.-Y., Gao, L., Fan, X.-L., Zeng, R.-C., Chen, D.-C., & Zhi, K.-Q. (2020). *Bioactive Materials, 5*, 364–376.
30. Li, J., He, N., Li, J., Fu, Q., Feng, M., Jin, W., Li, W., Xiao, Y., Yu, Z., & Chu, P. K. (2022). *Surface and Coating Technology, 439*, 128442.
31. Jiao, Z.-J., Yu, C., Wang, X.-M., Zhou, Y.-F., Guo, L., Xia, Y., Zhang, B.-C., & Zeng, R.-C. (2024). *Ceramics International, 50*, 5541–5551.
32. Chen, Y., Lu, X., Lamaka, S. V., Ju, P., Blawert, C., Zhang, T., Wang, F., & Zheludkevich, M. L. (2020). *Applied Surface Science, 504*, 144462.
33. Samadianfard, R., Seifzadeh, D., & Habibi-Yangjeh, A. (2021). *Surface and Coating Technology, 419*, 127292.
34. Farshid, S., Kharaziha, M., & Atapour, M. (2023). *Journal of Magnesium and Alloys, 11*, 592–606.
35. Cui, L.-Y., Gao, S.-D., Li, P.-P., Zeng, R.-C., Zhang, F., Li, S.-Q., & Han, E.-H. (2017). *Corrosion Science, 118*, 84–95.
36. Chen, Y., Wu, L., Yao, W., Chen, Y., Zhong, Z., Ci, W., Wu, J., Xie, Z., Yuan, Y., & Pan, F. (2022). *Corrosion Science, 194*, 109941.
37. Liu, S., Li, Z., Yu, Q., Qi, Y., Peng, Z., & Liang, J. (2021). *Chemical Engineering Journal, 424*, 130551.
38. Chen, J., Fang, L., Wu, F., Xie, J., Hu, J., Jiang, B., & Luo, H. (2019). *Progress in Organic Coating, 136*, 105234.
39. Huangfu, H., Guo, X., Li, N., Xiong, Y., Huang, Y., Zhang, J., & Wang, L. (2023). *Progress in Organic Coating, 181*, 107598.
40. Jiang, D., Xia, X., Hou, J., Cai, G., Zhang, X., & Dong, Z. (2019). *Chemical Engineering Journal, 373*, 285–297.
41. Chen, Y., Wu, L., He, X., Yao, W., Wu, J., He, Y., Zhou, Y., Yuan, Y., Wang, J., Xie, Z., Atrens, A., & Pan, F. (2024). *Journal of Alloys and Compounds, 989*, 174278.
42. Hiromoto, S. (2015). *Corrosion Science, 100*, 284–294.
43. Telmenbayar, L., Gopal Ramu, A., Yang, D., & Choi, D. (2023). *Chemical Engineering Journal, 458*, 141397.
44. Chen, Y., Wu, L., Wu, M., Yao, W., Wu, J., Zhou, Y., Yuan, Y., Xie, Z., & Pan, F. (2024). *Surface and Coating Technology, 493*, 131246.

45. Chen, P., Wu, Z., Huang, Q., Ji, S., Weng, Y., Wu, Z., Ma, Z., Chen, X., Weng, M., Fu, R. K. Y., Lin, H., Tian, X., Pan, F., & Chu, P. K. (2019). *Tribology International, 138*, 157–165.
46. Yang, S., Sun, R., & Chen, K. (2022). *Chemical Engineering Journal, 428*, 131198.
47. Qi, X., Gao, H., He, Y., Su, X., & Song, R. (2024). *Surface and Coating Technology, 484*, 130836.
48. Chen, Q., Lu, X., Serdechnova, M., Wang, C., Lamaka, S., Blawert, C., Zheludkevich, M. L., & Wang, F. (2022). *Corrosion Science, 209*, 110785.
49. Calado, L. M., Taryba, M. G., Carmezim, M. J., & Montemor, M. F. (2018). *Corrosion Science, 142*, 12–21.
50. Jia, Z., Xiong, P., Shi, Y., Zhou, W., Cheng, Y., Zheng, Y., Xi, T., & Wei, S. (2016). *Journal of Materials Chemistry B, 4*, 2498–2511.
51. Miyata, S. (1983). *Clays and Clay Minerals, 31*, 305–311.
52. Zahedi Asl, V., Chini, S. F., Zhao, J., Palizdar, Y., Shaker, M., & Sadeghi, A. (2021). *Surface and Coatings Technology, 426*, 127764.
53. Yang, S., Wang, J., Mao, W., Zhang, D., Guo, Y., Song, Y., Wang, J.-P., Qi, T., & Li, G. L. (2018). *Colloids and Surfaces, A: Physicochemical and Engineering Aspects, 555*, 18–26.
54. Wang, R., Zhao, C., Peng, Z., Yan, X., Sun, Y., Yu, Q., Yu, B., Cai, M., & Zhou, F. (2024). *Progress in Organic Coating, 186*, 108016.
55. Li, B., Zhou, Z., Wei, Y., Yu, Y., Huang, G., Qiu, R., & Ouyang, Y. (2024). *Surface and Coating Technology, 491*, 131146.
56. Liu, D., Han, E. H., Song, Y., & Shan, D. (2019). *Electrochimica Acta, 323*, 134796.
57. Yao, W.-H., Zhan, G.-X., Chen, Y.-H., Qin, J., Wu, L., Chen, Y.-N., Wu, J.-H., Jiang, B., Atrens, A., & Pan, F.-S. (2023). *Transactions of Nonferrous Metals Society of China, 33*, 3309–3318.
58. Tian, G., Wang, J., Su, H., Xue, C., Yang, X., Li, Q., Li, X., Miao, Y., & Yang, Z. (2024). *Progress in Organic Coating, 196*, 108696.
59. Toorani, M., Aliofkhazraei, M., Mahdavian, M., & Naderi, R. (2021). *Corrosion Science, 178*, 109065.
60. Zeng, R.-C., Liu, Z.-G., Zhang, F., Li, S.-Q., Cui, H.-Z., & Han, E.-H. (2014). *Journal of Materials Chemistry A, 2*, 13049–13057.
61. Trinh, D., Dauphin Ducharme, P., Mengesha Tefashe, U., Kish, J. R., & Mauzeroll, J. (2012). *Analytical Chemistry, 84*, 9899–9906.
62. Liu, C., Qiu, S., Du, P., Zhao, H., & Wang, L. (2018). *Nanoscale, 10*, 8115–8124.
63. Wu, J., Wu, L., Yao, W., Zhou, Y., Wu, M., Yuan, Y., Xie, Z., Atrens, A., Wang, J., & Pan, F. (2024). *Electrochimica Acta, 491*, 144358.
64. Akbarzadeh, S., Coelho, L. B., Dangreau, L., Lanzutti, A., Fedrizzi, L., & Olivier, M.-G. (2023). *Corrosion Science, 222*, 111424.
65. Liu, C., Wu, H., Qiang, Y., Zhao, H., & Wang, L. (2021). *Corrosion Science, 184*, 109355.
66. Wang, Y., Yu, D., Ma, K., Dai, C., Wang, D., & Wang, J. (2023). *Corrosion Science, 212*, 110927.
67. Sun, C., Liu, C., Wang, Y., Jiao, D., Zhang, A.-M., Sui, P.-S., Cui, L.-Y., & Zeng, R.-C. (2024). *Progress in Organic Coating, 186*, 108043.
68. Li, N., Ling, N., Fan, H., Wang, L., & Zhang, J. (2023). *Surface and Coating Technology, 454*, 129146.

Chapter 17
High-Temperature Thermal Protection by Plasma Electrolytic Oxidation Coatings

Yaming Wang, Shuqi Wang, and Zhiyun Ye

17.1 Introduction

Thermal protection for aerospace components, such as thrust cones, gas turbine engines, and jet engines, involves a series of disciplines including thermal insulation, thermal radiation, ablative resistance, hot corrosion resistance, high-temperature oxidation resistance, etc. [1–3]. A common dilemma faced by all these hot-end components is the unbearable surface temperature, which applies a huge thermal load to structural material, reducing the strength [4]. Notably, passive radiation heat dissipation is undoubtedly an optimized selection due to its advantages of not requiring additional energy consumption and being lightweight, especially suitable for the aerospace field at high-temperature environments [5]. Based on the Stefan-Boltzmann law [6], the improvement of the infrared emissivity of material surfaces is a necessary condition for enhancing thermal radiation capability. Therefore, high emissivity thermal protection coating plays an increasingly important role in the aerospace vehicles field, which can significantly reduce the thermal load of the alloy surface, especially at higher temperatures.

Titanium alloys, niobium alloys, and their composites are increasingly used in the aerospace field, due to their moderate density, excellent creep properties, and superior mechanical strength at high temperatures [7–9]. However, the surface temperatures of the aircraft skin and hot-end components of the turbine even exceed the working temperature limit of titanium alloys/niobium alloys, with the increased thrust-to-weight ratio of hypersonic vehicles. On the one hand, the surface emissivity

Y. Wang (✉) · S. Wang · Z. Ye
Institute for Advanced Ceramics, Harbin Institute of Technology, Harbin, China

Key Laboratory of Advanced Structure-Function Integrated Materials and Green Manufacturing Technology, Harbin Institute of Technology, Harbin, China
e-mail: wangyaming@hit.edu.cn; wang-shuqi@hit.edu.cn

V. S. Saji, M. Zheludkevich (eds.), *Plasma Electrolytic Oxidation – Fundamentals, Advances and Applications*,
https://doi.org/10.1007/978-3-031-92429-3_17

of titanium alloys (~0.2–0.3) and niobium alloys (~0.2) is low, showing poor radiation heat dissipation performance. On the other hand, the hot-end components of hypersonic vehicles usually face multiple extreme work environment challenges, such as high-temperature oxidation, hot corrosion, thermal shock, and complex gas scours, accelerating the failure of alloy [10]. Therefore, there is a critical need for thermal protection coatings that possess low thermal conductivity, high emissivity, and multiple thermal protection characteristics, which can unlock the full potential of these alloys and enable them to perform reliably under high-temperature extremes.

Plasma electrolytic oxidation (PEO) can achieve an in situ formation of protective ceramic coating and has emerged as an effective approach to address the above challenges [11]. It can produce a dense inner layer and porous ceramic oxide coatings, simultaneously providing low thermal conductivity and high emissivity, significantly enhancing the thermal protection and stability of the alloy. Moreover, it is proven that a novel coating with altered chemical compositions and functionalities can be prepared by the PEO technique with the functional particle addition into the electrolyte, which can be tailored for specific thermal protection function needs [12].

Recently, most of the research about PEO thermal protection coatings has focused on optimizing process parameters to tailor coatings for low thermal conductivity [13]. Moreover, researchers have explored the incorporation of nanoparticles to form the dual-layer composite coating, further improving the thermal insulation and infrared radiation performance [14]. Additionally, researchers have made great efforts to develop the composite method (PEO + other methods) for further improving the high-temperature protection properties while facing multiple service environments, such as PEO + magnetron sputtering [15], PEO + HAPC [16], and PEO + thermally sprayed [17], significantly broaden the applicability of substrate alloys in extreme environments.

In this chapter, the recent advancements in PEO thermal protection coatings on titanium alloys and niobium alloys are comprehensively reviewed, with particular emphasis on their thermal conductivity, emissivity, and multiple protection characteristics. The thermal protection mechanisms of PEO coatings are explored, highlighting significant innovations in coating design and preparation, and pointing out the emerging trends in the field. The aim is to provide a guide on how to design and prepare high-performance PEO thermal protection coatings on the metal surface, thereby promoting the development of thermal protection coating for the application on the hot-end components in important industries such as rockets, spacecraft, missiles, and high-speed aircraft.

17.2 PEO Coatings Design of High-Temperature Thermal Protection

Thermal protection for aerospace components, such as thrust cones, gas turbine engines, and jet engines, involves a series of disciplines including thermal insulation, thermal radiation, ablative resistance, hot corrosion resistance, high-temperature oxidation resistance, etc. [18–23]. Therefore, in addition to the requirement of high-temperature resistance, the multifunctional design of PEO thermal protection coatings plays an increasingly important role in the aerospace field.

17.2.1 PEO Coatings of High Emissivity

High-performance metals used for aerospace vehicle metal thermal protection systems (e.g., titanium, aluminum, magnesium, and niobium alloys) usually possess a low surface emissivity (about 0.1) [24], resulting in the heat conduction of most of the aerodynamic heat to the interior of the metal compartment (Fig. 17.1a). Conversely, 8–9 times of the aerodynamic heat can be radiated to the external cooling space when the metal surface is coated with high emissivity ceramic coating (emissivity is ~0.9) [25], effectively reducing the heat input to the aerospace vehicle compartment (Fig. 17.1b). As a result, it can reduce the heat input by up to 10–20%, significantly improving the metal's heat-bearing capacity [26]. Undoubtedly, PEO is a promising method for preparing high emissivity ceramic coating by multiphase doping and multi-chemical bonding broad-spectrum responsive absorption, which provide a new approach to the design of radiant heat protection coatings for aerospace vehicles. Based on the above considerations, the design and preparation of high emissivity PEO coatings may comply with the following design principles: (1)

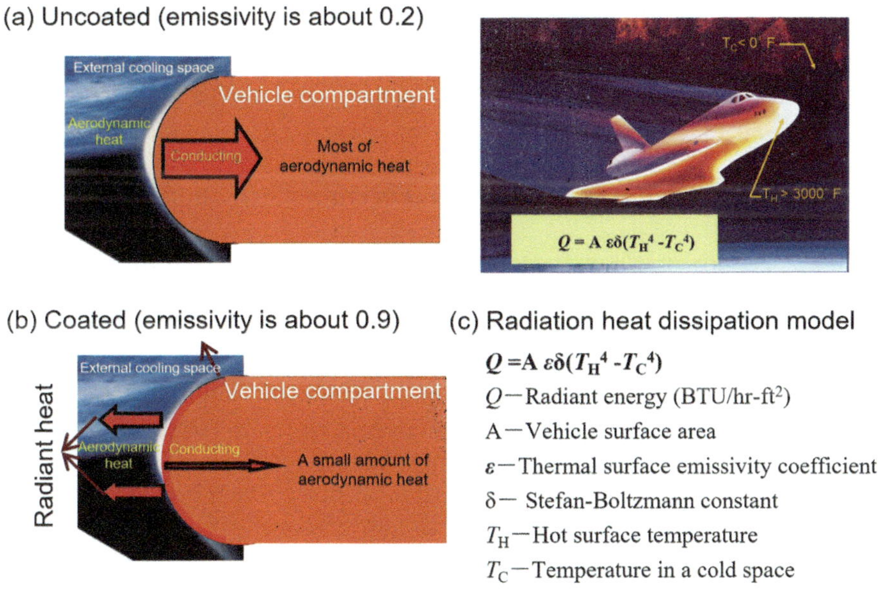

Fig. 17.1 (**a**, **b**) Schematic diagram of radiative cooling with high emissivity ceramic coatings on metallic parts of the aircraft

achieving the strengthening of polar bond vibrational absorption by multiphase and multipolar chemical bond matching to improve the emissivity at different bands (especially the low band) in specific electrolyte systems; (2) achieving the strong absorption in the optical localization of the matched "microporous resonance cavity" by controllable micropores (roughness) for improving the emissivity; (3) achieving the regulation of specific band emissivity by introducing high emissivity non-oxide components (such as SiC, $MoSi_2$, ZrC, and HfC) into the growth of PEO coatings.

17.2.2 PEO Coatings of Thermal Insulation

High-temperature resistance and thermal insulation coating with low thermal conductivity are necessary for high-speed aircraft or engine hot end parts, which can effectively block the heat transfer process and increase the temperature drop between the two sides of the coating, further reducing the surface temperature of the metal substrate for achieving the purpose of thermal insulation. The design and preparation of PEO thermal insulation coatings with low thermal conductivity should meet the following principles: (1) preparation of the oxide phases with low thermal conductivity by the special electrolyte; (2) formation of multistage holes within PEO coating by modulation of the electrical parameters for reducing thermal conductivity; (3) formation of large-thickness multilayer coating by introducing the low thermal conductivity components (e.g., ZrO_2) into the growth of the PEO coating. For example, mullite-rich PEO coating with microporous and amorphous phases can be prepared on aluminum alloys in a silicate electrolyte, showing a low thermal conductivity of 0.5 W m^{-1}·K^{-1}.

17.2.3 PEO Coatings of Multiple Thermal Protection

With the rapid development of aerospace technology, the hot end components (e.g., high-speed aircraft shells, aero-engine blades, rocket engine nozzles, etc.) usually face multiple extreme work environment challenges such as high-temperature oxidation, hot corrosion, thermal shock, and complex gas scours [27–30]. Therefore, developing an advanced thermal protection system (TPS) with multiple thermal protection functions is urgently needed, but it remains a challenge to date. PEO is a potential way to effectively integrate multiple thermal protection performances into one material, and the prepared PEO thermal protection coatings usually possess the following characteristics: (1) a dense barrier layer that prevents oxygen diffusion is formed on the metal substrate surface, which can be optimized through the electrolyte system and corresponding electrical parameters; (2) a large-thickness multilayer coating with multifunctional features is fabricated by introducing multifunctional components (e.g., SiC, $MoSi_2$, ZrC, etc.) into the growth of PEO coating, showing excellent thermal protection performance. For instance, to address the problem that the bottom of the bullet shedding has occurred during the target test of a certain type of aluminum alloy submunition. The anodic oxide protection coating is not resistant to gunpowder gas high-temperature ablation, while the PEO thermal protection coating shows excellent scouring resistance and ablation resistance, which can take a short period of time to withstand the high-temperature gas flow impact (2000 °C).

17.3 High Emissivity PEO Coatings on Ti_2AlNb Alloy

Titanium alloys and TiAl alloys are potential thermal structure materials for aerospace vehicles due to their lightweight and high strength. To further improve their heat-resistant temperature, a high emissivity ceramic coating can be prepared on metal surfaces and can reduce the surface thermal equilibrium temperature by 10–20% by radiation heat dissipation, enabling them to be used above their limiting service temperature. The high-temperature resistant TiAl alloy (Ti2AlNb), for instance, the effects of ionic additives (e.g., VO_3^{-1}) and particulate additives (e.g., SiC and Cr_2O_3) in the Na_2SiO_3 electrolyte on the microstructure and emissivity of PEO coatings were investigated. The electrolyte composition for high emissivity coating preparation on the surface of Ti2AlNb alloy is listed in Table 17.1. In addition, the exact electrical parameters, i.e., a voltage of 550 V, a frequency of 600 Hz, and a duty cycle of 8% were used during the PEO process.

17.3.1 PEO Coatings Prepared in Electrolyte with Different Additives

Figure 17.2 depicts the XRD patterns of PEO coatings with different additives in Na_2SiO_3 electrolyte, and the corresponding phase composition, thickness, and surface roughness of the coatings are listed in Table 17.2. Noticeably, SiC particles can enter into the discharge channel during the PEO process and participate in the

Table 17.1 Physical and chemical properties of silicon carbide matrix composites

Code	Na_2SiO_3/g·L^{-1}	SiC/g·L^{-1}	NH_4VO_3/g·L^{-1}	Cr_2O_3/g·L^{-1}	SDBS/g·L^{-1}
BS	25	0	0	0	0
S-5SiC		5	0	0	8
S-10SiC		10	0	0	8
S-3V		0	3	0	0
S-6V		0	6	0	0
S-5Cr_2O_3		0	0	5	8
S-10Cr_2O_3		0	0	10	8
S-10Cr_2O_3-6V		0	6	10	8

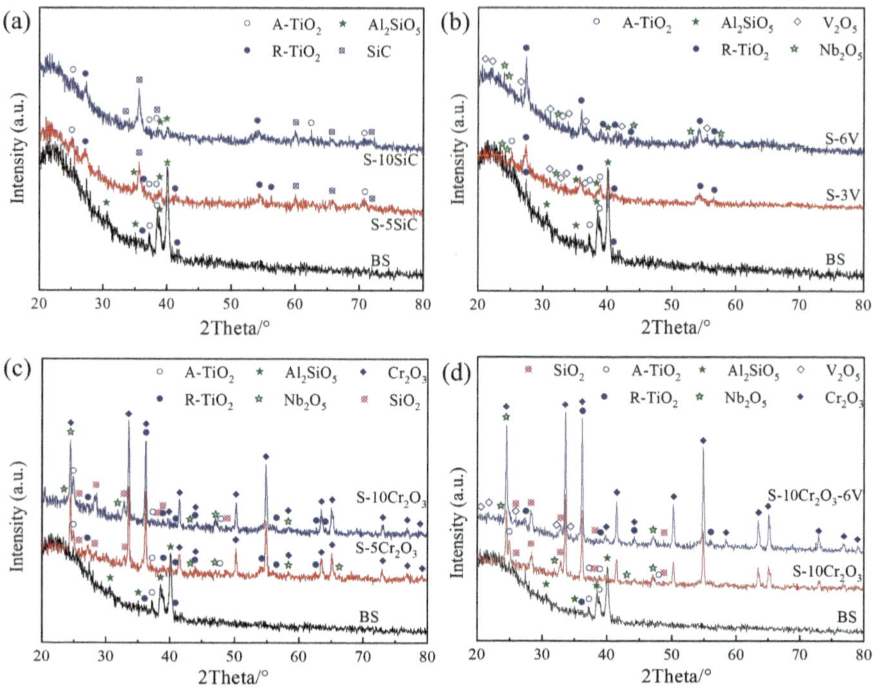

Fig. 17.2 XRD patterns of PEO coatings with the addition of different additives in Na_2SiO_3 electrolytes. (**a**) SiC; (**b**) NH_4VO_3; (**c**) Cr_2O_3; (**d**) Na_2SiO_3-NH_4VO_3-Cr_2O_3

growth of the coatings thus forming a SiC-doped TiO_2 coating. In addition, V_2O_5 phase can be formed after the addition of NH_4VO_3 in the electrolyte, as well as promote the formation of R-TiO_2 high-temperature stabilized phase. With increasing concentration of NH_4VO_3, more R-TiO_2 and V_2O_5 phases are produced while the content of A-TiO_2 and Al_2SiO_5 phases gradually decreases. After the addition of Cr_2O_3 in electrolyte, Cr_2O_3 dominates the coating and Al_2SiO_5/amorphous phase is replaced by the SiO_2 crystalline phase. The simultaneous addition of two additives of NH_4VO_3 and Cr_2O_3, not only introduces the V_2O_5 into the coating, but also further increases the content of the Cr_2O_3 within coatings, promoting the conversion of A-TiO_2 into R-TiO_2. As for the roughness, both the addition of NH_4VO_3 and Cr_2O_3 can increase the roughness of coating surface, and NH_4VO_3-rich coating shows a largest coating thickness and roughness.

17.3.2 Thermal Radiation Performance of Coatings with Different Additive

Figure 17.3 presents the emissivity spectrum of PEO coating prepared in different electrolytes at 600 °C. S-5SiC coating and S-10SiC coating possess a high emissivity in a wide wavelength range of 3~20 μm, which shows an enhanced trend as the

Table 17.2 Phase composition, coating thickness, and surface roughness of PEO coatings with addition of different additives in Na_2SiO_3 electrolyte

Code	Phase composition	The function of additives	Thickness/μm	Roughness/μm
BS	Al_2SiO_5 and a small amount of A-TiO_2 phase, as well as a large amount of amorphous phase	–	34 ± 2	2.21 ± 0.08
S-5SiC S-10SiC	R-TiO_2 and SiC in addition to A-TiO_2, Al_2SiO_5, and amorphous phases	Promoting the transition from nonsteady state A-TiO_2 to thermally stable R-TiO_2	34 ± 5 30 ± 3	5.77 ± 0.41 4.60 ± 0.13
S-3V S-6V	R-TiO_2, Nb_2O_5, V_2O_5, and amorphous SiO_2	NH_4VO_3 not only participates in the growth of PEO coatings to form V_2O_5 but also promotes the formation of the high-temperature stabilizing R-TiO_2	17 ± 2 44 ± 7	3.49 ± 0.30 12.35 ± 1.33
S-5Cr_2O_3 S-10Cr_2O_3	R-TiO_2, A-TiO_2, SiO_2, Nb_2O_5, and Cr_2O_3	Al_2SiO_5 and amorphous phases are replaced by SiO_2 crystalline phase, and Cr_2O_3 dominates the coating	37 ± 5 35 ± 5	10.53 ± 1.34 4.54 ± 0.31
S-10Cr_2O_3-6V	R-TiO_2, SiO_2, Nb_2O_5, Cr_2O_3, and V_2O_5	Introducing V_2O_5 into the coating further increases the content of Cr_2O_3 and promotes the conversion of A-TiO_2 to R-TiO_2	33 ± 4	9.06 ± 20.75

content of SiC increases. Especially in the short wavelength range of 3–9 μm, the emissivity of S-5SiC coating and S-10SiC coating can reach 0.7 and 0.8, respectively, significantly higher than that of BS coating (~0.5). As for the PEO coating prepared in Na_2SiO_3-NH_4VO_3 electrolyte, S-6 V coating possesses the highest emissivity in a range of 3–8 μm, while exhibiting the lowest emissivity in a range of 8–20 μm, and the average emissivity in full band of 3–20 μm is ~0.7. On the contrary, S-3V coating shows a similar law with that of BS coating, i.e., maintaining a constant emissivity value in the wavelength range of 3–8 μm, suddenly increasing from 8 μm to the highest at 10 μm, and then maintaining a high emissivity in the wavelength range of 10–20 μm.

For the PEO coating prepared in Na_2SiO_3-Cr_2O_3 electrolyte, the emissivity within the wavelength range of 3–8 μm is S-5Cr_2O_3 > S-10Cr_2O_3 > BS. In the wavelength range of 10–20 μm, the coating emissivity increases with the addition of more Cr_2O_3, and S-10Cr_2O_3 coating shows the highest surface emissivity. Overall, the introduction of nano-Cr_2O_3 can improve the thermal radiation performance of PEO coating, especially in the long wavelength range. When the Cr_2O_3 and NH_4VO_3 are added to the electrolyte simultaneously, S-10Cr_2O_3-6V coating exhibits high emissivity in the entire wavelength range of 3–20 μm. Among them, S-10Cr_2O_3-6V

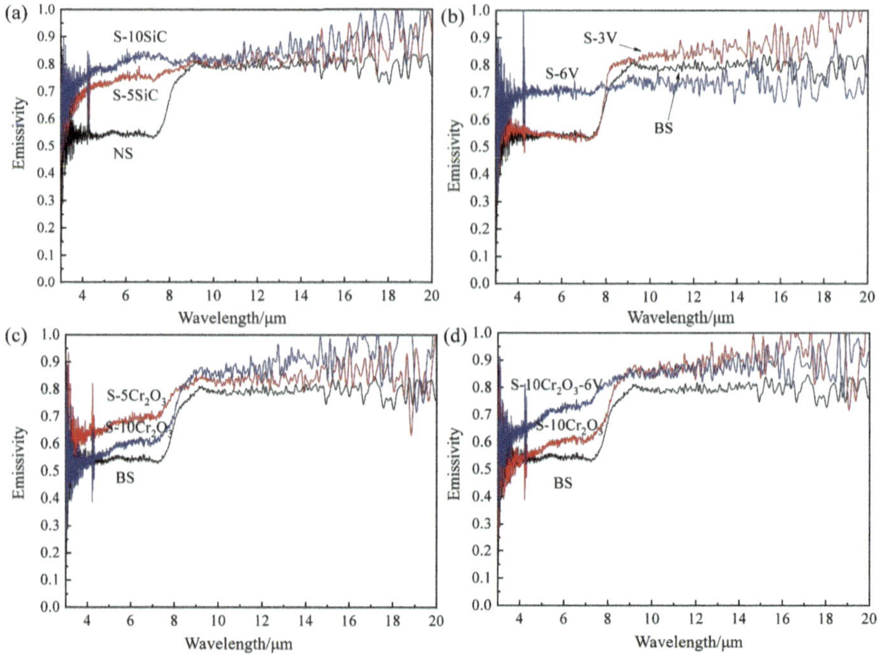

Fig. 17.3 The emissivity spectrum of PEO coating prepared in different electrolytes at 600 °C. (**a**) Na_2SiO_3-SiC, (**b**) Na_2SiO_3-NH_4VO_3, (**c**) Na_2SiO_3-Cr_2O_3, and (**d**) Na_2SiO_3-Cr_2O_3-NH_4VO_3

shows the highest in the wavelength range of 3–10 μm, and the emissivity gradually increases from 0.65 to 0.9 with the increase of wavelength. In the wavelength range of 10–20 μm, both S-10Cr_2O_3-6V and S-10Cr_2O_3 coatings possess a high emissivity that exceeds 0.9. As a result, the simultaneous addition of NH_4VO_3 and Cr_2O_3 not only maintains the high emissivity in the 10–20 μm wavelength range, but also improves the emissivity of PEO coating in the 3–10 μm wavelength range.

17.3.3 Thermal Radiation Mechanism of PEO Coatings on Titanium Alloy

Emissivity presents the ratio of the radiation energy absorbed from an object and a blackbody, which is affected by many factors, such as chemical composition, free carrier concentration, coating thickness, roughness, and microstructure [31–33]. It is well known that a PEO coating with high emissivity can be formed by optimizing the preparation process or introducing additives into the electrolyte. When the radiation energy is projected onto the coating surface and reaches a steady state, a portion is reflected, a portion attenuates due to absorption and backscattering when passing through the material, and the remaining portion passes through the material. Based on the law of energy conservation, the sump of absorption rate, reflectivity, and transmittance is 1. When the coating has sufficient thickness (transmittance is zero), the sump of absorption rate and reflectivity is 1. That is the emissivity of an object is equal to its absorption based on the Kirchhoff's law, i.e., high absorption coating is also known as high emissivity coating [5].

The reasons that high emissivity coating can be obtained by introducing the additive into the electrolyte are ascribed to two aspects: (1) the microstructure of PEO coating can be controlled to acquire high absorption; (2) high emissivity components can be directly introduced and involve the growth of PEO coating. Moreover, a rough and porous surface structure is beneficial for obtaining sufficient scattering, due to the multiple scattering and absorption of electromagnetic waves incident on porous structures, thereby improving the emissivity of coating [34]. For another, the emissivity of ceramic coating is mainly affected by the absorption of free charge carriers and lattice absorption, and doping high emitting phases in composite coatings can enhance the absorption performance of the ceramic coating [6]. Based on above theory analysis, the reasons that composite coating prepared in Na_2SiO_3-SiC electrolyte shows a high emissivity in the entire wavelength range of 3–20 μm are ascribed to the increased surface roughness, the formed porous structure, as well as the introduction of high emissivity SiC. In addition, noting that the PEO coating prepared in Na_2SiO_3-NH_4VO_3 electrolyte has a higher emissivity in the low wavelength range, while PEO coating prepared in Na_2SiO_3-Cr_2O_3 electrolyte has a higher emissivity in the high wavelength range. The possible reason is that the introduction of Cr_2O_3 provide extra absorption in the high wavelength range expect for the rough surface. Therefore, the PEO coating (S-10Cr_2O_3-6V) prepared in Na_2SiO_3-NH_4VO_3-Cr_2O_3 electrolyte shows a high emissivity in the full wavelength range, which combines the advantage both of NH_4VO_3 and Cr_2O_3.

17.4 Design and Preparation of PEO-Modified Thermal Protection Coatings on Niobium Alloy

As one of the superalloys, niobium alloys possess the minimum density and are increasingly being viewed as the promising structural material of rocket engines for high-temperature applications ascribed to a series of excellent performances, such as a high melting point (~ 2630 °C of Nb521), excellent thermal conductivity, and superior mechanical strength at high temperatures [8–10, 35, 36]. Nevertheless, the "pest" catastrophic oxidation of niobium alloy in an oxygen-containing environment at high temperatures (above 600 °C) is a fatal problem that greatly restricts its further application, despite it shows high chemical stability at room temperature [4, 37]. Besides, rocket engines face multiple harsh service environments except for high-temperature oxidation, accelerating the failure of niobium alloy (see Fig. 17.4). For instance, thrust extension section components are usually subjected to hot corrosion due to the deposition of solid fuel particles [39], the surface of the burner will face the large temperature gradient thermal cycle [25], as well as the nozzle will be subjected to strong thermal ablation from the flame flow [5]. In addition, a common dilemma faced by all rocket engine hot end components is the unbearable surface temperature, which applies a huge thermal load to niobium alloy structural material, reducing the strength of niobium alloy [26]. Therefore, high emissivity thermal protection coating plays an increasingly important role in the aerospace vehicles field (the emissivity of niobium alloy is only ~0.2 [40], which can significantly reduce the thermal load of the niobium alloy surface, especially at higher temperatures.

Among various coatings on niobium alloys, $NbSi_2$ coating is considered the most essential and effective coating to prevent further destruction caused by the external environment, since it can form a dense glass protective film layer characterized by a thermodynamically stable, low oxygen diffusion coefficient, and self-sealing [41–43]. Unfortunately, the large growth stress caused by the accumulation of high-temperature products results in cracks within the coating and destroys the integrality of the coating during the cooling process, which implies that the improved high-temperature protective performance of the single $NbSi_2$ coating reaches a bottleneck [44]. Besides, the emissivity of $NbSi_2$ coating is low of ~0.4 [45]. Therefore, the modified $NbSi_2$-based composite coating has attracted numerous attention for improved thermal radiation performance, high-temperature oxidation resistance, hot corrosion resistance, anti-ablation resistance, etc. In this section, the modified $NbSi_2$-based composite coatings were prepared on niobium alloys (Nb521) via halide-activated pack cementation (HAPC) and PEO methods, and the effects of components on the microstructure and high-temperature thermal protection performance were investigated. The electrolyte systems for the prepared high-temperature thermal protection coating on niobium alloys are presented in Table 17.3. During the PEO process, Na_2SiO_3 of 15 g/L, $(NaPO_3)_6$ of 6 g/L, NaOH of 1.2 g/L, a frequency of 600 Hz, and a duty cycle of 10% were fixed.

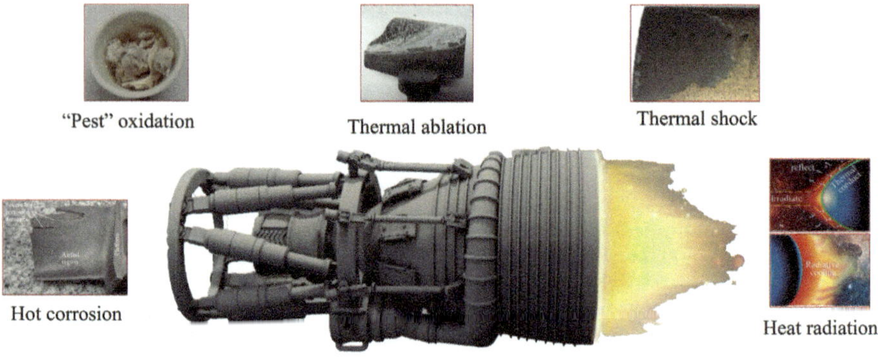

Fig. 17.4 Multiple service environments of thermal protection coating on niobium alloys. Reproduced with permission from [38] © 2024 Elsevier Ltd.

Table 17.3 Electrolyte composition design of thermal protective coating on Nb521 alloys

Coating systems	SiC ($g \cdot L^{-1}$)	$MoSi_2$ ($g \cdot L^{-1}$)	HfC ($g \cdot L^{-1}$)	Yb_2O_3 ($g \cdot L^{-1}$)	voltage/V	Duration/min
$NbSi_2/Nb_2O_5$-SiO_2 (PEO-$NbSi_2$)	0	0	0	0	450	11
$NbSi_2/Nb_2O_5$-SiO_2/SiC (PEO-SiC)	30	0	0	0	600	11
$NbSi_2/Nb_2O_5$-$SiO_2/MoSi_2$ (PEO-$MoSi_2$)	0	30	0	0	600	11
$NbSi_2/Nb_2O_5$-SiO_2/HfC (PEO-HfC)	0	0	30	0	600	11
$NbSi_2/Nb_2O_5$-SiO_2/ HfC-$MoSi_2$-Yb_2O_3 (PEO- HfC-$MoSi_2$-Yb_2O_3)	0	10	15	5	600	11

17.4.1 Phase Composition and Microstructures of Coatings Prepared in Different Electrolyte

Figure 17.5 depicts the XRD patterns of $NbSi_2$ coating and different PEO coatings on niobium alloys. During the PEO process, the $NbSi_2$ coating acts as the anode, and the in suit oxidation of $NbSi_2$ is achieved, forming the Nb_2O_5 and SiO_2 (see Fig. 17.5b). Moreover, SiO_2 phase in the PEO coating is obtained through two aspects: in situ oxidation of $NbSi_2$ and the high-temperature sintering of SiO_3^{2-} in the electrolyte, and it is mainly an amorphous glass phase confirmed by the halo peak in XRD pattern. With the addition of nanoparticles (SiC, $MoSi_2$, HfC, or Yb_2O_3), the obvious peaks of added nanoparticles are detected compared with the $NbSi_2/Nb_2O_5$-SiO_2 coating (PEO-$NbSi_2$), implying that these nanoparticles are involved in the growth of PEO coating. Besides, $NbSi_2/Nb_2O_5$-SiO_2/-SiC (PEO-SiC) and $NbSi_2/Nb_2O_5$-$SiO_2/MoSi_2$ (PEO-$MoSi_2$) coatings show an increased halo peak, and $NbSi_2/Nb_2O_5$-SiO_2/HfC (PEO-HfC) and $NbSi_2/Nb_2O_5$-SiO_2/HfC-$MoSi_2$-Yb_2O_3 (PEO-HfC-$MoSi_2$-Yb_2O_3) show a significant peak of HfO_2, indicating that the nanoparticles are easy to be oxidized due to the high applied voltage. Therefore, functional additives can be effectively deposited and sintered on $NbSi_2$ coating surface assisted by PEO technique with particle addition, which can achieve the construction of multilayer composite coating with tailored properties.

The surface and cross-sectional morphologies of $NbSi_2$ coating and different PEO coatings are shown in Fig. 17.6. The $NbSi_2$ coating surface is gray without obvious defects such as pores and micro-cracks, and it consists of many fine grains with an average size of 3–5 μm. Moreover, the $NbSi_2$ coating exhibits a structure with a double layer; the outer layer is $NbSi_2$ layer with a thickness of ~100 μm and the inner layer is a transition layer with a thickness of ~3–5 μm. After the PEO process in the electrolyte without the addition of nanoparticles, the typical volcanic-like surface with micropores is obtained, and a dense $Nb_2O_5 + SiO_2$ film layer with a thickness of 12~15 μm is produced on the $NbSi_2$ layer, resulting in a double-layer structure. With the addition of different nanoparticles, PEO coatings show clearly

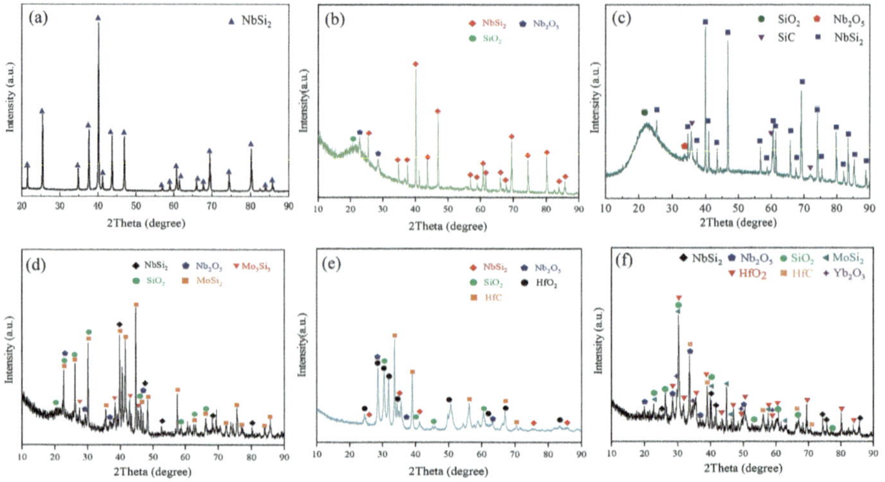

Fig. 17.5 XRD patterns of $NbSi_2$ coating and PEO coatings on niobium prepared in different electrolytes. (**a**) $NbSi_2$ coating; (**b**) $NbSi_2/Nb_2O_5$-SiO_2; (**c**) $NbSi_2/Nb_2O_5$-SiO_2/SiC; (**d**) $NbSi_2/Nb_2O_5$-$SiO_2/MoSi_2$; (**e**) $NbSi_2/Nb_2O_5$-SiO_2/HfC; (**f**) $NbSi_2/Nb_2O_5$-SiO_2/HfC-$MoSi_2$-Yb_2O_3. Reproduced (**a**), (**b**) and (**e**) with permission from [46] © 2024 Elsevier Ltd; (**c**) from [40] © 2025 Published by Elsevier Ltd on behalf of the editorial office of Journal of Materials Science & Technology; (**f**) from [47] © 2024 Elsevier B.V.

distinguished morphologies compared with $NbSi_2$ coating and Nb_2O_5-SiO_2 coating. As shown in Fig. 17.6e, f, PEO-SiC coating presents a gray intrinsic color of SiC nanoparticle, and the multilayer coating surface is completely filled with a large number of SiC nanoparticles, accompanied by a small number of holes. PEO-SiC coating possesses a multilayer structure with a ~100 μm $NbSi_2$ inner layer and ~80 μm PEO-SiC ceramic outer layer. Likewise, PEO-$MoSi_2$, PEO-HfC, and PEO-HfC-$MoSi_2$-Yb_2O_3 show similar morphologies, as well as the large thickness of the ceramic outer layer. Besides, it can be seen that the porosity of the PEO coating with the addition of SiC and $MoSi_2$ is lower than that of the PEO coating with the HfC addition. The reason is that SiC and $MoSi_2$ can provide more Si sources to form the extra SiO_2 glass during the PEO process, filling the portion of discharge channels. Therefore, different nanoparticles can be incorporated into the substrate coating by a special PEO method, forming a multilayer ceramic protective coating on Nb alloys, which is a promising strategy for designing and preparing the thermal protection performance with tailored properties.

17.4.2 High-Temperature Oxidation Behavior

Silicide coatings have proven to be promising for improving the high-temperature oxidation resistance of niobium alloy. However, the long-term protective property of single silicide coating remains a long-term endeavor due to the deficiency of

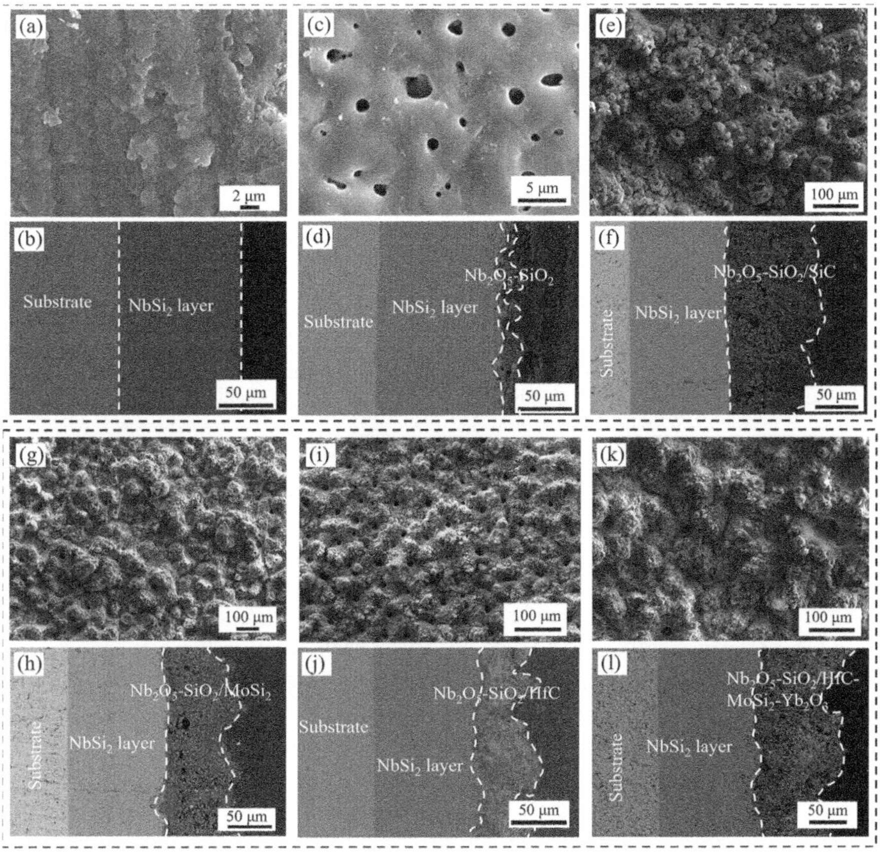

Fig. 17.6 Surface and cross-sectional morphologies of $NbSi_2$ coating and different PEO coatings. (**a, b**) $NbSi_2$ coating; (**c, d**) $NbSi_2/Nb_2O_5$-SiO_2; (**e, f**) $NbSi_2/Nb_2O_5$-SiO_2/SiC; (**g, h**) $NbSi_2/Nb_2O_5$-$SiO_2/MoSi_2$; (**i, j**) $NbSi_2/Nb_2O_5$-SiO_2/HfC; (**k, l**) $NbSi_2/Nb_2O_5$-SiO_2/HfC-$MoSi_2$-Yb_2O_3. Reproduced (**a, b**), (**c, b**) and (**i, j**) with permission from [46] © 2024 Elsevier Ltd.

oxygen-consuming phases, as well as the self-healing ability of the protective layer. Herein, a silicide-based composite coating is constructed on niobium alloy by incorporation of nano-SiC particles for enhancing the high-temperature oxidation resistance. Isothermal oxidation test for 1–50 h in the air at 1250 °C is conducted to accurately evaluate the high-temperature oxidation behavior of our designed multilayer coating, as shown in Fig. 17.7. For comparison, the oxidation results of the single $NbSi_2$ coated sample and $NbSi_2/Nb_2O_5$-SiO_2-coated sample at the same condition were provided as references. After oxidation, the mass gains of the $NbSi_2$ coated sample and the $NbSi_2/Nb_2O_5$-SiO_2-coated sample are 6.49 mg/cm^2 and 4.90 mg/cm^2 exposed to air for 50 h, respectively, and the oxide layer on the surface and edge is severely peeled off. On the contrary, the $NbSi_2/Nb_2O_5$-SiO_2/SiC-coated sample shows the least mass gain (2.49 mg/cm^2) with slightly damaged coating, even when the oxidation time extended to 50 h, exhibiting the best oxidation

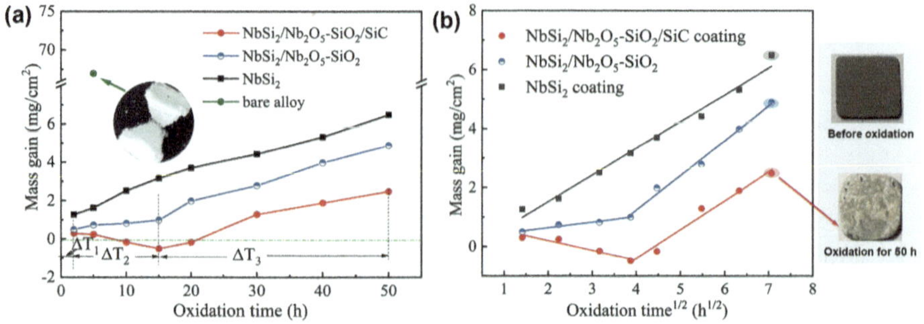

Fig. 17.7 Isothermal oxidation kinetic curves of $NbSi_2$ substrate coating, $NbSi_2/Nb_2O_5$-SiO_2 and $NbSi_2/Nb_2O_5$-SiO_2/SiC at 1250 °C in the air (**a**) and its corresponding fitting curve (**b**). Reproduced with permission from [40] © 2025 Published by Elsevier Ltd on behalf of the editorial office of Journal of Materials Science & Technology

resistance performance compared to that of $NbSi_2$-coated and $NbSi_2/Nb_2O_5$-SiO_2-coated samples.

As for $NbSi_2$-coated sample, the fitting curve shows a single linear relationship with a slope of 0.906, implying the continuous consumption of $NbSi_2$ layer. $NbSi_2/Nb_2O_5$-SiO_2-coated sample shows a similar oxidation behavior compared with that of $NbSi_2$. However, due to the existence of the pre-oxidation layer, it shows a segmented linear oxidation behavior, where the growth rate is 0.189 mg·cm^{-2} h$^{-1/2}$ during the first stage (1~15 h) and 1.179 mg·cm^{-2} h$^{-1/2}$ during the second stage (15~50 h). The $NbSi_2/Nb_2O_5$-SiO_2/SiC-coated sample shows a behavior of oxidation rate decreasing first and then increasing. Notably, $NbSi_2/Nb_2O_5$-SiO_2/SiC coating shows a negative gain with increasing oxidation time to 2~15 h. It can be ascribed to the domination by the volatilization of SiO_2 phase in this stage. The growth rate can be expressed as $h_1 = -0.334t^{1/2} + 0.879$ (1–15 h), $h_2 = 0.977t^{1/2} - 4.308$ (15–50 h). Meanwhile, no peeling and failure of the multilayer coating were observed. Summarily, $NbSi_2/Nb_2O_5$-SiO_2/SiC coating exhibits the best oxidation protective performance than those of substrate coating and $NbSi_2/Nb_2O_5$-SiO_2 coating at 1250 °C in the air, which is mainly ascribed to the continually supplemental SiO_2 protective layer.

Figure 17.8a describes the surface morphology after the oxidation time of 2 h, and the oxide layer surface exhibits a porous structure with uniform and fine pore sizes. Moreover, there exist many sub-micron-sized particles on the surface, which are SiO_2 phase identified by the XRD spectrum. As the oxidation time prolongs, SiO_2 particles are agglomerated and sintered, forming a micron-scale island structure (Fig. 17.8b). Meanwhile, the pore size of the surface micropores increases with the oxidation time and gradually transforms into a network structure (Fig. 17.8d). It can be ascribed to the sintering behavior at high temperatures, i.e., diffusing atoms always tend to migrate from small grains to large grains with lower surface energy. At the later stage of oxidation, the columnar Nb_2O_5 grains grow on the surface (Fig. 17.8e). It is the result that Nb atoms are diffused into the oxide layer surface

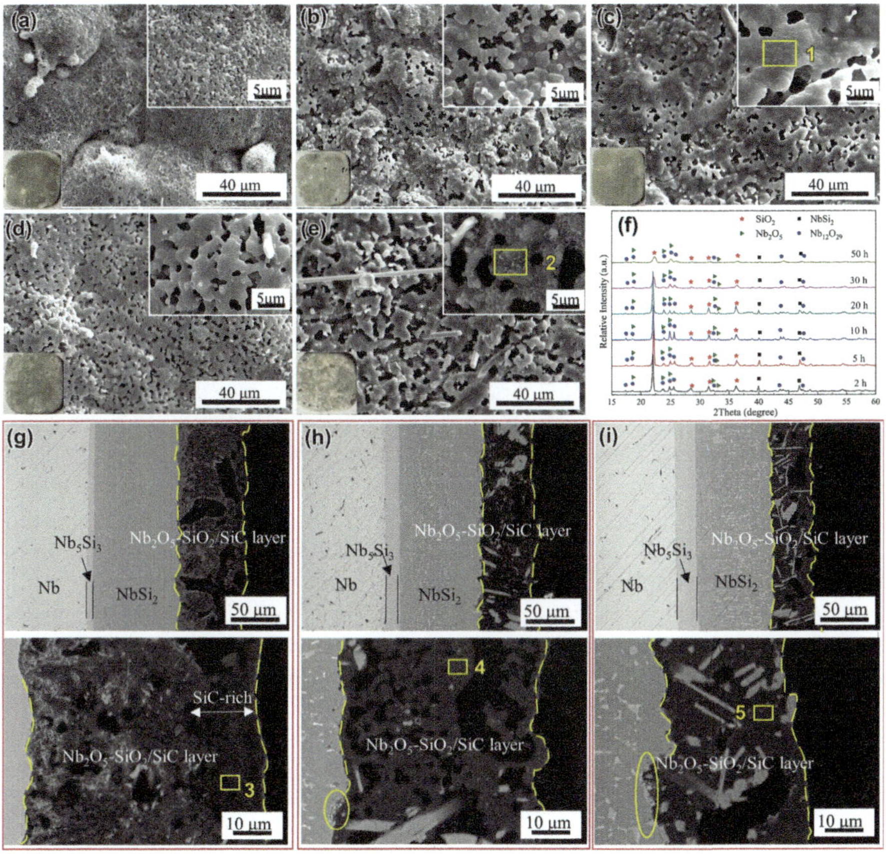

Fig. 17.8 $NbSi_2/Nb_2O_5$-SiO_2/SiC oxidized at 1250 °C for different times. Surface morphologies of 2 h (**a**), 5 h (**b**), 10 h (**c**), 20 h (**d**), and 50 h (**e**); XRD patterns (**f**); cross-sectional morphologies of 2 h (**g**), 20 h (**h**), and 50 h (**i**). Reproduced with permission from [40] © 2025 Published by Elsevier Ltd on behalf of the editorial office of Journal of Materials Science & Technology

and lead to Nb_2O_5 being produced in the external environment. These columnar Nb_2O_5 grains contact directly with the external environment and provide more oxygen diffusion channels due to their high oxygen diffusion coefficients, accelerating the coating's failure.

It can be seen from Fig. 17.8g that the cross-sectional morphologies show that an oxide layer with large voids is formed by transforming from Nb_2O_5-SiO_2/SiC ceramic layer. The SiC-rich outer ceramic layer and the dispersedly distributed SiC nanoparticles are oxidized, effectively reducing the inward oxygen flux by the formation of a protective oxygen diffusion barrier layer. In addition, some white columnar crystals (see Fig. 17.8i) are observed from the inner oxide layer, which is inferred to be niobium oxides ($Nb_2O_5/Nb_{12}O_{29}$). In substance, there are few effects on the protective properties of the oxide film caused by Nb_2O_5 grain growth, even the purity of SiO_2 in the oxide layer can be improved and is conducive to improving the SiO_2 mobility, resulting in the oxide layer with an enhanced self-healing property. As the oxidation procedure prolongs to 50 h (Fig. 17.8i), the oxide layer shows a decreased thickness of ~39 μm, while the diameter of the internal hole decreases and the compactness of the outer layer increases. Moreover, Fig. 17.8i shows that Nb_2O_5 columnar crystals can be up to 10–30 μm in length and are mainly distributed in the interior of the oxide layer, partial them grow to the surface coating. Notably, there exists some Nb_5Si_3 + SiO_2 mixing zone at the interface between $NbSi_2$ and oxide layer, and an interesting phenomenon that the thickness of the $NbSi_2$ layer with a little change after high-temperature oxidation is observed, as shown in Fig. 17.8h, i. It could be ascribed to the oxidation behavior of $NbSi_2$ under the condition of lower oxygen partial pressure.

In summary, Nb_2O_5-SiO_2/SiC ceramic layer is able to reduce the oxidative loss of $NbSi_2$ substrate coating, significantly improving the service life of the constructed $NbSi_2$/Nb_2O_5-SiO_2/SiC. On the one hand, SiC acts as the oxygen-consuming phase and reduces the huge amounts of the inward oxygen flux, and a dense SiO_2 self-healing film layer is formed with the extra Si source provided by SiC. On the other hand, the continuous decomposition of SiO_2 phase keeps the oxide layer's thickness in a stable range, avoiding the oxide layer's flaking and cracking caused by the large growth stress. As well as the high oxidation resistance Nb_5Si_3 transition layer grows, free of significant thickness decreased of $NbSi_2$ layer, which improves the reliability of the high-temperature protection of the coating. Our novel strategy sheds light on a prominent pathway of designing a large-thickness SiO_2 self-healing film for advanced thermal protective coating, which is beneficial for the service of silicide coating in more complex thermal environments.

17.4.3 Hot Corrosion Behavior

For the hot-end components of shipboard aircraft engines and gas turbines, molten salt mixtures containing Na, S, and Cl are usually deposited on the surface of the component due to the influence of the marine environment and fuel impurities, accelerating corrosion damage to the material. The accelerated oxidation caused by molten salt is called hot corrosion, and its harm far exceeds that of high-temperature oxidation. Herein, a novel HfC-$MoSi_2$-Yb_2O_3 ceramic outer layer has been fabricated on the $NbSi_2$ layer surface via PEO method with the addition of multiple nanoparticles, endowing the modified silicide-based multilayer composite coating with excellent hot corrosion resistance.

The hot corrosion behavior of $NbSi_2$/Nb_2O_5-SiO_2/HfC-$MoSi_2$-Yb_2O_3 (PEO-HfC-$MoSi_2$-Yb_2O_3) coating is investigated in a corrosive medium of 75%Na_2SO_4 + 25%NaCl. Figure 17.9 describes the corrosion kinetics curve and macroscopic morphologies of different coatings after hot corrosion at 900 °C in the air. Notably, the single $NbSi_2$ coating is partially peeling due to the accelerated corrosion by corrosive salt after 100 h and complete failure after 150 h of corrosion

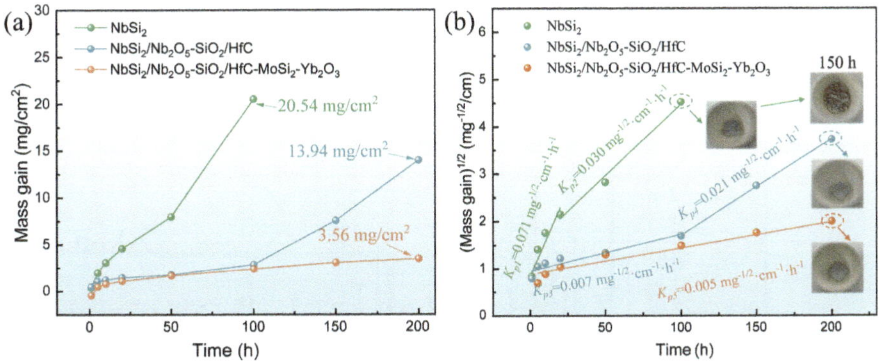

Fig. 17.9 Corrosion kinetics analysis of $NbSi_2/Nb_2O_5$-SiO_2/HfC-$MoSi_2$-Yb_2O_3 coating after hot corrosion at 900 °C in the air. (**a**) Curves of the mass gain vs. the hot corrosion time, (**b**) linear fit when (mass gain)1/2 (mg1/2/cm) is plotted vs. *t* (h). Reproduced with permission from [38] © *2024 Elsevier Ltd*

time, leading to the catastrophic destruction of the substrate. On the contrary, the coatings with the addition of HfC or HfC, $MoSi_2$, and Yb_2O_3 nanoparticles retain the intact morphologies after the corrosion time of 200 h, implying an improved hot corrosion resistance. The corrosion gain of a single $NbSi_2$ coating is 20.54 mg·cm^{-2} after 100 h, while the corrosion gain decreases to 13.94 mg·cm^{-2} with the addition of the HfC nanoparticle, even if the corrosion time extends to 200 h. As for the PEO-HfC-$MoSi_2$-Yb_2O_3 coating, the corrosion gain further decreases and is only 3.56 mg·cm^{-2} after 200 h, showing the best hot corrosion resistance. Moreover, the corrosion rate curves depict that the sample coated with PEO-HfC-$MoSi_2$-Yb_2O_3 coating always possesses the lowest corrosion rate during the whole hot corrosion process. Both the samples coated with $NbSi_2$ coating and $NbSi_2/Nb_2O_5$-SiO_2/HfC (PEO-HfC) coating show the segmented corrosion rate, while PEO-HfC-$MoSi_2$-Yb_2O_3 coating maintains a dynamic balance during the whole hot corrosion process. Undoubtedly, the addition of HfC, $MoSi_2$, and Yb_2O_3 nanoparticles is beneficial to improve the hot corrosion resistance of $NbSi_2$-based coating under the corrosive molten salt.

As demonstrated in Fig. 17.10, the corrosion products of PEO-HfC-$MoSi_2$-Yb_2O_3 coating are complex, mainly including SiO_2, $NaNbO_3$, $Na_2Nb_4O_{11}$, $Na_2Nb_8O_{21}$, $Na_2Si_2O_5$, Yb_2SiO_5, $NaHfSiO_5$, and Nb_2O_5, as well as the oxidation products of HfO_2 and $HfSiO_4$. In the initial stage of molten salt corrosion, Na_2SO_4 decomposes into Na_2O and SO_x, and Cl_2 is produced from the reaction between NaCl salt and SO_x byproduct. After the volatilization of Cl_2 and SO_x, an alkaline corrosion environment is formed at the corrosion salt/coating interface, dominated by the residual Na_2O. Subsequently, the reactions between Na_2O and the produced oxides are achieved, leading to the degradation and failure of coatings. During the hot corrosion process, HfC and $MoSi_2$ phases are consumed continuously with increasing corrosion time, and HfC phase is completely transformed into HfO_2 after 5 h of corrosion, while the peak of $MoSi_2$ phase cannot be detected after 50 h of corrosion.

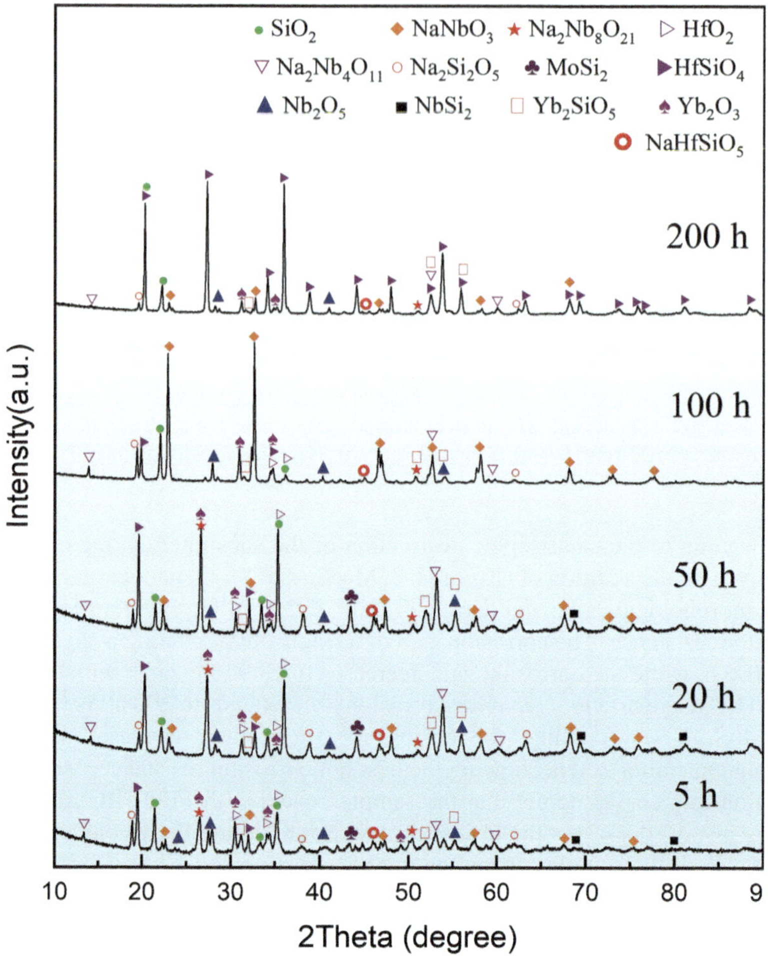

Fig. 17.10 XRD patterns of $NbSi_2/Nb_2O_5$-SiO_2/HfC-$MoSi_2$-Yb_2O_3 coating subjected to varying corrosion times. Reproduced with permission from [38] © 2024 Elsevier Ltd.

Moreover, the intensity of the $NaNbO_3$ diffraction peaks increased slightly with the corrosion time. After the corrosion time of 200 h, the intensity of the $NaNbO_3$ and Nb_2O_5 diffraction peaks increased and the intensity of SiO_2 decreased. Besides, the peak intensity of the $HfSiO_4$, Yb_2SiO_5 byproduct reaches a maximum after 200 h of corrosion, since the oxidation of the coating dominated the later hot corrosion process. The formation of Hf-Yb-Si-O film layer with a low oxygen diffusion coefficient significantly enhances the resistance to corrosive medium and oxygen.

To accurately analyze the hot corrosion protection mechanism of PEO-HfC-$MoSi_2$-Yb_2O_3 coating at 900 °C, macro-microstructure, and thickness evolution after varying corrosion times are necessary. Figure 17.11 illustrates the macroscopic and microscopic morphology evolution of the PEO-HfC-$MoSi_2$-Yb_2O_3 coating subjected to corrosion for 50 h, 100 h, and 200 h. As shown in Fig. 17.11a1, b1, and c1, all the coating surfaces show intact corrosion surfaces without obvious damage such as cracks and spalling. It is worth noting that the coating surface subjected to corrosion for 50 h and 100 h still remains a thin corrosive medium film covering the center coating, which is different from the coating surface after 200 h corrosion, implying the corrosive medium is not enough in the later stages of the corrosion. As shown in Fig. 17.11a2, the typical porous characteristic of PEO coating disappears and the coating surface is covered by the glass film layer, since the $MoSi_2$ provides a sufficient Si source in the early stage of corrosion. When the corrosion time extends to 100 h, the coating surface is dominated by the agglomerated $NaNbO_3$ and Nb_2O_5 particles and the coating porosity increases (Fig. 17.11b2), since the $MoSi_2$ is depleted after 50 h corrosion, and the produced SO_x/Cl_2 break the integrity of oxide film. After the corrosion of 200 h (Fig. 17.11c2), the corroded surface exhibits a loose stacked structure and the coating porosity further increases, which provides more reaction channels to the $NbSi_2$ inner layer for the corrosion media and oxygen. Moreover, the hot corrosion process during the later stage is dominated by the oxidation reaction of $NbSi_2$ due to the pores/cracks caused by molten salt corrosion as well as the insufficient corrosive medium of Na_2SO_4/NaCl.

From the cross-sectional morphologies after 50 h, 100 h, and 200 h corrosion (Fig. 17.11a3, b3, and c3), the corroded layer thickness increases from 47 μm (50 h) to 92 μm (200 h), while the residual decreases from 116 μm (50 h) to 98 μm (200 h),

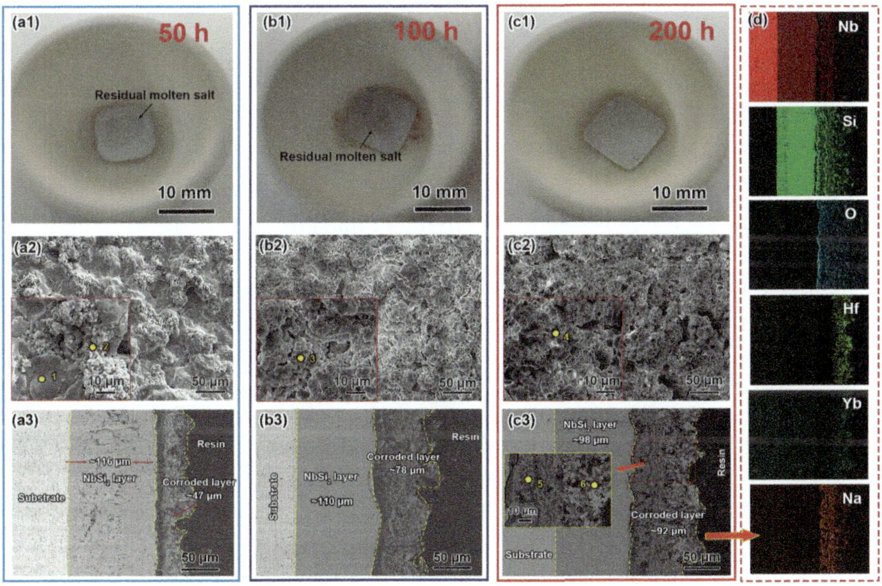

Fig. 17.11 Morphologies of $NbSi_2$/Nb_2O_5-SiO_2/HfC-$MoSi_2$-Yb_2O_3 coating after the corrosion times at 50 h, 100 h, and 200 h. (**a1**, **b1**, **c1**) macro morphologies; (**a2**, **b2**, **c2**) surface SEM images; (**a3**, **b3**, **c3**) cross-section SEM images; (**d**) the EDS mappings of the (**c3**). Reproduced with permission from [38] © 2024 Elsevier Ltd.

implying that PEO-HfC-$MoSi_2$-Yb_2O_3 coating still remains the protection capability for isolating the corrosive medium/oxygen. After a corrosion time of 50 h, the cross-sectional morphology shows a compact corroded layer due to the produced SiO_2 glass phase filling in the corroded pores and cracks. With the hot corrosion time extended to 200 h, visible corroded pores are observed except for the thickened corroded layer. Meanwhile, the EDS mappings (Fig. 17.11d) of the corroded layer demonstrate that the corroded layer can be divided into the inner oxidation layer including Nb_2O_5, SiO_2, and $NaNbO_3$, as well as the outer corroded skeleton layer mainly composed of the Hf-Yb-rich corroded products. Undoubtedly, the formation of the special corroded structure has proven our speculation that the later hot corrosion process is dominated by the oxidation of $NbSi_2$, due to the insufficient corrosive medium and source of SiO_2. As a result, the corroded layer is transformed into the visible dual-layer structure with increasing corrosion time (Fig. 17.11c3), including an Nb_2O_5-SiO_2 oxidation inner layer and the $HfSiO_4$-Yb_2SiO_5-SiO_2 corroded skeleton layer filled with the growing $NaNbO_3$. In summary, the PEO-HfC-$MoSi_2$-Yb_2O_3 coating shows the lowest corrosion mass gain compared with the single $NbSi_2$ coating and still possesses ~98 μm of the residual thickness of the $NbSi_2$ layer. The enhanced hot corrosion resistance provided by the added nanoparticles is mainly attributed to these two aspects. On the one hand, a dense barrier layer is formed during the early hot corrosion process and significantly reduces the inward oxygen/corrosive medium flux. On the other hand, the Hf-Yb-rich corroded skeleton layer with high hot corrosion stability is formed and makes the corroded layer maintain an intact structure even when the corrosion time extends to 200 h, greatly increasing the service life of the coating.

17.4.4 Ablation Behavior

When a high Mach number vehicle is flying in the atmosphere, the ultra-high temperature ablation and high-speed scouring environment caused by the aerodynamic heating put more demanding requirements on the engine hot end component materials. In this section, we demonstrate an $NbSi_2$/Nb_2O_5-SiO_2/HfC multilayer protective coating via PEO method, which is expected to have excellent anti-ablation performance. On the one hand, HfC acted as an oxygen-consuming phase which was introduced into multilayer coating to mitigate the generation of Nb_2O_5. On the other hand, the oxidation products of HfC eventually formed the $HfSiO_4$-SiO_2 skeleton with a low oxygen diffusion can improve the stability of the oxide scale and inhabit the inward diffusion of oxygen, retarding the growth of niobium oxide and the crack initiation. Therefore, the multilayer thermal protective coating is promising to serve in hot-end components in important industries such as rockets, spacecraft, missiles, and high-speed aircraft.

Oxyacetylene ablation (oxygen flow rate of 15.0 L/min and acetylene flow rate of 8.5 L/min to ensure a heat flow density of 1.5 MW/m^2) is employed to simulate the operating environment of vehicles and evaluate the high temperature ablation

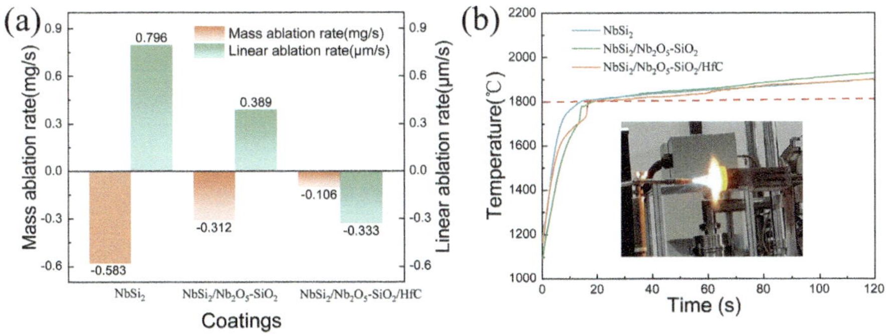

Fig. 17.12 (**a**) The mass ablation rate and linear ablation rate of $NbSi_2$, $NbSi_2/Nb_2O_5$-SiO_2, and $NbSi_2/Nb_2O_5$-SiO_2/HfC coatings after oxyacetylene ablation for 120 s; (**b**) the surface temperature curves of the coatings during ablation. Reproduced with permission from [46] © 2024 Elsevier Ltd.

resistance of $NbSi_2/Nb_2O_5$-SiO_2 coating. As shown in Fig. 17.12a, PEO-HfC coating shows the lowest mass ablation rate of −0.106 mg/s and a linear ablation rate of −0.333 μm/s, while the mass/line ablation rate of the $NbSi_2$ coating and $NbSi_2/Nb_2O_5$-SiO_2 coating are (−0.583 mg/s, 0.796 μm/s) and (−0.312 mg/s, 0.389 μm/s), respectively. Meanwhile, the surface temperature curves of the coatings during ablation depict that all three coatings possess a high surface temperature above 1800 °C with increasing the ablation time to 20 s (see Fig. 17.12b). Thereby, all three coatings show a negative mass ablation rate, due to the high-temperature oxidation reactions of the coatings. Among them, PEO-HfC coating possesses the lowest mass ablation rate, implying an excellent ablation resistance.

The macroscopic morphologies, as shown in Fig. 17.13a, demonstrate that PEO-HfC coating shows an intact macroscopic morphology after ablation, without obvious blisters, cracks, and flakes. Based on the EDS results (Fig. 17.13f) and the XRD pattern (Fig. 17.13c) of the surface coating after ablation, confirming that it contains mainly HfO_2, SiO_2, Nb_2O_5, and $HfSiO_4$. It can be verified that the oxidation of HfC dominates the short-time ablation process, as well as the formation of a $HfSiO_4$ new phase. Due to the deposition of the oxygen-consuming HfC phase on the surface coating, the inward oxygen flux is reduced significantly, thereby resulting in the lowest mass ablation rate. The cross-sectional morphology (Fig. 17.13d) of the PEO-HfC coating after ablation also depicts that an oxide scale is produced as similar to the oxidation test sample. However, this scale maintains a porous nature of $NbSi_2/Nb_2O_5$-SiO_2/HfC coating after a short time ablation, even when the scale thickness is above 50 μm, which corresponds to the negative line ablation rate (Fig. 17.12a). According to the EDS points/maps results (Fig. 17.13f, g), it can be seen that Hf anchors the whole scale, thereby resulting in a Hf-rich barrier layer. It should be mentioned that $NbSi_2$ coating and $NbSi_2/Nb_2O_5$-SiO_2 show a positive linear ablation rate after the ablation, which is ascribed to the violent flushing of the oxyacetylene flame. Conversely, PEO-HfC coating with the negative line ablation rate indicates that the produced scale possesses high stability to resist the

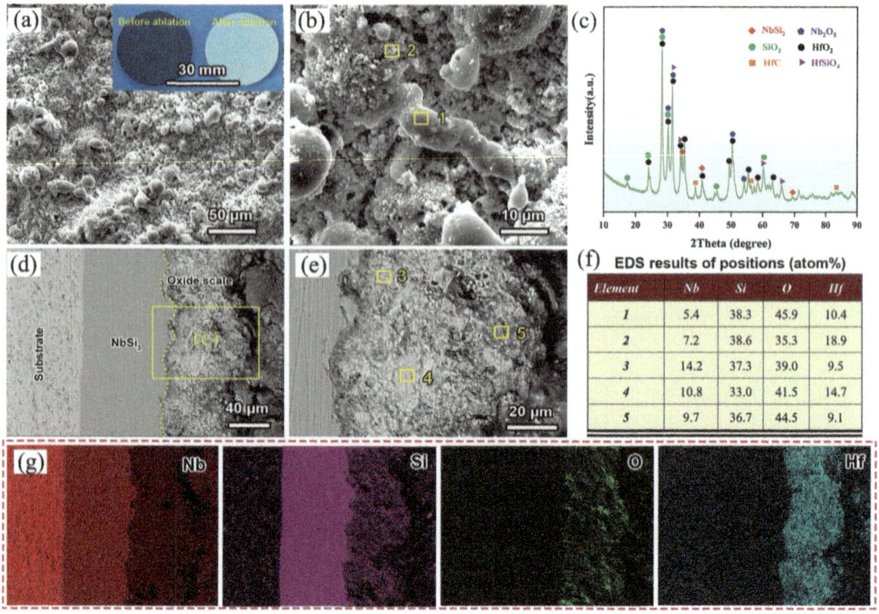

Element	Nb	Si	O	Hf
1	5.4	38.3	45.9	10.4
2	7.2	38.6	35.3	18.9
3	14.2	37.3	39.0	9.5
4	10.8	33.0	41.5	14.7
5	9.7	36.7	44.5	9.1

Fig. 17.13 The morphology and XRD patterns of $NbSi_2/Nb_2O_5$-SiO_2/HfC coating after oxyacetylene ablation for 120 s. (**a**) and (**b**) surface morphologies; (**c**) XRD pattern; (**d**) and (**e**) cross-sectional morphologies; (**f**) and (**g**) corresponding EDS results. Reproduced (**a**–**g**) with permission from [46] © 2024 Elsevier Ltd.

high-speed airflow erosion of the oxyacetylene flame. Summarily, PEO-HfC coating exhibits an enhanced oxyacetylene ablation resistance compared with $NbSi_2$ coating and $NbSi_2/Nb_2O_5$-SiO_2, which is attributed to the formed Hf-rich scale with high stability, thereby isolating the $NbSi_2$ and oxyacetylene flame flow.

17.4.5 *High Emissivity*

Passive radiative heat dissipation with the characteristic of free additional energy consumption, noise-free, energy conservation, and environmental protection, etc., is a promising thermal protective method. Moreover, passive radiative heat dissipation dominates the heat exchange in high-temperature environments, especially in rapidly rising temperature enclosed environments. In the current niobium alloys/$NbSi_2$ coating system, the room temperature emissivity of the bare niobium alloys and $NbSi_2$ coating is ~0.21 and ~0.48 (see Fig. 17.14) sequentially across the band range of 3~20 μm, which is seriously insufficient for heat dissipation via passive radiative. Therefore, developing a high emissivity coating on niobium alloy surfaces undoubtedly can radiate the heat flux from the surface of hot end components to the external cold environment, resulting in a significant passive cooling at the substrate alloy/

coating interface, thereby breaking through the operation temperature limit of the niobium alloys.

PEO treatment is a promising surface modification method, which can improve the emissivity, especially in wavelength 8–20 μm, due to the treated coating having a loose porous surface. As shown in Fig. 17.14a, the emissivity in wavelength 3–20 μm can be enhanced to 0.80 (PEO-$NbSi_2$) through the PEO treatment, and the emissivity in wavelength 8–20 μm even reaches 0.95. To further improve the infrared thermal radiation property of niobium alloys, abundant high emissivity components such as $MoSi_2$, SiC, and HfC are incorporated into the substrate coating via a PEO method. It predicts that the emissivity of the multilayer coating will further enhance under the combined action of surface structure and phase constituent. As shown in Fig. 17.14b, the emissivity increases significantly as expected, especially in wavelength 3–8 μm. The PEO coatings with the addition of nanoparticles exhibit enhanced emissivity above 0.9 across the infrared wavelength (3–20 μm), which makes up for the insufficient emissivity of PEO-$NbSi_2$ coating in wavelength 3–8 μm.

In substance, the emissivity of the material is dominated by the intrinsic absorption [48] and the surface morphology [49]. Besides the lattice vibration absorption

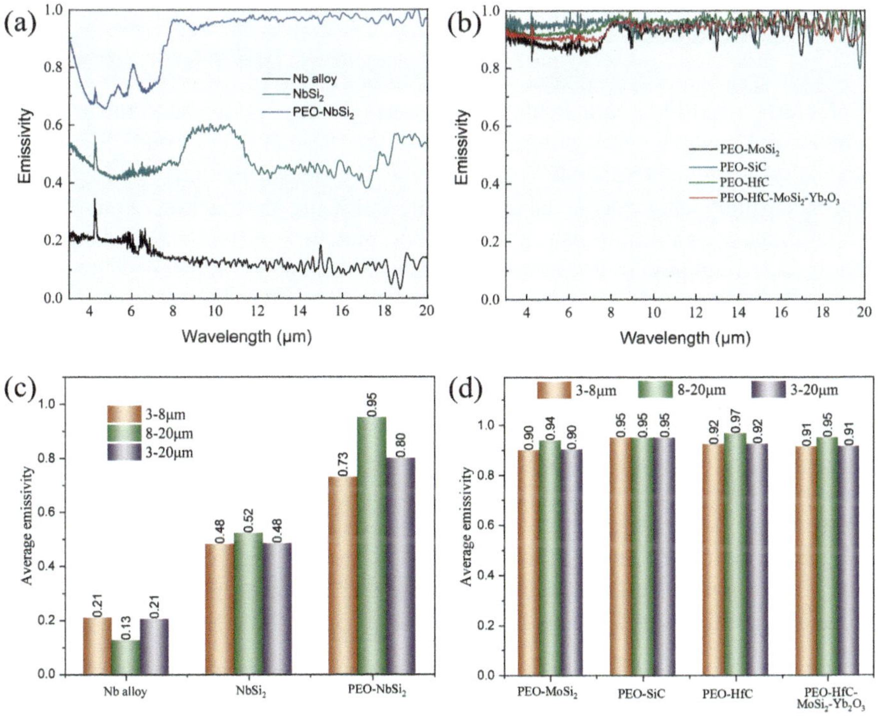

Fig. 17.14 The emissivity of various coatings. (**a**, **b**) spectral emissivity and (**c**, **d**) average emissivity at different wavelength ranges. Reproduced (**a**, **c**) with permission from [38] © 2024 Elsevier Ltd.

matching caused by the addition of nanoparticles, lattice distortion is also an important factor that can improve the absorption and radiation capabilities. It is noted that lattice distortion could reduce the degree of intrinsic lattice symmetry and enhance the crystal vibrational activity, thereby enhancing the non-simple harmonic effects of lattice polar vibrations [50]. During the PEO process, there exists a large degree of lattice distortion caused by the fast cooling of the molten oxide, as well as the production of amorphous regions within the coating, enhancing the absorptive capacity of PEO coating. For another, PEO method endows the coating to a porous surface structure with huge amounts of bulges, which significantly increases the BET surface area of coating surface. Thus, the probability of multiple reflections and absorptions of infrared waves increases, enhancing the thermal radiation performance of coatings [51]. To sum up, vibration-matching absorption of multiple chemical bonds, lattice distortion, and surface roughness are the main control factors of the high emissivity PEO coating. The above evidence has proven that our design and prepared composite is an expected high emissivity thermal protection coating on niobium alloys, which provides the material foundation for radiation heat dissipation multifunctional integrated thermal protection coating.

17.5 Conclusions and Outlook

High emissivity PEO coatings play an increasingly important role in the aerospace vehicles field, which can significantly reduce the thermal load of the alloy surface, especially at higher temperatures. Until now, the design of high emissivity PEO coatings usually can be achieved through the following approaches: (1) introducing the multiphase and multipolar chemical bond matching; (2) adjusting the porosity and surface roughness of PEO coatings; and (3) introducing high emissivity components. The PEO coating on titanium alloys can achieve a high emissivity in the entire wavelength range of 3–20 μm while using the Na_2SiO_3-SiC electrolyte. In addition, the hot-end components of hypersonic vehicles usually face multiple extreme work environment challenges. Achieving the high emissivity and multiple protection characteristics of thermal protection coatings on niobium alloys simultaneously is the future research objective, and the corresponding coating preparation method should attract our attention. The PEO method provides a new path for preparing the modified composite coating via the composite processes with other technologies, which can achieve tailored properties and has been shown to broaden the tunability of multifunctional characteristics in thermal protection coatings.

Future research should focus on the following aspects:

1. Current researches on PEO thermal protection coatings mainly focus on the regulation of low thermal conductivity and high emissivity, and the intrinsic properties of coatings such as bonding strength and thermal stability are often ignored. Therefore, while improving thermal radiation performance is still crucial, it is

equally important to conduct systematic and comprehensive research on the overall performance of PEO coating.
2. The characteristic of PEO coating is a loose and porous structure, which is unfavorable for high-temperature thermal protection performance. Therefore, reducing the porosity and pore size of the coating by process control, or forming the composite coatings with dense layers, can further improve the thermal protection performance of PEO coatings.
3. The preparation of PEO coatings with large thickness and high bond strength is difficult, which poses more stringent requirements for long-term high-temperature thermal protection performance. Therefore, combining PEO method with other technologies to prepare large-thickness composite coatings is the future development trend to further improve the thermal protection performance of metal surfaces.
4. Current research on the failure behavior of PEO thermal protection coatings in different service environments is fragmented. Therefore, investigating the coupled failure mechanism of PEO coatings under different harsh service environments is an important basis for guiding the design of high-performance thermal protection coatings.

Summarily, the PEO method is a promising way to prepare multifunctional thermal protection coating on metal surfaces due to its strong designability, despite the thermal protection performance of single PEO coatings reaching a bottleneck. In the future, integrating other advanced technologies with the PEO method will likely further improve thermal protection performance and fully unblock the potential of high-performance metals (such as titanium alloy, niobium alloy, and aluminum alloy) in the aerospace field.

Acknowledgments The partial support from the NSFC Grant nos. 523B2010, U21B2053, 52071114, 52301084, and 52001100, China Postdoctoral Science Foundation (2023M730843), Heilongjiang Touyan Team Program, Heilongjiang Postdoctoral Science Foundation (No. AUGA4110004823), and Research start-up Fund by HIT are gratefully acknowledged.

References

1. Li, Z. B., & Peng, L. M. (2007). *Acta Materialia, 55*, 6573–6585.
2. Su, L., Jia, L., Jiang, K., & Zhang, H. (2017). *International Journal of Refractory Metals and Hard Materials, 69*, 131–137.
3. Li, L., & Guo, X. (2023). *Corrosion Science, 220*, 111293.
4. Błachnio, J., Spychała, J., & Zasada, D. (2021). *Engineering Failure Analysis, 127*, 105554.
5. Zhang, G., Jiang, S., Zhang, H., Yao, W., & Liu, C. (2016). *RSC Advances, 6*, 61686–61694.
6. Wang, S., Wang, Y., Zhang, H., Zou, Y., Chen, G., Ouyang, J., Jia, D., & Zhou, Y. (2022). *Chemical Engineering Journal, 427*, 131701.
7. Yao, W., Li, X., Wei, Q., Fu, M., Wang, B., & Pan, S. (2024). *Journal of Alloys and Compounds, 983*, 173707.
8. Wang, Q., Wang, Q., Chen, R., Wang, X., Yongchun, Z., Su, Y., & Fu, H. (2024). *Composites Part B: Engineering, 283*, 111634.

9. Li, Y., Lin, X., Hu, Y., Fan, W., Gao, X., Yu, J., Dong, H., & Huang, W. (2024). *Composites. Part B, Engineering, 278*, 111427.
10. Zhang, W., Qiao, Y., & Guo, X. (2022). *Corrosion Science, 199*, 110201.
11. Apelfeld, A. V., Borisov, A. M., Krit, B. L., Ludin, V. B., Polyansky, M. N., Romanovsky, E. A., Savushkina, S. V., Suminov, I. V., Tkachenko, N. V., Vinogradov, A. V., & Vostrikov, V. G. (2015). *Surface and Coating Technology, 269*, 279–285.
12. Shen, D. J., Wang, Y. L., Nash, P., & Xing, G. Z. (2007). *Journal of Materials Processing Technology, 205*, 477–481.
13. Du, W., Zhang, S., Luo, X., Tao, X., Fang, C., & Yao, Z. (2017). *Journal of Alloys and Compounds, 729*, 970–977.
14. Ge, Y., Wang, Y., Chen, J., Zou, Y., Guo, L., Ouyang, J., Jia, D., & Zhou, Y. (2018). *Journal of Alloys and Compounds, 745*, 271–281.
15. Kang, S., Tu, W., Han, J., Li, Z., & Cheng, Y. (2018). *Surface and Coating Technology, 358*, 879–890.
16. Xie, T., Jiang, C., Ning, B., Qian, W., Shi, L., & Chen, Y. (2024). *Surface and Coating Technology, 489*, 131161.
17. López-Ortega, A., Arana, J. L., Rodríguez, E., & Bayón, R. (2018). *Corrosion Science, 143*, 258–280.
18. Wang, L., Li, J., Li, K., Wang, Y., & Ma, C. (2023). *Chinese Journal of Aeronautics, 37*, 471–481.
19. Ni, D. W., Cheng, Y., Zhang, J., Liu, J., Zou, J., Chen, B., Wu, H., Li, H., Dong, S., Han, J., Zhang, X., Fu, Q., & Zhang, G. (2021). *Journal of Advanced Ceramics, 11*, 1–56.
20. Sun, J., Liu, S. B., Li, W., Yu, H. J., Jiang, S. M., Gong, J., & Sun, C. (2019). *Corrosion Science, 149*, 207–217.
21. Tsakiropoulos, P. (2022). *Progress in Materials Science, 123*, 100714.
22. Ma, R., & Guo, X. (2023). *Journal of Materials Science and Technology, 132*, 27–41.
23. Kaseem, M., Fatimah, S., Nashrah, N., & Ko, Y. G. (2020). *Progress in Materials Science, 117*, 100735.
24. Chen, G., Wang, Y., Zou, Y., Jia, D., & Zhou, Y. (2019). *Chemical Engineering Journal, 374*, 231–241.
25. Lee, J. S., Kim, T., Lee, H. H., & Kim, Y. H. (2018). *Journal of Materials Chemistry C, 6*, 2515–2521.
26. Yu, W., Duan, Z., Zhang, G., Liu, C., & Fan, S. (2018). *Nano Letters, 18*, 1770–1776.
27. Padture, N. P. (2016). *Nature Materials, 15*, 804–809.
28. Squire, T. H., & Marschall, J. (2010). *Journal of the European Ceramic Society, 30*, 2239–2251.
29. Mungiguerra, S., Cecere, A., Savino, R., Saraga, F., Monteverde, F., & Sciti, D. (2021). *Corrosion Science, 178*, 109067.
30. Xiao, Y. Q., Liu, Z. Y., Zhu, W., & Peng, X. M. (2021). *Surface and Coating Technology, 423*, 12757.
31. Zou, Y., Wang, Y., Wei, D., Ge, Y., Ouyang, J., Jia, D., & Zhou, Y. (2019). *ACS Applied Electronic Materials, 1*, 1527–1537.
32. Hsu, P. C., Song, A. Y., Catrysse, P. B., Liu, C., Peng, Y., Xie, J., Fan, S., & Cui, Y. (2016). *Science, 353*, 1019–1023.
33. Xia, Z., Fang, Z., Zhang, Z., Shi, K., & Meng, Z. (2020). *ACS Applied Electronic Materials, 12*, 27241–27248.
34. Li, Z., Zhang, X., Yan, L., Du, H., Guo, A., Mu, Y., & Liu, J. (2024). *Ceramics International, 50*, 28857–28866.
35. Li, B. Q., Beyerlein, I. J., Shinzato, S. H., Ogata, S., & Han, W. Z. (2024). *Journal of Materials Science and Technology, 203*, 167–179.
36. Wang, Q., Wang, Q., Chen, R., Wang, X., Su, Y., & Fu, H. (2024). *Corrosion Science, 227*, 11170.
37. Zhou, X., Xu, J., Zha, Y., Zhang, Y., Deng, G., Zhao, X., Cai, Z., Liu, S., & Xiao, L. (2024). *Journal of the European Ceramic Society, 44*, 4425–4436.

38. Ye, Z., Wang, S., Zou, Y., Chen, G., Yu, S., Wen, L., Zhao, L., Zhang, G., Wang, Y., Jia, D., & Zhou, Y. (2024). *Composites. Part B, Engineering, 288*, 111905.
39. Wang, Z., Wang, S., Wang, Y., Zou, Y., Chen, G., Wen, L., Zhang, G., Zhao, L., Ouyang, J., Jia, D., & Zhou, Y. (2024). *Ceramics International, 50*, 10497–10514.
40. Wang, S., Ye, Z., Ge, Y., Zou, Y., Zhang, T., Zhao, X., Wang, M., Song, C., Wang, Y., & Zhou, Y. (2025). *Journal of Materials Science and Technology, 210*, 159–169.
41. Ji, X., Chen, Y., Yao, L., Zhang, Y., Ren, X., Wang, P., Kiryukhantsev-Korneev, P. V., Levashov, E. A., Shi, J., Kang, X., Zhang, B., Zhang, P., & Feng, P. (2024). *Corrosion Science, 233*, 112070.
42. Yoon, J. K., & Kim, G. H. (2018). *Corrosion Science, 141*, 97–108.
43. Kuang, J., Zhang, P., Wang, Q., Hu, Z., Liang, X., & Shen, B. (2022). *Corrosion Science, 198*, 110134.
44. Wang, C., Shao, W., Wang, W., & Zhou, C. (2020). *Corrosion Science, 163*, 108249.
45. Li, L., Guo, X., & Qiao, Y. (2023). *Journal of Alloys and Compounds, 938*, 168456.
46. Ye, Z., Zou, Y., Wang, S., Chen, G., Wang, Z., Wen, L., Zhang, G., Zhao, L., Wang, Y., & Zhou, Y. (2024). *Corrosion Science, 230*, 111908.
47. Ye, Z., Wang, S., Zou, Y., Chen, G., Zhao, X., Wen, L., Zhang, G., Zhao, L., Wang, Y., & Zhou, Y. (2024). *Chemical Engineering Journal, 496*, 153556.
48. Zhu, M., Song, Y., Liu, Z., Xu, D., Dong, K., & Han, E. H. (2022). *Surface and Coating Technology, 446*, 128797.
49. Yao, R., Yao, Z., Li, Y., Zhang, P., Lu, S., & Wu, X. (2023). *Surface and Coating Technology, 454*, 129152.
50. Park, K., Kim, H., Kim, D. H., & Gwon, S. Y. (2023). *Ceramics International, 49*, 15176–15182.
51. Chen, G., Fu, H., Zou, Y., Wang, S., Gao, Y., Yue, T., Cao, J., Wang, Y., Qiu, J., Zhao, J., Ouyang, J., Jia, D., Shuai, Y., & Zhou, Y. (2023). *Advanced Functional Materials, 33*, 2305650.

Chapter 18
Advances in Thermal Control Plasma Electrolytic Oxidation Coatings for Lightweight Materials

Guoliang Chen and Yaming Wang

18.1 Introduction

Lightweight alloys, including aluminum (Al), titanium (Ti), and magnesium (Mg) alloys, are increasingly integral to industries that require materials with high performance and low weight, such as aerospace, automotive, and energy sectors. The low density of these alloys plays a critical role in reducing energy consumption, which is essential for improving fuel efficiency and minimizing emissions. Consequently, lightweight alloys are particularly well suited for applications that necessitate effective thermal management, especially in high-power electronics and advanced aerospace systems subjected to extreme thermal conditions.

Despite their advantages in weight and strength, lightweight alloys often suffer from limited thermal stability and poor resistance to both high and low temperatures. These limitations restrict their use in applications, such as aerospace and automotive components, which must withstand extreme thermal conditions—ranging from intense heat to severe cold—while maintaining mechanical integrity. Therefore, there is a critical need for thermal control coatings that can enhance the thermal properties of these alloys, enabling them to perform reliably under both high and low temperature extremes.

Plasma electrolytic oxidation (PEO), also known as micro-arc oxidation (MAO), has emerged as an effective approach to address these challenges. This technique produces ceramic oxide coatings on the surface of light alloys, significantly enhancing their thermal protection and stability. The high-energy plasma discharges generated during the PEO process lead to the formation of dense, hard ceramic layers that

G. Chen · Y. Wang (✉)
Harbin Institute of Technology, Harbin, China
e-mail: chenguoliang@hit.edu.cn; wangyaming@hit.edu.cn

V. S. Saji, M. Zheludkevich (eds.), *Plasma Electrolytic Oxidation – Fundamentals, Advances and Applications*,
https://doi.org/10.1007/978-3-031-92429-3_18

not only provide protection to the underlying metal but also improve thermal management performance. PEO coatings can be tailored for specific thermal control needs, offering a versatile approach to extending the functional lifespan of light alloys in demanding thermal environments [1, 2].

Recent advances in PEO thermal control technology have centered on optimizing process parameters to tailor coatings for specific thermal control requirements. Researchers have explored the incorporation of nanoparticles, rare earth elements, and composite materials to enhance the thermal insulation, conductivity, and overall performance of PEO coatings [2–5]. Additionally, efforts have been made to improve coating uniformity, adhesion, and mechanical properties to broaden the applicability of light alloys in extreme environments [6].

This chapter presents a comprehensive review of recent advancements in PEO coatings for lightweight materials, with particular emphasis on their thermal control properties. It explores the fundamental mechanisms underlying the PEO process, highlights significant innovations in material modification and coating design, and examines emerging trends in the field. The aim is to provide a thorough understanding of how PEO coatings enhance the thermal performance and durability of lightweight alloys, thereby facilitating their application across a range of industries, including aerospace, automotive, and energy storage.

18.2 Theory and Classification of PEO TCCs

18.2.1 Theory of TCCs

Thermal control coatings (TCCs) are designed to achieve effective thermal management of objects, particularly in space or high-performance applications. Their primary function is to regulate the absorption and emission of thermal energy, ensuring precise temperature control under varying environmental conditions.

The two critical properties of TCCs are emissivity (ε) and absorptivity (α). Absorptivity is the fraction of incident energy absorbed by a surface, while emissivity refers to the efficiency with which a surface emits thermal radiation relative to a blackbody at the same temperature. These properties determine the thermal balance of the surface.

In space, an object achieves thermal equilibrium when the energy absorbed from solar radiation is balanced by the energy emitted as infrared thermal radiation. This relationship is governed by the energy balance equation:

$$Q_{\text{in}} = Q_{\text{out}} \tag{18.1}$$

where Q_{in} represents the incoming solar energy (I_{sol}) absorbed by the surface, given by:

$$Q_{\text{in}} = \alpha_{\text{s}} \cdot I_{\text{sol}} \cdot A \tag{18.2}$$

Here, α_s is the solar absorptivity, and A is the surface area. The outgoing energy via emitting thermal radiation is expressed using the Stefan-Boltzmann law:

$$Q_{\text{out}} = \varepsilon \cdot \sigma \cdot A \cdot T^4 \tag{18.3}$$

where ε is the emissivity of the surface, σ is the Stefan-Boltzmann constant (5.67×10^{-8} W/$m^2 \cdot$ K^4), and T is the surface temperature.

At equilibrium, the object reaches a temperature where the absorbed solar energy is balanced by the emitted thermal radiation, as shown in Eq. (18.1). Rearranging this equation yields the equilibrium temperature:

$$T_{\text{eq}} = \left(\frac{\alpha_s \cdot I_{\text{sol}}}{\varepsilon \cdot \sigma} \right)^{1/4} \tag{18.4}$$

This equation defines the equilibrium temperature as a function of the surface's solar absorptivity (α_s), emissivity (ε), the incident solar radiation (I_{sol}), and the Stefan-Boltzmann constant (σ). Since I_{sol} and σ are constants, the temperature of a spacecraft can be controlled by adjusting the surface's radiative properties, specifically the absorption-emission ratio (α_s/ε).

The solar absorptivity (α_s) of thermal control coatings can be measured using a UV-VIS-NIR spectrophotometer, which operates over the wavelength range of 300–2500 nm, encompassing the full solar radiation spectrum. Thus, α_s is derived from the reflectance spectrum ($R(\lambda)$) and the solar spectral intensity ($I_{\text{sol}}(\lambda)$) using the following formula [7]:

$$\alpha_s = \frac{\int_{300\,\text{nm}}^{2500\,\text{nm}} I_{\text{sol}}(\lambda)\left(1 - R(\lambda)\right) d\lambda}{\int_{300\,\text{nm}}^{2500\,\text{nm}} I_{\text{sol}}(\lambda)\, d\lambda} \tag{18.5}$$

where $I_{\text{sol}}(\lambda)$ and $R(\lambda)$ are the intensities of the incident and reflection spectrum at a given wavelength (λ).

The infrared emissivity (ε) of TCCs is determined using a Fourier transform infrared (FT-IR) spectrometer over the wavelength range of 2.5–25 μm. The ε is calculated as [8]:

$$\varepsilon = \frac{\int_{2.5\,\mu m}^{25\,\mu m} \rho(\lambda)\left(1 - R(\lambda)\right) d\lambda}{\int_{2.5\,\mu m}^{25\,\mu m} \rho(\lambda)\, d\lambda} \tag{18.6}$$

where $\rho(\lambda)$ represents the blackbody spectral radiance and $R(\lambda)$ is the reflectance spectrum of the coating.

According to the absorption-emission ratio, thermal control coatings can be broadly categorized into two main types:

1. **Infrared-band thermal control coatings** characterized by a high absorption-emission ratio. These coatings exhibit strong solar absorption and high emissiv-

ity in the infrared spectrum, meaning they possess high absorption across the 0.25–25 μm spectrum.
2. **Solar diffuse reflection thermal control coatings** defined by a low absorption-emission ratio. These coatings demonstrate low absorption within the 0.25–2.5 μm range while maintaining high emissivity in the infrared band (2.5–25 μm).

Historically, radiative heat dissipation technologies were primarily applied in high-temperature environments and often overlooked in power devices. However, for high-power electronic devices, significant temperature rise occurs, particularly in enclosed environments with limited convection, making radiative heat dissipation essential. The emissivity value is a critical parameter that quantifies a material's thermal radiation performance.

Key factors influencing emissivity include the intrinsic properties of the material (such as phase composition, texture orientation, and doping), coating thickness, and surface structure. Typically, the surface emissivity of a metallic heat sink is around 0.1–0.2; however, increasing the emissivity to 0.8–0.9 can enhance radiative heat dissipation by nearly a factor of nine. This results in significantly improved thermal management. Optimizing the heat dissipation structure and designing radiation cooling coatings for devices like Light Emitting Diodes (LEDs) and Insulated Gate Bipolar Transistors (IGBTs) are effective strategies to address the thermal challenges of high-power electronic devices.

18.2.2 Design Principles of PEO TCCs

The PEO technique offers a highly effective method for fabricating thermal control coatings on lightweight materials such as aluminum (Al), magnesium (Mg), titanium (Ti), and their respective alloys. These materials are widely used in aerospace, satellite, and high-power electronic applications due to their advantageous mechanical and thermal properties. However, the efficiency of thermal control for these materials in extreme environments, such as outer space, is often constrained by the need to manage heat transfer through radiation and absorption, especially when exposed to solar radiation. To address these challenges, the design of PEO TCCs must be meticulously designed to achieve specific absorption-emission characteristics, tailored to the operational requirements of the system.

1. **Design of high absorption-emission ratio coatings:** High solar absorptivity and high infrared emissivity thermal control coatings are essential for applications requiring efficient heat absorption from solar radiation and rapid heat dissipation through radiation, such as in spacecraft, high-power electronic devices, and systems operating in extreme thermal environments. The foundational PEO electrolyte system, typically composed of silicate, phosphate, or aluminate chemistries, serves as a base matrix for functionalization. Ion doping, using species such as tungstate (WO_4^{2-}), vanadate (VO_3^-), chromium (Cr^{3+}), molybdate

(MoO_4^{2-}), ferrous (Fe^{2+}), copper (Cu^{2+}), and manganese (Mn^+), introduces chemical interactions within the PEO layer, forming complex bonds that enhance solar absorptivity and emissivity across a broad wavelength range.

In addition to ion doping, particle doping offers another route to improve the absorption and emission properties of the coating. Particles such as copper oxide (CuO), carbon nanotubes (CNT), silicon carbide (SiC), iron oxide (Fe_3O_4), and carbon black are incorporated into the electrolyte to form composite phases within the coating. A combination of these dopants can be strategically used to maximize absorption in the solar spectrum (0.25–2.5 μm) and increase emissivity in the infrared range (2.5–25 μm), thereby producing a coating with an optimized high solar absorption-emission ratio. These modifications can be precisely tuned to meet the thermal regulation needs of satellites, spacecraft, and other equipment that operate in direct exposure to solar radiation.

2. **Design of Low Absorption-Emission Ratio Coatings:** In scenarios requiring minimal solar absorption, such as spacecraft exposed to direct sunlight or systems designed to maintain lower temperatures, coatings with low solar absorption-emission ratios are critical. Similar to coatings with high α_s/ε, the PEO process starts with a silicate, phosphate, or aluminate base electrolyte but employs different dopants to achieve low absorption properties. Ion doping with zinc (Zn^{2+}), titanium (Ti^{4+}), and zirconium (Zr^{2+}) can help produce coating phases that exhibit low solar absorption while maintaining high infrared emissivity.

   Furthermore, particle doping with materials such as zinc oxide (ZnO), aluminum oxide (Al_2O_3), yttrium oxide (Y_2O_3), zirconium oxide (ZrO_2), boron nitride (BN), titanium dioxide (TiO_2), and polytetrafluoroethylene (PTFE) enhances the coating's capacity to reflect incoming solar radiation, while still emitting heat effectively. These dopants contribute to forming a thermally stable coating with low solar absorption and efficient radiative cooling in the infrared range, making them ideal for thermal control applications.
3. **Optimization of Process Parameters:** The performance of PEO coatings is heavily influenced by precise control of processing parameters, including electrolyte conductivity, applied current density, and treatment time. Adjusting these parameters can modify the coating's thickness (h), surface roughness (Ra), and quality. For instance, increasing current density and treatment duration enhances coating thickness and surface roughness, improving absorption in specific wavelength ranges for high-emissivity applications. These optimizations not only improve radiative heat dissipation but also ensure long-term durability in harsh conditions such as the vacuum of space or high-temperature environments.
4. **Multi-step composite design:** To further enhance the thermal control properties, multi-step composite approach can be employed. Combining PEO with additional surface modification techniques, such as dip coating, magnetron sputtering, sol-gel processes, or vapor deposition. These post-PEO treatments allow for the creation of finely controlled microstructures on the coating surface, which can be designed to match specific wavelengths of the incident solar spectrum. The introduction of multi-component composite materials in these secondary

layers further improves the thermal control properties of the coating by enhancing its ability to reflect or emit heat rays.

For example, combining PEO with magnetron sputtering produces highly stable coatings with superior wear and oxidation resistance, while sol-gel processes facilitate the introduction of nano-sized particles for enhanced spectral performance. These multilayered composite coatings provide advanced thermal regulation capabilities, tailored to meet the rigorous demands of aerospace and high-power electronic applications.

In conclusion, the PEO technique, combined with precise material doping and multistep composite designs, enables the development of advanced thermal control coatings with high or low solar absorption-emission ratios. These coatings play a pivotal role in improving thermal management for spacecraft, satellites, and high-power electronics, ensuring operational stability and efficiency in extreme environments.

18.3 Design and Preparation of PEO TCCs

18.3.1 PEO TCCs on Aluminum Alloy

18.3.1.1 High Absorption-Emission Ratio Coating

The electrolyte systems used for plasma PEO thermal control coatings on aluminum alloys primarily consist of silicates, phosphates, and aluminates [6, 9]. To produce coatings with high solar absorptivity and infrared emissivity, ion doping is utilized to incorporate species such as WO_4^{2-}, VO_3^- Cr^{3+}, MoO_4^{2-}, Fe^{2+}, Cu^{2+}, and Mn^+ into the electrolyte [4, 10–13]. Additionally, particle doping with materials such as CuO, CNT, SiC, Fe_3O_4, and carbon black is employed to further enhance the thermal control properties of the coatings [14, 15]. Table 18.1 summarizes the compositions of the electrolyte, electrical parameters (current density, oxidation time, frequency, and duty), and the thermal control performance of PEO coatings with high absorptance-emittance ratios applied to aluminum alloy surfaces.

1. **Matrix Composition**: The 2024 aluminum alloy contains transition metal elements such as Fe, Cu, and Mn. During the PEO process, these elements form their respective oxides in situ via spark discharges, which are subsequently incorporated into the coating. These oxides act as black colorants within the alumina ceramic matrix, enhancing visible light absorption and imparting a black appearance to the coating. The PEO process employed a 0.1 mol/L sodium aluminate electrolyte with a frequency of 50 Hz, a current density of 8 A/dm^2, and an oxidation duration of 180 min [16].
2. **Ion Doping Modification**: PEO coatings were initially prepared using a base electrolyte of 3 g/L Na_2SiO_4 and 6 g/L KOH, with a current density of 15 A/dm^2, a frequency of 1000 Hz, and a duty cycle of 20%, referred to as the B coating. Further modification was performed by adding 4 g/L ammonium molybdate,

Table 18.1 Electrolyte system, process parameters, and thermal control performance of PEO coatings with high absorption-emission ratio on aluminum alloy surfaces

Electrolyte	Additives	Electrical parameters				h, μm	α_s	ε
		I, A/dm^2	t, min	f, Hz	D, %			
0.1 mol/L $NaAlO_2$	–	8	180	50	45	101	>0.9	>0.77
16 g/L $NaAlO_2$	1 g/L Na_2MoO_4	15	120	100	45	–	0.91	0.8
10 g/L KH_2PO_4; 3 g/L $K_4[Fe(CN)_6]$	5 g/L Na_2WO_4	10	20	500	45	77	0.93	0.87
3 g/L Na_2SiO_4; 6 g/L KOH	–	15	10	1000	20	20	0.53	0.7
	4 g/L $(NH_4)_6Mo_7O_{24}$					28	0.84	0.73
	4 g/L Na_2WO_4					26	0.75	0.76
	4 g/L K_2TiF_6					31	0.9	0.79
35 g/L $(NaPO_3)_6$; 1.5 g/L NaOH	$Sm_{0.5}Sr_{0.5}CoO_3$	1.6	15	500	–	75/11	0.89	0.9
0.08 mol/L Na_2SiO_3	50 mL/L CNT	10	15	–	–	10	–	0.89

sodium tungstate, and potassium titanium fluoride into the electrolyte, producing the BM, BW, and BT coatings, respectively. The results indicated that the B coating exhibited an absorptivity of 0.53 and an emissivity of 0.7; the BM coating had an absorptivity of 0.84 and emissivity of 0.73; the BW coating had an absorptivity of 0.75 and emissivity of 0.76; and the BT coating displayed an absorptivity of 0.9 and an emissivity of 0.79. The enhanced absorptivity of the BM, BW, and BT coatings is attributed to the presence of Mo, W, and Ti elements, which absorb visible light and increase the porosity of the coatings [17].

3. **Particle Doping Modification:** The introduction of $Sm_{0.5}Sr_{0.5}CoO_3$ microparticles into an electrolyte containing 35 g/L $(NaPO_3)_6$ and 1.5 g/L NaOH electrolyte, under a current density of 1.6 A/dm^2, a frequency of 500 Hz, a pulse width of 60 μs, and an oxidation time of 15 min, resulted in a PEO coating with a high absorptivity of 0.89 and an emissivity of 0.9 [18]. In another study, a thermal control coating with a thickness of approximately 10 μm was fabricated by adding 50 mL/L of 1 wt.% carbon nanotubes suspension to a 0.08 mol/L sodium silicate electrolyte. Under a current density of 100 mA/cm^2 and a treatment time of 900 s, the coating exhibited an emissivity of 0.89 in the infrared range from 5 μm to 20 μm, indicating that carbon nanotube doping effectively enhances the heat dissipation capabilities of the ceramic coating.

18.3.1.2 Low Absorption-Emission Ratio Coating

A series of low absorption-emission ratio PEO coating was developed on the surface of aluminum alloy [19]. The base electrolyte consisted of phosphates, aluminates, and silicates. Ion doping was carried out using Zn^{2+}, Ti^{4+}, and Zr^{2+}, etc. Besides, particle doping was also employed with materials such as ZnO, Al_2O_3, Y_2O_3, ZrO_2, BN, TiO_2, and PTFE to enhance the coating's thermal control properties.

Here, a phosphate-based electrolyte was formulated with 30 g/L $Na_5P_3O_{10}$, 1 g/L Na_2EDTA, 5 g/L KOH, 0.5 g/L NaF, and applied under a voltage of 500 V, a duty cycle of 15%, and a frequency of 450 Hz for PEO treatment. During the process, ZnO, TiO_2, and ZrO_2 were synthesized in situ within the coating by introducing Zn^{2+}, Ti^{4+}, and Zr^{2+} ions. The effect of ion doping on the coating's microstructure and thermal control performance were investigated.

Furthermore, ZnO, TiO_2, and ZrO_2 nanoparticles were directly added to the electrolyte, and the influence of particle type and concentration on the coating's structural and thermal properties was systematically studied. The particle doping type and concentration were optimized to achieve desired thermal control properties. A multiparticle co-doping strategy was also explored to enhance the coating's thickness, reduce interface absorption, improve light scattering, and minimize light absorption. This approach further optimized the thermal control performance of PEO coatings for advanced applications.

The selection of nanoparticles is based on their specific properties that enhance the performance of PEO coatings. TiO_2 is chosen for its energy level structure, which matches well with ultraviolet (UV) light, making it resistant to ionization and decomposition under UV irradiation thus ensuring long-term stability. Additionally, TiO_2 exhibits high reflectivity and excellent thermal stability, further improving the coating's performance. ZrO_2 and ZnO are selected for their superior reflectivity, contributing to a reduced overall absorptivity of the coating. The incorporation of these nanoparticles enables the creation of a coating with enhanced thermal control properties, particularly in applications requiring low absorptivity.

1. **Basic electrolyte system**: As detailed in Table 18.2, the thickness and roughness of the PEO coating are influenced by the applied voltage, which in turn impacts the coating's optical properties, including absorptivity and emissivity. With increasing voltage, the absorptivity of the coating decreases, while the emissivity increases. For instance, at a voltage of 600 V, the coating exhibits a high emissivity of 0.721 and a low absorptivity of 0.32, indicating its suitability for applications requiring efficient radiative cooling and minimal solar heat absorption.
2. **$ZnSO_4$ doping**: As shown in Table 18.3, increasing the concentration of $ZnSO_4$ in the electrolyte enhances both the thickness and roughness of the coating, resulting in a decrease in absorptivity and an increase in emissivity. At a $ZnSO_4$ concentration of 15 g/L, the coating achieves a high emissivity of 0.768 and a low absorptivity of 0.23. Macroscopic observations (Fig. 18.1) reveal that as the $ZnSO_4$ concentration increases, the coating exhibits uniform growth and progressively turns white in appearance, indicating improved optical performance.
3. **Potassium titanium oxalate doping**: In an electrolyte system containing 10 g/L $K_2TiO(C_2O_4)_2$, the PEO process was carried out at various voltages.

   At 450 V, the coating exhibited an off-white color, but as the voltage increased to 500 V, dark blue streaks appeared at the edges, and at 550 V, the entire coating turned dark blue. At 600 V, the coating reverted to an off-white color (Fig. 18.2). The dark blue coloration is attributed to the valence change of Ti^{4+}, leading to the

Table 18.2 PEO TCCs on Al alloys with different applied voltages and dopants

Sample	Electrolyte	Dopant content (g/L)	Voltage(V)	Parameters	h (μm)	Ra (μm)	α	ε
J-500	30 g/L $Na_5P_3O_{10}$; 1 g/L Na_2EDTA; 5 g/L KOH; 0.5 g/L NaF	–	500	Duty:15% Frequency:450 Hz Time:15 min	35	1.35	0.34	0.715
J-550			550		42	1.55	0.32	0.719
J-600			600		45	2.21	0.32	0.721
Zn-5	30 g/L $Na_5P_3O_{10}$; 1 g/L Na_2EDTA; 5 g/L KOH; 0.5 g/L NaF; **$ZnSO_4$(dopant)**	5	550	Duty:15% Frequency:450 Hz Time:15 min	42	3.322	0.32	0.703
Zn-10		10			46	3.652	0.29	0.761
Zn-15		15			50	3.712	0.23	0.768
T-5-550	30 g/L $Na_5P_3O_{10}$ 1 g/L Na_2EDTA 5 g/L KOH 0.5 g/L NaF **$K_2TiO(C_2O_4)_2$(dopant)**	5	550	Duty:15% Frequency:450 Hz Time:15 min	50	6.213	0.33	0.707
T-10-450		10	450		24	1.556	0.28	0.764
T-10-500			500		36	5.326	0.41	0.774
T-10-550			550		54	4.121	0.53	0.775
T-10-600			600		58	5.740	0.34	0.780
T-15-550		15	550		56	5.679	0.30	0.782
Zr-5	30 g/L Na_5P_3O 1 g/L Na_2EDTA 5 g/L KOH 0.5 g/L NaF **$Zr(NO_3)_4$(dopant)**	5	550	Duty:15% Frequency:450 Hz Time:15 min	45	1.844	0.26	0.783
Zr-10		10			51	3.541	0.31	0.802
Zr-15		15			55	4.772	0.25	0.814

Table 18.3 Process parameters and thermal control properties of PEO coating on aluminum alloy with ZnO, TiO_2, and ZrO_2 nanoparticle-doped electrolyte system

Sample	Electrolyte	Nanoparticle dopants	Parameters	h (μm)	Ra (μm)	α	ε
nZn-20	30 g/L $Na_5P_3O_{10}$ 1 g/L Na_2EDTA 5 g/L KOH 0.5 g/L NaF	20 g/L ZnO	Voltage: 550 V Duty: 15% Frequency: 450 Hz Time:15 min	46	2.527	0.27	0.806
nZn-30		30 g/L ZnO		50	2.822	0.27	0.819
nTi-20	30 g/L $Na_5P_3O_{10}$ 1 g/L Na_2EDTA 5 g/L KOH 0.5 g/L NaF	20 g/L TiO_2	Voltage: 550 V Duty: 15% Frequency: 450 Hz Time:15 min	32	3.225	0.25	0.755
nTi-30		30 g/L TiO_2		36	3.447	0.27	0.769
nZr-20	30 g/L $Na_5P_3O_{10}$ 1 g/L Na_2EDTA 5 g/L KOH 0.5 g/L NaF	20 g/L ZrO_2	Voltage: 550 V Duty: 15% Frequency: 450 Hz Time:15 min	29	1.921	0.25	0.871
nZr-30		30 g/L ZrO_2		40	2.221	0.25	0.876
nZr-40		40 g/L ZrO_2		60	2.312	0.23	0.893

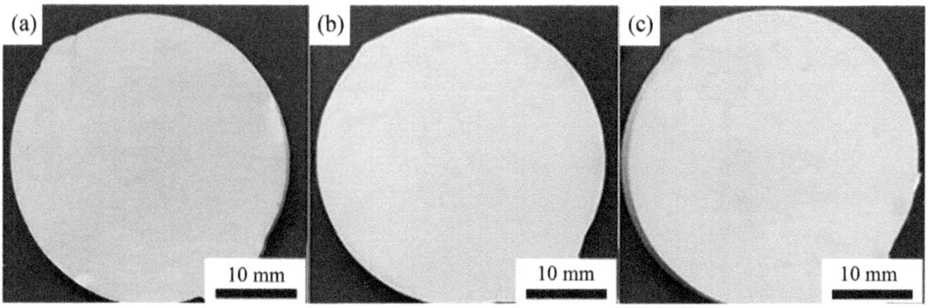

Fig. 18.1 Macroscopic morphology of thermal control coatings prepared by micro-arc oxidation in electrolyte systems doped with different contents of $ZnSO_4$. (**a**) 5 g/L, (**b**) 10 g/L, (**c**) 15 g/L

Fig. 18.2 Macroscopic morphology of micro-arc oxidation thermal control coatings at different voltages in 10 g/L $K_2TiO(C_2O_4)_2$-doped electrolyte system. (**a**) 450 V, (**b**) 500 V, (**c**) 550 V, (**d**) 600 V

formation of TiO. Table 18.3 shows that as voltage increases, both the thickness and roughness of the coating increase, accompanied by an enhancement in emissivity. Besides, higher concentrations of potassium titanium oxalate in the electrolyte result in increased emissivity. The absorptivity varies based on the coating color: off-white coating has an absorptivity of 0.28, while the dark blue coating has a higher absorptivity of approximately 0.53.

4. **$Zr(NO_3)_4$ doping**: Increasing the concentration of $Zr(NO_3)_4$ in the electrolyte, as detail in Table 18.3, leads to progressive increase in coating thickness and roughness, with the coating color gradually shifting toward white. At a $Zr(NO_3)_4$ concentration of 15 g/L, the coating achieves a high emissivity of 0.804 and a low absorptivity of 0.25, demonstrating its potential for thermal control applications requiring low solar absorption and high radiative cooling efficiency.
5. **Particle addition**: The addition of nanoparticles to the electrolyte enables the formation of composite PEO coatings through synchronous deposition and sintering of particles [5]. When 30 g/L ZnO is added to the base electrolyte, the coating exhibits a high emissivity (0.819) and a low absorptivity (0.27); Similarly, adding 30g/L TiO_2 resulting in a high emissivity (0.769) and a low absorptivity (0.27) (Table 18.3). When ZrO_2 nanoparticles are added at concentrations of 20 g/L, 30 g/L, and 40 g/L, the composite coating grows uniformly and exhibits enhanced deposition with increasing particle concentration. The coating color progressively turns white, and the thickness increases significantly. At a ZrO_2 concentration of 40 g/L, the coating achieves a high emissivity of 0.893 and a low absorptivity of 0.23, indicating superior thermal control properties.

In summary, for ion-doped in situ conversion coatings, cations in the electrolyte react with O^{2-} ions generated during the PEO process, forming specific oxidation products. The discharge channels created during PEO allow ions to penetrate the coating and undergo chemical reactions, resulting in the formation of low-absorptivity ZnO, TiO_2, and ZrO_2 phases. The addition of nanoparticles also enhances coating performance, with optimized deposition and sintering processes enabling the development of coatings with high emissivity and low absorptivity, suitable for advanced thermal management applications.

18.3.1.3 Radiation Heat Dissipation Coatings on Aluminum Alloy

Lightweight aluminum alloys with high thermal conductivity, such as 6061Al, are critical materials for heat dissipation substrates in high-power electronic devices. However, these alloys exhibit low surface emissivity and suboptimal radiation heat dissipation capacity, especially in enclosed environments with limited convective heat transfer. This drawback restricts their broader utilization in high-reliability, long-lifespan, and high-power electronic systems. To address this issue, PEO coatings were developed with optimized compositions and structures.

Two types of coatings—a standard PEO radiation heat dissipation coating and a composite PEO coating incorporating nano-sized SiC and hexagonal boron nitride (*h*-BN) particles—were fabricated on the aluminum alloy surfaces. The thermal radiation performance and cooling efficiencies of these coatings were systematically assessed. Table 18.4 provides detailed data on the electrolyte compositions,

Table 18.4 Electrolyte system, process parameters, emissivity, and radiation cooling effect of micro-arc oxidation and nano-modified composite coatings

Sample	Electrolyte	SiC /g·L[-1]	h-BN /g·L[-1]	Parameter V	Parameter Time (min)	Parameter f Duty	T (°C)	h (μm)	Ra (μm)	ε_{avg} 3~8	ε_{avg} 8~20	Chip IGBT radiation cooling /°C 1 W	Chip IGBT radiation cooling /°C 5 W
PEO-4	6 g/L (Na_2SiO_3) 35 g/L ($NaPO_3)_6$ 1.2 g/L NaOH			400	30	600H 8%	30~50	4	0.64	0.3	0.78	6.8	9.7
PEO-8				500	30		30~50	8	1.05	0.32	0.80	7.3	9.8
PEO-12				600	30		30~50	12	1.65	0.35	0.87	7.9	10.5
SiC-0	6 g/L (Na_2SiO_3) 35 g/L ($NaPO_3)_6$ 1.2 g/L NaOH	0		550	10	600 Hz 8%	70~90	7	0.5	0.31	0.76	7.1	9.8
SiC-2		2		550	10		70~90	6	0.9	0.5	0.85	7.5	10.9
SiC-4		4		550	10		70~90	6	1.3	0.72	0.85	8.9	14
SiC-8		8		550	10		70~90	75	3.41	0.86	0.88	10	15.8
BN-0	6 g/L (Na_2SiO_3) 35 g/L ($NaPO_3)_6$ 1.2 g/L NaOH		0	600	10	600 Hz 8%	70~90	12	0.5	0.32	0.78	7.1	–
BN-2			2	600	10		70~90	9	0.62	0.72	0.8	7.4	–
BN-8			8	600	10		70~90	7	0.74	0.73	0.8	8.2	–
BN-40			40	600	10		70~90	40	2.76	0.78	0.85	9.6	13
SiC/BN		8	40	600	10		70~90	78	3.89	0.86	0.91	10.4	16.8

processing parameters, emissivity, and radiation cooling performance of the PEO and nano-modified composite coatings.

1. **Basic electrolyte system:** The PEO process was conducted in an electrolyte consisting of 6 g/L Na_2SiO_3, 35 g/L $(NaPO_3)_6$, and 1.2 g/L NaOH, under a frequency of 600 Hz, an 8% duty cycle, and an oxidation duration of 30 min. The effects of varying voltages (400 V, 500 V, and 600 V) on the coating's structure and infrared emissivity were investigated. The results revealed that as the voltage increases, both the coating thickness and surface roughness progressively increase, leading to a continuous rise in emissivity. As shown in Table 18.4, the average emissivity of the MAO-12 coating, primarily composed of the γ-Al_2O_3 phase, is 0.67 in the 3–20 μm infrared range, but only 0.35 in the 3–8 μm range. This suggests that the potential for significantly improving emissivity through adjustments in PEO process parameters is somewhat limited.
2. **SiC additive**: To address the emissivity limitations of single-layer PEO coatings, nano-sized SiC particles (~30 nm) were introduced into the PEO process. A high-emissivity composite coating was developed by introducing SiC into the electrolyte under the following conditions: voltage of 550 V, electrolyte temperature of 70–90 °C, treatment duration of 10 min, with continuous stirring. The unmodified PEO coating exhibited a white surface, whereas the addition of 8 g/L SiC resulted in a gray-green appearance, signifying significant microstructural changes at critical SiC concentrations. With increasing SiC content, the surface density of nanoparticles progressively increased, as shown in Fig. 18.3b–

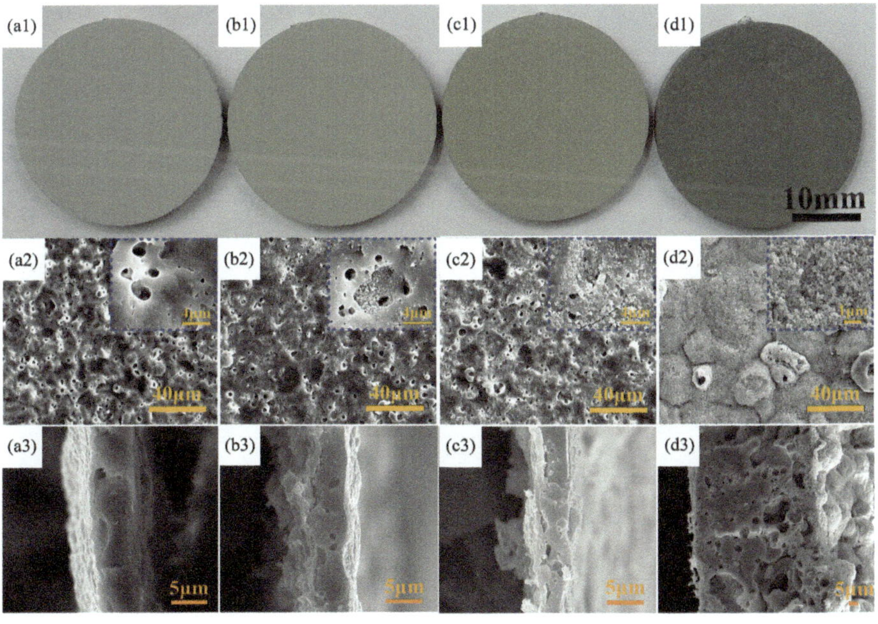

Fig. 18.3 PEO coating modified by SiC particles with different contents. (**a1–a3**) 0 g/L;(**b1–b3**) 2 g/L; (**c1–c3**) 4 g/L; (**d1–d3**) 8 g/L

d, where SiC particles are embedded within the micropores of the PEO coating. The SiC-8 composite coating (Fig. 18.3d) exhibited a surface densely populated with nanoparticles, accompanied by a notable increase in thickness to 75 μm. This structure consisted of a dense barrier layer, a middle layer (~55 μm thick), and an outer SiC-nanoparticle composite layer (~15 μm thick). Additionally, surface roughness increased with SiC content, reaching 3.41 μm for the SiC-8 coating. The incorporation of SiC significantly enhanced the average emissivity to 0.88 in the 3–20 μm wavelength range, with emissivity reaching 0.86 in the 3–8 μm range, as detailed in Table 18.4. These results demonstrate the efficacy of SiC-modified PEO coatings in enhancing thermal radiation properties and cooling performance.

3. ***h*-BN additive:** The influence of *h*-BN particle incorporation on the properties of PEO coatings was also systematically evaluated by introducing *h*-BN powder (particle size 100–300 nm) with different concentrations of 0 g/L, 20 g/L, 40 g/L, and 80 g/L into the base electrolyte. The PEO process was conducted under the following conditions: a loading voltage of 600 V, electrolyte temperature maintained between 70 °C and 90 °C, an oxidation duration of 10 min, and continuous stirring. As shown in Fig. 18.4, the surface color of the coating gradually lightens with increasing *h*-BN particle concentration, with the 40 g/L *h*-BN modified coating appearing completely white (Fig. 18.4d). This observation indicates a significant accumulation of h-BN particles on the coating surface. Table 18.4 illustrates that as the *h*-BN concentration increases from 0 g/L to 8 g/L, the surface density of nanoparticles rises, while the coating thickness decreases. However, at a concentration of 40 g/L, a substantial number of nanoparticles are observed on the coating surface, and the thickness reaches 42 μm. The surface roughness also increases with rising particle concentration, with the roughness of the BN-40 coating reaching 2.76 μm. Correspondingly, the emissivity of the coating increases with higher particle content, although the enhancement was

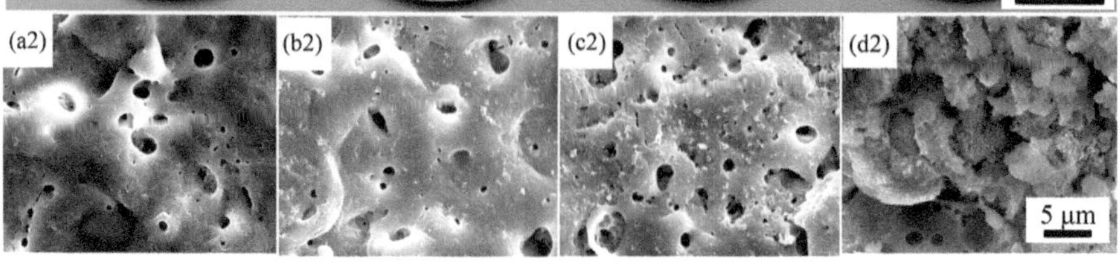

Fig. 18.4 PEO coatings modified by *h*-BN particles with different contents. (**a1**, **a2**) 0 g/L; (**b1**, **b2**) 2 g/L; (**c1**, **c2**) 8 g/L; (**d1**, **d2**) 40 g/L

minimal at lower concentrations. The BN-40 composite coating demonstrated an emissivity of 0.8 across the 3–20 μm wavelength range (Table 18.4), highlighting its excellent thermal radiation performance [20].

4. **Co-doping with SiC and *h*-BN:** The co-addition of SiC and *h*-BN particles was optimized to produce a synergistically enhanced PEO composite coating with superior thermal radiation properties. The optimized process parameters included a loading voltage of 600 V, a frequency of 600 Hz, a duty cycle of 8%, an electrolyte temperature of 70–90 °C, and an oxidation duration of 10 min. The concentrations of SiC and h-BN particles were set at 8 g/L and 40 g/L, respectively. The resulting composite coating exhibited a thickness of 78 μm, consisting of a three-layer structure: a dense barrier layer (0.6 μm), an intermediate layer (~50–60 μm), and an outer nanocomposite layer (20 μm). The surface roughness was measured at 3.89 μm, with SiC and *h*-BN nanoparticles uniformly distributed across the surface. The emissivity of the co-doped SiC/h-BN composite coating was 0.86 in the 3–8 μm wavelength range and 0.91 in the 8–20 μm range (Table 18.4), demonstrating significant improvements in thermal radiation performance.

The radiative cooling performance of the PEO and composite coatings was assessed by measuring the junction temperature of LED lamp beads under constant heat sources of 1 W and 5 W in an enclosed space (volume of 1.3 dm^3, heat dissipation surface area of 7 cm^2). For aluminum alloy substrates without coatings, the junction temperatures of 1 W and 5 W LEDs were 82.4 °C and 105.4 °C, respectively. Coated substrates exhibited a pronounced cooling effect, with the performance improving as the coating thickness increased. The PEO-12 coating reduced the junction temperature of the 5 W LED by 10.5 °C, while the SiC-8, BN-40, and SiC/h-BN composite coatings reduced the temperature by 15.8 °C, 13 °C, and 16.8 °C, respectively (Table 18.4).

The heat dissipation efficiency offers a direct measure of the radiative cooling performance of the coatings and is calculated using the following formula:

$$\delta = \frac{\Delta T}{\Delta T_c} \tag{18.7}$$

where δ represents the heat dissipation efficiency, ΔT is the temperature difference between the coated sample and the aluminum substrate, and ΔT_c is the temperature difference between the aluminum substrate and the test environment.

For the 5 W LED heat source, the heat dissipation efficiencies of the PEO-12, SiC-8, BN-40, and SiC/h-BN coatings were 13%, 21%, 17.2%, and 22.3%, respectively. The trend in heat dissipation efficiency was consistent with the emissivity values, particularly for the nanoparticle-modified composite coatings, which showed marked improvements in radiative cooling performance. The superior performance of the SiC/*h*-BN composite coating was attributed to its high emissivity across the 3–20 μm wavelength range, demonstrating that radiative heat dissipation plays a critical role in the thermal management of high-power electronic devices.

This highlights the potential of the co-doped SiC/*h*-BN composite coatings for applications in advanced thermal management systems.

18.3.2 PEO TCCs on Magnesium Alloy

Magnesium alloys, particularly magnesium-lithium (Mg-Li) alloys, are highly regarded for their low density, high strength, excellent specific stiffness, large elastic modulus, and superior processing performance. These alloys also exhibit a higher capacity for withstanding impact loads compared to aluminum alloys, making them ideal materials for lightweight aerospace structural applications. Table 18.5 presents a detailed summary of the electrolyte systems, process parameters, and thermal control performance of PEO TCCs on magnesium alloys. This comprehensive data provides critical insights into the influence of different PEO treatment conditions on the thermal properties and efficiency of these coatings, underscoring their potential for advancing thermal management in lightweight aerospace structures.

18.3.2.1 High Absorption-Emission Ratio Coating

Similar to aluminum alloys, the development of high absorption-emission ratio PEO coatings on magnesium alloys commonly employs phosphate- and silicate-based electrolyte systems [21]. To enhance the thermal control properties of these coatings, various ion doping strategies have been explored, including Fe^{2+}, Fe^{3+}, Cu^{2+}, Ni^{+}, Mn^{2+}, VO_3^{-}, VO_4^{3-}, WO_4^{2-}, and $Cr_2O_7^{2-}$ ions [22]. Additionally, particle doping is utilized to optimize the coating's structure and functionality, with common dopants including CNT, copper oxide (CuO), silicon carbide (SiC), and iron oxide (Fe_3O_4). These strategies effectively improve the absorption-emission ratio of the coatings, significantly enhancing their potential for thermal control applications [23].

1. **Process parameters**: In a basic electrolyte comprising potassium sodium tartrate ($C_4H_4KNaO_6$), NaF, $CuSO_4$, $NaAlO_2$ and Na_2WO_3, it was observed that the coating thickness increased with higher oxidation voltages. Concurrently, the micropores and cracks in the outer coating layer enlarged with increasing voltage and treatment time. Notably, the trends in hemispherical emissivity and surface roughness of the black MAO coating were consistent when the coating compositions were comparable. Under optimized conditions of 1.5 g/L CuSO4, an oxidation voltage of 450 V, and a treatment duration of 20 min, the resulting coating achieved the highest hemispherical emissivity of 0.84 [24].
2. **Ion doping**: High absorption-emission ratio coatings were developed using an electrolyte system containing can also be developed using an electrolyte system containing 25 g/L Na_3PO_4 and 25 g/L EDTA, with varying concentrations of

Table 18.5 Electrolyte system, process parameters, and thermal control performance of thermal control coating on magnesium alloy surface

Electrolyte	Additives	Parameter				h, Ra	α_s	ε
		Electric	t/min	f/Hz	D/%	/μm		
$(C_4H_4KNaO_6)$ + NaF, $NaAlO_2$ + Na_2WO_3	1.5 g/L $CuSO_4$	450 V	20	–	–	–/2.41	–	0.84
25 g/L Na_3PO_4 + 25 g/L EDTA-2Na	8 g/L $Fe_2(SO_4)_3$	5A/dm^2	10	500	10	28	0.94	0.83
10 g/L Na_2SiO_3 + 1 g/L KOH	8 g/L Na_3VO_4	500 V4.5A/dm^2	40	–	–	–	0.918	0.84
$(NaPO_3)_6$ + $NiSO_4$	1.5 g/L $CuSO_4$	5A/dm^2	20	1000	20	57.7/3.1	0.9	0.82
15 g/L Na_3PO_4 + 3 g/L NaF + 5.6 g/L KOH	10 g/L $FeC_6H_5O_7$	2A/dm^2	–	–	20~30	–	0.79	0.82
(Mg alloy) 22 g/L Na_2SiO_3 + 25 g/L $(NaPO_3)_6$	10 g/L NH_4VO_3	2A/dm^2	10	600	20	–	0.95	0.94
(Mg-Li alloy) 24 g/L Na_2SiO_3 + 25 g/L $(NaPO_3)_6$	10 g/L NH_4VO_3	3A/dm^2	10	600	20	–	0.964	0.951
20 g/L $(NaPO_3)_6$ + 8 g/L Na_2SiO_3 + 2 g/L NaOH	10 g/L SiC	550 V	10	600	10	–	0.76	0.89
6 g/L Na_2SiO_3 + 2 g/L KOH + 2 g/L KF	0~10 g/L CNT	10A/dm^2	–	–	–	–	–	0.87
10 g/L Na_2SiO_3 + 1 g/L KOH	–	500 V4.5A/dm^2	45	–	–	–	0.44	0.80
30 g/L $Na_5P_3O_{10}$ + 5 g/L KOH + 0.5 g/L NaF + 1 g/L EDTA-2Na	10 g/L $Zr(NO_3)_4$	10A/dm^2	15	50	45	–	0.41	0.87
30 g/L $Na_5P_3O_{10}$ + 5 g/L NaOH	4 g/L $ZnSO_4$	15A/dm^2	10	50	45	–	0.35	0.88
(Mg-Li alloy) 14 g/L Na_2SiO_3 + 1 g/L KOH + 1 g/L NaF	–	10A/dm^2	20	500	20	–	0.35	0.82
(Mg-Li alloy) 8.25 g/L [$(NaPO_3)_6$ + $(NaPO_3)_n$] + 3 g/L NaOH+1 g/L NaF	–	10A/dm^2	30	50	20	–	0.33	0.85

$Fe_2(SO_4)_3$ (0 g/L, 4 g/L, 8 g/L, and 12 g/L). The process parameters were set at a current density of 5 A/dm^2, frequency of 500 Hz, a duty cycle of 10%, and an oxidation time of 10 min. The results demonstrated that as the $Fe_2(SO_4)_3$ content increased, the coating thickness slightly decreased while roughness increased. The solar absorptance of the coating improved from 0.28 for the magnesium alloy substrate to 0.43, 0.93, 0.94, and 0.94, respectively, as $Fe_2(SO_4)_3$ concentration increased. Similarly, the infrared emissivity increased from 0.06 for the substrate to 0.78, 0.82, 0.83, and 0.83. The enhancement in emissivity and solar absorptance is attributed to two key factors: (1) the in situ formation of Fe_3O_4, which exhibits high solar absorption and infrared emissivity, and (2) the incorporation of iron ions into the MgO crystal structure, resulting in lattice distortion. This distortion reduces the symmetry of the crystal structure, enhancing polarization and promoting non-resonant lattice vibrations, phonon coupling, and phonon combination radiation, thus improving the infrared radiation and absorption characteristics [25].

Besides, by introducing different coloring salts ($Cu(Ac)_2$, Na_2WO_4, Na_3VO_4) onto the surface of magnesium alloys, four distinct colored coatings with white, reddish-brown, gray, and black appearances were successfully prepared (Fig. 18.5). These coatings exhibited solar absorption rates ranging from 0.439 to 0.918, while the emissivity remained consistently high at approximately 0.85. Notably, the black coating, prepared using Na_3VO_4 as the coloring agent, achieved the highest solar absorption and emissivity values of 0.918 and 0.84, respectively. The specific electrolyte compositions, process parameters, and corresponding thermal control performances are summarized in Table 18.6 [26].

In addition to the above commonly used additives, colorants for preparing high absorption-emission ratio black ceramic coatings also include copper tartrate, copper acetate, copper pyrophosphate, ferric citrate, ferric oxalate, nickel acetate, nickel oxalate, cobalt sulfate, cobalt oxalate, manganese sulfate, manganese acetate, etc.

3. **Particle Doping**: In addition to ion doping, particle doping plays a pivotal role in enhancing the performance of high absorption-emission ratio ceramic coatings on magnesium alloys [27]. For instance, by adding 10 g/L of nano-sized SiC particles (30 nm) into an electrolyte containing 20 g/L $(NaPO_3)_6$, 8 g/L Na_2SiO_3,

Fig. 18.5 Macroscopic morphology of coatings prepared in different electrolyte systems (from white to black) corresponds to (Si, Cu0.5, W2, V8). Reproduced with permission from [26] © 2013 Elsevier B.V.

Table 18.6 Electrolyte composition and process parameters for preparing thermal control coatings by adding different ions. Reproduced with permission from [26] © 2013 Elsevier B.V.

Sample	Electrolyte	Bipolar pulse V/V	Positive I/A·dm^{-2}	Negative I/A·dm^{-2}	t/min	α_s	ε
Si	10 g/L Na_2SiO_3 + 1 g/L KOH	500	4.5	2.3	45	0.439	0.82
Cu0.5	10 g/L Na_2SiO_3 + 8 g/L KOH + 20 g/L KF + 0.5 g/L $Cu(Ac)_2$ + 100 mL $NH_3{\cdot}H_2O$	400	4.5	0	10	0.49	0.78
W2	10 g/L Na_2SiO_3 + 1 g/L KOH + 2 g/L Na_2WO_4	490	4.5	2.3	30	0.7	0.82
V8	10 g/L Na_2SiO_3 + 1 g/L KOH + 8 g/L Na_3VO_4	500	4.5	2.3	40	0.918	0.84

and 2 g/L NaOH, a ceramic coating with high absorption and emissivity can be prepared under a voltage of 550 V, duty cycle of 10%, frequency of 600 Hz, and reaction time of 10 min. The resulting coating exhibited an absorption rate of 0.76 and an emissivity of 0.89. Similarly, in an electrolyte composed of 6 g/L Na_2SiO_3, 2 g/L KOH, and 2 g/L KF, different concentrations of CNTs (0 g/L, 2.5 g/L, 5 g/L, and 10 g/L) were added to produce CNT-doped PEO ceramic coatings on magnesium alloys, using a current density of 10 A/dm^2. The coating without CNT was white, and as the CNT content increased, the coating gradually darkened. At a CNT concentration of 10 g/L, the surface became entirely black, with CNTs primarily distributed around the pores, resulting in high absorption and emissivity values of 0.87.

Magnesium-lithium alloys, due to their lower density compared to magnesium alloys, have gained increasing attention in the aerospace field. When processed in a 10 g/L NH_4VO_3 electrolyte, the PEO coating on the surface of magnesium-lithium alloys consists of MgO and partially amorphous phases. The coating is porous with a large specific surface area, exhibiting a high absorption rate of 0.964 and a high emissivity of 0.951.

18.3.2.2 Low Absorption-Emission Ratio Coating on Magnesium Alloy

In the development of low absorption-emission ratio PEO coatings for magnesium alloys, both ion and particle doping play critical roles. These coatings are essential in applications such as thermal control for aerospace structures, where optimizing the balance between solar absorption and infrared emission is crucial for maintaining ideal thermal conditions. The basic electrolyte systems typically consist of sodium phosphate and silicate, into which a variety of ion and particle dopants can be introduced to tailor the coating's properties [28, 29]. Common ion dopants include Zr^{2+}, Zn^{2+}, and Ti^{4+}, while particle dopants often include ZnO, Al_2O_3, Y_2O_3, ZrO_2, BN, and TiO_2, each contributing unique structural and thermal attributes to the coatings.

1. **Zr^{2+} ion doping:** The use of Zr^{2+} ion doping in PEO coatings has demonstrated significant potential for enhancing emissivity while maintaining controlled absorptivity. In this process, an electrolyte composed of 30 g/L $Na_5P_3O_{10}$, 5 g/L KOH, 0.5 g/L NaF, and 1 g/L Na_2EDTA is used, with varying concentrations of $Zr(NO_3)_4$ added to tailor the coating properties. The concentrations of $Zr(NO_3)_4$ were varied at 0 g/L, 5 g/L, 10 g/L, and 15 g/L, with the process parameters set at a current density of 10 A/dm^2, frequency of 50 Hz, and a duty cycle of 45%.

   Increasing $Zr(NO_3)_4$ concentrations reduced the pore size and number, leading to smoother coatings and enhanced optical properties. The formation of ZrO_2 crystals within the coating was particularly effective in increasing emissivity due to their excellent infrared radiation properties. At an optimal $Zr(NO_3)_4$ concentration of 10 g/L, the coating achieved a solar absorption rate of 0.405 and an emissivity of 0.873, demonstrating superior thermal control performance [30].
2. **Zn^{2+} ion doping:** Zn^{2+} ion doping presents another effective approach for reducing solar absorptivity while enhancing infrared emissivity in PEO coatings. The process involves using an electrolyte consisting of 30 g/L $Na_5P_3O_{10}$ and 5 g/L NaOH, with $ZnSO_4$ concentrations set at 0 g/L, 2 g/L, 4 g/L, and 6 g/L. The process was conducted at a current density of 15 A/dm^2, frequency of 50 Hz, and a duty cycle of 45%, with an oxidation time of 10 min.

   As the concentration of $ZnSO_4$ increased, the coatings exhibited a gradual rise in emissivity, with average values of 0.68, 0.79, 0.88, and 0.88 for $ZnSO_4$ concentrations of 0 g/L, 2 g/L, 4 g/L, and 6 g/L, respectively. Interestingly, the solar absorptivity decreased with increasing Zn^{2+} content, reaching values of 0.58, 0.46, 0.35, and 0.36, respectively. This reduction in absorption is primarily attributed to the high refractive index of ZnO, which reflects a significant portion of incoming solar radiation, thereby lowering the absorptivity. This characteristic makes Zn^{2+}-doped coatings particularly suitable for spacecraft thermal control applications, where low solar absorption is desired to minimize heat buildup. The white coating produced with 4 g/L $ZnSO_4$ demonstrated a particularly favorable balance, with a solar absorption rate of 0.35 and an emissivity of 0.88, making it an ideal candidate for thermal control coatings in aerospace [31].
3. **Particle doping**: Beyond ion doping, the incorporation of specific particles into the PEO coating further enables the tuning of the absorption-emission ratio. Particle doping not only affects the optical properties but also influences the microstructure, surface morphology, and thermal stability of the coatings [32]. For example, by utilizing an electrolyte containing 15 g/L Na_2SiO_3, 8 g/L KF, and 4 g/L NaOH, PEO coatings primarily composed of MgO were produced on magnesium alloy surfaces under a voltage of 500 V, duty cycle of 8%, and frequency of 600 Hz, at an electrolyte temperature of 50 °C–70 °C, for a duration of 20 min. These coatings exhibited moderate thermal properties, with a solar absorptivity of 0.4 and an infrared emissivity of 0.7.

In contrast, introducing 11 vol% PTFE emulsion into the basic electrolyte led to the formation of an MgO/PTFE double-layer coating. PTFE's excellent reflectivity across the visible spectrum significantly reduced the solar absorption rate to 0.12

while increasing emissivity to 0.92. This dual-layer structure improved both optical properties and environmental resistance, making it highly suitable for radiative cooling applications. The combination of MgO's thermal stability and PTFE's reflectivity presents a robust solution for advanced thermal management in radiative cooling environments.

Through optimized ion and particle doping strategies, PEO coatings on magnesium alloys can be precisely tailored to achieve desired thermal control properties. Ion dopants such as Zr^{2+} and Zn^{2+} effectively modulate emissivity and absorption rates, while particle doping, particularly with materials like PTFE, further enhances reflectivity and thermal performance. These advancements underscore the potential of PEO coatings as effective thermal control solutions for lightweight, high-performance aerospace materials.

18.3.3 PEO TCCs on Titanium Alloy

Lightweight titanium alloys have emerged as a key material for the fabrication of spacecraft, aircraft, and missile components, owing to their exceptional properties, such as structural stability, high specific strength, superior corrosion resistance, and excellent thermal performance. The application of PEO technology for the development of TCCs on titanium alloy surfaces further expands the potential utility of these materials in advanced engineering applications [33, 34].

18.3.3.1 High Absorption-Emission Ratio Coatings

High absorption-emission ratio coatings are widely used in optical remote sensing systems, infrared guidance, and laser communication technologies. PEO coatings are particularly well suited for these demanding environments due to their excellent weather resistance and strong adhesion. The development of high absorption-emission ratio PEO coatings on titanium alloys commonly involves the use of electrolytes based on silicates, phosphates, and aluminates, with performance enhancements achieved through ion or particle doping. Table 18.7 provides a comprehensive overview of the electrolyte systems, process parameters, and thermal control properties associated with these coatings. The preparation of high absorption and high emission black coatings can be accomplished using the following methodologies:

1. **Ion doping modification**: Ion-type additives such as Cu^{2+}, WO_4^{2-}, VO_3^-, Co^{2+}, Ni^{2+}, and Fe^{2+} with complexing agents like EDTA stabilizer are introduced into the basic electrolyte [35–37]. For example, using a base electrolyte composed of 1 g/L $Na_5P_3O_{10}$, 4.5 g/L $(NaPO_3)_6$, 6 g/L Na_2SiO_3, and 20 g/L EDTA-2Na, increasing the concentrations of NH_4VO_3, $FeSO_4$, and $Ni(CH_3COO)_2{\cdot}4H_2O$, results in coatings with a darker appearance. Increasing the NH_4VO_3 concentration significantly enhances the roughness and thickness of the coating, leading

Table 18.7 Electrolyte system, process parameters, and thermal control performance of PEO coating with high absorption-emission ratio on titanium alloy surface

Electrolyte	Additive/ Post-processing	Parameters Electric	t/min	f/Hz	D/%	h/Ra /μm	α_s	ε
10 g/L Na_3PO_4 + 5 g/L $NaAlO_2$	–	400 V	1/6	1000	30	–	0.78	0.28
10 g/L Na_3PO_4 + 2 g/L $NaAlO_2$	Sol-gel TiO_2 500 °C-1 h	450 V	2	250	5	–	0.99	0.06
18.6 g/L $(NaPO_3)_6$ + 3.09 g/L $(NaPO_3)_n$ + 10 g/L EDTA-2Na	–	3A/cm^2	20	200	45	–	0.797	0.793
	10 g/L $Co(Ac)_2$	3A/cm^2	20	200	45	17/1.4	0.82	0.78
	10 g/L Ni salts	3A/cm^2	20	1000	45	27/2.2	0.93	0.84
	10 g/L $FeSO_4$	3A/cm^2	20	1000	45	22/1.7	0.9	0.84
1 g/L $(NaPO_3)_n$ + 4.5 g/L $(NaPO_3)_6$ + 6 g/L Na_2SiO_3 + 20 g/L EDTA-2Na	3 g/L NH_4VO_3	400 V	10	1000	30	44.2/3.5	0.962	0.95
	0~5 g/L $FeSO_4$	400 V	10	1000	30	35~44/ 3.5~4.2	0.96	0.95
	5 g/L $Ni(Ac)_2$	400 V	10	1000	30	44.2/3.45	0.962	0.95
70 g/L NaP_3O_{10} + 10 g/L $FeSO_4$ + 10 g/L $Co(Ac)_2$ + 10 g/L $Ni(Ac)_2$ + 30 g/L EDTA-2Na	7.5 g/L Na_2WO_4	1.5A/ cm^2	25	1000	–	44.1/1.3	0.93	0.88

to an increase in the absorption rate from 0.94 to 0.96 and in the infrared emissivity from 0.90 to 0.95. Similarly, elevating the concentration of $Ni(CH_3COO)_2·4H_2O$ causes minor increases in roughness and thickness, raising the absorption rate from 0.92 to 0.96 and the emissivity from 0.93 to 0.95. By contrast, $FeSO_4$ concentration has minimal impact on the coating's roughness, thickness, and morphology, with only slight variations in absorption and emissivity. While higher voltages increase the coating's roughness and thickness significantly, they have little effect on absorption and only slightly increase emissivity. Prolonged oxidation times also minimally affect roughness and thickness, yielding slight improvements in both absorption and emissivity. The optimized process includes 3 g/L NH_4VO_3, 5 g/L $Ni(CH_3COO)_2·4H_2O$, and 2–5 g/L $FeSO_4$ in the base electrolyte, with an applied voltage of 400 V, oxidation time of 10 min, duty cycle of 30%, and frequency of 1000 Hz. This process achieves a ceramic coating with a high absorption rate of 0.96 and a high emissivity of 0.95.

2. **Multi-salt co-doping modification**: Multi salt co-doping, utilizing $(NaPO_3)_6$ and $(NaPO_3)_n$ as primary salts with the addition of 30 g/L EDTA-2Na, 10 g/L $Ni(Ac)_2$, $Co(Ac)_2$, and $FeSO_4$, significantly impacts the thermal properties of the coating. Varying concentrations of Na_2WO_4 (5 g/L, 10 g/L, 15 g/L) were shown to enhance performance. Adding Na_2WO_4 increases the absorptivity from 0.81 to 0.92 with minimal changes in emissivity. Further increases in Na_2WO_4 concentration do not affect absorptivity but raise emissivity to 0.87.

The optical properties of the coating are closely linked to its composition and surface state. Higher oxide content increases emissivity, while the absorptivity is strongly influenced by coating thickness and surface condition. To further enhance absorption, particularly for solar radiation that includes ultraviolet light, UV absorption should be considered. Since UV radiation penetrates deeper layers, a multilayer coating design may be advantageous. The inner layer can incorporate materials optimized for UV absorption, while the outer layer is tailored for visible and infrared light absorption, enabling the development of an optimized high absorption and high emission coating design.

18.3.3.2 Low Absorption-Emission Ratio Coatings

The preparation of PEO coatings with low absorption-emission ratios typically involves electrolyte systems containing phosphates, silicates, aluminates, zirconates, and titanates. Table 18.8 presents a summary of the various electrolyte systems, processing parameters, and thermal control properties associated with PEO coatings exhibiting low absorption-emission ratios on titanium alloys.

1. **Phosphate-based electrolyte coatings**: Coatings prepared in phosphate-based electrolyte systems exhibit a darker color, with relatively low thickness and surface roughness. These coatings demonstrate an absorption ratio of 0.9 and an emissivity of 0.85. Upon annealing at 500 °C for 2 h, the absorption ratio decreases from 0.92 to 0.85, while the emission ratio remains virtually unchanged.
2. **Aluminate-based electrolyte coatings**: Coatings prepared in aluminate-based electrolyte systems are characterized by a grayish-white color, with slightly

Table 18.8 Electrolyte system, process parameters, and thermal control performance of PEO coating with low absorption-emission ratio on titanium alloy surface

Electrolyte	Additive/ post-processing	Parameters Electric	t/min	f/Hz	D/%	h/Ra /μm	α_s	ε
10 g/L Na_2SiO_3 + 1 g/L NaH_2PO_2	–	10Adm^{-2}	–	500	45	28/2.1	0.37	0.93
6 g/L K_2ZrF_6 + 0.5 g/L NaH_2PO_4 + 2 mL/L H_3PO_4	–	12Adm^{-2}	20	50	45	130	0.32	0.96
6 g/L K_2ZrF_6 + 2 mL/L H_3PO_4	–	12Adm^{-2}	30	100	30	170/11	0.237	0.99
10 g/L Na_2SiO_3 + 1 g/L NaH_2PO_2	–	9Adm^{-2}	30	500	–	93/8.5	0.41	0.92
9 g/L Na_2SiO_3 + 3 g/L Na_3PO_4 + 3 g/L $NaAlO_2$	–	400 V	20	600	8	15/1.2	–	>0.8
	Fluorosilane modification	400 V	20	600	8	15/2.0	–	>0.8
12 g/L Na_2SiO_3 + 3 g/L NaH_2PO_2	PTFE modification	500 V	15	600	8	–	0.19	0.85

higher thickness and roughness compared to phosphate-based coatings. Composed primarily of TiO_2 and Al_2TiO_5, these coatings exhibit an absorption ratio of 0.6 and an emission ratio of 0.8. After annealing at 500 °C for 2 h, the absorption ratio increases, while the emission ratio decreases.
3. **Silicate-based electrolyte coatings**: Coatings prepared in silicate-based electrolyte systems are white, with larger thickness and roughness. Following process optimization (10 g/L Na_2SiO_3, 1 g/L NaH_2PO_2, frequency 500 Hz, current density 10 A/dm^2, time 45 min), the coatings achieve an absorptivity of 0.37 and an emissivity of 0.93. After annealing, the absorptivity slightly increases, while the emissivity slightly decreases.
4. **Zirconate-based electrolyte coatings**: In zirconate-based electrolyte systems, white coatings with low absorption-emission ratios are prepared using an electrolyte composed of 6 g/L K_2ZrF_6 and 1 mL/L H_3PO_4 (85%). These coatings are produced at a current density of 12 A/dm^2, a duty cycle of 30%, and a frequency of 100 Hz. The resulting coatings exhibit a remarkably low absorptivity of 0.24 and a high emissivity of 0.99 [38].

From the above analysis, it can be concluded that the zirconate-based electrolyte system produces superior coatings with low absorption-emission ratios. To further explore the impact of various parameters on the microstructure and thermal control properties of PEO ceramic coatings, additional studies were conducted using a potassium fluozirconate-phosphate-sodium hypophosphite electrolyte system (pH = 1). The effects of potassium fluozirconate concentration, current density, treatment time, duty cycle, and frequency on coating characteristics were systematically investigated.

1. **Potassium fluozirconate concentration:** PEO ceramic coatings were prepared in potassium fluozirconate electrolytes at concentrations of 4 g/L, 6 g/L, and 8 g/L, with a current density of 10 A/dm^2, frequency of 50 Hz, duty cycle of 45%, and oxidation time of 30 min. As the K_2ZrF_6 concentration increased, both coating thickness and roughness initially increased, followed by a decrease at higher concentrations. The absorptivity exhibited a trend of first decreasing and then increasing, while the emissivity gradually decreased.
2. **Treatment time:** With an increase in treatment time, both coating thickness and roughness progressively increased, while the absorptivity accordingly decreased. The emissivity increased rapidly initially, but changed little after 10 min. This behavior is attributed to the high rate of film formation within the first 10 min, leading to significant changes in the emission ratio. After 10 min, the coating surface became more uniform, resulting in smaller variations in the emissivity. In this system, longer treatment times tend to produce coatings with higher emissivity and lower absorption characteristics.
3. **Current density:** As the current density increased, both the coating thickness and roughness increased, while the absorptivity decreased. The emissivity

showed only a slight increase, indicating that the current density has minimal effect on the emissivity under these conditions.

4. **Power frequency and duty cycle:** These parameters primarily affect the coating thickness and surface roughness, which in turn influence the absorptivity and emissivity. For instance, as the power frequency increased, the coating thickness decreased, roughness increased, and the absorptivity increased, while the emissivity remained nearly unchanged. This suggests that frequency has a limited effect on the emissivity (with lower power frequencies being more favorable for producing coatings with high emissivity and low absorptivity). Increasing the duty cycle resulted in an increase in both coating thickness and roughness, a decrease in the absorptivity, and minimal changes in the emissivity.

18.3.3.3 High Emissivity Thermal Protection Coatings

When titanium alloys are used in the metal shells of aerospace vehicles, the high temperatures generated by aerodynamic heating or combustion can cause rapid surface heating, leading to a significant reduction in the mechanical properties of high-strength titanium alloys. To mitigate this, thermal protection systems are typically applied to the vehicle's surface, with emissivity being a critical factor. Higher high-temperature emissivity of the surface enhances thermal radiation performance. For instance, Plasma Electrolytic Oxidation (PEO) coatings with different compositions were applied to Ti_2AlNb alloy, and their spectral emissivity was evaluated at 600 °C. The findings were as follows:

1. **Silicate-based electrolyte system**: PEO coating prepared in this electrolyte system exhibits emissivity >0.5 in the 3–8 μm wavelength range and >0.8 in the 10–20 μm range.
2. **Ion doping**: The incorporation of NH_4VO_3 and Na_2CrO_4 into the electrolyte increased the emissivity of the coating to 0.7 in the 3–8 μm range and 0.9 in the 10–20 μm range.
3. **Particle addition**: A composite coating prepared with the addition of SiC nanoparticles demonstrated emissivity >0.8 across the entire 3–20 μm wavelength range

These results suggest that the high-temperature emissivity of PEO ceramic coatings is primarily governed by their phase composition and surface roughness. Phases with high intrinsic absorption within the coating enhance the absorption of incident electromagnetic waves, while a rough surface structure increases the scattering path of these waves, dissipating more energy. Therefore, optimizing the microstructure or incorporating specific dopants can effectively enhance the thermal radiation properties of PEO ceramic coatings.

18.4 Structure and Thermal-Optical Properties of PEO Coatings

18.4.1 Coating Thickness

The thickness of PEO coatings can be effectively modulated by adjusting parameters such as applied current/voltage and oxidation duration. This parameter is intrinsically linked to key functional properties, including thermal management, corrosion resistance, and wear resistance. For coatings with similar compositions, the thermal performance is primarily governed by their structural characteristics, notably macroscale features such as thickness and surface roughness. Several studies have demonstrated the correlation between coating thickness and thermal control properties in light alloys.

It is widely known that the infrared emissivity of the MAO coating is proportional to the coating thickness, and this relationship can be described by the following equation [39]:

$$\varepsilon = \frac{(1-\rho_c)(e_s e^4 \beta d)}{(e^4 \beta d - \rho_c^S)} = (1-\rho_c)\left[1 - \frac{(\rho_s(1-\rho_c))}{(e^4 \beta d - \rho_c \rho_s)}\right] \tag{18.8}$$

where ρ_c represents the reflectance of the coating, ρ_s is the reflectance of the substrate, d denotes the coating thickness, and ρ is the Beer-Lambert absorption coefficient. Consequently, the emissivity increases with coating thickness, a trend consistent with previous experimental findings. Although it is not feasible to produce coatings of infinite thickness in practice, the penetration depth of light is typically limited. As a result, the emissivity of the coating stabilizes once it reaches a critical thickness.

As for the solar absorptivity, some researchers found that thicker, nonmetallic PEO coatings are capable of reducing solar energy absorption at the surface. However, some studies report contrasting results, showing that absorptivity increases with greater thickness. This discrepancy is likely due to the fact that absorptivity is more significantly influenced by the coating composition than by its thickness.

18.4.2 Surface Roughness

Surface roughness plays a critical role in determining the thermal control properties of PEO coatings. The interaction of electromagnetic waves with coatings of varying surface roughness is schematically illustrated in Fig. 18.6. For a smooth surface, most of the electromagnetic waves are reflected, with minimal absorption (Fig. 18.6a). In contrast, surfaces with higher roughness demonstrate significantly enhanced absorptivity and reduced reflectance (Fig. 18.6b). According to Kirchhoff's

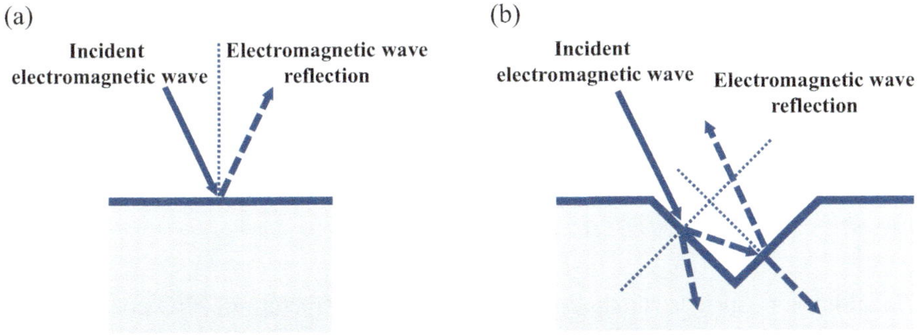

Fig. 18.6 Schematic view of electromagnetic wave acting on coatings at different surface roughness: (**a**) smooth and (**b**) rough surface. Reproduced with permission from [40] © 2009 Elsevier B.V.

law, in an isothermal system at thermal equilibrium, absorptivity is equal to emissivity. Therefore, PEO coatings with high surface roughness exhibit reduced reflectance and increased absorptivity, resulting in enhanced emissivity/absorptivity.

These principles are consistently supported by findings in the literature, where increased surface roughness correlates with higher absorptivity. The enhanced roughness in PEO coatings is typically associated with the formation of surface features such as bulges and micropores, which act as optical traps, promoting multiple reflections of incident light and thereby reducing overall reflectivity. Additionally, the larger surface area created by increased roughness facilitates greater radiative energy dissipation, further contributing to elevated emissivity. It is well documented that surface roughness tends to increase with prolonged PEO processing, leading to a corresponding rise in the actual surface area. This increase in surface area enhances both reflection and emissivity effects. Simultaneously, the pore size of the coatings also grows with extended processing time. However, as these pores fall outside the infrared spectral range, their light-trapping effectiveness diminishes, resulting in reduced solar absorption and infrared emissivity.

The correlation between emissivity/absorptivity and surface roughness can be described by the following equation [41]:

$$\varepsilon = 1 - \rho_{\mathrm{P}} = 1 - \rho_{\mathrm{R}} \exp\left(-\frac{4\pi R}{\lambda}\right)^2 \quad (18.9)$$

where ρ_{p} and ρ_R represent the reflectance of polished and rough surfaces, respectively, R is the roughness, and λ is the wavelength. According to this equation, higher roughness leads to higher emissivity, consistent with previous studies. However, some studies report opposite findings, where both absorptivity and emissivity are positively correlated with roughness. This discrepancy may arise because the effect of roughness is smaller than that of coating composition within certain ranges.

In summary, increased surface roughness enhances the actual surface area, promoting greater absorption and emissivity. Coatings with higher roughness benefit from an enlarged radiative surface area, positively influencing emissivity. However, the relationship between absorptivity and roughness is more complex, as coating composition and thickness typically exert a more significant influence on absorptivity than surface roughness alone.

18.4.3 Surface Structure

In addition to surface roughness, the surface microstructure of PEO coatings plays a critical role in determining their thermal control performance. The interaction between incident light and the micropores and micro-protrusions on the coating surface directly influences the absorptivity and emissivity of the coating [42].

A typical microporous morphology of PEO coatings is illustrated in Fig. 18.7a, showing a surface dominated by volcano-like structures formed through localized micro-discharges. The rapid solidification of molten material around discharge channels results in the formation of microcracks and particle accumulation. Chen et al. demonstrated this phenomenon by fabricating textured TiO_2 coatings on titanium alloys via MAO, systematically varying applied voltages (300, 400, and

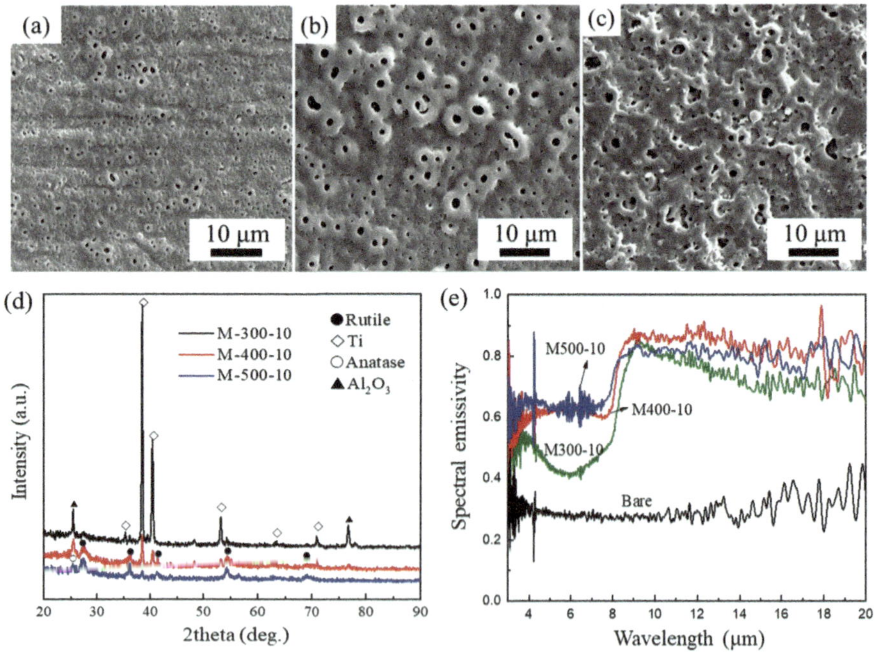

Fig. 18.7 (**a–c**) SEM images, (**d**) XRD patterns, and (**e**) corresponding spectral emissivity of the PEO coatings subjected to different oxidation times. Reproduced with permission from [42] © 2024 Published by Elsevier B.V.

500 V) and oxidation durations (10, 20, and 30 min). The applied voltage predominantly influenced surface morphology. At 300 V, the process generated small micropores (~1 μm in diameter) due to random white sparks. Increasing the voltage to 400 V produced uniform discharges, resulting in pores expanding to ~3 μm after 10 min of oxidation. At 500 V, the process intensified, creating an undulating surface with fewer but larger and irregularly distributed pores (Fig. 18.7a–c).

The emissivity of the coatings across the infrared spectrum under different voltages is depicted in Fig. 18.7e. Higher voltages, particularly within the 3–9 μm wavelength range, resulted in increased emissivity, correlating with larger micropore sizes.

To elucidate the enhancement of emissivity through structural optimization, the electric field intensity distributions of flat and textured TiO_2 surfaces were analyzed across infrared wavelengths (Fig. 18.8). In the 3–7 μm range, the flat TiO_2 coating exhibited a uniform electric field distribution, leading to minimal light absorption due to strong reflection at the TiO_2-Ti interface. In contrast, the textured surface allowed a gradual variation in refractive index, facilitating impedance matching and enhancing emissivity, which exceeded 0.6 in this spectral band. In the 8–5 μm range, strong surface phonon polaritons (SPhPs) in titania caused attenuation of the electric field in the flat coating, resulting in low emissivity. The presence of

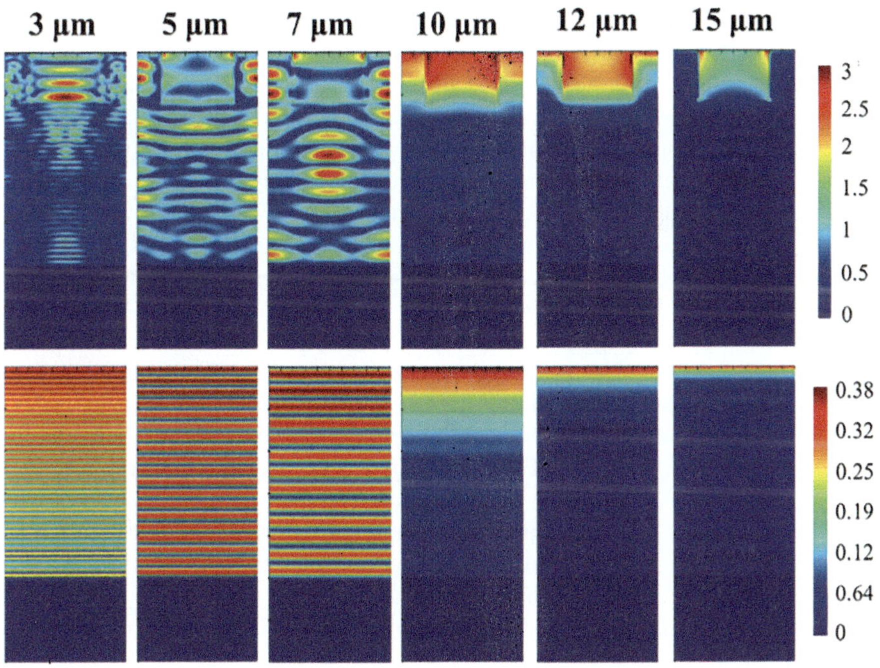

Fig. 18.8 Comparison of the electric field intensity distribution of TiO_2 coating with micropore (upper panel) or planar (lower panel) structure in the x–z plane at six specific wavelengths. Reproduced with permission from [42] © 2024 Published by Elsevier B.V.

micropores, however, activated localized SPhPs and diffracted the excitation to the far field, maintaining high emissivity.

In conclusion, surface structure, especially the presence of micropores and micro-protrusions, plays a critical role in the thermal control properties of PEO coatings. By optimizing coating structure and processing parameters, it is possible to significantly enhance emissivity, particularly in the infrared range, improving the coating's thermal control performance.

18.5 Application Examples of PEO TCCs

MAO thermal control coatings are widely used in aviation, aerospace, and many other fields.

18.5.1 Black PEO TCCs

Black PEO TCCs, developed on aluminum, magnesium, and titanium alloys, have emerged as highly effective inorganic ceramic solutions for extreme thermal environments. These coatings exhibit exceptional stability at both high and low temperatures, addressing the rigorous demands of space applications, where temperatures can reach up to 180 °C. Unlike conventional organic coatings, which are prone to aging, degradation, or evaporation under such conditions, black PEO coatings are resistant to thermal stress, effectively mitigating risks of contamination to optical components and preventing failure of critical payloads.

The black surface of PEO coatings enhances light absorption, minimizing reflection and suppressing stray light from sources such as sunlight or ground illumination. This property is particularly advantageous for satellite optical systems, enabling improved detection capabilities for observing faint targets. Additionally, these coatings exhibit a high and tunable solar absorptance-emittance ratio, providing precise thermal regulation tailored to specific mission requirements. Such thermal control is critical for maintaining stable operational conditions in space, where rapid and extreme temperature fluctuations are commonplace.

Beyond space applications, PEO technology has been successfully implemented in high-performance engineering components, including optical structural parts for satellites, aluminum alloy mounting boxes in fighter aircraft cockpits, and magnesium alloy optical sights for sniper rifles (Fig. 18.9a, b). These examples underscore the versatility of PEO coatings in aerospace and defense, where they enhance thermal control, optical performance, and environmental resistance, ensuring reliability under demanding conditions.

18.5.2 White PEO TCCs

White PEO coatings have been applied to aluminum alloy components, such as regulators and hand wheels, used in extravehicular space suits. These coatings provide crucial thermal control and enhanced reflectivity, contributing to the functional and esthetic requirements of space equipment. In these fields, PEO coatings are utilized not only for their thermal control properties but also for their ability to regulate light reflection and provide decorative finishes, highlighting their multifunctionality across different sectors.

18.5.3 PEO TCCs for Passive Radiative Cooling

As illustrated in Fig. 18.9c, radiation heat dissipation coatings have been successfully extended to a variety of applications, including air conditioners, laptops, transformers, motors, and engines. These coatings achieve a temperature reduction of 8–25%, significantly reducing the overall weight and cost of the systems while enhancing their operational lifespan and reliability. The integration of these coatings into such devices demonstrates their potential for widespread use in thermal management, offering both performance improvements and economic benefits across multiple industries.

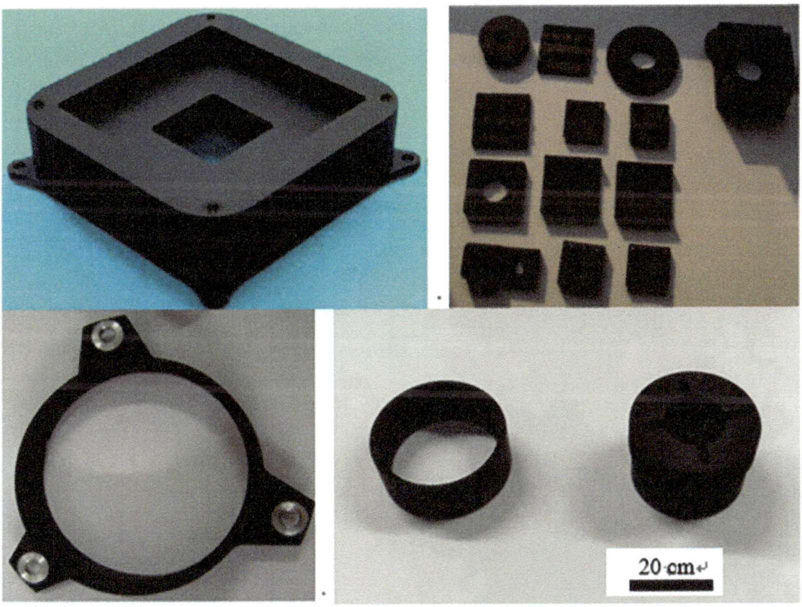

Fig. 18.9 (**a, b**) Magnesium alloy optical sight components for sniper rifles to eliminate stray light and thermal control coatings. (**c**) Low radiation absorption ratio, heat dissipation, and cooling coating on the surface of aluminum alloy housing of electronic equipment

18.6 Conclusions and Outlook

Compared to thermal control coatings applied to aluminum and titanium alloys, those with high and low solar absorptance-emittance ratios on magnesium and magnesium-lithium alloys exhibit slightly inferior thermal control performance. The PEO coatings on magnesium and magnesium-lithium alloys are generally characterized by reduced thickness and surface roughness compared to those on aluminum and titanium alloys. This observation highlights that surface morphology and microstructure are crucial factors influencing absorptance and emittance. Besides, employing techniques such as ion doping, particle doping, and composite processes with other technologies has been shown to broaden the tunability of absorptivity and emissivity in thermal control coatings.

Future research should address the following critical areas:

1. Current studies primarily focus on regulating absorptivity and emissivity, often neglecting key practical properties such as adhesion strength, thermal stability, and corrosion resistance. Therefore, while enhancing thermal control performance remains essential, systematic and comprehensive investigations into the overall performance of these coatings are equally important.
2. The effects of space radiation environments on the thermal control performance of PEO coatings remain underexplored. Given the indispensable role of these coatings in space applications, it is imperative to study their performance degradation under space radiation conditions. This can be achieved by combining simulation modeling with experimental validation, providing both theoretical insights and empirical data to guide practical applications.
3. Thermal control coatings with variable emissivity hold significant research and development potential. Leveraging the liquid-phase-assisted doping and sintering mechanisms inherent in the PEO process, the development of variable-emissivity materials through ion or micro-nanoparticle doping warrants further investigation.
4. Relying solely on the PEO technique often falls short of meeting both the thermal control performance requirements and the comprehensive performance needs of coatings. Therefore, integrating PEO with other advanced technologies offers a promising pathway for improving overall performance. Such hybrid approaches are expected to unlock broader application prospects for thermal control coatings.

In conclusion, addressing the challenges associated with magnesium and magnesium-lithium alloy thermal control coatings requires a multidimensional approach that balances thermal regulation, material stability, and durability under space conditions. Integrating advanced technologies with PEO will likely expand its applicability and deliver innovative solutions for thermal management across various critical fields.

Acknowledgments The partial supports from the NSFC grant nos. 52301085 and U21B2053, Heilongjiang Postdoctoral Science Foundation (No. AUGA5630105223), Heilongjiang Natural Science Foundation Joint Guidance Project, and Research Start-Up Fund by HIT are gratefully acknowledged.

References

1. Li, Z., Zhou, Z., Wang, Q., & Lu, S. (2023). *Current Applied Physics, 55*, 44–52.
2. Wang, Y., Ba, F., Chai, Z., & Zhang, Z. (2024). *International Journal of Electrochemical Science, 19*, 100514.
3. Yao, R., Yao, Z., Lu, S., Zhang, P., Xu, Y., & Wu, X. (2024). *Thin Solid Films, 797*, 140335.
4. Liang, J., Peng, Z., Cui, X., Li, R., & Wang, B. (2024). *Journal of Materials Research and Technology, 30*, 110–119.
5. Liang, J., Peng, Z., Li, R., & Wang, B. (2023). *Ceramics International, 49*, 29133–29140.
6. Liang, J., Peng, Z., Cui, X., Li, R., & Wang, B. (2024). *Ceramics International, 50*, 6349–6357.
7. Tachikawa, S., Nagano, H., Ohnishi, A., & Nagasaka, Y. (2022). *International Journal of Thermophysics, 43*, 91.
8. Cockeram, B. V., & Hollenbeck, J. L. (2002). *Surface and Coatings Technology, 157*, 274–281.
9. Zhu, M., Song, Y., Liu, Z., Xu, D., Dong, K., & Han, E. H. (2022). *Surface and Coatings Technology, 446*, 128797.
10. Pan, J., Wen, Y., Wang, L., Wu, Z., Dong, H., & Ye, Z. (2022). *Surface and Coatings Technology, 431*, 128035.
11. Pillai, A. M., Rajendra, A., Sharma, A. K., Bera, P., Poornima, S., & Sampath, S. (2021). *Surface and Coatings Technology, 410*, 126891.
12. Du, C., Zhao, H., Dai, Z., Tian, Z., Wang, J., & Wang, Z. (2019). *Materials Letters, 236*, 723–726.
13. Hakimizad, A., Raeissi, K., Santamaria, M., & Asghari, M. (2018). *Electrochimica Acta, 284*, 618–629.
14. Nagumothu, R. B., Thangavelu, A., Nair, A. M., Sukumaran, A., & Anjilivelil, T. (2018). *Transactions of the Indian Institute of Metals, 72*, 47–53.
15. Tu, C., Chen, X., Liu, C., & Li, D. (2023). *Materials, 16*, 4142.
16. Wu, X., Qin, W., Cui, B., Jiang, Z., & Lu, W. (2007). *International Journal of Applied Ceramic Technology, 4*, 269–275.
17. Arunnellaiappan, T., Anoop, S., & Rameshbabu, N. (2016). *Surface and Coatings Technology, 307*, 735–746.
18. Ma, D., Lu, C., Fang, Z., Yan, W., Wei, L., Ni, Y., & Xu, Z. (2016). *Applied Surface Science, 389*, 874–881.
19. Pillai, A. M., Rajendra, A., Sharma, A. K., & Sampath, S. (2019). *Journal of Applied Electrochemistry, 49*, 1239–1254.
20. Chen, G., Wang, Y., Zou, Y., Wang, H., Qiu, J., Cao, J., Wang, S., Jia, D., & Zhou, Y. (2021). *Chemical Engineering Journal, 421*, 127802.
21. Tu, W., Zhu, Z., Zhuang, X., Cheng, Y., & Skeldon, P. (2019). *Surface and Coatings Technology, 372*, 34–44.
22. Li, X., Xia, Q., Chen, C., & Yao, Z. (2019). *Materials Research Express, 6*, 106428.
23. Zhang, Z., He, F., Huang, C., Song, Z., Yang, J., & Wangm, X. (2022). *International Journal of Applied Ceramic Technology, 19*, 2203–2212.
24. Li, H., Wang, Y., Geng, J., Li, S., & Chen, Y. (2022). *Materials, 15*, 8755.
25. Lu, S., Qin, W., Wu, X., Wang, X., & Zhao, G. (2012). *Materials Chemistry and Physics, 135*, 58–62.

26. Wang, L., Zhou, J., Liang, J., & Chen, J. (2013). *Applied Surface Science, 280*, 151–155.
27. Wang, X., Ju, P., Lu, X., Chen, Y., & Wang, F. (2022). *Ceramics International, 48*, 3615–3627.
28. Wang, X., Lu, X., Ju, P., Chen, Y., Zhang, T., & Wang, F. (2020). *Surface and Coatings Technology, 393*, 125709.
29. Xia, Q., Zhang, D., Li, D., Jiang, Z., & Yao, Z. (2019). *Surface and Coatings Technology, 369*, 252–256.
30. Li, H., Lu, S., Qin, W., & Wu, X. (2017). *Acta Astronautica, 136*, 230–235.
31. Li, H., Lu, S., Wu, X., & Qin, W. (2015). *Surface and Coatings Technology, 269*, 220–227.
32. Wang, X., Lu, X., Ju, P., Chen, Y., Zhang, T., & Wang, F. (2021). *Surface and Coatings Technology, 409*, 126905.
33. Han, J., Chen, Y., Tu, W., Zhan, T., & Cheng, Y. (2018). *Applied Surface Science, 428*, 684.
34. Yao, Z., Xia, Q., Shen, Q., Ju, P., Su, P., Hu, B., & Jiang, Z. (2015). *Solar Energy Materials and Solar Cells, 143*, 236–241.
35. Yao, R., Li, Y., Yao, Z., Zhang, P., Lu, S., & Wu, X. (2022). *Surface and Coatings Technology, 429*, 127934.
36. Madhuri, D., Ghosh, R., Hasan, M. A., Dey, A., Pillai, A. M., Angamuthu, M., Anantharaju, K. S., & Rajendra, A. (2022). *Ceramics International, 48*, 35906–35914.
37. Yao, Z., Hu, B., Shen, Q., Niu, A., Jiang, Z., Su, P., & Ju, P. (2014). *Surface and Coatings Technology, 253*, 166–170.
38. Xia, Q., Wang, J., Liu, G., Wei, H., Li, D., Yao, Z., & Jiang, Z. (2016). *Surface and Coatings Technology, 307*, 1284–1290.
39. Wu, G., & Yu, D. (2013). *Progress in Organic Coating, 76*, 107–112.
40. Yu, H., Shen, X., Yan, X., & Cheng, C. (2009). *Applied Surface Science, 255*, 6077–6081.
41. Wen, C. D., & Mudawar, I. (2006). *International Journal of Heat and Mass Transfer, 49*, 4279–4289.
42. Chen, G., Xie, E., Zhao, Q., Ye, Z., Wang, S., Zou, Y., Wang, Y., Qiu, J., Wang, J., Shuai, Y., & Zhou, Y. (2024). *Applied Surface Science, 673*, 160818.

Chapter 19
Plasma Electrolytic Oxidation in Catalytic and Battery Materials Applications

Viswanathan S. Saji

19.1 Introduction

Plasma electrolytic oxidation (PEO) is an intricate electrochemical conversion approach that applies high voltages (~150–600 V) to fabricate a thick and adherent oxide layer on metals and alloys. This is achieved through complex thermal, electrochemical, and plasma-chemical reactions at the metal/electrolyte interface. The various process stages include the initial growth of thin passive surface oxides, dielectric breakdown, formation of discharge channels, generation of plasma, plasma-assisted reactions, emission of gases, and creation of crater-like surface oxides [1–6]. The basic details of PEO are described in the previous chapters of this book, highlighting the significance of process parameters.

Catalysts are vital in many industrial processes. Photocatalysts have been widely explored to address severe water pollution from industrial waste and human activities. They are reliable for air treatment, water treatment, green hydrogen production, carbon dioxide reduction, and various photocatalyzed chemical processes [7–10]. The photocatalysis process hinges on the fascinating power of light irradiation (UV/visible) on a semiconductor surface. When the photon energy surpasses the semiconductor's bandgap energy, electrons become excited and shift to the conduction band (CB), generating excitons (electron-hole pairs). However, the crucial step in this process is the delayed electron-hole recombination. If this recombination is delayed, they will participate in the desired reactions on the semiconductor's surface [11]. Conversely, electrocatalysts depend entirely on electrochemical processes facilitated by exchanging electrons and charge carriers between reactants.

V. S. Saji (✉)
Interdisciplinary Research Center for Advanced Materials, King Fahd University of Petroleum & Minerals, Dhahran, Saudi Arabia
e-mail: saji.viswanathan@kfupm.edu.sa

V. S. Saji, M. Zheludkevich (eds.), *Plasma Electrolytic Oxidation – Fundamentals, Advances and Applications*,
https://doi.org/10.1007/978-3-031-92429-3_19

Electrocatalysts are significant in various applications ranging from electrolyzers and fuel cells for energy storage devices to electro-organic synthesis [10]. Photoelectrocatalysis combines photocatalytic and electrochemical approaches. In this method, the photocurrent responses of the electrodes are measured with an applied potential [12].

Compared to powdered photocatalysts, metallic-supported catalysts are attractive due to their enhanced mechanical and physical properties and easy shape adjustment. It is more reliable than coating the active component on a conducting support. PEO is an excellent approach in this line. The porous outer layer of the PEO layer could help improve the effective active surface area. The developed PEO oxide layer itself could function as a catalyst. The catalytic activity can be further boosted by various approaches, such as doping and composite formation via in situ integration or post-impregnation of active components. For in situ incorporation, the required chemicals/particles are added to the PEO electrolyte, which is subsequently assimilated via electro/electrophoretic deposition during PEO. Various post-impregnation methods are also attempted to decorate the PEO layer with additional active components [3, 13–17].

Only a few works have been reported on the electrocatalytic applications of PEO compared to photocatalytic studies. A few reports focused on photoelectrochemical studies (Fig. 19.1). The most significant contribution to photo/photoelectro-catalysis is obvious as researchers attempted to develop suitable semiconductor oxides on various valve metals known for their superior photocatalytic activity [3, 13, 18, 19]. This chapter provides a concise account of PEO-derived electro (Sect. 19.2.1), photo (Sect. 19.2.2), and photoelectro (Sect. 19.2.3) catalysts in different applications. Also provided is an account of PEO in battery materials applications (Sect. 19.3).

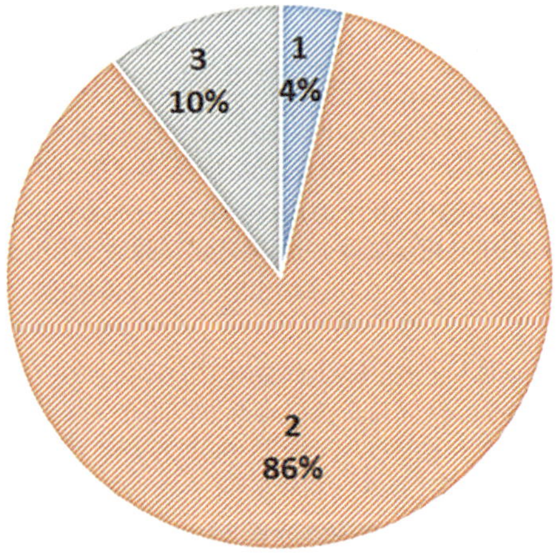

Fig. 19.1 A Pie chart on the extent of reported studies on PEO in catalytic applications: (1). Electrocatalysis, (2). Photocatalysis, (3) Photoelectrocatalysis (Source: Scopus/various sources)

19.2 Catalytic Applications

19.2.1 Electrocatalysis

PEO technology could fabricate electrocatalyst electrodes with large-area adhesion and uniformity at a reduced cost. Only a few recent works have explored PEO in electrocatalytic applications, which include Pd catalyst supported on porous TiO_2 for CO oxidation [8], Ni and Cu hydroxide suspensions on TiO_2 for urea oxidation reaction (UOR) [20], MoTiN and NiMoTiN coated on Ti for hydrogen evolution reaction (HER) [21], and Zn_2SnO_4-Ti for water oxidation reaction (WOR) [22]. Among them, a few works attempted post-impregnation [8] of the active material into the PEO support, whereas others attempted in situ amalgamation [20] of the active material during the PEO process.

Samadi et al. explored a Pd electrocatalyst supported by TiO_2/Ti for CO conversion into CO_2. The CO oxidation capability of the Pd catalyst strongly depended on the nature of the support. The TiO_2 was fabricated by PEO of Ti, and the Pd catalyst was impregnated by an adsorption method. The optimum concentration of Pd was ~0.14 wt%. The catalyst reached a 100% conversion of CO at 200 °C and, importantly, was found to be stable under the reaction conditions [8]. Wala et al. explored in situ electrocatalyst (Cu or Ni)-doped TiO_2 for UOR for application as anode materials in urea-fueled fuel cells. The PEO of Ti was conducted in an alkaline silicate solution containing Ni and Cu hydroxide suspensions. The micro/mesopores on the PEO surface provided a highly active area. PEO in electrolytes containing the additives led to reduced discharges, leading to smaller diameter pores, which benefited the electrochemical activity. Their studies in 1 M KOH + 0.15 M urea solution showed that adding Cu and Ni hydroxides to the PEO electrolyte has risen the UOR current densities. However, the performance was inferior compared to the reports on TiO_2 photocatalyst for UOR [20].

PEO electrodes have shown promising potential to boost the mechanical strength of HER catalysts. Wei et al. prepared MoTiN and NiMoTiN-coated Ti via PEO. Na_2MoO_4 and $NiAc_2$ were introduced into a phosphate electrolyte, and MoTiN and NiMoTiN coatings were developed using PEO and NH_3 annealing treatments. The electrochemically active phases toward HER were MoN and NiMoN. In 1.0 M KOH, the overpotentials (at 10 mA/cm^2) required for NiMoTiN and MoTiN electrodes were 121 and 126 mV, respectively, which were comparable to the Pt electrode. The MoTiN coating fabricated with 8 g/L Na_2MoO_4 yielded the best HER activity. Higher concentrations caused intense sparking and substantial voltage fluctuations during PEO, which drastically reduced the quality and adhesion of the coating [21]. Another study utilized PEO to fabricate Zn_2SnO_4-Ti anodes for two-electron WOR for H_2O_2 as potent in situ oxidants for the elimination of carcinogenic compounds in aqueous solutions [22].

The instant synthesis of Ru nanoparticles entrenched on a porous MgO support was realized by PEO. The developed Ru/MgO catalyst presented high hydrogen generation via hydrolysis of sodium and ammonium borohydrides [23]. These reports indicate that suitably optimized supported electrocatalysts can be fabricated under optimized PEO conditions. More studies are required in this area, and their long-term durability needs to be verified.

19.2.2 Photocatalysis

TiO_2 is one of the most effective photocatalysts due to its low cost, chemical stability, and ability to absorb photons and generate rapid excitons. The band gap energies of the rutile and anatase phases of TiO_2 are 3.0 and 3.2 eV, respectively [24, 25]. Irrespective of the smaller bandgap, anatase TiO_2 exhibits better photocatalytic activity than rutile, owing to the lesser electron-hole recombination and larger active material volume [26, 27]. One advantage of the PEO process is that the anatase content within pristine TiO_2 films can be manipulated by simply adjusting PEO parameters [28, 29].

However, the wide bandgap (limited visible light absorption) and fast electron-hole recombination reduce the quantum yield [26, 27]. The primary approaches attempted to shift the bandgap of the TiO_2 to lower values, where visible light would be adequate, are doping and heterojunctions [11]. Earlier studies showed that nanocrystalline titania films synthesized by PEO of Ti or its alloys exhibit better photocatalytic activity [30, 31], where the anatase to rutile phase ratio was decisive [28, 29].

Typically, organic dyes (methylene blue, rhodamine B, etc.) are used for photocatalytic studies. The dye degradation is a first-order kinetic reaction, where the rate constant (k) is calculated using the equation $\ln (C_0/C_t) = k$, where C_0 signifies the initial concentration and C_t denotes the residual concentration after irradiation time t [32, 33].

Several studies attempted to fabricate TiO_2 coatings with a pure anatase phase. Lin et al. have shown that PEO in optimized NaOH + $(NaPO_3)_6$ electrolytes helped to transform rutile-anatase mixed phases into pure anatase. The photocatalytic activity of the PEO layer developed variedly with the PEO voltage (Fig. 19.2a) and the electrolyte concentration (Fig. 19.2b). The linear relation supports the first-order reaction kinetics (Fig. 19.2a). The rate constants k, calculated from the slopes of the plots (Fig. 19.2b), were in accordance with the extent of anatase phase formed [34].

The crystal phase of TiO_2 can be significantly influenced by the PEO processing time and the voltage/current applied. The PEO layer with a 100% anatase phase will likely be formed at lower voltages and processing times. Coto et al. showed that at 1–8 min of processing, 100% anatase phase was detected, whereas the sample processed for 40 min had approximately a 70/30% of anatase/rutile ratio. This variation was attributed to the improved coating thickness and more energetic discharges, favoring rutile formation. The optimal PEO structure, observed with the sample processed for 8 min, gave the maximum photocatalytic effect [35].

Doping with nonmetals, noble metals, and transition metals could significantly boost the photocatalytic activity of PEO-derived TiO_2. Transition metals with

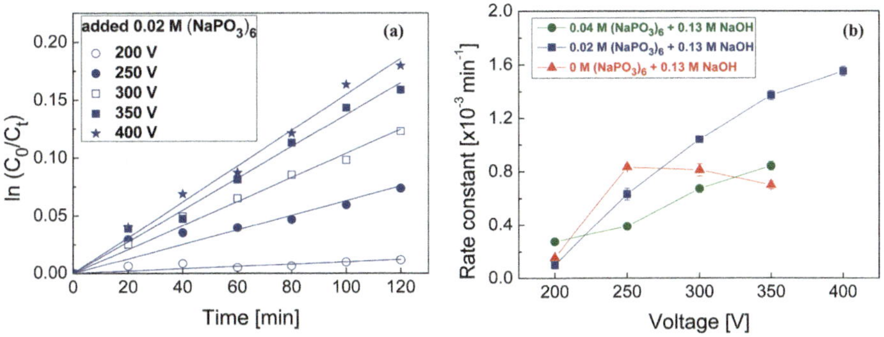

Fig. 19.2 (**a**) Photodegradation of TiO_2 coatings fabricated at different PEO voltages (10 min, 0.13 M NaOH + 0.02 M $(NaPO_3)_6$ electrolyte) vs irradiation time. (**b**) Rate constants vs applied voltage plots with various electrolyte concentrations. Reproduced with permission from [34] © 2018 Elsevier B.V.

atomic radii similar to Ti^{4+} (0.68 Å) can favor their facile integration into the TiO_2 crystal lattice without persuading substantial crystal structure deformation, which allows novel intra-bandgap states [36, 37]. The nonmetal (such as N, F, B, S, and C) doping of TiO_2 has gathered significant research interest, as they could reduce the band gap and recombination centers. Co-doping of more than one nonmetal/metal was explored to further extend the photocatalytic activity into the visible region [38].

Nitrogen is one of the most efficient non-metal dopants. N can be doped from the substrate alloy itself, or a suitable electrolyte additive (e.g., NH_3) can be used. Sridhar et al. fabricated a TiN coating on Ti by plasma nitriding and further subjected it to PEO. The band gap of TiO_2 was reduced to 2.92 eV after N-doping. Photoluminescence investigation showed that doping reduced the recombination kinetics. The higher photocatalytic activity was attributed to the proper anatase/rutile ratio, lowered band gap energy, lesser recombination kinetics and greater active surface area [39]. Another study reported that N-doped anatase TiO_2 was fabricated using TiN seeding layers (on Ti) and an NH_4OH-added PEO electrolyte. The band gap of anatase TiO_2 developed by PEO was reduced from 3.01 to 2.76 eV after N-doping, suggesting a facile method of preparing N-doped TiO_2 by PEO [40]. The enhanced photocatalytic efficiency of a B and N co-doped TiO_2 fabricated on a TiN/Ti sample in H_2SO_4/HBO_3 electrolyte was endorsed to the combined effect of interstitial N, chemisorbed NO_x, and the occurrence of Ti-O-B-N structure. The co-doped TiO_2 demonstrated significantly greater photocurrent response compared to the N-doped or non-doped counterparts (Fig. 19.3a). The photocatalytic mechanism was suggested to involve different processes (Fig. 19.3b), including (1) photo-generated electron transfer from the valence band (VB) to conduction (CB) of pure TiO_2, (2) and substitutional B ions, (3, 4) partial reduction of Ti^{4+} to Ti^{3+} by the interstitial B leading to charge separation and photo-generated electron traps, (5) those associated with substitutional N ions, (6) new state energy level by the

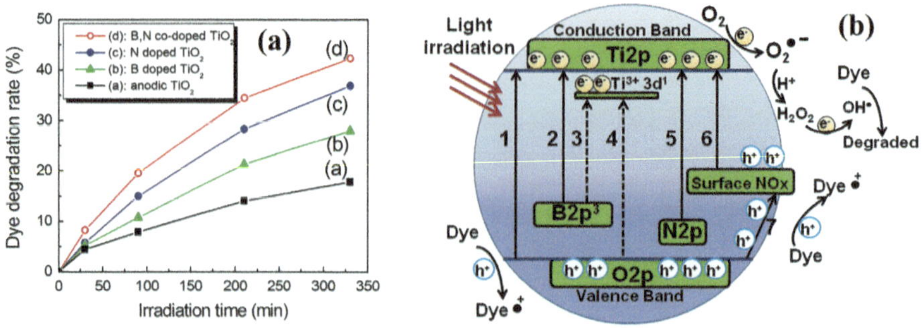

Fig. 19.3 (**a**) Photocatalytic degradation rate of aniline blue with differently doped samples. (**b**) Photocatalytic mechanism for the B, N co-doped TiO_2. Reproduced with permission from [38] © 2015 Elsevier Ltd. and Techna Group S.r.l.

interstitial N, and (7) photo-induced hole transfer from the VB of TiO_2 to the energy level of NO_x species [38].

A study on transition metals (Mn, Ni, and Co)-doped TiO_2 demonstrated that the PEO coatings formed in 0.75 g/L MnO, 1.5 g/L NiO, and 2.0 g/L Co_3O_4 added electrolyte had the highest photocatalytic activity. The primary reason for the enhanced activity was correlated to the reduced recombination rate of electron-hole pairs [41]. Fe is an appropriate candidate for doping TiO_2. Zhang et al. fabricated PEO on Ti in a phosphate electrolyte and further infused it with a $Fe(NO_3)_3$ solution. The photocatalytic degradation of the anatase-rich composite film prepared after 24 h of $Fe(NO_3)_3$ impregnation has reached ~60% (180 min of irradiation), much better than that of the plain TiO_2 film [42]. A study on three TiO_2 films with different anatase/rutile compositions showed that the Fe concentration integrated into the TiO_2 can be simply regulated by suitably altering the Fe salt concentration in the electrolyte. The film with 4% Fe concentration exhibited an optimum decomposition rate of methylene blue (MB) under visible light irradiation, owing to an improved light absorption and reduced recombination [29]. Similarly, doping with Ni [43], V [44], W [45], and rare earths [46, 47], as well as their co-doping [41], have been proven to enhance the photocatalytic activity of PEO-derived TiO_2.

The strategy of coupling TiO_2 with suitable semiconductor oxides has been explored with PEO-made coatings. The Bi_2O_3-modified TiO_2 coating developed by PEO of Ti in Bi_2O_3 particles-added phosphate electrolyte, showed a remarkable improvement in photoactivity. The highest activity was detected for the coating developed in 1.2 g/L Bi_2O_3 added electrolyte. This improvement was attributed to the reduced electron-hole recombination rate due to TiO_2-Bi_2O_3 coupling [48]. CdS (band gap ~2.3 eV) mixing enhanced the photocatalytic activity of PEO-TiO_2. CdS-modified anatase TiO_2 was developed on Ti by PEO in 10 g/L Na_3PO_4 + 0.4 g/L CdS electrolyte [49]. SnO_2-TiO_2 films prepared by PEO of Ti from electrolytes containing sodium stannate [50] and SnO_2 particles [51] yielded better photocatalytic activities. Stojadinovic et al. showed that the photocatalytic activity of TiO_2/SnO_2 coating greatly depends on the SnO_2 particle's concentration in the electrolyte, i.e., the maximum activity was recorded with 0.3 g/L SnO_2. The coating was partially crystallized and comprised mainly of anatase TiO_2. The SnO_2 particles suppressed the electron-hole recombination. The enhanced photocatalytic activity was correlated to the improved charge separation due to coupling semiconductors with discrete energy levels [51].

Several reports showed that coupling Fe_2O_3 with PEO-TiO_2 has a significant effect. Fe_2O_3 is abundant, environmentally compatible, and economical [52–54]. The PEO layer developed in phosphate electrolyte carrying 3 g/L Fe_2O_3 nanoparticles significantly enhanced the dye degradation. With the embedment of Fe_2O_3 in the TiO_2 layer, the absorption band-edge was shifted to longer wavelengths and the optical band gap was reduced from 3.12 to 2.0 eV. The corresponding reaction rate constants were 0.0135 and 0.0770 h^{-1} [52]. A study on Fe-TiO_2 made in the $FePO_4$ electrolyte via PEO yielded a reduced band gap of 2.57 eV. 92% of MB decomposition was attained during 270 min of visible illumination [54]. Similar enhancements were reported for other composites, such as TiO_2-CuO [55], TiO_2-V_2O_3 [56, 57], and TiO_2-WO_3 [58–61]. Photocatalytic activity of TiO_2-WO_3 coatings developed in 12-tungstosilicic acid altered with PEO duration (30–600 s), and greater photoactivity was achieved for shorter times. The oxide coatings were partially crystallized and were mainly comprised of WO_3 and anatase TiO_2 (band gap ~2.6 eV). Decreased microdischarge channels, particle agglomeration, and increased recombination kinetics reduced the photocatalytic activity [58]. Figure 19.4 shows the effect of electrolyte concentration, current density, and duration on the methyl orange photodecolorization efficiency of the TiO_2-WO_3 PEO layer by Petrović et al. [61]. Bayati et al. showed that WO_3 was distributed in the TiO_2 matrix and doped into the TiO_2 lattice. Band gap energy was ~2.87 eV [60].

Wu et al. prepared TiO_2-WO_3/TiO_2 film via a two-step methodology. The WO_3/TiO_2 was created by PEO, followed by depositing N, C co-doped TiO_2 by magnetron sputtering. The substantially porous microstructure and well-crystallized N, C-TiO_2 top-layer synergistically contributed to the photocatalytic activity [59]. Budnikova et al. reported $Fe_xCo_{1-x}WO_4$ film on Ti developed in a tungstate

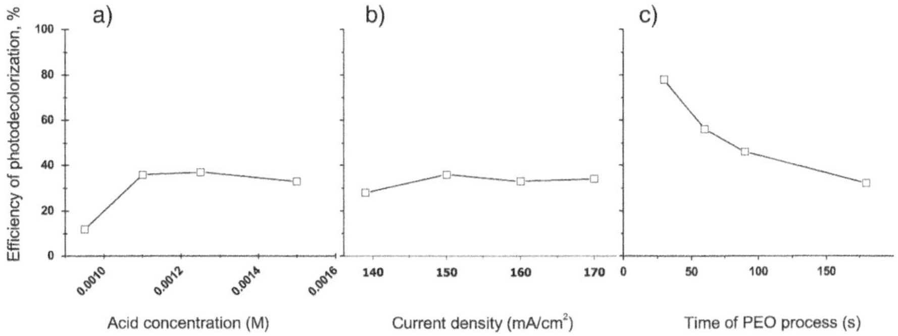

Fig. 19.4 PEO process parameters vs. photodecolorization efficiency at different (**a**) electrolyte concentration (150 mA/cm^2, 90 s), (**b**) current densities (1.1 × 10^{-3} M, 90 s), and (**c**) time (150 mA/cm^2, 1.1 × 10^{-3} M) [61]. Reproduced with permission from [61] © 2014 Elsevier B.V.

electrolyte containing Co(II)-EDTA and Fe(II)-EDTA anions. The coating contained crystalline anatase TiO_2 and orthorhombic WO_3. The band gap determined was in the range of 2.5–2.9 eV. The higher methyl orange degradation obtained for the Ti-W-Co sample in the presence of H_2O_2 was explained by the n-p heterojunction formation between TiO_2-WO_3 and $CoWO_4$ and the activation of H_2O_2 by Co and Fe via Fenton-like processes [62]. ZnO-WO_3-TiO_2 composite coating was prepared on Ti in Na_2PO_3 + $H_3O_{40}PW_{12}$ electrolyte having 1.5 g/L ZnO nanoparticles by PEO. The coating formed at 360 V yielded the maximum activity; about 80% of the MB was removed during UV irradiation for 120 min. ZnO doping significantly enhanced the activity; the best activity was achieved by the coating with the minimal W:Zn ratio [63]. Lukiyanchuk et al. fabricated a PEO layer on Ti in $Na_6P_6O_{18}$ + $NaVO_3$ + Na_2MoO_4 electrolyte. The results suggested that V substituted Ti while Mo was concentrated in the surface layers [64].

ZrO_2 has excellent mechanical properties, low friction coefficient, high biocompatibility, and good thermal and chemical stability. The presence of ZrO_2 could be beneficial for avoiding the anatase to rutile phase transformation and reducing the recombination kinetics. ZrO_2-TiO_2 film was fabricated on Ti using in situ incorporation of $Zr(OH)_4$ colloidal particles. The film had a porous and lamellar structure consisting of ZrO_2 and anatase-rutile phases. The composite and pure film's optical absorption edges were about 421 and 412 nm, respectively. The dye degradation rate constants for the two cases were 0.0442 and 0.0186 h^{-1}, respectively [65]. The photocatalytic activity varied with the ZrO_2 content in the electrolyte and the duty cycle. The enhancement was ascribed to the decreased electron-hole recombination in TiO_2 due to the ZrO_2 incorporation [66]. A study on PEO on Ti-6Al-4 V alloy in an electrolyte containing ZrO_2 nanoparticles showed that the composite coated at 2 A/dm^2 exhibited the best photocatalytic activity [67].

Rare earth oxides are a promising choice [68, 69]. A CeO_2-Nb_2O_5 coating was successfully applied to Ti via PEO in ammonium niobium oxalate + CeO_2 electrolyte. The band gap of the composite TiO_2 was ~3.3 eV. The sample fabricated after 600 s of PEO demonstrated the most substantial MB degradation under UV irradiation. The study also revealed that the activity of $AlNbO_4$-ZnO and $AlNbO_4$-Eu_2O_3 coatings, formed in $NaAlO_2$ electrolyte containing ZnO and Eu_2O_3 particles, was significantly improved compared to an $AlNbO_4$ counterpart [68]. Orsetti et al. showed that CeO_2-Nb_2O_5 coating made via 600 s of PEO yielded 58% MB degradation after 210 min of UV exposure [69].

Coupling TiO_2 with nanocarbons such as carbon nanotubes (CNT) and graphene is promising due to the large surface area, high charge carrier mobility, and electron-trapping ability of nanocarbons. The high photocatalytic dye degradation efficiency (93%) obtained for a W-doped TiO_2-reduced graphene oxide (rGO) coating was ascribed to the synergistic effect of 2D rGO and the doped W. Here, Na_2WO_4 (0–8 g/L) was added to a 5 g/L $Na_3(PO_4)_6$ + 2 g/L KOH + 0.5 g/L rGO electrolyte. The study suggested that the W concentrates on the TiO_2 grains, impedes crystal growth and TiO_2 phase transformation, and decreases recombination kinetics. rGO provided enhanced specific surface area [70]. Yu et al. developed a GO-doped PEO coating using GO (100 mg/L) and sodium dodecyl benzene sulfonate (SDS, 2 g/L) added sodium silicate electrolyte. The composite coating outperformed the plain PEO coating, achieving a maximum degradation efficiency of 95% after 5 h of simulated sunlight exposure. The anatase/rutile ratios in the GO and GO-SDS coatings were 1.85 and 1.62, respectively. The increased specific surface area due to GO was a critical factor in enhancing the photocatalytic activity [71].

These reports stress the potential of PEO layers developed on Ti using optimized PEO parameters and electrolyte compositions for industrial photocatalytic applications. Compared to Ti, reported works on other metals and alloys are few. The second most investigated are Mg [72–76] and Al [77–79]. Recent reports are also available on other metals, including Nb [80], Ta [81], Zr [82], Sn [83], and Zn [84].

19.2.3 Photoelectrochemical Catalysis

TiO_2 has been significantly researched for various photoelectrocatalytic applications, mainly photoelectrochemical water splitting for green hydrogen production [85]. The photocurrent (difference of current responses under irradiation and dark) and the incident photon-to-current conversion efficiency ($IPCE = j/\lambda P$) are primarily used to evaluate the effectiveness, where j is short circuit current density, λ is monochromatic light wavelength, and P is incident radiant power [12, 86].

A study on PEO-made TiO_2 in 0.5 M H_2SO_4 electrolyte at 160 V showed that lowering the electrolyte temperature (20 °C to −3.5 °C) can increase the anatase weight fraction up to 92%. The highest photocurrent density (0.18 mA/cm^2) was obtained on TiO_2 coatings developed in refrigerated electrolytes. The IPCE recorded at 254 nm irradiation was nearly 100% [86]. The anatase/rutile relative amount varied from 100% anatase at low potentials (110–130 V) to 100% rutile at higher potentials (180–200 V). Accordingly, the band gap changed from ~3.20 to 2.94 eV. There was a marginal red shift of ~0.1 eV as the electrolyte temperature was reduced from 25 °C to 0 °C. The TiO_2 photoanode synthesized at 140 V reached the highest IPCE of 88%. Correspondingly, the fastest dye decolorization kinetics were observed [87].

S-doped TiO_2 coatings developed by PEO in 0.1 M H_2SO_4 showed that the duty cycle strongly influenced the crystalline/amorphous and anatase/rutile ratios. The sample developed with a 50% duty cycle had the lowest bandgap and highest photoelectrochemical activity [88]. A study on Ni-doped TiO_2 films fabricated in 1.5 M $H_2SO_4 + NiSO_4$ electrolyte demonstrated that the rutile phase stabilizes as the $NiSO_4$ concentration in the electrolyte increases. The photocurrent varied with the $NiSO_4$ concentration in the electrolyte. Maximum IPCE values of 55% were obtained for 0.2 mM $NiSO_4$ [89]. Tucci et al. on Fe, Zn, and Cu-doped TiO_2 fabricated in refrigerated electrolytes showed that metal doping had a minor influence on the anatase/rutile crystal structure. Doped oxides made via low metal precursor concentration had higher IPCE values, which were in the order Zn > Cu > Fe-doped TiO_2 [37].

The larger photocurrents obtained during photoelectrochemical water splitting for TiO_2/Cu_xO photoanodes compared to TiO_2 were attributed to the improved crystallinity and reduced band gap. The photoanode performance was improved by adding methanol (hole scavenger) due to the reduced surface electron-hole recombination. The peak photocurrent was attained rapidly with the illumination of the composite TiO_2 photoanode compared to the approximately 2 s delay of the plain TiO_2 signaling improved photoexcited charge transfer (Fig. 19.5a, b). The respective IPCE values confirmed that the heterostructures generated greater photocurrents (Fig. 19.5c). Longer duration studies (25 mW/cm^2 illumination at 1.23 V for 2 h) displayed better stability (Fig. 19.5d) [27].

Bilayer MoS_2 (light-absorbing) and rGO (conductive layer) sensitized TiO_2 was fabricated. MoS_2 has a lower bandgap and matching energy band structure with TiO_2. The 2D structure of MoS_2 could provide a high specific surface area. The band gap was reduced from 3.07 to 2.55 eV. The recombination kinetics is also reduced. The bilayer photocatalyst enhanced the degradation kinetics of MB by 1.6 times under irradiation and 5 V applied bias. The electrode revealed better photo-corrosion stability [90]. Another study utilized a Ti/TiO_2-$ZrTiO_4$-ZrO_2 composited electrode. The composite electrode generated photocurrents of 230 μA in the potential range of 0.2–1.2 V, which is favorable for use as photoanodes for water splitting. The high photoelectrochemical activities were attributed to the effective photogenerated

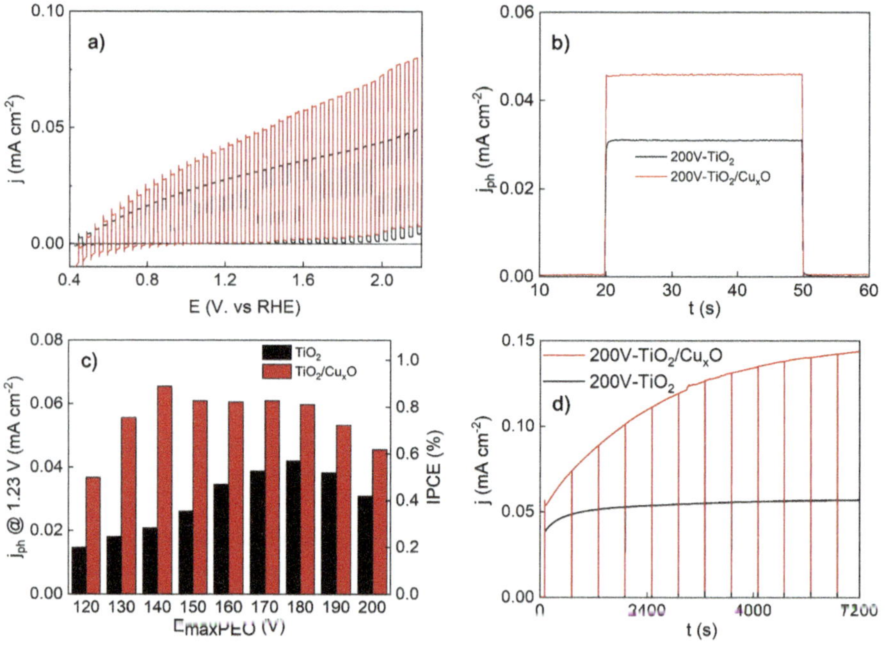

Fig. 19.5 Photoelectrochemical properties of TiO_2/Cu_xO and TiO_2 films: (**a**) linear sweep voltammetry at 5 mV/s with chopped illumination, (**b**) steady-state photocurrent pulses at 1.23 V, (**c**) photocurrents at 1.23 V and IPCE, (**d**) longer duration study. Reproduced with permission from [27] © 2024 The Author(s). Published on behalf of The Electrochemical Society by IOP Publishing Limited

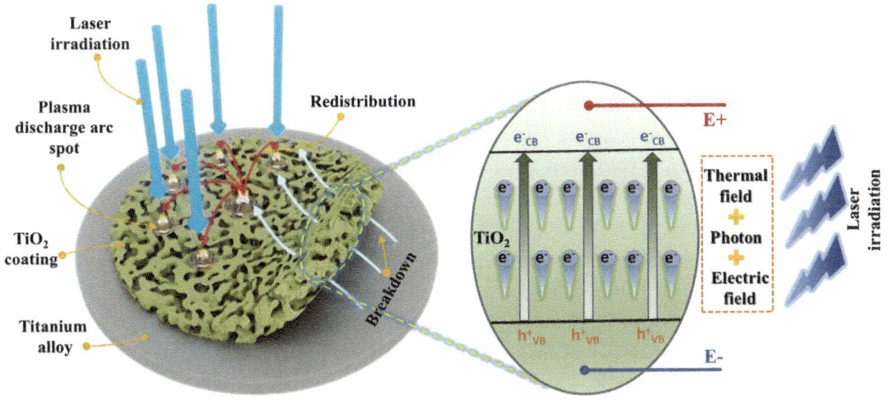

Fig. 19.6 A schematic of the laser irradiation effect on the PEO process [93]. Reproduced with permission from [93] © 2024 Elsevier B.V.

electron-hole separation and heterojunction-assisted improved electron transfer [91].

Nanoscale noble metal could reduce electron-hole recombination kinetics and favor efficient charge separation. A study on a 10 nm gold film (by electron beam evaporation) deposited Ti/TiO_2 composite electrode showed that the Au deposition was beneficial for better conductivity and photocurrent for water electrolysis. The photocurrent density of the composite was three orders of magnitude higher [92].

Laser irradiation can be used along with the PEO process. Laser-assisted irradiation could promote PEO, accelerate anatase-rutile conversion, reduce oxygen vacancy defects, and create a unique porous structure. Figure 19.6 shows a schematic of the mechanism of the effect of laser irradiation on PEO. The synergistic effect between the thermal and electric field and the photoelectric effect during laser-assisted plasma electrolysis processing could significantly reduce the conditions for generating electron avalanches, resulting in the induction of more plasma discharges on the anode surface [93]. The photoelectrochemical activity of the laser-assisted PEO coating was better than the plain PEO coating [93, 94].

19.3 Battery Materials Applications

Several studies have delved into the potential of PEO-based anode materials for lithium-ion batteries (LIBs), offering a promising alternative to conventional graphite anodes. The low theoretical specific capacity of graphite-based electrodes (372 mAh/g) is not suitable for achieving the energy density required for large-capacity energy storage systems and electric vehicles. Transition metal oxides, including TiO_2, have attracted significant interest as anode materials due to their suitable discharge potential (~1.7 V), negligible volume expansion ratio, and exceptional cycling stability. However, the limitations of TiO_2 anodes, such as low electrical conductivity and capacity, are significant hurdles in the field [95–97].

Transition metal dichalcogenides have been widely studied as a high-capacity anode (~670 mAh/g); however, they have a high-volume expansion ratio during lithiation/delithiation, leading to rapid capacity fading. The TiO_2-MoS_2 composite anode, fabricated by a two-step approach using PEO of Ti and magnetron sputtering of MoS_2, has been shown to be an effective anode. The composite can use the high specific capacity of MoS_2 while avoiding significant volume change. TiO_2 could buffer the volume change. A constructed battery demonstrated a specific capacity >400 mAh/g at 100 μA/cm^2, much greater than TiO_2 anode. The specific capacity remained constant after 100 cycles (Fig. 19.7a), highlighting the stability of the composite anode over repeated use. The discharge capacity is significantly greater than the capacity of TiO_2 nanotubes and TiO_2/Au anodes but lower than the initial capacity of pure MoS_2 [97].

The theoretical capacity (1494 mAh/g) of SnO_2 is approximately four times that of graphite. As discussed above, its large volume change (> 300%) is the drawback. SnO_2-TiO_2/graphene composite anode was fabricated using graphene as a conductive additive. Here, the composite electrode was prepared by PEO in Na_2SnO_3 + Na_3PO_4 electrolyte with or without added graphene. After 200 cycles, the composite electrode delivered a high reversible capacity of 482 mAh/g at 1.0 C. The result demonstrated the significant role of SnO_2 and graphene in the composite electrode's improved specific capacity and cycle stability. The cycling performance varied with PEO time; the best performance was obtained for a 120 s sample (Fig. 19.7b) [96].

Silicon-based materials are attractive anode materials owing to their high theoretical capacity. A porous SiO_2-TiO_2 composite prepared via PEO using Ti covered with polydimethylsiloxane resulted in a SiO_2 integrated oxide layer with larger pores. The pores facilitated Li-ion transportation, and the electrode attained a higher

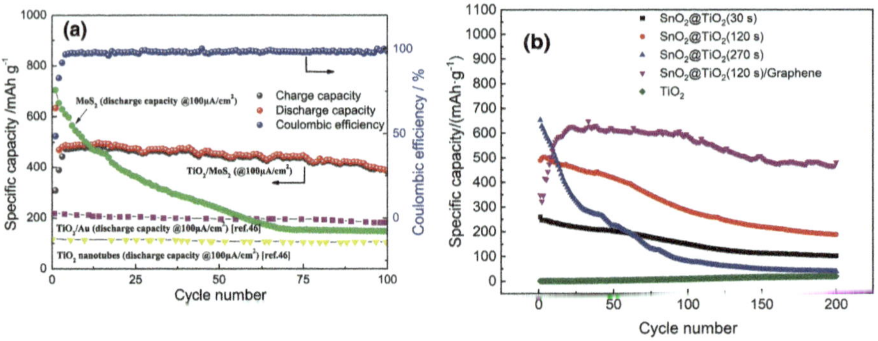

Fig. 19.7 (**a**) Capacity retention of TiO_2/MoS_2 composite anode at 100 μA/cm^2 [97]. (**b**) Cycling performance of SnO_2-TiO_2-based electrodes [96]. Reproduced (**a**) from [97] © 2021 Elsevier B.V. and (**b**) [96] © 2022 Elsevier B.V.

areal capacity of 1505.35 μAh/cm^2 at 500 μA/cm^2. The electrode's capacity retained a high value of 1071.84 μAh/cm^2 after 200 cycles [98].

19.4 Conclusions and Outlook

This chapter succinctly explains the catalytic and battery materials applications of plasma electrolytic oxidized (PEO) metals and alloys. PEO-derived electro-, photo-, and photoelectro-catalysts, with their adherent oxide layers, have recently garnered significant attention in catalysis research. These metallic-supported catalysts, prepared by PEO, are desirable for practical applications due to their self-supporting structure and enhanced mechanical and physical properties. The porous outer layer further enhances the active surface area. The catalytic activity of the developed oxide layer can be significantly improved by various approaches, such as doping and heterojunction formation via in situ integration or post-impregnation of the active components. The TiO_2 layers developed on titanium and its alloys have been the most extensively studied, but PEOs of aluminum, magnesium, zirconium, and various other metals and alloys have also been explored for catalytic applications.

Only a few works are explored for electrocatalytic applications compared to the photocatalytic field. Only a few of them addressed photoelectrochemical studies. Considering the advantages of robust and conducting metallic support of PEO-derived catalysts, more studies on electrocatalytic and electrophotocatalytic applications are in demand. Despite the significant reports on photocatalysts, the exact correlation between PEO parameters and the photocatalytic behavior of oxide coatings has not been well addressed. A detailed understanding of the mechanism of photocatalytic improvement needs to be addressed further so that precise optimization of preparation techniques and testing protocols can be derived, which could pave the way for the practical application of the PEO-derived photocatalysts. This chapter also provides an account of PEO-derived anode electrode materials for applications in lithium-ion batteries.

Acknowledgements The authors acknowledge INAM 2411 project, IRC-AM, KFUPM.

References

1. Yerokhin, A. L., Nie, X., Leyland, A., Matthews, A., & Dowey, S. J. (1999). *Surface and Coating Technology, 122*, 73–93.
2. Kaseem, M., Fatimah, S., Nashrah, N., & Ko, Y. G. (2021). *Progress in Materials Science, 117*, 100735.
3. Karbasi, M., Nikoomanzari, E., Hosseini, R., Bahramian, H., Chaharmahali, R., Giannakis, S., Kaseem, M., & Fattah-alhosseini, A. (2023). *Journal of Environmental Chemical Engineering, 11*, 110027.

4. Saji, V. S., Sankara Narayanan, T. S. N., & Chen, X. B. (2022). *Conversion coatings for magnesium and its alloys*. Springer Nature.
5. Saji, V. S., Ogunlakin, N., & Madhan Kumar, A., (2024). *Applied Surface Science, 669*, 160476.
6. Saji, V. S. (2024). *Chemical Engineering Journal, 498*, 155066.
7. Hassaan, M. A., El-Nemr, M. A., Elkatory, M. R., Ragab, S., Niculescu, V. C., & Nemr, A. E. (2023). *Topics in Current Chemistry, 381*, 31.
8. Samadi, P., Binczarski, M. J., Pawlaczyk, A., Rogowski, J., Szynkowska-Jozwik, M. I., & Witonska, I. A. (2022). *Materials, 15*, 4301.
9. Saji, V. S. (2024). *Corrosion and degradation in fuel cells, supercapacitors and batteries*. Springer Nature.
10. Saji, V. S., & Pillai, V. K. (2024). *Multi-functional electrocatalysts: Fundamentals and applications*. Royal Society of Chemistry.
11. Aliofkhazraei, M., Macdonald, D. D., Matykina, E., Parfenov, E. V., Egorkin, V. S., Curran, J. A., Troughton, S. C., Sinebryukhov, S. L., Gnedenkov, S. V., Lampke, T., Simchen, F., & Nabavi, H. F. (2021). *Applied Surface Science Advances, 5*, 100121.
12. Grishina, E. P., Kudryakova, N. O., Rumyantsev, P. A., Zhirov, A. V., Gladiy, Y. P., & Belkin, P. N. (2013). *Surface Engineering and Applied Electrochemistry, 49*, 83–90.
13. Lu, X., Blawert, C., Zheludkevich, M. L., & Kainer, K. U. (2015). *Corrosion Science, 101*, 201–207.
14. Krishnan, A., Swarnalal, A., Das, D., Krishnan, M., Saji, V. S., & Shibli, S. M. A. (2024). *Journal of Environmental Sciences, 139*, 389–417.
15. Saji, V. S. (2025). *Next Energy, 8*, 100259.
16. Saji, V. S. (2021). *Journal of Industrial and Engineering Chemistry, 103*, 358–372.
17. Saji, V. S. (2025). *Applied Surface Science Advances, 26*, 100700.
18. Lukiyanchuk, I. V., Rudnev, V. S., & Tyrina, L. M. (2016). *Surface and Coating Technology, 307*, 1183–1193.
19. Samadi, P., & Witonska, I. A. (2023). *Catalysis Communications, 181*, 106722.
20. Wala, M., Łubiarz, D., Waloszczyk, N., & Simka, W. (2023). *Materials, 16*, 2191.
21. Wei, F., Li, J., Lu, W., An, V., Tang, J., Zhao, L., Hua, Z., Yang, L., Yao, J., & Liu, B. (2024). *International Journal of Hydrogen Energy, 68*, 798–806.
22. Huang, W., Yu, Y., Wang, D., Ma, H., & Wang, G. (2024). *Journal of Environmental Management, 356*, 120724.
23. Jun, Z., Cao, X., Yang, X., Wang, Z., Wu, X., Chen, P., & Chen, J. (2023). *Materials Chemistry and Physics, 301*, 127654.
24. Fujishima, A., Hashimoto, K., & Watanabe, T. (2001). *TiO_2 photocatalysis: Fundamentals and applications*. Tokyo Bkc.
25. Friedemann, A. E. R., Thiel, K., Gesing, T. M., & Plagemann, P. (2018). *Surface and Coating Technology, 344*, 710–721.
26. Luttrell, T., Halpegamage, S., Tao, J., Kramer, A., Sutter, E., & Batzill, M. (2014). *Scientific Reports, 4*, 4043.
27. Levinas, R., Pakstas, V., Selskis, A., Murauskas, T., Viter, R., Anspoks, A., Pudza, I., Kuzmin, A., Tamasauskaite-Tamasiunaite, L., Cesiulis, H., & Norkus, E. (2024). *Journal of the Electrochemical Society, 171*, 036501.
28. Mirelman, L. K., Curran, J. A., & Clyne, T. W. (2012). *Surface and Coating Technology, 207*, 66–71.
29. Cai, N., Mai, Y., Su, R., & Lv, D. (2024). *Materials Letters, 365*, 136464.
30. Shin, Y. K., Chae, W. S., Song, Y. W., & Sung, Y. M. (2006). *Electrochemistry Communications, 8*, 465–470.
31. Bayati, M. R., Moshfegh, A. Z., & Golestani-Farda, F. (2010). *Electrochimica Acta, 55*, 2760–2766.
32. An, T. C., Zhu, X. H., & Xiong, Y. (2002). *Chemosphere, 46*, 897–903.
33. Nagaveni, K., Sivalingam, G., Hegde, M. S., & Madras, G. (2004). *Environmental Science & Technology, 38*, 1600–1604.

34. Lin, G. W., Chen, J. S., Tseng, W., & Lu, F. H. (2019). *Surface and Coating Technology, 357*, 28–35.
35. Coto, M., Troughton, S. C., Knight, P., Joshi, R., Francis, R., Kumar, R. V., & Clyne, T. W. (2021). *Surface and Coating Technology, 411*, 127000.
36. Jiang, D., Otitoju, T. A., Ouyang, Y., Shoparwe, N. F., Wang, S., Zhang, A., & Li, S. (2021). *Catalysts, 11*, 1039.
37. Tucci, A. P., Fumagalli, S., Livolsi, S., Chiarello, G. L., Selli, E., Ruiz-Delgado, A., Malato, S., Bestetti, M., & Franz, S. (2024). *Catalysis Today, 441*, 114851.
38. Lee, J. H., Youn, J. I., Kim, Y. J., Kim, I. K., Jang, K. W., & Oh, H. J. (2015). *Ceramics International, 41*, 11899–11907.
39. Sridhar, S., Arunnellaiappan, T., Rameshbabu, N., Mika, S., & Viswanathan, A. (2017). *Surface Engineering, 33*, 779–786.
40. Lin, G. W., Huang, Y. H., Tseng, W., & Lu, F. H. (2019). *Ceramics International, 45*, 22506–22512.
41. Stojadinović, S., Radić, N., Vasilić, R., Tadić, N., & Tsanev, A. (2022). *Solid State Sciences, 129*, 106896.
42. Zhang, Y., Fana, W., Dua, H. Q., Zhao, Y. W., Song, R. G., & Xiang, N. (2017). *Surface and Coating Technology, 315*, 196–204.
43. Yao, Z., Jia, F., Tian, S., Li, C. X., Jiang, Z., & Bai, X. (2010). *ACS Applied Materials & Interfaces, 2*, 2617–2622.
44. Yang, W., Xu, D., Chen, T., Wang, J., & Chen, J. (2020). *Science and Engineering of Composite Materials, 27*, 19–27.
45. Tadić, N., Stojadinović, S., Radić, N., Grbić, B., & Vasilić, R. (2016). *Surface and Coating Technology, 305*, 192–199.
46. Stojadinović, S., Radić, N., Grbić, B., Maletić, S., Stefanovc, P., Pačevski, A., & Vasilić, R. (2016). *Applied Surface Science, 370*, 218–228.
47. Stojadinović, S., Tadić, N., Radić, N., Grbić, B., & Vasilić, R. (2018). *Surface and Coating Technology, 337*, 279–289.
48. Stojadinović, S., Radić, N., Tadić, N., Vasilić, R., & Tsanev, A. (2022). *Journal of Materials Science: Materials in Electronics, 33*, 4467–4481.
49. Stojadinović, S., Tadić, N., Radić, N., Grbić, B., & Vasilić, R. (2018). *Surface and Coating Technology, 344*, 528–533.
50. He, J., Cai, Q. Z., Xiao, F., Li, X. W., Sun, W., & Zhao, X. (2011). *Journal of Alloys and Compounds, 509*, L11–L13.
51. Stojadinović, S., Tadić, N., Radić, N., Grbić, B., & Vasilić, R. (2017). *Materials Letters, 196*, 292–295.
52. Hoseini, A., & Yarmand, B. (2020). *Journal of Nanoparticle Research, 22*, 312.
53. Vasilyeva, M. S., Rudnev, V. S., Zvereva, A. A., Ustinov, A. Y., Arefieva, O. D., Kuryavyi, V. G., & Zverev, G. A. (2018). *Journal of Photochemistry and Photobiology A: Chemistry, 356*, 38–45.
54. Kim, Y. S., Shin, K. R., Kim, G. W., Ko, Y. G., & Shin, D. H. (2016). *Surface Engineering, 32*, 443–447.
55. Bahramian, H., Fattah-alhosseini, A., Karbasi, M., Nikoomanzari, E., & Giannaki, S. (2023). *Chemical Engineering Journal, 476*, 146588.
56. Bayati, M. R., Golestani-Fard, F., & Moshfegh, A. Z. (2010). *Catalysis Letters, 134*, 162–168.
57. Vasilyeva, M. S., Lukiyanchuk, I. V., Yarovaya, T. P., Ustinov, A. Y., Nedozorov, P. M., Fedorets, A. N., & Arefieva, O. D. (2021). *Surface and Coating Technology, 410*, 126898.
58. Stojadinović, S., Tadić, N., Radić, N., Petković, M., Stefanov, P., Zeković, L., & Grbić, B. (2012). *Applied Catalysis. B, Environmental, 126*, 334–341.
59. Wu, K. R., Hung, C. H., Yeh, C. W., & Wu, J. K. (2012). *Applied Surface Science, 263*, 688–695.
60. Bayati, M. R., Moshfegh, A. Z., Golestani-Fard, F., & Molaeia, R. (2010). *Materials Chemistry and Physics, 124*, 203–207.

61. Petrović, S., Stojadinović, S., Rožić, L., Radić, N., Grbić, B., & Vasilić, R. (2015). *Surface and Coating Technology, 269*, 250–257.
62. Budnikova, Y. B., Vasilyeva, M. S., Lukiyanchuk, I. V., Egorkin, V. S., Ustinov, A. Y., Kuryavyi, V. G., & Shlyk, D. H. (2023). *Journal of Materials Science: Materials in Electronics, 34*, 1973.
63. Qin, H., Chen, L., Yu, X., Wu, M., & Yan, Z. (2018). *Journal of Materials Science: Materials in Electronics, 29*, 2060–2071.
64. Lukiyanchuk, I. V., Tarkhanova, I. G., Vasilyeva, M. S., Yarovaya, T. P., Ustinov, A. Y., Vyaliy, I. E., & Kuryavyi, V. G. (2024). *Materials Chemistry and Physics, 311*, 128520.
65. Luo, Q., Cai, Q., Li, X., Pan, Z., Li, Y., Chen, X., & Yan, Q. (2013). *Transactions of the Nonferrous Metals Society of China, 23*, 2945–2950.
66. Babaei, M., Dehghanian, C., Taheri, P., & Babaei, M. (2016). *Surface and Coating Technology, 307*, 554–564.
67. Nikoomanzari, E., Fattah-alhosseini, A., Karbasi, M., & Nourian, A. (2022). *Surfaces and Interfaces, 32*, 102128.
68. Stojadinović, S., Radić, N., & Perković, M. (2024). *Journal of Materials Science: Materials in Electronics, 35*, 410.
69. Orsetti, R., Bukman, L., Santos, J. S., Nagay, B. E., Rangel, E. C., & Cruz, N. C. (2021). *Applied Surface Science Advances, 6*, 100143.
70. Manojkumar, P., Lokeshkumar, E., Premchand, C., Saikiran, A., Rama Krishna, L., & Rameshbabu, N. (2022). *Physica B, 631*, 413680.
71. Gao, Y., Yang, W., Xu, D., Chen, J., & Jiang, B. (2018). *Journal of Wuhan University of Technology Materials Science Edition, 33*, 1524–1529.
72. Tang, M., Li, G., Li, W., Liu, H., & Zhu, L. (2013). *Journal of Alloys and Compounds, 562*, 84–89.
73. Chen, W. H., Huang, S. Y., Chu, Y. R., Yang, S. H., Cheng, I. C., Jian, S. Y., & Lee, Y. L. (2024). *Surface and Coating Technology, 476*, 130238.
74. Stojadinović, S., Radić, N., & Vasilić, R. (2022). *Coatings, 12*, 1830.
75. Liu, M., Yang, X., Wang, Z., Wu, Y., Zhou, J., Cao, X., & Chen, P. (2024). *Materials Letters, 363*, 136321.
76. Safira, A. R., Fattah-alhosseini, A., & Kaseem, M., (2024). *Journal of Magnesium and Alloys, 12*, 2413–2432.
77. Ignjatović, S., Blawert, C., Serdechnov, M., Karpushenkov, S., Damjanović, M., Karlova, P., Wieland, D. C. F., Starykevich, M., Stojanović, S., Damjanović-Vasilić, L., & Zheludkevich, M. L. (2021). *Applied Surface Science, 544*, 148875.
78. Stojadinović, S., Vasilić, R., Radić, N., Tadić, N., Stefanovc, P., & Grbić, B. (2016). *Applied Surface Science, 377*, 37–43.
79. Mojsilović, K., Božović, N., Stojanović, S., Damjanović-Vasilić, L., Serdechnova, M., Blawert, C., Zheludkevich, M. L., Stojadinović, S., & Vasilić, R. (2021). *Surfaces and Interfaces, 26*, 101307.
80. Wu, T., Blawert, C., Serdechnova, M., Karlova, P., Dovzhenko, G., Wieland, D. C. F., Stojadinović, S., Vasilić, R., Mojsilovic, K., & Zheludkevich, M. L. (2022). *Applied Surface Science, 573*, 151629.
81. Stojadinović, S., Radić, N., & Vasilić, R. (2023). *Micromachines, 14*, 701.
82. Stojadinović, S., Radić, N., & Perković, M. (2023). *Optical Materials, 146*, 114608.
83. Luo, Q., Cai, Q. Z., He, J., Li, X. W., & Chen, X. D. (2014). *Advances in Applied Ceramics, 113*, 228–233.
84. Stojadinović, S., & Radić, N. (2024). *Solid State Sciences, 153*, 107578.
85. Hisatomi, T., Kubota, J., & Domen, K. (2014). *Chemical Society Reviews, 43*, 7520–7535.
86. Franz, S., Perego, D., Marchese, O., Lucotti, A., & Bestetti, M. (2016). *Applied Surface Science, 385*, 498–505.
87. Franz, S., Arab, H., Lucotti, A., Castiglioni, C., Vicenzo, A., Morini, F., & Bestetti, M. (2020). *Catalysts, 10*, 325.

88. Vargas-Villanueva, S., Velásquez-Tamayo, J. P., Torres-Cerón, D. A., Mercado, D. F., Torres-Palma, R. A., Riassetto, D., Riva, J. S., Amaya-Roncancio, S., Castilla-Acevedo, S. F., & Restrepo-Parra, E. (2023). *Journal of Environmental Chemical Engineering, 11*, 110246.
89. Arab, H., Chiarello, G. L., Selli, E., Bomboi, G., Calloni, A., Bussetti, G., Albani, G., Bestetti, M., & Franz, S. (2020). *Surfaces, 3*, 168–181.
90. Hoseini, A., & Yarmand, B. (2024). *Journal of Alloys and Compounds, 984*, 173976.
91. Vasilyeva, M. S., Lukiyanchuk, I. V., Shchitovskaya, E. V., Golushko, A. D., & Kondrikov, N. B. (2022). *Russian Journal of Inorganic Chemistry, 67*, 1460–1464.
92. Vasilyeva, M. S., Lukiyanchuk, I. V., Sergeeva, K. A., Sergeev, A. A., Shchitovskaya, E. V., Egorkin, V. S., Kuryavii, V. G., Ustinov, A. Y., Sinebryukhov, S. L., & Gnedenkov, S. V. (2022). *Journal of Alloys and Compounds, 913*, 165320.
93. Yin, Y., Li, L., Wu, G., Wang, Y., Yang, Z., Wen, C., & Yao, J. (2024). *Journal of Alloys and Compounds, 1002*, 175501.
94. Wu, G., Yin, Y., Zhu, L., Wang, Y., Wang, R., Yao, Z., & Yao, J. (2024). *Ceramics International, 50*, 6229–6240.
95. Lu, P., Wu, J., Zhao, H., Zhao, M., Dong, L., Xue, W., & Li, D. (2021). *Surface Engineering, 37*, 918–925.
96. Wei, B., Yan, S., Jia, D., Feng, C., Yin, J., & Wang, Z. (2022). *Inorganic Chemistry Communications, 145*, 110048.
97. Sun, M., Wu, J., Lu, P., Zhang, Z., Zhang, Y., & Li, D. (2021). *Journal of Alloys and Compounds, 892*, 162075.
98. Gim, H. G., Kim, Y. T., & Choi, J. (2023). *Electrochemistry Communications, 148*, 107455.

Chapter 20
Plasma Electrolytic Oxidation Coatings for Photoluminescent Applications

Stevan Stojadinović

20.1 Introduction

Luminescence in solids is the phenomenon in which electronic states of solids are excited by an external energy source and the excitation energy is released by the emission of electromagnetic radiation, i.e., photons. If the excitation of solid materials is triggered by electromagnetic radiation, the phenomenon is called photoluminescence (PL). Solid materials that exhibit PL are called phosphors [1, 2]. Phosphors are generally divided into three groups based on their performance: (1) host phosphor, (2) host phosphor + activator, and (3) host phosphor + sensitizer + activator [3]. The first type of phosphor consists of active centres that absorb excitation energy and emit radiation, the second type consists of an inactive host with luminescent active centres, and the third type of phosphor consists of a sensitizer that absorbs the excitation energy and transfers it to activators in the host, revealing the enhancement of the luminescence properties of the material.

According to the distinction between the phenomenon of radiant energy absorption and emission, PL is widely divided into two mechanisms, down-conversion and up-conversion emission. Down-conversion PL is a phenomenon in which the absorption of a single high-energy photon is followed by the emission of lower-energy photons, but in most cases, the emission of only one low-energy photon occurs [4]. This process obeys the Stokes law with a wavelength change known as the Stokes shift. Conversely, up-conversion PL is a nonlinear process in which the absorption of two or more low-energy photons leads to the emission of a higher-energy photon, i.e., the wavelength of the emitted radiation is shorter than that of the excitation radiation [5]. Most up-conversion materials transform near-infrared

S. Stojadinović (✉)
Faculty of Physics, University of Belgrade, Belgrade, Serbia
e-mail: sstevan@ff.bg.ac.rs

V. S. Saji, M. Zheludkevich (eds.), *Plasma Electrolytic Oxidation – Fundamentals, Advances and Applications*,
https://doi.org/10.1007/978-3-031-92429-3_20

radiation into visible or ultraviolet radiation through multiple absorptions or energy transfer processes.

Doped materials with lanthanide (Ln) ions exhibit the down- and up-conversion PL playing an important role in modern lighting and display fields due to the abundant emission colors based on their *4f–4f* or *5d–4f* transitions [6, 7]. The Ln elements are lanthanum (La), cerium (Ce), praseodymium (Pr), neodymium (Nd), promethium (Pm), samarium (Sm), europium (Eu), gadolinium (Gd), terbium (Tb), dysprosium (Dy), holmium (Ho), erbium (Er), thulium (Tm), ytterbium (Yb), and lutetium (Lu), a group of elements of very similar chemical properties determined by their electronic structure and exceptional spectroscopic properties.

Table 20.1 shows the atomic number (Z) and the electronic configuration of Ln atoms and ions. The general electronic configuration of Ln atoms is usually referred to as $[Xe]4f^n5d^16s^2$, where [Xe] is the electronic configuration of the noble gas xenon and n is the number of electrons from 0 to 14 (0 for La and 14 for Lu). Ln atoms have two electronic configurations based on their relative energy levels: $[Xe]4f^n6s^2$ and $[Xe]4f^{n-1}5d^16s^2$ (n = 1–14). La, Ce, Gd, and Lu belong to the latter type, while the rest of Ln belongs to the former (for Tb, both configurations are energetically close to each other, so both can be used). After ionization, the most common oxidation state of Ln ions is Ln^{3+}, which is usually also the most energetically stable. All trivalent Ln ions have the electronic configuration $[Xe]4f^n$, which means that the two 6*s* electrons and possibly the 5*d* electron have been lost. Some Ln ions, including Nd, Sm, Eu, Dy, Tm, and Yb, may be present as Ln^{2+} ions. However, some Ln ions, including Ce, Pr, Nd, Tb, and Dy, can also occur as Ln^{4+} ions.

Table 20.1 Electronic configurations of the Ln atoms and their common ions

Z	Elements	Electronic structure			
		Ln	Ln^{3+}	Ln^{4+}	Ln^{2+}
57	La	$[Xe]5d^16s^2$	[Xe]		
58	Ce	$[Xe]4f^15d^16s^2$	$[Xe]4f^1$	[Xe]	
59	Pr	$[Xe]4f^36s^2$	$[Xe]4f^2$	$[Xe]4f^1$	
60	Nd	$[Xe]4f^46s^2$	$[Xe]4f^3$	$[Xe]4f^2$	$[Xe]4f^4$
61	Pm	$[Xe]4f^56s^2$	$[Xe]4f^4$		
62	Sm	$[Xe]4f^66s^2$	$[Xe]4f^5$		$[Xe]4f^6$
63	Eu	$[Xe]4f^76s^2$	$[Xe]4f^6$		$[Xe]4f^7$
64	Gd	$[Xe]4f^75d^16s^2$	$[Xe]4f^7$		
65	Tb	$[Xe]4f^96s^2$	$[Xe]4f^8$	$[Xe]4f^7$	
66	Dy	$[Xe]4f^{10}6s^2$	$[Xe]4f^9$	$[Xe]4f^8$	$[Xe]4f^{10}$
67	Ho	$[Xe]4f^{11}6s^2$	$[Xe]4f^{10}$		
68	Er	$[Xe]4f^{12}6s^2$	$[Xe]4f^{11}$		
69	Tm	$[Xe]4f^{13}6s^2$	$[Xe]4f^{12}$		$[Xe]4f^{13}$
70	Yb	$[Xe]4f^{14}6s^2$	$[Xe]4f^{13}$		$[Xe]4f^{14}$
71	Lu	$Xe]4f^{14}5d^16s^2$	$[Xe]4f^{14}$		

For a given $[Xe]4f^n$ electron configuration, it can be assumed that the n $4f$ electron spins s_i and the n $4f$ electron orbital angular momentum l_i are composed separately to form a total spin angular momentum $S = \Sigma s_i$ and a total orbital angular momentum $L = \Sigma l_i$. L stands for the letters $S, P, D, F, G, H, I, K, L, M, N$... corresponding to the resulting orbital quantum number of the $4f$ electrons $L = 0, 1, 2, 3, 4, 5, 6, 7, 8$..., respectively. The total angular momentum (J) can be represented as follows: $J = L - S$ (number of $4f$ electrons is <7) and $J = L + S$ (number of $4f$ electrons is >7). When the Ln ions are doped into a solid, the degeneracy of the levels of the free ions is removed by coulombic, spin-orbit, and crystal field interactions. Coulombic (or electrostatic) interactions result from the existence of repulsions between the $4f$ electrons of the Ln ion under consideration. This leads to a splitting of the energy levels of the free ion into ^{2S+1}L energy levels, each of which is $(2L + 1)(2S + 1)$-fold degenerate. The electrostatic energy splitting of the ^{2S+1}L terms is typically around 10^4 cm^{-1}. Spin-orbit interactions (arising from the electromagnetic interaction between the electron spin and the magnetic field generated by the electron motion) split each ^{2S+1}L term into $(2J + 1)$ states labelled $^{2S+1}L_J$. The typical splitting of the $^{2S+1}L_J$ terms is of the order of 10^3 cm^{-1}. Finally, the interaction of the $4f$ electrons with the surrounding crystal field leads to a Stark splitting of the $^{2S+1}L_J$ terms into $(2J + 1)$ terms (if the number n of $4f$ electrons is even) or $(J + 1/2)$ terms (if n is odd). The Stark splitting is typically of the order of 10^2 cm^{-1}.

When a photon interacts with $4f$ electrons, three kinds of transitions can exist in Ln ions based on the local environment of Ln ions: sharp intra-configurational $4f$–$4f$ transitions, broad $4f^n$–$4f^{n-1}5d^1$ (f–d) transitions, and broad charge transfer (CT) transitions [7, 8]. Since $4f$ electrons are shielded by $5s^2$ and $5p^6$ electrons, the $4f$–$4f$ transitions are sharp and are typically interpreted using the electric dipole (ED), magnetic dipole (MD), and electric quadrupole (EQ) transitions. The parity selection rule, also known as the Laporte rule, states that MD and EQ transitions are allowed in a centrosymmetric environment, while ED transitions between states with the same parity are forbidden. The ED transition is partially allowed when the Ln ion is under the influence of a crystal field, since non-centrosymmetric interactions allow the mixing of electronic states of opposite parity into the $4f$ wave functions, somewhat relaxing the selection rules. The Laporte rule allows for interconfigurational f–d transitions, which involve promoting one electron from the $4f$ energy state to the higher $5d$ energy state and vice versa. These transitions produce intense and broad bands in the absorption and emission spectra. The most commonly observed f–d transition Ln ions are trivalent ions Ce^{3+}, Pr^{3+}, and Tb^{3+}, as well as divalent ions Eu^{2+}, Sm^{2+}, Yb^{2+}, Tm^{2+}, Dy^{2+}, and Nd^{2+}. Most Ln^{3+} ions have too high-energy f–d transitions, which typically occur in the vacuum UV or UV region of the electromagnetic spectrum. Ce^{3+} undergoes the lowest energy f–d transition. Ln^{2+} ions have lower energy f–d transitions than Ln^{3+} ions and exhibit f–d transitions from the ultraviolet to near-infrared region of the electromagnetic spectrum. The selection rules allow interconfigurational CT transitions in which $2p$ electrons are transferred from neighboring anions (such as O^{2-} in oxides) into the $4f$ shell. This leads to a broad excitation band known as the charge transfer band (CTB).

Metal oxides are excellent host matrices for Ln ions due to their outstanding physicochemical properties such as good thermal, chemical, and mechanical stability, large band gap, wide optical transparency, and low phonon frequency. In recent years, the author and coworkers have demonstrated that the plasma electrolytic oxidation (PEO) process in direct current (DC) regime is a very practical way to form Ln ions doped metal oxide semiconductor coatings such as ZrO_2 [9–22], Al_2O_3 [23–34], TiO_2 [35–38], ZnO [39], Nb_2O_5 [17, 40], HfO_2 [41, 42], Ta_2O_5 [43, 44], Y_2O_3 [45], and Gd_2O_3 [46, 47] by PEO of pure Zr, Al, Ti, Zn, Nb, Hf, Ta, Y, and Gd metals, respectively, in an electrolyte containing suitable lanthanide oxide (Ln_2O_3) particles. PEO on Al in electrolytes containing transition metal oxide particles such as Cr_2O_3, MnO, and TiO_2 is also appropriate for the formation of $Al_2O_3{:}Cr^{3+}$ [48], $Al_2O_3{:}Mn^{2+}$ [33], and $Al_2O_3{:}Ti^{3+}$ [49] coatings.

PEO is an electrochemical process based on the formation of oxide coatings on the surfaces of some metals and their alloys (mostly Al, Mg, Ti, Zr, Ta, Nb, and Zn) when a high voltage is applied above the dielectric breakdown, leading to the formation of micro-discharges that propagate continuously over the coating surface [50, 51]. PEO is usually carried out in an alkaline electrolyte without the use of hazardous chemicals, making the process environmentally friendly. The high pressures and temperatures that develop in the micro-discharge channels during PEO lead to a variety of complex processes, including heat release and thermal, electrochemical, and plasma chemical reactions that produce dense, thick coatings. The coatings formed consist of base metal oxides that may be doped with impurities added to the electrolyte. Ln_2O_3 particles have a negative zeta potential in alkaline media, i.e., their particles are negatively charged. Due to the electrophoretic effect, the Ln_2O_3 particles move toward the metal anode during the DC PEO. The melting point of the Ln_2O_3 particles (~2400 °C) is much lower than the local high temperatures (~10^4 °C) generated at the micro-discharge sites [52], so the molten Ln_2O_3 particles react with the metal oxide and form coatings doped with Ln ions.

In this chapter, we review the results of investigations down-conversion PL properties of Eu^{3+}-doped ZrO_2, HfO_2, Nb_2O_5, TiO_2, Gd_2O_3, and Y_2O_3 coatings formed by PEO and their application in luminescence thermometry. The PL properties of Al_2O_3 coatings doped with Ln ions and the up-coverage properties of Ta_2O_5 coatings doped with Er^{3+} and Ho^{3+} are also presented.

20.2 PL of Eu^{3+} in Different Metal Oxide Coatings

One of the most attractive and extensively used PL red activators in phosphors is Eu^{3+}, which has a characteristic 4*f*–4*f* absorption in the ultraviolet region and sharp red emission peaks resulting from $^5D_0 \rightarrow {}^7F_J$ ($J = 0, 1, 2, 3, 4$) transitions [53]. Due to the crystal field effect generated by the surrounding ions of Eu^{3+} in the host lattice, the $^5D_0 \rightarrow {}^7F_J$ transitions can have different components depending on the site symmetry. The $^5D_0 \rightarrow {}^7F_0$ transition is strictly forbidden according to the Judd–Ofelt theory. The observation of this transition is an indication that the Eu^{3+} ion occupies a site with C_{nv}, C_n, or C_s symmetry. The ${}^5D_0 \rightarrow {}^7F_1$ MD transition is almost independent of the local environment of the Eu^{3+} ion, while the hypersensitive ${}^5D_0 \rightarrow {}^7F_2$ ED transition is greatly dependent on the local symmetry of the Eu^{3+} ion and the coupling strength of the host. The ${}^5D_0 \rightarrow {}^7F_3$ transition is forbidden and dependent on the environment of the Eu^{3+} ion. ${}^5D_0 \rightarrow {}^7F_4$ is ED transition dependent on the local environment but to a lesser extent than the ${}^5D_0 \rightarrow {}^7F_2$. The ${}^5D_0 \rightarrow {}^7F_1$ and ${}^5D_0 \rightarrow {}^7F_2$ transitions are dominant in most Eu^{3+}-doped phosphor materials. The emission intensity ratio R between ${}^5D_0 \rightarrow {}^7F_2$ and ${}^5D_0 \rightarrow {}^7F_1$ transitions (also called the asymmetric ratio) can be used as an effective spectroscopic technique to explain the symmetry of the local environment of the Eu^{3+} ion. The R is more prominent when the intensity of the ED transition is higher than the MD transition. CT and 4*f*–4*f* transitions can be observed in the PL excitation spectra of Eu^{3+} in different materials, but unlike 4*f*–4*f* absorptions, CT transition enables very efficient excitation of Eu^{3+} ions in most metal oxides. CTB is observed in the UV region [22].

Figure 20.1 shows typical potential-time curves for the anodization of Zr, Hf, Nb, Ti, Gd, and Y at 150 mA/cm² for 10 min in a water solution containing 10 g/L $Na_3PO_4 \cdot 12H_2O$ + 2 g/L Eu_2O_3 particles, as well as the top view and the cross-sectional morphology, the chemical content of Eu and the phase composition of the

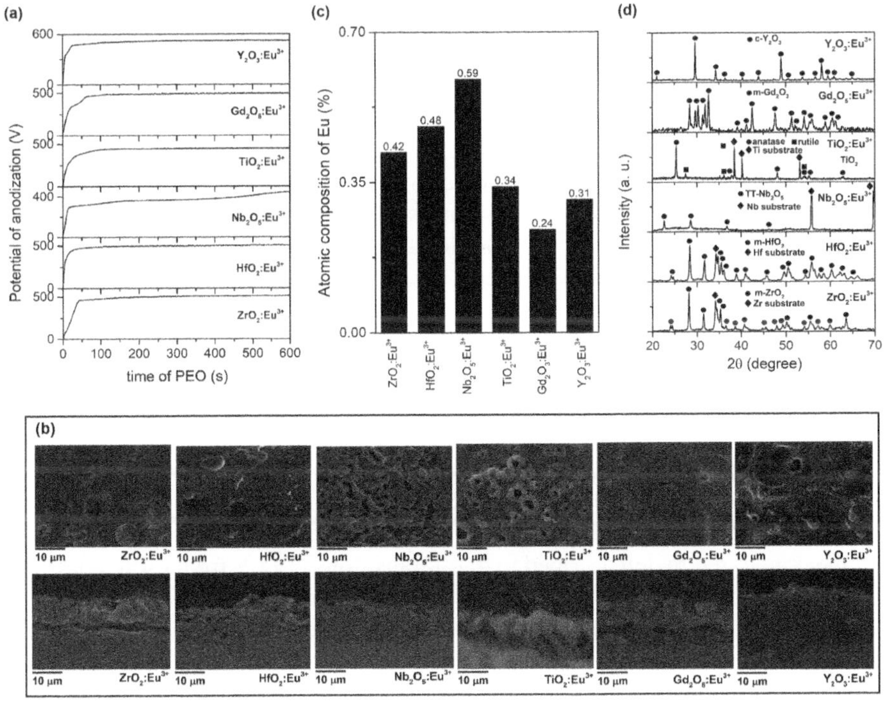

Fig. 20.1 (**a**) Potential–time curve; (**b**) top view and cross-section micrographs; (**c**) EDS Eu content; (**d**) XRD patterns; of coatings formed by PEO of Zr, Hf, Nb, Ti, Gd, and Y at 150 mA/cm² for 10 min in a water solution containing 10 g/L $Na_3PO_4 \cdot 12H_2O$ + 2 g/L Eu_2O_3 particles

coatings formed after 10 min. Depending on the increasing trend of the potential, at least three regions can be identified (Fig. 20.1a). From the beginning of anodization, the potential increases linearly with time, resulting in a constant rate of formation of a thin, compact barrier oxide layer on the substrate. The oxide layer forms at the metal/oxide and oxide/electrolyte interfaces due to the migration of O^{2-}/OH^- anions and metal cations Me^{n+} (n = 2, 3, 4, or 5) through the oxide, supported by a high electric field (~10^7 V/cm) [52]. During anodization, small amounts of anionic electrolyte components are also incorporated into the oxide at the oxide/electrolyte interface. The barrier oxide layer is amorphous and has a high ionic conductivity and a high electrical resistance. The thickness of the barrier oxide layer is limited to a few hundred nanometers due to the dielectric breakdown under high electric fields. This stage of anodization is followed by a deviation from the linearity of the potential-time curves, which begins at the so-called breakdown voltage (beginning of PEO). After breakdown, the potential continues to increase, but the slope of the potential-time curve decreases and a large number of small micro-discharges appear uniformly over the entire sample surface. Further anodizing leads to a relatively stable anodizing potential.

The top view and cross-sectional SEM micrographs of formed coatings are shown in Fig. 20.1b. The morphology and thickness of the coatings vary depending on the substrate on which they were formed. However, they all exhibit a structure typical of PEO coatings, which includes a variety of differently sized and shaped pores, cracks, and areas created by the rapid cooling of the molten material. The main elements of the coatings are Zr, Hf, Nb, Ti, Gd, or Y from substrates and O from the electrolyte [20, 35, 40, 41, 45, 46]. The Eu content in the formed coatings is low, <0.6 atomic percent (Fig. 20.1c).

The XRD patterns of the formed coatings are shown in Fig. 20.1d. The $ZrO_2:Eu^{3+}$, $HfO_2:Eu^{3+}$, and $Gd_2O_3:Eu^{3+}$ coatings consist of the monoclinic (m) phase of ZrO_2 (ICDD JCPDS 37-1484), HfO_2 (ICDD JCPDS 78-0049), and Gd_2O_3 (ICCD JCPDS 43-1015), respectively. The $Nb_2O_5:Eu^{3+}$ consists of a pseudohexagonal (TT) Nb_2O_5 phase (ICDD JCPDS 028-0317). The anatase (PDXL DB Card No. 9008213) and rutile (PDXL DB Card No. 9009083) phases of TiO_2 are observed in the XRD pattern of the $TiO_2:Eu^{3+}$ coating, while the cubic (c) phase of Y_2O_3 is detected in the XRD pattern of the $Y_2O_3:Eu^{3+}$ coating (ICDD JCPDS 41-1105). The structure of ZrO_2, HfO_2, Nb_2O_5, TiO_2, Gd_2O_3, and Y_2O_3 is not affected by the presence of Eu^{3+} ions, which may be due to the low concentration of Eu uniformly distributed over the entire surface of the coatings [20, 35, 40, 41, 45, 46].

PL emission spectra excited at 260 nm are characterized by sharp emission bands in the orange-red region ascribed to 4*f*–4*f* transitions of Eu^{3+} from excited level $^5D_0$ to lower levels ^7F_J (J = 0, 1, 2, 3, and 4) (Fig. 20.2a) [53]. The most intense band corresponds to the ED transition $^5D_0 \rightarrow {}^7F_2$. Also, bands originating from Gd_2O_3 and HfO_2 can be observed in the PL emission spectra of $Gd_2O_3:Eu^{3+}$ and $HfO_2:Eu^{3+}$, respectively [41, 46]. The excitation PL spectra monitored at the wavelength of the most intense peak in emission PL spectra ($^5D_0 \rightarrow {}^7F_2$ transition) are shown in Fig. 20.2b. The excitation PL spectra are featured by two overlapping regions: a strong PL band in the middle UV region, and a series of weak, sharp peaks that

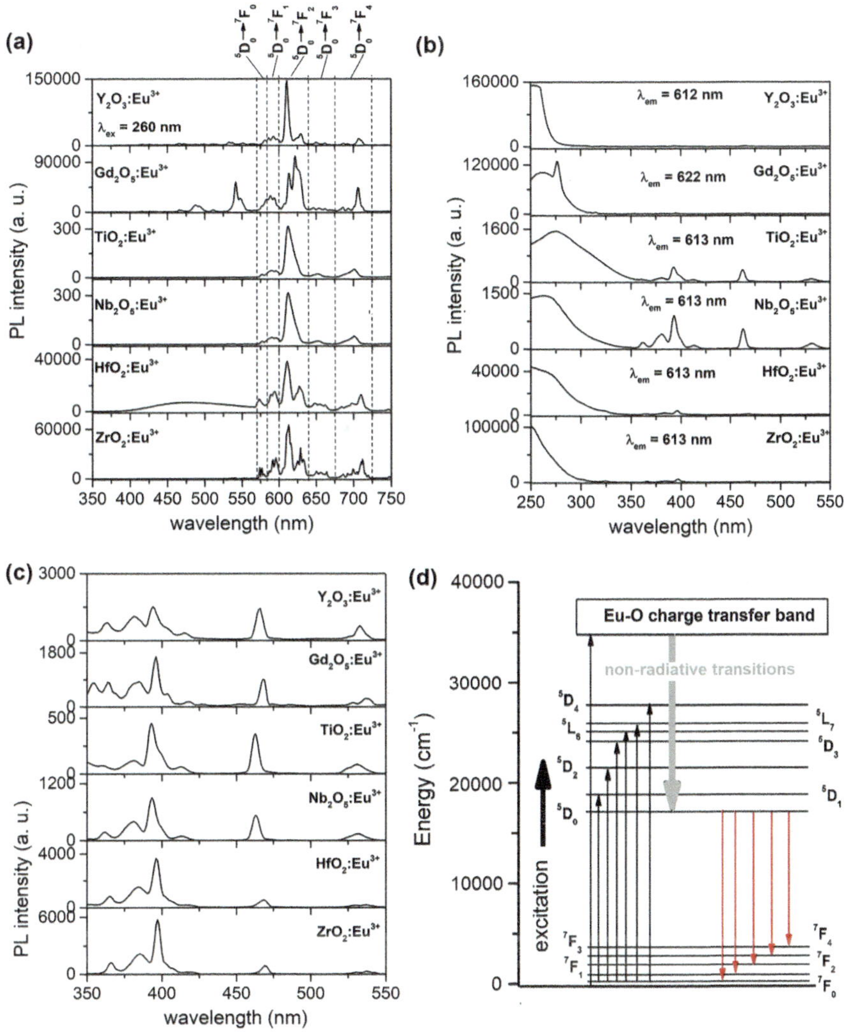

Fig. 20.2 (**a**) PL emission spectra excited at 260 nm; (**b**) PL excitation spectra monitored at the wavelength of the peak of $^5D_0 \rightarrow {}^7F_2$ transition in the range 250–550 nm; (**c**) PL excitation spectra monitored at the wavelength of the peak of $^5D_0 \rightarrow {}^7F_2$ transition in the range 350–550 nm; (**d**) Energy level diagram mechanism of charge transfer pathways involved in the PL of Eu^{3+}

correspond to direct excitation of the Eu^{3+} ground state $^7F_0$ into higher levels of the 4*f*-manifold. A strong PL band in the middle UV region is attributed to the CTB caused by the electron transfer between the completely filled 2*p* orbital of O^{2-} ions and the partially filled 4*f* orbital of the Eu^{3+} ions [20, 35, 40, 41, 45, 46]. The band at about 276 nm observed in the PL excitation spectrum of Gd_2O_3:Eu^{3+} originated from Gd^{3+} and Gd_2O_3 [46]. The main 4*f*–4*f* excitation bands are assigned to the electronic transition $^7F_0 \rightarrow {}^5D_4$ at 364 nm, $^7F_0 \rightarrow {}^5L_7$ at 384 nm, $^7F_0 \rightarrow {}^5L_6$ at 395 nm, $^7F_0 \rightarrow {}^5D_3$ at 416 nm, $^7F_0 \rightarrow {}^5D_2$ at 465 nm, $^7F_0 \rightarrow {}^5D_1$ at 527 nm, and $^7F_1 \rightarrow {}^5D_1$ at 534 nm (Fig. 20.2c).

Figure 20.2d shows the energy level diagram mechanism of charge transfer pathways involved in the PL of Eu^{3+}-doped ZrO_2, HfO_2, Nb_2O_5, TiO_2, Gd_2O_3, and Y_2O_3 coatings. Excitation may take place in two processes, either by transferring electrons to the CTB or by electron jumping to the higher levels of Eu^{3+}. After the excitation, excited electrons (due to charge transfer) are transferred to the lowest excited level of Eu^{3+}, i.e., $^5D_0$ by the non-radiative transition. Only upon non-radiative transfer electrons can make transitions to the lower levels ^7F_J (J = 0, 1, 2, 3, and 4) of Eu^{3+}.

The low phonon energy of the Y_2O_3 (~380 cm^{-1}), ZrO_2 (~470 cm^{-1}), HfO_2 (~470 cm^{-1}), and Gd_2O_3 ZrO_2 (~600 cm^{-1}) increases the probability of radiative transitions between the electronic energy levels of the incorporated Ln^{3+} ions and makes these metal oxides the ideal matrix component for the preparation of highly efficient PL materials by doping with Ln^{3+} ions [54–57]. Eu^{3+}-doped Y_2O_3, ZrO_2, HfO_2, and Gd_2O_3 PEO coatings emit intense PL under UV excitation, which makes these coatings suitable for red solid-state lighting.

20.3 Application of Eu^{3+}-Doped Metal Oxide Coatings in Luminescence Thermometry

Luminescence thermometry is a noncontact temperature measurement technique with high accuracy and sensitivity. Almost any property related to PL emission bands, such as the intensity of PL emission bands, the ratio of intensities of PL emission bands, the width of emission bands, the lifetimes of excited states, and the emission rise times, can be used for temperature measurement [58]. The luminescence intensity ratio (LIR) is most commonly used to determine temperature by comparing the relative PL intensities of two radiative transitions typically emitted from Ln ion-doped materials [59]. The changes in the intensity ratio as a function of temperature are not affected by the power of the source, as variations in excitation power affect both levels equally. This improves the measurement sensitivity and the stability of the sensor.

LIR utilizes the ratio of emission intensities of two thermally coupled levels. If the energy gap is <2000 cm^{-1}, then the thermal energy is sufficient to occupy the higher (H) from the lower (L) excited level [60]. The relative populations are given by the Boltzmann distribution: $N_H/N_L = \exp.(-\Delta E/kT)$, where $\Delta E = E_H - E_L$ and $k = 0.695\ cm^{-1}\ K^{-1}$ is the Boltzmann constant. The intensity of the transition is given by: $I = h\nu Ag$, where $h\nu$ is the energy of the resulting photon, A is the spontaneous emission probability and $g = 2J + 1$ is the degeneracy of the emitting level. Then, the LIR is given by [61]:

$$\mathrm{LIR}(T) = \frac{I_{\mathrm{H}}}{I_{\mathrm{L}}} = B\exp\left(-\frac{\Delta E}{kT}\right) \tag{20.1}$$

where $B = \nu_{\mathrm{H}} A_{\mathrm{H}} g_{\mathrm{H}}/(\nu_{\mathrm{H}} A_{\mathrm{H}} g_{\mathrm{H}})$ represents the material-dependent constant, independent of the temperature. The performance variables that describe the performance of the temperature sensor are the absolute and relative sensitivities given by Eqs. (20.2) and (20.3), respectively:

$$S(T) = \frac{d[\mathrm{LIR}(T)]}{dT} = \frac{\Delta E}{kT^2}\mathrm{LIR}(T), \tag{20.2}$$

$$S_{\mathrm{R}}(T) = \frac{1}{\mathrm{LIR}(T)} \cdot \frac{d[\mathrm{LIR}(T)]}{dT} = \frac{\Delta E}{kT^2} \tag{20.3}$$

PEO-based coatings of Eu^{3+}-doped ZrO_2 [17], Nb_2O_5 [17], Gd_2O_3 [46], and TiO_2 [38] are perspective for LIR thermometry. This section will illustrate the properties of $TiO_2:Eu^{3+}$ coatings. For Eu^{3+}, the $^5D_1$ and $^5D_0$ levels are separated by approximately 1750 cm^{-1} [62], and any combination of emissions $^5D_1 \rightarrow {}^7F_J$ and $^5D_0 \rightarrow {}^7F_J$ can be used for LIR. It is ideal to have an emission that is less temperature-dependent and serves as an internal reference. The hypersensitive transition $^5D_1 \rightarrow {}^7F_1$ is used as an internal reference due to its constant peak intensity at 543 nm across all temperature ranges. The hypersensitive transition $^5D_0 \rightarrow {}^7F_2$ is used to evaluate the LIR (Fig. 20.3a). Fitting the ratio of integrated emission intensities of $^5D_1 \rightarrow {}^7F_1$ and

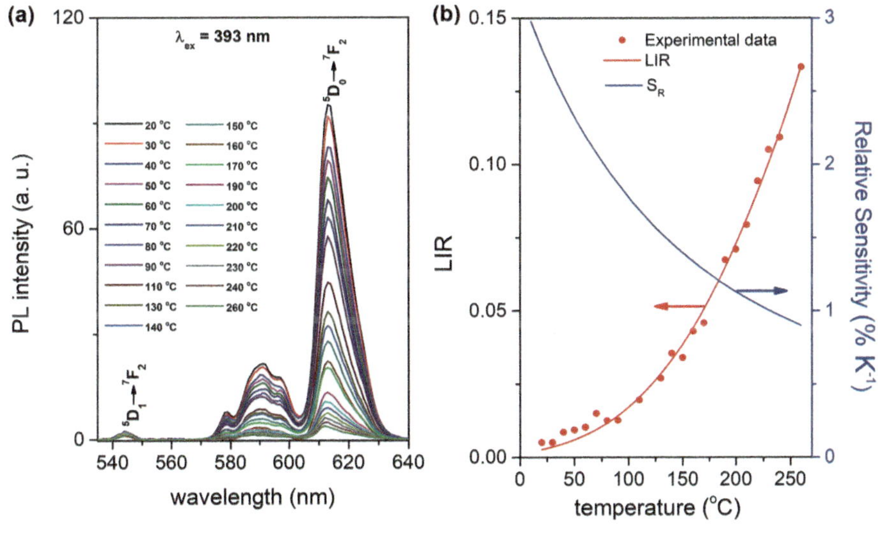

Fig. 20.3 (**a**) PL emission spectra of the $TiO_2:Eu^{3+}$ coating at temperatures ranging from 20 °C to 260 °C; (**b**) LIR experimental data, fit, and calculated sensitivity

$^5D_0 \rightarrow {}^7F_2$ results in $B = 16.2$ and $\Delta E = 1777$ cm^{-1}. Figure 20.3b shows the theoretical fit and the relative sensitivity.

20.4 PL of Al_2O_3 Coatings Doped with Ln Ions

In this subsection, the PL properties of Al_2O_3 coatings doped with Ln ions formed by PEO of aluminum for 10 min in an electrolyte containing 0.1 M H_3BO_3 + 0.05 M $Na_2B_4O_7 \cdot 10\,H_2O$ (supporting electrolyte—SE) with the addition of Ln_2O_3 (Ln = Eu, Sm, Pr, Tm, Dy), CeO_2 and Tb_3O_4 particles at different concentrations are presented. Al_2O_3 is a material with excellent properties for Ln ion host matrices, including high transparency from ultraviolet to near-infrared, high refractive index, good chemical, photochemical, and mechanical properties, high melting point, and so on [63, 64].

Characterization of coatings formed in SE containing Eu_2O_3 particles in different concentrations using SEM/EDS, XRD, and XPS is presented in Fig. 20.4. Generally independent of the Ln oxide particles in the SE, the formed coatings have similar morphology, phase, and chemical compositions. Characterization of Al_2O_3 coatings formed in SE with the addition of Ln_2O_3 (Ln = Sm, Pr, Tm, Dy), CeO_2, and Tb_3O_4 particles at different concentrations can be found in Refs. [23, 25–32].

Figure 20.4a shows that the surface morphology of the coatings is not significantly influenced by the concentration of Eu_2O_3 particles in SE. The main elements of the coatings are Al, O, and Eu. The EDS elemental mapping of the coating formed in the SE with an addition of 4.0 g/L Eu_2O_3 shows that all elements are rather evenly distributed in the PEO coatings (Fig. 20.4a). The Eu content in the formed coatings is low and increases with the concentration of Eu_2O_3 particles in the SE (Fig. 20.4b). The formed coatings are crystallized, and the main phase is gamma-Al_2O_3 (ICDD JCPDS 00-10-0173). In addition, the alpha phase of Al_2O_3 (ICDD JCPDS 00-10-0425) is observed in coatings formed in SE without and with the addition of Eu_2O_3 particles up to 1.0 g/L. It has been reported that the incorporation of Eu ions into Al_2O_3 prevents the formation of the alpha-Al_2O_3 phase and promotes the thermal stability of the gamma-Al_2O_3 phase [65, 66], which is consistent with the XRD measurements. The diffraction peaks attributed to metallic aluminum originate from the substrate. The lack of visible peaks of Eu species in the XRD patterns could be a consequence of the low concentration of uniformly distributed Eu_2O_3 particles in the Al_2O_3 coatings or the substitution of Eu ions into the crystal lattice of Al_2O_3.

The XPS study was conducted to further investigate the oxidation state of Eu in the coatings. Figure 20.4d shows the high-resolution Eu 3*d* spectrum of a Eu-doped Al_2O_3 coating formed in the SE with 4.0 g/L Eu_2O_3 particles. The Eu 3*d* core level spectrum shows the presence of Eu^{3+} and Eu^{2+} oxidation states. The spectrum exhibits two doublets: one with the Eu $3d_{5/2}$ and Eu $3d_{3/2}$ peaks at about 1135 eV and 1165 eV, corresponding to the Eu^{3+} oxidation state, and a second doublet at lower binding energies (around 1125 eV and 1155 eV) assigned to the Eu^{2+} oxidation state

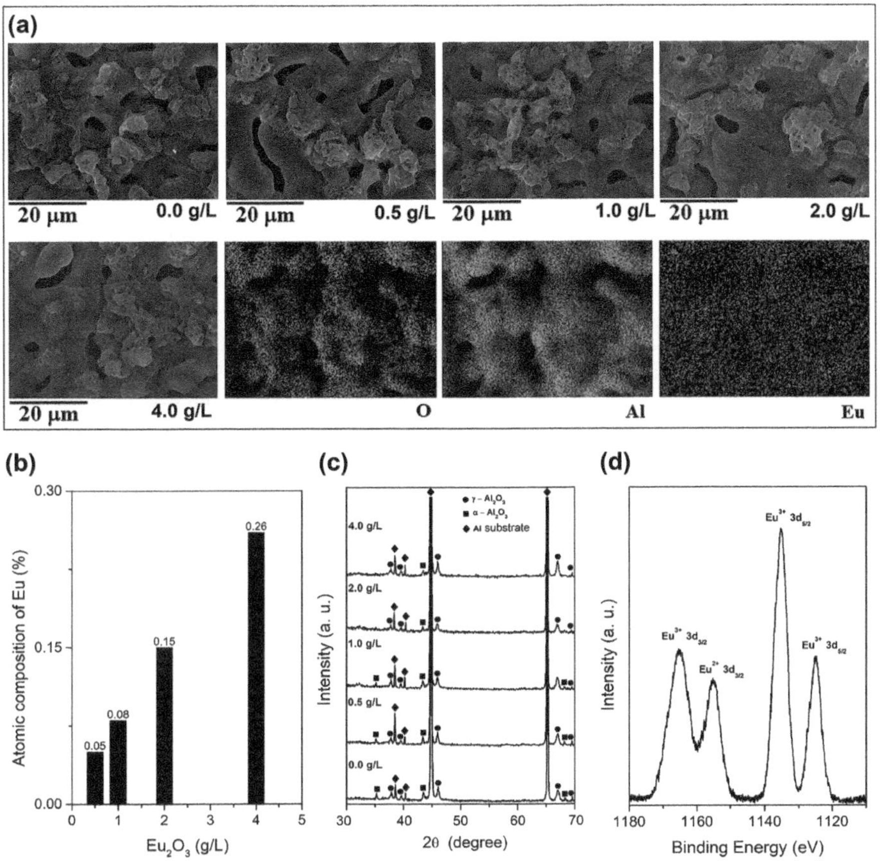

Fig. 20.4 (**a**) SEM micrographs and EDS mapping; (**b**) EDS Eu content; (**c**) XRD patterns; (**d**) high-resolution Eu 3d XPS spectrum; of coatings formed in SE with the addition of Eu_2O_3 particles in different concentrations

[67]. This result suggests that some Eu^{3+} ions are reduced to Eu^{2+} ions during the PEO process.

Due to the large discrepancy in ionic radii and charge imbalance between Al^{3+} (53 pm) and Eu^{3+} (95 pm), oxygen vacancies occur when Al^{3+} ions are substituted by Eu^{3+} ions, causing lattice distortion. It has been suggested that in similar systems, positively charged oxygen vacancies remain in the lattice while the corresponding negatively charged oxygen ion is released into the lattice. This negatively charged oxygen ion diffuses randomly through the lattice and releases its electrons with the release of oxygen ($2O^{2-} - 4e \rightarrow O_2$). The released electrons can be captured by Eu^{3+} ions, leading to the reduction of Eu^{3+} to Eu^{2+} ($Eu^{3+} + e \rightarrow Eu^{2+}$) [68, 69].

20.4.1 PL of Al_2O_3:Eu^{2+} Coatings

Eu^{2+} ions show intense excitation and strong emission band due to the parity allowed (Laporte's selection rule is obeyed) *5d–4f* electronic transitions [70]. The positions of excitation and emission PL bands of Eu^{2+} ions are strongly dependent on the crystal field splitting of the *5d* excited level. The *5d* electrons are located in the outermost orbitals, which are not shielded by outer orbital electrons. Therefore, the electrons in the *5d* orbitals are significantly influenced by the crystal fields of the host lattice [71]. As a result, Eu^{2+} ions emit light in a wide wavelength range, from ultraviolet to red.

Eu^{2+} does not occur in any naturally occurring material. Raw materials for Eu compounds are usually Eu_2O_3 or $EuCl_3$, where Eu ions are present in the trivalent oxidation state. Eu^{2+}-doped phosphors are produced by reducing Eu^{3+} to Eu^{2+}. Generally, if preparation is done in reducing atmospheres like H_2, H_2/N_2, or CO, the reduction of Eu^{3+} to Eu^{2+} in the matrix could be realized [72]. In some cases, this reduction occurs during the preparation in air at high temperatures [73]. Eu^{2+}-doped Al_2O_3 coatings can be formed by PEO of aluminum in the electrolyte containing Eu_2O_3 particles [24].

The PL of the Eu-doped Al_2O_3 coating formed by PEO is derived from the Al_2O_3 coating and the Eu ions incorporated into it. Figure 20.5a depicts the PL emission spectra excited at 260 nm by pure Al_2O_3 coatings formed in SE and Eu-doped Al_2O_3 coatings formed in SE with varying concentrations of Eu_2O_3 particles. The PL of

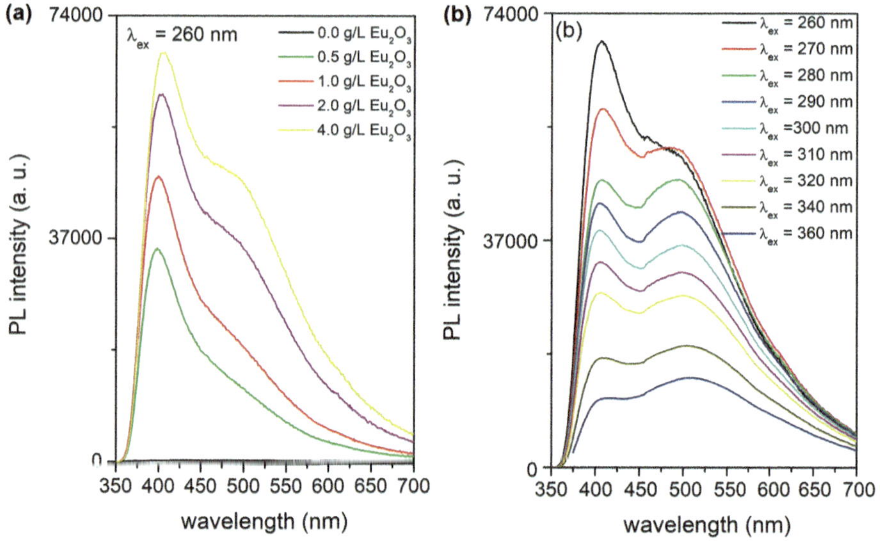

Fig. 20.5 (**a**) The effect of Eu_2O_3 particle concentration in SE on PL emission spectra excited at 260 nm; (**b**) PL emission spectra of the coating formed in SE with the addition of 4.0 g/L Eu_2O_3 particles

pure Al_2O_3 coating is caused by optical transitions in defect centers associated with oxygen vacancies containing a single electron (F^+ centers) and two electrons (F centers) [74]. The PL of pure Al_2O_3 coating is more than two orders of magnitude lower than that of Eu-doped Al_2O_3 coatings. This result indicates that Eu ions contribute most to the PL of Eu-doped Al_2O_3 coatings. The intensity of PL increases with the amount of Eu incorporated into the coatings, which is determined by the concentration of Eu_2O_3 particles in the SE (Fig. 20.4b).

Figure 20.5b shows the PL emission spectra of the Eu-doped Al_2O_3 coating formed in SE by adding 4.0 g/L Eu_2O_3 particles and excited at different excitation wavelengths. The PL emission spectra consist of two distinct bands with maxima at about 410 nm and 500 nm. During the PEO process, most Eu^{3+} ions are reduced to Eu^{2+} ions, as evidenced by the absence of sharp emission bands in the orange-red region (600–700 nm) originating from $4f$–$4f$ transitions of Eu^{3+} from the excited levels $^5D_0$ to the lower levels ^7F_J [75]. The PL emission is caused by the electronic transition $4f^65d^1 \rightarrow 4f^7$of Eu^{2+} ions in the low-symmetry coordination environment of the Al_2O_3 host. Two PL bands show that the Eu^{2+} ions occupy two different lattice sites in the Al_2O_3 host [76]. The excitation wavelength of 260 nm gives the most intense PL. The relative ratio between the first (410 nm) and the second (500 nm) PL maxima varies with the excitation wavelength (Fig. 20.5b). The 410 nm PL emission band of Eu^{2+} is dominant at shorter excitation wavelengths, while the 500 nm PL emission band dominates at longer excitation wavelengths. As the concentration of Eu^{2+} increases, the PL band at 500 nm becomes more pronounced (Fig. 20.5a).

20.4.2 PL of Al_2O_3:Sm^{2+}/Sm^{3+} Coatings

The PL emission spectra of the Sm-doped Al_2O_3 coating formed in SE with the addition of 4.0 g/L Sm_2O_3 particles are shown in Fig. 20.6a. The PL emission spectra clearly show three distinct regions. The first region consists of the characteristic broad PL band of Al_2O_3 with a maximum at about 410 nm. The second region in the range from 540 nm to 670 nm exhibits sharp emission bands related to the $4f$–$4f$ transitions of Sm^{3+}: $^4G_{5/2} \rightarrow {}^6H_{5/2}$, $^4G_{5/2} \rightarrow {}^6H_{7/2}$, and $^4G_{5/2} \rightarrow {}^6H_{9/2}$ positioned at about 563 nm, 599 nm, and 645 nm, respectively [77]. The third region in the range from 670 nm to 750 nm is characterized by sharp emission bands related to $^5D_0 \rightarrow {}^7F_0$, $^5D_0 \rightarrow {}^7F_1$, and $^5D_0 \rightarrow {}^7F_2$ transitions of Sm^{2+} positioned at about 688 nm, 703 nm, and 728 nm, respectively [78].

Figure 20.6b shows PL excitation spectra monitored at 410 nm, 599 nm, and 688 nm, i.e., at the wavelength of the most intense PL in each of the three regions in emission PL spectra. PL excitation spectrum of Al_2O_3 monitored at 410 nm exhibits a maximum at about 260 nm. The PL excitation spectrum of Sm^{3+} in Al_2O_3 monitored at 599 nm (transition $^4G_{5/2} \rightarrow {}^6H_{5/2}$) can be divided into two regions: the broad band region from 240 nm to 300 nm with a maximum at about 250 nm associated with charge CTB of Sm^{3+}, which originates from the electronic transition between

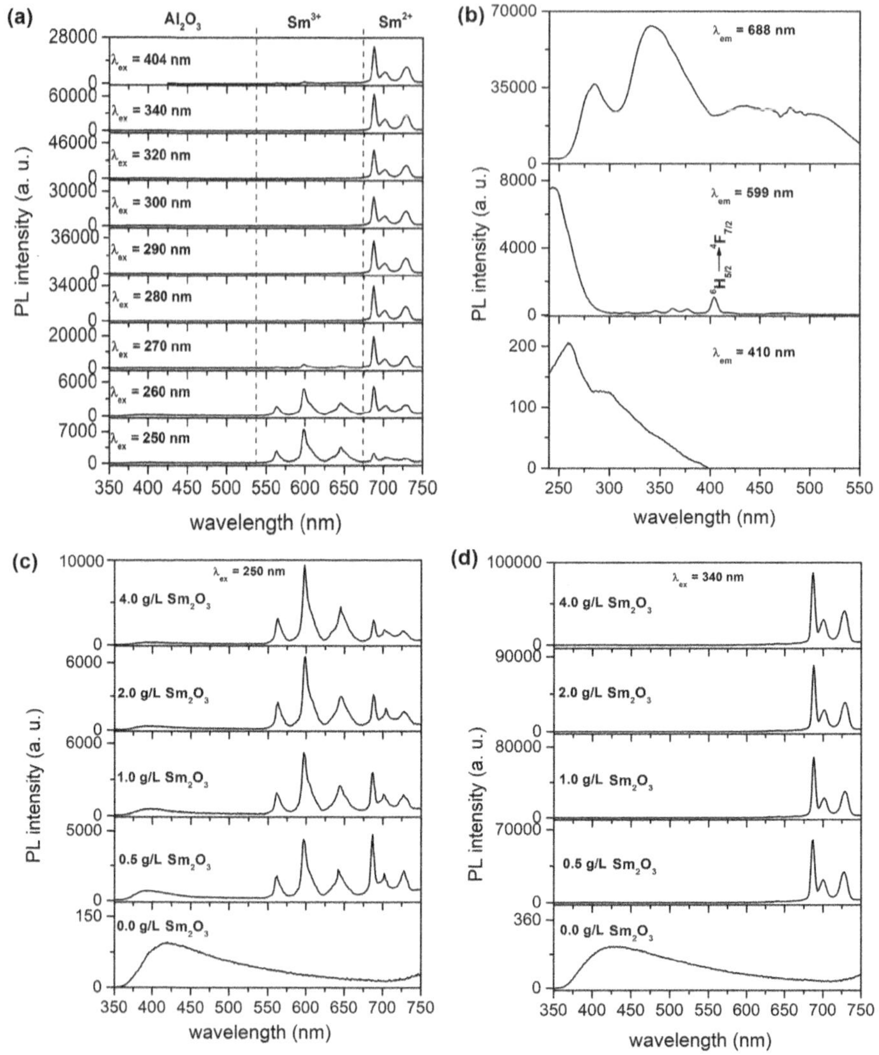

Fig. 20.6 (**a**) PL emission spectra and (**b**) PL excitation spectra of the coating formed in SE with the addition of 4.0 g/L Sm_2O_3 particles. The influence of concentration of Sm_2O_3 particles in SE on PL emission spectra: (**c**) excited at 250 nm; (**d**) excited at 340 nm

the completely filled $2p$ orbital of O^{2-} ions ($2p^6$) and empty $4f$ orbital of Sm^{3+} ions ($4f^6$) [79], and the series of sharp peaks in the range from 300 nm to 550 nm correspond to direct excitation of Sm^{3+} ground state $^6H_{5/2}$ to higher levels, where the most intense peak at 404 nm is assigned to the electronic transition $^6H_{5/2} \rightarrow {}^4F_{7/2}$ [80]. PL excitation spectrum of Sm^{2+} in Al_2O_3 monitoring at 688 nm shows three broad bands centered around 285 nm, 340 nm, and 480 nm originating from the $4f^6 \rightarrow 4f^5 5d^1$ transitions of Sm^{2+} ions [81].

Figure 20.6c, d illustrates how the concentration of Sm_2O_3 particles in SE affects PL emission. The PL intensity of Sm^{3+} emission bands (Fig. 20.6c) and Sm^{2+} emission bands (Fig. 20.6d) increases with the concentration of Sm_2O_3 particles in the SE as the Sm content of the coatings increases.

20.4.3 PL of Al_2O_3: Ce^{3+} Coatings

Figure 20.7a shows the PL emission spectra excited at 285 nm of Ce-doped Al_2O_3 coatings formed in SE with the addition of different concentrations of CeO_2 particles. Since the PL of the pure Al_2O_3 coating is almost three orders of magnitude lower than that of the counterpart formed in SE with CeO_2 particles, this result indicates that the main contribution to the PL comes from the Ce^{3+} ions incorporated into the Al_2O_3 coatings. Since the Ce ion in CeO_2 is in the +4 oxidation state, the XPS result (Fig. 20.7b) suggests that the interaction between CeO_2 particles and Al_2O_3 under PEO conditions promotes the reduction of Ce^{4+} to Ce^{3+} ions in the Al_2O_3 host [25]. The deconvolution of the high-resolution Ce $3d$ XPS peaks shows that the level of the Ce ion consists of four peaks, i.e., two groups of doublets ($3d_{5/2}$ at 881.5 eV and 885.3 eV; $3d_{3/2}$ at 899.5 eV and 903.9 eV), which are characteristic of the Ce^{3+} oxidation state [82].

The characteristic broad emission band of Ce^{3+}-doped Al_2O_3 coatings exhibits a peak at about 345 nm at an excitation of 285 nm (Fig. 20.7a), which could be attributed to the transitions of Ce^{3+} ions from the excited $5d$ state to the split ground state, into their $^2F_{5/2}$ and $^2F_{7/2}$ components [83]. The intensity of PL emission increases with the concentration of Ce incorporated in the Al_2O_3 coatings. When the excitation wavelength increases from 260 nm to 320 nm, the corresponding emission spectra change their shape and their maximum positions shift to longer wavelengths

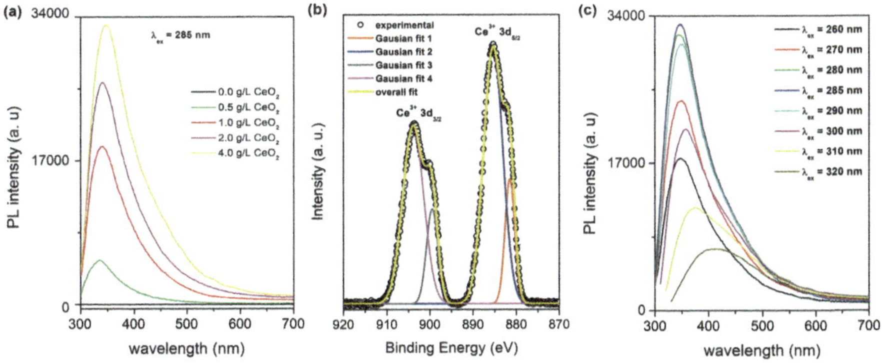

Fig. 20.7 (**a**) The influence of concentration of CeO_2 particles in SE on PL emission spectra excited at 285 nm; (**b**) high-resolution Ce $3d$ XPS spectrum; (**c**) PL emission spectra of coating formed in SE with the addition of 4.0 g/L CeO_2 particles excited with different wavelengths

(345–410 nm) (Fig. 20.7b). This red shift is thought to be due to a 5*d* centroid shift of Ce^{3+} ions [84, 85].

20.4.4 Sensitization of Sm^{2+} Emission by Eu^{2+} Under UV Excitation in Al_2O_3 Coatings

A typical PL emission spectrum excited at 260 nm of the Al_2O_3:Eu^{2+} coating and a PL excitation spectrum monitored at 687 nm of the Al_2O_3:Sm^{2+}/Sm^{3+} coating formed in SE with the addition of 4.0 g/L Eu_2O_3 and 4.0 g/L Sm_2O_3 particles, respectively, are shown in Fig. 20.8a. Due to the overlap between the Eu^{2+} emission band and Sm^{2+} excitation, an efficient energy transfer from Eu^{2+} ions to Sm^{2+} is expected.

Figure 20.8b shows the PL emission spectra of Sm^{2+}/Sm^{3+}, Eu^{2+}, and Sm^{2+}/Sm^{3+}/Eu^{2+}-doped Al_2O_3 coatings excited by 260 nm radiation. The Al_2O_3:Sm^{2+}/Sm^{3+} coating shows a weak PL emission spectrum. The PL emission intensity of Sm^{2+} ions in Al_2O_3:Sm^{2+}/Sm^{3+}/Eu^{3+} is about one order of magnitude higher than that in Al_2O_3:Sm^{2+}/Sm^{3+}, while the PL of the bands related to Sm^{3+} ions is almost unchanged. This result confirms that Eu^{2+} is a very efficient sensitizer for Sm^{2+} to enhance the emission of Sm^{2+} ions under UV excitation in Al_2O_3 coatings. Moreover, it is obvious that the PL emission of Eu^{2+} decreases, indicating that a non-radiative energy transfer from Eu^{2+} ions to Sm^{2+} ions occurs.

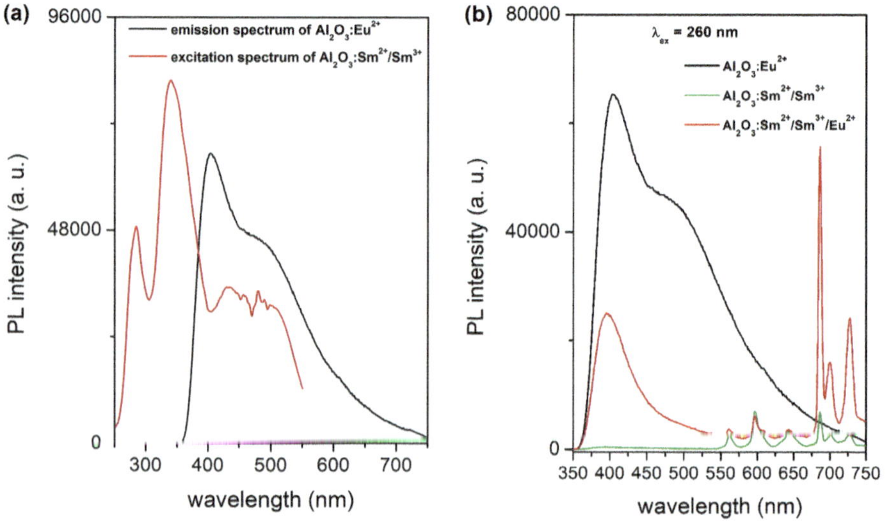

Fig. 20.8 (**a**) Spectral overlap between the PL emission spectrum of Al_2O_3:Eu^{2+} excited at 260 nm and the PL excitation spectrum of Al_2O_3:Sm^{2+}/Sm^{3+} monitored at 687 nm. (**b**) PL emission spectra of Eu^{2+}, Sm^{2+}/Sm^{3+}, and Sm^{2+}/Sm^{3+}/Eu^{2+}-doped Al_2O_3 coatings excited by 260 nm radiation

Based on previous PL spectra analysis, the charge transfer pathways involved in PL of Al_2O_3:Sm^{2+}/Sm^{3+}/Eu^{3+} coatings at 260 nm excitation can be hypothesized as follows. Eu^{2+} ions are excited directly from the ground state $4f^7$ to the $4f^65d^1$ levels before relaxing to the lowest $5d$ crystal field splitting state. Part of the absorption energy of Eu^{2+} ions is released to the ground level via broadband emission, while another portion is transferred to the $5d$ state of Sm^{2+} ions before relaxing to the $^5D_0$ level of Sm^{2+} [86]. After $^5D_0$ level excitation, the excited electrons are transferred to the ^7F_J ($J = 0, 1, 2$) ground state of Sm^{2+}, emitting intensive red PL.

20.4.5 Sensitizing Effect of Ce^{3+} on Eu^{2+} Emission in Al_2O_3:Ce^{3+}/Eu^{2+} Coatings

The overlap of Ce^{3+} PL emission with Eu^{2+} PL excitation in Al_2O_3 between 300 nm and 400 nm (Figs. 20.5 and 20.7) indicates that energy transfer from a sensitizer Ce^{3+} to an activator Eu^{2+} is possible. To verify $Ce^{3+}\rightarrow Eu^{2+}$ energy transfer in Al_2O_3:Ce^{3+}/Eu^{2+} coatings, the PL emission spectra excited at 260 nm and 285 nm from coatings formed in SE with the addition of 4.0 g/L CeO_2, 0.5 g/L Eu_2O_3, and 4.0 g/L CeO_2 + 0.5 g/L Eu_2O_3 are shown in Fig. 20.9. The intensity of the PL band of Ce^{3+} decreases while the intensity of the PL bands of Eu^{2+} increases in the Al_2O_3:Ce^{3+}/Eu^{2+} coating compared to their PL intensities in Al_2O_3:Ce^{3+} and Al_2O_3:Eu^{2+} coatings, respectively. These results indicate that the energy transfer from Ce^{3+} to Eu^{2+} takes place.

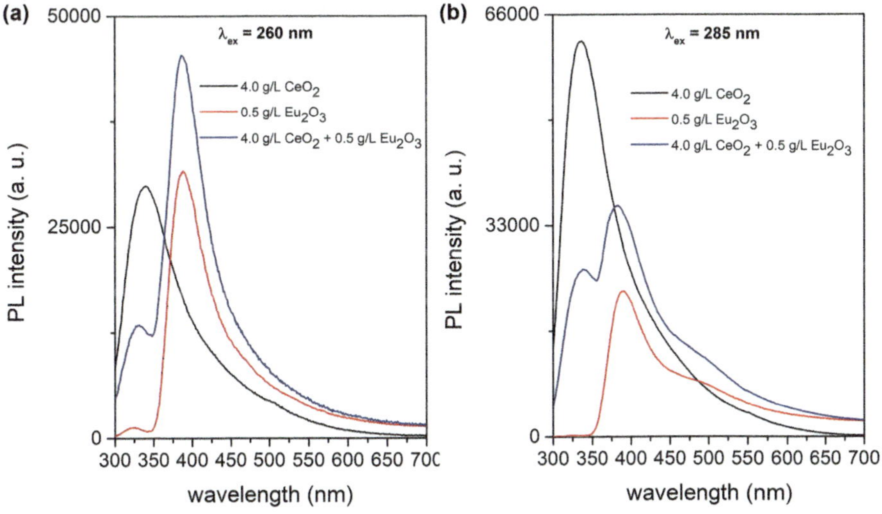

Fig. 20.9 PL emission spectra of Al_2O_3:Ce^{3+}, Al_2O_3:Eu^{2+}, and Al_2O_3:Ce^{3+}/Eu^{2+} coatings excited at: (**a**) 260 nm; (**b**) 285 nm

20.4.6 PL of Al_2O_3:Ln^{3+} (Ln = Dy, Tb, Pr, Tm) Coatings

Figure 20.10 shows the PL excitation and emission spectra of Al_2O_3:Dy^{3+}, Al_2O_3:Tb^{3+}, Al_2O_3:Pr^{3+}, and Al_2O_3:Tm^{3+} coatings formed in SE with the addition of 4.0 g/L Dy_2O_3, Tb_3O_4, Pr_2O_3, and Tm_2O_3 particles, respectively. The PL excitation spectrum of Al_2O_3:Dy^{3+} observed at 577 nm ($^4F_{9/2}\rightarrow{}^6H_{13/2}$ transition of Dy^{3+}) consists of several bands ranging from 250 to 500 nm, which correspond to the direct excitation from the $^6H_{15/2}$ ground state to higher energy levels of Dy^{3+} [87]. The band at 355 nm ($^6H_{15/2}\rightarrow{}^6P_{7/2}$ transition) has a maximum intensity and is used as the excitation wavelength for recording the PL emission spectrum. Under the excitation at 355 nm, the PL emission spectrum of the Al_2O_3:Dy^{3+} coating consists of three emission bands located in the blue-yellow-red region between 460 nm and 700 nm. The bands are centered at about 484 nm, 577 nm, and 668 nm and are assigned to the $^4F_{9/2}\rightarrow{}^6H_{15/2}$, $^4F_{9/2}\rightarrow{}^6H_{13/2}$, and $^4F_{9/2}\rightarrow{}^6H_{11/2}$ transitions, respectively [87].

The PL excitation spectrum of Al_2O_3:Tb^{3+} monitored at 541 nm ($^5D_4\rightarrow{}^7F_5$ transition of Tb^{3+}) can be divided into two regions: the broad band in the range from 250 to 300 nm which mainly originates from the $4f^8\rightarrow 4f^7 5d^1$ transition of Tb^{3+} [88], and several small PL bands in the range from 300 to 550 nm, which correspond to the $4f$–$4f$ transitions of Tb^{3+}. The PL emission spectra excited at 260 nm feature several bands attributed to $4f$–$4f$ transitions of Tb^{3+} from the excited level $^5D_4$ to lower levels ^7F_J (J = 3,4,5, and 6) [89]. The PL bands are centered at about 486 nm, 541 nm, 585 nm, and 621 nm, and are assigned to $^5D_4\rightarrow{}^7F_6$, $^5D_4\rightarrow{}^7F_5$, $^5D_4\rightarrow{}^7F_4$, and $^5D_4\rightarrow{}^7F_3$ transitions, respectively.

The PL excitation spectrum of Al_2O_3:Pr^{3+} monitored at 644 nm ($^3P_0\rightarrow{}^3F_2$ transition of Pr^{3+}) consists of three broad bands centered at about 446 nm, 471 nm, and

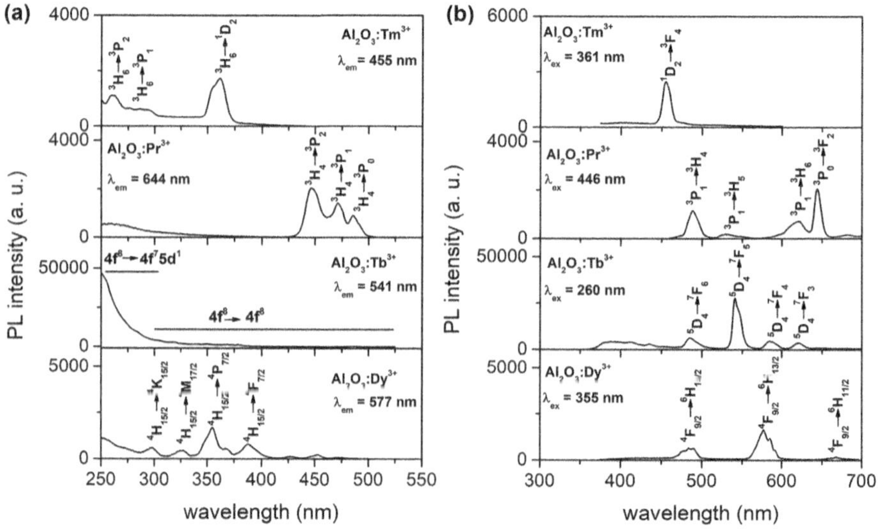

Fig. 20.10 (**a**, **b**) PL excitation and emission spectra of Al_2O_3:Ln^{3+} (Ln = Dy, Tb, Pr, Tm) coatings

485 nm, corresponding to the direct excitation of the Pr^{3+} ground state $^3H_4$ to higher levels ^3P_J ($J = 0, 1, 2$) [16]. The PL emission spectrum excited at 446 nm features bands corresponding to 4*f*–4*f* transitions of Pr^{3+}. The most intense bands are assigned to the $^3P_1 \rightarrow {}^3H_J$ ($J = 4, 5, 6$) and $^3P_0 \rightarrow {}^3F_2$ transitions [90].

The PL excitation spectrum of $Al_2O_3:Tm^{3+}$ monitored at 455 nm ($^1D_2 \rightarrow {}^3F_4$ transition of Tm^{3+}) consists of one intense and sharp band with maximum at 361 nm assigned to direct excitation from ground state $^3H_6$ to excited state $^1D_2$ of Tm^{3+} and several weak bands corresponding to 4*f*–4*f* transitions of the Tm^{3+} from $^3H_6$ to higher levels [91]. Upon 361 nm excitation, the PL emission spectrum of $Al_2O_3:Tm^{3+}$ consists of a strong blue emission band, corresponding to $^1D_2 \rightarrow {}^3F_4$ transition of Tm^{3+} [91].

20.5 Up-Conversion PL Properties of $Ta_2O_5:Ln^{3+}/Yb^{3+}$ (Ln = Er, Ho) Coatings

Ta_2O_5 has attracted much attention in the field of optics and electronics due to its remarkable physical and chemical properties, which include low optical absorption coefficient, high dielectric constant, and refractive index, good photoelectric performance, good chemical stability, low photon energy (strong phonon modes are at 200–300 cm^{-1}) and high energy band gap [92–94]. In this section, the up-conversion PL properties of the $Ta_2O_5:Ln^{3+}/Yb^{3+}$ (Ln = Er, Ho) coatings formed by PEO on the Ta substrate in 10 g/L $Na_3PO_4 \cdot 10H_2O$ + 4 g/L Ln_2O_3 with the addition of Yb_2O_3 particles in concentrations ranging from 1 g/L to 4 g/L for 15 min at a current density of 150 mA/cm^2 are presented. The main crystalline phase in $Ta_2O_5:Ln^{3+}/Yb^{3+}$ coatings is orthorhombic Ta_2O_5 [43, 44].

20.5.1 Up-Conversion PL of $Ta_2O_5:Er^{3+}/Yb^{3+}$ Coatings

Up-conversion PL spectra under 980 nm excitation of formed coatings can be recognized as characteristic 4*f*–4*f* transitions of Er^{3+} as shown in Fig. 20.11a [13]. The shape of the spectra is the same for $Ta_2O_5:Er^{3+}$ and $Ta_2O_5:Er^{3+}/Yb^{3+}$, while the PL intensities of $Ta_2O_5:Er^{3+}/Yb^{3+}$ coatings are significantly higher. Er^{3+} has a very narrow absorption cross-section in the near-infrared region, and only low PL efficiency could be achieved in Er^{3+}-doped materials. Yb^{3+} has a strong and broad absorption band in the near-infrared region (between 850 nm and 1100 nm) and is used as a sensitizer that efficiently transfers its energy to Er^{3+} thus increasing the up-conversion PL emission of Er^{3+}/Yb^{3+}-doped materials [5].

The $Ta_2O_5:Er^{3+}$ coating shows a significantly higher intensity of the $^4S_{3/2} \rightarrow {}^4I_{15/2}$ transition compared to other Er^{3+} transitions (Fig. 20.11a). By excitation at 980 nm, Er^{3+} at the intermediate $^4I_{11/2}$ level is populated by ground state absorption (GSA)

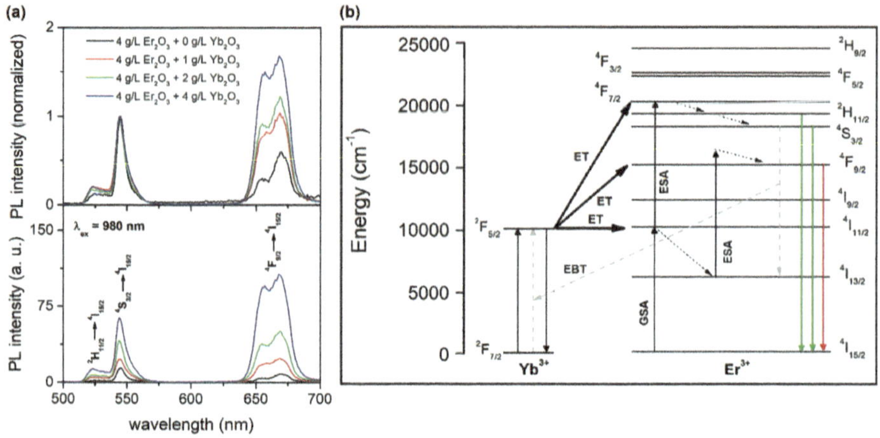

Fig. 20.11 (**a**) Up-conversion PL spectra of Ta_2O_5:Er^{3+}/Yb^{3+} coatings under 980 nm excitation; (**b**) energy level diagram with corresponding transitions

from the $^4I_{15/2}$ level (Fig. 20.11b). Er^{3+} at the $^4I_{15/2}$ level is excited to the $^4F_{7/2}$ level via the excited state absorption (ESA) process, in which incident 980 nm photons are reabsorbed. The ions in the $^4F_{7/2}$ level quickly non-radiatively relax to the $^2H_{11/2}$ and $^4S_{3/2}$ levels due to a small energy gap between $^4F_{7/2}$ and $^2H_{11/2}$, $^4S_{3/2}$ levels, after which they return to the ground state and emit green light ($^2H_{11/2}\rightarrow{}^4I_{15/2}$ and $^4S_{3/2}\rightarrow{}^4I_{15/2}$ transitions). The transition probability of Er^{3+} from the $^2H_{11/2}$ and $^4S_{3/2}$ levels to the $^4F_{9/2}$ level is low, since the energy gap between the $^4S_{3/2}$ and $^4F_{9/2}$ levels is about 3100 cm^{-1}. The $^4F_{9/2}$ level can therefore be populated by the following mechanism: Er^{3+} rapidly relaxes from the intermediate $^4I_{11/2}$ level to the long-lived $^4I_{13/2}$ level, which is further excited to the $^4F_{9/2}$ level by the ESA process. The transition from the $^4F_{9/2}$ to the $^4I_{15/2}$ ground state leads to a red emission.

The increase in up-conversion PL intensity of Ta_2O_5:Er^{3+}/Yb^{3+} coatings is due to the efficient ET of Yb^{3+} to Er^{3+}. Er^{3+} can be directly excited by the above processes, i.e., via ESAs. However, in Ta_2O_5:Er^{3+}/Yb^{3+} coatings, the ESAs are significantly smaller than ET, since Yb^{3+} has a larger absorption cross-section in the near-infrared region than Er^{3+}. Yb^{3+} has only two energy levels, $^4F_{7/2}$ and $^4F_{5/2}$, and the energy gap between these levels corresponds to the energy of 980 nm laser radiation. Yb^{3+} absorbs most of the excitation energy and is excited from the ground state $^2F_{7/2}$ to the state $^2F_{5/2}$ via the GSA (Fig. 20.11b), with a radiative lifetime of about 1 ms [95]. With a radiative lifetime of about 10 ms, Er^{3+} is excited from the ground state $^4I_{15/2}$ to the state $^4I_{11/2}$ by the GSA under the same excitation process. In the $^2F_{5/2}$ state, the excited Yb^{3+} either relaxes to the ground state $^2F_{7/2}$ or uses the ET process to transfer its energy to the nearest Er^{3+}, causing Er^{3+} to transition from the $^4I_{15/2}$ state to the $4I_{11/2}$ state [$^2F_{5/2}$ (Yb^{3+}) + $^4I_{15/2}$ (Er^{3+}) → $^2F_{7/2}$ (Yb^{3+}) + $^4I_{11/2}$ (Er^{3+})] [96]. Following that, the second ET process from Yb^{3+} to Er^{3+} [$^2F_{5/2}$ (Yb^{3+}) + $^4I_{11/2}$ (Er^{3+}) → $^2F_{7/2}$ (Yb^{3+}) + $^4F_{7/2}$ (Er^{3+})] or ESA occurs, and some electrons at the $^4I_{11/2}$ level are excited to the $^4F_{7/2}$ level. Additional processes have already been reported in Er^{3+}-doped

Ta_2O_5 coatings. Additional processes have already been described in Er^{3+}-doped Ta_2O_5. ESA or ET can populate the $^4F_{9/2}$ level of Er^{3+} from the $^4I_{13/2}$ level for red emission [$^2F_{5/2}$ (Yb^{3+}) + $^4I_{13/2}$ (Er^{3+}) → $^2F_{7/2}$ (Yb^{3+}) + $^4F_{9/2}$ (Er^{3+})].

Increasing the Yb^{3+} concentration in the coatings leads to an increased PL intensity of the red transition $^4F_{9/2}$→$^4I_{15/2}$ compared to the green transition $^4S_{3/2}$→$^4I_{15/2}$ (Fig. 20.11a). A phonon-assisted energy back-transfer (EBT) process could promote the red transition: $^4S_{3/2}$ (Er^{3+}) + $^2F_{7/2}$ (Yb^{3+}) → $^4I_{13/2}$ (Er^{3+}) + $^2F_{5/2}$ (Yb^{3+}), which is dependent on the Yb^{3+} concentration [97]. In the EBT process, the energy of the mismatch is about 300 cm^{-1}, which can be easily dissipated by low-energy phonons in the Ta_2O_5. The ETB process reduces the $^4S_{3/2}$ population (which reduces the green emissions) while increasing the population at the $^4I_{13/2}$ level of Er^{3+}. Since the population of the $^4F_{9/2}$ level results from the $^4I_{13/2}$ level of Er^{3+} via the ET, this process leads to an increase in red emission.

20.5.2 Up-Conversion PL of Ta_2O_5:Ho^{3+}/Yb^{3+} Coatings

Figure 20.12a shows the PL up-conversion spectra of the Ta_2O_5:Ho^{3+}/Yb^{3+} coatings when excited at 980 nm. Ta_2O_5:Ho^{3+} coatings do not produce detectable up-conversion PL emission because there are no Ho^{3+} energy levels that match the 980 nm irradiation. The up-conversion PL spectra show three bands centered around 542 nm, 649 nm, and 775 nm. These bands correspond to the transitions of Ho^{3+}: $^5F_4$, $^5S_2$→$^5I_8$, $^5F_5$→$^5I_8$, and $^5F_4$, $^5S_2$→$^5I_5$, respectively [43]. The PL intensity of these bands increases with increasing Yb^{3+} content in the Ta_2O_5:Ho^{3+}/Yb^{3+} coatings, suggesting that the energy is efficiently transferred from excited Yb^{3+} to Ho^{3+}.

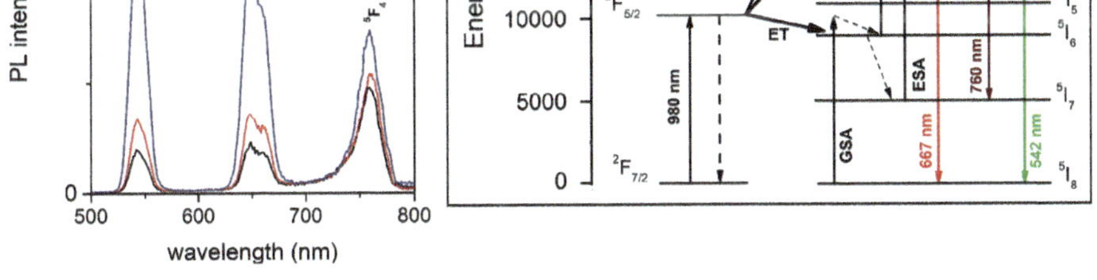

Fig. 20.12 (**a**) Up-conversion PL spectra of Ta_2O_5:Ho^{3+}/Yb^{3+} coatings under 980 nm excitation; (**b**) energy level diagram with corresponding transitions

Figure 20.12b shows an energy level diagram of Ho^{3+} and Yb^{3+} ions with potential excitation and emission pathways that illustrate the mechanisms of PL up-conversion. Yb^{3+} absorbs 980 nm photons by GSA corresponding to the $^2F_{7/2}\rightarrow^2F_{5/2}$ absorption transition. Then these Yb^{3+} transfer energy to the ground state $^5I_8$ of Ho^{3+}, which fills up the excited state $^5I_6$ via GSA. This is a phonon-assisted energy transfer process due to an energy mismatch between the $^2F_{5/2}$ level of Yb^{3+} and the $^5I_6$ level of Ho^{3+} [98]. After reabsorption of the second 980 nm photon, Ho^{3+} ions in the $^5I_6$ level are excited to the $^5F_4$ and $^5S_2$ levels via ESA. The transitions from the $^5F_4$ and $^5S_2$ levels to the $^5I_8$ ground level produce green emissions at around 540 nm, while the transition to the $^5I_7$ level produces emissions at 760 nm. Two proposed pathways for red $^5F_5\rightarrow^5I_8$ emission involve the excitation of Ho^{3+} to the $^5F_5$ excited state [99, 100]. In the first mechanism, Ho^{3+} ions at the $^5F_4$ and $^5S_2$ levels can relax to the $^5F_5$ level by phonon relaxation without radiation and then transition to the $^5I_8$ ground level. The second mechanism is associated with the population of $^5I_7$ level via non-radiative relaxation from the excited state $^5I_6$. ESA excites Ho^{3+} ions in the $^5I_7$ level to the $^5F_5$ level. Finally, the electrons in the $^5F_5$ level jump to the ground state of Ho^{3+}, leading to emission at 667 nm.

20.6 Conclusions and Outlook

This chapter presents the main findings on PEO of various metals (Al, Ti, Zn, Nb, Hf, Ta, Y, and Gd) in electrolytes containing Ln_2O_3 (Ln = Eu, Sm, Pr, Tm, Dy, Er, Ho), CeO_2 and Tb_3O_4 particles to form oxide coatings that can be applied as photoluminescent (PL) materials. Both down-conversion and up-conversion PL emissions are observed with the typical *4f–4f* transition of Ln^{3+} ions. In addition, PEO of Al in an electrolyte containing Eu_2O_3, CeO_2, and Sm_2O_3 particles is suitable for the formation of $Al_2O_3:Eu^{2+}$, $Al_2O_3:Ce^{3+}$, and $Al_2O_3:Sm^{2+}/Sm^{3+}$ coatings. $Al_2O_3:Eu^{2+}$ and $Al_2O_3:Ce^{3+}$ coatings exhibit intense continuous PL emission in the visible and UV regions, respectively, due to *d-f* transitions in Eu^{2+} and Ce^{3+}. $Al_2O_3:Sm^{2+}/Sm^{3+}$ coatings emit intense red PL emission when excited by the $4f^6 \rightarrow 4f^55d^1$ transitions of Sm^{2+}, which correspond to the $^5D_0\rightarrow^7F_J$ ($J = 0, 1, 2$) transitions of Sm^{2+}.

PEO coatings doped with Ln ions have the potential for use in solid-state lighting and luminescence thermometry. Under UV light excitation, $Y_2O_3:Eu^{3+}$, $ZrO_2:Eu^{3+}$, $HfO_2:Eu^{3+}$, and $Gd_2O_3:Eu^{3+}$ coatings emit intense PL, making them good candidates for red phosphors. Eu^{3+}-doped ZrO_2, Nb_2O_5, Gd_2O_3, and TiO_2 coatings are promising for luminescence intensity ratio thermometry. $Al_2O_3:Eu^{2+}$ coatings are suitable for a continuous violet-blue-green light source, while $Al_2O_3:Ce_{3+}$ coatings are suitable for a near-UV light source. The up-conversion emission of Er^{3+}/Yb^{3+} and Ho^{3+}/Yb^{3+}-doped PEO coatings (Ta_2O_5, ZrO_2, etc.) has the potential to be used in optoelectronic and photonic applications.

References

1. Gupta, S. K., Kadam, R. M., & Pujari, P. K. (2020). *Coordination Chemistry Reviews, 420*, 213405.
2. Luo, X., & Xie, R.-J. (2020). *Journal of Rare Earths, 38*, 464–473.
3. Mehare, M. D., Mehare, C. M., Swart, H. C., & Dhoble, S. J. (2023). *Progress in Materials Science, 133*, 101067.
4. Yu, D., Yu, T., Lin, H., Zhuang, S., & Zhang, D. (2022). *Advanced Optical Materials, 10*, 2200014.
5. Zhou, J., Liu, Q., Feng, W., Sun, Y., & Li, F. (2015). *Chemical Reviews, 115*, 395–465.
6. Gupta, I., Singh, S., Bhagwan, S., & Singh, D. (2021). *Ceramics International, 47*, 19282–19303.
7. Kaur, S., Kaur, H., Rao, A. S., & Prakash, G. V. (2024). *Physica B, 690*, 416224.
8. Ansari, A. A., Muthumareeswaran, M. R., & Lv, R. (2022). *Coordination Chemistry Reviews, 466*, 214584.
9. Stojadinović, S., Tadić, N., & Vasilić, R. (2016). *Materials Letters, 164*, 329–332.
10. Stojadinović, S., & Vasilić, R. (2016). *Journal of Luminescence, 176*, 25–31.
11. Stojadinović, S., Tadić, N., & Vasilić, R. (2018). *Journal of Luminescence, 197*, 83–89.
12. Stojadinović, S., Tadić, N., & Vasilić, R. (2018). *Materials Letters, 219*, 251–255.
13. Stojadinović, S., Tadić, N., & Vasilić, R. (2019). *Journal of Luminescence, 208*, 296–301.
14. Stojadinović, S., Tadić, N., & Vasilić, R. (2019). *Journal of Alloys and Compounds, 785*, 1222–1232.
15. Ćirić, A., & Stojadinović, S. (2019). *Journal of Luminescence, 214*, 116568.
16. Ćirić, A., & Stojadinović, S. (2019). *Journal of Alloys and Compounds, 803*, 126–134.
17. Ćirić, A., Stojadinović, S., & Dramićanin, M. D. (2019). *Measurement Science and Technology, 30*, 045001.
18. Ćirić, A., & Stojadinović, S. (2020). *Journal of Alloys and Compounds, 832*, 154907.
19. Ćirić, A., & Stojadinović, S. (2023). *Journal of Luminescence, 257*, 119665.
20. Ćirić, A., & Stojadinović, S. (2024). *Journal of Luminescence, 266*, 120268.
21. Ćirić, A., & Stojadinović, S. (2021). In S. Thomas, K. Upadhyay, R. K. Tamrakar, & N. Kalarikkal (Eds.), *Upconversion nanophosphors* (pp. 103–115). Elsevier.
22. Ćirić, A., & Stojadinović, S. (2023). In P. M. Visakh, A. Semkin, R. Balakrishnan, & S. Lazović (Eds.), *Nanotechnology in electronics* (pp. 279–300). Wiley-VCH GmbH.
23. Stojadinović, S., Tadić, N., & Vasilić, R. (2017). *Journal of Luminescence, 192*, 110–116.
24. Stojadinović, S., & Vasilić, R. (2018). *Journal of Luminescence, 199*, 240–244.
25. Stojadinović, S., & Vasilić, R. (2018). *Journal of Luminescence, 203*, 576–581.
26. Stojadinović, S., & Vasilić, R. (2019). *Materials Letters, 234*, 9–12.
27. Stojadinović, S., & Ćirić, A. (2019). *Coatings, 9*, 819.
28. Stojadinović, S., & Ćirić, A. (2020). *Journal of Luminescence, 226*, 117403.
29. Stojadinović, S., & Ćirić, A. (2020). *Journal of Luminescence, 226*, 117407.
30. Stojadinović, S., & Ćirić, A. (2021). *Journal of Materials Science: Materials in Electronics, 32*, 7841–7847.
31. Ćirić, A., Stojadinović, S., Ristić, Z., Zeković, I., Kuzman, S., Antić, Ž., & Dramićanin, M. D. (2021). *Advanced Materials Technologies, 6*, 2001201.
32. Stojadinović, S., & Ćirić, A. (2021). *Journal of Luminescence, 239*, 118323.
33. Stojadinović, S., & Ćirić, A. (2021). *Journal of Materials Science: Materials in Electronics, 32*, 23472–23485.
34. Ćirić, A., Stojadinović, S., & Dramićanin, M. (2023). *Advanced Materials Technologies, 8*, 2300607.
35. Stojadinović, S., Radić, N., Grbić, B., Maletić, S., Stefanov, P., Pačevski, A., & Vasilić, R. (2016). *Applied Surface Science, 370*, 218–228.
36. Stojadinović, S., Tadić, N., Radić, N., Grbić, B., & Vasilić, R. (2018). *Surface and Coating Technology, 337*, 279–289.

37. Torres-Ceron, D. A., Stojadinović, S., Radić, N., Amaya-Roncancio, S., Velasquez-Tamayo, J. P., Benavides-Palacios, V., & Restrepo-Parra, E. (2024). *Applied Surface Science, 665*, 160351.
38. Ćirić, A., Stojadinović, S., & Dramićanin, M. D. (2018). *Optical Materials, 85*, 261–266.
39. Stojadinović, S., & Ćirić, A. (2021). *Journal of Luminescence, 235*, 118022.
40. Stojadinović, S., & Vasilić, R. (2016). *Journal of Alloys and Compounds, 685*, 881–889.
41. Stojadinović, S., Tadić, N., Ćirić, A., & Vasilić, R. (2018). *Optical Materials, 77*, 19–24.
42. Ćirić, A., Stojadinović, S., & Dramićanin, M. D. (2019). *Optical Materials, 88*, 392–395.
43. Stojadinović, S., Perković, M., & Ćirić, A. (2023). *Journal of Materials Science: Materials in Electronics, 34*, 1658.
44. Stojadinović, S., & Ćirić, A. (2023). *Journal of Electronic Materials, 52*, 7384–7392.
45. Ćirić, A., & Stojadinović, S. (2020). *Journal of Luminescence, 217*, 116762.
46. Ćirić, A., & Stojadinović, S. (2020). *Optical Materials, 99*, 109546.
47. Ćirić, A., Stojadinović, S., & Dramićanin, M. D. (2020). *Ceramics International, 46*, 23223–23231.
48. Ćirić, A., Stojadinović, S., Ristić, Z., Antić, Ž., & Dramićanin, M. D. (2021). *Sensors Actuators A, 331*, 112987.
49. Ćirić, A., Stojadinović, S., & Dramićanin, M. D. (2022). *Journal of Alloys and Compounds, 900*, 163544.
50. Kaseem, M., Fatimah, S., Nashrah, N., & Ko, Y. G. (2020). *Progress in Materials Science, 117*, 100735.
51. Clyne, T. W., & Troughton, S. C. (2019). *International Materials Review, 64*, 127–162.
52. Stojadinović, S., Vasilić, R., & Perić, M. (2014). *RSC Advances, 4*, 25759.
53. Binnemans, K. (2015). *Coordination Chemistry Reviews, 295*, 1–45.
54. Kostyukov, A. I., Panchenko, V. N., Nashivochnikov, A. A., Rakhmanova, M. I., Cherepanova, S. V., Suprun, E. A., & Antonova, O. (2023). *Ceramics International, 49*, 41158–41166.
55. García Ramírez, V. M., García Murillo, A., Romo, F. D. J. C., Tobón, A. C., & Castro, N. C. (2023). *Bulletin of Materials Science, 46*, 75.
56. Liu, D., Wang, M., Gong, L., Zhao, J., Zhu, M., & Wang, X. (2021). *Journal of Alloys and Compounds, 864*, 158781.
57. Aguilar-Castillo, A., Aguilar-Hernández, J. R., García-Hipólito, M., López-Romero, S., Swarnkar, R. K., Báez-Rodríguez, A., Fragoso-Soriano, R. J., & Falcony, C. (2017). *Ceramics International, 43*, 355–362.
58. Dramićanin, M. D. (2020). *Journal of Applied Physics, 128*, 040902.
59. Jia, M., Chen, X., Sun, R., Wu, D., Li, X., Shi, Z., Chen, G., & Shan, C. (2023). *Nano Research, 16*, 2949–2967.
60. Dramićanin, M. D. (2016). *Methods and Applications in Fluorescence, 4*, 042001.
61. Collins, S. F., Baxter, G. W., Wade, S. A., Sun, T., Grattan, K. T. V., Zhang, Z. Y., & Palmer, A. W. (1998). *Journal of Applied Physics, 84*, 4649–4654.
62. Carnall, W. T., Goodman, G. L., Rajnak, K., & Rana, R. S. (1989). *The Journal of Chemical Physics, 90*, 3443–3457.
63. de Azevedo, W. M., de Carvalho, D. D., de Vasconcelos, E. A., & da Silva, E. F. (2004). *Applied Surface Science, 234*, 457–461.
64. Nayar, P., Zhu, X.-Y., Yang, F., Lu, M., Lakshminarayana, G., Liu, X. P., Chen, Y.-F., & Kityk, I. V. (2016). *Optical Materials, 60*, 57–61.
65. Ozuna, O., Hirata, G. A., & McKittrick, J. (2004). *Applied Physics Letters, 84*, 1296–1298.
66. Loong, C. K., Richardson, J. W., Jr., & Ozawa, M. (1997). *Journal of Alloys and Compounds, 250*, 356–359.
67. Rogers, J. J., MacKenzie, K. J. D., Rees, G., & Hanna, J. V. (2018). *Ceramics International, 44*, 1110–1119.
68. Xie, H., Lu, J., Guan, Y., Huang, Y., Wei, D., & Seo, H. J. (2014). *Inorganic Chemistry, 53*, 827–834.

69. Zhai, B., Liu, D., He, Y., Yang, L., & Huang, Y. M. (2018). *Journal of Luminescence, 194*, 485–493.
70. Dorenbos, P. (2003). *Journal of Luminescence, 104*, 239–260.
71. Chen, M., Xia, Z., Molokeev, M. S., Lin, C. C., Su, C., Chuang, Y. C., & Liu, Q. (2017). *Chemistry of Materials, 29*, 7563–7570.
72. Budde, B., Luo, H., Dorenbos, P., & van der Kolk, E. (2017). *Optical Materials, 69*, 378–381.
73. Peng, M., & Hong, G. (2007). *Journal of Luminescence, 127*, 735–740.
74. Stojadinovic, S., Vasilic, R., Nedic, Z., Kasalica, B., Belca, I., & Zekovic, L. (2011). *Thin Solid Films, 519*, 3516–3521.
75. Onishi, Y., Nakamura, T., & Adachi, S. (2016). *Journal of Luminescence, 176*, 266–271.
76. Pillonnet, A., Pereira, A., Marty, O., & Champeaux, C. (2011). *Journal of Physics D: Applied Physics, 44*, 375402.
77. Haritha, P., Dwaraka Viswanath, C. S., Linganna, K., Babu, P., Jayasankar, C. K., Lavín, V., & Venkatramu, V. (2016). *Journal of Luminescence, 179*, 533–538.
78. Sakirzanovas, S., Katelnikovas, A., Dutczak, D., Kareiva, A., & Justel, T. (2011). *Journal of Luminescence, 131*, 2255–2262.
79. Yang, L., Yu, X., Yang, S., Zhou, C., Zhou, P., Gao, W., & Ye, P. (2008). *Materials Letters, 62*, 907–910.
80. David, A. D. J., Muhammad, G. S., & Sivakumar, V. (2016). *Journal of Luminescence, 177*, 104–110.
81. Zeng, Q., Kilah, N., & Riley, M. (2003). *Journal of Luminescence, 101*, 167–174.
82. Yagoub, M. Y. A., Swart, H. C., Noto, L. L., Bergman, P., & Coetsee, E. (2015). *Maternité, 8*, 2361–2375.
83. Martínez-Martínez, R., García-Hipólito, M., Ramos-Brito, F., Hernández-Pozos, J. L., Caldiño, U., & Falcony, C. (2005). *Journal of Physics. Condensed Matter, 17*, 3647–3656.
84. Wang, F., Wang, W., Zhang, L., Zheng, J., Jin, Y., & Zhang, J. (2017). *RSC Advances, 7*, 27422–27430.
85. Dorenbos, P. (2013). *Journal of Luminescence, 135*, 93–104.
86. Wang, Y., Ding, J., Li, Y., & Wang, Y. (2015). *RSC Advances, 5*, 88477–88484.
87. Wang, C., Jiang, J., Zhu, G., & Xin, S. (2018). *Optical Materials, 86*, 66–70.
88. Martínez-Martínez, R., Álvarez, E., Speghini, A., Falcony, C., & Caldiño, U. (2010). *Thin Solid Films, 518*, 5724–5730.
89. Chen, Y., Luo, D., Luo, L., Wang, X., Tang, T., & Luo, W. (2014). *Journal of Non-Crystalline Solids, 386*, 124–128.
90. Rao, V. H., Prasad, P. S., & Babu, K. S. (2020). *Optical Materials, 101*, 109740.
91. Li, Y., Zheng, J., Li, Z., Yang, X., Chen, J., & Chen, C. (2018). *Optik, 169*, 257–263.
92. Bright, T. J., Watjen, J. I., Zhang, Z. M., Muratore, C., Voevodin, A. A., Koukis, D. I., Tanner, D. B., & Arenas, D. J. (2013). *Journal of Applied Physics, 114*, 083515.
93. Zhang, H.-F., Ning, B.-Y., Weng, T.-C., & Ning, X. J. (2021). *Journal of the American Ceramic Society, 104*, 6413–6423.
94. Sharma, N., Kumar, V. S. R. S. P., Kumar, M., Kumari, N., Karar, V., Goswamy, J. K., & Sharma, A. L. (2020). *Optik, 202*, 163697.
95. Salas, P., Angeles-Chávez, C., Montoya, J. A., De la Rosa, E., Diaz-Torres, L. A., Desirena, H., Martínez, A., Romero-Romo, M. A., & Morales, J. (2005). *Optical Materials, 27*, 1295–1300.
96. Xie, J., Bin, J., Guan, M., Liu, H., Yang, D., Xue, J., Liao, L., & Mei, L. (2018). *Journal of Luminescence, 200*, 133–140.
97. Zhang, X., Xu, D., Zhou, G., Wang, X., Liu, H., Yu, Z., Zhang, G., & Zhu, L. (2016). *RSC Advances, 6*, 103973–103980.
98. Pandey, A., Kumar, V., Kroon, R. E., & Swart, H. C. (2017). *Journal of Luminescence, 192*, 757–760.

99. Zhang, M., Wen, H., Yu, H., Ai, F., Shao, H., Pan, X., Tang, M., Yu, J., Gai, L., & Liu, Y. (2016). *Journal of Alloys and Compounds, 672*, 7–12.
100. Li, N., Wang, W., Duan, P., Wang, Y., Sun, X., Di, J., Li, W., Chu, B., & He, Q. (2016). *Chemical Physics Letters, 644*, 152–156.

Chapter 21
Plasma Electrolytic Oxidation Coatings for Antibacterial and Antibiofouling Applications

Gurleen Singh Sandhu, Md Julker Nine, and Dusan Losic

21.1 Introduction

Plasma electrolytic oxidation (PEO) process has become one of the very promising techniques in the fabrication of robust porous coatings applied in biomedical applications [1]. In the process, ceramic oxide coatings are formed on different valve metals such as aluminum (Al), magnesium (Mg), titanium (Ti), tantalum (Ta), and niobium (Nb) and their alloys, which are widely used in medical implant materials because of their biocompatibility and mechanical properties. Features such as porous structure, possibility of bioactive compound incorporation, and advanced surface properties of PEO coatings make them especially suitable in applications that require bioactive antimicrobial properties. For example, coatings developed on Ti alloys using PEO facilitate hydroxyapatite (HAp) incorporation, resulting in enhanced biocompatibility or bioactivity of the implant surfaces. This is particularly important in orthopedic and dental applications, in which the integration of implants with bone tissue is so crucial [2]. The incorporation of bioactive materials in PEO coatings can enhance osteoconductive properties, hence enhancing the general process of healing, which makes such coatings highly desirable in biomedical applications [3]. PEO also enables the incorporation of various antibacterial agents directly into the coating matrix. For example, silver nanoparticles (Ag NPs) can be entrapped within the porous structure of the PEO coatings to ensure sustained antibacterial activity against a wide spectrum of pathogens, including *methicillin-resistant Staphylococcus aureus* (*S. aureus*) (MRSA) [4]. The released silver ions (Ag^{2+}) from the coating have been reported to successfully inhibit bacterial growth, hence minimizing the chances of infection caused by the implanted medical devices [5]. It

G. S. Sandhu · M. J. Nine (✉) · D. Losic (✉)
School of Chemical Engineering, The University of Adelaide, Adelaide, SA, Australia
e-mail: mdjulker.nine@adelaide.edu.au; dusan.losic@adelaide.edu.au

V. S. Saji, M. Zheludkevich (eds.), *Plasma Electrolytic Oxidation – Fundamentals, Advances and Applications*,
https://doi.org/10.1007/978-3-031-92429-3_21

is also possible to modify the chemical and physical characteristics of these coatings with the use of different electrolytes, allowing their antibacterial properties to be optimized. The addition of nanoparticles (NPs), like zinc oxide (ZnO) to calcium phosphate (CaP), increases its antibacterial efficacy without compromising its biocompatibility [3, 6]. This possibility of customization makes PEO to be considered as an adaptable technique in developing coatings for various biomedical applications. Apart from improving biocompatibility and antibacterial properties, PEO coatings excel in corrosion resistance, which is one of the most essential concerns for biomedical implants to ensure longevity and reliability [7]. The nature of PEO coatings being dense and compact decreases the permeability of corrosive agents markedly and protects the degradation of the underlying metal substrate from physiological environments [8, 9].

This chapter is organized into several sections: the general impression and significance of PEO coatings for antibacterial properties are highlighted at the beginning of the introductory section. In Sect. 21.2, different influential factors of PEO process have been explained showing the effect of electrolyte composition, structural properties, and wettability on the performance of antibacterial/antibiofouling properties. Section 21.3 discusses how antibacterial coatings are used in different areas like biomedical devices, surgical instruments and tools, and biofouling in marine structures. Finally, the conclusions and outlook are briefed in Sect. 21.4.

21.2 Influential Factors of PEO Coatings to Control Bacterial Proliferation

Modification of surface properties in several ways during the PEO process enables the fabrication of antibacterial coatings. The incorporation and deposition of various elements with the choice of different electrolytes, the introduction of micro/nanostructures, alteration of wettability of the coatings and fabrication of porous surface coatings offer active pockets to load antibacterial agents, as shown in Fig. 21.1. A few of the influential factors towards enhanced antibacterial PEO coatings are discussed below.

21.2.1 Effect of Elemental Incorporation on Antibacterial PEO Coatings

The electrolyte plays a significant role in bringing elemental variation during PEO process enhancing biocompatibility and antibacterial properties in PEO coatings. The addition of bioactive elements, such as calcium (Ca) and phosphorus (P) from phosphoric acid-based (H_3PO_4) electrolytes, triggers the formation of HAp known

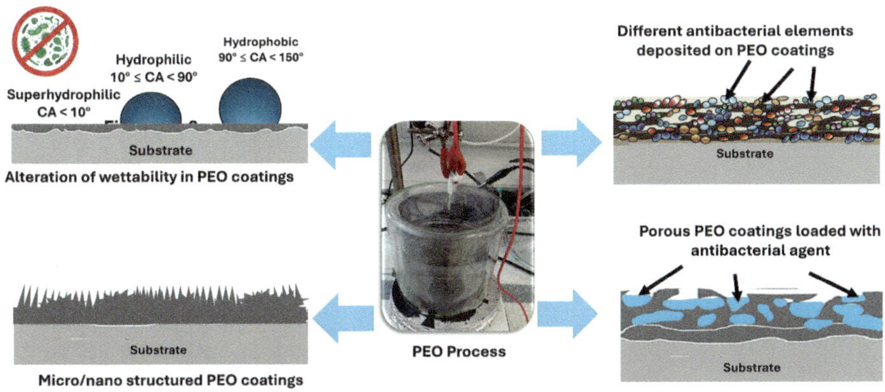

Fig. 21.1 Schematic of influential factors of PEO coatings to control bacterial proliferation

for its biocompatibility and reduction of bacterial adhesion [10]. Incorporating different NPs can also add to the antibacterial properties of the PEO coatings as well. For instance, the doped addition of Ag NPs to the electrolyte offers huge antibacterial effects due to the release of Ag^{2+} ions inhibiting microbial growth during long-term use [4, 10]. The porous properties and other structural features of the PEO coatings are also related to the electrolyte composition itself and account for better nutrient exchange and cell attachment required for osseointegration [11]. Further, the corrosion resistance of the coatings can be improved when silicate (SiO_3^{2-}) and fluoride (F^-) ions are added to the electrolyte, which indirectly enhances the antibacterial efficiency of coatings during service by sustaining the surface integrity [12]. Different prominent works showing the effect of electrolyte composition in fabricating antibacterial PEO coatings have been discussed in Table 21.1.

21.2.2 Effect of Modified Structures and Porosity on Antibacterial PEO Coating

The changes in the structural properties of PEO coatings may provide a real opportunity to improve their antibacterial efficiency for biomedical applications. Increased porosity within PEO coatings allows better ion release, which can actively kill bacteria and improve bioactivity due to nutrient exchange and cell attachment. The antibacterial activity of coatings with enhanced porosity can release more antibacterial agents [21–23]. On the other hand, surface roughness plays a dual role; while excessive roughness can promote bacterial adhesion, optimal roughness can mechanically disrupt bacterial membranes, enhancing antibacterial efficacy. This has been observed in studies concerning coatings with different surface roughness and their adhesion properties [24, 25]. While surface topography, at the nanoscale,

Table 21.1 Summary of the influence of different electrolyte compositions on the antibacterial activity and bacterial inhibition efficiencies of PEO coatings

S. no.	Electrolyte	Surface composition	Bacterial strain	Bacterial inhibition rate (%)	Ref
1	NaH_2PO_4, NaOH	TiO_2 + Ag NPs	*E. coli*	99.9	[13]
			S. aureus	90	
2	$HF_{(aq)}$	TiO_2 + Ag NPs	*E. coli*	99.85	[14]
			S. aureus	99.85	
3	Na_2SiO_3, KF, NaOH	Ag NPs	*S. aureus*	99	[15]
4	$Na_3PO_4.12H_2O$, KOH, $Ca_3(PO_4)_2$,	ZrO_2 NPs	*S. aureus*	94.37	[16]
5	$Na_3PO_4.12H_2O$, KOH, and $Ca_3(PO_4)_2$	TiO_2 NPs	*S. aureus*	94.47	[17]
6	$Zn(NO_3)_2{\cdot}4\ H_2O$ and NaH_2PO_4	Zn/P-Si and Zn/P-P	*E. coli*	94.58	[18]
			S. aureus	96.04	
7	$C_3H_7Na_2O_6P$, KOH	FeOOH nanosheet	*S. aureus*	86.7	[19]
8	Na_2SiO_3, H_3PO_4, KOH	Graphene oxide (GO)	*E. coli*	80	[20]
			S. aureus	100	

mostly, may result in changes in wettability and surface energy of the coatings, thus affecting antibacterial behavior, nanoscale features can enhance the release of antibacterial ions and promote biofilm resistance. It has also been proved that micro/nanostructured surfaces can enhance ion release and, thus, provides improved antibacterial properties [4, 26]. Finally, surface patterning of micro and nanostructures can build physical barriers to restrict bacterial colonization and facilitate the release of highly reactive oxygen species (ROS) to elicit bacterial death. The lack of available literature concerning the direct linking of surface patterning with enhanced antibacterial properties in PEO coatings justifies the demand for more research in the field. Accordingly, PEO coatings can be designed for effective antibacterial surfaces by making strategic modifications in the properties of these structures for various biomedical applications and a comparative study analysis is enlisted in Table 21.2.

21.2.3 Effect of Wettability on Bioactivity and Antibacterial PEO Coatings

Wettability is one critical aspect that affects the performance of PEO coatings especially in biomedical applications where interaction with biological entities at surfaces is important. Wettability, described by the contact angle that a liquid

Table 21.2 Summary of the effect of porosity, surface roughness, and surface topography on a range of bacteria with inhibition rates

S. no.	Porosity (%)	Surface roughness (RA, μm)	Surface topography	Bacterial strain	Bacterial inhibition rate (%)	Ref
1	1.4–5.1	0.6–1.5	Rough	*E. coli*	100	[27]
				S. epidermidis	100	
2	2.1–7.3	0.27–0.83	Rough	*S. aureus*	94	[16]
3	–	0.65–0.85	Micro-nano porous	*E. coli*	80	[20]
				S. aureus	100	
4	3.4–4.3	0.04–0.06	Rough	*S. aureus*	>90	[28]
5	–	0.9–1.4	Nodular	*E. coli*	23	[29]
				S. aureus	45	
6	8	2.5–2.6	–	*P. aeruginosa*	82	[30]
				S. aureus	64	
				E. faecium	57	
7	1.1–8.7	0.15–2.7	Porous	*E. coli*	87	[31]
8	–	1.1–2.0	Microporous	*E. coli*	83	[32]

droplet assumes upon a surface, has a strong influence on cell adhesion and, subsequently, biofilm formation, which is very important regarding the performance of medical implants and devices. PEO coatings are known to improve wettability, which is helpful in enhancing biocompatibility and antibacterial properties. For example, PEO-treated magnesium (Mg) alloys showed increased hydrophilicity, with contact angles reducing to about 55.2° from the uncoated surfaces, thus improving biological performance [33]. Hydrophilic surfaces favor the adsorption of proteins and subsequent cell attachment hence promoting osseointegration processes [34]. This hydrophilicity is essential for the effective functioning of antibacterial agents incorporated within the coatings, as it promotes the release of these agents into the surrounding environment. The increased surface roughness associated with PEO coatings also contributes to bacterial adhesion; though rough surfaces can enhance adhesion; the right balance of wettability and roughness can minimize bacterial colonization. Further surface treatment on PEO coatings with the resultant sharp-edged spike morphology plays a significant role in killing bacteria by mechanical piercing [16, 35]. While microstructurally featured surfaces may enhance cellular adhesion, a combination of micro/nanostructure and superwetting properties could effectively disrupt bacterial membranes. This duality highlights careful design in PEO coatings for the optimization of both wettability and antibacterial properties, and some of the results from recent studies are tabulated in Table 21.3.

Table 21.3 Summary of the effect of wettability on the antibacterial effect of PEO coatings

S. no.	Material	Wettability classification	Contact angle (°)	Bacteria studied	Inhibition rate (%)	Ref
1	Ta_2O_5	Hydrophilic	20	*S. mutans*	38	[36]
2	GO	Superhydrophilic	4.3	*S. aureus*	–	[37]
3	Ag	Hydrophilic	20	*S. aureus*	–	[13]
				E. coli	–	
4	GO	Hydrophilic	16	*S. aureus*	100	[20]
				E. coli	80	
5	ZrO_2	Superhydrophilic	10	*S. epidermidis*	100	[27]
				E. coli	100	
6	TiO_2	Hydrophilic	31	*S. aureus*	95	[16]
7	TiO_2	Superhydrophilic	5	*S. aureus*	97	[38]
8	$ZnO–ZrO_2$	Hydrophilic	55	*S. aureus*	90	[28]
9	MXene	Hydrophilic	43	*S. aureus*	97	[39]

21.3 Applications of Antibacterial and Antibiofouling PEO Coatings

Antibacterial PEO coatings have gained a lot of interest due to their efficiency in preventing bacterial colonization and biofilm formation, hence highly suitable for different critical applications (shown in Fig. 21.2). The enhanced antibacterial properties are being exploited in dental implants and prosthetics, which lead to reduced infection risks and improve the longevity of implants. In addition, surgical tools coated with antibacterial PEO layers maintain sterility during procedures and minimize postoperative infections. Besides, the performance and potential of the coatings in combating aquatic biofouling offer promising avenues to solve problems of microbial buildup on surfaces of marine structures.

21.3.1 Antibacterial PEO Coatings for Implants and Prosthetics

PEO is a superior surface modification process that has attained a great deal of interest within the biomedical device area, mainly in implants and prosthetics. The technique realizes the fabrication of hard, bioactive coatings onto popular metallic implants, essentially made of Ti and its alloys which are extensively used in orthopedic and dental applications [35]. The coatings made by PEO not only improve the mechanical properties of implants but also provide antibacterial features, which are very important to avoid post-operative infections.

The application of PEO coatings in biomedical devices is very much related to the incorporation of antibacterial agents within the coatings. For example, Ag-doped TiO_2 coatings fabricated by PEO showed excellent antibacterial properties against

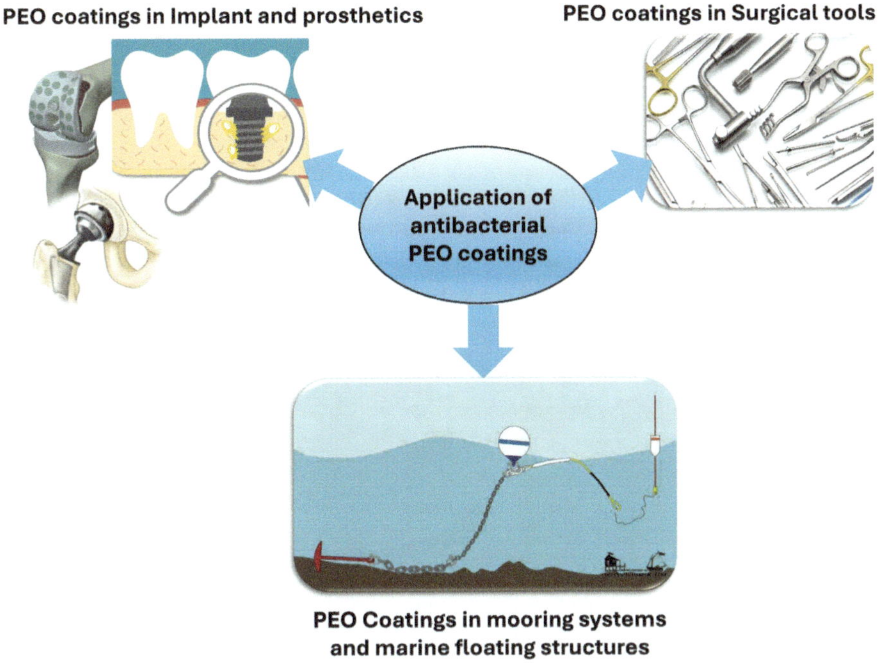

Fig. 21.2 Applications of the PEO coatings in (1) implant and prosthetics, (2) surgical tools, and (3) marine structures (e.g., mooring systems, grounded structures of floating devices)

the most common pathogens, such as *S. aureus* and *E. coli*, which are often responsible for implant-related infections [13, 40]. The antibacterial action of such coatings is determined by the release of active agents (e.g., Ag^+ ions), which disturb the transport of substances in and out of the bacterial cells and block bacterial growth [13]. Moreover, due to the porous structure of PEO coatings, embedding a variety of NPs, such as ZnO and HAp significantly enhances their bioactivity and antibacterial efficiency simultaneously, as depicted in Fig. 21.3 [2, 41]. PEO coatings not only have antibacterial properties but also greatly increase implant biocompatibility. By simulating the natural bone environment, the coatings can encourage osteoconductive and osteointegration. For instance, it has been demonstrated that adding calcium (Ca^{2+}) and phosphate ions to PEO coatings improves osteoblast adhesion and proliferation, which are critical for bone integration and healing [2, 38, 40]. The success of orthopedic implants depends on improved cell adhesion, and nutrition exchange made possible by PEO coatings' porosity nature [13].

Moreover, PEO coatings have very good corrosion resistance, which is important for the implant life span due to the very aggressive physiological environment in the human body. The dense layer formed does not allow the underlying metal to corrode, hence reducing the failure of implants [22, 42]. It has been proved in the studies that PEO-treated Ti alloys exhibit higher corrosion resistance compared with untreated ones and, therefore, are more suitable for enhanced service life of implants

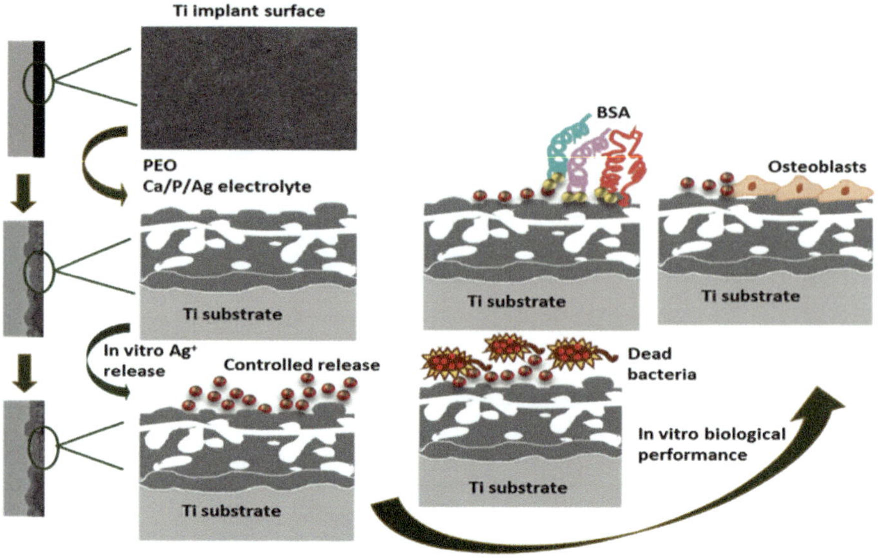

Fig. 21.3 Schematic illustration of antibacterial bioactive PEO coatings with embedded Ca/P/Ag elements. Reproduced with permission from Ref. [40] © 2020 American Chemical Society

[43]. In another study, addition of zirconium dioxide (ZrO_2) to fabricate PEO coatings has been proven for long-term corrosion resistance for orthopedic and dental implants [16]. Further, the same nanocomposite coatings were found to be efficient in inhibiting *S. aureus* bacteria up to ~94% within 24 h as shown in Fig. 21.4.

21.3.2 Mechanical Robust PEO Coatings for Surgical Instruments and Tools

The PEO process allows the possibility of tailoring coatings to specific clinical needs. Researchers can control mechanical, chemical, and biological properties by manipulating the electrolyte composition and processing parameters of PEO coatings [44, 45]. This ability to adapt is of critical importance for the development of the next-generation biomedical devices that must prevent infection. Moreover, the mechanical properties of PEO coatings make them appropriate for surgical instruments. The coatings exhibit high resistance to wear and high hardness, which are very important properties in order to retain the integrity of the surgical tools upon timely use [43, 46]. For instance, it was reported that PEO-treated Ti surfaces show better wear resistance compared to untreated ones, which is an important feature for instruments subjected to repeated sterilization and mechanical stressing [44]. In a different work, the PEO process was used to oxidize Ti6Al4V scaffolds that were

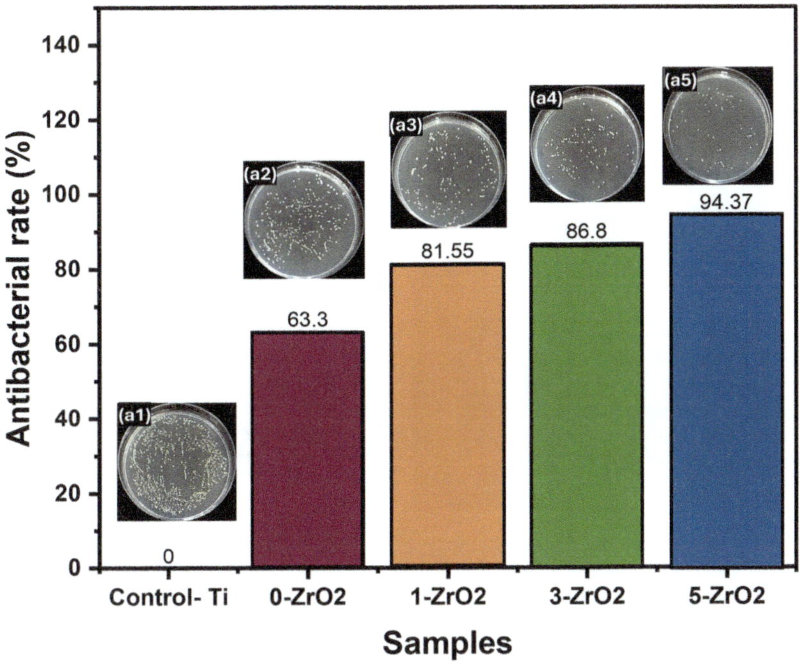

Fig. 21.4 Bacterial colonies (inset image) of *S. aureus* bacteria at 37 °C (a1, a2, a3, a4, and a5) after 24 h with examined antibacterial rate of coated and uncoated samples. Reproduced with permission from Ref. [16] © 2022 Elsevier B.V.

additively fabricated using the cubic unit cell and had four distinct relative densities ranging from 13 to 37% [47]. The scaffolds' mechanical characteristics (elastic modulus, yield stress, and plateau stress) were not significantly altered by the oxidation process. However, even after 106 loading cycles at 70% of their yield strength, the fatigue tests showed that none of the scaffolds under any of the oxidation conditions failed. Similarly, the improved corrosion resistance of PEO-coated surfaces guarantees functionality and safety for longer periods in surgical instruments, even in aggressive environments like those found in surgical settings [48].

21.3.3 Antibiofouling PEO Coatings in Marine Structures

Biofouling represents one of the most serious problems in many industries, such as biomedical, marine, and oil/gas industries, causing serious economic problems and environmental consequences [49, 50]. So far researchers have applied different methods to address this issue, but one promising process could be to employ the PEO coatings to a few selective marine structures to function against fouling

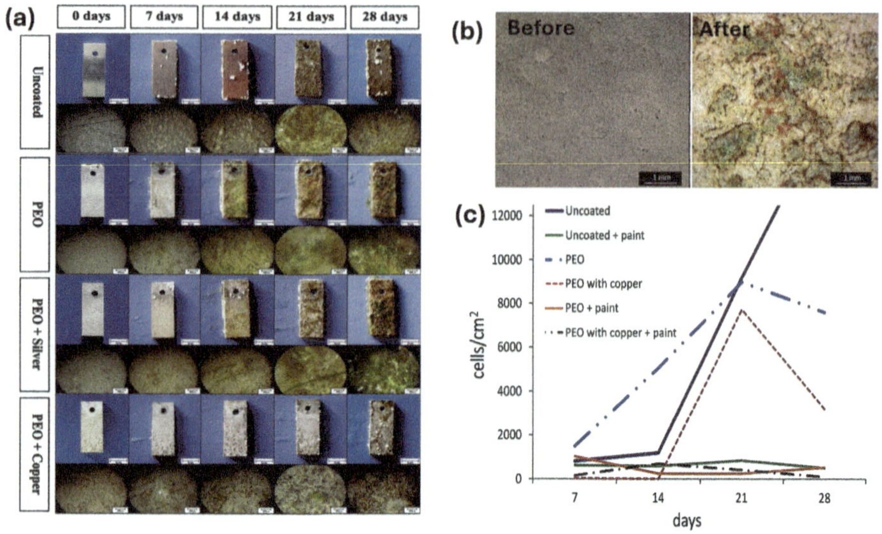

Fig. 21.5 (**a**) Digital images of uncoated and PEO-treated Al samples. Reproduced with permission from Ref. [51]© 2018 Elsevier B.V. (**b**) Stereomicroscopic images of PEO surface before and after PEO treatment with copper, (**c**) Results of quantitative analyses about the treatment effects on bacterial colonization. Reproduced with permission from Ref. [52] © 2020 Elsevier B.V.

deposition. The efficacy of the technique has shown that the PEO-treated Al samples exhibit good antifoulant properties [51]. Particularly, Ag and Cu NPs have been incorporated into PEO electrolytes to impart antibiofouling properties, and among them, the Cu NPs have shown enhanced barriers toward antibiofouling and anticorrosion. The analysis of the antibiofouling test conducted and observed under a stereomicroscope has been shown in Fig. 21.5a. It is distinguishable that the PEO-treated Al samples with the incorporation of Cu NPs exhibited excellent antibiofouling in comparison to incorporation of Ag NPs.

Apart from biofouling, metallic substrates used for different applications in seawater are always prone to the problem of biocorrosion. PEO surface treatment permits the addition of NPs to the coating and significantly improves the corrosion and wear resistance of light metals compared to conventional anodizing. A study examined the antibiofouling characteristics of Cu-containing PEO coatings fabricated on AA 7075 in seawater (Fig. 21.5b) [52]. They have introduced Cu NPs in the PEO electrolyte and consequently have successfully incorporated Cu NPs on the surface of PEO treated sample. Further, they evaluated the PEO + Cu treated sample for anticorrosion and antibiofouling in the seawater. Further, Fig. 21.5c presents a comparative analysis of the effect of PEO-treated samples, showing very low fouling of the PEO + Cu with paint as well as a low number of colonies formed. The uncoated sample demonstrated a slow colonization by the microalgal population throughout the course of the four-week test.

21.4 Conclusions and Outlook

PEO coatings with their distinct capability of combining customized bioactive and antibacterial attributes made them suitable for biomedical applications. By incorporating NPs (e.g., Ag, Cu, ZnO), bioinspired compounds, and hybrid additives as bioactive agents, researchers have upscaled the antibacterial capacity of these coatings. These improvements are even more significant when applied in dental implants, orthopedic devices, and surgical tools, which are all about biocompatibility, infection prevention, and material durability. Despite the positive developments occurring in the biomedical sector, advances in marine applications remain underlying. Marine coatings offer solutions to the problems caused by the specific obstacles triggered by biofouling and tough climate conditions. The future track for fouling-resistant PEO coatings should focus on widening the range of their applicability to mitigate biofouling of sophisticated metallic devices such as sensors, which are prone to biological and marine environments. In order to overcome the existing gaps such as long-term antibacterial performance, scalability, and environmental compatibility, collaboration across a range of disciplines will be crucial. The antibacterial PEO coatings are extremely promising as a next-generation solution to sustainable protection if the research on them continues exploring their full potential in both biomedical and marine environments.

References

1. Saji, V. S. (2020). *Advances in Colloid and Interface Science, 283*, 102245.
2. Kazek-Kęsik, A., Kalemba-Rec, I., & Simka, W. (2019). *Materials, 12*, 3002.
3. Chebodaeva, V., Sedelnikova, M., Khimich, M., Bakina, O., Tolmachev, A., Miller, A., Golohvast, K., Zakharenko, A., Egorkin, V., Vyaliy, I., & Sharkeev, Y. (2023). *Coatings, 13*, 1921.
4. Wu, Z., Luo, J., Zhang, J., Huang, H., Xie, Z., & Xie, X. (2021). *Coatings, 11*, 716.
5. Sobolev, A., Valkov, A., Kossenko, A., Wolicki, I., Zinigrad, M., & Borodianskiy, K. (2019). *ACS Applied Materials & Interfaces, 11*, 39534–39544.
6. Fathyunes, L., Hosseini, M., & Khalil-Allafi, J. (2024). *Materials Today Communications, 40*, 109443.
7. Saji, V. S. (2024). *Chemical Engineering Journal, 498*, 155066.
8. Lara Rodriguez, L., & Sundaram, P. A. (2016). *Materials Chemistry and Physics, 181*, 67–77.
9. Monetta, T., Parnian, P., & Acquesta, A. (2020). *Metals, 10*, 907.
10. Oleshko, O., Liubchak, I., Husak, Y., Korniienko, V., Yusupova, A., Oleshko, T., Banasiuk, R., Szkodo, M., Matros-Taranets, I., & Kazek-Kęsik, A. (2020). *Materials, 13*, 4359.
11. Sowa, M., & Simka, W. (2018). *Materials, 11*, 723.
12. Mashtalyar, D. V., Nadaraia, K. V., Imshenetskiy, I. M., Belov, E. A., Gerasimenko, M. S., Sinebryukhov, S. L., & Gnedenkov, S. V. (2023). *Molecules, 28*, 465.
13. Thukkaram, M., Cools, P., Nikiforov, A., Rigole, P., Coenye, T., Van Der Voort, P., Du Laing, G., Vercruysse, C., Declercq, H., & Morent, R. (2020). *Applied Surface Science, 500*, 144235.
14. Yang, Y., Zhang, Y., Hu, R., Huang, Q., Wu, K., Zhang, L., Tang, P., & Lin, C. (2017). *RSC Advances, 7*, 38434–38443.

15. Han, Q., Li, Y., Lu, X., Mei, D., Chen, Q., Zhang, T., & Wang, F. (2021). *Materials Letters, 293*, 129731.
16. Molaei, M., Fattah-alhosseini, A., Nouri, M., Mahmoodi, P., Navard, S. H., & Nourian, A. (2022). *Surfaces and Interfaces, 30*, 101967.
17. Molaei, M., Fattah-alhosseini, A., Nouri, M., Mahmoodi, P., & Nourian, A. (2022). *Ceramics International, 48*, 21005–21024.
18. Qian, K., Wang, C., Yang, M., Zou, X., Zhang, Y., Dong, Q., Wang, X., Shao, Y., Cheng, Z., Ju, J., Chu, C., Chen, Q., Xue, F., & Bai, J. (2024). *Corrosion Science, 226*, 111674.
19. Zhang, D., Peng, F., Tan, J., Zhang, Y., Wang, F., Xie, J., Xu, R., Du, H., Qian, S., Qiao, Y., Li, M., & Liu, X. (2022). *Chemical Engineering Journal, 437*, 135257.
20. Mazinani, A., Nine, M. J., Chiesa, R., Candiani, G., Tarsini, P., Tung, T. T., & Losic, D. (2021). *Materials & Design, 200*, 109443.
21. Fattah-alhosseini, A., Molaei, M., Attarzadeh, N., Babaei, K., & Attarzadeh, F. (2020). *Ceramics International, 46*, 20587–20607.
22. Rokosz, K., Hryniewicz, T., & Dudek, Ł. (2020). *Materials, 13*, 2468.
23. Husak, Y., Olszaniecki, J., Pykacz, J., Ossowska, A., Blacha-Grzechnik, A., Waloszczyk, N., Babilas, D., Korniienko, V., Varava, Y., & Diedkova, K. (2024). *Applied Surface Science, 654*, 159387.
24. Leśniak-Ziółkowska, K., Brodacz, K., Babilas, D., Dulski, M., Blacha-Grzechnik, A., Lu, X., Kazek-Kęsik, A., & Simka, W. (2023). *Applied Surface Science, 615*, 156285.
25. Makurat-Kasprolewicz, B., & Ossowska, A. (2023). *Materials Today Communications, 34*, 105425.
26. Lokeshkumar, E., Kashimbetova, A., Shishir, R., Manojkumar, P., Ravisankar, B., Montufar, E., Čelko, L., & Rameshbabu, N. (2023). *Transactions of the Indian Institute of Metals, 76*, 511–518.
27. Nikoomanzari, E., Fattah-alhosseini, A., Alamoti, M. R. P., & Keshavarz, M. K. (2020). *Ceramics International, 46*, 13114–13124.
28. Nadimi, M., & Dehghanian, C. (2021). *Ceramics International, 47*, 33413–33425.
29. Seyfi, M., Fattah-alhosseini, A., Pajohi-Alamoti, M., & Nikoomanzari, E. (2021). *Journal of Asian Ceramic Societies, 9*, 1114–1127.
30. Parfenova, L. V., Galimshina, Z. R., Gil'fanova, G. U., Alibaeva, E. I., Danilko, K. V., Pashkova, T. M., Kartashova, O. L., Farrakhov, R. G., Mukaeva, V. R., & Parfenov, E. V. (2022). *Surfaces and Interfaces, 28*, 101678.
31. Shishir, R., Nasiruddin, U., Manojkumar, P., Ponnilavan, V., Lokeshkumar, E., Krishna, L. R., & Rameshbabu, N. (2024). *Ceramics International, 50*, 15538–15550.
32. Usmaniya, N., Shishir, R., Ponnilavan, V., & Rameshbabu, N. (2024). *Journal of Alloys and Compounds, 1004*, 175723.
33. Daavari, M., Atapour, M., Mohedano, M., Matykina, E., Arrabal, R., & Nesic, D. (2023). *Journal of Functional Biomaterials, 14*, 475.
34. Caetano, G. L., dos Santos Junior, J., Pereira, B. L., & Benegra, M. (2021). *Surface Engineering, 37*, 1467–1475.
35. Mazinani, A., Nine, M. J., Rastin, H., Chiesa, R., Candiani, G., Tarsini, P., Ghomashchi, R., & Losic, D. (2024). *Advanced Engineering Materials, 26*, 2401305.
36. Alam, K., Jo, Y. Y., Park, C.-K., & Cho, H. (2020). *International Journal of Nanomedicine, 15*, 5813–5824.
37. Huang, H.-L., Tsai, M.-T., Lin, Y.-J., & Chang, Y.-Y. (2019). *Thin Solid Films, 688*, 137268.
38. Mozafarnia, H., Fattah-Alhosseini, A., Chaharmahali, R., Nouri, M., Keshavarz, M. K., & Kaseem, M. (2022). *Coatings, 12*, 1967.
39. Mazinani, A., Rastin, H., Nine, M. J., Lee, J., Tikhomirova, A., Tung, T. T., Ghomashchi, R., Kidd, S., Vreugde, S., & Losic, D. (2021). *Journal of Materials Chemistry B, 9*, 6412–6424.
40. Thukkaram, M., Coryn, R., Asadian, M., Esbah Tabaei, P. S., Rigole, P., Rajendhran, N., Nikiforov, A., Sukumaran, J., Coenye, T., & Van Der Voort, P. (2020). *ACS Applied Materials & Interfaces, 12*, 30155–30169.

41. Chebodaeva, V., Sedelnikova, M., Khimich, M., Bakina, O., Tolmachev, A., Miller, A., Golohvast, K., Zakharenko, A., Egorkin, V., & Vyaliy, I. (2023). *Coatings, 13*, 1921.
42. Rizwan, M., Alias, R., Zaidi, U. Z., Mahmoodian, R., & Hamdi, M. (2018). *Journal of Biomedical Materials Research Part A, 106*, 590–605.
43. Wang, P., Liu, D. X., Li, J. P., Guo, Y. C., & Yang, Z. (2012). *Advanced Materials Research, 479*, 178–181.
44. Guo, P. Y., Wang, N., & Fan, P. (2012). *Applied Mechanics and Materials, 174*, 596–599.
45. Krząkała, A., Kazek-Kęsik, A., & Simka, W. (2013). *RSC Advances, 3*, 19725–19743.
46. Hussein, R., Nie, X., & Northwood, D. (2016). *WIT Transactions on the Built Environment, 166*, 439–454.
47. Karaji, Z. G., Hedayati, R., Pouran, B., Apachitei, I., & Zadpoor, A. A. (2017). *Materials Science and Engineering: C, 76*, 406–416.
48. Fernández-Hernán, J. P., López, A. J., Torres, B., Martínez-Campos, E., Matykina, E., & Rams, J. (2021). *ACS Biomaterials Science & Engineering, 7*, 5861–5877.
49. Banerjee, I., Pangule, R. C., & Kane, R. S. (2011). *Advanced Materials, 23*, 690–718.
50. Kamjunke, N., Spohn, U., Morig, C., Wagner, G., & Neu, T. R. (2020). *Water, 12*, 1597.
51. Cerchier, P., Pezzato, L., Moschin, E., Coelho, L. B., Olivier, M. G. M., Moro, I., & Magrini, M. (2018). *International Biodeterioration & Biodegradation, 133*, 70–78.
52. Cerchier, P., Pezzato, L., Gennari, C., Moschin, E., Moro, I., & Dabalà, M. (2020). *Surface and Coatings Technology, 393*, 125774.

Chapter 22
Advances in Biomedical Applications of Plasma Electrolytic Oxidation Coatings

J. P. Fernández-Hernán

22.1 Introduction

Some metallic materials have a series of properties that make them ideal for use in biomedical applications. For example, they are used in the manufacture of stents and implants for the treatment of cardiovascular diseases, or in orthopaedic and dental osteosynthesis treatments. Metals have high mechanical strength, making them very interesting for orthopaedic and dental applications, acting as scaffolds for osteosynthesis, where they support damaged bone tissue while it regenerates, or as a physical medium to fill critical-sized bone defects. Similarly, they can be used to manufacture containment elements for organs and tissues, such as metal meshes or stents for the treatment of cardiovascular diseases like atherosclerosis, where they are used to open blood vessels occluded by the formation of atheroma plaques, increasing the internal lumen of the vessel and providing rigidity to prevent it from collapsing again.

In any case, whenever we talk about biomedical applications in which the material will come into contact with the tissues and organs of the patient in whom it is implanted, it is essential that the material presents excellent biological properties. In fact, the classification of biomaterials into different generations is based on these properties (Table 22.1). Thus, first-generation biomaterials, which have been used since the second half of the twentieth century, are biomaterials that are not toxic to the body and also exhibit bioinert behaviour. This means that once they come into contact with the living tissues of the patient in whom they are implanted, they do not

J. P. Fernández-Hernán (✉)
Departamento de Matemática Aplicada, Ciencia e Ingeniería de los Materiales y Tecnología Electrónica, Escuela Superior de Ciencias Experimentales y Tecnología, Universidad Rey Juan Carlos, Móstoles (Madrid), Spain
e-mail: juanpablo.fernandez@urjc.es

V. S. Saji, M. Zheludkevich (eds.), *Plasma Electrolytic Oxidation – Fundamentals, Advances and Applications*,
https://doi.org/10.1007/978-3-031-92429-3_22

Table 22.1 Biomaterial generations through history

Biomaterials generation	Year	Properties	Biomaterials
First	~1960	Non-toxic, bioinert	Stainless Steel 316L, Cr-Co-Mo, Ti6Al4V, PMMA
Second	1970–1990	Bioactive, biodegradable	Hydroxyapatite, bioactive ceramics, Mg alloys, biopolymers (PLA, PCL)
Third	2000~	Bioactive, biodegradable, promote specific cellular response	Bioglass®, biosilicates
Fourth	2016~	Interactions with cellular signals. Biological medium adaptability	Carbon-based materials, electrical-conductive polymers

PMMA Poly(methyl methacrylate), *PLA* Polylactic acid, *PCL* Polycaprolactone

provoke any adverse response from the body beyond being surrounded by a fibrous tissue capsule that is naturally generated, isolating it from the surrounding environment, as the body recognises it as an exogenous and foreign element.

The second generation of biomaterials consists of those that are not toxic to the body but are also bioactive. This means that once in contact with biological tissues, they promote cell colonisation and proliferation on their surface. Similarly, they can favour various cellular processes that culminate in the formation of new biological tissue on and around the implant material. This behaviour makes materials of this generation optimal for tissue engineering applications.

In practice, the classification of biomaterials extends beyond these first two generations, reaching a third and fourth generation of biomaterials. These include those that can provoke specific cellular responses or interact with cellular signals. However, these biomaterials are based on biosilicates, conductive polymers, and carbon-based materials, whose study is beyond the scope of this chapter.

In practice, it is common that no single material has properties sufficient to meet all the requirements for an optimal material for a certain biomedical application. Metals are widely used in implantology mainly due to their mechanical properties. However, many of them exhibit high reactivity when in contact with a physiological environment. Similarly, at the surface level, not all metals used in biomedical applications have adequate cytocompatibility or biological properties.

In materials science and engineering, there are numerous strategies that can be followed to modify the properties of a material solely at the surface level. The surface properties and the properties of the interior of the material do not have to be the same. However, it is clear that when we talk about materials for biomedical applications, the surface must have excellent biocompatibility or bioactivity, depending on the application, as it is the surface that comes into contact with the biological tissues of the implant host. Among all the possible options for treating materials to achieve optimal surface properties, the generation of coatings is a technique in constant development and evolution. When it comes to metals, the use of the PEO technique is a versatile method that allows the modification of numerous surface properties:

cytocompatibility, antibacterial behaviour, increased surface hardness, control of the degradation rate of biodegradable implants, etc.

The PEO technique is a cost-effective method that allows for the creation of durable coatings with significant thicknesses and good adhesion to the substrate surface. Additionally, it can be applied to the surface of a variety of metallic materials of interest in the biomedical field, such as steels, titanium alloys, or magnesium alloys, and it allows for coating samples with complex morphologies, such as those found in many implantable devices like stents, screws, or osteosynthesis plates [1].

All these characteristics make PEO coatings particularly interesting as protective systems against corrosion degradation in biodegradable metallic materials, which are of interest for the development of temporary bioresorbable implants. This is the case with magnesium and its alloys, which are biocompatible and biodegradable materials with mechanical properties close to those of cortical bone tissue, making them very interesting for osteosynthesis and implantology applications in traumatology. However, the main disadvantage of these materials, and the reason their use is not more widespread, is that they are highly reactive, with a high tendency to undergo corrosion degradation processes once introduced into an aqueous environment, such as the extracellular medium. This would imply a significant loss of material and a rapid decrease in the mechanical integrity of the implant, which would not align with the necessary timeframes to allow for the regeneration of the bone tissue supported by the implant before the onset of degradation and its progressive disappearance, leading to the failure of the medical treatment. However, the generation of PEO coatings on the surface of these materials could contribute to better control of the degradation rate of bioresorbable materials, allowing it to be adapted to the bone regeneration times so that the implant does not suffer a significant loss of mechanical strength until the newly generated bone tissue can replace the implant in providing mechanical support, at which point the implant could gradually degrade until it completely disappears from the body, thus avoiding the need for post-treatment surgical interventions that are necessary when using non-biodegradable materials to remove the implant from the patient once the osteosynthesis process is complete.

Another advantage of the PEO technique that can be utilised in the biomedical field is the flexibility in the composition of the electrolytes used for coating generation. Thus, depending on the final application of the coated material, the electrolyte can be designed to include a series of chemical elements, ions, or molecules (P, Ca^{2+}, Mg^{2+}, proteins, growth factors, etc.) that, by interacting with the different cell lines expected depending on the implantation site, can enhance their ability to adhere to and expand on the implant surface, promote cell growth and differentiation, and participate in various cellular metabolic processes. Ultimately, the presence of these elements would be reflected in an increase in cell viability on the implant surface, reducing the possibility of rejection and increasing its biocompatibility. Furthermore, during the coating generation process, numerous particles (carbon nanotubes, graphene, CaO nanoparticles, etc.) can be added to the electrolyte to endow the coating with certain properties provided by the inclusion of these nano- and microparticles within and on the surface of the coating. Finally, the morphology of PEO coatings, particularly due to their high surface porosity, makes them potential drug delivery systems, in which the pharmacological substance can be introduced into the coating pores to subsequently allow for controlled release into the surrounding medium, where the substance can carry out its function.

Thus, research and development in the use of the PEO technique can lead to the development of medical devices with improved antimicrobial behaviour, biodegradation, and biocompatibility properties, contributing to advances in biomedical techniques and the improvement of people's quality of life.

22.2 PEO Coatings for Biocompatibility Improvement

Previously, different generations of biomaterials have been discussed, and how they would be classified into one generation or another based on their properties. Biomaterials can be bioinert, biodegradable, bioactive, etc., but among all the characteristics that biomaterials can have, one is essential for their safe use in biomedical applications, and that characteristic is biocompatibility. A material is considered biocompatible when, upon being implanted in an organism and coming into contact with various biological tissues, it does not provoke an adverse response from them. This could be due to the action of elements that are part of the material's composition and can be released into the extracellular medium, or due to surface parameters of the material such as porosity, roughness, surface tension, etc., which significantly influence the ability of cells to adhere, proliferate, and differentiate on its surface [2–4].

At this point, it is necessary to make a brief clarification to distinguish between biocompatibility and cytocompatibility. Biocompatibility has already been defined, and it has been explained that to verify the biocompatibility of a material, it must come into contact with the biological tissues of an organism, which in practice translates to performing in vivo implants. However, it is common to find scientific literature claiming to study the biocompatibility of a material without conducting any in vivo tests, exposing the material under study only to the action of certain cell lines cultures in a controlled environment that simulates a biological medium in vitro (Fig. 22.1). Regulations for conducting animal testing are very strict and require a series of preliminary studies whose data show the potential biocompatibility of the material. This is why cytocompatibility studies (in vitro) are carried out in a preliminary stage before biocompatibility studies (in vivo).

Having clarified this point, the generation of coatings on metallic substrates using the PEO technique offers the possibility of increasing the cytocompatibility/biocompatibility of metallic materials whose properties are of interest for biomedical applications. This increase in cytocompatibility and/or biocompatibility can occur for several reasons. The first is that the presence of the coatings can prevent the degradation of the material's surface by protecting it against corrosion and wear, thus preventing the release of particles from the implant surface during wear, or the liberation of metal ions or molecules as corrosion products, which could provoke a

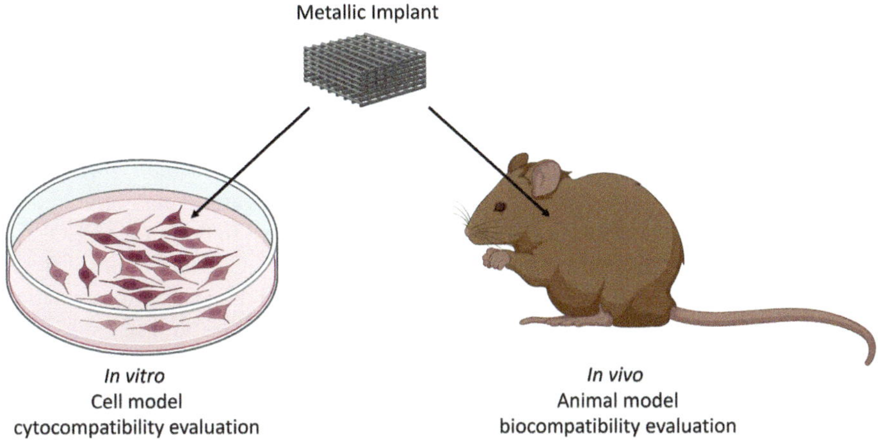

Fig. 22.1 Differences between cytocompatibility and biocompatibility tests of metallic implants

rejection response from the body by causing changes in certain physiological parameters such as the pH of the extracellular medium [5–7].

Moreover, considering that the surface roughness of the implant significantly affects the ability of cells to adhere to its surface, the intrinsic porosity of the coatings generated by PEO could be considered a beneficial factor for increasing the biocompatibility of the implant. This porosity can be controlled according to the parameters used during the coating generation process to adapt it to the ideal surface roughness values depending on the cell lines intended to colonise and proliferate on the implant's surface [8].

Similarly, the porosity of PEO coatings can be utilised to use the coating as a controlled release system for pharmacologically active molecules, such as growth factors, proteins, etc., which could promote and enhance the ability of cells to adhere, proliferate, and differentiate on the implant's surface [9].

Another aspect that can improve biocompatibility through the use of PEO coatings is their composition. By controlling the presence of certain chemical elements and molecules in the coatings, cell growth and development on their surface can be promoted. Similarly, the addition of various nano- and microparticles to the coating can improve the biocompatibility of implants. One of the advantages of coatings generated by PEO is that it is relatively easy to control their composition by controlling the composition of the electrolytes used during the coating generation process. For example, in the case of implants for osteosynthesis, that is, for medical treatments that allow the generation of bone tissue for the healing of bone fractures, the incorporation of calcium and phosphorus compounds provides improved biocompatibility of the implant and can also accelerate osteoconductive behaviour, increasing the rate at which bone cells colonise the implant's surface to speed up the bone regeneration process [10].

As exposed, there are many possibilities that coatings generated by PEO offer to increase the cytocompatibility and, ultimately, the biocompatibility of metallic implants. Numerous examples can be found in the scientific literature on the study and development of PEO coatings to achieve improved biocompatibility.

Focusing solely on the increase in the cytocompatibility of metallic biomaterials through the application of PEO coatings on their surface, there are numerous studies in the scientific literature demonstrating that this type of treatment significantly improves the ability of implants to allow cell proliferation on their surface. For example, A. Santos-Coquillat et al. [11] generated PEO coatings on commercially pure grade I Ti substrates of interest for dental applications. In this study, PEO coatings were generated from a base electrolyte composed of Ca and P, with different compositions depending on the presence of Si, Mg, Zn, and F. The cytocompatibility of the coatings was studied in vitro in terms of osteoblastic and osteoclastic genesis after 7 and 5 days of cell culture, using the MC3T3 preosteoblastic cell line. The results of this study show that all the generated PEO coatings favour the proliferation and differentiation of osteoblasts on their surface. Similarly, the presence of PEO coatings allows for up to three times more collagen secretion in the cell culture compared to Ti substrates, demonstrating a significant improvement in the cytocompatibility of the biomaterial. In another study, A. Santos-Coquillat et al. [12] generated four different types of PEO coatings from an electrolyte with Ca, P, Si, and F compounds on Mg0.8Ca magnesium alloy substrates of interest for the design of cardiovascular stents and bone regeneration applications. In this study, the cytocompatibility of the generated coatings was studied against various cell lines: C2C12-GFP premyoblastic cells, C166-GFP endothelial cells, MC3T3-E1 preosteoblastic cells, and RAW 264.7 cells as a model for osteoclast differentiation. Direct cytocompatibility studies were conducted by culturing the C166-GFP and C2C12-GFP cell lines on samples with different coating conditions, and indirect cytocompatibility studies of the generated coatings were conducted using extracts, where there is no direct contact between the cell cultures and the sample but contact with a medium in which the samples treated with PEO coatings had previously been. The formation of compounds such as hydroxyapatite and fluorapatite was observed during the coating generation, which influences their cytocompatibility. However, it was found that the presence of certain elements in the PEO coating, such as F, can negatively affect certain cell lines, such as the C166-GFP endothelial cell line, making proper control of the electrolyte composition critical. On the other hand, the presence of Si resulted in the coatings with the best cytocompatibility of all those studied. The results show an improvement in the processes of osteoblast and osteoclast generation in co-culture of MC3T3-E1 and RAW 264.7 cells for samples treated with PEO coatings compared to uncoated Mg0.8Ca alloy substrates, making this type of treatment a very promising option for use in bone regeneration applications.

In their research, P. Tian et al. [13] generated hydrothermally treated PEO coatings on biodegradable magnesium alloy AZ31 substrates. The coatings were generated from an electrolyte composed of 0.04 M $Na_2SiO_3{\cdot}9H_2O$, 0.1 M KOH, and 0.2 M $KF{\cdot}2H_2O$. Subsequently, a hydrothermal treatment was performed by immersing the coated samples for 4 h in a solution heated to 90 °C and composed of 0.1 M Ca-EDTA and 0.1 M KH_2PO_4. In this study, the MC3T3-E1 preosteoblastic cell line was used to culture the samples with different coating conditions and evaluate the cytocompatibility of the generated coatings in terms of cell adhesion and proliferation. The hydrothermally treated PEO coatings showed a significant increase in cytocompatibility compared to uncoated samples and PEO-coated samples without hydrothermal treatment (Fig. 22.2). This increase in cytocompatibility was associated with the formation of a hydroxyapatite layer on the PEO coating, which was generated during the hydrothermal treatment.

There are fewer studies that have gone a step further in studying the biocompatibility of PEO coatings, but scientific literature includes research that has successfully demonstrated the improvement in the biocompatibility of metallic implants through the application of PEO coatings, having conducted in vivo studies with animal models. For example, Anastasiya D. Popova et al. [9] generated PEO coatings on grade 4 Ti substrates, composed of electrolytes containing Na_2HPO_4,

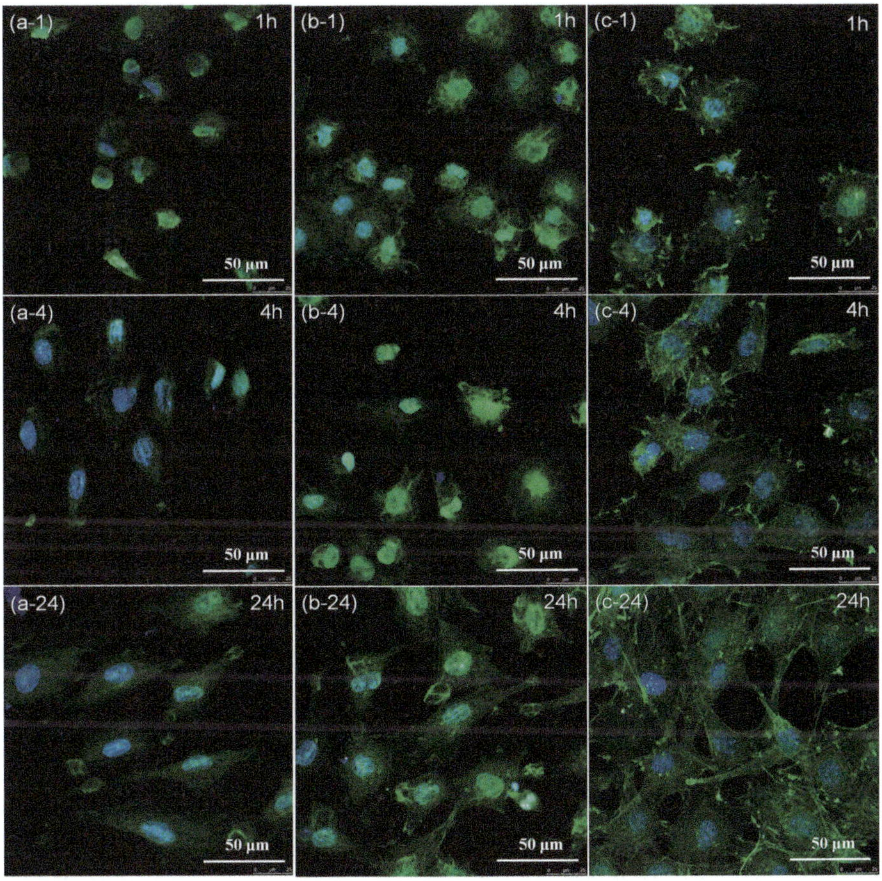

Fig. 22.2 Fluorescence microscopy images of MC3T3-E1 cells cultured on AZ31 alloy (**a**), PEO (**b**) and PEO/HT (**c**) for 1 h (i-1), 4 h (i-4) and 24 h (i-24) (i stands for **a**, **b** and **c**) with actin stained with FITC (green) and the nucleus stained with PDAI (blue). Reproduced with permission from [13] © 2015 Elsevier B.V.

$Ca(OH)_2$, $(NH_2)_2CO$, Na_2SiO_3, $CuSO_4$, and KOH. Additionally, the generated PEO coatings were loaded with bone morphogenetic protein-2 (BMP-2) to study the biocompatibility and the ability of these coatings to promote bone remodelling. The protein was loaded into the PEO coatings by adding a 20 mM sodium phosphate buffer solution containing BMP-2. They conducted in vivo studies using mice that underwent a craniotomy. They found that PEO coatings loaded with BMP-2 significantly improved osteoconductive behaviour and osteogenesis in in vivo models compared to PEO coatings without the protein, promoting a significant improvement in the biocompatibility and bioactivity of these types of implants.

22.3 PEO Coatings for Biodegradability Control

When discussing the use of implants in traumatology, depending on the type of condition the patient has, it may be necessary to use either permanent or temporary implants to correct it. Permanent implants are those that are implanted in the body, generally to serve as a support or to join tissues, and are expected to fulfil this function throughout the patient's life. Examples of such implants include hip or knee implants. On the other hand, temporary implants are those that are implanted in the patient's body for a limited time, so their presence is required only for the duration of the medical treatment, after which they are usually removed from the patient. Examples of this type of implant include those used in osteosynthesis procedures for the treatment of bone fractures.

Thus, temporary implants present a series of challenges that should be addressed to improve patients' quality of life and to reduce healthcare costs. The main challenge with temporary implants is that, in many cases, at the end of the medical treatment, the implant must be removed from the patient's body, which involves subjecting the patient to another surgical intervention, with the associated risks of any medical procedure, as well as an increase in the cost of medical treatment. Therefore, research is currently being conducted into the development of biodegradable biomaterials for the manufacture of temporary implants. The main advantage of using biodegradable biomaterials is that once implanted in the body, these materials begin to degrade under the action of the extracellular medium and various molecules and ions present in the body (H_2O, O_2, Cl^-, etc.). This degradation process ends with the complete disappearance of the material within the patient's body. If these biodegradable biomaterials also have good bioactivity, they will be able to promote certain cellular processes to facilitate tissue regeneration, actively participating in the healing process and not just serving as support during the treatment period.

There are several challenges in designing biodegradable biomaterials: Firstly, biocompatible materials must be used, and their degradation products, which will be released into the patient's body, must be easily assimilated or eliminated, so they do not cause any pathology due to accumulation. Secondly, they must not be cytotoxic to avoid adverse reactions from the tissues near the implant. Additionally, and

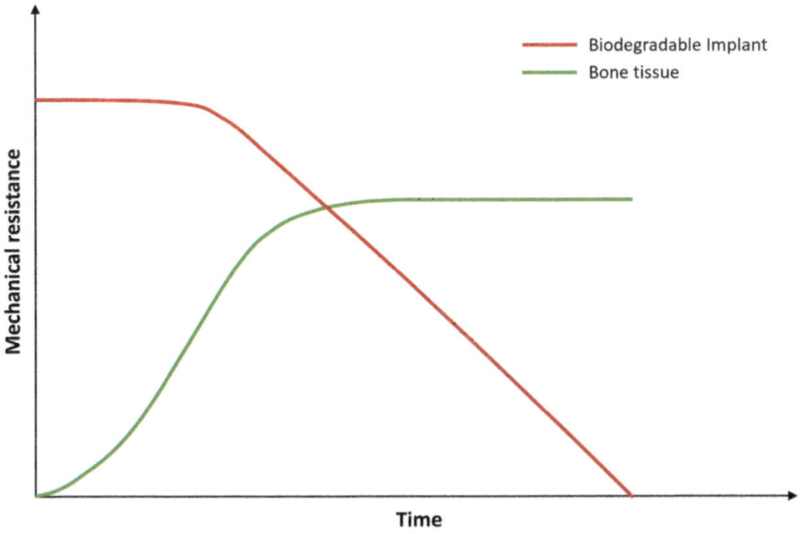

Fig. 22.3 Diagram showing the relation between biodegradable implant loss of mechanical resistance and bone tissue increment in mechanical resistance during osteosynthesis treatment

especially when these materials are designed for the manufacture of temporary implants for osteosynthesis, they must have a degradation rate that matches the bone regeneration rate during the bone fracture healing process, so that the implant provides mechanical support to the bone during its regeneration, then gradually loses mass, and therefore mechanical strength, as the bone regenerates and can support itself (Fig. 22.3). Otherwise, if the degradation rate of the biodegradable implant is not synchronised with the regeneration rate, for example, if it is higher than the bone tissue regeneration rate, the implant's integrity and mechanical strength could critically decrease before the bone has regenerated and can provide support, causing treatment failure and necessitating another surgical intervention to correct the situation.

Currently, research lines in the development of biodegradable metallic biomaterials are based on the use of Mg, Zn, Fe, and their alloys. These metallic materials are of interest for the development of biodegradable temporary implants for several reasons: Firstly, all of them are elements naturally present in the human body, as their ions actively participate in numerous biological processes at the cellular level within the organism. Therefore, their presence is not foreign to the body, and they exhibit high biocompatibility. All of these materials degrade through oxidation processes when exposed to an environment containing water and oxygen, such as the interior of the human body, making them biodegradable. Lastly, being metals, they possess sufficient mechanical strength to be used in a multitude of biomedical applications. At first glance, these materials seem ideal for the manufacture of biodegradable temporary implants. However, there are several aspects that need to be considered, which is why their use is not yet widespread.

Firstly, although these materials are biocompatible in principle, the degradation process generates a series of products that can negatively affect the ability of cells to adhere to and grow on their surface. For example, focusing on magnesium, which is one of the most studied and promising biodegradable materials, Eq. (22.1), which describes its degradation process in an aqueous medium, shows the generation of degradation products such as gaseous hydrogen, which is released from the surface of the metallic substrate, as well as magnesium hydroxide, which forms a low-strength, low-adherence coating layer on the surface of the metallic substrate. Therefore, the presence of these products will negatively affect the ability of cells to adhere to the substrate and proliferate on it.

$$Mg + 2H_2O \rightarrow Mg^{2+} + 2OH^- + \uparrow H_{2(g)} \tag{22.1}$$

At this point, PEO technology is very interesting as a tool to help control the degradation rate of materials that could be used to manufacture biodegradable implants. PEO coatings could contribute to this objective in several ways: Firstly, the generation of coatings has long been used to increase corrosion resistance and prevent the degradation of metallic materials. This effect is mainly achieved due to the barrier effect provided by the coating generated on the surface of the substrate, so that the metallic substrate is isolated from the aggressive environment or medium that, without the presence of the coating, would attack its surface. However, for the specific application of developing biodegradable implants, this corrosion protection is sought only temporarily, as the ultimate goal is that, once the implant's function is completed, it degrades slowly so that the degradation products can be easily assimilated and eliminated by the body, ending in the complete disappearance of the implant. This is where PEO coatings are interesting, as they are metallic oxide coatings that exhibit some porosity and, therefore, are susceptible to gradually degrading, which means that their protective capacity decreases over time. Consequently, in a controlled manner, the substrate is allowed to start corroding at a much lower degradation rate than it would without the presence of the coating. This behaviour of the coating can be modified depending on the composition of the electrolyte used, as well as the presence or absence of particles of different natures suspended in the electrolyte during the coating generation process [14].

In cases where the presence of the coating alone is not sufficient to reduce the degradation rate to the level required by the medical treatment, it is possible to make certain modifications to the PEO coatings to provide them with greater protective capacity and thus further reduce the degradation rate of the biodegradable metal on which they have been generated. The most common method to reduce the porosity of PEO coatings is to deposit a new material, generally some biopolymer, on them to fill and occlude the pores, thereby increasing the protective capacity of the coating. Examples of polymers used for this purpose include poly(tetrafluoroethylene) (PTFE), poly(L-lactic acid) (PLLA), or polycaprolactone) (PCL) [15, 16].

On the other hand, instead of considering the intrinsic porosity of PEO coatings as a disadvantage in providing corrosion protection, this characteristic can be exploited to enhance their protective capacity. The pores of the coating can be sealed using biocompatible substances that offer corrosion-inhibiting behaviour. As the coating degrades, these substances can be released and act to slow down the degradation of the metallic substrate on which the coating has been generated. In this way, degradation rates of biomaterials can be achieved that are tailored to each treatment and the time required for the complete healing of bone tissue.

In the scientific literature, various examples can be found of the use of PEO coatings to control the degradation rate of biomaterials, as well as examples of the use of biopolymers and corrosion-inhibiting substances in combination with coatings generated by PEO, to increase the protective capacity of the coatings and obtain better control of the degradation rate of the metallic substrates on which they are generated. For instance, M. Daavari et al. [17] generated PEO coatings reinforced with multi-walled carbon nanotubes (MWCNTs) on AZ31 magnesium alloy substrates. The reinforced PEO/MWCNT coatings were obtained by incorporating the nanoparticles into the electrolyte used for generating the PEO coatings at a concentration of 5 g/L, ensuring their dispersion through mechanical stirring. The evaluation of the coatings showed that the presence of carbon nanoparticles generated larger pores in the outer layers of the PEO coating but also allowed for more uniform and thicker layers in the denser part of the PEO coating, located in contact with the metallic substrate. This characteristic of the reinforced PEO/MWCNT coatings allows them to offer better corrosion protection compared to unreinforced PEO coatings. Additionally, they observed that the presence of MWCNTs within the pores of the reinforced coating improved pitting corrosion resistance, relating this to the nanoparticles' ability to enhance the adhesion and cohesion of the PEO coating.

A.S. Gnedenkov et al. [18] used the electrophoretic deposition (EPD) technique to modify PEO coatings generated on MA8 magnesium alloy substrates composed of magnesium oxides and silicates and hydroxyapatite. Through EPD, they managed to deposit a coating of polytetrafluoroethylene (PTFE) particles that sealed the surface porosities of the initial PEO coating, endowing the combined PEO–PTFE coating with significantly greater protective capacity than the PEO coating alone, achieving significantly lower and more consistent corrosion rates of the substrate over time.

Z. Wei et al. [16] generated composite PEO/poly(L-lactide acid) (PLLA) coatings on AZ31 magnesium alloy substrates. In this research, the deposition of PLLA on the PEO coating was carried out by immersing the PEO-coated samples in a dichloromethane solution with a 6% weight content of PLLA. The evaluation of the results showed a significant reduction in the corrosion rate of the magnesium alloy substrate in the composite PEO/PLLA coating compared to the uncoated or PEO-only coated sample, which also implies an improvement in the corrosion rate of the metallic substrate (Fig. 22.4).

In their research, A. Bordbar-Khiabani et al. [19] considered the composition and pH conditions of the surrounding medium after implant placement due to the inflammatory reaction that occurs in the body following such interventions. These composition and pH conditions can negatively affect the degradation rate of the biodegradable implant. To prevent the adverse effect that the inflammation of

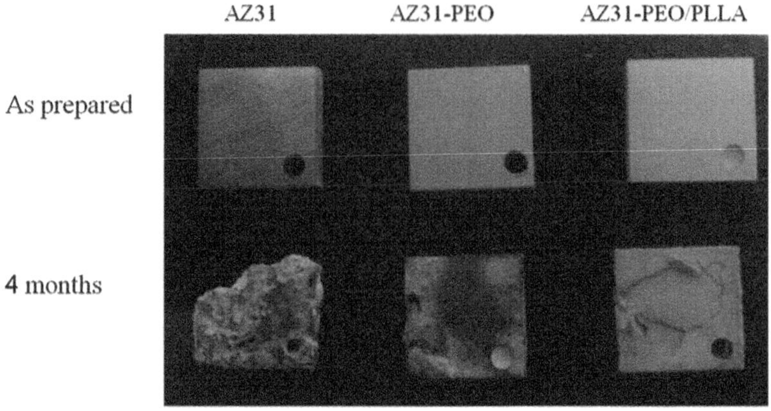

Fig. 22.4 Surface appearances of AZ31, AZ31-PEO, and AZ31-PEO/PLLA before and after immersion in SBF for 4 months. Reproduced with permission from [16] © 2014 Wiley Periodicals, Inc.

tissues surrounding the implant can have on its degradation rate, they developed composite PEO/betamethasone sodium phosphate (BSP) coatings deposited on AZ91 magnesium alloy, with this compound being a drug with anti-inflammatory properties. In this case, the BSP coating was deposited on the AZ91 substrate previously coated by PEO using the dip-coating technique, immersing the PEO-coated metallic substrates for 1 h in a commercial solution with a BSP concentration of 3 mg/mL. They evaluated the ability of these composite coatings to reduce the degradation rate under simulated postoperative inflammation conditions, and their results show that, under acute inflammatory reaction conditions, the presence of BSP significantly improves corrosion resistance and, therefore, the degradation rate of the metallic substrate.

22.3.1 Antimicrobial PEO Coatings

During a surgical procedure for the placement of an implant, one of the most important aspects to control is to follow strict hygiene and asepsis measures for all personnel and elements involved in the surgical process. Inadequate or insufficient control of these measures can lead to bacterial colonisation of the area where the intervention is carried out, as well as the tools used and the surface of the element to be implanted in the patient. This increases the likelihood of nosocomial infections, also known as hospital-acquired infections, which can be acquired during the patient's hospital stay and were not present before admission [20].

The development of infections after implant placement can have a series of potentially dangerous consequences for the patient: increased recovery times, complications such as bone tissue infections or infections around the implant that compromise its stability, and even the possibility of sepsis, leading to potentially life-threatening situations for the patient. All of this can result in the failure of the surgical treatment and the need for a new intervention to remove the implant and treat the infection, which also increases medical costs.

There are various approaches that can help reduce the likelihood of nosocomial infections following implantology and trauma treatments. From the implant's perspective, different surface treatments can be applied to provide it with antibacterial properties, both by preventing bacterial adhesion to its surface and by promoting the bacterial death of those bacteria that manage to reach and adhere to the implant surface, such as antibiotic impregnation [21].

In this regard, PEO technology can be used to generate coatings on metallic implants that can provide them with the ability to prevent and combat bacterial colonisation and infection of their surface. In this case, PEO-generated coatings can act against bacteria due to the presence of certain chemical elements in their composition, as well as molecules with pharmacological behaviour, drugs, or nano- and microparticles with known bactericidal effects. Additionally, due to their nature, PEO coatings exhibit high porosity, especially in the outer layers, which allows the coating itself to act as a drug delivery system if the pores are used as reservoirs for antimicrobial molecules or particles that can be released in a controlled manner from the coating into the physiological environment, eliminating the bacterial threat even before adhesion to the implant surface occurs [22, 23].

In the scientific literature, studies can be found on various strategies for generating PEO coatings applied to metallic substrates of different natures, although mainly Ti and its alloys, to provide the material with antibacterial and antimicrobial properties. Ti and its alloys are widely used in dental and orthopaedic implantology applications due to their excellent properties, such as bio-inertness, mechanical strength, and corrosion resistance [22]. However, they are not materials that inherently possess antibacterial properties, so it is highly recommended to perform surface treatments that add this characteristic to avoid the possibility of infections and improve the likelihood of successful surgical treatment.

From a biological perspective, microbial adhesion and accumulation are the processes that have the greatest influence on the development of nosocomial infections [22]. Therefore, the first step to control the adhesion of microorganisms to the surface of an implant is precisely to design and modify the surface morphologically to control how microorganisms interact with it [24]. The next step would be to add particles and/or substances with proven antimicrobial efficacy so that, acting alone or synergistically with an appropriate surface design, they allow a significant reduction in the number of microorganisms capable of reaching and adhering to the surface. In case this happens, they promote the death of the adhered bacteria before they can multiply and develop a biofilm on the implant surface that complicates their elimination. The PEO technique allows for these two possibilities, as the coating generation process itself causes a morphological change in the surface of the coated material. Additionally, by modifying the composition of the electrolyte used, the composition of the coating can be controlled, allowing the presence of elements such as Cu, Zn, or Ag, which have demonstrated bactericidal effects [25, 26].

However, nano- or microparticles of Ag, Zn, Mg, Cu, with various morphologies, can also be added to the electrolyte. During the coating process, the particles are kept in suspension in the electrolyte so that they are incorporated into the coating.

The way in which different elements, molecules, or particles attack bacteria and prevent their proliferation can vary: from causing bacterial and microbial death by direct contact with microorganisms, causing the rupture of their cytoplasmic membrane [27], modifying the permeability of the bacterial membrane to Na^{+}, K^{+}, Mg^{2+}, or Ca^{2+} cations, interfering with bacterial metabolic processes and causing microbial death [28, 29], as well as the assimilation of these substances by microorganisms, interfering with their metabolic processes, causing oxidative stress, and preventing their viability [30–32]. It should be noted that the antibacterial capacity of these elements, molecules, and particles depends on their final concentration in the coating, linked to their minimum inhibitory concentration (MIC), which in microbiology refers to the lowest concentration of a substance with antimicrobial effect that inhibits the growth of the microorganism.

On the other hand, instead of adding the elements, molecules, or particles that will provide the bactericidal behaviour to the coating of the electrolyte, various studies show the possibility of using PEO coatings as drug delivery systems, taking advantage of their intrinsic porosity (Fig. 22.5). Through post-PEO coating generation treatments, substances with pharmacological and antibiotic effects can be loaded into the pores of the coating, allowing their release from the coating once the coated material is implanted to prevent bacterial adhesion and to decrease bacterial viability in the medium surrounding the implant [33–35].

In the scientific literature, a multitude of elements, substances, and particles are used as antimicrobial agents [36, 37]. In the specific case of PEO coatings, the most common approach is the use of molecules from which metal cations with well-established bactericidal power can be obtained, such as copper (Cu^{2+}), zinc (Zn^{2+}), or silver (Ag^{+}) [25, 26]. Similarly, there are different ways to incorporate antibacterial agents into PEO coatings, such as impregnating the coatings with antibiotic substances through immersion or vacuum application as a post-coating treatment, using techniques like physical vapour deposition (PVD) to deposit thin layers of bactericidal elements like silver on PEO coatings [38]. Another technique is the

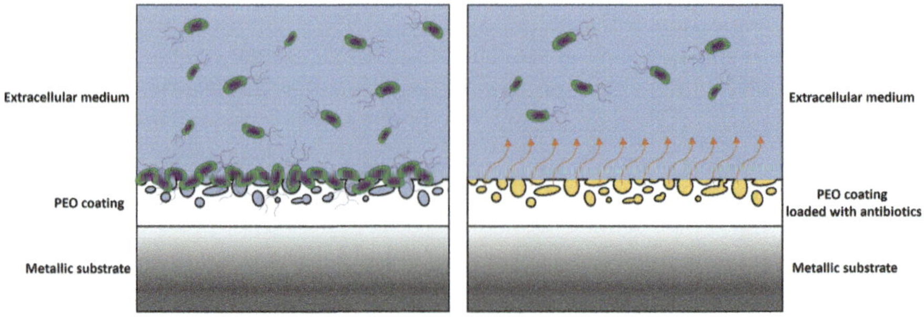

Fig. 22.5 Schematic effect of antibiotics been delivered from porous PEO coating

addition of molecules or particles of Cu, Zn, Ag, Zr to the electrolyte, allowing their incorporation into the coating during its generation and the subsequent release of ions after immersing the coated sample in the physiological medium [27, 34]. These antimicrobial agents act in various ways to promote bacterial death: breaking the bacterial cytoplasmic membrane, interfering with the normal permeability of the membrane, causing metabolic disorders, or altering what is known as cellular respiration [27]. It should be noted that the antibacterial capacity of the aforementioned elements depends on their concentration, but similarly, certain concentration levels can affect the different cell types intended to colonise and grow on the material's surface. Therefore, it is necessary to find a balance to attack bacteria while minimally affecting the implant's biocompatibility.

In their research, R.C. Costa et al. [22] highlight some of the main metallic elements used as antibacterial agents in PEO coatings, especially elements like Ag, Cu, and Zn due to their broad-spectrum antibacterial behaviour. These elements can act in combination with PEO coatings to provide antibacterial properties. However, it should be noted that depending on the concentration of these elements in the coating, some cytotoxicity and decreased cytocompatibility could occur. Therefore, it is necessary to study the minimum inhibitory concentration of each compound that prevents bacterial proliferation while minimally affecting cell proliferation.

Janaina S. Santos et al. [39] evaluated the antibacterial behaviour of PEO coatings treated with Cu as an antibacterial agent. They developed two different types of coatings: PEO coatings decorated with Cu and PEO coatings anodically doped with Cu ions. They exposed the coatings to *E. coli* and *S. aureus* strains, observing that the presence of Cu-treated PEO coatings showed marked antibacterial behaviour after 24 h of testing. Cu-decorated coatings were more effective against *S. aureus*, while Cu-doped PEO coatings were more effective against *E. coli*.

Maryam Molaei et al. [40] incorporated three different concentrations (1, 3, and 5 g/L) of ZrO_2 nanoparticles into the electrolyte used to generate PEO coatings on titanium substrates to evaluate the antibacterial behaviour of these coatings. The nanoparticles were incorporated into the surface and pores of the PEO coating. They exposed the coatings to *S. aureus* bacterial strains, finding a significant increase in antibacterial activity during the first 24 h of testing.

On the other hand, P. Cerchier et al. [41] investigated the bactericidal effect against *E. coli* and *S. aureus* of PEO coatings doped with Ag microparticles. During the PEO coating generation process, Ag microparticles were added directly to the electrolyte in two different concentrations. Additionally, they performed a sealing treatment of the PEO coatings to enhance their corrosion protection effect by immersing the previously coated samples in boiling water for 15 min, having added more Ag microparticles during the sealing process. To conduct this study, the bacterial medium was in contact with the surface of the samples with different coating conditions for 3 h. Subsequently, the medium was collected and spread on agar plates to incubate for 24–48 h and evaluate bacterial colony growth, thus estimating the bactericidal capacity by counting bacterial colonies grown on the agar plate. The results showed that exposing the PEO coatings with different Ag concentrations to

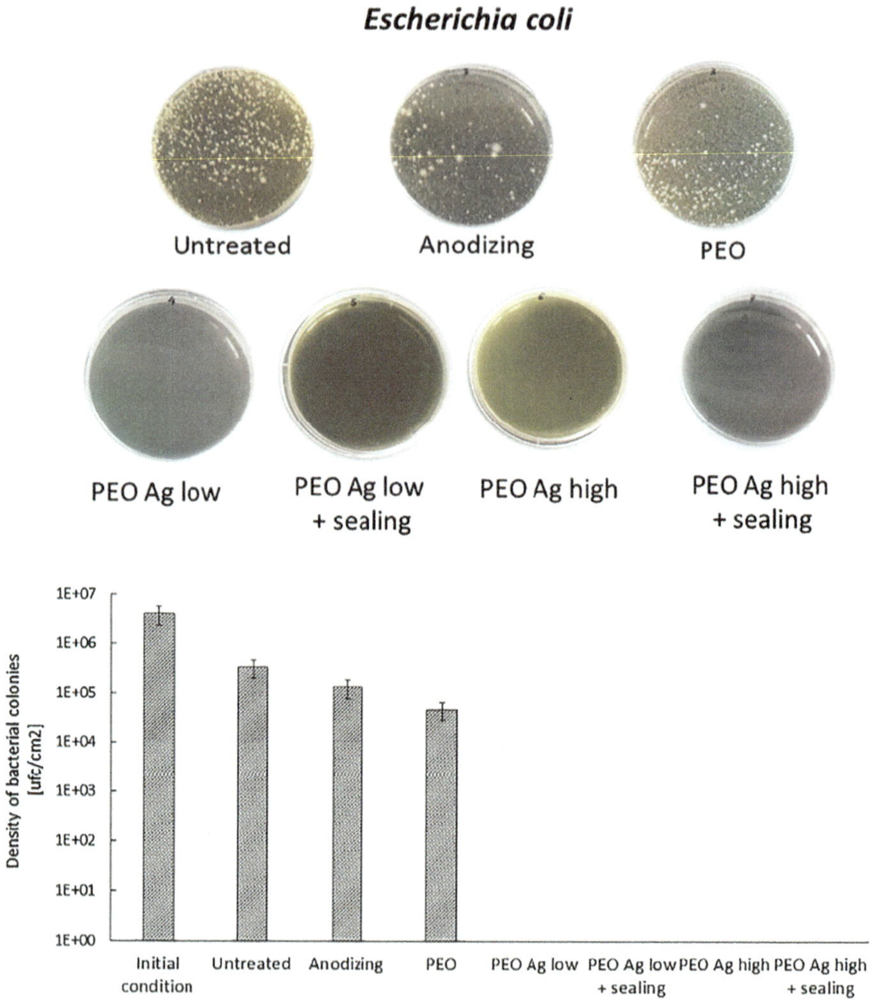

Fig. 22.6 Counting of *E. coli* and images of the slides with the bacteria colonies after 3 h of exposition to the different surfaces. Reproduced with permission from [41] © 2017 Elsevier B.V.

the bacteria eliminated bacterial presence in the case of *E. coli* (Fig. 22.6) and significantly reduced it in the case of *S. aureus* (Fig. 22.7).

In their research, E. Nikoomanzari et al. [42] used ZrO_2 nanoparticles in different concentrations in the phosphate-based electrolyte (0, 1, 3, and 5 g/L) added to PEO coatings developed on Ti6Al4V titanium alloy substrates. They studied the bactericidal effect of these coatings against four types of bacteria: *S. aureus* (ATCC 25923) and *S. epidermidis* (ATCC 12228) as gram-positive pathogenic microorganisms, and *E. coli* (ATCC 25922) and *P. aeruginosa* (ATCC 15442) as gram-negative pathogenic microorganisms. They found that the presence of ZrO_2 nanoparticles allowed

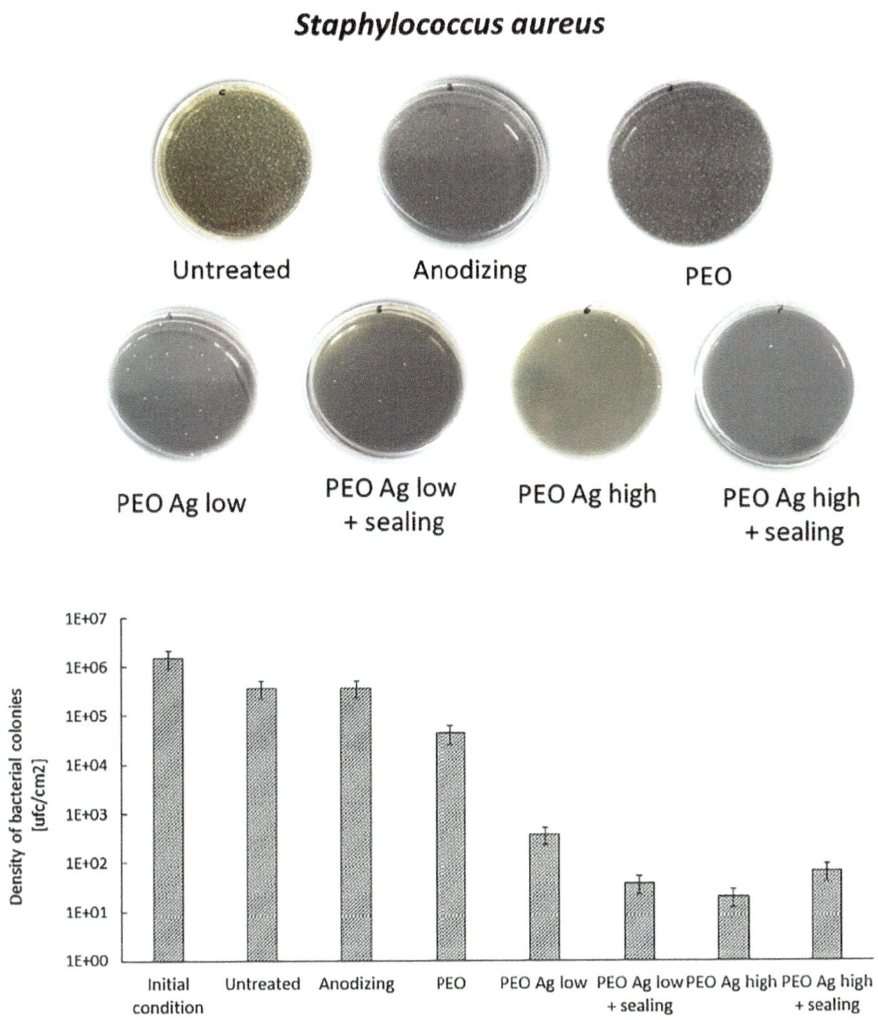

Fig. 22.7 Counting of *S. aureus* and images of the slides with the bacteria colonies after 3 h of exposition to the different surfaces. Reproduced with permission from [41] © 2017 Elsevier B.V.

the elimination of gram-positive bacteria with increasing nanoparticle concentrations up to 3 g/L in the electrolyte. However, the antibacterial effectiveness against gram-negative bacteria decreased with increasing nanoparticle concentration.

D.V. Mashtalyar et al. [34] generated PEO coatings with a composition based on Ca/P compounds. Additionally, they endowed the coatings with antibacterial behaviour by impregnating the PEO coating with a Ringer's solution containing vancomycin, an antibiotic substance for treating infections caused by gram-positive bacteria, using vacuum application to effectively incorporate the substance into the coating, allowing it to penetrate the pores. They tested the antibacterial power of the

PEO coating impregnated with vancomycin against *S. aureus* PS84 and MRSA (methicillin-resistant *Staphylococcus aureus*) bacterial strains, achieving good inhibitory activity against both bacterial strains.

B. Zhang et al. [43] generated TiO_2 coatings on Ti6Al4V alloy doped with yttrium by adding yttrium acetate to the electrolyte and demonstrated significant antibacterial effects against both *E. coli* and *S. aureus*.

Not only antibiotics or other specific elements can be used to provide antibacterial effects to PEO coatings. The use of carbon nanoparticles like graphene oxide (GO) can be found in the literature for this purpose. Thus, A. Mazinani et al. [44] generated TiO_2 PEO coatings on pure Ti substrates and decorated the coatings with graphene oxide via electrophoretic deposition. The antibacterial behaviour was evaluated against *E. coli* and *S. aureus*, obtaining a reduction of 80% and 100% against *E. coli* and *S. aureus*, respectively, compared with non-treated substrates.

Y. Husak et al. [23] generated antibacterial PEO coatings on magnesium substrates by adding different concentrations of CuO nanoparticles dispersed into the electrolyte by magnetic stirring. The antibacterial behaviour of the coatings was evaluated against *E. coli*, *S. aureus* bacterial strains, and the pathogenic yeast *C. albicans*. The antimicrobial evaluation showed dependence on the concentration of CuO particles, exhibiting different behaviour depending on their concentration.

22.4 Conclusions and Outlook

Throughout this chapter, the versatility and interest of PEO technology for generating coatings with biomedical applications in the fields of orthopaedic, cardiology and dental implantology have been demonstrated. The various possibilities that this technique offers for generating coatings that can improve the cytocompatibility and biocompatibility of implants made from metallic biomaterials have been explored. Advanced coatings that reduce the likelihood of postoperative infections following implant surgery and coatings that control the degradation rate of an implant made from biodegradable metal to increase the success rate of medical treatments have been discussed. These advancements improve patients' quality of life and reduce medical costs associated with surgical interventions, which can be enhanced through the use of technologies such as those presented in this chapter. Thanks to the efforts of many researchers today and due to the promising results obtained so far, it is clear that PEO technology will play a significant role in the biomedical field. It will enable the generation of advanced multifunctional coatings that will take drug delivery systems, biodegradation protection systems, and biocompatibility in implantology to the next level.

References

1. Babaei, K., Fattah-alhosseini, A., & Chaharmahali, R. (2020). *Surfaces and Interfaces, 21*, 100719.
2. Rosales-Leal, J. I., Rodríguez-Valverde, M. A., Mazzaglia, G., Ramón-Torregrosa, P. J., Díaz-Rodríguez, L., García-Martínez, O., et al. (2010). *Colloids and Surfaces A: Physicochemical and Engineering Aspects, 365*, 222–229.
3. Ranella, A., Barberoglou, M., Bakogianni, S., Fotakis, C., & Stratakis, E. (2010). *Acta Biomaterialia, 6*, 2711–2720.
4. Lampin, M., Warocquier-Clérout, R., Legris, C., Degrange, M., & Sigot-Luizard, M. F. (1997). *Journal of Biomedical Materials Research, 36*, 99–108.
5. Zhang, L., Haddouti, E. M., Welle, K., Burger, C., Wirtz, D. C., Schildberg, F. A., et al. (2020). *Frontiers in Cell and Developmental Biology, 8*, 1–17.
6. Lohmann, C. H., Schwartz, Z., Köster, G., Jahn, U., Buchhorn, G. H., MacDougall, M. J., et al. (2000). *Biomaterials, 21*, 551–561.
7. Rose, S. F., Weaver, C. L., Fenwick, S. A., Horner, A., & Pawar, V. D. (2012). *Journal of Biomedical Materials Research – Part B: Applied Biomaterials, 100 B*, 1359–1368.
8. Shin, K. R., Kim, Y. S., Jeong, J. H., Ko, Y. G., & Shin, D. H. (2016). *Surface Engineering, 32*, 418–422.
9. Popova, A. D., Sheveyko, A. N., Kuptsov, K. A., Advakhova, D. Y., Karyagina, A. S., Gromov, A. V., et al. (2023). *ACS Applied Materials & Interfaces, 15*, 37274–37289.
10. Esmaily, M., Svensson, J. E., Fajardo, S., Birbilis, N., Frankel, G. S., Virtanen, S., et al. (2017). *Progress in Materials Science, 89*, 92–193.
11. Santos-Coquillat, A., Mohedano, M., Martinez-Campos, E., Arrabal, R., Pardo, A., & Matykina, E. (2019). *Materials Science and Engineering: C, 97*, 738–752.
12. Santos-Coquillat, A., Esteban-Lucia, M., Martinez-Campos, E., Mohedano, M., Arrabal, R., Blawert, C., et al. (2019). *Materials Science and Engineering: C, 105*, 110026.
13. Tian, P., Liu, X., & Ding, C. (2015). *Colloids and Surfaces. B, Biointerfaces, 128*, 44–54.
14. Heimann, R. B. (2021). *Surface and Coatings Technology, 405*, 126521.
15. Gnedenkov, S. V., Sinebryukhov, S. L., Mashtalyar, D. V., & Imshinetskiy, I. M. (2015). *Surface and Coatings Technology, 283*, 347–352.
16. Wei, Z., Tian, P., Liu, X., & Zhou, B. (2015). *Journal of Biomedical Materials Research – Part B: Applied Biomaterials, 103*, 342–357.
17. Daavari, M., Atapour, M., Mohedano, M., Arrabal, R., Matykina, E., & Taherizadeh, A. (2021). *Surfaces and Interfaces, 22*, 100850.
18. Gnedenkov, A. S., Lamaka, S. V., Sinebryukhov, S. L., Mashtalyar, D. V., Egorkin, V. S., Imshinetskiy, I. M., et al. (2021). *Corrosion Science, 182*, 109254.
19. Bordbar-Khiabani, A., Yarmand, B., Sharifi-Asl, S., & Mozafari, M. (2020). *Materials Chemistry and Physics, 239*, 122003.
20. Raoofi, S., Kan, F. P., Rafiei, S., Hosseinipalangi, Z., Mejareh, Z. N., Khani, S., et al. (2023). *PLoS One, 18*, 1–17.
21. Mishra, A., Aggarwal, A., & Khan, F. (2024). *Antibiotics, 13*, 1–16.
22. Costa, R. C., Nagay, B. E., Dini, C., Borges, M. H. R., Miranda, L. F. B., Cordeiro, J. M., et al. (2023). *Advances in Colloid and Interface Science, 311*, 102805.
23. Husak, Y., Ma, J., Wala-Kapica, M., Leśniak, K., Babilas, D., Blacha-Grzechnik, A., et al. (2024). *Materials Chemistry and Physics, 323*, 129627.
24. Souza, J. G. S., Bertolini, M., Costa, R. C., Cordeiro, J. M., Nagay, B. E., De Almeida, A. B., et al. (2020). *ACS Applied Materials & Interfaces, 12*, 10118–10129.
25. Liu, Y., Chen, C., Liang, T., Wang, Y., Zhao, R., Li, G., et al. (2024). *Colloids and Surfaces. B, Biointerfaces, 233*, 113634.
26. Vasiliev, G., Kubo, A. L., Vija, H., Kahru, A., Bondar, D., Karpichev, Y., et al. (2023). *Scientific Reports, 13*, 1–15.

27. Fattah-alhosseini, A., Molaei, M., Nouri, M., & Babaei, K. (2022). *Journal of Magnesium and Alloys, 10*, 81–96.
28. Díaz, P., D'Suze, G., Salazar, V., Sevcik, C., Shannon, J., Sherman, N., et al. (2009). *Toxicon, 54*, 802–817.
29. Xin, J., Pu, Q., Wang, R., Gu, Y., He, L., Du, X., et al. (2024). *Frontiers in Veterinary Science, 11*.
30. Albesa, I., Cecilia Becerra, M., Battán, P., & Páez, P. (2004). *Biochemical and Biophysical Research Communications, 317*, 605–609.
31. Mourenza, A., Gil, J., Mateos, L., & Letek, M. (2020). *Antioxidants, 9*, 361.
32. Gurunathan, S., Han, J. W., Dayem, A. A., Eppakayala, V., & Kim, J. (2012). *International Journal of Nanomedicine, 7*, 5901–5914.
33. Leśniak, K., Płonka, J., Śmiga-Matuszowicz, M., Brzychczy-Włoch, M., & Kazek-Kęsik, A. (2020). *Journal of Biomedical Materials Research – Part B: Applied Biomaterials, 108*, 1167–1753.
34. Mashtalyar, D. V., Nadaraia, K. V., Plekhova, N. G., Imshinetskiy, I. M., Piatkova, M. A., Pleshkova, A. I., et al. (2022). *Materials Letters, 317*, 132099.
35. Moon, S., Kang, M., Im, S., Kim, K., & Kwon, J. S. (2012). *Journal of the Korean Physical Society, 60*, 954–958.
36. Huang, S., Song, Y., Zhang, J. R., Chen, X., & Zhu, J. J. (2023). *Small, 19*, 1–28.
37. Badoni, A., & Prakash, J. (2024). *Micro and Nano Engineering, 22*, 100239.
38. Aktug, S. L., Durdu, S., Aktas, S., Yalcin, E., & Usta, M. (2019). *Surface and Coatings Technology, 375*, 46–53.
39. Santos, J. S., Márquez, V., Buijnsters, J. G., Praserthdam, S., & Praserthdam, P. (2023). *Applied Surface Science, 607*, 155072.
40. Molaei, M., Fattah-alhosseini, A., Nouri, M., Mahmoodi, P., Navard, S. H., & Nourian, A. (2022). *Surfaces and Interfaces, 30*, 101967.
41. Cerchier, P., Pezzato, L., Brunelli, K., Dolcet, P., Bartolozzi, A., Bertani, R., et al. (2017). *Materials Science and Engineering: C, 75*, 554–564.
42. Nikoomanzari, E., Fattah-alhosseini, A., Pajohi Alamoti, M. R., & Keshavarz, M. K. (2020). *Ceramics International, 46*, 13114–13124.
43. Zhang, B., Li, B., Gao, S., Li, Y., Cao, R., Cheng, J., et al. (2020). *Materials and Design, 192*, 1–12.
44. Mazinani, A., Nine, M. J., Chiesa, R., Candiani, G., Tarsini, P., Tung, T. T., et al. (2021). *Materials and Design, 200*, 109443.

Chapter 23
Posttreatment Sealing Strategies for Plasma Electrolytic Oxidation Coatings

You Zhang, Chenkai Xu, Zheng Zhang, Zhe Zhang, and Dahai Gao

23.1 Introduction

This chapter presents the latest techniques and progress in posttreatment sealing strategies for plasma electrolytic oxidation (PEO) coatings. It covers three main categories: (1) physical posttreatment methods, such as polishing and mechanical processing, which enhance surface properties through physical means [1]; (2) chemical posttreatment sealing strategies, including organic and inorganic coating sealing, conversion coating sealing, electrodeposition, and electrophoretic deposition [2], etc., which improve performance by altering the chemical composition and microstructure of the coating; and (3) methods that involve both structural and performance modifications, such as heat treatment, hydrothermal sealing, electroless plating, etc., which enhance coating quality and performance through a combination of physical and chemical mechanisms [3].

The primary function of posttreatment processes for PEO coatings is to further improve their performance. These processes typically aim to enhance adhesion, improve aesthetics, and optimize specific physical and chemical properties. The selection of these posttreatment steps is carefully tailored to the unique characteristics of the PEO coating.

The PEO process is well known for generating ceramic coatings on metal surfaces, which exhibit excellent wear resistance, corrosion resistance, and high-temperature performance. The formation of PEO coatings is based on the theory of dielectric breakdown, wherein localized dielectric breakdown and plasma generation continuously produce the coatings. However, during the arc discharge process,

Y. Zhang (✉) · C. Xu · Z. Zhang · Z. Zhang · D. Gao (✉)
Beijing Key Lab of Special Elastomeric Composite Materials, College of New Materials and Chemical Engineering, Beijing Institute of Petrochemical Technology, Beijing, China
e-mail: youzhang@bipt.edu.cn; gaodh@bipt.edu.cn

V. S. Saji, M. Zheludkevich (eds.), *Plasma Electrolytic Oxidation – Fundamentals, Advances and Applications*,
https://doi.org/10.1007/978-3-031-92429-3_23

local regions experience extremely high temperatures and pressures, resulting in rapid thermal shock. This shock leads to gas expansion, which forms pores within the coating [4]. Additionally, the oxidation reaction releases oxygen and hydrogen gases, which can become trapped within the coating, leading to the formation of bubbles that can eventually grow into pores. Furthermore, during the cooling process, the rapid solidification of molten oxides in the electrolyte solution induces significant thermal stress, resulting in the formation of microcracks within the oxide coating structure [4].

These pores provide "fast paths" for the ingress of corrosive media, which significantly reduce the long-term corrosion resistance of the PEO coating. Therefore, sealing treatment of the PEO coatings is crucial prior to their application. This chapter will focus not only on the parameter adjustments made during the formation of PEO coating but also on the posttreatment processes that are essential for achieving effective sealing.

23.2 Mechanical Surface Refinement Sealing

The posttreatment of PEO coatings is a critical step in enhancing their overall performance, particularly by refining surface morphology, roughness, and porosity. Mechanical surface refinement sealing is an effective approach for optimizing the coating's surface density and uniformity. Given the changes in the surface structure and morphology, these strategies enhance key functional properties, including corrosion resistance, wear resistance, and hardness. This is because the outer layer of the coating directly interacts with the external environment, and its surface topography plays a critical role in determining the coating's performance and durability. Techniques such as polishing, surface laser process, and shot peening [5–7] not only improve the mechanical properties of the coating but also enhance its resistance to harsh environmental conditions. These methods are essential components of the posttreatment process, contributing to the optimization of PEO coatings for various applications.

23.2.1 Polishing

Polishing is a widely adopted mechanical refinement technique for PEO coatings, offering a combination of simplicity and efficacy in improving surface characteristics. The process effectively removes most micro-scale imperfections, contributing to a smoother and more uniform surface [5, 8]. This reduction in surface roughness not only enhances the coating's appearance but also plays a critical role in mitigating potential sites that could trigger corrosion initiation. Furthermore, polishing minimizes localized stress concentrations, thereby preventing structural degradation in corrosive environments [9].

Polishing offers versatility and precision, making it suitable for a variety of substrates and coating systems under mild conditions, which is crucial for applications requiring high surface quality. It is particularly beneficial in industries such as automotive manufacturing, where a fine, attractive gloss is essential. Fernandez-López et al. [5] demonstrated polishing of the porous layer of PEO coatings on the surface of Al-Si cast alloys. The PEO coating exhibited a rough and porous structure (Fig. 23.1a). After polishing the outer PEO coating, the surface cracks and roughness were decreased, while the overall elemental distribution of the coating remained unchanged (Fig. 23.1b). Results showed that the polishing applies mechanical forces to the surface of the PEO coating, causing the irregularly shaped porous regions to undergo plastic deformation. This results in the mechanical sealing of the pores, which effectively reduces surface roughness and improves both wear and corrosion resistance of the samples.

However, polishing has limitations, as it requires careful control to prevent excessive material removal, which could compromise the thickness or integrity of the PEO coating. Additionally, the process is considered time-consuming and energy-intensive, particularly for large-scale or intricate components. Despite these challenges, advancements in polishing technology, including automated systems and precision abrasives, have enhanced both its efficiency and consistency.

In summary, polishing is an effective method for improving surface smoothness and reducing defect sites, making it an indispensable posttreatment strategy for PEO coatings, particularly in applications where precision and performance are critical.

23.2.2 Surface Laser Process

Surface laser processing (SLP) has emerged as a versatile and effective method for refining the properties of PEO coatings. By employing high-energy laser beams, SLP can modify the surface morphology, densify the coating structure, and reduce the inherent porosities and microcracks typically found in PEO layers. Additionally, laser processing achieves these modifications without direct physical contact, minimizing the mechanical damage to the coating [3].

SLP techniques, such as laser surface melting (LSM), laser shock adhesion (LSA), laser surface texturing (LST), and laser shock peening (LSP) [3], offer unique advantages for enhancing the properties of PEO coatings. These methods utilize low-energy, high-power density laser beams to improve both corrosion resistance and mechanical performance [10].

LSM effectively eliminates surface defects like microcracks and seals porosities by melting and resolidifying a thin surface layer (Fig. 23.2) [11]. This process creates a dense, uniform structure that enhances both corrosion resistance and wear performance [12]. In addition, LSA enriches the surface layer with alloying elements by melting a shallow layer of the material and introducing desired elements into the melt pool. This results in a metallurgically bonded layer with improved

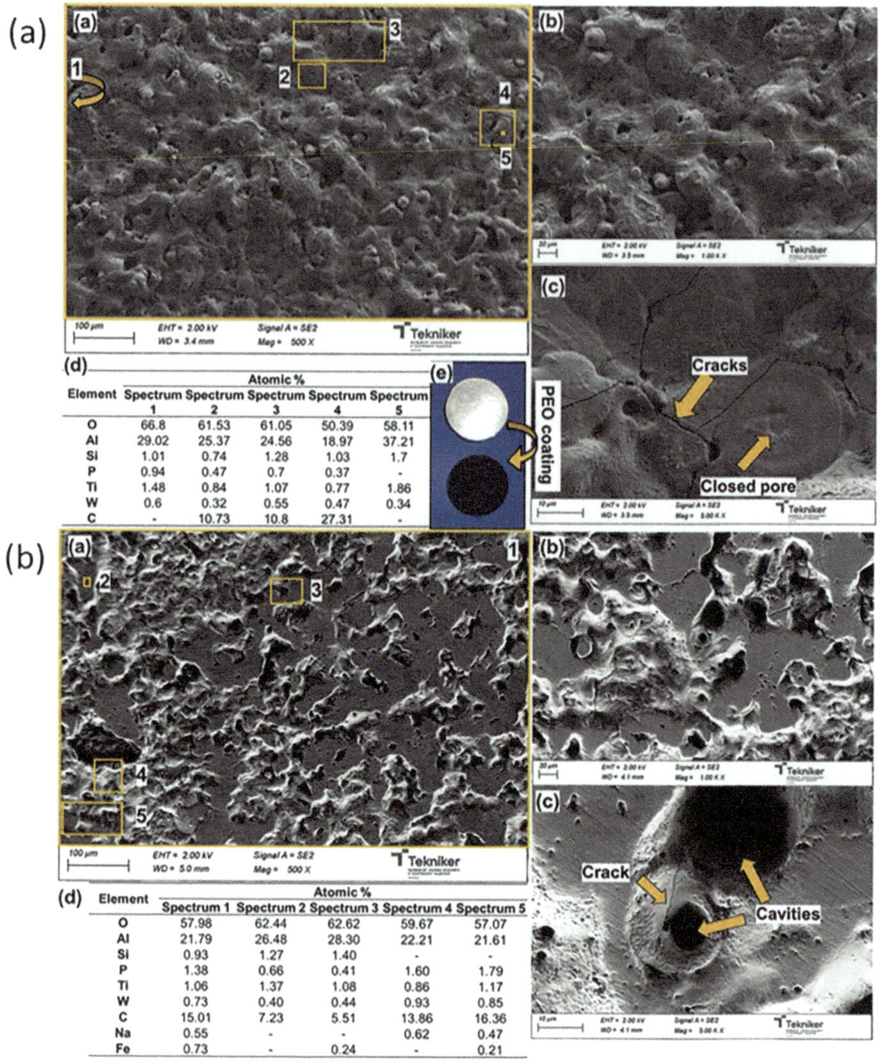

(a) (d)

Element	Atomic %				
	Spectrum 1	Spectrum 2	Spectrum 3	Spectrum 4	Spectrum 5
O	66.8	61.53	61.05	50.39	58.11
Al	29.02	25.37	24.56	18.97	37.21
Si	1.01	0.74	1.28	1.03	1.7
P	0.94	0.47	0.7	0.37	-
Ti	1.48	0.84	1.07	0.77	1.86
W	0.6	0.32	0.55	0.47	0.34
C	-	10.73	10.8	27.31	-

(b) (d)

Element	Atomic %				
	Spectrum 1	Spectrum 2	Spectrum 3	Spectrum 4	Spectrum 5
O	57.98	62.44	62.62	59.67	57.07
Al	21.79	26.48	28.30	22.21	21.61
Si	0.93	1.27	1.40	-	-
P	1.38	0.66	0.41	1.60	1.79
Ti	1.06	1.37	1.08	0.86	1.17
W	0.73	0.40	0.44	0.93	0.85
C	15.01	7.23	5.51	13.86	16.36
Na	0.55	-	-	0.62	0.47
Fe	0.73	-	0.24	-	0.21

Fig. 23.1 SEM micrographs and EDS distribution of elements of (**a**) PEO coating and (**b**) PEO coating after the polishing of its porous layer. Reproduced with permission from [5] © 2022 Elsevier Ltd.

surface properties [13]. LST creates micro- or nanoscale surface patterns, which increase the surface area and improve coating functionalities such as mechanical interlocking and subsequent layer adhesion [14]. LSP enhances the mechanical properties of PEO coatings by inducing compressive residual stresses, improving fatigue resistance and durability under dynamic loading conditions [15]. These laser-based techniques, each with distinct mechanisms, offer specialized solutions for controlling the morphology and properties of PEO coatings, thereby expanding their potential applications in demanding environments.

Despite its advantages, SLP requires precise control over parameters such as laser power, scanning speed, and pulse duration to avoid overheating of the substrate or degradation of the coating. Furthermore, the integration of SLP into industrial-scale applications increases both cost and process complexity. Nevertheless, SLP represents a promising approach to enhancing the durability and multifunctionality of PEO-treated materials, with potential applications spanning aerospace to biomedical devices.

23.2.3 *Shot Peening*

Shot peening, a technique traditionally used for surface preparation, has been explored to a limited extent as a posttreatment for PEO coatings [6]. This process introduces compressive stresses to the surface through the controlled impact of high-velocity media, such as steel or ceramic beads [16]. While shot peening is primarily used in pre-treatment to enhance the fatigue resistance of substrates, its potential to compact surface porosities and refine coating texture makes it an intriguing candidate for PEO posttreatment under specific conditions.

In the context of PEO coatings, shot peening could facilitate surface densification, improving resistance to crack initiation and propagation under alternating mechanical loads. Additionally, it may enhance the adhesion of subsequent protective layers or topcoats [6]. However, its practical implementation faces challenges, including the risk of surface damage due to the brittle nature of some PEO coatings, as well as the need for precise parameter control to avoid exposing the substrate.

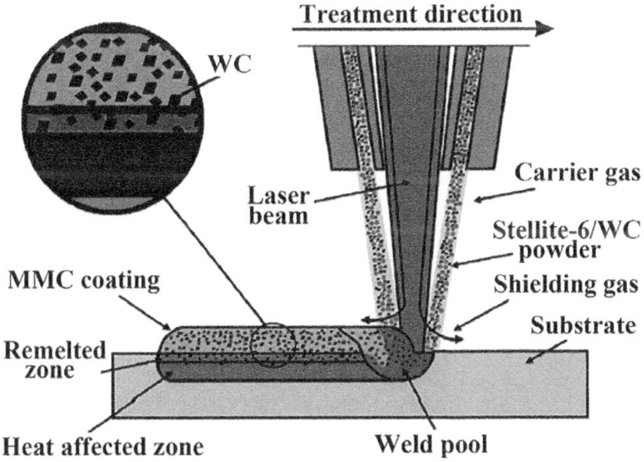

Fig. 23.2 Laser surface melting. Reproduced with permission from [11] © 2014 Elsevier Ltd.

This section highlights the pivotal role of mechanical surface refinement in enhancing the performance of PEO coatings. By addressing challenges such as surface roughness and porosity, these techniques significantly improve the mechanical strength, corrosion resistance, and adhesion of the coatings. However, the selection of appropriate posttreatment methods depends on the specific application requirements and performance objectives. In practice, a combination of multiple posttreatment techniques is often employed to achieve optimal results. For instance, polishing and shot peening can improve surface morphology and mechanical properties, while laser treatment further enhance hardness and wear resistance. In contrast, chemical treatments are primarily beneficial for enhancing corrosion resistance and adhesion. The integration of these techniques not only demonstrates the versatility of mechanical surface refinement but also expands the potential applications of PEO coatings in various engineering fields. Subsequent sections will explore these complementary techniques, examining their mechanisms, advantages, and performance in practical applications, providing a comprehensive overview of posttreatment strategies for PEO coatings.

23.3 Chemical Treatment Methods

Chemical treatment methods for PEO coatings are designed not only to address structural defects but also to meet specific functional demands for different applications [17]. These methods do more than merely fill pores or seal micro-cracks; they play a vital role in modifying the surface properties of PEO coatings, enhancing their corrosion resistance, wear resistance, and multifunctionality [18].

The primary goal of chemical treatments is to refine the coating's microstructure and seal surface defects, thereby significantly improving its protective capabilities. Organic and inorganic coating sealing strategies, as well as conversion coating treatments [19], are particularly effective in this regard, offering solutions for pores filling and improving the overall coating density. Additionally, other advanced surface treatment techniques have been developed to further increase the durability of coatings by modifying surface composition and imparting specialized properties.

This section examines various chemical treatment methods, highlighting their roles in enhancing PEO coatings and expanding their functionality to meet the evolving demands of modern engineering.

23.3.1 Organic and Inorganic Coating Sealing Strategies

The integration of organic and inorganic coatings as sealing layers for PEO coatings offers an effective approach to addressing the inherent defects of PEO coating, such as porosity and susceptibility to environmental degradation. The cross-sectional micrographs of samples treated with these sealing strategies are presented in

Fig. 23.3. It was clearly found that a continuous interface between the sol-gel layer and PEO coating, can effectively sealing original pores developed during the preparation of PEO. A comparison of different immersion times reveals that prolonged immersion in either organic or inorganic coating solutions contributes to an increase in the thickness of the coating deposited above the PEO surface. Organic coatings can cover the surface of PEO to form a thick membrane, thereby showing more effective micropore sealing.

Organic coatings based on polymeric matrix, such as epoxy resins, polyacrylate, and polyurethane [21], provide excellent barrier properties and hydrophobicity, both of which are essential for corrosion resistance by preventing moisture ingress. On the other hand, inorganic sealing methods, such as sol-gel coatings, and oxide thin films, form chemically stable and dense barriers over the PEO surface [22]. These coatings create continuous, adherent layers that chemically bond to the PEO surface, blocking pathways for corrosive media and ensuring long-term durability, even in aggressive environments.

Moreover, these coating matrices can be functionalized, enabling the incorporation of corrosion inhibitors or self-healing agents [23], further enhancing the active protective capabilities of the system. For instance, Yang et al. [24] developed a hybrid PEO-epoxy coating for magnesium alloy, incorporating sodium salt of

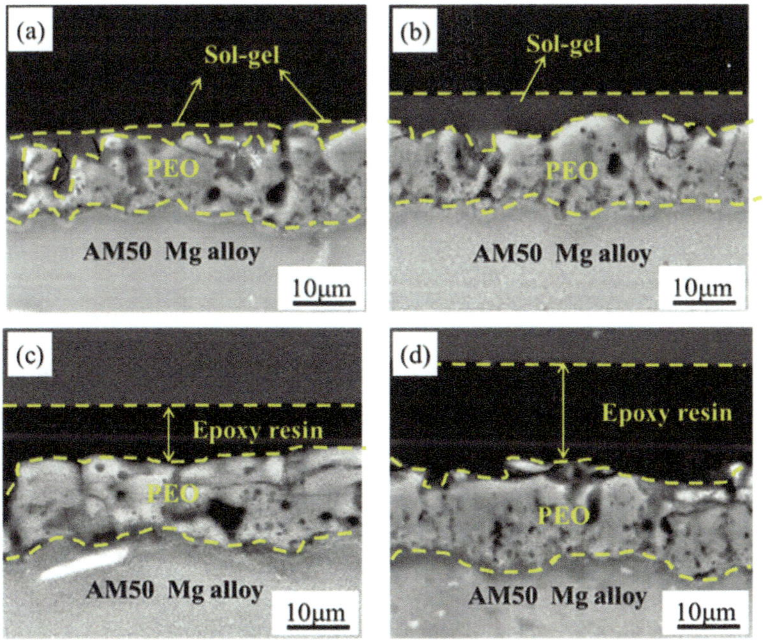

Fig. 23.3 The cross-sectional micrographs of samples treated with organic or inorganic coatings sealing strategies. (**a**) sealing by sol-gel coating, dipped for 5 s, (**b**) sealing by sol-gel coating, dipped for 60 s, (**c**) sealing by epoxy resin coating, dipped for 5 s, (**d**) sealing by epoxy resin coating, dipped for 60 s. Reproduced with permission from [20] © 2021 Elsevier B.V.

3-methylsalicylate as an inhibitor. This system demonstrated enhanced corrosion resistance and active protection through controlled dip-coating parameters and inhibitor release.

These sealing strategies not only improve the intrinsic properties of PEO coatings but also extend their applicability to high-performance industrial sectors. By varying the composition and structure of the sealing layer, it is possible to tailor the coating to meet the specific demands of applications in aerospace, marine, and automotive industries, where resistance to harsh environmental conditions is critical. Furthermore, these methods offer practical advantages, such as scalability and cost-effectiveness, as they generally do not require complex equipment or highly specialized conditions for application.

However, these sealing methods also have limitations. While organic coatings provide excellent corrosion resistance, they may underperform in extreme environmental conditions. Inorganic coatings, though thermally and chemically stable, tend to be more brittle under impact and thermal expansion [25], which could limit their applicability under certain mechanical stresses. Additionally, both organic and inorganic sealants often require posttreatment processes, such as curing, which can add extra production time and costs. Furthermore, the sealing layers may restrict the "breathing" ability of the PEO coating, potentially affecting its performance in applications where gas exchange is necessary.

23.3.2 Conversion Coating Sealing

The application of conversion coatings as sealing layers for PEO coatings represents an advanced strategy to address the inherent porosity and environmental susceptibility of PEO layers. Common types of conversion coatings include metal oxide conversion films, layered double hydroxides (LDHs), metal-organic frameworks (MOFs), phosphates, and hydroxyapatite films [26–30]. In addition to enhancing corrosion resistance, certain types, such as LDH and MOF-based films, also impart self-healing properties [31]. These coatings provide a versatile platform for controlling surface functionalities to meet the demands of diverse applications. This section will primarily focus on the first three conversion coating sealing strategies.

Metal oxide conversion coatings enhance the protective properties of PEO layers by forming chemically stable and dense films through reactions with the substrate. These coatings effectively seal the intrinsic porosity of PEO coatings with a continuous layer, thereby improving their barrier performance. Among these conversion coatings, chromate-based conversion coatings, despite their excellent corrosion resistance, are not widely accepted due to concerns over their toxicity and environmental impact. In contrast, cerium-based conversion coatings are particularly notable for their robust adhesion and superior corrosion resistance, making them highly suitable for use in aggressive environments [26].

LDH coatings provide advanced protection by utilizing their layered structure to load and release corrosion inhibitors, which results from their responsive defense mechanism. This intercalation allows for the gradual release of active species in response to environmental triggers, thereby imparting self-healing capabilities [32, 33]. Since Serdechnova et al. reported the in situ growth of LDH on PEO-coated AA2024-T3 aluminum alloy in 2017 [34], LDH coatings have been regarded as an effective posttreatment method for sealing PEO coatings [35–37]. As demonstrated in Fig. 23.4a–d [38], the LDH film is in situ grown on the surface of the PEO coating, effectively sealing the pores of the PEO layer. This results in a significant alteration of the surface morphology. Notably, the characteristic honeycomb-like structure was maintained even after the exchange of VO_3^- and MoO_4^{2-} (Fig. 23.4e–h). Furthermore, the inherent ability of the LDH to capture Cl^- ions, coupled with the corrosion inhibitors intercalated through the anion exchange process, is expected to further enhance the corrosion resistance of the treated samples. Both the two types of ions were reduced into metal oxide and were loaded onto the substrates, as shown in the following equations.

$$2Al + 3H_2O \rightarrow Al_2O_3 + 6H^+ + 6e^- \quad (23.1)$$

$$VO_3^- + 2H^+ + e^- \rightarrow VO_2 + H_2O \quad (23.2)$$

$$4VO_2 + O_2 \rightarrow 2V_2O_5 \quad (23.3)$$

$$MoO_4^{2-} + 4H^+ + 2e^- \rightarrow MoO_2 + 2H_2O \quad (23.4)$$

$$2MoO_2 + O_2 \rightarrow 2MoO_3 \quad (23.5)$$

Bouali et al. [39] report a novel approach for growing LDH films on any PEO coating on AA2024 using an Al_2O_3 sol-gel layer, independent of the nature of the PEO coating. Taking LDH-NO_3 as an example, its growth mechanism can be divided into two steps:

1. Al_2O_3 gel formation:

$$Al(NO_3)_3 \cdot 9H_2O \leftrightharpoons \left[Al(H_2O)_6\right]^{3+} + 3NO_3^- \quad (23.6)$$

$$\left[Al(H_2O)_6\right]^{3+} \leftrightharpoons \left[Al(H_2O)_6\right](OH)_a^{(3-a)+} \rightarrow Al_2O_3(OH)_z \cdot nH_2O \quad (23.7)$$

2. LDH film growth:

$$Al_2O_3(OH)_z \cdot nH_2O + H_2O + OH^- \rightarrow xAl(OH)_4^- \quad (23.8)$$

$$Zn^{2+} + (OH)^- \rightarrow Zn(OH)^+ \quad (23.9)$$

$$Zn(OH)^+ + xAl(OH)_4^- + NO_3^- + H_2O \rightarrow LDH - NO_3 \quad (23.10)$$

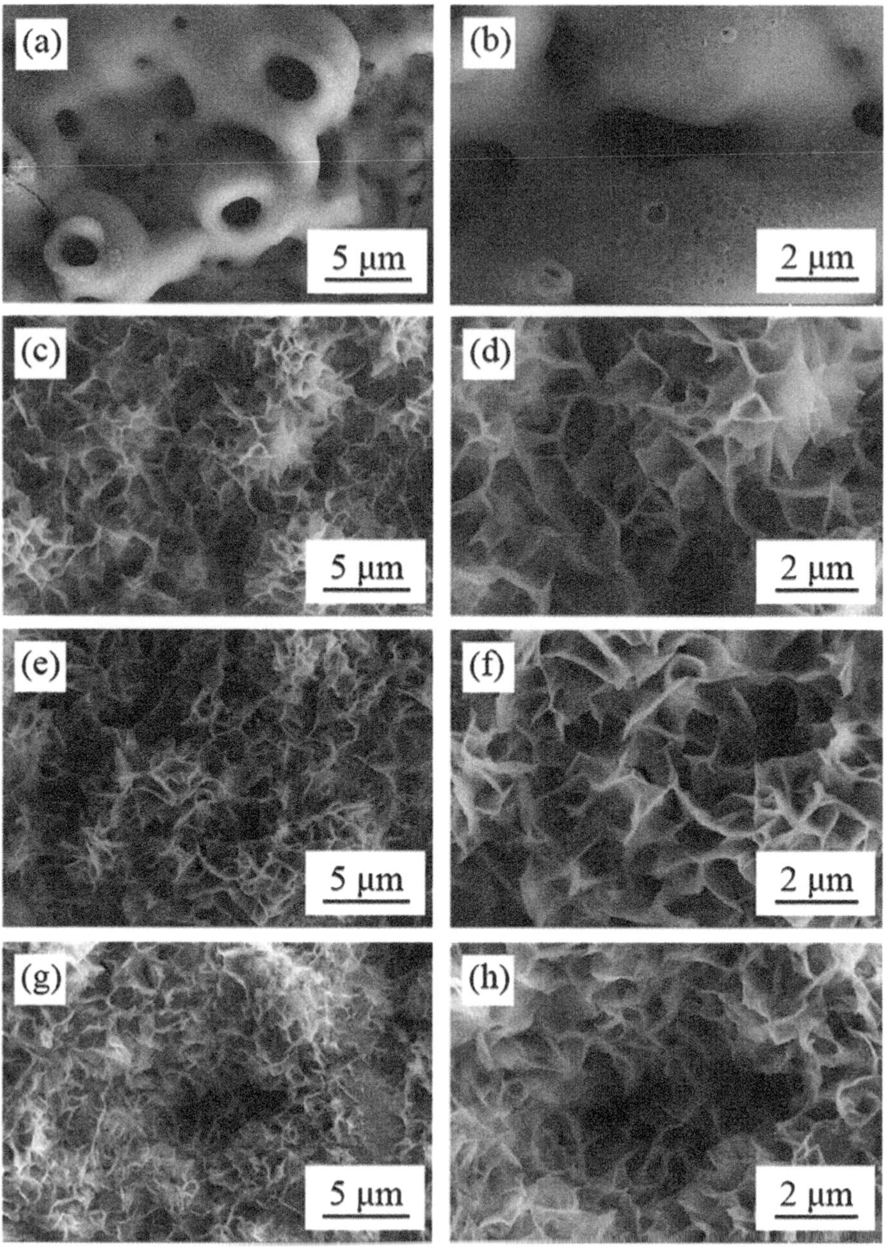

Fig. 23.4 Surface morphology of PEO coatings (**a** and **b**) PEO, (**c** and **d**) PEO-LDH-NO_3, (**e** and **f**) PEO-LDH-VO_3 and (**g** and **h**) PEO-LDH-MoO_4. Reproduced with permission from [38] © 2022 Elsevier B.V.

As shown in Fig. 23.5a, the intrinsic pores on the PEO surface of the PEO-X-LDH-VO_x samples were filled by deposits, which exhibit a strong vanadium signal. This suggests that the anion inhibitor (VO_3^-) was incorporated onto LDH layers via ion exchange, and transformed into VO_x to repair the porous surface. Scanning vibrating electrode technique (SVET) test results demonstrated that after artificial drilling and immersion for a period of time, the surface of the PEO-X-LDH-VO_x sample did not exhibit the zone of high corrosion current, as observed in the other samples. This behavior can be ascribed to the increase in pH and the presence of corrosive species such as Cl^-, which facilitate the release of vanadate corrosion inhibitors from the LDH nanocontainers near the active regions. The self-healing performance of LDH film (Fig. 23.5b). The LDH sealing strategy significantly reduces the porosity of the PEO surface. Meanwhile, vanadate-loaded LDH releases corrosion inhibitors when corrosion occurs, endowing the coating with self-healing abilities.

Additionally, LDH coatings offer excellent environmental stability, mechanical strength, and adaptability to different metal substrates [40]. The ability to provide both barrier and active protection, combined with their multifunctionality, makes LDH coatings generally applicable for enlarging the durability of PEO-treated materials. Furthermore, LDH-based conversion coatings can be tailored for specific applications by adjusting the composition and release profiles of the intercalated inhibitors, offering a customizable approach to corrosion protection [41].

Metal-organic frameworks further enhance the functionality of conversion coatings by utilizing their nanostructured, highly porous frameworks [42]. MOF-based coatings can encapsulate and release corrosion inhibitors or self-healing agents in a controlled manner, offering a dynamic and responsive defense against corrosion [43]. This unique combination of high porosity and functional versatility makes MOFs a promising strategy for the development of advanced protective systems for PEO coatings [27, 44, 45]. During the self-assembly process of the MOF film on the surface of the PEO coating, a dense top layer and stable composition are formed. This structure effectively seals defects and prevents the infiltration of corrosive ions into the metallic substrate [27]. However, MOF coatings face challenges with adhesion to the substrate, often requiring an additional precursor layer to improve bonding strength. This not only complicates the coating process but also leads to higher material consumption. As demonstrated in Fig. 23.6, Jiang et al. [45] synthesized a ZIF-8 flowerlike film on the surface of a PEO coating, utilizing flower-like ZnO nanorods as a pre-treating layer. This nanoflower structure imparts superhydrophobicity to the film, thereby enhancing its corrosion resistance. The resulting composite coating exhibited excellent corrosion resistance and outstanding characterization properties.

In summary, conversion coating sealing strategies effectively enhance the performance of PEO coatings by filling their inherent porosity and environmental vulnerabilities. Meanwhile, due to the high designability of conversion coatings, numerous studies in recent years have reported on the application of conversion coating sealing strategies on the surface of PEO coating in fields such as photocatalysis and antibacterial protection [46, 47]. However, several challenges remain in

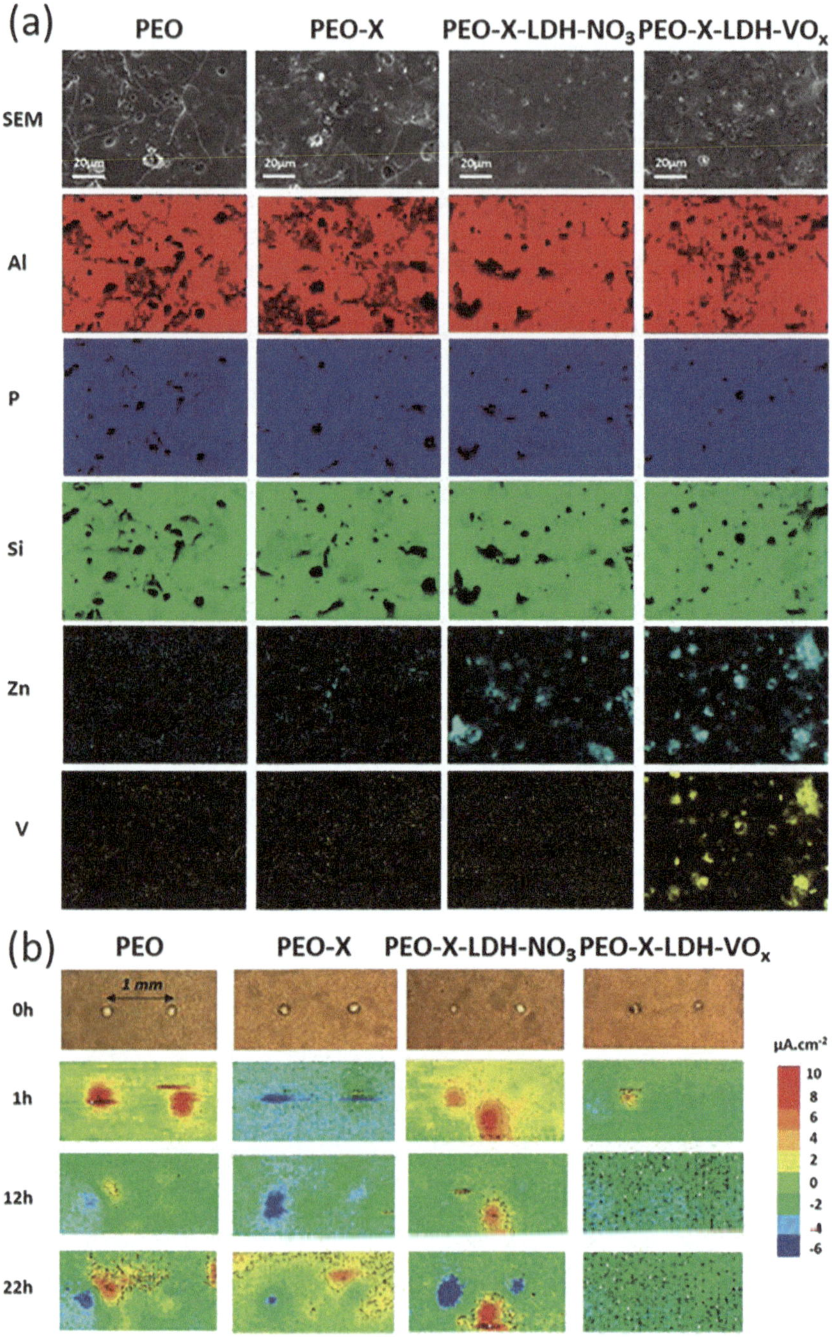

Fig. 23.5 (**a**) SEM and EDS elemental distribution for samples, (**b**) Micrographs and SVET maps of samples. Reproduced with permission from [39] © 2019 The Authors. Published by Elsevier B.V.

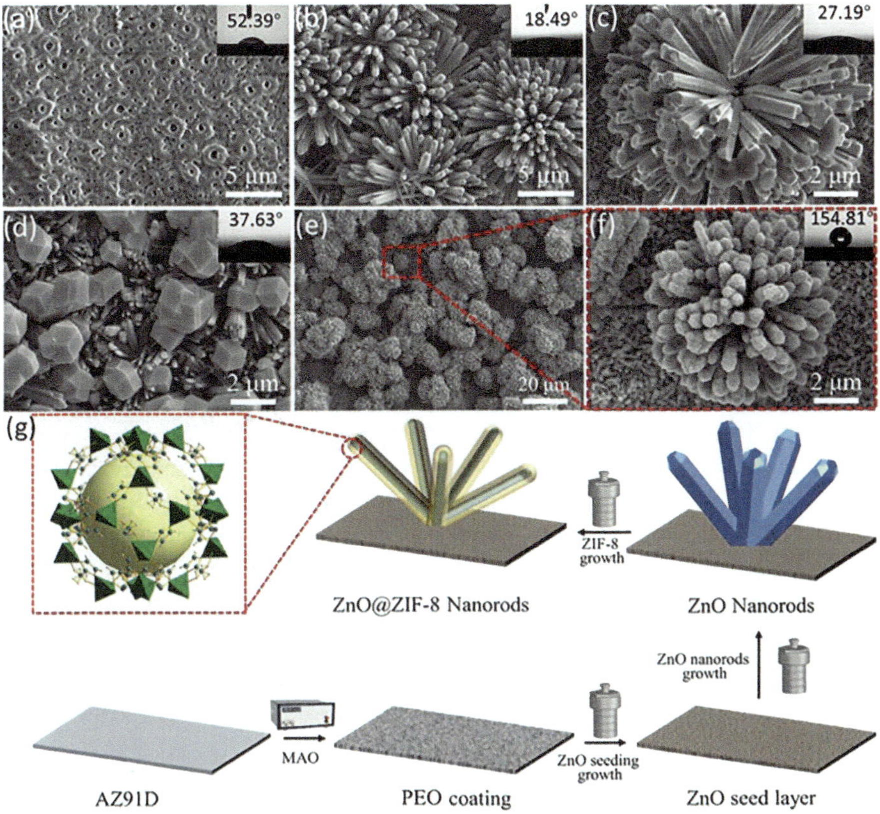

Fig. 23.6 SEM images of (**a**) PEO coating, (**b**) ZnO array, and samples with different ligand concentrations in (**c**) 0.1 mol/L, (**d**) 0.2 mol/L, (**e**, **f**) 0.5 mol/L. The insets of the images of water drop on the surface, (**g**) schematic illustration of preparation of ZnO@ZIF-8 nanorods array. Reproduced with permission from [45] © 2022 Chongqing University. Publishing services provided by Elsevier B.V. on behalf of KeAi Communications Co. Ltd.

implementing these sealing approaches. Traditional metal-oxide conversion coatings often suffer from poor adhesion, limited wear resistance, and frequent defects. Similarly, LDH and MOF conversion coatings face issues such as inadequate adhesion, complex processing methods, and high costs, which restrict their practical applications. Therefore, future research should focus on optimizing the adhesion strength of these coatings, simplifying processing techniques, and enhancing the long-term stability and performance of conversion coatings to meet the evolving demands of industrial applications.

23.3.3 Other Chemical Sealing Methods

In addition to organic/inorganic coating sealing and conversion coating sealing, researchers have investigated several other chemical treatment methods, including electrodeposition, electrophoretic deposition (EPD), and chemical vapor deposition (CVD) [48, 49]. These approaches can effectively address the inherent porosity of PEO coatings while enhancing their functional performance.

Electrodeposition and electrophoretic deposition have gained attention as effective posttreatment techniques for enhancing the properties of PEO coatings. These methods leverage the structural features of PEO layers, such as porosity and surface roughness [50], to create coatings with enhanced corrosion resistance.

The electrodeposition process involves deposition of materials, such as metals, ceramics, and composites, onto PEO-coated surfaces under an applied electric field. This process effectively seals micro-pores and enhances the corrosion resistance of the coating. For instance, hydroxyapatite and calcium phosphate have been electrodeposited onto PEO-treated magnesium [48, 49], forming hybrid coatings that significantly improve both biodegradability and corrosion resistance. In addition, materials like graphene oxide [51] can be deposited to create laminar structures that can block pores and provide opportunities for further functionalization (Fig. 23.7).

In contrast, EPD is a versatile technique that deposits charged particles—ranging from ceramics to polymers or composites—onto PEO-treated surfaces [52, 53].

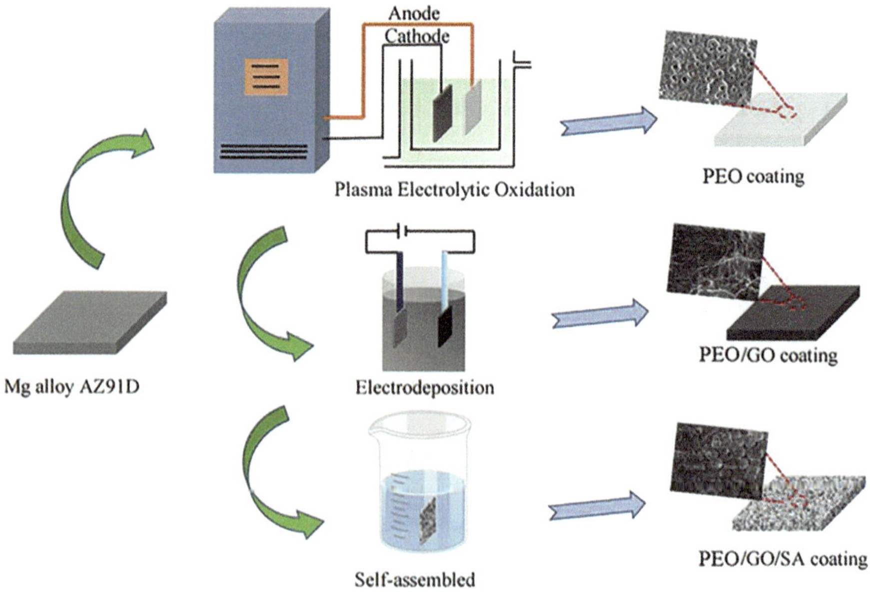

Fig. 23.7 Flow chart of the preparation of the PEO/GO/SA coating on a Mg alloy. Reproduced with permission from [51] © 2023, University of Science and Technology Beijing

This method optimizes the special functions of coatings, such as wear resistance and adhesion strength, by adding specific substances to the solution. Achieving uniform and defect-free coatings through EPD, however, requires careful control of the suspension's particle size, charge, and dispersion stability.

Chemical vapor deposition is a widely applied process that deposits thin, conformal layers onto substrates through chemical reactions from vapor-phase precursors. Renowned for producing coatings with excellent uniformity, high purity, and superior interface bonding, CVD is commonly employed to create dense protective layers. When applied to PEO-treated surfaces, CVD effectively seals micro-pores and further enhances the resistance of the coating [54, 55]. The conformal nature of CVD ensures that the deposited layer uniformly covers the surface, thereby improving the barrier performance of PEO coatings against environmental factors. Certain studies have also indicated that CVD, as a sealing strategy for PEO coatings, can enhance the wear resistance or biocompatibility of the coatings [55–57]. Hao et al. [55]. prepared a bilayer organic–inorganic functional hybrid coating with excellent tribological properties. This was achieved by depositing a dense poly(tetrafluoro-*p*-xylylene) layer onto the PEO coating on the surface of tantalum-tungsten alloys using the CVD process. As shown in Fig. 23.8, the deposition of poly(tetrafluoro-p-xylylene) (PF) formed a cauliflower-like polymer layer on the PEO coating surface, with an effective sealing of the pore structures. The deposited polymer layer

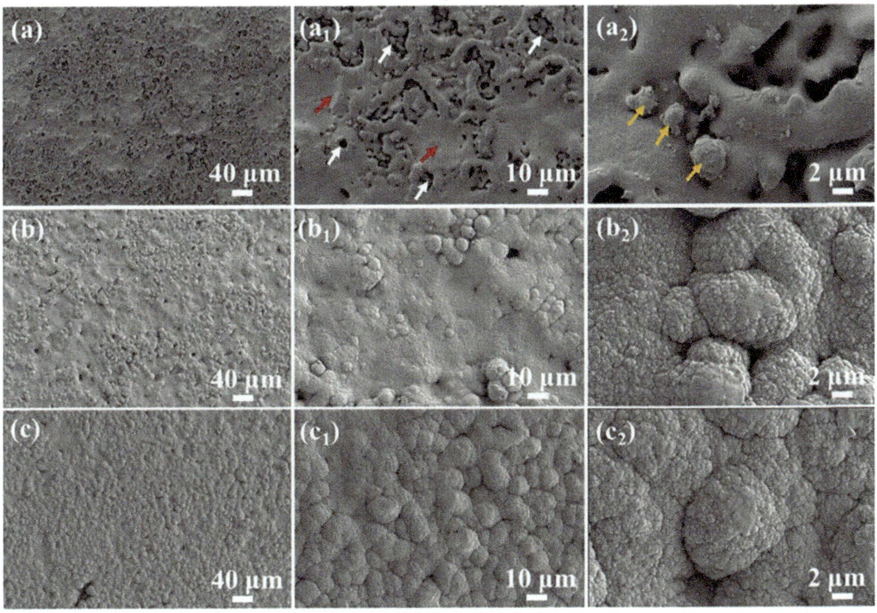

Fig. 23.8 Surface morphology of the coatings at different magnifications: (**a**, **a**$_1$, **a**$_2$) PEO; (**b**, **b**$_1$, **b**$_2$) PEO-PF-4 μm; (**c**, **c**$_1$, **c**$_2$) PEO-PF-8 μm. Reproduced with permission from [55] © 2024, The Author(s), under exclusive license to Springer Science Business Media, LLC

enhances the wear resistance of the sample, attributed to the advanced low friction of fluorinated polymers.

However, the process typically requires high temperatures and specialized equipment, which can increase production costs and limit its compatibility with certain substrates. Nonetheless, CVD remains a promising posttreatment technique for enhancing the protective capabilities of PEO coatings, especially when precise dimension control is required.

Posttreatment sealing strategies play a critical role in enhancing the performance of PEO coatings by addressing inherent porosity and surface irregularities. This section discusses three major categories of surface morphology improvement methods: organic and inorganic coating sealing, conversion coating sealing, and other chemical sealing methods. Each approach demonstrates unique advantages in modifying PEO-coated surfaces for improved corrosion resistance, mechanical durability, and functionality. For instance, organic and inorganic coatings provide flexibility in tailoring protective layers, while conversion coatings offer strong interfacial bonding and self-healing properties. Advanced techniques, such as electrodeposition, electrophoretic deposition, and chemical vapor deposition, further expand the potential of PEO coatings by introducing functional enhancements.

Future research should focus on integrating these methods to develop multifunctional coatings with optimized performance. Additional emphasis should also be placed on improving the adhesion and long-term stability of the sealed layers, as well as simplifying the processing steps and reducing material consumption, to enable broader industrial applications.

23.4 Structure and Performance Comprehensive Adjustment Methods

In the posttreatment process of PEO coating, there are some comprehensive adjustment strategies to improve its structure and performance [58]. These strategies often integrate various techniques to improve the overall properties of the PEO coating [58, 59]. Common posttreatment methods include heat treatment, hydrothermal sealing, physical vapor deposition, electroless plating, and so on. Each of these methods is discussed in detail below.

23.4.1 Heat Treatment

Heat treatment is a widely used process that alters the material structure by controlling the heating and cooling conditions. It can improve the hardness, wear resistance, strength, and other properties of PEO coating on a large scale. The primary function of heat treatment is grain refinement: it refines the grains in the PEO coating, enhancing its hardness and wear resistance [60]. For instance, the pore structure of the PEO coating changes at different temperatures, as shown in Fig. 23.9 [59]. This research reveals that thermal treatment generally enhances the integrity of PEO coatings because when the treatment temperature increases, the pore size and surface porosity decrease. The morphological transformation contributes to the strengthening of corrosion resistance.

Currently, the most common heat treatment processes are solid solution treatment and annealing. In solid solution treatment, the coating is heated to a specific temperature, held for a designated period, and then rapidly cooled to improve hardness and wear resistance. In annealing, the coating is heated to an appropriate temperature and cooled slowly, which contributes to reduce the brittleness and improve toughness. A representative of heat treatment process is shown in Fig. 23.10 [58].

Generally, heat treatment is carried out within a temperature range of 500–700 °C. It should be emphasized that the choice of temperature is not only dependent on the coating but also on the properties of the substrate. While higher temperatures can improve the coating performance, they may also cause damage to the substrate. Heat treatment can recrystallize the grains in the coating or form new phases [58], thereby improving the mechanical properties and wear resistance of the coating.

Heat treatment also promotes densification via the combination of various metal oxides. This phase transformation mechanism leads to a reduction in pores and cracks, contributing to a continuous coating with improved corrosion resistance, as illustrated in Fig. 23.11 [61]. When comparing the surface morphology of PEO coatings treated under the air and vacuum conditions, the oxygen atmosphere helps to thicken the surface oxides layers, which significantly enlarged the thickness of coatings. However, after treating under vacuum, the thickness coating decreases due to the oxygen atom diffusion and incorporation into the metal bulk. This mechanism also induces permeable cracks on the surface, which significantly undermine the corrosion protection capabilities.

23.4.2 Hydrothermal Sealing

Hydrothermal sealing is a process used to treat PEO coatings by immersing them in water or an aqueous solution under high temperature and pressure conditions [62]. Hydrothermal treatment (HT) sealing layer can effectively fill the micropores in the PEO coating (Figs. 23.12 and 23.13 [62]), thereby improving the sealing and corrosion resistance of the coating. In Fig. 23.12, the PEO coating (a) presents crater-like micro-pores and micro-cracks, which are caused by the PEO process and will affect the protective performance. The PEO/HT coating (Fig. 23.12b) shows Mg $(OH)_2$ crystal after HT treatment. Figure 23.12c shows the influence of chitosan (CS) layer on PEO coating. The micropores are effectively sealed, resulting in a smooth, uniform surface. The CS layer in PEO/CS coating (Fig. 23.12d), effectively covers and fills the pores and cracks of the PEO coating. The difference between PEO/CS and

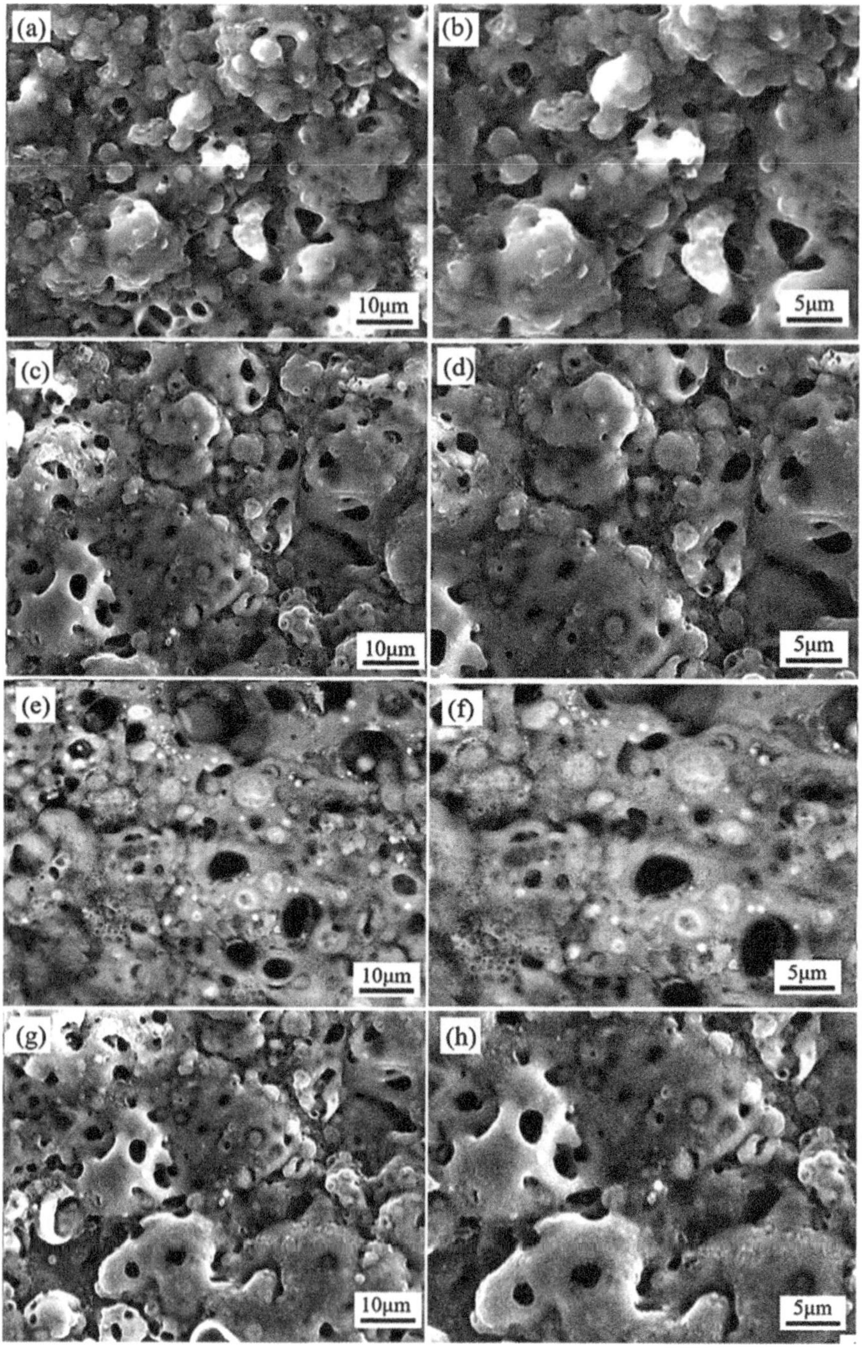

Fig. 23.9 The pore diameter changes of PEO coatings at different temperatures, (**a**, **b**) 0 °C, (**c**, **d**) 150 °C, (**e**, **f**) 200 °C, (**g**, **h**) 250 °C. Reproduced with permission from [59] © 2023 Elsevier B.V.

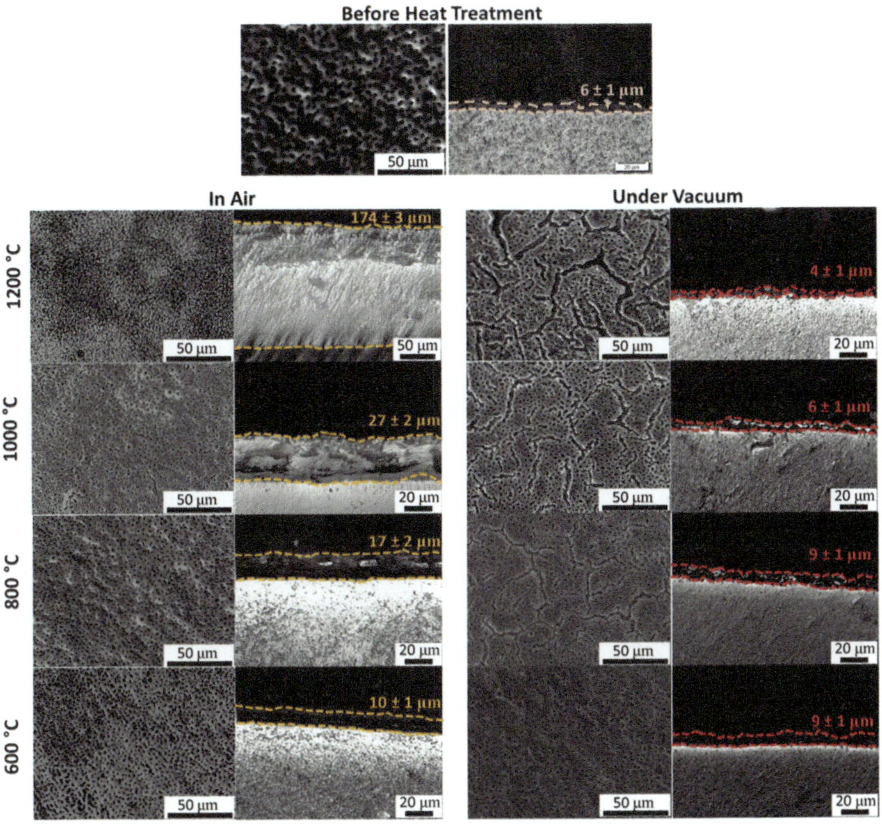

Fig. 23.10 (**a**, **b**) A way of heat treatment process. Reproduced with permission from [60] © 2024 The Authors. Published by Elsevier B.V.

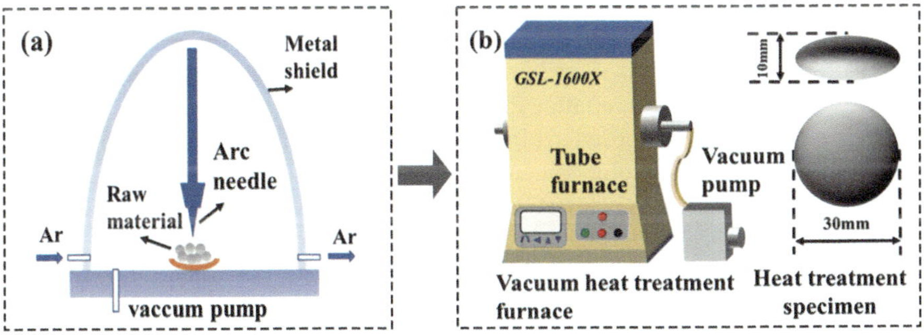

Fig. 23.11 Surface and cross section morphologies of PEO coatings at different temperatures. Reproduced with permission from [61] © 2024 Elsevier B.V.

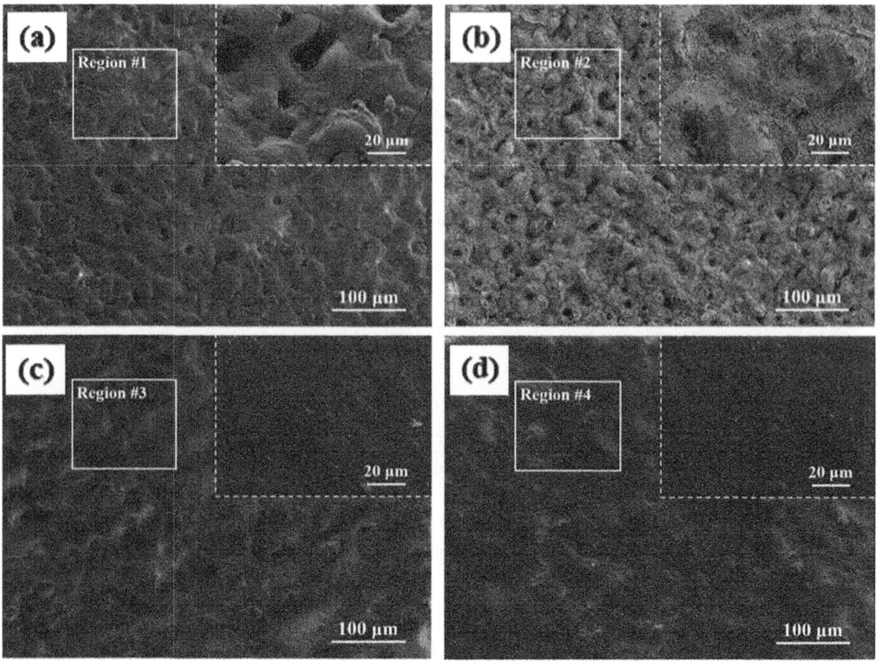

Fig. 23.12 The surface morphologies of the (**a**) PEO, (**b**) PEO/HT, (**c**) PEO/CS, and (**d**) PEO/HT/CS samples with individual enlarged parts. Reproduced with permission from [62] © 2022 Elsevier B.V.

PEO/HT/CS could be investigated from the cross-section of coating in Fig. 23.13. Obviously, the combination of HT and organic coatings almost resulted in filling the whole permeable tube in the coatings, which ensure the satisfactory performance in corrosion resistance. Significant structural changes inside the PEO coating after HT treatment, particularly at the interface tightness between the CS layer and the PEO layer, which improves the overall compactness. The primary purpose of this process is to increase the compactness and improve the corrosion resistance of the coating.

In the standard procedure, after PEO treatment, the coated sample is immersed in a liquid medium and exposed to high temperature and pressure conditions. During this process, the water or solution penetrates the coating, allowing the micropores to form a denser, more compact layer. Zhang et al. [63] have combined LDH with PEO coating using the hydrothermal method to prepare a denser coating. This process increases the coating thickness, thereby improving its corrosion resistance, as shown in Fig. 23.14. Maryam et al. [64] found that combining the hydrothermal method with PEO improves both the photocatalytic performance and corrosion resistance of the coating.

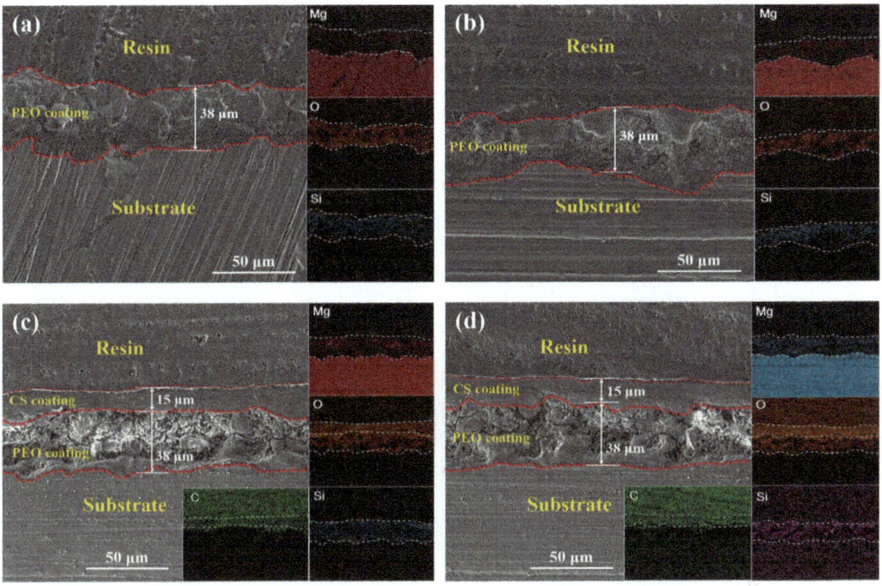

Fig. 23.13 The cross-sectional morphologies and element mappings of Mg, O, Si, and C for (**a**) PEO, (**b**) PEO/HT, (**c**) PEO/CS, and (**d**) PEO/HT/CS samples. Reproduced with permission from [62] © 2022 Elsevier B.V.

23.4.3 *Physical Vapor Deposition*

Physical vapor deposition is a widely used technique that involves the physical transformation of a material from its vapor phase to a solid state, which is then deposited onto the surface of a substrate [65, 66]. The PVD process can enhance the properties of PEO coatings by producing a denser and more uniform coating, thereby further improving its performance [67]. The primary objective of PVD in this context is to improve the hardness, wear resistance, and corrosion resistance of the coating [66]. An effective strategy is to deposit metal, ceramic, or composite films by PVD after the PEO treatment. Common PVD techniques include evaporation deposition [68], sputtering deposition [69] and others. Sidra et al. [65] obtained a coating with satisfactory corrosion resistance and biocompatibility by applying shot blasting to the PEO coating on a Ti-6Al-4V alloy (Figs. 23.15 and 23.16).

These posttreatment processes significantly improve the overall performance of PEO coatings, particularly in terms of wear resistance, corrosion protection and high-temperature stability. Additionally, PVD can provide a rigid protective layer for the coating. These processes are often combined based on specific application requirements to achieve optimal performance enhancement.

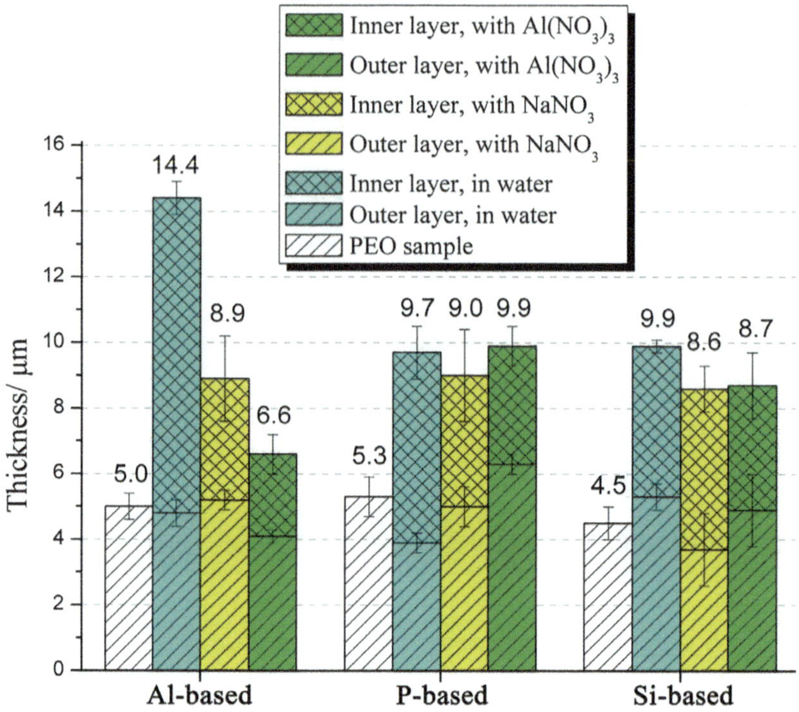

Fig. 23.14 Evolution of total coating thickness. Reproduced with permission from [63] © 2021 Chongqing University. Publishing services provided by Elsevier B.V. on behalf of KeAi Communications Co. Ltd.

23.4.4 Electroless Plating

After the PEO coating is formed, residual electrolytes or impurities may remain on the substrate surface. Electroless plating is a process in which metal ions are reduced and deposited onto the substrate surface, forming a coating. Electroless nickel plating (Ni-P), as the most common type, creates a uniform nickel-phosphorus alloy layer on the metal surface [70]. Therefore, it is essential to clean the PEO coating prior to electroless plating. Typically, a weak acid or neutral solution is used to remove contaminants, salts, and oxides, ensuring proper adhesion of the chemical coating. After electroless plating, posttreatment processes such as drying and curing are applied to enhance the wear and corrosion resistance of the coating. Additionally, heat treatment can be used to further improve the hardness and corrosion resistance of the nickel-plated layer. The combination of electroless plating and the PEO process significantly enhances the properties of metal substrates, improving corrosion resistance, surface hardness, wear resistance, and coating uniformity. It also adds a metal layer to the PEO surface to enhancing adhesion, especially in applications involving high loads or friction.

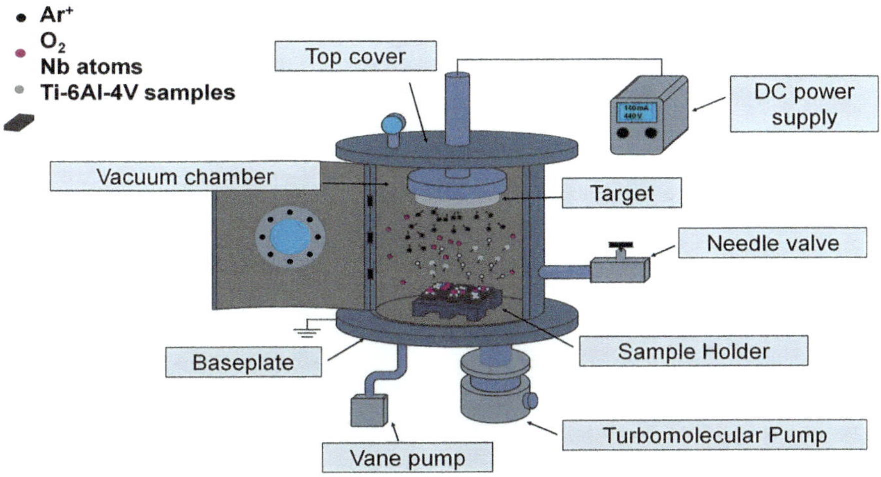

Fig. 23.15 The invention relates to a physical vapor deposition process. Reproduced with permission from [66] © 2021 Elsevier B.V.

Li et al. [70] studied the impact of nickel electroless plating. The morphology of PEO coating with micropores, mesopores, and macropores is prepared using NaF, Na_2SiO_3, and Na_3PO_4 as electrolytes, which are shown in Fig. 23.17a–c. After electroless plating, the pores with various sizes are no longer clearly visible and are replaced by a regular, spherelike microstructure (Fig. 23.17d, e), resulting in a reduced corrosion current density of ~10^{-8} A/cm^2.

Zhou et al. [71] also applied nickel electroless plating over PEO coating. As shown in Fig. 23.16a, the preparation scheme describes the entire process, including PEO treatment and electroless nickel plating. Figure 23.16b, c show the surface morphology and elemental distribution before and after electroless plating, to reveal the interaction of PEO coating and plating layer. The sample treated by electroless plating exhibited an amorphous outer layer composed of Ni and P, as shown by the element distribution. The researchers pointed out that a small amount of Ni was detected under the outer layer and suggested that pores and cracks of PEO coating were filled with the product of electroless plating, which could be verified by element mapping in microscopic analysis. After electroless plating, the corrosion current density was reduced by 97%. Figure 23.16f illustrates the interfacial state of electroless plating and PEO coating, which helps to explain the microstructure of the coating and sealing mechanism.

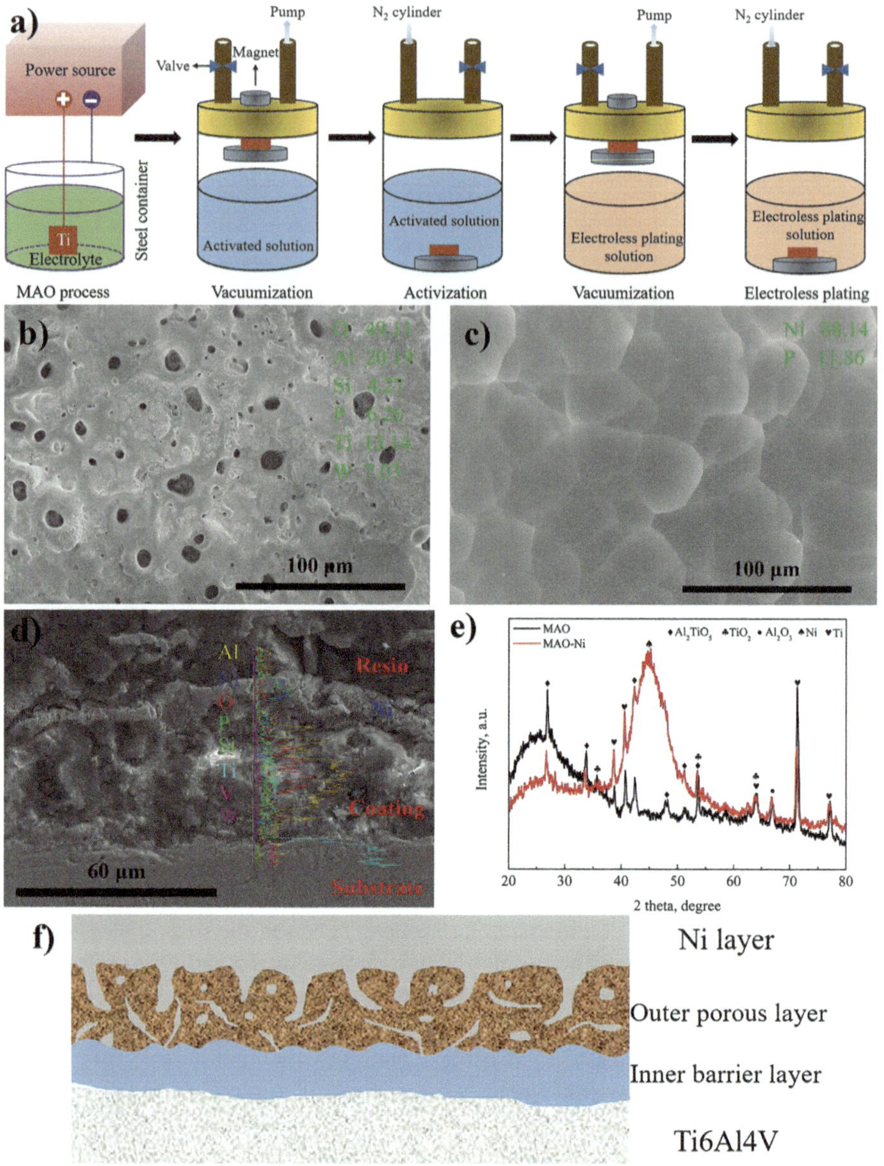

Fig. 23.16 (**a**) Schematic illustration of preparation process, surface topography of PEO (**b**) and PEO-Ni (**c**), (**d**) cross-section morphology of PEO-Ni, (**e**) XRD patterns of PEO and PEO-Ni, (**f**) structural representation of PEO-Ni. Reproduced with permission from [70] © 2022 Elsevier B.V.

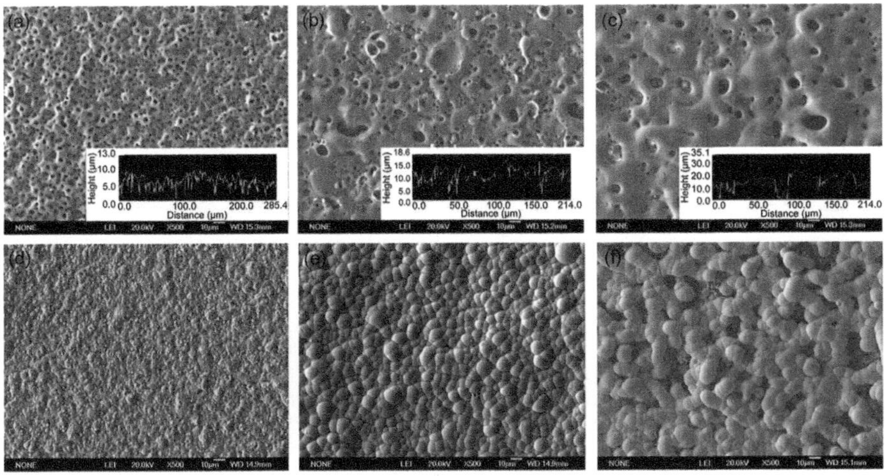

Fig. 23.17 (**a–c**) SEM images of porous structures on magnesium substrates formed by PEO pretreatment in NaF, Na_2SiO_3, and Na_3PO_4 solutions, respectively, and the insets show the CLSM-measured height profiles; (**d–f**) SEM images of the corresponding EN layer deposited on above pretreated substrates. Reproduced with permission from [71] © 2012 Elsevier B.V.

23.5 Conclusions and Outlook

This chapter presents an overview of various posttreatment sealing strategies for PEO coatings, with a focus on the development and current status of these techniques. The diversity of posttreatment methods is discussed in terms of their underlying mechanisms and applications. A review of the current literature reveals that mechanical surface refinement, chemical sealing strategies, and conversion coating sealing are key strategies for enhancing the corrosion resistance, mechanical strength, and adhesion properties of PEO coatings. Additionally, current posttreatment sealing strategies focus on enhancing the self-healing, superhydrophobicity, biocompatibility, and photocatalytic sterilization capabilities of the coatings. These enhancements aim to meet the increasingly complex and diverse application scenarios and performance requirements of PEO coatings.

As these methods continue to evolve, an increasing variety of techniques and approaches are emerging to optimize the performance of PEO coatings. Among the most promising developments are the integration of corrosion inhibitors, self-healing agents, and functionalized nanomaterials, which hold great potential for strengthening the long-term durability and efficiency of PEO coatings. The versatility and adaptability of these posttreatment strategies underscore the significance of PEO technology in addressing the complex challenges posed by harsh environments.

From an industrial perspective, however, the main challenges in developing these strategies are related to scaling up for large-scale applications. Issues such as coating adhesion, cracking, and process optimization must be addressed to ensure the reliable and cost-effective implementation of posttreatment sealing strategies. Additionally, the establishment of a robust, systematic detection and evaluation system will be critical for assessing the long-term performance and stability of these coatings.

References

1. Fattah-alhosseini, A., Fardosi, A., Karbasi, M., & Kaseem, M. (2024). *Journal of Magnesium and Alloys, 12*, 465–489.
2. Hosseini, R., Fattah-alhosseini, A., Karbasi, M., & Giannakis, S. (2024). *Applied Catalysis B: Environmental, 340*, 123197.
3. Fattah-alhosseini, A., & Chaharmahali, R. (2023). *Journal of Magnesium and Alloys, 11*, 4390–4406.
4. Lu, X., Blawert, C., Huang, Y., Ovri, H., Zheludkevich, M. L., & Kainer, K. U. (2016). *Electrochimica Acta, 187*, 20–33.
5. Fernández-López, P., Alves, S. A., Azpitarte, I., San-José, J. T., & Bayón, R. (2022). *Corrosion Science, 207*, 110548.
6. Martin, J., Teriokhina, E., Marcos, G., Czerwiec, T., Henrion, G., Novelli, M., & Grosdidier, T. (2024). *Materials Letters, 370*, 136840.
7. Wang, M., Wang, Y. L., Liu, Z. D., & Ding, H. (2013). *Applied Mechanics and Materials, 365–366*, 1110–1113.
8. Shirani, A., Joy, T., Rogov, A., Lin, M., Yerokhin, A., Mogonye, J.-E., Korenyi-Both, A., Aouadi, S. M., Voevodin, A. A., & Berman, D. (2020). *Surface and Coating Technology, 397*, 126016.
9. Fernández-López, P., Alves, S. A., López-Ortega, A., San José-Lombera, J. T., & Bayón, R. (2021). *Ceramics International, 47*, 31238–31250.
10. Kannatey-Asibu, E., Jr. (2009). Laser surface modification. In *Principles of laser materials processing* (pp. 568–615). Wiley.
11. Bartkowski, D., Młynarczak, A., Piasecki, A., Dudziak, B., Gościański, M., & Bartkowska, A. (2015). *Optics and Laser Technology, 68*, 191–201.
12. Wang, L., Zhou, J., Liang, J., & Chen, J. (2012). *Surface and Coating Technology, 206*, 3109–3115.
13. Rapheal, G., Kumar, S., Blawert, C., & Dahotre, N. B. (2011). *Wear, 271*, 1987–1997.
14. Li, Y., Guan, Y., Zhang, Z., & Ynag, S. (2019). *Applied Surface Science, 478*, 866–871.
15. Xiong, Y., Hu, Q., Song, R., & Hu, X. (2017). *Materials Science and Engineering: C, 75*, 1299–1304.
16. Xie, X., Ye, Y., Zou, Z., Mo, Y., Liang, Z., & Tang, G. (2024). *Journal of Materials Processing Technology, 332*, 118557.
17. Yao, W., Tan, Y., Lu, Q., Yi, H., Cheng, C., Wu, L., Saji, V. S., & Pan, F. (2024). *Journal of Materials Research and Technology, 31*, 3238–3254.
18. Mingo, B., Arrabal, R., Mohedano, M., Llamazares, Y., Matykina, E., Yerokhin, A., & Pardo, A. (2018). *Applied Surface Science, 433*, 653–667.
19. Pezzato, L., Babbolin, R., Cerchier, P., Marigo, M., Dolcet, P., Dabalà, M., & Brunelli, K. (2020). *Corrosion Science, 173*, 108741.
20. Liu, C., Lu, X., Li, Y., Chen, Q., Zhang, T., & Wang, F. (2021). *Journal of Alloys and Compounds, 870*, 159462.
21. Xue, S., Li, B., Mu, P., & Li, J. (2022). *Progress in Organic Coating, 170*, 106966.
22. Ostapiuk, M., Taryba, M. G., Calado, L. M., Bieniaś, J., & Montemor, M. F. (2021). *Corrosion Science, 186*, 109470.

23. Parichehr, R., Dehghanian, C., & Nikbakht, A. (2021). *Journal of Alloys and Compounds, 876*, 159995.
24. Yang, J., Blawert, C., Lamaka, S. V., Snihirova, D., Lu, X., Di, S., & Zheludkevich, M. L. (2018). *Corrosion Science, 140*, 99–110.
25. Tiwari, A., & Hihara, L. H. (2012). *Surface and Coating Technology, 206*, 4606–4618.
26. Hu, Y., Du, H., Lu, J., Li, Z., Ma, Y., & Song, S. (2023). *Fuel, 354*, 129417.
27. Khan, M. A., Safira, A. R., Aadil, M., & Kaseem, M. (2024). *Journal of Magnesium and Alloys, 12*, 586–607.
28. Shulha, T., Shikun, M., Serdechnova, M., Naacke, T., Kasneryk, V., Heitmann, V., Karlova, P., Davydok, A., Karpushenkov, S. A., Blawert, C., & Zheludkevich, M. L. (2024). *Applied Surface Science, 670*, 160707.
29. Qian, K., Zhang, Y., Dong, Q., Shao, Y., Cheng, Z., Ju, J., Xue, F., Chu, C., Xia, D., & Bai, J. (2023). *Corrosion Science, 219*, 111218.
30. Nisar, S. S., & Choe, H.-C. (2024). *Ceramics International, 50*, 17703–17719.
31. Wu, J., Wu, L., Yao, W., Zhou, Y., Wu, M., Yuan, Y., Xie, Z., Atrens, A., Wang, J., & Pan, F. (2024). *Electrochimica Acta, 491*, 144358.
32. Chen, J., Lin, W., Liang, S., Zou, L., Wang, C., Wang, B., Yan, M., & Cui, X. (2019). *Applied Surface Science, 463*, 535–544.
33. Jiang, D., Xia, X., Hou, J., Cai, G., Zhang, X., & Dong, Z. (2019). *Chemical Engineering Journal, 373*, 285–297.
34. Serdechnova, M., Mohedano, M., Bouali, A. C., Höche, D., Kuznetsov, B., Karpushenkov, S., Blawert, C., & Zheludkevich, M. L. (2017). *Coatings, 7*, 190.
35. Zhang, Y., Yu, P., Wang, J., Li, Y., Chen, F., Wei, K., & Zuo, Y. (2018). *Applied Surface Science, 433*, 927–933.
36. Chen, Q., Lu, X., Serdechnova, M., Wu, T., Kasneryk, V., Naacke, T., Wieland, D. C. F., Blawert, C., Zhang, Y., Xu, Y., Zheludkevich, M. L., & Wang, F. (2024). *Surface and Coating Technology, 485*, 130907.
37. Chen, F., Yu, P., & Zhang, Y. (2017). *Journal of Alloys and Compounds, 711*, 342–348.
38. Liu, G., Lu, X., Zhang, X., Zhang, T., & Wang, F. (2022). *Surface and Coating Technology, 441*, 128528.
39. Bouali, A. C., Straumal, E. A., Serdechnova, M., Wieland, D. C. F., Starykevich, M., Blawert, C., Hammel, J. U., Lermontov, S. A., Ferreira, M. G. S., & Zheludkevich, M. L. (2019). *Applied Surface Science, 494*, 829–840.
40. Peng, F., Wang, D., Zhang, D., Yan, B., Cao, H., Qiao, Y., Liu, X., & Biomater, A. C. S. (2018). *The Sciences and Engineering, 4*, 4112–4121.
41. Hou, L., Li, Y., Sun, J., Zhang, S. H., Wei, H., & Wei, Y. (2019). *Applied Surface Science, 487*, 101–108.
42. Wei, K., Wei, Y., Zhang, Y., Kasneryk, V., Serdechnova, M., Wang, H., Zhang, Z., Yuan, Y., Blawert, C., Zheludkevich, M. L., & Chen, F. (2024). *Corrosion Science, 227*, 111731.
43. Zhang, Z., Wang, J., Serdechnova, M., Kasneryk, V., Zhang, Z., Blawert, C., Wang, H., Zheludkevich, M. L., Chen, F., & Zhang, Y. (2024). *ACS Applied Materials & Interfaces, 16*, 11944–11956.
44. Xie, Z.-H., Luo, Y., Liu, R., Long, J., Wu, L., Wang, K., & Zhong, C.-J. (2024). *Chemical Engineering Journal, 500*, 156627.
45. Jiang, S., Li, W., Liu, J., Jiang, J., Zhang, Z., Shang, W., Peng, N., & Wen, Y. (2023). *Journal of Magnesium and Alloys, 11*, 3287–3301.
46. Aadil, M., Safira, A. R., Fattah-alhosseini, A., Alkaseem, M., & Kaseem, M. (2024). *Journal of Magnesium and Alloys, 12*, 3729–3743.
47. Yang, E., Liang, F., Yang, R., Mo, Q., Jia, N., & Li, W. (2024). *Surface and Coating Technology, 494*, 131292.
48. Shahin, N., Shamanian, M., & Kharaziha, M. (2022). *Diamond and Related Materials, 130*, 109405.

49. Han, J., Blawert, C., Tang, S., Yang, J., Hu, J., & Zheludkevich, M. L. (2021). *Surface and Coating Technology, 412*, 127030.
50. Zhao, C., Wang, X., Yu, B., Cai, M., Yu, Q., & Zhou, F. (2023). *Coatings, 13*, 1189.
51. Wang, D., Ma, C., Liu, J., Li, W., Shang, W., Peng, N., & Wen, Y. (2023). *International Journal of Minerals, Metallurgy, and Materials, 30*, 1128–1139.
52. Wang, C., Jiang, B., Liu, M., & Ge, Y. (2015). *Journal of Alloys and Compounds, 621*, 53–61.
53. Gnedenkov, S. V., Sinebryukhov, S. L., Mashtalyar, D. V., & Imshinetskiy, I. M. (2015). *Surface and Coating Technology, 283*, 347–352.
54. Wang, G., Guo, L., Ruan, Y., Zhao, G., Zhang, X., Liu, Y., & Kim, D.-E. (2024). *Surface and Coating Technology, 479*, 130556.
55. Hao, Y., Wang, L., Cao, Y., Du, Y., Ye, Z., & Wang, C. (2024). *Journal of Materials Science, 59*, 14028–14045.
56. Yang, W., Gao, Y., Guo, P., Xu, D., Hu, L., & Wang, A. (2020). *Journal of the Mechanical Behavior of Biomedical Materials, 101*, 103448.
57. Hao, Y., Du, Y., Wang, L., Cao, Y., Ye, Z., & Wang, C. (2024). *Ceramics International, 50*, 53835–53850.
58. Wang, S., Liu, X., Yin, X., Zhang, B., & Du, N. (2020). *Surface and Coating Technology, 404*, 126473.
59. Zhao, Y., Li, Y., Yin, T., Zhang, T., & Wang, F. (2023). *Surface and Coating Technology, 472*, 129873.
60. Li, B., Tian, M., Feng, L., Wu, D., Guo, H., Gao, Y., Xu, Y., An, C., & Liang, C. (2024). *Journal of Materials Research and Technology, 33*, 8545–8554.
61. Kuroda, P. A. B., Cardoso, G. C., Rossi, M. C., Afonso, C. R. M., & Grandini, C. R. (2024). *Materials Letters, 373*, 137155.
62. Xu, L., Fu, X., Su, H., Sun, H., Li, R., & Wan, Y. (2022). *Progress in Organic Coating, 170*, 107002.
63. Zhang, G., Wu, L., Serdechnova, M., Tang, A., Wang, C., Blawert, C., Pan, F., & Zheludkevich, M. L. (2023). *Journal of Magnesium and Alloys, 11*, 2376–2391.
64. Sharifiyan, M. S., Fattah-alhosseini, A., & Karbasi, M. (2023). *Applied Surface Science Advances, 18*, 100541.
65. Nisar, S. S., Arun, S., Toan, N. K., Ahn, S.-G., & Choe, H.-C. (2024). *Journal of Materials Research and Technology, 31*, 1282–1303.
66. do Nascimento, J. P. L., Ferreira, M. O. A., Gelamo, R. V., Scarmínio, J., Steffen, T. T., da Silva, B. P., Aoki, I. V., dos Santos, A. G., Jr., de Castro, V. V., de Fraga Malfatti, C., & Moreto, J. A. (2021). *Surface and Coating Technology, 428*, 127854.
67. Wei, B., Cheng, Y., Liu, Y., Zhu, Z., & Cheng, Y. (2021). *Transactions of the Nonferrous Metals Society of China, 31*, 2287–2306.
68. Bamunuarachchi, B. A. H. M., Jin, J., Lee, H. J., Choi, C. K., & Lee, S. H. (2024). *International Communications in Heat and Mass Transfer, 159*, 108292.
69. Byun, J. S., Song, Y., Montero, J., Yoon, J.-H., Österlund, L., Kim, S., & Song, P. (2024). *Vacuum, 230*, 113679.
70. Li, J., Zhang, Q., Cai, H., Wang, A., Zhang, J., & Hua, X. (2013). *Materials Letters, 93*, 263–265.
71. Zhou, T., Zhang, X., Wang, R., Long, F., Li, X., & Liu, L. (2022). *Materials Letters, 315*, 131944.

Index

V. S. Saji, M. Zheludkevich (eds.), *Plasma Electrolytic Oxidation – Fundamentals, Advances and Applications*,
https://doi.org/10.1007/978-3-031-92429-3

M

N

O

P

S

GPSR Compliance

The European Union's (EU) General Product Safety Regulation (GPSR) is a set of rules that requires consumer products to be safe and our obligations to ensure this.

If you have any concerns about our products, you can contact us on ProductSafety@springernature.com

In case Publisher is established outside the EU, the EU authorized representative is:

Springer Nature Customer Service Center GmbH
Europaplatz 3
69115 Heidelberg, Germany

Batch number: 10755458

Printed by Printforce, the Netherlands